T0328856

LONDON MATHEMATICAL SOCIETY LECTURE NOTE SERIES

Managing Editor: Professor J.W.S. Cassels, Department of Pure Mathematics and Mathematical Statistics, University of Cambridge, 16 Mill Lane, Cambridge CB2 1SB, England

The titles below are available from booksellers, or, in case of difficulty, from Cambridge University Press.

46 *p*-adic Analysis: a short course on recent work, N. KOBLITZ
50 Commutator calculus and groups of homotopy classes, H.J. BAUES
59 Applicable differential geometry, M. CRAMPIN & F.A.E. PIRANI
66 Several complex variables and complex manifolds II, M.J. FIELD
69 Representation theory, I.M. GELFAND *et al*
86 Topological topics, I.M. JAMES (ed)
87 Surveys in set theory, A.R.D. MATHIAS (ed)
88 FPF ring theory, C. FAITH & S. PAGE
89 An F-space sampler, N.J. KALTON, N.T. PECK & J.W. ROBERTS
90 Polytopes and symmetry, S.A. ROBERTSON
92 Representation of rings over skew fields, A.H. SCHOFIELD
93 Aspects of topology, I.M. JAMES & E.H. KRONHEIMER (eds)
96 Diophantine equations over function fields, R.C. MASON
97 Varieties of constructive mathematics, D.S. BRIDGES & F. RICHMAN
98 Localization in Noetherian rings, A.V. JATEGAONKAR
99 Methods of differential geometry in algebraic topology, M. KAROUBI & C. LERUSTE
100 Stopping time techniques for analysts and probabilists, L. EGGHE
104 Elliptic structures on 3-manifolds, C.B. THOMAS
105 A local spectral theory for closed operators, I. ERDELYI & WANG SHENGWANG
107 Compactification of Siegel moduli schemes, C.-L. CHAI
109 Diophantine analysis, J. LOXTON & A. VAN DER POORTEN (eds)
113 Lectures on the asymptotic theory of ideals, D. REES
114 Lectures on Bochner-Riesz means, K.M. DAVIS & Y.-C. CHANG
116 Representations of algebras, P.J. WEBB (ed)
119 Triangulated categories in the representation theory of finite-dimensional algebras, D. HAPPEL
121 Proceedings of *Groups - St Andrews 1985*, E. ROBERTSON & C. CAMPBELL (eds)
128 Descriptive set theory and the structure of sets of uniqueness, A.S. KECHRIS & A. LOUVEAU
129 The subgroup structure of the finite classical groups, P.B. KLEIDMAN & M.W. LIEBECK
130 Model theory and modules, M. PREST
131 Algebraic, extremal & metric combinatorics, M.-M. DEZA, P. FRANKL & I.G. ROSENBERG (eds)
132 Whitehead groups of finite groups, ROBERT OLIVER
133 Linear algebraic monoids, MOHAN S. PUTCHA
134 Number theory and dynamical systems, M. DODSON & J. VICKERS (eds)
135 Operator algebras and applications, 1, D. EVANS & M. TAKESAKI (eds)
137 Analysis at Urbana, I, E. BERKSON, T. PECK, & J. UHL (eds)
138 Analysis at Urbana, II, E. BERKSON, T. PECK, & J. UHL (eds)
139 Advances in homotopy theory, S. SALAMON, B. STEER & W. SUTHERLAND (eds)
140 Geometric aspects of Banach spaces, E.M. PEINADOR & A. RODES (eds)
141 Surveys in combinatorics 1989, J. SIEMONS (ed)
144 Introduction to uniform spaces, I.M. JAMES
146 Cohen-Macaulay modules over Cohen-Macaulay rings, Y. YOSHINO
148 Helices and vector bundles, A.N. RUDAKOV *et al*
149 Solitons, nonlinear evolution equations and inverse scattering, M. ABLOWITZ & P. CLARKSON
150 Geometry of low-dimensional manifolds 1, S. DONALDSON & C.B. THOMAS (eds)
151 Geometry of low-dimensional manifolds 2, S. DONALDSON & C.B. THOMAS (eds)
152 Oligomorphic permutation groups, P. CAMERON
153 L-functions and arithmetic, J. COATES & M.J. TAYLOR (eds)
155 Classification theories of polarized varieties, TAKAO FUJITA
156 Twistors in mathematics and physics, T.N. BAILEY & R.J. BASTON (eds)
158 Geometry of Banach spaces, P.F.X. MÜLLER & W. SCHACHERMAYER (eds)
159 Groups St Andrews 1989 volume 1, C.M. CAMPBELL & E.F. ROBERTSON (eds)
160 Groups St Andrews 1989 volume 2, C.M. CAMPBELL & E.F. ROBERTSON (eds)
161 Lectures on block theory, BURKHARD KÜLSHAMMER
162 Harmonic analysis and representation theory, A. FIGA-TALAMANCA & C. NEBBIA
163 Topics in varieties of group representations, S.M. VOVSI
164 Quasi-symmetric designs, M.S. SHRIKANDE & S.S. SANE
166 Surveys in combinatorics, 1991, A.D. KEEDWELL (ed)
168 Representations of algebras, H. TACHIKAWA & S. BRENNER (eds)
169 Boolean function complexity, M.S. PATERSON (ed)
170 Manifolds with singularities and the Adams-Novikov spectral sequence, B. BOTVINNIK
171 Squares, A.R. RAJWADE
172 Algebraic varieties, GEORGE R. KEMPF
173 Discrete groups and geometry, W.J. HARVEY & C. MACLACHLAN (eds)
174 Lectures on mechanics, J.E. MARSDEN
175 Adams memorial symposium on algebraic topology 1, N. RAY & G. WALKER (eds)
176 Adams memorial symposium on algebraic topology 2, N. RAY & G. WALKER (eds)

177 Applications of categories in computer science, M. FOURMAN, P. JOHNSTONE & A. PITTS (eds)
178 Lower K- and L-theory, A. RANICKI
179 Complex projective geometry, G. ELLINGSRUD *et al*
180 Lectures on ergodic theory and Pesin theory on compact manifolds, M. POLLICOTT
181 Geometric group theory I, G.A. NIBLO & M.A. ROLLER (eds)
182 Geometric group theory II, G.A. NIBLO & M.A. ROLLER (eds)
183 Shintani zeta functions, A. YUKIE
184 Arithmetical functions, W. SCHWARZ & J. SPILKER
185 Representations of solvable groups, O. MANZ & T.R. WOLF
186 Complexity: knots, colourings and counting, D.J.A. WELSH
187 Surveys in combinatorics, 1993, K. WALKER (ed)
188 Local analysis for the odd order theorem, H. BENDER & G. GLAUBERMAN
189 Locally presentable and accessible categories, J. ADAMEK & J. ROSICKY
190 Polynomial invariants of finite groups, D.J. BENSON
191 Finite geometry and combinatorics, F. DE CLERCK *et al*
192 Symplectic geometry, D. SALAMON (ed)
193 Computer algebra and differential equations, E. TOURNIER (ed)
194 Independent random variables and rearrangement invariant spaces, M. BRAVERMAN
195 Arithmetic of blowup algebras, WOLMER VASCONCELOS
196 Microlocal analysis for differential operators, A. GRIGIS & J. SJÖSTRAND
197 Two-dimensional homotopy and combinatorial group theory, C. HOG-ANGELONI, W. METZLER & A.J. SIERADSKI (eds)
198 The algebraic characterization of geometric 4-manifolds, J.A. HILLMAN
199 Invariant potential theory in the unit ball of $\mathbf{C}^n$, MANFRED STOLL
200 The Grothendieck theory of dessins d'enfant, L. SCHNEPS (ed)
201 Singularities, JEAN-PAUL BRASSELET (ed)
202 The technique of pseudodifferential operators, H.O. CORDES
203 Hochschild cohomology of von Neumann algebras, A. SINCLAIR & R. SMITH
204 Combinatorial and geometric group theory, A.J. DUNCAN, N.D. GILBERT & J. HOWIE (eds)
205 Ergodic theory and its connections with harmonic analysis, K. PETERSEN & I. SALAMA (eds)
206 An introduction to noncommutative differential geometry and its physical applications, J. MADORE
207 Groups of Lie type and their geometries, W.M. KANTOR & L. DI MARTINO (eds)
208 Vector bundles in algebraic geometry, N.J. HITCHIN, P. NEWSTEAD & W.M. OXBURY (eds)
209 Arithmetic of diagonal hypersurfaces over finite fields, F.Q. GOUVÊA & N. YUI
210 Hilbert C*-modules, E.C. LANCE
211 Groups 93 Galway / St Andrews I, C.M. CAMPBELL *et al* (eds)
212 Groups 93 Galway / St Andrews II, C.M. CAMPBELL *et al* (eds)
214 Generalised Euler-Jacobi inversion formula and asymptotics beyond all orders, V. KOWALENKO, N.E. FRANKEL, M.L. GLASSER & T. TAUCHER
215 Number theory 1992–93, S. DAVID (ed)
216 Stochastic partial differential equations, A. ETHERIDGE (ed)
217 Quadratic forms with applications to algebraic geometry and topology, A. PFISTER
218 Surveys in combinatorics, 1995, PETER ROWLINSON (ed)
220 Algebraic set theory, A. JOYAL & I. MOERDIJK
221 Harmonic approximation, S.J. GARDINER
222 Advances in linear logic, J.-Y. GIRARD, Y. LAFONT & L. REGNIER (eds)
223 Analytic semigroups and semilinear initial boundary value problems, KAZUAKI TAIRA
224 Computability, enumerability, unsolvability, S.B. COOPER, T.A. SLAMAN & S.S. WAINER (eds)
225 A mathematical introduction to string theory, S. ALBEVERIO, J. JOST, S. PAYCHA, S. SCARLATTI
226 Novikov conjectures, index theorems and rigidity I, S. FERRY, A. RANICKI & J. ROSENBERG (eds)
227 Novikov conjectures, index theorems and rigidity II, S. FERRY, A. RANICKI & J. ROSENBERG (eds)
228 Ergodic theory of $\mathbf{Z}^d$ actions, M. POLLICOTT & K. SCHMIDT (eds)
229 Ergodicity for infinite dimensional systems, G. DA PRATO & J. ZABCZYK
230 Prolegomena to a middlebrow arithmetic of curves of genus 2, J.W.S. CASSELS & E.V. FLYNN
231 Semigroup theory and its applications, K.H. HOFMANN & M.W. MISLOVE (eds)
232 The descriptive set theory of Polish group actions, H. BECKER & A.S. KECHRIS
233 Finite fields and applications, S. COHEN & H. NIEDERREITER (eds)
234 Introduction to subfactors, V. JONES & V.S. SUNDER
235 Number theory 1993–94, S. DAVID (ed)
236 The James forest, H. FETTER & B. GAMBOA DE BUEN
237 Sieve methods, exponential sums, and their applications in number theory, G.R.H. GREAVES, G. HARMAN & M.N. HUXLEY (eds)
238 Representation theory and algebraic geometry, A. MARTSINKOVSKY & G. TODOROV (eds)
239 Clifford algebras and spinors, P. LOUNESTO
240 Stable groups, FRANK O. WAGNER
241 Surveys in combinatorics, 1997, R.A. BAILEY (ed)
242 Geometric Galois actions I, L. SCHNEPS & P. LOCHAK (eds)
243 Geometric Galois actions II, L. SCHNEPS & P. LOCHAK (eds)
244 Model theory of groups and automorphism groups, D. EVANS (ed)
245 Geometry, combinatorial designs and related structures, J.W.P. HIRSCHFELD, S.S. MAGLIVERAS & M.J. DE RESMINI (eds)
246 *p*-Automorphisms of finite *p*-groups, E.I. KHUKHRO
247 Analytic number theory, Y. MOTOHASHI (ed)
251 Gröbner bases and applications, B. BUCHBERGER & F. WINKLER (eds)

London Mathematical Society Lecture Note Series. 251

Gröbner Bases and Applications

Edited by

B. Buchberger & F. Winkler
Johannes Kepler University of Linz

CAMBRIDGE UNIVERSITY PRESS
Cambridge, New York, Melbourne, Madrid, Cape Town, Singapore,
São Paulo, Delhi, Dubai, Tokyo, Mexico City

Cambridge University Press
The Edinburgh Building, Cambridge CB2 8RU, UK

Published in the United States of America by
Cambridge University Press, New York

www.cambridge.org
Information on this title: www.cambridge.org/9780521632980

First published 1998

A catalogue record for this publication is available from the British Library

ISBN 978-0-521-63298-0 Paperback

Table of Contents

Preface

Gröbner bases were introduced in 1965 by the first editor (Buchberger) in his Ph.D. thesis. It took some years before the concept became known to the research communities in Mathematics and Theoretical Computer Science, but about ten years later many other research groups around the world started working on or using this concept. The theory of Gröbner bases has become an important subarea in computer algebra, it is included in all the major program systems of symbolic computation, and it is being fruitfully applied in a variety of seemingly unrelated research areas.

We at RISC-Linz wanted to celebrate the fact that Gröbner bases have been around now for a third of a century, and in doing so provide a snapshot of the state of the art. For this reason we organized a Special Year on Gröbner Bases, culminating in an intensive course for young researchers (January 1998) and in the conference "33 Years of Gröbner Bases", held at RISC-Linz in February 2–4, 1998. The conference included both tutorials on various aspects of Gröbner bases and their application, and the presentation of original research papers on new developments in the theory of Gröbner bases.

The present book is the outcome of these activities. It contains a short introduction to the theory of Gröbner bases, tutorial papers on the interaction between Gröbner bases and various other mathematical theories, and the original research papers presented at the conference. Finally, an English translation by Michael Abramson and Robert Lumbert of the journal version of Buchberger's Ph.D. thesis is included.

We want to thank all the people who have contributed to this book: the members of the program committee, who did a great job organizing the refereeing process for the conference, the authors of tutorial papers and research papers, who demonstrate the wide range usefulness and interconnections of Gröbner bases, and the publisher, who managed to print and distribute the book in exceptionally short time. Many thanks also to Daniela Vasaru for her organizational help in preparing the camera-ready manuscript of the book. Without the help of all these people we could never have succeeded in this project.

Bruno Buchberger
Franz Winkler

Program Committee of the Conference "33 Years of Gröbner Bases"

W.W. Adams (College Park, USA)
B. Buchberger (Linz, Austria)
J.H. Davenport (Bath, England)
R. Fröberg (Stockholm, Sweden)
A. Galligo (Nice, France)
V.P. Gerdt (Dubna, Russia)
M. Giusti (Palaiseau, France)
M. Kalkbrener (Zürich, Switzerland)
W.W. Küchlin (Tübingen, Germany)
Y.N. Lakshman(Philadelphia, USA)
A.H.M. Levelt (Nijmegen, Netherlands)
B. Mishra (New York, USA)
T. Mora (Genova, Italy)
T. Recio (Santander, Spain)
H.J. Stetter (Vienna, Austria)
M. Sweedler (Ithaca, USA)
C. Traverso (Pisa, Italy)
V. Weispfenning (Passau, Germany)
F. Winkler (Linz, Austria)

Tutorials

Introduction to Gröbner Bases[1]

Bruno Buchberger

Research Institute for Symbolic Computation
Austria-4232, Schloss Hagenberg
Bruno.Buchberger@RISC.uni-linz.ac.at

Outline

A comprehensive treatment of Gröbner bases theory is far beyond what can be done in one article in a book. Recent text books on Gröbner bases like (Becker, Weispfenning 1993) and (Cox, Little, O'Shea 1992) present the material on several hundred pages. However, there are only a few key ideas behind Gröbner bases theory. It is the objective of this introduction to explain these ideas as simply as possible and to give an overview of the immediate applications. More advanced applications are described in the other tutorial articles in this book.

The concept of Gröbner bases together with the characterization theorem (by "S-polynomials") on which an algorithm for constructing Gröbner bases hinges has been introduced in the author's PhD thesis (Buchberger 1965), see also the journal publication (Buchberger 1970). In these early papers we also gave some first applications (computation in residue class rings modulo polynomial ideal congruence, algebraic equations, and Hilbert function computation), a computer implementation (in the assembler language of the ZUSE Z23V computer), and some first remarks on complexity. Later work by the author and by many other authors has mainly added generalizations of the method and quite a few new applications for the algorithmic solution of various fundamental problems in algebraic geometry (polynomial ideal theory, commutative algebra). Also, complexity issues have been treated extensively. The field is still under active development both into the direction of improving the method by new theoretical insights and by finding new applications.

This article is structured as follows:

In the first section we give a variety of examples demonstrating the versatility of the method of Gröbner bases for problems that involve finite sets of multivariate polynomials.

In the second section, the main idea contained in the notion of Gröbner bases and the main theorem about them, which also leads to an algorithmic

[1]An earlier version of this paper appeared in the Proceedings of the Marktoberdorf Summer School 1995, published by Springer Heidelberg, 1997.

construction of Gröbner bases, is explained. The proof of the main theorem is spelled out in detail.

The third section systematically summarizes the most important immediate applications of Gröbner bases.

1 Gröbner Bases at Work

1.1 Example: Fermat Ideals

The following polynomials are called Fermat polynomials:

$$F_n := x^n + y^n - z^n \ (n \geq 1).$$

Question: Can, from some k on, F_n be expressed as a linear combination

$$F_n = \sum_{1 \leq i \leq k} h_{n,i} \cdot F_i$$

with $h_{n,i} \in \mathbb{Q}[x, y, z]$? In other words: Is F_n in Ideal$(F_1, \ldots, F_k)$, the ideal generated by $F_1, \ldots, F_k$? (This question was raised in connection with possible approaches to solving the Fermat problem. An affirmative but inconstructive answer, i.e. an answer that did not explicitly construct the $h_{n,i}$, was given in (Elias 88). This answer used quite heavy machinery from algebraic geometry.)

Solution by the Gröbner bases method: We compute a Gröbner basis G for Ideal(F_1, F_2, F_3) and check, by "reduction of F_4 modulo G", whether or not $F_4 \in$ Ideal(F_1, F_2, F_3). It turns out the answer is "yes", which can be seen from the fact that the reduction of F_4 modulo G yields 0. During the reduction we "collect the cofactors", which yields the representation

$$F_4 = S_0 \cdot F_1 - S_1 \cdot F_2 + S_2 \cdot F_3,$$

where the S_i are the elementary symmetric polynomials in x, y, z. ($S_0 := x\,y\,z$, $S_1 := x\,y + x\,z + y\,z$, $S_2 := x + y + z$.)

By the same method, we can now check whether $F_5 \in$ Ideal(F_1, F_2, F_3, F_4). It turns out that, even, $F_5 \in$ Ideal(F_2, F_3, F_4) and, surprisingly, again

$$F_5 = S_0 \cdot F_2 - S_1 \cdot F_3 + S_2 \cdot F_4.$$

This leads immediately to the conjecture that, for arbitrary $n \geq 1$,

$$F_{n+3} = S_0 \cdot F_n - S_1 \cdot F_{n+1} + S_2 \cdot F_{n+2}. \qquad \text{(identity}_3\text{)}$$

This conjecture can be verified easily by elementary formula manipulation. (One may want to use a symbolic computation software system for this verification!)

This identity yields the by-product that the "Fermat ideal" generated by the infinitely many F_n $(n \geq 1)$ is already generated by the first three Fermat polynomials. Of course, one can now go immediately one step further and may conjecture that, for the "generalized Fermat polynomials"

$$F_{m,n} := x_1{}^n + \ldots + x_{m-1}{}^n - x_m{}^n \quad (m, n \geq 1)$$

the following identity holds

$$F_{m,n+m} = \sum_{0 \leq k < m} (-1)^{m+k+1} \cdot S_{m,k} \cdot F_{m,n+k} \qquad (\text{identity}_m)$$

where the $S_{m,k}$ are the elementary symmetric polynomials in $x_1, \ldots, x_m$. This formula can be proved by straight-forward induction on m or by generating functions. The details can be found in (Buchberger, Elias 1992). Note that the same identity holds for the symmetric "exponential sums". However, the $F_{m,n}$ are not symmetric and this could be the reason why (identity_m) seems to have gone unnoticed in the literature.

1.2 Example: Geometry Theorem Proving

(This example is taken from (Buchberger, Kutzler 1986).)

Question: Is the following proposition true?

"In an arrangement of the form shown in the Figure 1, K, L, M are collinear."

(In the drawing, A$< 0, y_1 >$ etc. denotes the point A with coordinates 0 and y_1, etc.)

(The above statement is *Pappus' Theorem* and it is well known that it is true. The point is that one can ask this question about any geometric proposition whose premises and conclusion, after being described in Cartesian coordinates, can be expressed by multivariate polynomials, and the method given below will answer the question automatically.)

Solution by the Gröbner basis algorithm: An algebraic formulation of the problem is as follows:

$$\forall y_1, \ldots, y_{12}$$
$$(p_1(y_1, \ldots, y_{12}) = 0 \wedge \ldots \wedge p_6(y_1, \ldots, y_{12}) = 0 \Longrightarrow c(y_1, \ldots, y_{12}) = 0)$$

where $p_1, \ldots, p_6$, and c are non-linear polynomials in the variables $y_1, \ldots, y_{12}$ that express the premises and the conclusion of the theorem. For example, p_1 expresses the condition that K is on the line $\overline{AE}$ and has the following form

$$p_1(y_1, \ldots, y_{12}) := (y_7 - y_1)y_5 + y_8 y_1.$$

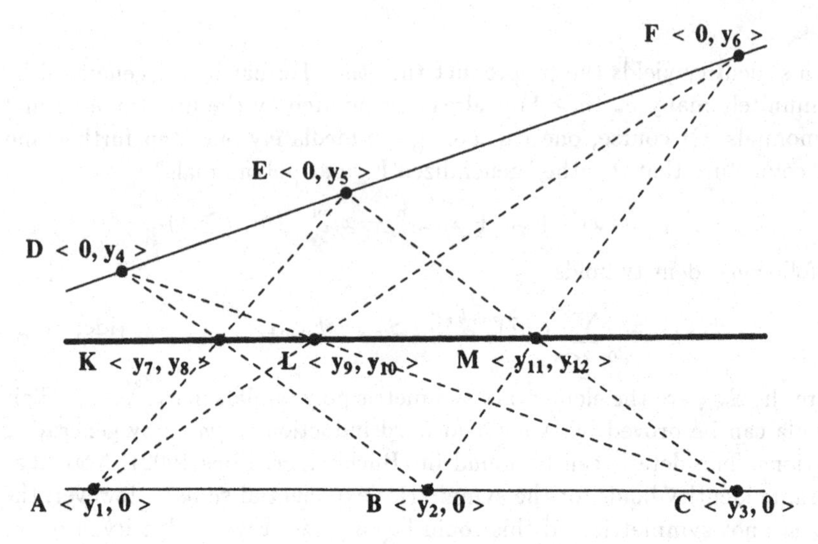

Figure 1: Pappus' Theorem

In our example, the other polynomials are

$$\begin{aligned}
p_2(y_1,\ldots,y_{12}) &:= (y_7 - y_2)y_4 + y_8\,y_2,\\
p_3(y_1,\ldots,y_{12}) &:= (y_9 - y_1)y_6 + y_{10}\,y_1,\\
p_4(y_1,\ldots,y_{12}) &:= (y_9 - y_2)y_4 + y_{10}\,y_3,\\
p_5(y_1,\ldots,y_{12}) &:= (y_{11} - y_2)y_6 + y_{12}\,y_2,\\
p_6(y_1,\ldots,y_{12}) &:= (y_{11} - y_3)y_5 + y_{12}\,y_3,\\
c(y_1,\ldots,y_{12}) &:= (y_9 - y_7)(y_{12} - y_8) + (y_{10} - y_8)(y_{11} - y_7).
\end{aligned}$$

We now input the following system of polynomials to the Gröbner basis algorithm:

$$\{p_1,\ldots,p_6, c\cdot y - 1\},$$

where y is a new variable. It can be shown that a theorem of the above form is true iff the Gröbner basis produced for the above input contains the polynomial 1. This is the case in our example and, hence, we know that the theorem is true.

1.3 Example: Invariant Theory

(This example is taken from (Sturmfels 1993).)

Question: Compute all algebraic relations between the fundamental invariants for the invariant ring of the cyclic group Z_4 of order 4, i.e. a set of generators for the ring

$$\{f \in \mathbb{C}[x_1, x_2] \mid f(x_1, x_2) = f(-x_2, x_1)\}$$

and represent the invariant $x_1^7 x_2 - x_1 x_2^7$ by the fundamental invariants.

Solution by the Gröbner basis method: The following polynomials

$$I_1 := x_1^2 + x_2^2,\ I_2 := x_1^2 x_2^2,\ I_3 := x_1^3 x_2 + x_1 x_2^3$$

form a system of fundamental invariants for Z_4. Now we compute the Gröbner basis of

$$\{-I_1 + x_1^2 + x_2^2, -I_1 + x_1^2 x_2^2, -I_3 + x_1^3 x_2 + x_1 x_2^3\}$$

(in the polynomial ring with added slack variables I_1, I_2, I_3) with respect to the lexical ordering determined by $I_1 < I_2 < I_3 < x_1 < x_2$. In our case this yields the set

$$\{I_1^2 I_2 - 4I_2^2 - 4I_3^2, I_2 - I_1 x_2^2 + x_2^4,$$
$$\text{... (6 other polynomials in which } x_1 \text{ and } x_2 \text{ occur)...}\}$$

Now, those polynomials in this Gröbner basis that depend only on I_1, I_2, and I_3 generate the ideal of all algebraic relations between I_1, I_2, and I_3. In our case this ideal is, hence, generated by

$$I_1^2 I_2 - 4I_2^2 - 4I_3^2.$$

Furthermore, by reducing any given polynomial g in x_1, x_2 modulo $\{I_1^2 I_2 - 4I_2^2 - 4I_3^2, \ldots\}$ one can check whether or not g is invariant (iff the reduction yields a polynomial that does not contain x_1, x_2 anymore) and, if it is invariant, this reduction yields a representation of g in terms of the fundamental invariants. In our example, the reduction of $x_1^7 x_2 - x_1 x_2^7$ yields $I_1^2 I_3 - I_2 I_3$.

1.4 Example: Systems of Polynomial Equations

(This example is taken from (Buchberger, Kutzler 1986)).

Systems of multivariate polynomial equations are pervasive in all areas of engineering. For example, consider the simple robot from Figure 2.

After appropriate coordinatization (using the Denavit-Hartenberg approach), the relation between the angles d_1 and d_2 at the links of the robot and the position of the gripper (described by the coordinates p_x, p_y, p_z) and its orientation (described, for example, by the Euler angles φ, θ, ψ) can be characterized

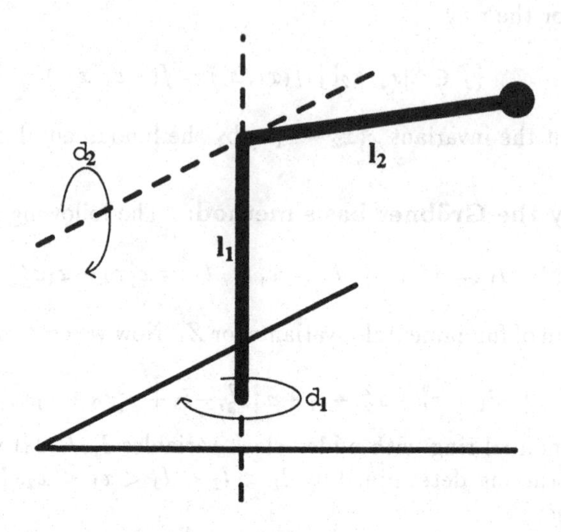

Figure 2: Simple Robot.

by the following system of polynomial equations:

$$
\begin{aligned}
c_1c_2 - c_fc_tc_p - s_fs_p &= 0,\\
s_1c_2 - s_fc_ts_p - c_fs_p &= 0,\\
s_2 + s_tc_p &= 0,\\
-c_1s_2 - c_fc_ts_p + s_fc_p &= 0,\\
-s_1s_2 + s_fc_ts_p - c_fc_p &= 0,\\
c_2 - s_ts_p &= 0,\\
s_1 - c_fs_t &= 0,\\
c_1 + s_fs_t &= 0,\\
c_t &= 0,\\
l_2c_1c_2 - p_x &= 0,\\
l_2s_1c_2 - p_y &= 0,\\
l_2s_2 + l_1 - p_z &= 0,\\
c_1^2 + s_1^2 - 1 &= 0,\\
c_2^2 + s_2^2 - 1 &= 0,\\
c_f^2 + s_f^2 - 1 &= 0,\\
c_t^2 + s_t^2 - 1 &= 0,\\
c_p^2 + s_p^2 - 1 &= 0.
\end{aligned}
$$

Here, $c_1, s_1, c_2, s_2, c_f, s_f, c_t, s_t, c_p, s_p$ are the cosines and sines of the angles d_1, d_2, φ, θ, ψ, respectively. These values are algebraically related by the last five additional equations. The arm lengths l_1 and l_2 are parameters. The kinematics problem asks for finding the value of some of these variables if the value of the rest of the variables is given. Because of the limited degree of freedom, in this example, we can only give the value of two of these variables. The others will then be determined. For example, we may fix p_x and p_y and ask for suitable values for the other variables.

Solution by the Gröbner basis algorithm: If we input this system of polynomials to the Gröbner basis algorithm (setting thc "ordering parameter" to the "lexical ordering" determined by $c_1 < c_2 < s_1 < s_2 < p_y < c_f < c_t < c_p < s_f < s_t < s_p$ and taking p_x, p_z, l_1, and l_2 as parameters), we obtain the following output:

$$\begin{aligned}
c_1^2 + Q_1 &= 0,\\
c_2 + Q_2 c_1 &= 0,\\
s_1^2 + Q_3 &= 0,\\
s_2 + Q_4 &= 0,\\
p_y + Q_5 c_1 s_1 &= 0,\\
c_f^2 + Q_6 &= 0,\\
c_t &= 0,\\
c_p + Q_7 s_1 c_f &= 0,\\
s_f + Q_8 c_1 s_1 c_f &= 0,\\
s_t + Q_9 s_1 c_f &= 0,\\
s_p + Q_{10} c_1 s_1 c_f &= 0,
\end{aligned}$$

where the Q_i are rational functions in p_x, p_z and the parameters l_1, l_2. The Gröbner basis produced has the remarkable and useful property that it is "triangularized", i.e. its first equation is univariate in the lowest variable c_1, i.e. all the possible values for c_1 can be determined from this equation. The second equation contains c_1 and c_2 and, in fact, c_2 is "explicit". Thus, for each value of c_1 a corresponding value for c_2 can be determined and so on. Also, the Gröbner basis still contains p_x, p_z and the parameters l_1, l_2 in "symbolic form".

Of course, in this simple example, the "symbolic solution" could also be derived by a reasonably skillful analysis of the drawing. However, the Gröbner basis algorithm works in all situations and always results in a "triangularized" system.

2 The Main Theorem on Gröbner Bases

2.1 Polynomials

Let $\mathbb{N}$ be the set of natural numbers including zero. The variables i ,j, k, l, m, n will range over $\mathbb{N}$. Let $(\mathbf{K}, +, 0, -, \cdot, 1, /)$ be a field, let $n \in \mathbb{N}$, and let $x_1, \ldots, x_n$ be indeterminates. By $(\mathbf{K}[x_1, \ldots, x_n], +, 0, -, \cdot, 1)$ we denote (any of the infinitely many isomorphic representations of) the ring of polynomials over $\mathbf{K}$ with indeterminates $x_1, \ldots, x_n$. Furthermore, $[x_1, \ldots, x_n]$ will denote the set of power products (i.e. monomials with coefficient 1) over the indeterminates $x_1, \ldots, x_n$. Throughout this paper, $(\mathbf{K}, +, 0, -, \cdot, 1, /)$, n, and $x_1, \ldots, x_n$ will be fixed and we will also use the abbreviations

$$\begin{aligned} \mathbf{T} &:= [x_1, \ldots, x_n], \\ \mathbf{P} &:= \mathbf{K}[x_1, \ldots, x_n]. \end{aligned}$$

Note that, in this paper, we use the symbols "+", "0" etc. both for the operations in the original field and for the operations in the polynomial ring. In addition, we will use "$\cdot$" also for scalar multiplication between field elements and polynomials. (In fact, with this additional operation, the polynomial ring becomes a vector space and even an associative algebra over the field.) This overloading of operation symbols will not cause any confusion since we will stick to the following additional type convention: The variables a, b, c will range over $\mathbf{K}$; p, q, r, but also f, g, h will range over $\mathbf{P}$; and t, u, v will range over $\mathbf{T}$. The variables F and G will be used for subsets of $\mathbf{P}$.

With some care, all these variables will also be used for ranging over finite sequences of elements from the respective sets. For any set Y, Y^* will denote the set of finite sequences over Y. If $y \in Y^*$, y_i is the i-th element and $|y|$ is the length of y, respectively. Of course, if $y \in Y^*$ then $y_i \in Y$.

On $\mathbf{T}$, we consider the following three additional operations:

$$\begin{aligned} t|u &:\Longleftrightarrow && u \text{ is a multiple of } t, \\ t/u &:= && t \text{ divided by } u \text{ (in case } u|t), \\ \mathrm{LCM}(t,u) &:= && \text{the least common multiple of } t \text{ and } u. \end{aligned}$$

On $\mathbf{P}$ we introduce the following structural operations:

$$\begin{aligned} \mathrm{C}(p, t) &:= \text{the coefficient at } t \text{ in } p, \\ \mathrm{M}(p, t) &:= \mathrm{C}(p,t) \cdot t, \\ \mathrm{S}(p) &:= \{t \mid \mathrm{C}(p,t) \neq 0\}. \end{aligned}$$

(For $\mathrm{M}(p, t)$ and $\mathrm{S}(p)$ read "the monomial at t in p" and "the support of p", respectively.)

The theory of Gröbner bases will be formulated independently of any particular representation of the domain of polynomials. However, in our examples we will always use the "ordinary" representation of polynomials as arithmetical terms in "fully expanded form" as, for example, "$-3xy^2z + 3/2x^2y + 5/3yz^2$".

Formal text (definitions, theorems, proofs) that is followed by informal text will be terminated by the symbol □.

2.2 Polynomial Ideals

Definition (Congruence and Ideals):

$$g \equiv_F h \quad :\Longleftrightarrow \quad \exists p \in \mathbf{P}^* \, \exists f \in F^* (|p| = |f| \wedge g = h + \sum_{1 \leq i \leq |p|} p_i \cdot f_i).$$

$$\text{Ideal}(F) \quad := \quad \{g \mid g \equiv_F 0\}.$$

(For $f \equiv_F g$ read "f is congruent g modulo F". For Ideal(F) read "the ideal generated by F".)

2.3 Admissible Orderings on Power Products

Congruence modulo an F is a nonalgorithmic notion: For deciding whether or not $g \equiv_F h$, one could try to compare coefficients in the presentation $g = h + \sum p_i \cdot f_i$ and then use linear algebra. However, a priori, it is not clear how big the degrees of appropriate p_i might become. In (Hermann 1926) bounds for the degrees were derived and, thus, in principle, $g \equiv_F h$ could be decided algorithmically.

However, we are heading for a different approach which will allow us to solve a broader class of problems in polynomial ideal theory and yields algorithmic decidability of congruence modulo arbitrary F as a by-product. The first step towards this goal is to replace congruence by "reduction", which can be viewed as a kind of "directed congruence". For this purpose, we order the power products (and thereby also the polynomials) so that we will later be able to replace power products by "lower" polynomials modulo F. Certain special orderings ("total degree" orderings) on power products were used in algebra already at the beginning of this century, for example by F. S. Macaulay. Gröbner basis theory works, however, with respect to any "admissible" ordering, as has been noticed first in (Trinks 1978).

Definition (Admissible Ordering): Let $\prec$ be a total ordering on $\mathbf{T}$. Then,

$$\prec \text{ is admissible } :\Longleftrightarrow \forall t \neq 1 \; (1 \prec t),$$
$$\forall t, u, v \; (t \prec u \Longrightarrow t \cdot v \prec u \cdot v). \quad \text{(monotonicity)}$$

Examples of Admissible Orderings: The "lexical" ordering defined by $x \prec y$ orders the power products in $[x, y]$ in the following way: $1 \prec x \prec x^2 \prec x^3 \prec \ldots \prec y \prec xy \prec x^2y \prec \ldots \prec y^2 \prec xy^2 \prec x^2y^2 \prec \ldots$.

The "total degree" ordering defined by $x \prec y$ orders the power products in $[x, y]$ in the following way: $1 \prec x \prec y \prec x^2 \prec xy \prec y^2 \prec x^3 \prec x^2y \prec xy^2 \prec y^3 \prec x^4 \prec \ldots$.

Admissible orderings on **T** have two important properties:

Proposition (Properties of Admissible Orderings): Let $\prec$ be an admissible ordering on **T**. Then,

$$\forall t, u\ (t \mid u \Longrightarrow t \preceq u), \qquad (|\text{-compatibility})$$
$$\prec \text{ is Noetherian.}$$

(A relation is Noetherian iff there are no infinite descending chains w.r.t. the relation.)

Proof: The proof of |-compatibility is immediate using the definition of admissibility. The proof of Noetherianity can be given by using a combinatorial lemma known as Dickson's lemma introduced in (Dickson 1913), see for example (Becker, Weispfenning 1993), p. 163.

2.4 Order Dependent Decomposition of Polynomials

Given an admissible ordering $\prec$ on **T**, we can now introduce a couple of operations on **P** that decompose polynomials into various constituents:

$$\begin{aligned}
\mathrm{LPP}_{\prec}(p) &:= \max_{\prec}\mathrm{S}(p),\\
\mathrm{LC}_{\prec}(p) &:= \mathrm{C}(p, \mathrm{LPP}_{\prec}(p)),\\
\mathrm{LM}_{\prec}(p) &:= \mathrm{LC}_{\prec}(p)\cdot\mathrm{LPP}_{\prec}(p),\\
\mathrm{R}_{\prec}(p) &:= p - \mathrm{LM}_{\prec}(p),\\
\mathrm{H}_{\prec}(p,t) &:= \sum_{u\in \mathrm{S}(p)\wedge u\succ t} \mathrm{C}(p,u)\cdot u,\\
\mathrm{L}_{\prec}(p,t) &:= \sum_{u\in \mathrm{S}(p)\wedge t\succ u} \mathrm{C}(p,u)\cdot u,\\
\mathrm{B}_{\prec}(p,t_1,t_2) &:= \sum_{u\in \mathrm{S}(p)\wedge t_1\succ u\succ t_2} \mathrm{C}(p,u)\cdot u,
\end{aligned}$$

(If $\prec$ is clear from the context, we will omit the subscript $\prec$ at these operations. For $\mathrm{LPP}(p)$ etc. read "the Leading Power Product of p", "the Leading Coefficient of p", "the Leading Monomial of p", "the Remaining part of p", "the part of p Higher than t", "the part of p Lower than t", "the part of p Between t_1 and t_2", respectively.)

Of course, for any p and $t_1 \succ t_2$,

$$p = \mathrm{H}(p, t_1) + \mathrm{C}(p, t_1) \cdot t_1 + \mathrm{B}(p, t_1, t_2) + \mathrm{C}(p, t_2) \cdot t_2 + \mathrm{L}(p, t_2).$$

2.5 Admissible Orderings on Polynomials

Any admissible ordering $\prec$ on power products can be extended to a partial ordering on polynomials in the following way.

Definition (Extension of Admissible Ordering): Let $\prec$ be an admissible ordering on $\mathbf{T}$.

$$p \prec q :\Longleftrightarrow \exists t \; (\mathrm{H}_{\prec}(p,t) = \mathrm{H}_{\prec}(q,t) \wedge t \notin \mathrm{S}(p) \wedge t \in \mathrm{S}(q)). \; \square$$

In general, the extension of $\prec$ is not any more a total ordering. However, it is Noetherian, which is important for our algorithmic perspective:

Proposition (Properties of Admissible Orderings): Let $\prec$ be the extension of an admissible ordering to $\mathbf{P}$. Then,

$\prec$ is a partial ordering,

$\prec$ is Noetherian,

$\forall p \neq 0 \; (\mathrm{p} \succ 0 \;).$

Proof: Easy from the definitions. For proving Noetherianity of $\prec$ on $\mathbf{P}$, one uses Noetherian induction w.r.t. $\prec$ on $\mathbf{T}$.

2.6 Reduction Modulo Polynomials

From now on, let an admissible ordering $\prec$ on $\mathbf{T}$ (and, hence, on $\mathbf{P}$) be fixed. We will now define a binary relation "reduction modulo a set F of polynomials" and a corresponding reduction algorithm that reduces a given "reducible" polynomial, modulo F to a polynomial which is smaller w.r.t. $\prec$. It will turn out that the reflexive, symmetric, transitive closure of this reduction relation is identical to ideal congruence modulo F. However, reduction brings in an algorithmic flavor. Reduction modulo F can also be viewed as a sort of generalized polynomial division with respect to divisors in F.

Definition (Reduction Modulo Polynomials):

$$
\begin{aligned}
g \to_{f,t} h \;&:\Longleftrightarrow\; t \in \mathrm{S}(g) \wedge \mathrm{LPP}(f) \,|\, t \wedge h = g - (\mathrm{M}(g,t)/\mathrm{LM}(f)) \cdot f).\\
g \to_{f} h \;&:\Longleftrightarrow\; \exists t \in \mathrm{S}(g)\ (g \to_{f,t} h).\\
g \to_{F} h \;&:\Longleftrightarrow\; \exists f \in F\ (g \to_{f} h).\\
\underline{g}_F \;&:\Longleftrightarrow\; \neg \exists h\ (g \to_F h).
\end{aligned}
$$

(For $g \to_{f,t} h$, $g \to_f h$, $g \to_F h$ and $\underline{g}_F$ read "g reduces to h modulo f using t","g reduces to h modulo f", "g reduces to h modulo F" and "g is reduced modulo F", respectively.)

Example: Let $\prec$ be the total degree ordering defined by $x \prec y$ and let $g := x^2y^3 + 3xy^2 - 5x$, $f_1 := xy - 2y$, $f_2 := 2y^2 - x^2$. Then, for example

$$g \to_{f_1, xy^2} h_1 := g - (3xy^2/xy) \cdot f_1 = x^2y^3 + 6y^2 - 5x$$

but also

$$g \to_{f_1, x^2y^3} h_2 := g - (x^2y^3/xy) \cdot f_1 = 2xy^3 + 3xy^2 - 5x$$

and also

$$g \to_{f_2, x^2y^3} h_3 := g - (x^2y^3/2y^2) \cdot f_2 = 1/2x^4y + 3xy^2 - 5x.$$

We will now show that reduction is Noetherian, which is important for obtaining an algorithm that computes reduced polynomials modulo a polynomial set F.

Proposition (Noetherianity of Reduction Modulo Polynomials):

$$g \to_F h \Longrightarrow g \succ h.$$

$\to_F$ is Noetherian.

Proof: If $g \to_{f,t} h$, then $h = \mathrm{H}(g,t) + 0 \cdot t + r$, where $r := \mathrm{L}(g,t) - (\mathrm{M}(g,t)/\mathrm{LM}(f)) \cdot R(f)$. By $|$-compatibility and monotonicity of $\prec$, $\mathrm{LPP}(r) \prec t$. Hence, $\mathrm{H}(g,t) = \mathrm{H}(h,t)$. Furthermore, $t \in \mathrm{S}(g)$ but $t \notin \mathrm{S}(h)$ and, thus, $g \succ h$. Now, since $\prec$ is Noetherian, also $\to_F$ must be Noetherian.

Let $\overset{*}{\to}_F$ be the reflexive and transitive closure of $\to_F$. By the Noetherianity of $\to_F$ and by the fact that the existence and selection of suitable t and f in the definition of $\to_F$ can of course be handled algorithmically, "by iteration of this selection process", we can easily design an algorithm "RF" with the property stated in the following proposition. We omit the straight-forward details of this algorithm.

Proposition (Property of Reduction Algorithm):

$$g \to_F^* \mathrm{RF}(F,g),$$

$$\underline{\mathrm{RF}(F,g)}_F.$$

(For $\mathrm{RF}(F,g)$ read "a Reduced Form of g modulo F".) □

In fact, we can get out more information from these iterated selection steps. Namely, we can collect the appropriate multiples of the polynomials in F that were selected in the individual reduction steps so that, at termination of RF, we will also have accumulated polynomial "cofactors" available such that $\mathrm{RF}(F,g)$ can be represented as g plus a linear combination of the cofactors with the polynomials in F. More formally, we have an algorithm "Cofactors" that satisfies the following property:

Proposition (Property of Cofactor Algorithm):

$$\mathrm{RF}(F,g) = g + \sum_{f \in F} \mathrm{Cofactors}(F,g)_f \cdot f.$$

(For $\mathrm{Cofactors}(F,g)$ read "the cofactors of the reduced form of g modulo F".) □

Reduction modulo polynomials has a couple of useful elementary properties that will play a crucial role in Gröbner bases construction.

Proposition (Compatibility of Reduction):

$$a \neq 0 \wedge f_1 = a \cdot f_2 \implies \to_{f_1} = \to_{f_2}, \qquad \text{(monicity)}$$

$$g \to_f h \implies a \cdot t \cdot g \to_f a \cdot t \cdot h, \qquad \text{(product compatibility)}$$

$$g \to_f h \implies \exists q\,(g + p \to_f^* q \leftarrow_f^* h + p). \qquad \text{(sum semi-compatibility)}$$

Proof: Monicity and product compatibility are straight-forward from the definitions. Because of monicity, in the sequel we will be able to restrict our considerations to monic polynomials f, i.e. polynomials whose leading coefficient is 1. This will make the presentation slightly simpler.

Now assume that $g \to_{f,t} h$, and consider an arbitrary p. Define $u := t/\mathrm{LPP}(f)$. Of course, $h = g - \mathrm{C}(g,t) \cdot u \cdot f$. Now

$$\begin{array}{rllll} f = & \mathrm{LPP}(f) + & \mathrm{R}(f), & & \\ g = & \mathrm{H}(g,t) + & \mathrm{C}(g,t) \cdot t & + \mathrm{L}(g,t), & \\ h = & \mathrm{H}(g,t) + & 0 \cdot t & + \mathrm{L}(g,t) & - \mathrm{C}(g,t) \cdot u \cdot \mathrm{R}(f), \\ p = & \mathrm{H}(p,t) + & \mathrm{C}(p,t) \cdot t & + \mathrm{L}(p,t). & \end{array}$$

We now have three cases:

Case $C(p,t) = 0$: In this case, $g + p \rightarrow_{f,t} g + p - C(g,t) \cdot u \cdot f = h + p$.

Case $0 \neq C(p,t) = -C(g,t)$: In this case, $g+p = h+p-C(p,t)\cdot u\cdot f \leftarrow_{f,t} h+p$.

Case $0 \neq C(p,t) \neq -C(g,t)$: In this case,

$$g+p \rightarrow_{f,t} g+p-(C(g,t)+C(p,t))\cdot u\cdot f = h+p-C(p,t)\cdot u\cdot f \leftarrow_{f,t} h+p.\Box$$

Note that sum compatibility, i.e.

$$g \rightarrow_f h \Longrightarrow g + p \rightarrow_f h + p,$$

does not hold in general. This fact is the reason for additional technical difficulties encountered in the proof of the main theorem for Gröbner bases (in comparison with analogous situations in general rewriting).

Congruence and reduction are intimately related. For expressing the relation, let now $\longleftrightarrow_F^*$ denote the reflexive, symmetric, and transitive closure of $\rightarrow_F$.

Proposition (Relation Between Reduction and Congruence):

$$g \equiv_F h \Longleftrightarrow g \longleftrightarrow_F^* h.$$

Proof: "$\Longleftarrow$": This direction is easy. Just notice that, if $g \rightarrow_f h$, then h results from g by subtracting a multiple of f.
"$\Longrightarrow$": For this direction we observe, first, that $g \equiv_F h$ implies that, for certain $a \in \mathbf{K}^*$, $t \in \mathbf{T}^*$, $f \in F^*$ with $|a| = |t| = |F|$,

$$g = h + \sum_{1 \leq i \leq |a|} a_i \cdot t_i \cdot f_i.$$

Now one can proceed by induction on $|a|$ using sum semi-compatibility. $\Box$

(If you have the feeling that the direction "$\Longrightarrow$" in the above proposition, and sum semi-compatibility, is trivial then you should better go back to the definition of $\rightarrow_F$ and study it carefully. The point is that $g \rightarrow_f h$ entails that $h = g + a \cdot u \cdot f$ for some a and u with the additional property that a power product $t \in S(g)$ is "cancelled" and $g \succ h$. In contrast, if $h = g + a \cdot u \cdot f$ for some general a and u, we cannot conclude at all that any cancellation takes place nor that $g \succ h$. Therefore the above lemma is non-trivial.)

The above relation does not yet help us deciding whether or not $g \equiv_F h$. We were much better off if we could prove, for example,

$$g \equiv_F h \iff \exists p\, (g \rightarrow_F^* p \leftarrow_F^* h)$$

because this equivalence would allow a "directed" search for an appropriate p in order to decide whether or not $g \equiv_F h$.

However, this equivalence is not true in general.

Now, those sets F for which the above property is true are called *Gröbner bases*. Those are the sets for which $g \longleftrightarrow_F^* h$ (and, hence, $g \equiv_F h$) can be decided by a directed search. Fortunately, we will be able to prove that any F that does not satisfy the above property can be transformed (algorithmically) into an "equivalent" Gröbner basis G, i.e. into a Gröbner basis G for which $\equiv_G = \equiv_F$. This will provide a uniform methodology for tackling quite a few fundamental problems in polynomial ideal theory by structurally simple algorithms.

Before we go into the details of this program, we will summarize a few fundamental properties of general reduction relations that do not depend on the special context of polynomials.

2.7 Some General Properties of Noetherian Reduction Relations

In this subsection, the variables x, y, z, w range over an arbitrary set X. Let $\rightarrow$ be a binary relation on X. Let $\longleftrightarrow$, $\rightarrow^*$ and $\longleftrightarrow^*$ denote the symmetric, the reflexive-transitive, and the reflexive-symmetric-transitive closure of $\rightarrow$, respectively. Furthermore, $\underline{x} :\Longleftrightarrow \neg\exists y\,(x \rightarrow y)$. Furthermore, let NF be a function on X such that $\forall x\,(x \rightarrow^* \mathrm{NF}(x))$ and $\forall x\,(\underline{\mathrm{NF}(x)})$. For "NF($x$)" read "the Normal Form of x produced by $\rightarrow$". Finally, $x \downarrow^* y :\Longleftrightarrow \exists z\,(x \rightarrow^* z \leftarrow^* y)$. (For $x \downarrow^* y$ read "x and y have a common $\rightarrow$ successor".)

Proposition (Church-Rosser Property, Confluence, and Local Confluence): Let $\rightarrow$ be Noetherian. Then the following properties are equivalent:

$\forall x, y\,(x \longleftrightarrow^* y \Longrightarrow x \downarrow^* y)$, (Church-Rosser property)

$\forall x, y\,(x \longleftrightarrow^* y \Longrightarrow \mathrm{NF}(x) = \mathrm{NF}(y))$, (Church-Rosser normal form property)

$\forall x, y, z\,(x \leftarrow^* z \rightarrow^* y \Longrightarrow x \downarrow^* y)$, (confluence)

$\forall x, y, z\,(x \leftarrow z \rightarrow y \Longrightarrow x \downarrow^* y)$. (local confluence)

Proof: The equivalence of the first three properties is easy and, of course, confluence implies local confluence. The converse is the so-called Newman-Lemma introduced in (Newman 1942) whose proof, by Noetherian induction, can be found for example in (Becker, Weispfenning 1993). □

The test for checking the Church-Rosser property can be simplified further. For this, let $<$ be a partial ordering on X.

Definition (Connectibility):

$$x \overset{<w}{\longleftrightarrow}^* y \quad :\Longleftrightarrow \quad \exists z \in X^* \\ (x = z_1 < w \wedge \\ \forall 1 \leq i < |z| \, (z_i \longleftrightarrow z_{i+1} < w) \wedge \\ z_{|z|} = y < w).$$

(For $x \overset{<w}{\longleftrightarrow}^* y$ read "x can be connected with y by $\rightarrow$ staying $< w$".)

Proposition (Generalized Newman Lemma): Let $<$ be Noetherian and $\rightarrow \subseteq >$. Then the following two properties are equivalent:

$$\forall x, y, z \, (x \leftarrow z \rightarrow y \Longrightarrow x \downarrow^* y), \qquad \text{(local confluence)}$$

$$\forall x, y, z \, (x \leftarrow z \rightarrow y \Longrightarrow x \overset{<z}{\longleftrightarrow}^* y). \qquad \text{(local connectibility)}$$

Proof: This version of Newman's lemma and its proof, by Noetherian induction, is implicit in (Buchberger 1979). I formulated and proved the lemma explicitly in (Winkler, Buchberger 1983). □

When we apply the above proposition to the case of reductions modulo polynomial sets F, we see that "$g \longleftrightarrow_F^* h$" and, hence, "$g \equiv_F h$" could be decided algorithmically by checking whether or not $\mathrm{RF}(F, g) = \mathrm{RF}(F, h)$ if $\rightarrow_F$ had the Church-Rosser property. This motivates the following definition of the concept of *Gröbner basis*.

2.8 Gröbner Bases

Definition (Gröbner Basis):

F is a Gröbner basis $:\Longleftrightarrow$ $\rightarrow_F$ has the Church-Rosser property (i.e. $\forall g, h \, (g \longleftrightarrow_F^* h \Longrightarrow g \downarrow_F^* h)$. □

It is relatively easy to give an inconstructive proof that

$$\forall F \, \exists G \, (G \text{ is a Gröbner basis and } \longleftrightarrow_F^* = \longleftrightarrow_G^*).$$

However, what we want is an *algorithm* that constructs G from F. For this we first try to develop an algorithmic test for deciding whether or not a given F *is* a Gröbner basis. By Newman's lemma, for this it is sufficient to test, for all g, h, and q with $g \leftarrow_F q \rightarrow_F h$, whether or not there exists a p such that $g \rightarrow_F^* p \leftarrow_F^* h$. However, still, this requires infinitely many tests.

The crucial idea in Gröbner basis theory is the observation that these infinitely many tests can be replaced by the consideration of finitely many "critical situations" that can be characterized by the so-called "S-polynomials" of F.

2.9 S-Polynomials

Definition (S-Polynomials): Let f_1, f_2 be monic polynomials. Then,

$$\mathrm{SP}(f_1, f_2) := u_1 \cdot f_1 - u_2 \cdot f_2,$$

where $u_1 := w/\mathrm{LPP}(f_1)$, $u_2 := w/\mathrm{LPP}(f_2)$, $w := \mathrm{LCM}(\mathrm{LPP}(f_1), \mathrm{LPP}(f_2))$. (For $\mathrm{SP}(f_1, f_2)$ read "the S-polynomials of f_1 and f_2".) □

The intuition behind considering S-polynomials is that LCM($\mathrm{LPP}(f_1)$, $\mathrm{LPP}(f_2)$) is the first power product (in the divisibility ordering) that can be reduced both modulo f_1 and modulo f_2, i.e. where reduction may "diverge" and, hence, an injury of the Church-Rosser property may occur. In the proof of the main theorem we will then see the surprising fact that, fortunately, if for a given (finite) set F no divergence is detected at any of the finitely many $\mathrm{LCM}(\mathrm{LPP}(f_1), \mathrm{LPP}(f_2))$ $(f_1, f_2 \in F)$ then no divergence can occur at any point in the infinite "reduction graph" of $\rightarrow_F$, i.e. $\rightarrow_F$ has the Church-Rosser property or, in other words F is a Gröbner basis. Thus, by considering S-polynomials we obtain a finite algorithmic check for testing whether or not a given F is a Gröbner basis.

2.10 The Main Theorem: Algorithmic Characterization of Gröbner Bases by S-Polynomials

In the sequel, by the property (monicity), we may assume without loss of generality that all polynomials in F are monic.

Theorem (Main Theorem of Gröbner Basis Theory):

$$F \text{ is a Gröbner basis} \iff \forall f_1, f_2 \in F\, (\mathrm{RF}(F, \mathrm{SP}(f_1, f_2)) = 0).$$

Proof: The direction "$\Longrightarrow$" is easy. Namely, if $f_1, f_2 \in F$, then $\mathrm{SP}(f_1, f_2) \in \mathrm{Ideal}(F)$, i.e. $\mathrm{SP}(f_1, f_2) \equiv_F 0$. By the relation between reduction and congruence this implies that $\mathrm{SP}(f_1, f_2) \longleftrightarrow^*_F 0$. Hence, $\mathrm{RF}(\ F,\ \mathrm{SP}(f_1, f_2)\ ) = \mathrm{RF}(F, 0) = 0$ because F is a Gröbner basis (and because of the equivalence between the Church-Rosser property and the normal form Church-Rosser property).

For the direction "$\Longleftarrow$", by the generalized Newman lemma and the fact that $\rightarrow_F \subseteq \succ$, it suffices to prove local connectibility, i.e. it suffices to prove that under the assumption

$$g_1 \leftarrow_F h \rightarrow_F g_2$$

we have

$$g_1 \overset{\prec h}{\longleftrightarrow}{}^*_F\, g_2.$$

By the assumption, there exist $f_1, f_2 \in F$ and $t_1, t_2 \in \mathrm{S}(h)$ with $\mathrm{LPP}(f_1) \mid t_1$ and $\mathrm{LPP}(f_2) \mid t_2$, such that $h \rightarrow_{f_1,t_1} g_1$ and $h \rightarrow_{f_2,t_2} g_2$.

Now we have three cases.

Case $t_1 \succ t_2$: In this case,

$$g_1 = \mathrm{H}(h,t_1) + 0 \cdot t_1 + \mathrm{B}(h,t_1,t_2) + \mathrm{C}(h,t_2) \cdot t_2 + \mathrm{L}(h,t_2) - \\ -\mathrm{C}(h,t_1) \cdot u_1 \cdot \mathrm{R}(f_1)$$

and

$$g_2 = \mathrm{H}(h,t_1) + \mathrm{C}(h,t_1) \cdot t_1 + \mathrm{B}(h,t_1,t_2) + 0 \cdot t_2 + \mathrm{L}(h,t_2) - \\ -\mathrm{C}(h,t_2) \cdot u_2 \cdot \mathrm{R}(f_2),$$

where $u_1 := t_1/\mathrm{LPP}(f_1)$, $u_2 := t_2/\mathrm{LPP}(f_2)$.

Furthermore,

$$g_2 \rightarrow_{f_1} g_{1,2} := \mathrm{H}(h,t_1) + 0 \cdot t_1 + \mathrm{B}(h,t_1,t_2) + 0 \cdot t_2 + \mathrm{L}(h,t_2) - \\ -\mathrm{C}(h,t_1) \cdot u_1 \cdot \mathrm{R}(f_1) - \\ -\mathrm{C}(h,t_2) \cdot u_2 \cdot \mathrm{R}(f_2).$$

Now, $g_1 = h - \mathrm{C}(h,t_1) \cdot u_1 \cdot f_1$ and $g_{1,2} = g_2 - \mathrm{C}(h,t_1) \cdot u_1 \cdot f_1$ and, by assumption, $h \rightarrow_F g_2$. Hence, by sum semi-compatibility, $g_1 \downarrow_F^* g_{1,2}$ and, hence, $g_1 \overset{\prec h}{\longleftrightarrow}{}^*_F\, g_2$. (Note that, in general, $g_1 \rightarrow_{f_1} g_{1,2}$ need not be the case. Why not?)

Case $t_1 \prec t_2$: Analogous.

Case $t := t_1 = t_2$: In this case,

$$g_1 = \mathrm{H}(h,t) + 0 \cdot t + \mathrm{L}(h,t) - \\ -\mathrm{C}(h,t) \cdot u_1 \cdot \mathrm{R}(f_1)$$

and

$$g_2 = \mathrm{H}(h,t) + 0 \cdot t + \mathrm{L}(h,t) - \\ -\mathrm{C}(h,t) \cdot u_2 \cdot \mathrm{R}(f_2).$$

Hence,

$$\begin{aligned} g_1 - g_2 &= -\mathrm{C}(h,t) \cdot (u_1 \cdot R(f_1) - u_2 \cdot R(f_2)) = \\ &= -\mathrm{C}(h,t) \cdot (u_1 \cdot f_1 - u_2 \cdot f_2) = -\mathrm{C}(h,t) \cdot v \cdot SP(f_1, f_2), \end{aligned}$$

where $v := t/\ \mathrm{LCM}(\ \mathrm{LPP}(f_1),\mathrm{LPP}(f_2)\ )$.

We have assumed that RF(F,SP(f_1, f_2)) $= 0$, i.e. SP(f_1, f_2) $\rightarrow_F^* 0$. Hence, by product compatibility, $g_1 - g_2 = -\mathrm{C}(h,t) \cdot v \cdot SP(f_1, f_2) \rightarrow_F^* 0$. This means that there exists a sequence $p \in \mathbf{P}^*$ such that

$$p_1 = g_1 - g_2,$$
$$\forall 1 \leq i < |p| \; (p_i \rightarrow_F p_{i+1}), \tag{*}$$

and

$$p_{|p|} = 0.$$

Furthermore note that, because of $\rightarrow_F \subseteq \succ$,

$$\forall 1 \leq i \leq |p| \; (p_i \preceq g_1 - g_2 \prec h).$$

Thus, by sum semi-compatibility applied to (*),

$$g_1 = p_1 + g_2,$$
$$\forall 1 \leq i < |p| \; (p_i + g_2 \downarrow_F^* p_{i+1} + g_2),$$
$$g_2 = p_{|p|} + g_2.$$

Also, we have

$$\forall 1 \leq i \leq |p| \; (p_i + g_2 \prec h)$$

because

$$\forall 1 \leq i \leq |p| \; (\mathrm{H}(p_i + g_2, t) = \mathrm{H}(h,t) \wedge \mathrm{C}(p_i + g_2, t) = 0).$$

Thus, summarizing, $g_1 \overset{\prec h}{\longleftrightarrow}{}^*_F \, g_2$ also in this case.

2.11 An Algorithm for Constructing Gröbner Bases

The main theorem can immediately be read as an algorithm for testing whether or not a given *finite* F is a Gröbner basis. However, it can also be used to show that the following, structurally simple, algorithm "Gröbner-Basis" (formulated here in a functional style) meets the specification given in the proposition below.

Algorithm (Construction of a Gröbner Basis):

> Gröbner-Basis(F) := GB(F, $\{\{f_1, f_2\} \mid f_1, f_2 \in F\}$).
>
> GB(F, $\emptyset$) := F,
>
> GB(F, $\{f_1, f_2\} \smile B$) :=
>
>> GB(F, B), if $h = 0$,
>> GB($F \cup \{h\}$, $B \cup \{\{h, f\} \mid f \in F\}$), otherwise,
>> where h := RF(F, SP(f_1, f_2)).□

(Here, $\smile$ is a constructor for finite sets such that $x \smile X = Y$ iff $\{x\} \cup X = Y$ and $x \notin X$.)

Proposition (Correctness of the Gröbner-Basis Algorithm): For finite sets F:

Gröbner-Basis(F) is a Gröbner basis,

Ideal(F) = Ideal(Gröbner-Basis(F)).

Proof (Sketch): (The algorithm always terminates by Dickson's lemma because the leading power product of a polynomial h that gets adjoined to an intermediate set F is not a multiple of the leading power product of any $f \in F$.) The final set F clearly satisfies the condition in the main theorem and, hence, is a Gröbner basis. Furthermore, any h that gets adjoined to an intermediate F is in Ideal(F) and hence in the ideal generated by the input set. □

Gröbner bases can be made unique by "interreducing" them:

Definition (Reduced Gröbner Bases):

$$F \text{ is a reduced Gröbner basis} \quad :\Longleftrightarrow \quad F \text{ is a (monic) Gröbner basis, } \forall f \in F\, (\underline{f}_{(F-\{f\})}). \ \square$$

The above algorithm can be easily converted into an algorithm "Reduced-Gröbner-Basis" that yields a reduced Gröbner basis and it is easy to show that these bases are unique in the following sense:

Proposition (Canonicality):

$$\text{Ideal}(F) = \text{Ideal}(G) \Longrightarrow$$

$$\text{Reduced-Gröbner-Basis}(F) = \text{Reduced-Gröbner-Basis}(G). \ \square$$

The above crude form of the algorithm can be made much more efficient by introducing "criteria" by which one can predict that certain S-polynomials reduce to zero without actually carrying out the reduction. This idea was introduced in (Buchberger 1979) and is an easy consequence of the main theorem in the form given above using our generalized Newman lemma. The strategy of using "criteria" later was carried over and proved useful also in the Knuth-Bendix completion algorithm in the area of rewriting.

In fact, the above algorithms Gröbner-Basis and Reduced-Gröbner-Basis should be indexed by one more parameter, namely the admissible ordering $\prec$ used. When, in the applications below, indication of the ordering used is important we will therefore write "Gröbner-Basis$_\prec$" etc.

2.12 Other Characterizations of Gröbner Bases

Gröbner bases can also be characterized by quite a few other properties, see the text books on the subject. The equivalence proofs are quite simple except the one that relates Gröbner bases F to "syzygies", i.e. to the solutions of linear diophantine equations with coefficients from F.

3 Applications of Gröbner Bases

3.1 Overview

Over the years, literally dozens of applications have been found for the Gröbner bases algorithm, see the recent text books, for example (Becker, Weispfenning 1993), (Cox et al. 1992) and the papers on applications in this book. In this section, we only briefly summarize the most fundamental problems to which the Gröbner bases algorithm can be applied.

We present most of these problems and their algorithmic solutions by formulating a mathematical theorem whose left-hand side defines a problem and whose right-hand describes the algorithmic solution. The proofs of most of these theorems are easy consequences of the main theorem and the various equivalent forms of it. Some of the applications need additional knowledge for which we refer to the text books.

3.2 Ideal Membership, Canonical Simplification, Ideal Identity

$f \in \text{Ideal}(F) \Longleftrightarrow \text{RF}(\text{Gröbner-Basis}(F), f) = 0.$
(This problem is sometimes called the "main problem of polynomial ideal theory".)

$f \equiv_F g \Longleftrightarrow \text{RF}(\text{Gröbner-Basis}(F), f) = \text{RF}(\text{Gröbner-Basis}(F), g),$

$f \equiv_F \text{RF}(\text{Gröbner-Basis}(F), f).$

(The last two properties show that "RF" modulo a Gröbner-Basis(F) is a canonical simplifier for $\equiv_F$. For the notion of "canonical simplifier" see (Buchberger, Loos 1982).)

$\text{Ideal}(F) \subseteq \text{Ideal}(G) \Longleftrightarrow \forall f \in F\, (\text{RF}(\text{Gröbner-Basis}(G), f) = 0).$

$\text{Ideal}(F) = \text{Ideal}(G) \Longleftrightarrow$
$\text{Reduced-Gröbner-Basis}(F) = \text{Reduced-Gröbner-Basis}(G).$

Ideal(F) = Ideal(Reduced-Gröbner-Basis(F)).

(The last two properties show that "Reduced-Gröbner-Basis" is a canonical simplifier for the equivalence $\sim$ defined by $F \sim G :\Longleftrightarrow$ Ideal(F) = Ideal(G) on the set of subsets of $\mathbf{P}$.)

3.3 Radical Membership

f $\in$ Radical(F) $\Longleftrightarrow$ $1 \in$ Gröbner-Basis($F \cup \{y \cdot f - 1\}$) (where y is a new indeterminate). □

(On this test, a systematic method for deciding a large class of geometrical propositions can be based, see the example in the first section. It is much harder to compute, by Gröbner bases, a basis for the radical of an ideal because one must have a means of factorization in extension fields. Even harder is the determination of a complete "primary decomposition" of an ideal that, roughly, corresponds to a decomposition of algebraic varieties into irreducible varieties, see (Becker, Weispfenning 1993).)

3.4 Computation in Residue Class Rings Modulo Ideals

Let $(\mathbf{P}_F, +_F, 0_F, -_F, \cdot_F, 1_F)$ be the residue class ring of $(\mathbf{P}, +, 0, -, \cdot, 1)$ modulo Ideal(F).

Let $\underline{F}$:= Gröbner-Basis(F) and $(\underline{\mathbf{P}}, \underline{+}, \underline{0}, \underline{-}, \underline{\cdot}, \underline{1})$ be the following structure:

$$\begin{aligned}
\underline{\mathbf{P}} &:= \{f \in \mathbf{P} | \underline{f}_{\underline{F}}\}, \\
f \underline{+} g &:= \mathrm{RF}(\underline{F}, f + g), \\
\underline{0} &= 0, \\
\underline{-} f &:= \mathrm{RF}(\underline{F}, -f), \\
f \underline{\cdot} g &:= \mathrm{RF}(\underline{F}, f \cdot g), \\
\underline{1} &= 1,
\end{aligned}$$

Then $(\mathbf{P}_F, +_F, 0_F, -_F, \cdot_F, 1_F)$ is isomorphic to $(\underline{\mathbf{P}}, \underline{+}, \underline{0}, \underline{-}, \underline{\cdot}, \underline{1})$ by the following isomorphism i:

$$i(f) := \text{ the residue class of } f \text{ modulo Ideal}(F) \text{ (for all} f \in \underline{\mathbf{P}}).$$

$\mathbf{B} := \{ t \in \mathbf{T} \mid \neg\exists f \in \underline{F} \text{ (LPP}(f) \,|\, t) \}$ is a linearly independent basis for the vector space $(\mathbf{K}, \underline{\mathbf{P}}, \underline{+}, \underline{0}, \underline{-}, \underline{\cdot})$.

The following elements in $\mathbf{K}$

$$\mathrm{SC}_{t,u,v} := \mathrm{C}(\mathrm{RF}(\underline{F}, t \cdot u), v),$$

are the "structure constants" of the associative algebra $(\mathbf{K}, \underline{\mathbf{P}}, \underline{+}, \underline{0}, \underline{-}, \underline{\cdot})$, i.e., for all $t, u, v \in \mathbf{B}$,

$$t \underline{\cdot} u = \sum_{v \in \mathbf{B}} \mathbf{SC}_{t,u,v} \cdot v.$$

In other words, having computed a Gröbner basis $\underline{F}$ for a given F, one can master arithmetics in the residue class ring modulo Ideal(F) completely algorithmically. On this method, one can also base algorithms for computing in algebraic extension fields which can be considered as residue class rings modulo polynomial ideals.

3.5 Leading Power Products

$\{\text{LPP}(f) | \ f \in \text{Ideal}(F)\} = \{u \cdot \text{LPP}(f) \ | \ u \in \mathbf{T} \wedge f \in \text{Gröbner-Basis}(F)\}$.

Ideal(F) is a principal ideal $\Longleftrightarrow$ Reduced-Gröbner-Basis(F) has exactly one element.

3.6 Polynomial Equations

$$\begin{aligned} \text{Ideal}(F) = \mathbf{P} \quad &\Longleftrightarrow \quad 1 \in \text{Gröbner-Basis}(F) \\ &\Longleftrightarrow \quad \text{Reduced-Gröbner-Basis}(F) = \{\ 1\}. \end{aligned}$$

Let $\mathbf{K}$ be algebraically closed:
F is solvable (i.e. has a solution in $\mathbf{K}$) $\Longleftrightarrow 1 \notin$ Gröbner-Basis(F).

F has only finitely many solutions $\Longleftrightarrow$
$\forall 1 \le i \le n \ \exists f \in \text{Gröbner-Basis}(F) \ \exists j \ (\ \text{LPP}(f) = x_i^j\ )$.

For all F with finitely many solutions:
the number of solutions of F (counting multiplicities) =
$|\ \{t \in \mathbf{T} \mid \neg \exists f \in \text{Gröbner-Basis}(F)\ (\ \text{LPP}(f) \,|\, t\ )\ \}\ |$.

Let $U \subset \mathbf{T}$ be finite:
$\exists$ f $\in$ Ideal(F) (S(f) $\subseteq U$) $\Longleftrightarrow$
$\{$ RF(Gröbner-Basis(F), u) $|\ u \in U\}$ is linearly dependent over $\mathbf{K}$. □

By applying this property successively to the powers $1, x, x^2, x^3, \ldots$, one can algorithmically find, for example, the univariate polynomial in x of minimal degree in Ideal(F) if it exists. Such a polynomial exists iff F has only finitely many solutions, which can be checked algorithmically by the above method. On this algorithm a general method for solving arbitrary systems of polynomial equations can be based, see (Buchberger 1970), which works for arbitrary term orderings $\prec$ whereas the elimination method mentioned below works only for elimination orderings. Also, this algorithm can be used for transforming a Gröbner basis w.r.t. a given admissible ordering $\prec_1$ into a Gröbner basis w.r.t. $\prec_2$. This is sometimes helpful because the complexity of Gröbner bases computation depends strongly on the term ordering chosen. Basically this idea is used in the recent, most advanced, version of the Gröbner bases algorithm for improving efficiency, see (Collart et al.).

3.7 Linear Syzygies

Let $F \subseteq \mathbf{P}$ be finite and $h : F \to \mathbf{P}$ (i.e. a sequence of polynomials indexed by F).

> h is a (linear) syzygy for $(F, g) :\Longleftrightarrow \sum_{f \in F} h_f \cdot f = g$.
> (For "h is a linear syzygy for (F, g)" we also say "h is a solution of the inhomogeneous diophantine equation given by F and g.").
>
> h is a (linear) syzygy for $F :\Longleftrightarrow h$ is a syzygy for $(F, 0)$ (i.e. $\sum_{f \in F} h_f \cdot f = 0$).
> (For "h is a linear syzygy for F" we also say "h is a solution of the homogeneous diophantine equation given by F.").

Let F be a Gröbner basis:

> The diophantine equation given by (F, g) is solvable (i.e. $\exists h$ (h is a syzygy for (F, g)) $\Longleftrightarrow$ RF$(F, g) = 0$.
>
> The diophantine equation given by (F, g) is solvable $\Longrightarrow$ Cofactors(F, g) is a syzygy for (F, g).

The following set of sequences is a finite basis for the infinite module of all solutions of the homogeneous diophantine equation given by F:

$$\{S^{f,g} \mid f, g \in F\},$$

where, for arbitrary $f, g \in F$,

$$S^{f,g} : F \to \mathbf{P},$$

$$S_f^{f,g} := u - P_f^{f,g},$$

$$S_g^{f,g} := -v - P_g^{f,g}$$

$$S_h^{f,g} := -P_h^{f,g}, \text{ if } h \in F - \{f, g\},$$

$$u := \mathrm{LCM}(\mathrm{LPP}(f), \mathrm{LPP}(g))/\mathrm{LPP}(f),$$

$v := \mathrm{LCM}(\mathrm{LPP}(f), \mathrm{LPP}(g))/\mathrm{LPP}(g)$,

$P^{f,g} := \mathrm{Cofactors}(F, \mathrm{SP}(f,g))$.□

Thus, essentially, the reduction of all the S-polynomials of F establishes a finite basis for the module of homogeneous syzygies. By adding one solution of the inhomogeneous equations one obtains all the solutions of the inhomogeous diophantine equation.

For obtaining a basis for the syzygies for a diophantine equation with arbitrary F, one first computes $\underline{F}$ = Gröbner-Basis(F). Then one solves the problem for $\underline{F}$ as above. The solutions found can transformed back to solutions for F by multiplication with matrices with polynomial entries that can be obtained from expressing $\underline{F}$ in terms of F and F in terms of $\underline{F}$ using the Cofactor algorithm.

Systems of linear diophantine equations can be handled by reducing the problem recursively to the case of just one equation.

3.8 Hilbert Functions

Let the term ordering $\prec$ be a total degree ordering. The "Hilbert function" H is defined as follows:

H(d, F) := the number of modulo Ideal(F) linearly independent polynomials f with Degree(f) $\leq d$.

(The Hilbert function is important because it allows to read off various structural information on the variety of F.)

Now,

$$\mathrm{H}(d, F) = \binom{d+n}{n} -$$

$$-|\{u \in \mathbf{T} \mid \mathrm{Degree}(u) \leq d \wedge \neg\exists f \in \text{Gröbner-Basis}(F)\,(\mathrm{LPP}(f) \mid u)\}|.$$

3.9 Elimination Ideals

Let $\prec$ be the lexical term ordering defined by $x_1 \prec x_2 \prec \ldots \prec x_n$. Then,

Gröbner-Basis$_\prec(F) \cap \mathbf{K}[x_1, \ldots, x_i]$ is a Gröbner basis for (Ideal(F) $\cap\, \mathbf{K}[x_1, \ldots, x_i]$).

Reduced-Gröbner-Basis$_\prec(F) \cap \mathbf{K}[x_1, \ldots, x_i]$ =

= Reduced-Gröbner-Basis$_\prec$(Ideal(F) $\cap\, \mathbf{K}[x_1, \ldots, x_i]$).□

This property leads immediately to a general solution method, by "successive substitution", for arbitrary systems of polynomial equations with finitely many solutions, see the example in the first lecture.

Let $\prec$ be a term ordering in which power products containing no other indeterminates except $x_{i_1}, \ldots, x_{i_m}$ are $\prec$ then all the other power products.

$$\mathrm{Ideal}(F) \cap \mathbf{K}[x_{i_1}, \ldots, x_{i_m}] = \emptyset \iff$$
$$\text{Gröbner-Basis}_{\prec}(F) \cap \mathbf{K}[x_{i_1}, \ldots, x_{i_m}] = \emptyset. \square$$

(In case $\mathrm{Ideal}(F) \cap \mathbf{K}[x_{i_1}, \ldots, x_{i_m}] = \emptyset$ one says that the indeterminates x_{i_1}, $\ldots$, x_{i_m} are algebraically independent modulo Ideal(F). The above criterion for independence yields an algorithm for determining the dimension of Ideal(F), i.e. the maximal number of independent indeterminates modulo Ideal(F).)

3.10 Ideal Operations

Let $\prec$ be an admissible term ordering in which power products containing $x_1, \ldots, x_n$ are $\prec$ than any power product containing the new variable y:

Gröbner-Basis$_{\prec}(\{y \cdot f \mid f \in F\} \cup \{(1-y) \cdot g \mid g \in G\}) \cap \mathbf{K}[x_1, \ldots, x_n]$ is a Gröbner basis for Ideal(F) $\cap$ Ideal(G).

Reduced-Gröbner-Basis$_{\prec}$(Ideal(F) $\cap$ Ideal(G)) =
= Reduced-Gröbner-Basis$_{\prec}$($\{y \cdot f \mid f \in F\} \cup \{(1-y) \cdot g \mid g \in G\}$)
$\cap\ \mathbf{K}[x_1, \ldots, x_n\ ]$.

(This property yields also an algorithm for quotients of finitely generated ideals because the determination of such quotients can be reduced to the determination of intersections. Alternatively, quotients of ideals can be computed by using the algorithm for linear syzygies.)

3.11 Algebraic Relations and Implicitization

Let $F = \{f_1, \ldots, f_m\} \subseteq \mathbf{K}[x_1, \ldots, x_n]$, let $y_1, \ldots, y_m$ be new indeterminates and let $\prec$ be an admissible ordering in which power products containing only the y_i are $\prec$ any other power product. Then,

Gröbner-Basis$_{\prec}(\{y_1 - f_1, \ldots, y_m - f_m\}) \cap \mathbf{K}[y_1, \ldots, y_m]$ is a Gröbner basis for $\{g \in \mathbf{K}[y_1, \ldots, y_m] \mid g(f_1, \ldots, f_m) = 0\}$.

(The set $\{g \in \mathbf{K}[y_1, \ldots, y_m] \mid g(f_1, \ldots, f_m) = 0\}$ is, in fact, an ideal. It is called the "ideal of algebraic relations (or non-linear syzygies) over F".)

Let $R :=$ Gröbner-Basis$_\prec(\{y_1 - f_1, \ldots, y_m - f_m\}) \cap \mathbf{K}[y_1, \ldots, y_m]$ and let $\mathbf{K}$ be algebraically closed. Then the variety of R is the smallest variety in $\mathbf{K}$ that contains the set

$$\{(f_1(x_1, \ldots, x_n), \ldots, f_m(x_1, \ldots, x_n)) \mid x_1, \ldots, x_n \in \mathbf{K}\}.$$

(This means that R is an "implicit" representation of the set given in "parametric" representation by $f_1, \ldots, f_m$.)

3.12 Inverse Mappings

Let $F = \{f_1, \ldots, f_n\} \subseteq \mathbf{K}[x_1, \ldots, x_n]$, let $y_1, \ldots, y_n$ be new indeterminates and let $\prec$ be an admissible ordering with the property of the previous section. Then,

$$\begin{array}{rcl} \text{the mapping } M : \mathbf{K}^n & \rightarrow & \mathbf{K}^n \text{ is bijective} \\ (a_1, \ldots, a_n) & \mapsto & (f_1(a_1, \ldots, a_n), \ldots, f_n(a_1, \ldots, a_n)) \end{array}$$

$$\Longleftrightarrow$$

$$\begin{array}{c} \text{Reduced-Gröbner-Basis}_\prec(y_1 - f_1, \ldots, y_n - f_n) \text{ has the form} \\ \{x_1 - g_1, \ldots, x_n - g_n\} \text{ with } \{g_1, \ldots, g_n\} \subseteq \mathbf{K}[y_1, \ldots, y_n]. \end{array}$$

(Moreover, the g_i define the mapping which is inverse to M.)

3.13 Miscellaneous

In fact, by combining the above methods, a big number of particular problems in various areas of mathematics have been attacked in the literature. Some of these applications of Gröbner bases are quite unexpected. For example, geometrical theorem proving, integer programming, integration of rational functions, greatest common divisor and factorization of multivariate polynomials, bases for Bezier splines, bases for Runge-Kutta numerical integration formulae, interpretation of resolution theorem proving in boolean rings etc. have been studied successfully using Gröbner bases, see the other tutorial papers in this book.

Acknowledgement: My sincere thanks to Daniela Văsaru and Wolfgang Windsteiger for carefully checking the paper and preparing the Latex version. Some work on the paper was done during a stay of the author at the University of Tsukuba, chair of Professor Tetsuo Ida, in the frame of the TARA project and an invitation by the Japanese Society for the Promotion of Science.

References

An extensive bibliography on Gröbner bases giving credit also to the original papers on the various applications is contained in the book by Becker and Weispfenning (1993). Below we only list those papers that are explicitly referenced in the present paper.

Becker, T., Weispfenning, V. (1993): Gröbner Bases: A Computational Approach to Commutative Algebra. Springer-Verlag, New York

Buchberger, B. (1965): An Algorithm for Finding a Basis for the Residue Class Ring of a Zero-Dimensional Polynomial Ideal (German). PhD Thesis, University of Innsbruck, Institute for Mathematics

Buchberger, B. (1970): An Algorithmic Criterion for the Solvability of Algebraic Systems of Equations (German). Aequationes Mathematicae 4: 374-383

Buchberger, B. (1979): A Criterion for Detecting Unnecessary Reductions in the Construction of Gröbner bases. In: Ng (ed.): Proceedings of the EUROSAM '79 , Springer, pp. 3-21 (Lecture notes in computer science, vol. 72)

Buchberger, B. (1985): Gröbner Bases: An Algorithmic Method in Polynomial Ideal Theory. In: Bose, N. K. (ed): Multidimensional Systems Theory. D. Reidel, Dordrecht, pp. 184-232

Buchberger, B., Elias, J. (1992): Using Gröbner Bases for Detecting Polynomial Identities: A Case Study on Fermat's Ideal. J. Number Theory 41: 272-279

Buchberger, B., Kutzler, B. (1986): Computer-Algebra for the Engineer (German). In: Rechnerorientierte Verfahren, Teubner Verlag, Stuttgart, pp. 11-69

Buchberger, B., Loos, R. (1982): Algebraic Simplification. In: Buchberger, B., Collins, G.E., Loos, R. (eds): Computer Algebra: Symbolic and Algebraic Computation. Springer, Vienna, pp. 11-44

Collart, S., Kalkbrener, M., Mall, D. (1997): Converting Bases with the Gröbner Walk. J. Symb. Comp., to appear.

Cox, D., Little, J., O'Shea, D. (1992): Ideals, Varieties, and Algorithms: An Introduction to Computational Algebraic Geometry and Commutative Algebra. Springer-Verlag, New York

Dickson, L.E. (1913): Finiteness of the Odd Perfect and Primitive Abundant Numbers With n Distinct Prime Factors. American J. Math. 35: 413-422

Elias, J. (1988): On Fermat's Ideal. Technical Report, Univ. Barcelona, Dept. of Algebra and Geometry

Hermann, G. (1926): The Question of Finitely Many Steps in the Theory of Polynomial Ideals (German). Mathematische Annalen 95: 736-788

Newman, M.H.A. (1942): On Theories with a Combinatorial Definition of Equivalence. Annals of Mathematics 43: 233-243

Sturmfels, B. (1993): Algorithms in Invariant Theory. In: Buchberger, B., Collins, G. (eds.): Texts and Monographs in Symbolic Computation. Springer Verlag Wien

Trinks, W. (1978): On Buchberger's Method for Solving Systems of Algebraic Equations (German). J. Number Theory 10: 475-488

Winkler, F., Buchberger, B. (1983): A Criterion for Eliminating Unnecessary Reductions in the Knuth-Bendix Algorithm. Colloquia Math. Soc. J. Bolyai 42: 849-869

Gröbner Bases, Symbolic Summation and Symbolic Integration

Frédéric Chyzak

INRIA-Rocquencourt and École polytechnique (FRANCE)
Frederic.Chyzak@inria.fr

Abstract

The treatment of combinatorial expressions and special functions by linear operators is amenable to Gröbner basis methods. In this tutorial, we illustrate the applications of Gröbner bases to symbolic summation and integration.

Introduction

In the late 1960's, Risch (1969, 1970) developed an algorithm for symbolic indefinite integration. The approach followed there consists in computing a tower of *differential extensions* in order to determine if an indefinite integral can be expressed in terms of elementary functions. Risch's algorithm became very popular and is now at the heart of the integration routines of many computer algebra systems. In the early 1980's, Karr (1981, 1985) appealed to similar ideas, namely *difference extensions*, in order to develop an algorithm for symbolic indefinite summation. Despite its indisputable algorithmic interest, Karr's algorithm has unfortunately not received as much attention as it deserves yet, due to its complexity and the difficulty to implement it.

In the early 1990's, Zeilberger (1990b) initiated a different approach to symbolic summation and integration. As opposed to the approach by differential or difference extensions, Zeilberger studies the action of algebras of differential or difference *linear operators* in order to compute special operators that determine the sum or integral under consideration. The method is based on the theory of *holonomy* (Bernstein 1971, 1972; Björk 1979; Coutinho 1995; Ehlers 1987; Kashiwara 1978). Since then, Zeilberger has improved his first approach so much that his "fast algorithm" is the basis for the definite summation routines of many computer algebra systems (Petkovšek *et al.* 1996; Wilf and Zeilberger 1992a, 1992b; Zeilberger 1990a).

More specifically, the starting point of the operator approach is to consider which linear operators annihilate a given function or sequence. A function, respectively a sequence, is then described by a set of annihilating operators.

For example, the binomial coefficients $\binom{n}{j}$ satisfy linear recurrence equations. Let S_n denote the operator of *shift* with respect to n: this operator acts on a function f by $(S_n \cdot f)(n) = f(n+1)$. Shifts with respect to other variables are denoted analogously. Identifying a rational function r in (n, j) with the linear operator of multiplication by r, we get a non-commutative algebra whose product denotes the composition of operators. With this notation, the binomial coefficients are annihilated by each operator of the system

$$(n+1-j)S_n - (n+1), \qquad (j+1)S_j - (n-j). \tag{0.1}$$

These operators correspond to the vertical and horizontal recurrences in Pascal's triangle. Summation is then recast into a different form: informally, the *left ideal* spanned by the previous operator system encodes all the linear equations with rational function coefficients satisfied by the binomial coefficients; summation then reduces to deducing an operator of a special form starting from the description (0.1) of the ideal. More precisely, the summation over j requires an operator which does not involve j, i.e., an operator in n, S_n and S_j only. For instance, deriving Pascal's triangle rule, as encoded by the operator $P = S_n S_j - S_j - 1$, suffices to perform the summation over j. This derivation is based on the *elimination* of j, for which methods are detailed in the main sections of this tutorial. A rewritten form for P is $(S_j - 1)(S_n - 1) + (S_n - 2)$, which applied to $\binom{n}{j}$ yields

$$g_{n,j+1} - g_{n,j} + [(S_n - 2) \cdot f](n, j) = 0$$

for $f_{n,j} = \binom{n}{j}$ and $g_{n,j} = \frac{j}{n+1-j}\binom{n}{j}$. Then, summing over j in $\mathbb{Z}$ yields that the operator $S_n - 2$ annihilates the sum $\sum_{j=0}^{n} \binom{n}{j}$. This method of *creative telescoping* will be described in §2.1. Solving the corresponding recurrence, we deduce that $\sum_{j=0}^{n} \binom{n}{j} = 2^n$. More generally, the algorithms described in this tutorial input and output linear systems. Typically, the input is a partial differential or recurrence system, while the output is an ordinary differential or recurrence equation, whose resolution is left as a post-processing.

As another example, the Appell F_4 bivariate hypergeometric function

$$F_4(a, b, c, d; x, y) = \sum_{m,n=0}^{\infty} \frac{(a)_{m+n}(b)_{m+n}}{m!n!(c)_m(d)_n} x^m y^n, \tag{0.2}$$

where $(x)_n$ denotes the Pochhammer symbol $\Gamma(x+n)/\Gamma(x)$, satisfies linear differential equations (Erdélyi 1981). Let D_x denote the operator of *derivation* with respect to x: this operator acts on a function f by $(D_x \cdot f)(x) = f'(x)$. Denoting derivations with respect to other variables in analogously, the Appell F_4 function is annihilated by

$$\begin{cases} xD_x(xD_x + c - 1) - x(xD_x + yD_y + a)(xD_x + yD_y + b), \\ yD_y(yD_y + d - 1) - y(xD_x + yD_y + a)(xD_x + yD_y + b), \end{cases} \tag{0.3}$$

with a similar identification for rational functions in (x, y) as in the previous example. Here, determining an operator which does not involve y, i.e., in x, D_x and D_y only, is enough to obtain an ordinary differential equation satisfied by an integral of the Appell function.

More generally, the operator approach applies to mixed difference-differential systems. For example, the Jacobi orthogonal polynomials $P_n^{(a,b)}(z)$, viewed as a function P in (n, a, b, z), is annihilated by the classical linear difference-differential system (Erdélyi 1981)

$$\begin{cases} (1-z^2)D_z^2 + (b-a-(a+b+2)z)D_z + n(n+a+b+1), \\ (2n+a+b+2)(1-z^2)S_nD_z \\ \qquad -(n+1)(a-b-(2n+a+b+2)z)S_n \\ \qquad -2(n+a+1)(n+b+1), \\ 2(n+2)(n+a+b+2)(2n+a+b+2)S_n^2 \\ -(2n+a+b+3) \\ \qquad\qquad [(2n+a+b+2)(2n+a+b+4)z+(a^2-b^2)]S_n \\ \qquad\qquad\qquad +2(n+a+1)(n+b+1)(2n+a+b+4), \\ (2n+a+b+2)(1-z)S_a - 2(n+a+1) + 2(n+1)S_n, \\ (2n+a+b+2)(1+z)S_b - 2(n+b+1) - 2(n+1)S_n. \end{cases} \tag{0.4}$$

Again, deriving an operator that does not involve z suffices to perform integrations of P.

An abstract framework for linear operators is given by *Ore algebras*, which provide a unifying viewpoint on shift and derivation operators (Chyzak and Salvy 1996). They also provide a polynomial representation for the operators under consideration, generalizing the polynomial representation of the systems (0.1), (0.3) and (0.4). It turns out that the theory of Gröbner bases developed for commutative algebras of polynomials extends to this non-commutative setting. Early work in this area is due to Galligo (1985) in the differential case. Takayama (1989) used an analogous technique for difference-differential algebras. A general setting was introduced by Kandri-Rody and Weispfenning (1990) and Kredel (1993), and later adapted to Ore algebras (Chyzak and Salvy 1996). In this tutorial, we illustrate applications of non-commutative Gröbner bases to symbolic summation and integration. Only simple facts of the theory of commutative Gröbner bases will be required, for which we refer to (Buchberger 1998) and (Cox *et al.* 1992).

Gröbner bases are used for various purposes in the context of the symbolic manipulation of linear operators. First, by providing normal forms modulo an ideal, they are crucial to algorithms which compute in finite-dimensional quotient rings. This applies in particular to the algorithm for indefinite summation and integration which we present in Section 1, after setting up the algebraic framework of Ore algebras and recalling how Gröbner bases can be computed there. Next, the use of Gröbner bases for elimination purposes is the keystone of a general method of definite summation and integration called creative telescoping. An algorithm is presented in Section 2, together with an extension to multiple summations and integrations. A more efficient version in the case of natural boundaries is obtained by appealing to Gröbner bases for modules. We end the section by applying the algorithm of Section 1 to the definite case. In Section 3, we turn again to the use of Gröbner bases for computing normal forms and obtain algorithms for various other operations. We also present a second method for the indefinite case. Finally, Gröbner bases allow the calculation of ideal dimensions. This concept is related to the theory of holonomy, which we allude to in Section 4, and is used to explain the success of the various methods and to decide the termination of the corresponding algorithms.

1 Indefinite Summation and Integration

The algorithms which we illustrate simultaneously apply to summation and integration. Indeed, they share an interpretation in terms of *Ore operators* which generalize both derivation and difference operators. This algorithmically important fact stems from the essential property of Ore operators to satisfy a skew Leibnitz rule (see (1.1) below) which encompasses both the classical Leibnitz rule for derivations,

$$(fg)'(x) = f(x)g'(x) + f'(x)g(x),$$

and the skew Leibnitz rule for the finite difference operator Δ,

$$\Delta(fg)(x) = f(x+1)\Delta g(x) + (\Delta f(x))g(x),$$

where $\Delta f(x) = f(x+1) - f(x)$. Furthermore, Ore operators have a representation in special algebras, namely *Ore algebras*, which can be regarded as algebras of skew polynomials. Due to the nice properties of these polynomials, a theory of Gröbner bases is available in Ore algebras. To end the section, we present an algorithm for the indefinite case, which will be used for the definite case in §2.4.

1.1 Ore Operators, Ore Algebras, Annihilating Ideals

The purpose of this section is to provide a representation of linear operators which is suitable for symbolic manipulations, and in particular for Gröbner basis calculations. This encoding is in terms of *skew polynomials*, which encapsulate the difference and differential cases in a single algebraic framework (Cohn 1971). Considering the polynomial form of the operators in (0.1), (0.3) and (0.4), we introduce a generic indeterminate ∂ which may either represent a derivation operator, a shift operator, or more general *Ore operators* (Bronstein and Petkovšek 1994; Ore 1933). This ∂ inherits the action on functions of the original operator. Introducing several indeterminates ∂_i, we then describe multivariate cases. Imposing commutation on the ∂_i's, we are led to the definition of *Ore algebras* (Chyzak and Salvy 1996).

In view of the similarity between the treatments of functions and sequences, we henceforth indifferently use the word "function" for functions and sequences, and more generally for any object on which we apply linear operators. Let $\mathcal{F}$ be an algebra of functions over a commutative field $\mathbb{K}$ and denote the algebra of $\mathbb{K}$-linear endomorphisms of $\mathcal{F}$ by $\operatorname{End}_{\mathbb{K}} \mathcal{F}$.

By analogy with the prime notation for derivations, we denote by f^η the action of $\eta \in \operatorname{End}_{\mathbb{K}} \mathcal{F}$ on $f \in \mathcal{F}$. An endomorphism $\delta \in \operatorname{End}_{\mathbb{K}} \mathcal{F}$ which satisfies the Leibnitz law $(fg)^\delta = fg^\delta + f^\delta g$ for $f, g \in \mathcal{F}$ is called a *derivation* or *derivation operator*. On the other hand, an algebra endomorphism $\sigma \in \operatorname{End}_{\mathbb{K}} \mathcal{F}$ is called a *difference* or *difference operator*. Each difference σ induces an operator $\theta = \sigma - 1$ which satisfies the skew Leibnitz law $(fg)^\theta = f^\sigma g^\theta + f^\theta g$ for $f, g \in \mathcal{F}$. More generally, we are interested in pairs (σ, δ) of operators which satisfy such a skew Leibnitz law. Given a difference σ on $\mathcal{F}$, an endomorphism $\delta \in \operatorname{End}_{\mathbb{K}} \mathcal{F}$ is called an *Ore operator*, or a *σ-derivation* (Bronstein and Petkovšek 1994; Ore 1933), when

$$(fg)^\delta = f^\sigma g^\delta + f^\delta g \qquad \text{for all } f, g \in \mathcal{F}. \tag{1.1}$$

This includes classical derivations η, in which case one considers the pair $(1, \eta)$ where 1 denotes the identity.

We now proceed to describe a representation of subalgebras of $\operatorname{End}_{\mathbb{K}} \mathcal{F}$ generated by σ-derivations as skew polynomial rings. In order to allow linear operators over various domains of coefficients, we introduce a $\mathbb{K}$-algebra $\mathbb{A}$. In practice, this will always be a polynomial ring $\mathbb{A} = \mathbb{K}[\mathbf{x}]$ or a rational function field $\mathbb{A} = \mathbb{K}(\mathbf{x})$. In addition, we require $\mathcal{F}$ to include $\mathbb{A}$, so that $\mathbb{A}$ can also be viewed as a subalgebra of $\operatorname{End}_{\mathbb{K}} \mathcal{F}$ by the identification of $a \in \mathbb{A}$ with the operator of multiplication by a. In view of the operators that appear in the applications, we distinguish between the general case of a pair (σ, δ) and the special case (σ, θ) for $\theta = \sigma - 1$. In the general case, we put the emphasis on the σ-derivation δ, and a polynomial representation is obtained by encoding δ through $\partial \cdot f = f^\delta$ for $f \in \mathcal{F}$. When additionally σ and δ restrict to a difference and a σ-derivation on $\mathbb{A}$, the *skew polynomial ring* $\mathbb{S} = \mathbb{A}[\partial; \sigma, \delta]$ is defined as the set of polynomials in ∂ with coefficients in $\mathbb{A}$, with usual addition and a product defined by associativity from the commutation rule

$$\partial a = a^\sigma \partial + a^\delta \qquad \text{for } a \in \mathbb{A}$$

(Cohn 1971). This is the quotient of the free associative algebra in ∂ modulo the above relations. In the special case, the emphasis is put on the difference $\sigma = \theta + 1$ through the action $\partial \cdot f = f^\sigma$ for $f \in \mathcal{F}$. We then consider the skew polynomial ring $\mathbb{A}[\partial; \sigma, 0]$ governed by the commutation

$$\partial a = a^\sigma \partial \qquad \text{for } a \in \mathbb{A}.$$

We turn to the multivariate case. Given several pairs of operators (σ_i, δ_i) on the same algebra $\mathcal{F}$ with the property that σ_i and δ_j commute for $i \neq j$, we introduce indeterminates ∂_i's so as to consider the quotient of the free associative algebra in the ∂_i's modulo the relations $\partial_i a = a^{\sigma_i}\partial_i + a^{\delta_i}$ for $a \in \mathbb{A}$ and $i = 1, \ldots, r$, and $\partial_i\partial_j = \partial_j\partial_i$ for $i, j = 1, \ldots, r$. Such a skew polynomial ring is called an *Ore algebra*, denoted $\mathbb{A}[\boldsymbol{\partial}; \boldsymbol{\sigma}, \boldsymbol{\delta}]$. For convenience, we often denote an Ore algebra $\mathbb{A}[\boldsymbol{\partial}; \boldsymbol{\sigma}, \boldsymbol{\delta}]$ by the abusive iterated notation $\mathbb{A}[\partial_1; \sigma_1, \delta_1] \ldots [\partial_r; \sigma_r, \delta_r]$, which could be formalized. When $\mathbb{A}$ is of the form $\mathbb{K}[\mathbf{x}]$, we obtain the special case of *polynomial Ore algebras* $\mathbb{K}[\mathbf{x}][\boldsymbol{\partial}; \boldsymbol{\sigma}, \boldsymbol{\delta}]$. Consider the example of the $\mathbb{C}$-algebra $\mathcal{F} = \mathbb{C}(n, a, b)((z))$ of formal Laurent series. The operators (0.4) which annihilate the Jacobi polynomials can be considered in the Ore algebra

$$\mathbb{O}_{\mathrm{r}} = \mathbb{A}[S_n; S_n, 0][S_a; S_a, 0][S_b; S_b, 0][D_z; 1, D_z] \quad \text{for} \quad \mathbb{A} = \mathbb{C}(n, a, b, z), \tag{1.2}$$

but might as well be considered in the polynomial Ore algebra $\mathbb{O}_{\mathrm{p}}$ obtained when $\mathbb{A} = \mathbb{C}[n, a, b, z]$. This distinction will be crucial to treatments by Gröbner bases: the elimination of the indeterminates n, a, b or z is amenable to Gröbner basis methods in $\mathbb{O}_{\mathrm{p}}$ but not in $\mathbb{O}_{\mathrm{r}}$.

For a module $\mathcal{F}$ of functions over an Ore algebra $\mathbb{O}$ and a function $f \in \mathcal{F}$, the set of operators that annihilate the function f has the algebraic structure of a *left ideal*. It is classically called the *annihilating ideal* of f and is denoted $\operatorname{Ann}_{\mathbb{O}} f$. This algebraic structure is the basis for the application of Gröbner basis methods. Alternatively, the set $\mathbb{O} \cdot f$ obtained by the action of the Ore algebra $\mathbb{O}$ on the function f is a left $\mathbb{O}$-module, in fact, a left submodule of $\mathcal{F}$. The structure of the module $\mathbb{O} \cdot f$ keeps track of the dependencies between the derivatives $\boldsymbol{\partial}^\alpha \cdot f$ of f. More specifically, $\mathbb{O} \cdot f \simeq \mathbb{O} / \operatorname{Ann}_{\mathbb{O}} f$.

1.2 Gröbner Bases in Ore Algebras

Due to the encoding of Ore operators as skew polynomials, the theory of Gröbner bases extends to Ore algebras. Under mild conditions, the leading term of a product of skew polynomials is the product of the leading terms of the factors. This nice feature avoids any problem of non-noetherianity. The main result originates in works by Kandri-Rody and Weispfenning (1990) and Kredel (1993), which were adapted to Ore algebras by Chyzak and Salvy (1996). It states that polynomial Ore algebras $\mathbb{O} = \mathbb{K}[\mathbf{x}][\boldsymbol{\partial}; \boldsymbol{\sigma}, \boldsymbol{\delta}]$ such that $\boldsymbol{\partial}$, $\boldsymbol{\sigma}$, $\boldsymbol{\delta}$ and $\mathbf{x}$ satisfy relations of the type

$$\partial_i x_j = (a_{i,j} x_j + b_{i,j})\partial_i + c_{i,j}(\mathbf{x}), \qquad 1 \le i \le r, \quad 1 \le j \le s, \tag{1.3}$$

with $b_{i,j} \in \mathbb{K}$, $a_{i,j} \in \mathbb{K} \setminus \{0\}$ and $c_{i,j} \in \mathbb{K}[\mathbf{x}]$, are left noetherian and that a non-commutative version of Buchberger's algorithm terminates for term orders with respect to which all the ∂_i's are larger than the x_i's. When additionally all the $c_{i,j}$'s are of total degree at most 1 in $\mathbf{x}$, Buchberger's algorithm terminates for any term order on $\mathbf{x}$ and $\boldsymbol{\partial}$. In all cases of termination, Buchberger's algorithm computes a Gröbner basis with respect to the term order. Moreover, the main tools of computational commutative algebra extend to the skew setting as well, and calculations of Hilbert dimensions, polynomials and series, and of Gröbner bases for modules are also available in Ore algebras.

Contiguity Relations for the Appell F_4 Bivariate Hypergeometric Function

As an example, we borrow from Takayama (1989) the calculation of contiguity relations for the Appell F_4 bivariate hypergeometric function defined by Eq. (0.2). The method can be used to derive expressions for the shift and reverse shifts of $F_4(a,b,c,d;x,y)$ with respect to its parameters a, b, c and d in terms of its derivatives with respect to the variables x and y. Here, we get a differential expression for $F_4(a-1,b,c,d;x,y)$ which was first obtained by Takayama.

Consider the action of the Ore algebra $\mathbb{O} = \mathbb{K}(x,y)[D_x;1,D_x][D_y;1,D_y]$, where $\mathbb{K} = \mathbb{Q}(a,b,c,d)$, on the algebra $\mathcal{F} = \mathbb{K}((x,y))$ of Laurent power series. The Appell function is annihilated by the operators of (0.3). In fact, $\mathfrak{I} = \operatorname{Ann}_{\mathbb{O}} F_4$ is precisely the ideal spanned by these operators. Since

$$\frac{(a+1)_{m+n}}{(a)_{m+n}} = \frac{(m+n+a)}{a},$$

the operator $H_a = a^{-1}(xD_x + yD_y + a)$ satisfies

$$(H_a \cdot F_4)(a,b,c,d;x,y) = F_4(a+1,b,c,d;x,y);$$

for this reason, H_a is called a *step-up* operator. We proceed to compute a *step-down* operator B_a, i.e., an operator such that $B_{a+1}H_a = B_aH_{a-1} = 1$.

Computing a Gröbner basis for the ideal $\mathfrak{I}$ spanned by the operators (0.3) in $\mathbb{O}$ with respect to a total degree order on D_x and D_y such that $D_x \succ D_y$, we obtain

$$\begin{cases} 2xyD_xD_y + (xy+y^2-y)D_y^2 \\ \qquad + (a+b-c+1)xD_x + (dx+(a+b+1)y-d)D_y + ab, \\ xD_x^2 - yD_y^2 + cD_x - dD_y, \\ (2x^2y^2 - 4xy^3 + 2y^4 - 4xy^2 - 4y^3 + 2y^2)D_y^3 \\ \qquad + A_{0,2}D_y^2 + A_{1,0}D_x + A_{0,1}D_y + A_{0,0}, \end{cases} \tag{1.4}$$

where each $A_{i,j}$ is a large polynomial in x and y. It follows that the quotient ring $\mathbb{O}/\mathfrak{I}$ is a $\mathbb{Q}(a,b,c,d,x,y)$-vector space of dimension 4 which admits $\{1, D_x, D_y, D_y^2\}$ as a basis.

Thus, we set $B_{a+1} = c_0 + c_1D_x + c_2D_y + c_3D_y^2$ with unknowns c_i's. Reducing $B_{a+1}H_a - 1$ by the Gröbner basis (1.4) and identifying the coefficients of 1, D_x, D_y and D_y^2 to zero yields a linear system in (c_0, c_1, c_2, c_3) which we easily solve to obtain the step-down operator.

Note that in the previous calculation, $\mathfrak{I}$ is not required to be $\mathrm{Ann}_{\mathbb{O}} f$, but only a subideal of it such that $\mathbb{O}/\mathfrak{I}$ be finite-dimensional. In the present case, one could prove that $\mathfrak{I} = \mathrm{Ann}_{\mathbb{O}} f$. The problem of working with a subideal of $\mathrm{Ann}_{\mathbb{O}} f$ is that the returned step-down operator could be of higher order.

We refer to (Sturmfels and Takayama 1998) for further results relating to multivariate hypergeometric series.

1.3 ∂-Finite Functions

Many calculations described in this tutorial rely on a property of finiteness satisfied by the input of the algorithms. Informally, the requirement is that the calculations performed by the algorithms take place in a finite-dimensional vector space.

For instance, the key fact that makes the calculations on the example of the Appell function possible is that the quotient $\mathbb{O}/\mathfrak{I}$ is finite-dimensional. Such a situation is rather common. Consider for instance the Jacobi polynomials $P_n^{(a,b)}(z)$ regarded as a function $P \in \mathbb{F}((z))$ for $\mathbb{F} = \mathbb{C}(n, a, b)$ and the Ore algebra $\mathbb{O}_r$ defined by Eq. (1.2). It is obvious from (0.4) that the $\mathbb{F}(z)$-vector space $\mathbb{O}_r \cdot P$ is finite-dimensional, and is generated by $\{1, S_n, D_z\}$. In fact, computing a Gröbner basis for the set of operators (0.4) with respect to a total degree order on S_n, S_a, S_b and D_z such that $D_z \prec S_b \prec S_a \prec S_n$, we get:

$$\begin{cases} (n+a+b+1)S_b - (z-1)D_z - (n+a+b+1), \\ (n+a+b+1)S_a - (z+1)D_z - (n+a+b+1), \\ 2(n+1)(n+a+b+1)S_n - (z^2-1)(2n+a+b+2)D_z \\ \qquad -(n+a+b+1)((2n+a+b+2)z+a-b), \\ (z^2-1)D_z^2 + ((a+b+2)z+a-b)D_z - n(n+a+b+1). \end{cases}$$

This proves that $\mathbb{O}_r \cdot P = \mathbb{F}(z)P + \mathbb{F}(z)(D_z \cdot P)$ is a vector space of dimension at most 2. From the theory of orthogonal polynomials (Erdélyi 1981), we know that P and $D_z \cdot P$ cannot satisfy a linear dependency with rational function coefficients, so that the sum is indeed direct, $\mathbb{O}_r \cdot P$ has dimension 2 and $\mathrm{Ann}_{\mathbb{O}_r} P$ is spanned by the operators (0.4).

On the other hand, the function $f = e^{\sin z} \in \mathcal{F}$ is such that the module $\mathbb{O} \cdot f$ over $\mathbb{O} = \mathbb{C}(z)[D_z; 1, D_z]$ is an infinite-dimensional $\mathbb{C}(z)$-vector space that admits $\{(\cos z)^k f\}_{k \geq 0}$ as a basis. No algorithm of this tutorial applies to functions like f.

Formalizing, a function $f \in \mathcal{F}$ is called *∂-finite* with respect to an Ore algebra $\mathbb{O} = \mathbb{F}[\boldsymbol{\partial}; \boldsymbol{\sigma}, \boldsymbol{\delta}]$ over a field $\mathbb{F}$ when the vector space $\mathbb{O} \cdot f \simeq \mathbb{O}/\,\mathrm{Ann}_{\mathbb{O}} f$ is finite-dimensional (Chyzak and Salvy 1996). In the phrasing "∂-finite", ∂ is merely a generic symbol and bears no relation to the actual ∂_i's that appear in $\mathbb{O}$.

1.4 Indefinite Summation and Integration

We present an algorithm which was introduced in (Chyzak 1997) to compute indefinite sums and integrals of ∂-finite functions. More generally, the algorithm searches for particular solutions of certain integro-differential problems. The technique is restricted to single summations and integrations, but provides a fast algorithm in the definite case. This will be described in §2.4.

Consider functions in $(x_0, \mathbf{x})$; let $\mathbb{O} = \mathbb{K}(x_0, \mathbf{x})[\partial_0; \sigma_0, \delta_0][\boldsymbol{\partial}; \boldsymbol{\sigma}, \boldsymbol{\delta}]$ be an Ore algebra and f be a ∂-finite function with respect to this algebra. Given operators $P \in \mathbb{O}$ and $\Xi \in \mathbb{K}(x_0)[\partial_0; \sigma_0, \delta_0]$, we look for a function $F \in \mathbb{O} \cdot f$, if one exists, such that

$$\Xi(x_0, \partial_0) \cdot F = P(x_0, \mathbf{x}, \partial_0, \boldsymbol{\partial}) \cdot f. \tag{1.5}$$

The key idea of the algorithm is to write F in the form $Q \cdot f$, and then to solve for $Q \in \mathbb{O}$. A solution F found in this way is a ∂-finite particular solution of Eq. (1.5). In general, no such solution exists. In this case, the algorithm returns a proof that no solution exists in $\mathbb{O} \cdot f$.

As a special case, when ∂_0 is a derivation operator and $\Xi = \partial_0$, we get an algorithm for indefinite integration. When ∂_0 is a shift operator and $\Xi = \partial_0 - 1$, we get an algorithm for indefinite summation. In both cases, either we set $P = 1$ to integrate or sum the function f itself, or the algorithm applies to any element of the module $\mathbb{O} \cdot f$. When $\Xi = \partial_0 - 1$, $r = 0$, $\partial_0 = S_n$, $\mathbb{O} = \mathbb{Q}(n)[S_n; S_n, 0]$, and $\mathbb{O} \cdot f$ is of dimension 1, we get an algorithm for the indefinite summation of *hypergeometric terms*, that is based on Abramov's classical algorithm (Abramov 1989a, 1989b, 1995; Abramov and Kvashenko 1991; Abramov *et al.* 1995); this algorithm is an alternative to Gosper's algorithm (Gosper 1978).

Harmonic Summation

Harmonic summation identities like the indefinite summation

$$\sum_{k=1}^{n} \binom{k}{m} H_k = \binom{n+1}{m+1} \left(H_{n+1} - \frac{1}{m+1} \right)$$

are classically proved by summation by parts or by techniques of generating functions. Here, we prove the above identity following the operator approach.

Introducing $f_n = \binom{n}{m} H_n$, we show the equivalent form

$$\sum_{k=1}^{n} f_k = \frac{(n+1)^2}{(m+1)^2} f_n - \frac{(n-m)(n-m+1)}{(m+1)^2} f_{n+1}. \tag{1.6}$$

First, f satisfies the linear recurrence

$$\begin{aligned}(n-m+1)(n-m+2)f_{n+2}-(2n+3)(n-m+1)f_{n+1}\\ +(n+1)^2 f_n=0,\end{aligned} \tag{1.7}$$

which is obtained by homogenizing the relation

$$(n+1-m)f_{n+1}=(n+1)f_n+1,$$

allowing rational function coefficients in $\mathbb{Q}(n,m)$ only. In addition, one could prove that f cannot satisfy any such homogeneous relation of lower order. It follows that the sequence f is a ∂-finite function with respect to the Ore algebra $\mathbb{O} = \mathbb{Q}(n,m)[S_n; S_n, 0]$. We look for an indefinite sum F in the module $\mathbb{O}\cdot f$. Since the latter is a two-dimensional vector space with basis $\{f, S_n \cdot f\}$, we introduce a generic operator $Q=\alpha_n+\beta_n S_n$ such that $F = Q\cdot f$. We impose $(S_n-1)\cdot F = f$, so that we compute $Z=(S_n-1)Q-1$. Our goal is to find *rational functions* α and β such that Z is zero. We have

$$Z=\beta_{n+1}S_n^2+(\alpha_{n+1}-\beta_n)S_n-(\alpha_n+1),$$

which we reduce using Eq. (1.7). We obtain

$$\begin{aligned}Z=\left(\alpha_{n+1}+\frac{(2n+3)\beta_{n+1}}{n-m+2}-\beta_n\right)S_n\\ -\left(\alpha_n+\frac{(n+1)^2\beta_{n+1}}{(n-m+1)(n-m+2)}+1\right).\end{aligned} \tag{1.8}$$

Setting both terms in parentheses to zero, next uncoupling the recurrence system so as to get rid of α, yields the recurrence equation

$$\begin{aligned}(n+2)^2\beta_{n+2}-(2n+3)(n-m+3)\beta_{n+1}\\ +(n-m+2)(n-m+3)\beta_n+(n-m+3)(n-m+2)=0,\end{aligned} \tag{1.9}$$

which is solved for rational solutions by Abramov's algorithm (Abramov 1989a, 1989b, 1995; Abramov *et al.* 1995). In general, the uncoupling step can be performed using specialized algorithms (Abramov and Zima 1996; Barkatou 1993; Bronstein and Petkovšek 1996), or by a Gröbner basis calculation (see below). Substituting in the system and eliminating α_{n+1} between both equations yields an expression for α_n. The expressions found are

$$\alpha_n=\frac{(n+1)^2}{(m+1)^2} \quad \text{and} \quad \beta_n=-\frac{(n-m)(n-m+1)}{(m+1)^2},$$

which proves Eq. (1.6).

Particular Solutions

The classical method of variation of the constant to search for a particular solution of a non-homogeneous ordinary differential equation returns an output which involves an indefinite integral and a division. On the other hand, the method of the present section searches for explicit linear expressions in the class of ∂-finite functions.

As an example, let $\mathbb{O}$ be the Ore algebra $\mathbb{Q}(q,x)[D_x;1,D_x]$ and consider the equation:

$$\left(D_x^2 + \frac{1}{x}D_x - q^2\right) \cdot F(x) = (1-x^2)J_0(qx) - 2q^2xJ_1(qx),$$

where J_ν is the Bessel function of the first kind and order ν (Erdélyi 1981). This corresponds to the general setting for

$$f(x) = J_0(qx), \quad P = (1-x^2) - 2qxD_x \quad \text{and} \quad \Xi = D_x^2 + x^{-1}D_x - q^2.$$

Once again $\mathbb{O} \cdot f$ has dimension 2. Introducing $Q = \alpha(x) + \beta(x)D_x$ and reducing $Z = \Xi Q - P$ yields a linear differential system, which, once uncoupled, is solved by another algorithm of Abramov's (Abramov 1989a, 1989b; Abramov and Kvashenko 1991). The solution is

$$\alpha(x) = \frac{2q - 1 + x^2}{2q^2} \qquad \text{and} \qquad \beta(x) = \frac{(1-q^3)x}{q^2}.$$

Denoting by I_ν and K_ν the modified Bessel functions of the first and second kind and of order ν (Erdélyi 1981), the general solution to the differential equation above is:

$$\frac{2q-1+x^2}{2q^2}J_0(qx) + \frac{(q^3-1)x}{q^3}J_1(qx) + C_1I_0(qx) + C_2K_0(qx),$$

since both $I_0(qx)$ and $K_0(qx)$ are annihilated by Ξ.

Multivariate Extension

In other cases, we do not deal with a function f in a single variable x_0, but in variables $(x_0, \mathbf{x})$. We then consider the annihilating ideal of f with respect to an Ore algebra $\mathbb{O} = \mathbb{K}(x_0, \mathbf{x})[\partial_0; \sigma_0, \delta_0][\boldsymbol{\partial}; \boldsymbol{\sigma}, \boldsymbol{\delta}]$. Gröbner bases are then used for several purposes:

1. a Gröbner basis of $\mathfrak{I} = \mathrm{Ann}_{\mathbb{O}} f$ yields a basis of the $\mathbb{K}$-vector space $\mathbb{O}/\mathfrak{I}$ and determines how many undetermined coefficients have to be introduced in the generic operator Q;
2. the reduction of $\Xi Q - P$ is then performed using the same Gröbner basis;

3. the uncoupling step can be performed by computing the Gröbner basis of an $\mathbb{O}$-module.

Let us detail the last point. Assume that N undetermined coefficients η_i, for $i = 1, \dots N$, have been introduced in Q; in other words, we assume that $\mathbb{O} \cdot f$ has dimension N. We then introduce the free module $\mathfrak{M} = \bigoplus_{i=0}^{N} \mathbb{O} \cdot e_i$ over new indeterminates e_i and interpret any linear expression in the η_i's and their derivatives as an element of $\mathfrak{M}$ by mapping η_i onto e_i and making the non-homogeneous terms act on e_0. For instance, in the example of the harmonic summation above, we use e_1 and e_2 to keep track of the operators that act on α and β, respectively. An expression $U \cdot \alpha + V \beta + r$ for $U, V \in \mathbb{O}$ and $r \in \mathbb{Q}(n, m)$ is then encoded by $r \cdot e_0 + U \cdot e_1 + V \cdot e_2$.

In particular, any linear relation between the η_i's and their derivatives has its counterpart in the $\mathbb{O}$-module $\mathfrak{M}$. We encode the set of all such equations as the submodule $\mathfrak{N}$ of $\mathfrak{M}$ spanned by the polynomials obtained after encoding the coefficients of the reduced form of $Z = \Xi Q - P$ with respect to $(\partial_0, \boldsymbol{\partial})$. In order to uncouple the system corresponding to Z, we compute a Gröbner basis for $\mathfrak{N}$ with respect to a term order that sorts the e_i's lexicographically. We get a triangularized system, as required. Following up the example of the uncoupling step in the harmonic summation above, the expression Z that is defined by Eq. (1.8) is encoded by

$$\begin{cases} (n-m+2)S_n \cdot e_1 + [(2n+3)S_n - (n-m+2)] \cdot e_2, \\ (n-m+1)(n-m+2) \cdot (e_0 + e_1) + (n+1)^2 S_n \cdot e_2, \end{cases}$$

which generates a submodule $\mathfrak{N}$ of $\mathfrak{M} = \bigoplus_{i=0}^{2} \mathbb{O} \cdot e_i$. We compute a Gröbner basis for $\mathfrak{N}$ with respect to a term order that satisfies $e_2 \succ e_1 \succ e_0$. This eliminates e_2 and we obtain

$$\begin{aligned} \left[(n+2)^2 S_n^2 - (2n+3)(n-m+3)S_n + (n-m+2)(n-m+3)\right] \cdot e_1 \\ +(n-m+3)(n-m+2) \cdot e_0, \end{aligned}$$

which represents Eq. (1.9).

2 Definite Summation and Integration

A simple algorithm for definite summation and integration is based on a brute-force *elimination* by a Gröbner basis calculation and appeals to the method of *creative telescoping* (§2.1). Since several indeterminates can be eliminated simultaneously, this general method applies to multiple summations and integrations as well (§2.2). However, a partial elimination is always preferable. When in addition certain analytical conditions are satisfied, more specifically in the case of so-called *natural boundaries*, a partial elimination is made possible by the calculation of Gröbner basis for modules (§2.3). The corresponding algorithm also deals with the case of multiple summations and integrations. We finish the section with an algorithm that does not appeal to elimination by Gröbner bases, but is based on the algorithm of §1.4. It is also faster than the brute-force method, but restricted to single summations and integrations (§2.4).

2.1 Creative Telescoping and Elimination by Gröbner Bases

In this section, we turn to the problem of definite summation and integration by a method based on elimination (Almkvist and Zeilberger 1990). We exemplify the method by the integration of the function f defined by

$$f_n(x) = \frac{e^{-px}T_n(x)}{\sqrt{1-x^2}}, \quad \text{where} \quad T_n(x) = \frac{n}{2}\sum_{k=0}^{\lfloor n/2 \rfloor}(-1)^k\frac{(n-k-1)!}{k!(n-2k)!}(2x)^{n-2k}$$

denotes the Chebyshev orthogonal polynomials of the first kind (Erdélyi 1981). More specifically, we proceed to prove (Prudnikov *et al.* 1986: §2.18.1, Eq. (10)),

$$\int_{-1}^{+1}\frac{e^{-px}T_n(x)}{\sqrt{1-x^2}}dx = \pi(-1)^n I_n(p). \tag{2.1}$$

This identity appears for example in the study of the fundamental modes of vibration of a drum membrane in interaction with a mallet (Joly and Rhaouti 1997).

The most basic algorithm works as follows. We view f as a member of $\mathcal{F} = \mathbb{C}(p,n)((x))$ and make the algebra $\mathbb{O}_r = \mathbb{C}(p,n,x)[D_x;1,D_x][S_n;S_n,0]$ act on $\mathcal{F}$. Then, f is annihilated by the special operator

$$D_x(S_n^2-1) + pS_n^2 - 2(n+1)S_n - p, \tag{2.2}$$

and this operator suffices to obtain integrals of f with respect to x, as explained below. The peculiarity of the previous operator is that it does not involve the variable x of integration in its coefficients. Moreover, one can show within the theory of *holonomy* that such an eliminated operator exists and can be computed for a large class of functions (Zeilberger 1990b); see also Section 4. A similar phenomenon is exploited for *creative telescoping* in the summation case, where a special operator that does not involve the index of summation in its coefficients suffices to perform summation (Zeilberger 1991).

Indeed, applying the operator (2.2) to f and integrating between u and v yields the non-homogeneous recurrence

$$\left[(S_n^2-1)\cdot f\right]_u^v + (pS_n^2 - 2(n+1)S_n - p)\cdot\int_u^v f_n(x)dx = 0. \tag{2.3}$$

A series expansion indicates that the left term tends to 0 for any n when x goes to ± 1, so that for $u = -1$ and $v = +1$, the integral satisfies a homogeneous recurrence. Note that this recurrence could have been obtained directly by setting $D_x = 0$ in (2.2), if we had predicted the cancellation of the left term in Eq. (2.3) beforehand. Solving the second order recurrence can be done automatically. This yields two $\mathbb{C}$-independent solutions, $(-1)^n I_n(p)$ and $(-1)^n K_n(p)$. To complete the proof of Eq. (2.1), one needs to determine *which* solution of the two-dimensional vector space of solution the integral is. This is achieved by considering initial conditions at $n = 0$ and $n = 1$.

It remains to explain how an operator like (2.2) can be obtained. First, the polynomials $T_n(x)$ are ∂-finite; the following operators vanishing at f are variations on classical equations satisfied by $T_n(x)$ (Erdélyi 1981), obtained by the change of function $T_n(x) = \sqrt{1-x^2}e^{px} f_n(x)$:

$$\begin{cases} (x^2-1)D_x^2 + (2px^2+3x-2p)D_x + p^2x^2 + 3px - n^2 - p^2 + 1, \\ (x^2-1)D_xS_n + (px^2 - nx - p)S_n + n + 1, \qquad S_n^2 - 2xS_n + 1. \end{cases} \tag{2.4}$$

We could prove that these operators span the ideal $\mathrm{Ann}_{\mathbb{O}_r} f$. To get an operator that is free from x, we would like to compute the intersection of this ideal with the subalgebra $\mathbb{C}(p,n)[D_x][S_n; S_n, 0]$ of $\mathbb{O}_r$. To this end, it seems natural to perform this *elimination* by computing a Gröbner basis for a suitable term order. However, such an elimination is tractable by Gröbner basis calculations only if x appears in a polynomial way in the algebra. We therefore compute the Gröbner basis of the ideal spanned by the system (2.4) in the *polynomial* Ore algebra $\mathbb{O}_p = \mathbb{C}(p)[n,x][D_x; 1, D_x][S_n; S_n, 0]$ for a term order on n, x, D_x and S_n that sets x lexicographically greater than the other indeterminates (i.e., $x \succ n^i D_x^j S_n^k$ for any i, j and k) and breaks ties by a total degree order such that $S_n \succ D_x \succ n$. This yields

$$\begin{cases} S_n^2 D_x - D_x - 2S_n - 2nS_n + pS_n^2 - p, \\ xD_x - S_nD_x + px - pS_n + n + 1, \qquad 2xS_n - S_n^2 - 1. \end{cases}$$

The operator (2.2) is obtained as the only operator which does not involve x.

2.2 Multiple Summations and Integrations

The previous method by brute-force elimination also applies to multiple summations and integrations. For instance, we compute

$$\sum_{k=0}^{\infty}\sum_{\ell=0}^{\infty} \frac{(2k+2\ell+n+m)!}{(k+n)!(\ell+m)!k!\ell!}\left(\frac{x}{4}\right)^{k+\ell} \tag{2.5}$$

and prove that this *double* sum admits the (infinite) *single* sum representation

$$\frac{n!}{(n-m)!m!}\,{}_4F_3\left(\left.\begin{matrix} \frac{n+m}{2}+1, \frac{n+m}{2}+1, \frac{n+m+1}{2}, \frac{n+m+1}{2} \\ n+m+1, n+1, m+1 \end{matrix}\right| 4x\right), \tag{2.6}$$

in terms of the hypergeometric ${}_4F_3$ function. The double sum describes the probability of reaching the point (n, m) by a random walk on the integer lattice $\mathbb{N}^2$ starting from the origin, with probability $x/4$ to proceed in any of the four directions and probability $1 - x$ to stop the walk at each step (Trigg 1996).

Considering the Ore algebra $\mathbb{K}(k, \ell, x)[S_k; S_k, 0][S_\ell; S_\ell, 0][D_x; 1, D_x]$ over the field $\mathbb{K} = \mathbb{Q}(n, m)$, the following system of operators describes the summand s as a ∂-finite function:

$$\begin{cases} xD_x - (k + \ell), \\ 4(k+1)(k+1+n)S_k \\ \qquad - x(2(k+\ell) + n + m + 1)(2(k+\ell+1) + n + m), \\ 4(\ell+1)(\ell+1+m)S_\ell \\ \qquad - x(2(k+\ell) + n + m + 1)(2(k+\ell+1) + n + m). \end{cases} \tag{2.7}$$

These operators are obtained by computing normal forms for $(D_x \cdot s)/s$, $(S_k \cdot s)/s$ and $(S_\ell \cdot s)/s$ respectively, which turn out to be rational functions in $\mathbb{K}(k, \ell, x)$.

In view of the creative telescoping, we compute a Gröbner basis for the ideal spanned by (2.7) in the polynomial Ore algebra

$$\mathbb{K}(x)[k, \ell][S_k; S_k, 0][S_\ell; S_\ell, 0][D_x; 1, D_x].$$

To perform the elimination, we use a term order on the five polynomial indeterminates of the algebra so that k and ℓ are lexicographically greater than S_k, S_ℓ and D_x. Ties are broken by a total degree order on k and ℓ such that $k \prec \ell$ on the one hand, and by a total degree order on S_k, S_ℓ and D_x such that $S_k \prec S_\ell \prec D_x$ on the other hand. The Gröbner basis found has the following shape:

$$\begin{cases} D_x^5 S_k^2 S_\ell^2 + \langle 307 \text{ l.o.t.}\rangle, & D_x^4 S_k^3 S_\ell^3 + \langle 511 \text{ l.o.t.}\rangle, \\ k + \ell - xD_x, & kS_k^2 S_\ell^2 + \langle 192 \text{ l.o.t.}\rangle, \\ kS_k S_\ell + \langle 27 \text{ l.o.t.}\rangle, & kD_x^2 S_k + \langle 72 \text{ l.o.t.}\rangle, \\ kS_k S_\ell^3 + \langle 467 \text{ l.o.t.}\rangle, & kD_x^2 S_\ell^2 + \langle 103 \text{ l.o.t.}\rangle, \\ k^2 S_\ell + \langle 15 \text{ l.o.t.}\rangle, & k^2 S_k + \langle 14 \text{ l.o.t.}\rangle \end{cases} \tag{2.8}$$

where "p l.o.t." stands for "p lower order terms", and where each polynomial has been made monic. Both operators of the first line contain neither k nor ℓ. It has been noted in the previous section that a recurrence for the integral of Eq. (2.1) is obtained by setting $D_x = 0$ in the eliminated operator (2.2). In the present case, summing over k and ℓ in $(0, \infty)$ corresponds to setting $S_k = 1$ and $S_\ell = 1$ in the first two operators of (2.8). This yields two operators of $\mathbb{K}(x)[D_x; 1, D_x]$ of degree 4 and 5 in D_x, respectively. Computing a Gröbner basis with respect to a term order on D_x, we get their right gcd, namely the following operator of degree 4:

$$
\begin{aligned}
&4x^3(1-4x)D_x^4 + 8x^2((s+3)-2(2s+9)x)D_x^3 \\
&\quad + 4x((s^2+p+6s+7)-(6s^2+42s+77)x)D_x^2 \\
&\quad + 4((p+s+1)(s+1)-(2s+5)(s^2+5s+7)x)D_x - (s+1)^2(s+2)^2
\end{aligned}
$$

for $s = n+m$ and $p = nm$. The previous calculation reduces in fact to a skew Euclidean algorithm (Bronstein and Petkovšek 1996; Chyzak and Salvy 1996; Ore 1933). Solving the corresponding differential equation and taking into account four initial values at 0 yields the hypergeometric series representation of (2.6) for the double sum.

Following an approach that is not based on Gröbner bases, Wegschaider (1997) obtained a fast algorithm tailored for the calculation of binomial multiple sums like (2.5).

2.3 Takayama's Algorithm and Gröbner Bases of Modules

We proceed to describe how under certain analytical conditions, the brute-force elimination of the previous algorithms can be replaced by a moderated elimination.

Consider a function f in $(x_0, \mathbf{x})$ to be summed or integrated with respect to x_0, and let $\mathbb{O}_e$ be the Ore algebra $\mathbb{K}(\mathbf{x})[x_0][\partial_0; \sigma_0, \delta_0][\boldsymbol{\partial}; \boldsymbol{\sigma}, \boldsymbol{\delta}]$ that would be used by the algorithms of the previous sections. The key ingredient of the method is an annihilating operator like (2.2), of the form

$$\partial_0 Q(\mathbf{x}, \partial_0, \boldsymbol{\partial}) + P(\mathbf{x}, \boldsymbol{\partial}) \tag{2.9}$$

where x_0—the variable of integration when $\partial_0 = D_{x_0}$; the index of summation when $\partial_0 = S_{x_0} - 1$—neither appears in P nor in Q. It was observed by Almkvist and Zeilberger (1990) that the elimination of x_0 is more than needed. In fact, an annihilating operator of the form

$$\partial_0 Q(x_0, \mathbf{x}, \partial_0, \boldsymbol{\partial}) + P(\mathbf{x}, \boldsymbol{\partial}) \tag{2.10}$$

suffices for the next steps of creative telescoping. On the other hand, the term in Q results in a variational term $[Q \cdot f]_{x_0=u}^{x_0=v}$, like in Eq. (2.3). Cases where this term can be predicted to evaluate to zero are called summation and integration *over natural boundaries*. We then only need to compute P.

An algorithm to deal with this case in a differential framework is given in (Takayama 1990a, 1990b), and was further elaborated to accommodate difference operators in (Chyzak and Salvy 1996). The key idea is that the annihilating operators of the form (2.10) may be multiplied by $\mathbf{x}$, ∂_0 and $\boldsymbol{\partial}$, but not by x_0, and thus constitute a module over the algebra

$$\mathbb{O}_{\mathrm{m}} = \mathbb{K}(\mathbf{x})[\partial_0; \sigma_0, \delta_0][\boldsymbol{\partial}; \boldsymbol{\sigma}, \boldsymbol{\delta}].$$

The annihilating ideal $\mathfrak{I} = \mathrm{Ann}_{\mathbb{O}_{\mathrm{e}}} f$ of $\mathbb{O}_{\mathrm{e}}$ is then viewed as an $\mathbb{O}_{\mathrm{m}}$-module. As such, it is of infinite type and is generated by the products of the (finitely many) generators of $\mathfrak{I}$ with the (infinitely many) powers of x_0. The algorithm then proceeds by truncation of the module with respect to the degree in x_0: one computes an elimination Gröbner basis for the $\mathbb{O}_{\mathrm{m}}$-module spanned by all operators up to a given degree in x_0; if an operator of the type (2.10) is found, the algorithm returns the corresponding P; otherwise the maximum degree in x_0 is increased. The discussion of the termination of this algorithm is based on the concept of *holonomy* and *ideal dimension*, and is postponed until Section 4.

The algorithm also generalizes to multiple summations and integrations; in this case, one introduces an algebra $\mathbb{O}_{\mathrm{e}}$ is of the form

$$\mathbb{K}(\mathbf{x})[x_0, \ldots, x_s][\partial_0; \sigma_0, \delta_0] \ldots [\partial_s; \sigma_s, \delta_s][\boldsymbol{\partial}; \boldsymbol{\sigma}, \boldsymbol{\delta}]$$

to sum or integrate over $x_0, \ldots, x_s$ simultaneously. The form of the crucial operator (2.10) to look for then becomes

$$\partial_0 Q_0(x_0, \mathbf{x}, \partial_0, \ldots, \partial_s, \boldsymbol{\partial}) + \cdots + \partial_s Q_s(x_s, \mathbf{x}, \partial_0, \ldots, \partial_s, \boldsymbol{\partial}) + P(\mathbf{x}, \boldsymbol{\partial}). \quad (2.11)$$

We now proceed to exemplify the algorithm with a double summation.

Gordon's Generalization of the Rogers-Ramanujan Identities

An extension of the famous Rogers-Ramanujan identities (Rogers 1894) is due to Gordon (1961) and yields multiple summations identities. Gordon's theorem states that the partitions of n of the form $(\lambda_1, \ldots, \lambda_s)$ where $\lambda_i - \lambda_{i+k-1} \geq 2$ for each i and at most $r-1$ of the λ_i's equal 1 are equinumerous to the partitions of n into parts that are not congruent to 0, r or $-r$ modulo $2k+1$. We recover the Rogers-Ramanujan identities for $k = r = 2$ and $k = r + 1 = 2$. They possess an *infinite analytic* version due to Andrews (1974). For instance for $(k, r) = (3, 3)$, we have

$$\sum_{j=0}^{\infty} \sum_{i=0}^{\infty} \frac{q^{(i+j)^2+j^2}}{(q;q)_i (q;q)_j} = \frac{(q^3;q^7)_\infty (q^4;q^7)_\infty}{(q;q)_\infty},$$

where $(x;q)_n$ denotes the q-Pochhammer symbol $(1-x)(1-qx)\ldots(1-q^{n-1}x)$. More generally, each instance of the theorem above for a fixed k yields a k-fold summation identity which is in principle tractable by the algorithm of this section. Moreover, these identities can be obtained as a limiting case of a *finite analytic* version due to Paule (1985). For $k = r = 3$, the identity reads

$$\sum_{j=0}^{n}\sum_{i=0}^{n-j}\frac{q^{(i+j)^2+j^2}}{(q;q)_{n-i-j}(q;q)_i(q;q)_j}=\sum_{k=-n}^{n}\frac{(-1)^k q^{7/2k^2+1/2k}}{(q;q)_{n+k}(q;q)_{n-k}}. \tag{2.12}$$

To prove this double summation identity, we consider the algebra $\mathcal{F}$ of sequences from $\mathbb{N}^3$ to $\mathbb{Q}((q))$ indexed by (i, j, n) and view the summand f as an element of $\mathcal{F}$. We introduce the shift operators S_i, S_j and S_n, together with the operators of multiplication by q, q^i, q^j and q^n. Those operators generate the algebra $\mathbb{O} = \mathbb{Q}(q, q^i, q^j, q^n)[S_i; S_i, 0][S_j; S_j, 0][S_n; S_n, 0]$ of $\mathbb{Q}$-linear operators on $\mathcal{F}$. These are known as q-calculus operators, and satisfy the commutations

$$S_i q^i = qq^i S_i, \qquad S_j q^j = qq^j S_j, \qquad S_n q^n = qq^n S_n.$$

Then, f is a ∂-finite function since is vanishes at

$$\begin{cases}(1-qq^i)q^jS_i - qq^i(q^iq^j - q^n), \quad (1-qq^j)q^iS_j - qq^j(q^jq^i - q^n),\\ (q^iq^j - qq^n)S_n - q^iq^j.\end{cases} \tag{2.13}$$

This system is obtained by computing normal forms for the quotients $\frac{f_{i+1,j,n}}{f_{i,j,n}}$, $\frac{f_{i,j+1,n}}{f_{i,j,n}}$ and $\frac{f_{i,j,n+1}}{f_{i,j,n}}$ respectively, which turn out to be rational functions in q, q^i, q^j and q^n. We perform the algorithm described above. The ideal spanned by (2.13) is precisely $\mathrm{Ann}_{\mathbb{O}} f$ (otherwise f would have to be a constant). We compute the set of products of each operator p in (2.13) by $(q^i)^\alpha (q^j)^\beta$ for $\alpha + \beta \le 7 - d$, where d is the total degree of p in q^i and q^j. This yields 32 operators.

We then make S_i and S_j disappear from them. The point is that in the present case, the eliminated operator (2.11) takes the form

$$(S_i - 1)Q_0(q^i, q^j, q^n, S_i, S_j, S_n) + (S_j - 1)Q_1(q^i, q^j, q^n, S_i, S_j, S_n) + P(q^n, S_n),$$

where we need not compute the Q_i's, but only P. Furthermore, P can be obtained from the previous eliminated operator after two Euclidean divisions on the left by $S_i - 1$ next by $S_j - 1$. Performing these divisions *before* any Gröbner basis calculation, we are led to Gröbner basis calculations in an algebra over two indeterminates less, which speeds up calculations. The divisions are easily performed by following the rule

$$(q^i)^\alpha S_i^\beta = (-1)^\beta q^{-\alpha\beta}(q^i)^\alpha,$$

which holds modulo the *right* ideal spanned by $S_i - 1$. Next, a Gröbner basis is computed with respect to a term order on q^i, q^j and S_n that eliminates q^i and q^j, but with no multiplication by q^i and q^j allowed. In fact, this calculation takes place in the $\mathbb{Q}(n)[S_n; S_n, 0]$-module with the products $(q^i)^\alpha(q^j)^\beta$'s as a basis, and with an ordering on this basis which makes q^i and q^j disappear. We obtain the third-order operator

$$\begin{aligned}(q^3q^n - 1)S_n^3 \\ + \left[q^{10}(q^n)^4 + q - q^8(q^n)^3 + q^6(q^n)^2 + q^5(q^n)^2 - q^4q^n - q^3q^n + q^2 + 1\right] S_n^2 \\ -q(1 + q^5(q^n)^2 + q^4(q^n)^2 - q^3q^n + q^2 + q)S_n + q^3.\end{aligned}$$

Repeating the same process on the right-hand side, or using another summation algorithm (Paule and Riese 1997), we obtain the same operator. We complete the proof of Eq. (2.12) by checking that both sides agree for $n = 1$, 2 and 3.

2.4 Zeilberger's Fast Algorithm and its ∂-Finite Extension

The algorithm described in §1.4 for indefinite summation and integration also applies to definite summation and integration via creative telescoping (Chyzak 1997). Consider an Ore algebra $\mathbb{O}_r = \mathbb{K}(x_0, x_1)[\partial_0; \sigma_0, \delta_0][\partial_1; \sigma_1, \delta_1]$ and let f be a ∂-finite function with respect to $\mathbb{O}_r$. The method of creative telescoping relies on the search for a function $P \cdot f$ in the module $\mathbb{O}_r \cdot f$, which admits an indefinite sum or integral $F = Q \cdot f$ in the same module. This search corresponds to the search for an eliminated operator (2.10). In view of the summation or integration with respect to x_0, the method also requires P to be independent from (x_0, ∂_0). Assuming such a P is known, the algorithm for the indefinite case can be used to compute $F = Q \cdot f$. The idea of the algorithm in the definite case is therefore to introduce P in the undetermined form $P = \sum_{i=0}^{L} \eta_i(x_1)\partial_1^i$ and to solve for Q and *rational* η_i's simultaneously. Either a solution is found, or one gets a proof that no solution exists with P of degree L, in which case the search is repeated with a higher degree L of P. The discussion of the termination of this algorithm is similar to that of the algorithm of Section 2.3, and requires concepts to be introduced in Section 4.

Calkin's Curious Identity

Calkin (1994) proved the identity

$$\sum_{k=0}^{n}\left(\sum_{j=0}^{k}\binom{n}{j}\right)^3 = \frac{n}{2}8^n + 8^n - \frac{3n}{4}2^n\binom{2n}{n}$$

by manipulations of summation identities. This identity yields the expected value of the maximum of three independent Bernoulli random variables. Computing a closed form for such a left-hand side is now routine work for a computer. We now sketch the computer proof obtained by the algorithm described above.

Recall from the introduction that the binomial coefficients are annihilated by (0.1). It turns out that (0.1) already is is a Gröbner basis. Computing the inner indefinite sum by the algorithm of §3.2 yields the Gröbner basis

$$(n+1-k)S_n + kS_k + k - 2(n+1), \qquad (k+1)S_k^2 - (n+1)S_k + n - k$$

of the annihilating ideal of $\sum_{j=0}^{k}\binom{n}{j}$. Next, computing the cube of this sum by the algorithm of §3.1 yields a Gröbner basis of the following form:

$$S_nS_k + \langle 76 \text{ l.o.t.}\rangle, \qquad S_n^2 + \langle 249 \text{ l.o.t.}\rangle, \qquad S_k^3 + \langle 239 \text{ l.o.t.}\rangle \tag{2.14}$$

where "p l.o.t." stands for "p lower order terms". We introduce undetermined rational functions $\phi_i(k)$, $\psi(k)$ and η_i, each of which depends on n, and we set

$$P = \eta_0 + \eta_1 S_n + \eta_2 S_n^2, \quad \text{and} \quad Q = \phi_0(k) + \phi_1(k)S_k + \phi_2(k)S_k^2 + \psi(k)S_n.$$

Reducing $(S_k - 1)Q - P$ by the Gröbner basis (2.14) and extracting the coefficients of 1, S_n, S_n^2 and S_k yields a linear system of four first order recurrence equations in the ϕ_i's and ψ. Uncoupling the system by any method suggested in §1.4 yields a non-homogeneous linear recurrence of order 4 in $\psi(k)$, in which the η_i's appear linearly in the non-homogeneous part. All the coefficients of this equation are (large) polynomials of degree 19 or 20 in k with polynomial coefficients in n. We then solve this equation by an extension of Abramov's algorithm (Abramov 1989a, 1989b, 1995). This variant mimics the classical extension of Gosper's algorithm described in (Zeilberger 1991), and corresponds to searching for a rational solution ψ *together with* values of the η_i's which allow such a solution to exist. A solution is given by

$$\begin{cases} \eta_0 = -4(2n+1), \quad \eta_1 = -(7n+12), \quad \eta_2 = n+1, \quad \psi(k) = \frac{N(k)}{D(k)}, \\ D(k) = (k-2n-1)(k^2 + (1-2n)k + 2(n^2+n+1))(k-n-1)^3, \end{cases}$$

where $N(k)$ is a polynomial of degree 7 in k. Substituting back into the remaining equations yields values for the $\phi_i(k)$'s, each of which is a rational function with denominator $D(k)$. At this point, summing for k over the integer interval $(0, r)$ yields

$$[(Q(n,k,S_n,S_k)\cdot h)(n,k)]_{k=0}^{k=r+1} + \sum_{k=0}^{r}(P(n,S_n)\cdot h)(n,k) = 0$$

where $h_{n,k} = \left(\sum_{j=0}^{k}\binom{n}{j}\right)^3$, provided that $Q \cdot h$ may be evaluated on $(0, r)$. Actually, the previous equation holds for $r \leq n-3$ only, due to the singularity in $n-k-1$ in the denominator D of Q. We thus set $r = n-3$ to get the non-homogeneous recurrence equation

$$(P(n,S_n)\cdot H)(n) = \sum_{i=0}^{2}\eta_i(n)\sum_{k=n-2}^{n+i} h_{n+i,k} - [(Q(n,k,S_n,S_k)\cdot h)(n,k)]_{k=0}^{k=n-2}$$

where H_n is the sum $\sum_{k=0}^{n} h_{n,k}$ for which we are looking for a closed form. One easily evaluates the non-homogeneous term since $h_{n,n+i} = (2^n)^3$ for $i \geq 0$ and $h_{n,n-i} = (2^n - 1 - \cdots - \binom{n}{n-i})^3$ for small $i \geq 0$. This yields the following equation on the sum:

$$(n+1)H_{n+2} - (7n+12)H_{n+1} - 4(2n+1)H_n + 2(10-9n)8^n = 0.$$

Finally, solving the previous recurrence by Petkovšek's algorithm (Petkovšek 1992) computes the announced closed form evaluation as a linear combination of hypergeometric terms.

3 Closure Properties

The summands and integrands under consideration in the examples of §§1.4, 2.1 and 2.4 are products of ∂-finite functions. An important point is the possibility to generate the systems which define these functions as ∂-finite from a database of "elementary" ∂-finite functions. We begin the section by explaining how this can be performed via the closure properties of ∂-finite functions. To this end, we appeal once again to Gröbner bases to compute normal forms in quotient rings, and adapt the FGLM algorithm commonly used in the theory of commutative Gröbner bases (Faugère *et al.* 1993). Next, we illustrate another method for indefinite summation and integration based on a complete elimination and a Gröbner basis calculation for modules.

3.1 Addition, Product and Derivation of ∂-Finite Functions by the FGLM Algorithm

A nice property of ∂-finite functions—in fact, one of the main reasons for their introduction in (Chyzak and Salvy 1996)—is their closure under addition, product and derivation. More specifically, assume f and g to be two ∂-finite functions with respect to an Ore algebra $\mathbb{O} = \mathbb{F}[\partial; \sigma, \delta]$. Then $\mathbb{O} \cdot f$ and $\mathbb{O} \cdot g$ are two finite-dimensional $\mathbb{F}$-vector spaces which admit finite bases $\{b_i\}_{i=1,\dots,N}$ and $\{c_i\}_{i=1,\dots,M}$, respectively. Then, the derivatives of the sum $f+g$ can be reduced on the generating set constituted of the b_i's and the c_i's. The sum $f+g$ is thus ∂-finite. Similarly, the derivatives of the product fg can be reduced on the generating set constituted of the $c_i b_j$'s. This entails that the product is also ∂-finite. Finally, each derivative $\partial_k \cdot f$ is ∂-finite, as follows from the reduction of its derivatives on the basis of the b_i's.

Computationally, one usually chooses basis elements of the form ∂^α, and uses an extension of the FGLM algorithm (Faugère *et al.* 1993) to take mixed derivatives into account. This extension is described in (Chyzak and Salvy 1996), from which the following example is borrowed.

We compute annihilators for the sum of the exponential function $f(x,y) = \exp(\mu x+\nu y)$ and of the product of Bessel functions $g(x,y) = J_\mu(x)J_\nu(y)$. The functions f and g are defined by the systems

$$\begin{cases} f_x - \mu f = 0, \\ f_y - \nu f = 0 \end{cases} \quad \text{and} \quad \begin{cases} x^2 g_{x,x} + x g_x + (x^2-\mu^2)g = 0, \\ y^2 g_{y,y} + y g_y + (y^2-\nu^2)g = 0 \end{cases}$$

respectively (indices denoting differentiation). First, the algorithm reduces 1, D_y, D_x, D_y^2, D_xD_y and detects that they are independent. Then D_x^2 is reduced and found to satisfy a linear relation with the previous ones, expressed by the following operator:

$$\begin{aligned} p_1 = &-(x^2-\mu^2+x^2\mu^2+\mu x)y^2D_y^2 + x^2(y^2-\nu^2+y^2\nu^2+\nu y)D_x^2 \\ &-(x^2-\mu^2+x^2\mu^2+\mu x)yD_y + x(y^2-\nu^2+y^2\nu^2+\nu y)D_x \\ &-\mu^2y^2\nu^2+x^2\nu y+x^2y^2\nu^2-x^2\mu^2y^2+x^2\mu^2\nu^2-\mu x y^2+\mu x\nu^2-\mu^2\nu y. \end{aligned}$$

Next, the algorithm continues by reducing D_y^3 and finds another relation

$$\begin{aligned} p_2 = &y^2(y^2-\nu^2+\nu y+y^2\nu^2)D_y^3 - y(y^3\nu+y^3\nu^3-y^2-2\nu y-\nu^3y+3\nu^2)D_y^2 \\ &+(y^4+y^4\nu^2-y^3\nu^3-y^2-y^2\nu^4-4y^2\nu^2-\nu^2+\nu^4)D_y \\ &+\nu(-y^4-y^4\nu^2+y^2-\nu^4+2y^2\nu^2+\nu^2+y^2\nu^4-y^3\nu+3\nu^3y). \end{aligned}$$

Finally, the reduction of $D_xD_y^2$ produces the operator

$$p_3 = y^2D_xD_y^2 - \mu y^2D_y^2 + yD_xD_y - \mu yD_y + (y^2-\nu^2)D_x - \mu(y^2-\nu^2).$$

The system $\{p_1, p_2, p_3\}$ makes it possible to rewrite any derivative of the sum $f+g$ as a linear combination of 5 derivatives.

3.2 Indefinite Summation and Integration by Gröbner Bases

In §1.4, we looked for the indefinite sum or integral of a function in the module spanned by this function. More specifically, for a ∂-finite function f with respect to an Ore algebra $\mathbb{O}$, the method either returns an explicit indefinite sum or integral F in the form $F = Q\cdot f$ for $Q\in\mathbb{O}$ or a proof that no such F exists. In the present section, we turn to the unfavorable case. We no longer search for an explicit form for F, but for a system of operators that describes F as a ∂-finite function. This is performed by Gröbner basis calculations, as exemplified by the following example borrowed from (Chyzak and Salvy 1996). Informally, the idea is to apply creative telescoping by regarding an indefinite sum or integral as a definite one.

Let $f(x,y) = (1+xy+y^2)^{-2}$. We look for a specification of its indefinite integral $F(x,y) = -\int_y^{+\infty} f(x,t)\,dt$, which is defined for any x. Working in the

Ore algebra $\mathbb{O} = \mathbb{C}(x, y)[D_x; 1, D_x][D_y; 1, D_y]$, the algorithm of §1.4 returns a proof that no indefinite integral exists in $\mathbb{O} \cdot f = \mathbb{C}(x, y)$. We thus try another method. The function f is annihilated by both operators

$$p_x = (1 + xy + y^2)D_x + 2y \qquad \text{and} \qquad p_y = (1 + xy + y^2)D_y + 2x + 2y,$$

from which trivially follows that $p_x D_y$ and $p_y D_y$ annihilate the indefinite integral F. Our goal is to find other operators that annihilate F, so as to achieve a ∂-finite description. The elimination of y between the polynomials p_x and p_y in $\mathbb{C}(x)[y][D_x; 1, D_x][D_y; 1, D_y]$ yields

$$P = AD_y + B \quad \text{where} \quad \begin{cases} A = x(x-2)(x+2)D_x + xD_y + 2(x^2+1), \\ B = -x(x-2)(x+2)D_x^2 - 4(x^2+1)D_x. \end{cases}$$

Creative telescoping in the definite case would go on by the integration of the identity $P \cdot f = 0$. Similarly, $h = P \cdot F$ is an indefinite integral of $P \cdot f$ and we immediately have $D_y \cdot h = 0$, so that h is constant with respect to y. Since F and all its cross-derivatives tend to 0 when y tends to $+\infty$, so does h. Thus $h = 0$, i.e., $A \cdot f + B \cdot F = 0$. It then suffices to find a left annihilator for A modulo $\operatorname{Ann}_{\mathbb{O}} f$ to get an operator that annihilates F. To compute those C such that $CA = 0$ modulo the ideal generated by p_x and p_y in $\mathbb{O}$, we adapt a method used in the commutative case for the calculation of ideal quotients and based on the calculation of modules of syzygies (Becker and Weispfenning 1993: Algorithm IDEALDIV1). More precisely, we introduce new commutative indeterminates t, u, v and w, and eliminate t between the polynomials $u - tA$, $v - tp_x$ and $w - tp_y$, by computing a Gröbner basis in the algebra $\mathbb{C}(x, y, u, v, w)[t][D_x; 1, D_x][D_y; 1, D_y]$. In this Gröbner basis, those polynomials which do not involve t are of the form $uU + vV + wW$, where U, V and W are polynomials in $\mathbb{O}$ such that $UA + Vp_x + Wp_y = 0$. Consider the U's obtained in this way. Their right multiplication by B yields new operators that annihilate F:

$$\begin{cases} (y(x^3y + 4x^2 + 4 + 16xy + 4x^2y^2 + 4y^2) \\ \qquad + (1 + xy + y^2)(x^2y^2 + y^2 + 3xy + 1)D_y) \\ \qquad\qquad \times (4(x^2+1)D_x + x(x-2)(x+2)D_x^2), \\ (32y^2 + 32xy + 8 + 48x^2y^2 + 36xy^3 + 12x^3y^3)D_x \\ \qquad + x(15y^4 + 5x^2y^4 + 24xy^3 + 8x^3y^3 - 2y^2 + 32x^2y^2 + 28xy + 7)D_x^2 \\ \qquad\qquad + (x-2)(x+2)(1 + xy + y^2)(x^2y^2 + y^2 + 3xy + 1)D_x^3. \end{cases}$$

Computing a Gröbner basis from those polynomials adjoined to the ones known beforehand, namely $p_x D_y$ and $p_y D_y$, finally yields a basis of a subideal of $\operatorname{Ann}_{\mathbb{O}} F$ constituted of $p_x D_y$, $p_y D_y$ and a third polynomial

$$x(x^2-4)(1 + xy + y^2)D_x^2 + 4(x^2+1)(1 + xy + y^2)D_x \\ -(2x^2y^2 + 2y^2 + 6xy + 2)D_y.$$

This yields a description of F as a ∂-finite function.

Note that the method yields a subideal of $\mathfrak{I} = \mathrm{Ann}_{\mathbb{O}} F$ only. Due to the intrinsic weakness in the elimination step of the algorithm, the method may well fail to find the ideal $\mathfrak{I}$, and even a subideal of $\mathfrak{I}$ which describes F as ∂-finite. Reasons for this weakness are discussed in the next section.

4 Dimension and Holonomy

Intentionally, we have not discussed the *termination* of the algorithms yet, nor have we justified their *success* in case of termination. More specifically, all the algorithms for summation and integration in this tutorial are based on the existence of an eliminated polynomial of the form (2.9) or (2.10), and fail or loop forever if no such polynomial exists. In this concluding section, we interpret this existence in terms of *ideal dimension*, and apply tools based on Gröbner bases to predict the termination of the algorithms, and in some cases decide when to stop them.

The theory of ideal dimension in commutative polynomial algebras extends to a similar theory in Ore algebras. This is the basis for the theory of *holonomy* (Bernstein 1971, 1972; Björk 1979; Coutinho 1995; Ehlers 1987; Kashiwara 1978) in the differential case. Let $\mathbb{O}$ be an Ore algebra $\mathbb{K}[\mathbf{x}][\boldsymbol{\partial};\boldsymbol{\sigma},\boldsymbol{\delta}]$ which satisfies Eq. (1.3) for $c_{i,j}$'s of total degree at most 1 in $\mathbf{x}$. Let F_n be the $\mathbb{K}$-vector space in $\mathbb{O}$ spanned by all terms $\mathbf{x}^{\alpha}\boldsymbol{\partial}^{\beta}$ with total degree $|\boldsymbol{\alpha}| + |\boldsymbol{\beta}|$ at most n. Then, for an ideal $\mathfrak{I} \subseteq \mathbb{O}$, the *Hilbert function* $h(n) = \dim_{\mathbb{K}}(F_n/\mathfrak{I} \cap F_n)$ asymptotically agrees with a polynomial called *Hilbert polynomial*. The degree of this polynomial is called the *dimension* of the ideal $\mathfrak{I}$. These invariants can be attributed to the module $\mathbb{O}/\mathfrak{I}$ in view of the following generalization to any module. Let $\mathfrak{M}$ be an $\mathbb{O}$-module with generating set U, and let $\Gamma_n = F_n \cdot U$. Then all previous invariants are defined starting from $h(n) = \dim_{\mathbb{K}} \Gamma_n$. In the case of an ideal, U is the singleton consisting of the class of 1 modulo $\mathfrak{I}$, and $\Gamma_n = F_n/\mathfrak{I} \cap F_n$. All the previous invariants can be computed from a Gröbner basis for $\mathfrak{I}$ (Bayer and Stillman 1992; Becker and Weispfenning 1993).

In the differential case, the combinatorial description of the dimension of a module has several algebraic and analytic interpretations and is ruled by *Bernstein's inequality*: when the algebra is generated by r indeterminates x_i and r derivation operators D_{x_i}, Bernstein (1972) proved that the dimension d of a non-zero module $\mathfrak{M}$ satisfies $d \geq r$. A module over the polynomial algebra $\mathbb{O}_{\mathrm{p}} = \mathbb{K}[\mathbf{x}][\boldsymbol{\partial};\boldsymbol{\sigma},\boldsymbol{\delta}]$ is called *holonomic* when its dimension is precisely the number r of derivation operators in the algebra; a function f in $\mathbf{x}$ is called *holonomic* when its annihilating ideal $\mathrm{Ann}_{\mathbb{O}_{\mathrm{p}}} f$ is such that the module $\mathbb{O}_{\mathrm{p}}/\mathrm{Ann}_{\mathbb{O}_{\mathrm{p}}} f \simeq \mathbb{O}_{\mathrm{p}} \cdot f$ is holonomic. Moreover, a holonomic function f is ∂-finite with respect to the Ore algebra $\mathbb{O}_{\mathrm{r}} = \mathbb{K}(\mathbf{x})[\boldsymbol{\partial};\boldsymbol{\sigma},\boldsymbol{\delta}]$. The converse implication is a deep result due to Kashiwara (1978); see also (Takayama

1992). For further use of Gröbner bases in relation to holonomy, see (Sturmfels and Takayama 1998).

The motivation for appealing to the theory of holonomy in the context of symbolic summation and integration is that an eliminated operator of the form (2.9) exists as soon as the function is holonomic. More precisely, for *any* choice of $r-1$ indeterminates from the $2r$ indeterminates of the algebra, the holonomy of an ideal $\mathfrak{I} \subseteq \mathbb{O}_p$ ensures the existence of a non-zero operator in $\mathfrak{I}$ which does not involve these $r-1$ indeterminates. It follows that the multiple summation and integration of a holonomic function with respect to $r-1$ variables can be performed by a single elimination. See the examples of §§2.2 and 2.3.

To illustrate the combinatorial invariants described above, consider the example of the Appell F_4 bivariate hypergeometric function, which is obviously ∂-finite in view of (0.3). Computing a Gröbner basis for this set of operators in the polynomial algebra $\mathbb{O}_p = \mathbb{Q}(a,b,c,d)[x,y][D_x;1,D_x][D_y;1,D_y]$ with respect to a term order such that $D_x \prec D_y \prec x \prec y$ yields a system of 6 operators with respective leading terms $xy^2D_xD_y$, $x^2yD_x^2$, $x^3D_x^2$, $x^2y^2D_x^3$, $xy^3D_y^4$ and $y^4D_x^2D_y^3$. Computing the Hilbert polynomial of the corresponding ideal in $\mathbb{O}_p$ yields $\frac{21}{2}n^2 - \frac{85}{2}n + 73$. This entails that the module $\mathbb{O}_p \cdot F_4$ is of dimension 2, the number of derivation operators in the algebra. The module is thus holonomic, and any of the integration algorithms presented in this tutorial terminates on the input (0.3) computing an annihilating operator for the corresponding integral.

As another example, recall from §1.3 that the function $e^{\sin z}$ is not ∂-finite with respect to the algebra $\mathbb{O}_r = \mathbb{C}(z)[D_z;1,D_z]$. Correspondingly, the Hilbert polynomial of the module $\mathbb{O}_p \cdot f$ over $\mathbb{O}_r = \mathbb{C}[z][D_z;1,D_z]$ is $\frac{n^2}{2}+\frac{3n}{2}+1$ so that the module is of dimension 2, in particular more than 1, and the function is not holonomic. The integration of this function is therefore not directly tractable by the algorithms of creative telescoping presented in this tutorial.

As a non-differential example, the function $\binom{n}{k}$ viewed with respect to the algebra $\mathbb{O}_p = \mathbb{Q}[n,k][S_n;S_n,0][S_k;S_k,0]$ has an annihilating ideal $\mathfrak{I}$ with Hilbert polynomial $2n+1$. The corresponding dimension is 1, which shows that Bernstein's inequality does not hold in general in non-differential Ore algebras. However, the summation of this function is amenable to creative telescoping, provided a description of $\mathfrak{I}$ is given. Indeed, such a summation is always possible as soon as the dimension of the module $\mathbb{O}_p/\mathfrak{I}$ is less than or equal to 2.

¿From the examples of this tutorial, it might seem easy to integrate a function f as soon as it known to be holonomic. However, although a description of f as a ∂-finite function with respect to $\mathbb{O}_r$ is computationally easy to obtain, a holonomic ideal of annihilating operators in the polynomial Ore algebra $\mathbb{O}_p$ can be really hard to get. An example is provided by the function $f = 1/(y^2 - y + x)$. Computing integrals of f with respect to y

requires finding an operator free from y in $\mathbb{O}_p = \mathbb{Q}[x,y][D_x;1,D_x][D_y;1,D_y]$. The annihilating ideal of f in $\mathbb{O}_r = \mathbb{Q}(x,y)[D_x;1,D_x][D_y;1,D_y]$ is $\mathfrak{K} = \mathbb{O}_r P + \mathbb{O}_r Q$ where $P = D_y(y^2 - y + x) = (y^2 - y + x)D_y + (2y - 1)$ and $Q = D_x(y^2 - y + x) = (y^2 - y + x)D_x + 1$. Any larger ideal in $\mathbb{O}_r$ is $\mathbb{O}_r$ itself. The operator $U = D_y^2 + (4x-1)D_x^2 + 6D_x$ annihilates f, so that $U \in \mathfrak{K}$, hence $U \in \mathfrak{K} \cap \mathbb{O}_p$. However, U is *not* an element of $\mathfrak{J} = \mathbb{O}_p P + \mathbb{O}_p Q$ in $\mathbb{O}_p$. It follows that the algorithm based on a simple elimination of §2.1 fails to compute any integral, as long as it is given $\mathfrak{J}$ in input. Even worse, the algorithm of §2.3 fails to terminate. However, if one adjoins the operator $R = (y^2 - y + x)D_y D_x - 2D_y \in \mathfrak{K}$, the ideal $\mathbb{O}_p P + \mathbb{O}_p Q + \mathbb{O}_p R \subseteq \mathbb{O}_p$ contains the operator U and both algorithms find it. A similar phenomenon arises when integrating with respect to x. On the other hand, the algorithms of §1.4 and §2.4 only require the ∂-finite description and they succeed in finding the integrals.

Acknowledgements

Philippe Dumas, Philippe Flajolet, Peter Paule, Bruno Salvy and Berndt Sturmfels read and commented on preliminary versions of this tutorial. In particular, Philippe Dumas and Peter Paule provided me with extremely detailed and impressively accurate comments, which permitted me to improve on the paper significantly. I thank all these contributors most warmly for their precious help.

This work was supported in part by the Long Term Research Project Alcom-IT (#20244) of the European Union.

References

Abramov, S. A. (1989a): Problems in computer algebra that are connected with a search for polynomial solutions of linear differential and difference equations. *Moscow Univ. Comput. Math. Cybernet.*, **3**:63–68. Transl. from: *Vestn. Moskov. univ. Ser. XV Vychisl. mat. kibernet.*, **3**:56–60.

Abramov, S. A. (1989b): Rational solutions of linear differential and difference equations with polynomial coefficients. *USSR Computational Mathematics and Mathematical Physics*, **29**(11):1611–1620. Transl. from: Zh. vychisl. Mat. i mat. Fiz.

Abramov, S. A. (1995): Rational solutions of linear difference and q-difference equations with polynomial coefficients. In: Levelt, A. (ed.), *Symbolic and algebraic computation (ISSAC '95)*, Montreal, Canada. ACM Press, New York, pp. 285–289.

Abramov, S. A., Bronstein, M., Petkovšek, M. (1995): On polynomial solutions of linear operator equations. In: Levelt, A. (ed.), *Symbolic and algebraic computation (ISSAC '95)*, Montreal, Canada. ACM Press, New York, pp. 290–296.

Abramov, S. A., Kvashenko, K. Y. (1991): Fast algorithms for the search of the rational solutions of linear differential equations with polynomial coefficients. In: Watt, S. M. (ed.), *Symbolic and algebraic computation (ISSAC '91)*, Bonn, Germany. ACM Press, New York, pp. 267–270.

Abramov, S. A., Zima, E. V. (1996): A universal program to uncouple linear systems. Preprint.

Almkvist, G., Zeilberger, D. (1990): The method of differentiating under the integral sign. *J. Symbolic Comput.*, **10**:571–591.

Andrews, G. E. (1974): An analytic generalization of the Rogers-Ramanujan identities for odd moduli. *Proc. Nat. Acad. Sci. U.S.A.*, **71**:4082–4085.

Barkatou, M. A. (1993): An algorithm for computing a companion block diagonal form for a system of linear differential equations. *Appl. Algebra Engrg. Comm. Comput.*, **4**:185–195.

Bayer, D., Stillman, M. (1992): Computation of Hilbert functions. *J. Symbolic Comput.*, **14**:31–50.

Becker, T., Weispfenning, V. (1993): *Groebner bases.* Springer-Verlag *(Graduate Texts in Mathematics, vol. 141).* In cooperation with Heinz Kredel.

Bernstein, I. N. (1971): Modules over a ring of differential operators, study of the fundamental solutions of equations with constant coefficients. *Funct. Anal. Appl.*, **5**(2):1–16 (Russian); 89–101 (English translation).

Bernstein, I. N. (1972): The analytic continuation of generalized functions with respect to a parameter. *Funct. Anal. Appl.*, **6**(4):26–40 (Russian); 273–285 (English translation).

Björk, J. E. (1979): *Rings of Differential Operators.* North Holland, Amsterdam.

Bronstein, M., Petkovšek, M. (1994): On Ore rings, linear operators and factorisation. *Programmirovanie*, **1**:27–44. Also available as Research Report 200, Informatik, ETH Zürich.

Bronstein, M., Petkovšek, M. (1996): An introduction to pseudo-linear algebra. *Theoret. Comput. Sci.*, **157**(1).

Buchberger, B. (1998): Introduction to Gröbner bases. In the present volume.

Calkin, N. J. (1994): A curious binomial identity. *Discrete Math.*, **131**(1–3):335–337.

Chyzak, F. (1997): An extension of Zeilberger's fast algorithm to general holonomic functions. In: *Formal Power Series and Algebraic Combinatorics, 9th Conference (FPSAC '97)*, Vienna, Austria, pp. 172–183.

Chyzak, F., Salvy, B. (1996): Non-commutative elimination in Ore algebras proves multivariate holonomic identities. Research Report 2799, Institut National de Recherche en Informatique et en Automatique. Submitted to: *J. Symbolic Comput.*.

Cohn, P. M. (1971): *Free Rings and Their Relations.* Academic Press (*London Mathematical Society Monographs, number 2).*

Coutinho, S. C. (1995): *A Primer of Algebraic D-modules.* Cambridge University Press (*London Mathematical Society Student Texts, number 33).*

Cox, D., Little, J., O'Shea, D. (1992): *Ideals, Varieties, and Algorithms. An Introduction to Computational Algebraic Geometry and Commutative Algebra.* Springer-Verlag, New York.

Ehlers, F. (1987): *The Weyl Algebra*, volume 2 of *Perspectives in Mathematics*, chapter 5, pages 173–205. In: Borel, A. et al.: *Algebraic D-Modules.* Academic Press (*Perspectives in Mathematics, vol. 2).*

Erdélyi, A. (1981): *Higher Transcendental Functions*, vol. 1, 2, 3. R. E. Krieger Publishing Company, Malabar, Florida. Second edition.

Faugère, J., Gianni, P., Lazard, D., Mora, T. (1993): Efficient computation of zero-dimensional Gröbner bases by change of ordering. *J. Symbolic Comput.*, **16**:329–344.

Galligo, A. (1985): Some algorithmic questions on ideals of differential operators. In: Caviness, B. F. (ed.), *Proceedings EUROCAL '85*. Springer-Verlag, pp. 413–421 (*Lecture Notes in Computer Science, vol. 204)*.

Gordon, B. (1961): A combinatorial generalization of the Rogers-Ramanujan identities. *Amer. J. Math.*, **83**:393–399.

Gosper, R. W. (1978): Decision procedure for indefinite hypergeometric summation. *Proc. Nat. Acad. Sci. U.S.A.*, **75**(1):40–42.

Joly, P., Rhaouti, L. (1997): Private communication.

Kandri-Rody, A., Weispfenning, V. (1990): Non-commutative Gröbner bases in algebras of solvable type. *J. Symbolic Comput.*, **9**:1–26.

Karr, M. (1981): Summation in finite terms. *J. ACM*, **28**(2):305–350.

Karr, M. (1985): Theory of summation in finite terms. *J. Symbolic Comput.*, **1**:303–315.

Kashiwara, M. (1978): On the holonomic systems of linear differential equations II. *Invent. Math.*, **49**:121–135.

Kredel, H. (1993): *Solvable Polynomial Rings. (Reihe Mathematik)*, Verlag Shaker, Aachen, Germany, ISBN 3-86111-342-2. After: Kredel, H. (1992): *Solvable Polynomial Rings*. Doctoral Dissertation, Univ. Passau.

Ore, O. (1933): Theory of non-commutative polynomials. *Ann. of Math.*, **34**:480–508.

Paule, P. (1985): On identities of the Rogers-Ramanujan type. *J. Math. Anal. Appl.*, **107**(1):255–284.

Paule, P., Riese, A. (1997): A Mathematica q-analogue of Zeilberger's algorithm based on an algebraically motivated approach to q-hypergeometric telescoping. In: *Special functions, q-series and related topics*, Toronto, Ontario, 1995. Fields Inst. Commun. **14**:179–210. Amer. Math. Soc., Providence.

Petkovšek, M. (1992): Hypergeometric solutions of linear recurrences with polynomial coefficients. *J. Symbolic Comput.*, **14**:243–264.

Petkovšek, M., Wilf, H., Zeilberger, D. (1996): *A=B*. A. K. Peters, Wellesley, Massachusset, ISBN 1-56881-063-6.

Prudnikov, A. P., Brychkov, Y. A., Marichev, O. I. (1986): *Integrals and Series. Volume 2: Special functions*. Gordon and Breach. First edition by Nauka, Moscow (1983).

Risch, R. H. (1969): The problem of integration in finite terms. *Trans. Amer. Math. Soc.*, **139**:167–189.

Risch, R. H. (1970): The solution of the problem of integration in finite terms. *Bull. Amer. Math. Soc.*, **76**:605–608.

Rogers, L. J. (1894): Second memoir on the expansion of certain infinite products. *Proc. London Math. Soc.*, **25**:318–343.

Sturmfels, B., Takayama, N. (1998): Gröbner bases and hypergeometric functions. In the present volume.

Takayama, N. (1989): Gröbner basis and the problem of contiguous relations. *Japan J. Appl. Math.*, **6**(1):147–160.

Takayama, N. (1990a): An algorithm of constructing the integral of a module—an infinite dimensional analog of Gröbner basis. In: Watanabe, S. and Nagata, M. (eds.), *Symbolic and algebraic computation (ISSAC '90)*, Kyoto, Japan. ACM and Addison-Wesley, pp. 206–211.

Takayama, N. (1990b): Gröbner basis, integration and transcendental functions. In: Watanabe, S. and Nagata, M. (eds.), *Symbolic and algebraic computation (ISSAC '90)*, Kyoto. Japan, ACM and Addison-Wesley, pp. 152–156.

Takayama, N. (1992): An approach to the zero recognition problem by Buchberger algorithm. *J. Symbolic Comput.*, **14**:265–282.

Trigg, L. (1996): Private communication.

Wegschaider, K. (1997): *Computer generated proofs of binomial multi-sum identities. (Diploma Thesis)*, RISC, Univ. Linz.

Wilf, H. S., Zeilberger, D. (1992a): An algorithmic proof theory for hypergeometric (ordinary and "q") multisum/integral identities. *Invent. Math.*, **108**:575–633.

Wilf, H. S., Zeilberger, D. (1992b): Rational function certification of multisum/integral/"q" identities. *Bull. Amer. Math. Soc.*, **27**(1):148–153.

Zeilberger, D. (1990a): A fast algorithm for proving terminating hypergeometric identities. *Discrete Math.*, **80**:207–211.

Zeilberger, D. (1990b): A holonomic systems approach to special functions identities. *J. Comput. Appl. Math.*, **32**:321–368.

Zeilberger, D. (1991): The method of creative telescoping. *J. Symbolic Comput.*, **11**:195–204.

Gröbner Bases and Invariant Theory

Wolfram Decker and Theo de Jong

FB 9 Mathematik, Universität des Saarlandes, Saarbrücken

"As all the roads lead to Rome so I find in my own case at least that all algebraic inquiries, sooner or later, end at the Capitol of modern algebra over whose shining portal is inscribed the Theory Of Invariants." (Sylvester (1864, p. 380))

"Like the Arabian phoenix rising out of its ashes, the theory of invariants, pronounced dead at the turn of the century, is once again at the forefront of mathematics. During its long eclipse, the language of modern algebra was developed, a sharp tool now at last being applied to the very purpose for which it was invented." (Kung *et al.* (1984))

0 Introduction

Historically, one root of invariant theory can be found in a problem of number theory, namely the problem of representing integers by quadratic binary forms

$$F = ax^2 + 2bxy + cy^2,$$

where a, b, c are integer coefficients. The discovery by Lagrange (1773, 1775) that such a representation is far from being unique lead to the notion of equivalence of quadratic binary forms. This was introduced in a rigorous way by Gauss (1801) in his Disquisitiones Arithmeticae. Gauss called two forms F and $\tilde{F}$ equivalent (properly equivalent) if F can be transformed into $\tilde{F}$ by a change of coordinates with determinant ± 1 $(+1)$. A key point is that the discriminant $(b^2 - ac)$ is invariant under such coordinate transformations (invariant functions in a, b, c are tools to distinguish between inequivalent forms). Indeed, in the situation of Gauss, if a change of coordinates

$$x \to \alpha x + \beta y, \quad y \to \gamma x + \delta y$$

transforms F into the form

$$\tilde{F} = \tilde{a}x^2 + 2\tilde{b}xy + \tilde{c}y^2,$$

then there is the relation

$$\tilde{b}^2 - \tilde{a}\tilde{c} = (b^2 - ac)(\alpha\delta - \beta\gamma)^2.$$

This relation is of course not only true in the modular context, we may allow complex coefficients as well. We therefore arrive at the following result which we state in the modern language of representation theory. The special linear group $\mathrm{SL}_2(\mathbb{C})$ acts on the vector space V of quadratic binary forms over $\mathbb{C}$ by linear substitution. The dual action on the space of coordinates a, b, c gives rise to an action of $\mathrm{SL}_2(\mathbb{C})$ on the symmetric algebra

$$\mathbb{C}[a,b,c] = \mathbb{C}[V] = S(V^*) = \bigoplus_{d \geq 0} S^d V^*,$$

that is, on the algebra of polynomial functions on V. The result is that the discriminant $\delta = b^2 - ac$ is an *invariant* under this action. Even more is true: every invariant can be written as a polynomial in δ (see for example Sturmfels (1993)). In other words, the graded *ring of invariants* $\mathbb{C}[V]^{\mathrm{SL}_2(\mathbb{C})}$ is generated as a $\mathbb{C}$–algebra by δ:

$$\mathbb{C}[V]^{\mathrm{SL}_2(\mathbb{C})} = \mathbb{C}[\delta].$$

Algebraic invariants such as the discriminant naturally show up in algebraic geometry, when one asks for properties of geometric objects, which are invariant under certain classes of transformations. For example, the geometric significance of the discriminant is that a quadratic binary form defines two distinct points on the projective line $\mathbb{P}^1(\mathbb{C})$ if and only if its discriminant is non–zero. People became interested in such invariant properties especially after the introduction of homogeneous coordinates by Möbius (1827) and Plücker (1830). This was a major impetus for invariant theory.

In the first decades of invariant theory, say between 1840 and 1870, people were mainly concerned with the discovery of particular invariants. The major case of interest was that of forms of degree d in n variables, d and n arbitrary, with $\mathrm{SL}_n(\mathbb{C})$ acting by linear substitution. One way of producing new invariants is to apply certain *covariants* (classically called *concomitants*) like the Hessian to an invariant already known. Cayley's *Ω–process* (compare Sturmfels (1993)) generates invariants more systematically by means of a differential operator. Another important tool for computing and presenting invariants is the *symbolic method* developed by Aronhold, Clebsch and Gordan (compare Dieudonné *et al.* (1971) , Kung *et al.* (1984), Sturmfels (1993)), and referred to as "the great war–horse of nineteenth century invariant theory" by Weyl, and as "Formelngestrüpp" or "Rechnerei" by E. Noether.

As time went on people realized that they could find in many explicit cases such as the example above a *fundamental system of invariants*, that is, a minimal finite system of homogeneous generators of the ring of invariants under consideration. The question whether there always exists a fundamental system of invariants is nowadays also known as *Hilbert's 14th problem* (Hilbert 1900). Gordan (1868) gave a positive answer to this question in the case of binary forms of any given degree. The proof, which is rather complicated, is based on the symbolic method. It is constructive and can in principle be used to compute fundamental systems of invariants (compare also Kung *et al.* (1984), Sturmfels (1993)).

Classical invariant theory culminated in Hilbert's two landmark papers (1890) and (1893). In these papers Hilbert introduced stunning new ideas which have deeply influenced the development of modern algebra and algebraic geometry. As an application of his ideas Hilbert proved a finiteness theorem for forms in n variables, n arbitrary.

The first paper contains a non-constructive proof of this result which relies on the *Basis Theorem*, proved in the same paper, and the Ω–process as a tool to generate invariants. Hilbert remarked that his proof actually applies in a wider range of cases, namely in those cases in which there exists an analogue to the Ω–process for the generation of invariants (compare also Weyl (1946, Appendix C)). Thus, in modern terminology, it applies to rational representations of linearly reductive groups. In particular, it applies to the *classical groups*, to *tori*, and to *finite groups* in the *non–modular case*, that is, in the case where the group order $|G|$ is invertible in the field of coefficients k. Hilbert also studied the successive relations (*syzygies*) between the elements $f_1, \ldots, f_r$ of a fundamental system of invariants. This chain of syzygies is of finite length $\leq r$ by Hilbert's *syzygy theorem* which is proved in the same paper. As a corollary of the syzygy theorem Hilbert proved the *Hilbert–Serre Theorem* on the structure of the *Hilbert function.*

In the second paper Hilbert gave a constructive proof for his finiteness result. This paper was Hilbert's quick answer to those of his fellow mathematicians who harshly criticized the non–constructiveness of his first proof. The second paper contains the *Nullstellensatz*, the *Hilbert–Mumford criterion* for *Null-forms*, and the *"Noether normalization theorem (graded case)"*.

Two of E. Noether's papers on invariant theory are concerned with the finiteness theorem for finite groups. In the first one (1916) she gave two short constructive proofs which also yield an explicit bound for the degree of the generators. In this paper Noether worked in characteristic zero but, under some assumptions on the group order, her ideas also apply to the case of characteristic p (compare for example Smith (1995)). The second paper (1926) settles the general case of arbitrary characteristic with a non–constructive proof.

Highlights of post–Hilbert invariant theory include Weyl's work on semi–simple groups in characteristic zero, Nagata's counterexample to Hilbert's 14th problem, Mumford's geometric invariant theory, and work of Nagata, Haboush and others, who proved the finiteness theorem for rational representations of *reductive groups* (compare Newstead (1978, pp. 90–92) for a short history of Hilbert's 14th problem and precise references). For an overview concerning degree bounds we refer to Derksen *et al.* (1995) .

Nowadays, with powerful computers available, one is tempted to come back to the explicit computation of invariants in interesting examples coming from algebraic geometry, invariant theory, physics, combinatorics (see Stanley (1979), or coding theory (see Sloane (1977)). The key point is, as demonstrated in Sturmfels' book (1993), that the classical computational techniques and Gröbner bases (Buchberger 1965, 1970, 1985, 1998) fit nicely together, and yield modern algorithms for invariant theory. In this context Gröbner bases are mainly used in order to compute syzygies and Hilbert functions, and for elimination (compare Eisenbud (1995) for an introduction to Gröbner bases). Another application of Gröbner bases is the computation of normalization (Vasconcelos 1991), (Gianni *et al.* 1996), (de Jong 1997), which is needed in Hilbert's algorithm (1893).

In this note we review some of the algorithms. We focus on recent developments which are not yet covered by Sturmfels (1993). We discuss variants of Kemper's algorithm for finite groups in arbitrary characteristic (Kemper 1996), and the remarkable discovery of Derksen (1997), that the ideas of Hilbert (1890) can be refined and turned into an algorithm.

Some of the algorithms presented here are already implemented, others are about to be implemented. We mention the Maple package Invar for finite groups by Kemper (1993), the Singular library finvar.lib for finite groups by Heydtmann (1997), and the package for finite groups coming with Magma (compare Kemper *et al.* (1997)).

1 Basic definitions and problems

Our basic setting consists of a group G, a field k, a vector space V over k of finite dimension n, and a representation $\rho : G \to \mathrm{GL}(V)$ of G on V. We suppose that coordinates $x_1, \ldots, x_n \in V^*$ are chosen. The action of G on V induces an action of G on V^*,

$$(\pi f)(v) = f(\rho(\pi^{-1})v),$$

and thus on

$$k[V] := k[x] := k[x_1, \ldots, x_n] := S(V^*) = \bigoplus_{d \geq 0} S^d V^*,$$

that is, on the graded k–algebra of polynomial functions on V (as usual be careful with this notion over finite fields). The basic objects of study in invariant theory are the fixed point sets of such actions.

Definition 1.1. The *ring of invariants* in the above setting is the graded subalgebra

$$k[V]^G := k[x]^G := \{f \in k[V] \mid \pi f = f \text{ for any } \pi \in G\} \subset k[V]. \quad \square$$

$k[V]^G$ only depends on the image $\rho(G) \subset \mathrm{GL}(V)$. We therefore may and will if necessary suppose that G is given as a subgroup of $\mathrm{GL}(V)$.

Definition 1.2. If $k[V]^G$ is finitely generated as a k–algebra then any minimal (that is irredundant) finite system $f_1, \ldots, f_r$ of homogeneous generators is called a *fundamental system of invariants.* □

If $k[V]^G$ is finitely generated over k then it is a Noetherian ring. Homogeneous elements $f_1, \ldots, f_r \in k[V]^G$ of positive degree generate $k[V]^G$ as a k–algebra if and only if they generate the ideal $k[V]^G_+ = \bigoplus_{d \geq 1} k[V]^G_d \subset k[V]^G$, or in turn if and only if the residue classes $\bar{f}_1, \ldots, \bar{f}_r$ generate $k[V]^G_+/(k[V]^G_+)^2$ as a vector space over k. This follows from Nakayama's Lemma. Thus the number of elements (and their degrees) in a fundamental system of invariants is uniquely determined as the embedding dimension of $k[V]^G$. The system itself however, is in general not uniquely determined.

Example 1.3. The symmetric group S_n acts on $k[x_1, \ldots, x_n]$ by permuting the coordinates. Elements of the corresponding ring of invariants $k[x]^{S_n}$ are called *symmetric polynomials.* Examples are the *elementary symmetric polynomials*

$$\sigma_1(x_1, \ldots, x_n) := x_1 + \ldots + x_n,$$
$$\vdots$$
$$\sigma_n(x_1, \ldots, x_n) := x_1 \cdots x_n.$$

It is well-known that every symmetric polynomial can be written as a polynomial in the elementary symmetric polynomials. In other words, $\sigma_1, \ldots, \sigma_n$ is a fundamental system of invariants. Even more is true: the expression of a symmetric polynomial in the elementary symmetric polynomials is unique since $\sigma_1, \ldots, \sigma_n$ are algebraically independent. Thus

$$k[x]^{S_n} = k[\sigma_1, \ldots, \sigma_n]$$

is a polynomial algebra. In characteristic zero, another system of algebraically independent generators of $k[x]^{S_n}$ over k is provided by the power sums

$$p_l(x_1, \ldots, x_n) = x_1^l + \cdots + x_n^l, \quad l = 1, \ldots, n. \quad \square$$

In general $k[V]^G$ is not a polynomial algebra. (Compare Mumford *et al.* (1994, Appendix to Chapter 4, A) for a discussion of classification results.) This means that in general we expect algebraic relations among the elements of a fundamental system of invariants. Thus we face two fundamental problems when studying a ring of invariants $k[V]^G$.

- **First fundamental problem of invariant theory**: Find a fundamental system of invariants $f_1, \ldots, f_r$ of $k[V]^G$.

- **Second fundamental problem of invariant theory**: Find the *syzygies* of $k[V]^G$.

The second problem means to find the successive relations between $f_1, \ldots, f_r$. To be more precise, let $z_1, \ldots, z_r$ be new variables, $\deg(z_i) = \deg(f_i)$, and consider the graded morphism

$$\phi_0 : k[z_1, \ldots, z_r] \to k[V]^G, \quad z_i \mapsto f_i.$$

Then the kernel I of ϕ_0 describes the algebraic relations between $f_1, \ldots, f_r$. I is finitely generated by Hilbert's basis theorem (1890). (Compare Gordan (1899), Cox *et al.* (1997) for a proof involving Gröbner bases.) There is an obvious but rather costly algorithm for the computation of I:

Remark 1.4. The algebraic relations between $f_1, \ldots, f_r$ can be computed as follows. Consider the ideal

$$\tilde{I} = (f_1 - z_1, \ldots, f_r - z_r) \subset k[x_1, \ldots, x_n, z_1, \ldots, z_r].$$

Choose an elimination order on $k[x_1, \ldots, x_n, z_1, \ldots, z_r]$ with respect to $x_1, \ldots, x_n$. Compute a Gröbner basis $\mathcal{G}$ of $\tilde{I}$ with respect to this order. Then the elements of $\mathcal{G}$ which do not involve $x_1, \ldots, x_n$ generate the elimination ideal

$$I = \tilde{I} \cap k[z_1, \ldots, z_r],$$

that is, the ideal of algebraic relations between $f_1, \ldots, f_r$.

Notice that an expression of a given element $f \in k[V]^G$ as a polynomial in $f_1, \ldots, f_r$ can be found by computing a normal form reduction modulo the Gröbner basis $\mathcal{G}$. □

A minimal set of homogeneous generators of I gives rise to a graded morphism $\phi_1 : F_1 \to k[z_1, \ldots, z_r]$ where F_1 is a free $k[z_1, \ldots, z_r]$–module of finite rank. By applying the same arguments to $\operatorname{Ker} \phi_1$ and so on we obtain the minimal free resolution

$$\cdots \to F_2 \xrightarrow{\phi_2} F_1 \xrightarrow{\phi_1} k[z_1, \ldots, z_r] \xrightarrow{\phi_0} k[V]^G \to 0$$

of $k[V]^G$ as a graded $k[z_1, \ldots, z_r]$–module. The elements of $\operatorname{Ker} \phi_{i-1}$ are the ith *order syzygies* on $k[V]^G$. Hilbert's syzygy theorem (1890), compare Schreyer (1980), Eisenbud (1995) for a proof involving Gröbner bases, tells us that all syzygies of order $\geq r+1$ are zero.

Theorem 1.5 (Hilbert's Syzygy Theorem). *Let $k[z_1, \ldots, z_r]$ be a graded polynomial algebra and M a finitely generated graded $k[z_1, \ldots, z_r]$–module. Then the minimal free resolution of M is of type*

$$0 \to F_s \to \cdots \to F_1 \to F_0 \to M \to 0$$

for some $s \leq r$.

Applying for example Schreyer's variant of Buchberger's algorithm (Schreyer 1980), (Eisenbud 1995) for computing the syzygies of submodules of free modules over a polynomial algebra gives the higher order syzygies between $f_1, \ldots, f_r$.

Example 1.6. A beautiful interplay between geometry and invariant theory can be found in Klein's papers on modular curves. For example in Klein (1878/1879) he studies a natural embedding of the modular curve $X(7)$ of level 7 into the projective plane $\mathbb{P}^2(\mathbb{C})$. In fact, the embedded curve is defined by the quartic invariant

$$f = \lambda^3\mu + \mu^3\nu + \nu^3\lambda$$

of $G = \mathrm{PSL}_2(\mathbb{F}_7)$ acting on the homogeneous coordinate ring of $\mathbb{P}^2(\mathbb{C})$. G is a finite group of order 168, explicitly given as the subgroup of $\mathrm{GL}_3(\mathbb{C})$ generated by the matrices

$$\pi_1 = \begin{pmatrix} \gamma & 0 & 0 \\ 0 & \gamma^4 & 0 \\ 0 & 0 & \gamma^2 \end{pmatrix} \quad \text{and} \quad \pi_2 = \begin{pmatrix} A & B & C \\ B & C & A \\ C & A & B \end{pmatrix},$$

where $\gamma = e^{2\pi i/7}$ is a 7th root of unity, and where

$$A = \frac{(\gamma^5 - \gamma^2)}{\sqrt{-7}}, \quad B = \frac{(\gamma^3 - \gamma^4)}{\sqrt{-7}}, \quad C = \frac{(\gamma^6 - \gamma)}{\sqrt{-7}}.$$

¿From the single invariant f Klein successively derives further invariants by combining standard ad hoc techniques in invariant theory with geometric arguments. The second invariant he finds is the (suitably normalized) Hessian

$$\nabla = \frac{1}{54} \begin{vmatrix} \frac{\partial^2 f}{\partial \lambda^2} & \frac{\partial^2 f}{\partial \lambda \partial \mu} & \frac{\partial^2 f}{\partial \lambda \partial \nu} \\ \frac{\partial^2 f}{\partial \mu \partial \lambda} & \frac{\partial^2 f}{\partial \mu^2} & \frac{\partial^2 f}{\partial \mu \partial \nu} \\ \frac{\partial^2 f}{\partial \nu \partial \lambda} & \frac{\partial^2 f}{\partial \nu \partial \mu} & \frac{\partial^2 f}{\partial \nu^2} \end{vmatrix}$$

which has degree 6. The plane curves defined by ∇ and f intersect in the 24 flexes of the curve $\{f = 0\}$. Klein shows by using geometric arguments that the flexes form the only set of 24 points on $\{f = 0\}$ which is invariant under the action of G. Moreover, there is no such set of lower cardinality. Thus by Bezout's theorem f and ∇ are, up to scalars, the only invariants in degrees ≤ 6. Similar arguments show that the next possible degree for another generator of the invariant ring is 14 (recall that the curve $\{f = 0\}$ has 28 bitangents). Klein obtains an invariant of degree 14 which is different from $f^2\nabla$ by bordering the Hessian matrix:

$$C = \frac{1}{9} \begin{vmatrix} \frac{\partial^2 f}{\partial \lambda^2} & \frac{\partial^2 f}{\partial \lambda \partial \mu} & \frac{\partial^2 f}{\partial \lambda \partial \nu} & \frac{\partial \nabla}{\partial \lambda} \\ \frac{\partial^2 f}{\partial \mu \partial \lambda} & \frac{\partial^2 f}{\partial \mu^2} & \frac{\partial^2 f}{\partial \mu \partial \nu} & \frac{\partial \nabla}{\partial \mu} \\ \frac{\partial^2 f}{\partial \nu \partial \lambda} & \frac{\partial^2 f}{\partial \nu \partial \mu} & \frac{\partial^2 f}{\partial \nu^2} & \frac{\partial \nabla}{\partial \nu} \\ \frac{\partial \nabla}{\partial \lambda} & \frac{\partial \nabla}{\partial \mu} & \frac{\partial \nabla}{\partial \nu} & 0 \end{vmatrix}$$

Finally, he finds a fourth invariant by taking the Jacobian of f, ∇ and C:

$$K = \frac{1}{14}\begin{vmatrix} \frac{\partial f}{\partial \lambda} & \frac{\partial \nabla}{\partial \lambda} & \frac{\partial C}{\partial \lambda} \\ \frac{\partial f}{\partial \mu} & \frac{\partial \nabla}{\partial \mu} & \frac{\partial C}{\partial \mu} \\ \frac{\partial f}{\partial \nu} & \frac{\partial \nabla}{\partial \nu} & \frac{\partial C}{\partial \nu} \end{vmatrix}$$

Geometric arguments as above show that these four invariants generate the ring of invariants and that the relation

$$(-\nabla)^7 = (\frac{C}{12})^3 - 27(\frac{K}{216})^2.$$

generates the (first order) syzygies. □

Klein's arguments show that his ring of invariants is not a polynomial algebra. Finitely generated rings of invariants are, however, always integral over a polynomial algebra. This is due to a result of Hilbert (1893) which is nowadays known as the graded Noether normalization theorem (compare e.g. Bruns *et al.* (1993), Eisenbud (1995)). In the following all k–algebras will be associative, commutative, and with 1.

Theorem 1.7 (Graded Noether Normalization Theorem).
Let $A = \bigoplus_{d\geq 0} A_d$ be a finitely generated graded algebra over $A_0 = k$. Then there exist algebraically independent homogeneous elements $p_1, \ldots, p_m \in A$ such that A is integral over $k[p_1, \ldots, p_m]$.

It is a well-known fact that the number m does not depend on the particular Noether normalization. m is the *dimension* of A. Every system $p_1, \ldots, p_m$ as above is called a *homogeneous system of parameters.*

Theorem 1.8. *Let $A = \bigoplus_{d\geq 0} A_d$ be a finitely generated graded algebra of dimension m over $A_0 = k$.*

(1) The following are equivalent for homogeneous $p_1, \ldots, p_m \in A$:

  (a) $p_1, \ldots, p_m$ is a homogeneous system of parameters.

  (b) A is a finitely generated $k[p_1, \ldots, p_m]$–module.

  (c) $A/(p_1, \ldots, p_m)$ is a finite dimensional k–vector space.

(2) If A is an integral domain, then $m = \operatorname{trdeg}_k(A)$.

Here are two easy well-known criteria (compare e.g. Bruns *et al.* (1993)):

Lemma 1.9. *Let $A = \bigoplus_{d\geq 0} A_d$ be a finitely generated graded algebra over $A_0 = k$ of dimension m and $p_1, \ldots, p_i$ homogeneous elements of A. Then $\dim(A/(p_1, \ldots, p_i)) \geq m - i$ and equality holds if and only if $p_1, \ldots, p_i$ can be extended to a homogeneous system of parameters.*

Lemma 1.10. *Let $A = \bigoplus_{d \geq 0} A_d$ be a finitely generated graded algebra over $A_0 = k$ and let $p_1, \ldots, p_m$ be a homogeneous system of parameters. Then homogeneous $s_0, \ldots, s_l \in A$ form a minimal system of generators of A as a $k[p_1, \ldots, p_m]$–module if and only if the residue classes $\bar{s}_0, \ldots, \bar{s}_l$ form a basis of the k–vector space $A/(p_1, \ldots, p_m)$.*

Definition 1.11. If $A = k[V]^G$ is a finitely generated ring of invariants then any homogeneous system of parameters $p_1, \ldots, p_m$ is called a system of *primary invariants*. Any minimal system of homogeneous $s_0 = 1, s_1, \ldots, s_l$ generating $k[V]^G$ as a $k[p_1, \ldots, p_m]$–module is called a system of *secondary invariants*. □

Example 1.12. In 1.6 the invariants f, ∇, and C are algebraically independent. In particular, $\operatorname{trdeg}_{\mathbb{C}} \mathbb{C}[\lambda, \mu, \nu]^G = \operatorname{trdeg}_{\mathbb{C}} \mathbb{C}[\lambda, \mu, \nu] = 3$. In addition, Klein's geometric arguments show that f, ∇, and C cut out the empty subset of $\mathbb{P}^2(\mathbb{C})$. Therefore $\mathbb{C}[\lambda, \mu, \nu]/(f, \nabla, C)$ is a finite dimensional $\mathbb{C}$–vector space. So f, ∇, and C form a homogeneous system of parameters for $\mathbb{C}[\lambda, \mu, \nu]$ and thus also for $\mathbb{C}[\lambda, \mu, \nu]^G$. It is clear from the relation in 1.6 that 1 and K are corresponding secondary invariants. Of course, we could also take f, C, and K as primary invariants. Then $1, \nabla, \ldots, \nabla^6$ are corresponding secondary invariants. □

Primary and secondary invariants together (without $s_0 = 1$) provide a system of generators $f_1, \ldots, f_r$ of $k[V]^G$ as a k–algebra. But, as shown by the example, such a system might not be minimal. For each i one can check by linear algebra or via Gröbner bases whether f_i is contained in the $\deg(f_i)$–graded piece of the subalgebra generated by the remaining f_j. In this way superfluous generators can be detected and eliminated. Let us describe an alternative approach.

Remark 1.13. Suppose that

$$R := k[p_1, \ldots, p_m] \subset A = k[V]^G$$

is a Noether normalization. Let $s_0 = 1, s_1, \ldots, s_l$ be a minimal system of homogeneous generators of A as an R–module. Then one can describe A as a quotient of a polynomial ring in $m + l$ variables as follows. Let $y_1, \ldots, y_l$ be new variables, and consider the surjective map of k–algebras:

$$\psi : k[p_1, \ldots, p_m, y_1, \ldots, y_l] \longrightarrow A, \quad p_i \mapsto p_i, \; y_j \mapsto s_j.$$

The kernel of ψ describes the algebraic relations between the p_i and s_j. Let us single out two types of elements in $\operatorname{Ker} \psi$.

(1) Every R–module syzygy between the s_j, that is, every expression

$$\alpha_0 s_0 + \alpha_1 s_1 + \ldots + \alpha_l s_l = 0, \quad \alpha_i \in R,$$

gives an element $\alpha_0 + \alpha_1 y_1 + \ldots + \alpha_l y_l \in \operatorname{Ker} \psi$, which we call a *linear relation* .

(2) As A is also an R–algebra, one can form the products s_is_j, $1 \leq i \leq j \leq l$. Since A is generated as an R–module by $s_0, \ldots, s_l$ there exist $\beta_{ijk} \in R$ such that $s_is_j = \sum_k \beta_{ijk}s_k$. In this way we obtain $\binom{l+1}{2}$ elements $y_iy_j - \beta_{ij0} - \beta_{ij1}y_1 - \cdots - \beta_{ijl}y_l$ in $\operatorname{Ker}\psi$, which we call *quadratic relations.*

It is an easy fact that the kernel of ψ is generated by the linear and the quadratic relations. Superfluous generators among $p_1, \ldots, p_m, s_1, \ldots, s_l$ correspond to terms in these relations which are linear in either one of the p_i or one of the s_j. By eliminating these generators and the corresponding relations we eventually arrive at a fundamental system of invariants, and the remaining linear and quadratic relations generate the first order syzygies between these fundamental invariants. □

In those cases where A is a free R–module there are no linear relations. This property can be taken as a definition for A to be Cohen–Macaulay (compare Bruns *et al.* (1993), Springer (1989)):

Definition 1.14. Let $A = \bigoplus_{d \geq 0} A_d$ be a finitely generated graded algebra of dimension m over $A_0 = k$. Then A is *Cohen–Macaulay* if for some (and hence for any) homogeneous system of parameters $p_1, \ldots, p_m$ A is a free $k[p_1, \ldots, p_m]$–module. □

Rings of invariants are Cohen–Macaulay in a wide range of cases (compare the following section). In these cases one only has to compute the quadratic relations. This can be done by linear algebra or via Gröbner bases. Knowing the β_{ijk} one can write down a not necessarily minimal free resolution of $\operatorname{Ker}\psi$ as a $k[p_1, \ldots, p_m, y_1, \ldots, y_l]$–module without computing syzygies (Eisenbud *et al.* 1981).

For later use we recall some results on Hilbert functions. For details and proofs we refer to Bruns *et al.* (1993), Smith (1995) .

Definition 1.15. Let $A = \bigoplus_{d \geq 0} A_d$ be a finitely generated graded algebra over $A_0 = k$. The function

$$H(A, \_) : \mathbb{N} \to \mathbb{N}, \quad d \to \dim_k A_d$$

is the *Hilbert function* of A, and

$$H_A(t) = \sum_{d \geq 0} H(A, d)t^d$$

is the *Hilbert series* of A. □

Remark 1.16.

(1) If $A = k[f_1, \ldots, f_r]$ is a polynomial algebra, then the Hilbert series of A is of the form

$$H_A(t) = \frac{1}{\prod_{i=1}^{r}(1 - t^{\deg(f_i)})}.$$

(2) If A is Cohen–Macaulay, $p_1, \ldots, p_m$ form a homogeneous system of parameters, and $s_0, \ldots, s_l$ is a basis of A as a free $k[p_1, \ldots, p_m]$-module, then

$$H_A(t) = \frac{\sum_{j=0}^{l} t^{\deg(s_j)}}{\prod_{i=1}^{m}(1 - t^{\deg(p_i)})}. \qquad \square$$

Example 1.17. In 1.6, 1.12 the Hilbert series is

$$H_{\mathbb{C}[\lambda,\mu,\nu]^G}(t) = \frac{1+t^{21}}{(1-t)^4(1-t)^6(1-t)^{14}} = \frac{1+t^6+t^{12}+\ldots t^{36}}{(1-t)^4(1-t)^{14}(1-t)^{21}}$$

$$= 1 + t^4 + t^6 + t^8 + t^{10} + 2t^{12} + 2t^{14} + 2t^{16} + 3t^{18} + 3t^{20} + t^{21} + \ldots. \qquad \square$$

The additivity of the Hilbert function, Hilbert's syzygy theorem, and Remark 1.16, (1) yield the following

Theorem 1.18 (Hilbert–Serre). *Let $A = \bigoplus_{d \geq 0} A_d$ be a graded algebra which is finitely generated over $A_0 = k$ by homogeneous elements $f_1, \ldots, f_r$ of positive degrees. Then the Hilbert series of A is of the form*

$$H_A(t) = \frac{p(t)}{\prod_{i=1}^{r}(1 - t^{\deg(f_i)})}$$

for some $p(t) \in \mathbb{Z}[t]$.

2 Algorithms for finite groups

We start with E. Noether's finiteness theorem (Noether 1926). The non-constructive proof, which is in principle Noether's original proof, is taken from Smith (1995).

Theorem 2.1 (Noether's Finiteness Theorem). *Let $\rho : G \to \mathrm{GL}(V)$ be a representation of a finite group G on the n-dimensional k-vector space V. Then $k[V]$ is integral over $k[V]^G$. In particular,*

$$\mathrm{trdeg}_k\, k[V]^G = \mathrm{trdeg}_k\, k[V] = n.$$

Moreover, $k[V]^G$ is finitely generated as a k-algebra.

Proof. Let $f \in k[V] = k[x_1, \ldots, x_n]$. Then f is a root of the monic univariate polynomial

$$P_f(t) := \prod_{\pi \in G}(t - \pi f) \in k[V][t].$$

The coefficients of P_f belong to $k[V]^G$ since they are elementary symmetric polynomials in the elements πf, $\pi \in G$. It follows that $k[V]$ is integral over $k[V]^G$. In particular, $\operatorname{trdeg}_k k[V]^G = \operatorname{trdeg}_k k[V] = n$.

Let A be the subalgebra of $k[V]^G$ generated by the coefficients of the monic polynomials $P_{x_1}, \ldots, P_{x_n}$. Then $k[V]$ is integral over A as well, and hence $k[V]$ is finitely generated as an A–module. Since A is Noetherian also the submodule $k[V]^G$ of $k[V]$ is finitely generated as an A–module. It follows that $k[V]^G$ is finitely generated as a k–algebra. □

Next we introduce an important tool for the invariant theory of a finite group G. This tool is defined if the characteristic of k is zero, or a prime number which does not divide the group order $|G|$.

Remark and Definition 2.2. Let G be a finite group such that $|G|$ is invertible in k. Then

$$R : k[V] \to k[V], \quad f(x) \mapsto \frac{1}{|G|} \sum_{\pi \in G} (\pi f)(x)$$

is a well–defined k–linear graded map which projects $k[V]$ onto $k[V]^G$. Moreover, R is a $k[V]^G$–module homomorphism. R is called *Reynolds operator*. □

The following result first appeared in Hochster *et al.* (1971), but had been known before. The proof we present is taken from Stanley (1979).

Theorem 2.3. *Let G be a finite group such that $|G|$ is invertible in k. Then $k[V]^G$ is Cohen–Macaulay.*

Proof. Let

$$k[p_1, \ldots, p_n] \subset k[V]^G$$

be a Noether normalization. Since $k[V]$ is integral over $k[V]^G$ by Theorem 2.1, it is also integral over $k[p_1, \ldots, p_n]$. Therefore $p_1, \ldots, p_n$ is also a homogeneous system of parameters for $k[V]$. As $k[V]$ is Cohen–Macaulay, it follows that $k[V]$ is a free $k[p_1, \ldots, p_n]$–module. By Lemma 1.10 homogeneous $s_0, \ldots, s_r$ generate $k[V]$ as a $k[p_1, \ldots, p_n]$–module if and only if their residue classes generate $k[V]/(p_1, \ldots, p_n)$ as a k–vector space.
Now notice that $k[V] = k[V]^G \oplus \operatorname{Ker} R$ is a decomposition into $k[V]^G$–modules by 2.2. Choose homogeneous $s_0, \ldots, s_l \in k[V]^G$ and $s_{l+1}, \ldots, s_r \in \operatorname{Ker} R$ such that $\bar{s}_0, \ldots, \bar{s}_l$ and $\bar{s}_{l+1}, \ldots, \bar{s}_r$ are k–bases of $k[V]^G/(p_1, \ldots, p_n)$ and $\operatorname{Ker} R / \sum_{i=1}^n p_i \operatorname{Ker} R$ resp. This gives a decomposition

$$k[V]^G = \bigoplus_{j=0}^{l} s_j k[p_1, \ldots, p_n],$$

hence $k[V]^G$ is Cohen–Macaulay. □

This theorem holds more generally for rational representations of linearly reductive groups (see Hochster *et al.* (1974)).

Definition 2.4. Let G be a finite group. We speak of *non-modular* invariant theory if $|G|$ is invertible in k, and of *modular* invariant theory otherwise. □

In the following G will be a finite group explicitly given as a matrix group $G \subset \mathrm{GL}_k(V^*) \cong \mathrm{GL}_n(k)$ by a finite set of generating matrices, and acting on $k[V] \cong k[x]$ by linear substitution: $\pi f(x_1, \ldots, x_n) = f((x_1, \ldots, x_n) \cdot \pi)$.

How to compute a fundamental system of invariants? As in Klein's Example 1.6 the basic strategy is to proceed degree by degree. Thus the first task is to calculate for a given degree d a k–basis of the graded piece $k[x]_d^G$. There are two basic methods. To be invariant under just one element of G imposes a linear condition on the homogeneous polynomials of a given degree d. Therefore, for each d, we may compute a k–basis of $k[x]_d^G$ by solving a linear system of equations which depends on the given set of generators of G. In the non–modular case we may evaluate the Reynold's operator R in the monomials of degree d in order to calculate a k–basis of $k[x]_d^G$. If a priori information on the dimension of $k[x]_d^G$ is available we can stop evaluating R as soon as the right number of invariants has been computed. Therefore already at this point, it is useful to precompute the Hilbert series of $k[x]^G$. This is possible due to a result of Molien (1897) (compare Heydtmann (1996)).

Theorem 2.5 (Molien's Theorem). *In the non–modular case the Hilbert series of $k[x]^G$ is given as the Molien series of G which is defined as follows.*

(1) Let $\operatorname{char} k = 0$. *Then the Molien series of G is*

$$\frac{1}{|G|} \sum_{\pi \in G} \frac{1}{\det(1 - t\pi)}.$$

(2) Let $\operatorname{char} k = p > 0$, *$|G|$ not divisible by p. Fix primitive $|G|$th roots of unity over k and $\mathbb{Q}$ resp., say λ and $\tilde{\lambda}$. For $\pi \in G$ let $\lambda^{k_1(\pi)}, \ldots, \lambda^{k_n(\pi)}$ be the eigenvalues of π and*

$$\Phi_\pi(t) = \prod_{i=1}^{n} (1 - t\tilde{\lambda}^{k_i(\pi)}).$$

Then the Molien series of G is

$$\frac{1}{|G|} \sum_{\pi \in G} \frac{1}{\Phi_\pi(t)}.$$

Summing up what we have so far gives a first algorithm in the non–modular case: Compute invariants degree by degree until the Hilbert series of the k–subalgebra of $k[x]^G$ generated by the invariants computed so far equals the

Molien series. Or, again in the non–modular case, use degree bounds such as Noether's degree bound (1916) or that of Campbell *et al.* (1991) as criteria for termination.

Another idea, which yields more effective algorithms, is to compute a Noether normalization of $k[x]^G$ first. In the case of finite groups, by Theorem 2.1, every homogeneous system of parameters for $k[x]^G$ contains precisely n elements, and $p_1, \ldots, p_n \in k[x]^G$ are a homogeneous system of parameters for $k[x]^G$ if and only if they are a homogeneous system of parameters for $k[x]$. Lemma 1.9 suggests how to compute such a system $p_1, \ldots, p_n$ successively. If $p_1, \ldots, p_i$ with $\dim((p_1, \ldots, p_i)) = n - i$ are already constructed, then we need to find a homogeneous invariant q such that $\dim((p_1, \ldots, p_i, q)) = n - i - 1$. Refining this idea yields the

Algorithm 2.6 (Primary invariants). (Decker *et al.* 1997)
Input: Generators of G.
Output: Primary invariants for $k[x]^G$.
Procedure:

Initialize $d := i := 0$.
repeat
  Set $d := d + 1$.
  Compute a basis $b_1, \ldots, b_{c_d}$ of $k[x_1, \ldots, x_n]_d^G$.
  Compute $m = \dim((p_1, \ldots, p_i, b_1, \ldots, b_{c_d}))$ via Gröbner bases.
  while $n - i > m$ **do**
    if k infinite **then**
      set $i := i + 1$.
      Find a k–linear combination p_i of $b_1, \ldots, b_{c_d}$ such that
        $\dim((p_1, \ldots, p_{i-1})) > \dim((p_1, \ldots, p_i))$.
    else
      find a new k–linear combination q of $b_1, \ldots, b_{c_d}$.
      if $\dim((p_1, \ldots, p_i)) > \dim((p_1, \ldots, p_i, q)$ **then**
        set $i := i + 1$, $p_i := q$.
      if all linear combinations of $b_1, \ldots, b_{c_d}$ have been tested **then**
        break [of **while**–loop]
  end [of **while**–loop]
until $i = n$
return: $p_1, \ldots, p_n$
end [of **Procedure**].

¿From now on suppose that a system of primary invariants $p_1, \ldots, p_n$ and a Gröbner basis $\mathcal{G}$ of the ideal $(p_1, \ldots, p_n)$ are already given. The next step is to compute a system of secondary invariants, that is, a minimal system of homogeneous generators of $k[x]^G$ as a $k[p_1, \dots, p_n]$–module.

We first deal with the non–modular case. In this case the Reynolds operator R is available. By 2.2 R is a homomorphism of $k[x]^G$–modules and thus also of $k[p_1, \dots, p_n]$–modules. Hence homogeneous generators of $k[x]^G$ as a $k[p_1, \dots, p_n]$–module can be obtained by applying R to a (minimal) system of homogeneous generators of $k[x]$ as a $k[p_1, \dots, p_n]$–module, that is, to a system of homogeneous polynomials whose residue classes are a basis of the k–vector space $k[x]/(p_1, \dots, p_n)$ (compare Lemma 1.10). We may thus apply R to the finitely many monomials which are not in the initial ideal of $(p_1, \dots, p_n)$ (with respect to the chosen monomial order). In this way we obtain a not necessarily minimal system of generators of $k[x]^G$ as a $k[p_1, \dots, p_n]$–module. In order to avoid superfluous computations we refine the algorithm as follows. Write $(p_1, \dots, p_n)^{k[x]^G}$ for the ideal generated by $p_1, \dots, p_n$ in $k[x]^G$. Again by Lemma 1.10 homogeneous $s_0, \dots, s_j \in k[x]^G$ can be extended to a system of secondary invariants if and only if their residue classes modulo $(p_1, \dots, p_n)^{k[x]^G}$ are linearly independent. This in turn is equivalent to the fact that $s_0, \dots, s_j$ are linearly independent modulo $(p_1, \dots, p_n)$ (apply 2.2 to show that the natural map $k[x]^G/(p_1, \dots, p_n)^{k[x]^G} \to k[x]/(p_1, \dots, p_n)$ is injective). Since everything is homogeneous it is enough to check that for every d the normal forms with respect to $\mathcal{G}$ of those s_i having degree d are linearly independent. This check can be performed by solving a linear system of equations, or by a Gröbner basis calculation. The following numerical information provides a criterion for termination.

Remark 2.7. Let $|G|$ be invertible in k. Then $k[x]^G$ is Cohen–Macaulay, and we may compute the Hilbert series of $k[x]^G$ in two different ways. Comparing the formulae in 1.16, (2) and 2.5 gives the following result (see e.g. Heydtmann (1996)):

(1) The degrees d_j of the secondary invariants and their number k_j in each degree resp. are uniquely determined as the exponents and the coefficients resp. of the univariate polynomial

$$\sum_{j=0}^{b} m_j t^{d_j} := H_{K[x]^G}(t) \prod_{i=1}^{n} (1 - t^{\deg(p_i)}).$$

(2) The total number of secondary invariants equals

$$\frac{1}{|G|} \prod_{i=1}^{n} \deg(p_i). \qquad \square$$

Example 2.8. (1) In 1.6, 1.12, with primary invariants f, ∇ and C, the polynomial in 2.7, (1) reads

$$H_{\mathbb{C}[\lambda,\mu,\nu]^G}(t)(1-t^4)(1-t^6)(1-t^{14}) = 1 + t^{21}.$$

With primary invariants f, C and K we obtain

$$H_{\mathbb{C}[\lambda,\mu,\nu]^G}(t)(1-t)^4(1-t)^{14}(1-t)^{21} = 1 + t^6 + t^{12} + \cdots + t^{36}.$$

(2) The Heisenberg group H_5 of level 5 in its Schrödinger representation is the subgroup of $\mathrm{GL}_5(\mathbb{C})$ generated by the matrices

$$\pi_1 = \begin{pmatrix} 0 & 0 & 0 & 0 & 1 \\ 1 & 0 & 0 & 0 & 0 \\ 0 & 1 & 0 & 0 & 0 \\ 0 & 0 & 1 & 0 & 0 \\ 0 & 0 & 0 & 1 & 0 \end{pmatrix} \quad \text{and} \quad \pi_2 = \begin{pmatrix} 1 & 0 & 0 & 0 & 0 \\ 0 & \xi & 0 & 0 & 0 \\ 0 & 0 & \xi^2 & 0 & 0 \\ 0 & 0 & 0 & \xi^3 & 0 \\ 0 & 0 & 0 & 0 & \xi^4 \end{pmatrix},$$

where $\xi = e^{2\pi i/5}$ is a 5th root of unity. H_5 is a finite group of order 125. A partial expansion of the Molien series is

$$1 + 6t^5 + 41t^{10} + 156t^{15} + 426t^{20} + 951t^{25} + 1856t^{30} + \ldots \quad .$$

The invariants in degree 5 are the Horrocks–Mumford quintics (Horrocks *et al.* 1973). The Algorithm 2.6 finds three primary invariants in degree 5, and two in degree 10. The corresponding polynomial as in 2.7, (1) reads

$$1 + 3t^5 + 24t^{10} + 44t^{15} + 24t^{20} + 3t^{25} + t^{30}.$$

Thus, including 1, there are 100 secondary invariants of degrees up to 30. See Heydtmann (1996), where this example is worked out with the help of finvar.lib (Heydtmann 1997).

(3) The modular curve $X(11)$ is studied in Klein (1879). Klein discovered that X(11) is the singular locus of the cubic threefold in $\mathbb{P}^4(\mathbb{C})$ defined by

$$\nabla = y_1^2 y_9 + y_4^2 y_3 + y_5^2 y_1 + y_9^2 y_4 + y_3^2 y_5,$$

where y_1, y_4, y_5, y_9, y_3 are the homogeneous coordinates of $\mathbb{P}^4(\mathbb{C})$. ∇ is a cubic invariant of $G = \mathrm{PSL}_2(\mathbb{F}_{11})$ acting on the homogeneous coordinate ring of $\mathbb{P}^4(\mathbb{C})$. Explicitly, G is the subgroup of $\mathrm{GL}_5(\mathbb{C})$ generated by the matrices

$$\pi_1 = \begin{pmatrix} \rho & 0 & 0 & 0 & 0 \\ 0 & \rho^4 & 0 & 0 & 0 \\ 0 & 0 & \rho^5 & 0 & 0 \\ 0 & 0 & 0 & \rho^9 & 0 \\ 0 & 0 & 0 & 0 & \rho^3 \end{pmatrix} \quad \text{and} \quad \pi_2 = \begin{pmatrix} A & B & C & D & E \\ B & C & D & E & A \\ C & D & E & A & B \\ D & E & A & B & C \\ E & A & B & C & D \end{pmatrix},$$

where $\rho = e^{2\pi i/11}$ is a 11th root of unity, and where

$$A = \frac{(\rho^9 - \rho^2)}{\sqrt{-11}}, \quad B = \frac{(\rho^4 - \rho^7)}{\sqrt{-11}}, \quad C = \frac{(\rho^3 - \rho^8)}{\sqrt{-11}},$$

$$D = \frac{(\rho^5 - \rho^6)}{\sqrt{-11}}, \quad E = \frac{(\rho - \rho^{10})}{\sqrt{-11}}.$$

G is a finite group of order 660. Besides ∇ Klein only needs two more invariants for his purposes. One of them is the (suitably normalized) Hessian of ∇,

$$H = \begin{vmatrix} y_9 & 0 & y_5 & y_1 & 0 \\ 0 & y_3 & 0 & y_9 & y_4 \\ y_5 & 0 & y_1 & 0 & y_3 \\ y_1 & y_9 & 0 & y_4 & 0 \\ 0 & y_4 & y_3 & 0 & y_5 \end{vmatrix}.$$

A complete description of $\mathbb{C}[y]^G$ is given by Adler (1992) along the following lines. Adler starts his paper by quoting Klein (1879):

"Es ist nicht meine Absicht, alle ungeändert bleibenden gänzen Funktionen der y mitzuteilen; dies würde jedenfalls eine weitläufige und vielleicht eine schwierige Aufgabe sein."

The Molien series of $\mathbb{C}[y]^G$ is known to Adler from a letter of Edge:

$$H_{\mathbb{C}[y]^G}(t) = \frac{1 + t^7 + t^9 + t^{10} + t^{12} + 2t^{14} + t^{16} + t^{18} + t^{19} + t^{21} + t^{28}}{(1-t^3)(1-t^5)(1-t^6)(1-t^8)(1-t^{11})}$$

$$= 1 + t^3 + t^5 + 2t^6 + t^7 + 2t^8 + 3t^9 + 3t^{10} + 4t^{11} + \dots \quad .$$

Comparing with 2.7 one can hope to find primary invariants of degrees 3, 5, 6, 8, and 11 resp., and secondary invariants in the degrees prescribed by the numerator polynomial. Let us follow Adler's notation and provide every invariant with a subscript indicating the degree. Starting with the two invariants $f_3 := \nabla$ and $f_5 := H$, Adler derives two systems of invariants,

$$\mathcal{P} = \{f_3, f_5, f_6, f_8, f_{11}\},$$

and

$$\mathcal{S} = \{1, f_7, f_9, f_{10}, f_{12}, f_{14}, f_7^2, f_7 f_9, f_9^2, f_9 f_{10}, f_7^3, f_9^2 f_{10}\},$$

in the spirit of classical invariant theory. To be more precise, if $\Phi \in \mathbb{C}[y]^G$ is given, then define the differential operator D_Φ by

$$D_\Phi = \Phi(\frac{\partial}{\partial y_1}, \frac{\partial}{\partial y_4}, \frac{\partial}{\partial y_5}, \frac{\partial}{\partial y_9}, \frac{\partial}{\partial y_3}).$$

Then $D_\Phi(\Psi) \in \mathbb{C}[y]^G$ for every $\Psi \in \mathbb{C}[y]^G$. With $D = D_{f_3}$ Adler obtains e.g. the invariants

$$f_6 = \frac{1}{12}(D(f_3^3 - 168 f_3^2), \quad f_8 = \frac{1}{4}(D(f_3^2 f_5) - 150 f_3 f_5),$$

and

$$f_{11} = \frac{1}{2} D(f_3^2 f_8) - 219 f_3^2 f_5 + 56 f_5 f_6 - 150 f_3 f_8$$

(see Adler (1992) for $f_7, f_9, f_{10}, f_{12}, f_{14}$). Adler checks that the Jacobian matrix of the invariants in $\mathcal{P}$ does not vanish identically. Therefore these invariants are algebraically independent over $\mathbb{C}$. Now Adler proves by more or less explicit computations involving the quadratic relations as in 1.13 that $R_0 = \mathbb{Q}[f_3, f_5, f_6, f_8, f_{11}, f_7, f_9, f_{10}, f_{12}, f_{14}]$ is a free $A_0 = \mathbb{Q}[f_3, f_5, f_6, f_8, f_{11}]$–module with basis $\mathcal{S}$. The next step is to show that $R_0 \otimes \mathbb{C} \cong \mathbb{C}[y]^G$. This is a check on Hilbert functions. Altogether it follows that the elements of $\mathcal{P}$ are a homogeneous system of parameters, that $\mathbb{C}[y]^G$ is a free $\mathbb{C}[\mathcal{P}]$–module with basis $\mathcal{S}$, and that $f_3, f_5, f_6, f_7, f_8, f_9, f_{10}, f_{11}, f_{12}, f_{14}$ is a fundamental system of invariants with fifteen relations. □

The following algorithm singles out those secondary invariants which are not products of powers of secondary invariants of lower degree. Following Kemper *et al.* (1997) we call these secondary invariants *irreducible* (and the others, including 1, *reducible*). We write NF(f) for the normal form of a polynomial $f \in k[x]$ with respect to the fixed Gröbner basis $\mathcal{G}$.

Algorithm 2.9 (Secondary invariants in the non–modular case). (Kemper *et al.* 1997)

Input: Reynolds operator R, Hilbert series $H_{k[x]^G}(t)$, primary invariants $p_1, \ldots, p_n$ for $k[x]^G$, a Gröbner basis $\mathcal{G}$ of $(p_1, \ldots, p_n)$.

Output: Secondary invariants for $k[x]^G$ (irreducible and reducible).

Procedure:

Compute the polynomial $\sum_{j=0}^{b} m_j t^{d_j} = H_{K[x]^G}(t) \prod_{i=1}^{n}(1 - t^{\deg(p_i)})$.
Initialize $I := \emptyset$, $P := \{1\}$.
for $j = 1, \ldots, b$ **do**
  set i:=0.
  while $i < m_j$ **do**
    for all power products q of elements of I having degree d_j **do**
      if NF(s_1), ..., NF(s_i), NF(q) are linearly independent
        modulo $(p_1, \ldots, p_n)$ **then**
        set $i := i + 1$, $s_i := q$, $P := P \cup \{s_i\}$.
    end [of **for**–loop]
    Compute the monomials $m_1, \ldots, m_c$ of degree d_j which are
      not in the initial ideal of $(p_1, \ldots, p_n)$
    for $r = 1, \ldots, c$ **do**
      if NF(s_1), ..., NF(s_i), NF($R(m_r)$) are linearly independent
        modulo $(p_1, \ldots, p_n)$ **then**
        set $i := i + 1$, $s_i := R(m_r)$, $I := I \cup \{s_i\}$.
    end [of **for**–loop]
  end [of **while**–loop]

end [of **for**–loop]
return: I, P
end [of **Procedure**].

It is easy to see that for the calculation of fundamental invariants and syzygies as in Remark 1.13 it is enough to check the products of type $s_i s_j$, $s_i \in I$ and $s_j \in I \cup (P \setminus \{1\})$.

Remark 2.10. The computation of secondary invariants in the modular case can be reduced to the computation of secondary invariants in the non-modular case by an idea of Kemper (1996) (compare also Kemper *et al.* (1997)). Choose a subgroup $H \subset G$ such that $|H|$ is invertible in k (for example, the trivial subgroup will do). The given primary invariants $p_1, \ldots, p_n$ for G are also primary invariants for H. Write $A = k[p_1, \ldots, p_n]$. Compute secondary invariants $t_1, \ldots, t_m$ for H with respect to $p_1, \ldots, p_n$ (for example by Algorithm 2.9). Choose a subset $\Delta \subset G$ such that the elements of $\Delta \cup H$ generate G. Then there is a commutative diagram of graded $k[p_1, \ldots, p_n]$–modules with exact rows and columns,

$$\begin{array}{ccccccc}
 & & 0 & & 0 & & 0 \\
 & & \uparrow & & \uparrow & & \uparrow \\
0 & \longrightarrow & k[x]^G & \rightarrow & k[x]^H & \xrightarrow{\alpha} & \bigoplus_{\pi \in \Delta} k[x] \\
 & & \uparrow & & \beta \uparrow & & \uparrow \\
0 & \rightarrow & M & \longrightarrow & A^m & \xrightarrow{\tilde{\alpha}} & A^r \\
 & & \uparrow & & \uparrow & & \uparrow \\
 & & 0 & & 0 & & 0
\end{array}$$

where α is defined by $f \mapsto (\pi f - f)_{\pi \in \Delta}$, β is the map $(f_1, \ldots, f_m) \mapsto \sum_{j=1}^m f_j t_j$, and $r = |\Delta| \cdot \prod_{i=1}^n \deg(p_i)$. Indeed, $k[x]$ is a free $k[p_1, \ldots, p_n]$–module of rank $\prod_{i=1}^n \deg(p_i)$ with a basis $\mathcal{B}$ given by those monomials which are not in the initial ideal of $(p_1, \ldots, p_n)$ (with respect to the chosen monomial order). The map $\tilde{\alpha}$, that is, the representation of each $(\pi t_j - t_j)$ in terms of $\mathcal{B}$, can be computed by linear algebra or via Gröbner bases. In this way the computation of secondary invariants for G and of the $k[p_1, \ldots, p_n]$–module relations between the secondary invariants can be reduced to the computation of syzygies of submodules of free $k[p_1, \ldots, p_n]$–modules. □

Example 2.11. Kemper (1996) applies his algorithm to an example of Bertin (1967). The cyclic group G of order 4 acts on $\mathbb{F}_2[x_1, \ldots, x_4]$ by cyclic permutations of the variables. Kemper finds the primary invariants

$$p_1 = x_1 + x_2 + x_3 + x_4, \quad p_2 = x_1 x_3 + x_2 x_4,$$

$$p_3 = x_1 x_2 + x_2 x_3 + x_3 x_4 + x_4 x_1, \quad p_4 = x_1 x_2 x_3 x_4.$$

Choose the trivial subgroup H of G. The corresponding polynomial as in 2.7, (1) reads $1 + 3t + 4t^2 + 4t^3 + 3t^4 + t^5$. From the 16 secondary invariants for H Kemper obtains 5 secondary invariants for G,

$$s_0 = 1, s_1 = x_1^3 + x_2^3 + x_3^3 + x_4^3, \quad s_2 = x_1^2x_2 + x_2^2x_3 + x_3^2x_4 + x_4^2x_1,$$
$$s_3 = x_1^3x_2 + x_2^3x_3 + x_3^3x_4 + x_4^3x_1, \quad s_4 = x_1^3x_2^2 + x_2^3x_3^2 + x_3^3x_4^2 + x_4^3x_1^2,$$

and the relation

$$p_1(p_1^4 + p_2^2 + p_2p_3)s_1 + (p_1^2 + p_2 + p_3)s_2 + (p_1^2 + p_2)s_3 + p_1s_4 = 0.$$

It follows that $\mathbb{F}_2[x]^G$ is not Cohen–Macaulay. □

Remark 2.12.

(1) The following is easy to see (compare Kemper (1996)): let $p_1, \ldots, p_n$ be a system of primary invariants of degrees $d_1, \ldots, d_n$. Then $k[x]^G$ is Cohen–Macaulay if and only the number of secondary invariants equals $(\prod_{i=1}^n d_i)/|G|$.

(2) A check on the algorithms for secondary invariants shows that it is crucial to start with primary invariants $p_1, \ldots, p_n$ such that the product $\prod_{i=1}^n \deg(p_i)$ is as small as possible. Algorithm 2.6 does not always produce primary invariants with $\prod_{i=1}^n \deg(p_i)$ minimal. See Kemper (1997) for examples and a solution to this problem.

(3) For Adler's bicycles, which grew out of an attempt "to understand from a general point of view the results of Felix Klein on the equations defining modular curves of prime order" we refer to Adler (1997). □

3 Algorithms for linearly reductive groups

In this section our field of coefficients is $k = \mathbb{C}$, if not explicitly stated otherwise.

First we recall the notion of a linearly reductive group. The general linear group $\mathrm{GL}_m(k)$ is the Zariski open subset of k^{m^2} defined by $\det(x_{ij}) \neq 0$. We may identify $\mathrm{GL}_m(k)$ with the Zariski closed subset of k^{m^2+1} defined by the polynomial equation $z \cdot \det(x_{ij}) = 1$, where z is an additional coordinate. Thus $\mathrm{GL}_m(k)$ is an affine variety with coordinate ring

$$k[GL_m(k)] \cong k[x_{ij}, \det(x_{ij})^{-1}] \cong k[x_{ij}, z]/(z \cdot \det(x_{ij}) - 1).$$

A *linear algebraic group* is a subgroup $G \subset \mathrm{GL}_m(k)$ which is closed in the Zariski topology. Then G is itself an affine variety with coordinate ring $k[G] = k[GL_m(k)]/I$, where I is the defining ideal of G. A finite dimensional representation $\rho : G \to \mathrm{GL}(V) \cong \mathrm{GL}_n(k)$ of a linear algebraic group G is called *rational* if ρ is also a morphism between affine varieties.

Example 3.1. $\mathrm{SL}_2(\mathbb{C}) \subset \mathrm{GL}_2(\mathbb{C})$ is the Zariski closed subset defined by the equation $\det = 1$. The representation of $\mathrm{SL}_2(\mathbb{C})$ on the space V of quadratic binary forms is given by

$$\rho : \mathrm{SL}_2(\mathbb{C}) \to \mathrm{GL}(V) \cong \mathrm{GL}_3(\mathbb{C}), \quad \begin{pmatrix} \alpha & \beta \\ \gamma & \delta \end{pmatrix} \to \begin{pmatrix} \alpha^2 & 2\alpha\gamma & \gamma^2 \\ \alpha\beta & \alpha\delta + \beta\gamma & \gamma\delta \\ \beta^2 & 2\beta\delta & \delta^2 \end{pmatrix}.$$

Thus ρ is a polynomial map, hence ρ is a rational representation. □

Definition 3.2. An algebraic group G is said to be *linearly reductive*, if every finite dimensional rational representation of G is *completely reducible*, that is, a direct sum of irreducible representations. □

Since we are interested in actions of groups on polynomial rings we must consider infinite dimensional representations as well. If G is linear algebraic, a representation of G on an arbitrary vector space over k is called *rational* if every vector is contained in a finite dimensional G–invariant subspace W such that the induced representation on W is rational in the sense above. The following is easy to prove (compare Springer (1989)).

Proposition 3.3. *Let G be a linear algebraic group. The following are equivalent:*

(1) G is linearly reductive.

(2) The defining condition in 3.2 holds for any rational representation of G.

(3) For any rational representation $\rho : V \to \mathrm{GL}(V)$ there exists a k-linear map $p_V : V \to V$, with $p_V \circ \rho(\pi) = p_V$ for all $\pi \in G$, which projects V onto the fixed point set $V^G = \{v \in V \mid \pi v = v \text{ for any } \pi \in G\}$.

(4) There exists an invariant integral *on G, that is, a map $I : k[G] \longrightarrow k$ with $I(1) = 1$ and $I \circ \lambda(\pi) = I$ for all $\pi \in G$, where λ is the representation of G on $k[G]$ given by $(\lambda(\pi)f)(\xi) = f(\pi^{-1}\xi)$.*

Remark 3.4. If a rational representation of a linearly reductive group G on a finite dimensional vector space V is given, then the third part of 3.3 applies in particular to the induced rational representation on $k[V]$. As is easy to see, the corresponding map $R : k[V] \to k[V]$ is in addition a $k[V]^G$–module homomorphism. R is called *Reynolds operator.*

Example 3.5.

(1) Let G be a finite group. Then G has the structure of a linear algebraic group with coordinate ring $k[G]$, where $k[G]$ is the k-algebra of k-valued functions on G with pointwise defined multiplication. Suppose that the

group order $|G|$ is invertible in k (*non-modular case*). Then G is linearly reductive (Maschke's theorem, see (Maschke 1899)). Indeed, if V is a rational representation of G then averaging over the group gives a map as in 3.3, (3).

$$V \to V, \quad v \to \frac{1}{|G|} \sum_{\pi \in G} \pi v.$$

(2) The multiplicative group k^* is a linear algebraic group with coordinate ring $k[z, z^{-1}]$. The map $\sum a_i z^i \mapsto a_0$ is an invariant integral. Therefore k^* is linearly reductive. More generally, every torus $(k^*)^m$ is linearly reductive.

(3) For linear algebraic groups which are compact in the $\mathbb{C}$–topology one can construct an invariant integral by integrating functions with respect to the Haar measure. This measure is obtained from an arbitrary measure by averaging over the group.

(4) A linear algebraic group G, which has a subgroup that is compact in the $\mathbb{C}$–topology and dense in the Zariski topology, is linearly reductive (see Kraft (1985)). The linear reductivity of the classical groups can be deduced from this fact. □

The Reynolds operator is abstractly defined, and in general it is not easy to get a hand on it. But in most situations it suffices that one is able to generate all invariants of a given degree. There are several ways for producing the invariants in a given degree, each adapted to a particular situation. We sketch the Lie algebra method and refer to Sturmfels (1993) for Cayley's Ω–process and the symbolic method. Consider a rational representation $\rho : G \longrightarrow \mathrm{GL}(W)$ of a linear algebraic group on a finite dimensional vector space W, and the induced action on the Lie-algebras:

$$\rho_* : \mathrm{Lie}(G) \longrightarrow \mathrm{Lie}(\mathrm{GL}(W)).$$

If $w \in W$ is G–invariant, then differentiating the expression $\pi w = w$ with respect to π, shows that $\phi w = 0$ for any element ϕ in $\mathrm{Lie}(G)$. The converse holds, as soon as G is *connected.* Applying this to $W = S^d(V^*)$ shows that one may find all invariants of a given degree by solving linear equations. If G is not connected, one should first find the invariants for the connected component G^0 of the identity, and then pick out those invariants of G^0 which are invariant under the finite group G/G^0.

We now give a proof of Hilbert's finiteness theorem (Hilbert 1890).

Theorem 3.6 (Hilbert's Finiteness Theorem). *Let G be a linearly reductive group, acting rationally on a finite dimensional vector space V. Then the ring of invariants $k[V]^G$ is finitely generated as a k–algebra.*

Proof. Consider the ideal $I_{\mathcal{N}}$ in $k[V]$ generated by the homogeneous invariants of positive degree. Hilbert's basis theorem implies that already finitely many of those generate $I_{\mathcal{N}}$:

$$I_{\mathcal{N}} = (f_1, \ldots, f_r).$$

We claim that the invariants are generated as a k–algebra by $f_1, \ldots, f_r$. If f is any homogeneous invariant of positive degree, then $f \in I$, so we can write $f = \sum h_i f_i$. Applying the Reynolds operator R yields:

$$f = \sum R(h_i) f_i.$$

The $R(h_i)$ are invariants, which we may assume to be of lower degree than f. By induction, the $R(h_i)$ are in the subalgebra generated by $f_1, \ldots, f_r$. Hence so is f. □

The proof, as it stands, is non–constructive, because one has to find generators of $I_{\mathcal{N}}$. Because of this, Hilbert was harshly criticized. (Gordan: "Das ist Theologie und keine Mathematik"). Let us quote from a letter of Minkowski to Hilbert from 1892 (see Minkowski (1973) and compare Derksen *et al.* (1995) for a translation into English):

(...) Dass es nur eine Frage der Zeit sein konnte, wann Du die alten Invariantenfragen soweit erledigt haben würdest, dass kaum noch das Tüpfelchem auf dem i fehlt, war mir eigentlich schon seit lange nicht zweifelhaft. Dass es aber damit so schnell geht, und alles so überraschend einfach gelingt, hat mich aufrichtig gefreut, und beglückwünsche ich Dich dazu. Jetzt, wo Du in Deinem letzten Satze sogar das rauchlose Pulver gefunden hast, nachdem schon Theorem I nur noch vor Gordans Augen Dampf gab, ist es wirklich an der Zeit, dass die Burgen der Raubritter STROH, GORDAN, STEPHANOS und wie sie alle heissen mögen, welche die einzelreisenden Invarianten überfielen und in's Burgverlies sperrten, dem Erdboden gleichgemacht werden, auf die Gefahr hin, dass aus diesen Ruinen niemals wieder neues Leben spriesst. (...)

Later on Gordan admitted, that "theology also has its advantages" (compare Kline (1972)). In particular, he gave a new proof of Hilbert's basis theorem (Gordan 1899). In today's language, this proof consists of two parts:

- Every ideal generated by monomials is finitely generated (Dickson's Lemma).
- Every ideal has a finite *Gröbner basis.*

Taking this historical discussion into account, it came with a big surprise that Derksen (1997) could refine the ideas of Hilbert (1890) and turn them into an algorithm. In fact, Derksen came up with an algorithm for finding homogeneous generators for the ideal $I_{\mathcal{N}}$ by using Gröbner bases. First he remarks that it easily follows from the above proof of Hilbert's finiteness theorem, that for any set of homogeneous generators $f_1, \dots, f_r$ of $I_{\mathcal{N}}$, the ring of invariants is generated by $R(f_1), \dots, R(f_r)$ (see Derksen (1997)). As usual, the Reynolds operator can be replaced by any other means of generating invariants of a given degree. To find generators of $I_{\mathcal{N}}$, Derksen considers the map

$$\psi : G \times V \longrightarrow V \times V, \qquad (\pi, v) \mapsto (v, \pi v).$$

Let B be the Zariski closure of the image of ψ, with defining ideal $J \subset k[V \times V] \simeq k[x, y]$. Here $y = y_1, \dots, y_n$ is a new set of variables. Derksen proves that one can get generators for $I_{\mathcal{N}}$ by substituting $y = 0$ in a generating set of J, or to put it in another way:

Theorem 3.7.

$$((y) + J) \cap k[x] = I_{\mathcal{N}}.$$

Proof. " $\supset$ " Let $f(x) \in I_{\mathcal{N}}$ be an invariant of positive degree. Then $f(x) = f(y) + (f(x) - f(y)) \in ((y) + J)$ since $f(y)$ is in (y), and $(f(x) - f(y))$ is in J. Indeed, $(f(x) - f(y))(v, \pi v) = f(v) - f(\pi v) = 0$, as f is G–invariant.
" $\subset$ " To prove the other inclusion we look at the action of G on $V \times V$ given by $\pi(v, w) = (v, \pi w)$. The corresponding Reynolds operator

$$R_y : k[x, y] \longrightarrow k[y]^G[x]$$

has the following properties:

(1) R_y is $k[x]$-linear.

(2) $R_y(J) \subset J$. This is because B is G–invariant.

(3) The diagonal $\Delta \subset V \times V$ is a subset of B. Thus J is contained in the ideal of the diagonal, that is, in the kernel of $\phi : k[x, y] \to k[x], \quad p(x, y) \mapsto p(x, x)$.

(4) The restriction of R_y to $k[y]$ is the Reynolds operator for the given action of G on V, since this operator is uniquely determined by its properties as in 3.3.

Now take any homogeneous $f(x) = \sum c_i(x) f_i(y) + h(x, y) \in ((y) + J) \cap k[x]$, with $f_i(y) \in (y)$ homogeneous, and $h(x, y) \in J$. We have $R_y(h(x, y)) \in J$ by (2), hence $\phi(R_y(h(x, y))) = 0$ by (3). Therefore, by (1) and (4),

$$f(x) = \sum c_i(x) \phi(R_y(f_i(y))) = \sum c_i(x) R(f_i(x)).$$

The $R(f_i)$ are homogeneous invariants of positive degree. Hence $f \in I_{\mathcal{N}}$. □

Generators of J can be found by elimination, that is, via Gröbner bases. We consider an example.

Example 3.8. We take the action of $G = \mathrm{SL}_2(\mathbb{C})$ on the three dimensional vector space V of quadratic binary forms as in Example 3.1. Consider

$$\Gamma = \{(\pi, v, \pi v) : \pi \in G, v \in V\} \subset G \times V \times V.$$

Then B is the closure of $p(\Gamma)$, where p is the projection onto $V \times V$. The ideal

$$I \subset k[\alpha, \beta, \gamma, \delta, x_1, x_2, x_3, y_1, y_2, y_3]$$

of Γ is given by $(\alpha\delta - \beta\gamma - 1, -y_1 + \alpha^2 x_1 + 2\alpha\gamma x_2 + \gamma^2 x_3, -y_2 + \alpha\beta x_1 + (\alpha\delta + \beta\gamma)x_2 + \gamma\delta x_3, -y_3 + \beta^2 x_1 + 2\beta\delta x_2 + \delta^2 x_3)$. We eliminate $\alpha, \beta, \gamma, \delta$ via Gröbner bases. The result is the principal ideal

$$(x_2^2 - x_1 x_3 - y_2^2 + y_1 y_3).$$

Now we plug in $y_1 = y_2 = y_3 = 0$ and conclude that the discriminant $x_2^2 - x_1 x_3$, which is already invariant, generates the ring of invariants. □

Hilbert himself reacted to the criticism by writing a second paper (Hilbert 1893) in which he gave a constructive way of finding generators for the ring of invariants in the case of $\mathrm{SL}_n(\mathbb{C})$ acting on forms (or systems of forms) by linear substitution. Let us sketch this method. Instead of looking for generators of the ideal $I_{\mathcal{N}}$, Hilbert asks for homogeneous invariants cutting out the variety defined by $I_{\mathcal{N}}$. This variety is called the *Null-cone* by Hilbert. He first shows:

Theorem 3.9. *Let $f_1, \ldots, f_r$ be homogeneous invariants whose zero–set is the Null-cone. Then $K[V]^G$ is a finitely generated $K[f_1, \ldots, f_r]$-module.*

Proof. By the Nullstellensatz (proved in the same paper), there exists an $m > 0$, such that $I_{\mathcal{N}}^m \subset (f_1, \ldots, f_r)$. By applying the Reynolds operator and induction, it follows without too much difficulty that $k[V]^G$ is generated as a $K[f_1, \ldots, f_r]$–module by all invariants of degree $\leq mrd$. Here d is the maximum of the degrees of the f_i. □

How to compute $f_1, \ldots, f_r$ as above? Consider the maximal torus T in G, and compute its ring of invariants $k[V]^T$ via integer programming, see (Sturmfels 1993, section 1.4). The zero-set of the torus invariants of positive degree is called the *canonical cone.* The Hilbert–Mumford criterion (Hilbert 1893, Mumford *et al* 1994), compare also Kraft (1985), says that the Null-cone is the orbit under G of the canonical cone. It follows that one can find $f_1, \ldots, f_r$ as above from the torus invariants via Gröbner bases (see Sturmfels (1993)).

At this point, Hilbert gives two alternatives how to proceed. We sketch the first method.

(1) Compute a Noether normalization $k[p_1, \ldots, p_m]$ of $k[f_1, \ldots, f_r]$. Then $k[p_1, \ldots, p_m] \subset k[V]^G$ is also a Noether normalization, and therefore every element of $k[V]^G$ is integral over $k[p_1, \ldots, p_m]$.

(2) Compute the *quotient field* of $k[V]^G$.

(3) Via the theorem of the primitive element, find a further invariant p such that $k(p_1, \ldots, p_m, p)$ is equal to the quotient field of $k[V]^G$.

(4) It now follows easily that the *normalization* of $k[p_1, \ldots, p_m, p]$ is equal to $k[V]^G$.

For Hilbert's second method, which also reduces the problem to the computation of a normalization, we refer to Sturmfels (1993).

An efficient algorithm for computing the normalization has been recently described by de Jong (1997). This algorithm is implemented in Singular (Decker *et al.* 1997). See Greuel *et al.* (1997) for Singular.

References

Adler, A. (1992) 'Invariants of the Simple Group of Order 660 acting on $\mathbb{C}^5$', *Comm. Alg.* **22**, 2837–2862.

Adler, A. (1997) 'Invariants of $SL_2(\mathbb{F}_q) \cdot \mathrm{Aut}(\mathbb{F}_q)$ acting on $\mathbb{C}^n$', Preprint.

Bertin, M.–J. (1967) 'Anneaux d'invariants d'anneaux de polynomes en caractéristique p', *Comptes Rendus Acad. Sci. Paris (Serie A)* **264**, 653–656.

Bruns, W., Herzog, J. (1993) 'Cohen–Macaulay rings', Cambridge University Press, Cambridge.

Buchberger, B. (1965) 'On Finding a Vector Space Basis of the Residue Class Ring Modulo a Zero Dimensional Polynomial Ideal' (German). PhD Thesis, Univ of Innsbruck, Austria.

Buchberger, B. (1970) 'An Algorithmical Criterion for the Solvability of Algebraic Systems of Equations' (German), *Aequationes mathematicae* **4**, 374–383.

Buchberger, B. (1985) 'Gröbner-Bases: An Algorithmic Method in Polynomial Ideal Theory.' Chapter 6 in: Multidimensional Systems Theory (N.K. Bose ed.), Reidel Publishing Company, Dordrecht, 1985, 184-232.

Buchberger, B. (1998) 'Introduction to Gröbner Bases', this volume.

Campbell, H.E.A., Hughes, I.P., Pollack, R.D. (1991) 'Rings of Invariants and p-Sylow Subgroups', *Can. Math. Bull.* **34**, 42–47.

Cox, D., Little, J., O'Shea, D. (1997) 'Ideals, Varieties, and Algorithms, second ed.', Springer, New York.

Decker, W., Greuel, G.–M., de Jong, T., Pfister, G. (1997) 'The Normalization: a New Invented Algorithm, Implementation and Comparisons', Preprint, Kaiserslautern, Saarbrücken.

Decker, W., Heydtmann, A. E., Schreyer, F.-O. (1997) 'Generating a Noetherian Normalization of the Invariant Ring of a Finite Group', J. Symbolic Computation (to appear).

Derksen, H. (1997a) 'Constructive Invariant Theory and the Linearization Problem', Thesis, Basel.

Derksen, H. (1997b) 'Computation of Invariants for Reductive Groups', Preprint, Basel.

Derksen, H., Kraft, H. (1995) 'Constructive Invariant Theory', Preprint, Basel.

Dieudonné, J.A., Carrell, J.B. (1971) 'Invariant Theory, Old and New', Academic Press, New York, London.

Eisenbud, D. (1995) 'Commutative Algebra with a View Toward Algebraic Geometry', Springer, New York.

Eisenbud, D., Riemenschneider, O., Schreyer, F.-O. (1981) 'Projective Resolutions of Cohen-Macaulay Algebras', *Math. Ann.* **257**, 85–98.

Gauss, C.F. (1801) 'Disquisitiones arithmeticae', translation A.A. Clarke, Yale University Press, 1965.

Gianni, P., Trager, B. (1996) 'Integral Closure of Noetherian Rings', Preprint.

Gordan, P. (1868) 'Beweis, dass jede Covariante und Invariante einer binären Form eine ganze Funktion mit numerischen Coeffizienten einer endlichen Anzahl solcher Formen ist', *J. Reine Angew. Math.* **69**, 323–354.

Gordan, P. (1899) 'Neuer Beweis des Hilbertschen Satzes über homogene Funktionen', *Nachrichten König. Ges. der Wiss. zu Gött.*, 240–242.

Greuel, G.-M., Pfister, G., Schönemann, H. (1997) 'Singular Reference Manual', *Reports On Computer Algebra, number 12.* Centre for Computer Algebra, University of Kaiserslautern,

http://www.mathematik.uni-kl.de/~zca/Singular.

Heydtmann, A.E. (1996) 'Generating Invariant Rings of Finite Groups', Diplom Thesis, Saarbrücken.

Heydtmann, A.E. (1997) 'finvar.lib: A Singular Library to Compute Invariant Rings and more', http://www.mathematik.uni-kl.de/~zca/Singular.

Hilbert, D. (1890) 'Über die Theorie der algebraischen Formen', *Math. Ann.* **36**, 473–534.

Hilbert, D. (1893) 'Über die vollen Invariantensysteme', *Math. Ann.* **42**, 313–373.

Hilbert, D. (1900) 'Mathematische Probleme', *Nachr. v. d. Ges. d. Wiss. zu Göttingen*, 253–297.

Hochster, M., Eagon, J.A. (1971) 'Cohen–Macaulay Rings, Invariant Theory, and the Generic Perfection of Determinantal Loci', *Am. J. Math.* **93**, 1020–1058.

Hochster, M., Roberts, J. (1974) 'Rings of Invariants of Reductive Groups Acting on Regular Rings are Cohen-Macaulay', *Adv. in Math.* **13**, 115–175.

Horrocks, G., Mumford, D. (1973) 'A Rank 2 Vector Bundle on $\mathbb{P}^4$ with 15,000 Symmetries', *Topology* **12**, 63–81.

de Jong, T. (1997) 'An Algorithm for Computing the Integral Closure', Preprint, Saarbrücken.

Kemper, G. (1993) 'The Invar Package for Calculating Rings of Invariants', *IWR Preprint Universität Heidelberg*, 93-94.

Kemper, G. (1996) 'Calculating Invariant Rings of Finite Groups over Arbitrary Fields', *J. Symbolic Computation* **21**, 351–366.

Kemper, G. (1997) 'Calculating Optimal Homogeneous Systems of Parameters', *IWR Preprint Universität Heidelberg*, 97-08.

Kemper, G., Steel, A. (1997) 'Some Algorithms in Invariant Theory of Finite Groups', Preprint.

Klein, F. (1878/1879) 'Über die Transformationen siebenter Ordnung der elliptischen Funktionen', *Math. Ann.* **14**, 428–471.

Klein, F. (1879) 'Über die Transformationen elfter Ordnung der elliptischen Funktionen', *Math. Ann.* **15**, 533–555.

Kline, M. (1972) 'Mathematical Thought from Ancient to Modern Times', Oxford University Press, New York.

Kraft, H. (1985) 'Geometrische Methoden in der Invariantentheorie', Vieweg, Braunschweig.

Kung, J.P.S., Rota, G.–C. (1984) 'The Invariant Theory of Binary Forms', *Bull. Am. Math. Soc.* **10**, 27–85.

de Lagrange, J.–L. (1773, 1775) 'Recherches d'arithmétique', Œuvres, vol. 3, 695–795, Georg Olms Verlag, Hildesheim, New York 1973.

Maschke, H. (1899) 'Beweis des Satzes, dass diejenigen endlichen linearen Substitutionsgruppen, in welchen einige durchgehende verschwindende Coeffizienten auftreten, intransitiv sind', *Math. Ann.* **52**, 363–368.

Minkowski, H. (1973) ' Briefe an David Hilbert, mit Beiträgen und herausgegeben von L. Rüdenberg und H. Zassenhaus', Springer, Berlin, Heidelberg, New York.

Möbius, F. (1827) 'Der barycentrische Calcul', Gesammelte Werke, vol. 1, 1-388, Wiesbaden 1885.

Molien, T. (1897) 'Über die Invarianten der linearen Substitutionsgruppen', *Sitzungsber. König. Preuss. Akad. Wiss.*, 1152–1156.

Mumford, D., Fogarty, J., Kirwan, F. (1994) 'Geometric Invariant Theory', third ed., Springer, Berlin, Heidelberg, New York.

Newstead, P. (1978) 'Introduction to Moduli Problems and Orbit Spaces', Springer, Berlin, Heidelberg, New York.

Noether, E. (1916) 'Der Endlichkeitssatz der Invarianten endlicher Gruppen', *Math. Ann.* **77**, 89–92.

Noether, E. (1926) 'Der Endlichkeitssatz der Invarianten endlicher linearer Gruppen der Charakteristik p', *Nachr. v. d. Ges. d. Wiss. zu Göttingen*, 28–35.

Plücker, J. (1830) 'Über ein neues Coordinatensystem', *J. Reine Angew. Math.* **5**, 1–36.

Schreyer, F.–O. (1980) 'Die Berechnung von Syzygien mit dem verallgemeinerten Weierstrass'schen Divisionssatz', Diplom Thesis, Hamburg.

Sloane, N.J.A. (1977) 'Error Correcting Codes and Invariant Theory: New Applications of a Nineteenth Century Technique', *Am. Math. Monthly* **84**, 82–107.

Smith, L. (1995) 'Polynomial Invariants of Finite Groups', A.K. Peters, Wellesley.

Springer, T.A. (1989) 'Aktionen reduktiver Gruppen auf Varietäten', Algebraische Transformationsgruppen und Invariantentheorie, DMV-Seminar Notes 13, Birkhäuser, Basel, Boston.

Stanley, R. B. (1979) 'Invariants of Finite Groups and their Applications to Combinatorics', *Bull. Am. Math. Soc.* **1**, 475–511.

Sturmfels, B. (1993) 'Algorithms in Invariant Theory', Springer, Wien, New York.

Sylvester, J.J. (1904–1912) 'The Collected Mathematical Papers of James Joseph Sylvester', Cambridge University Press, Cambridge.

Vasconcelos, W. (1991) 'Computing the Integral Closure of an Affine Domain', *Proc. AMS* **113**, 633–638.

Weyl, H. (1946) 'The Classical Groups, their Invariants and Representations', second ed., Princeton Mathematical Series 1, Princeton University Press, Princeton.

A tutorial on generic initial ideals

Mark Green and Michael Stillman

Our goal in this tutorial is to give a quick overview of generic initial ideals. A more comprehensive treatment is in [Gr96]. We first lay out the basic facts and notations, and then deal with a few of the more interesting points in a question and answer format.

We would like to thank Bruno Buchberger for inviting us to contribute this paper, and also for having, through his fundamental contributions, made the work discussed in this tutorial possible.

For this tutorial we let $S = \mathbf{C}[x_1, \ldots, x_n]$. Some of what we do also works in characteristic $p > 0$, but the combinatorial properties of Borel fixed monomial ideals are more complicated. Later we give an indication of what is true in this case.

Let I be a homogeneous ideal in the polynomial ring S, and choose a monomial order. Throughout this tutorial, the only monomial orders that we consider satisfy $x_1 > x_2 > \cdots > x_n$. Any such order will do, but the most interesting from our present point of view are the lexicographic and reverse lexicographic orders. Given this monomial order, we may compute a Gröbner basis $\{g_1, \ldots, g_r\}$ of the ideal I, using Buchberger's algorithm. The **initial ideal** $\mathrm{in}(I)$ is the monomial ideal generated by the lead terms of $g_1, \ldots, g_r$. This monomial ideal has the same Hilbert function as I.

This procedure gives us a very rich combinatorial object to study containing a great deal of information about the original ideal I. In many situations—especially if I arises from geometry—the choice of a basis for the polynomial ring is arbitrary, and we thus may want to have an object that is independent of the choice of basis. One way to do this, of course, is to look at the action of $g \in \mathrm{GL}(n)$ on I by substitution, leading to an ideal $g(I)$ having the same Hilbert function as I. One might then study the map

$$\mathrm{GL}(n) \to \{\text{monomial ideals over } S\}$$

given by

$$g \mapsto \mathrm{in}(g(I)).$$

There are circumstances—for example, in questions of stability—when one needs all of this information, but for a wide variety of purposes the most important coordinate-free information contained in this map comes from the fact that on a non-empty Zariski open set $U \subseteq \mathrm{GL}(n)$, the map above is constant [Ga]:

THEOREM (Galligo). *Given a monomial order and an ideal $I \subseteq S$, there is a Zariski open subset $U \subseteq \mathrm{GL}(n)$, and a monomial ideal* gin(I) *(the* generic initial ideal*) such that for all $g \in U$, $\mathrm{in}(g(I)) = \mathrm{gin}(I)$. Furthermore,* gin$(I)$ *is a Borel-fixed monomial ideal (this is defined below).*

For example, if $I = (x_1^2, x_2^2 - x_3^2) \subset \mathbf{C}[x_1, x_2, x_3]$, then for the reverse lexicographic order, $\mathrm{gin}(I) = (x_1^2, x_1x_2, x_2^3)$, and for the lexicographic order (the "lex gin"), $\mathrm{gin}(I) = (x_1^2, x_1x_2, x_1x_3^2, x_2^4)$.

Generic initial ideals appear first in the work of Grauert and Hironaka; and there is a side of the theory dealing with singularities that we will be unable to cover due to lack of space and (more importantly) expertise. A nice survey of this side of the subject is [BR].

The first advantage that gin(I) has over in(I) is that it is a much more special kind of combinatorial object, namely a **Borel-fixed monomial ideal.** These are characterized in two equivalent ways: elegantly, if we adopt the notation

$$g(x_i) = \sum_j g_{ij} x_j,$$

and let B be the **Borel subgroup**

$$B = \{g \mid g_{ij} = 0 \text{ for } j > i\},$$

then a monomial ideal M is Borel-fixed if

$$g(M) = M \text{ for all } g \in B.$$

More computationally, define an **elementary move** e_k for $1 \leq k \leq n-1$ by

$$e_k(x^J) = x^{\hat{J}},$$

where, if $J = (j_1, \ldots, j_n)$ and $j_{k+1} > 0$, then

$$\hat{J} = (j_1, \ldots, j_{k-1}, j_k + 1, j_{k+1} - 1, j_{k+2}, \ldots, j_n).$$

If $j_{k+1} = 0$, we set $e_k(x^J) = 0$. Thus e_k increases the exponent of x_k by 1 and decreases the exponent of x_{k+1} by 1. Then a monomial ideal M is Borel-fixed if

$$x^J \in M \to e_k(x^J) \in M \quad \text{for all } k.$$

One almost-immediate payoff of the fact that gin(I) is Borel-fixed is:

THEOREM. *Every homogeneous ideal has the same Hilbert function as some Borel-fixed monomial ideal.*

With a non-trivial amount of work, one obtains [M], [Gr89]:

THEOREM (Macaulay). *A function $h: \mathbf{N} \to \mathbf{N}$ is the Hilbert function of some homogeneous ideal I over some $S = \mathbf{C}[x_1, \ldots, x_n]$ if and only if, for all d, if we expand*

$$h(d) = \binom{k_d}{d} + \binom{k_{d-1}}{d-1} + \cdots + \binom{k_\delta}{\delta},$$

where $k_d > k_{d-1} > \cdots > k_\delta > 0$, then

$$h(d+1) \leq \binom{k_d+1}{d+1} + \binom{k_{d-1}+1}{d} + \cdots + \binom{k_\delta+1}{\delta+1}.$$

This theorem can be viewed as a lower bound on the rate of growth of a homogeneous ideal. It has a wide range of applications, from commutative algebra to Hodge theory (see [Gr89b]).

Macaulay's theorem is the best possible general statement; its bound is realized by the **lexicographic ideal**. However, improvements exist, notably [Go], see also [Gr89]:

THEOREM (Gotzmann). *If I is a homogeneous ideal generated in degrees $\leq d$ for which equality holds in Macaulay's bound for $h_{S/I}(d+1)$, then equality holds in Macaulay's bound for $h_{S/I}(k)$ for all $k \geq d+1$.*

A second payoff of the fact that gin(I) is Borel-fixed is that the minimal free resolution of a Borel-fixed ideal is especially simple. The p-th syzygies of any monomial ideal M are generated by ones involving only terms of the form

$$x^{K_p} \otimes x^{K_{p-1}} \otimes \cdots \otimes x^{K_1} \otimes x^{K_0},$$

where

$$x^{K_0} \in M;$$

all terms occurring in the same syzygy have the same $K_0 + K_1 + \cdots + K_p$. We order such terms by the collection of rules that, when $K_0 + \cdots + K_p = L_0 + \cdots + L_p$,

$$x^{K_p} \otimes x^{K_{p-1}} \otimes \cdots \otimes x^{K_1} \otimes x^{K_0} > x^{L_p} \otimes x^{L_{p-1}} \otimes \cdots \otimes x^{L_1} \otimes x^{L_0}$$

if

(1) $|K_0| > |L_0|$, or $|K_0| = |L_0|$ and $|K_1| > |L_1|$, etc.;
(2) $|K_i| = |L_i|$ for all i and $x^{K_0} < x^{L_0}$ or $K_0 = L_0$ and $x^{K_1} < x^{L_1}$, etc.
It then follows that, although the minimal free resolution of a general monomial ideal can be quite complicated, for Borel-fixed monomial ideals the situation is very nice:[EK]

THEOREM (Eliahou-Kervaire). *For a Borel-fixed monomial ideal M, the initial terms of the p-th syzygies are precisely*

$$x_{k_p} \otimes x_{k_{p-1}} \otimes \cdots \otimes x_{k_1} \otimes x^J,$$

where $x^J \in M$ is a minimal generator, $J = (j_1, \ldots, j_n)$, and

$$k_p < k_{p-1} < \cdots < k_1 < \max\{m \mid j_m > 0\}.$$

Using this theorem, or a more direct argument, we may write down a formula for the graded betti numbers of a Borel-fixed monomial ideal M. Define $b_{p,d}(M)$ to be the number of minimal p-th syzygies of M of degree d. Thus $b_{0,d}(M)$ is the number of minimal generators of M of degree exactly d. For each (minimal) generator x^J of M, define $\operatorname{maxvar}(x^J) = \max\{m \mid j_m > 0\}$. We then conclude:

COROLLARY. *For a Borel-fixed monomial ideal M, the graded betti numbers satisfy*

$$b_{p,d+p}(M) = \sum_{x^J} \binom{\operatorname{maxvar}(x^J) - 1}{p},$$

where the sum is over all minimal generators x^J of M having degree exactly d.

The resolutions of I and $\operatorname{in}(I)$ are related: Since there is a flat family I_t for any I with $I_0 = \operatorname{in}(I)$ and $I_t = g_t(I)$ for a diagonal matrix g_t, we conclude rather easily that:

PROPOSITION. *The minimal free resolution of I is obtained from that of $\operatorname{in}(I)$ by cancelling some adjacent terms of the same degree.*

We remark that the **Hilbert syzygy theorem** is a simple corollary of this result and the Eliahou-Kervaire Theorem.

Combined with the Eliahou-Kervaire Theorem, this proposition gives a lot of information about the minimal free resolution of I that can be read off of $\operatorname{gin}(I)$; however, there is not enough information to reconstruct the minimal free resolution of I in most cases.

A lot of fundamental information about I can be read off from $\operatorname{gin}(I)$ if we use the **reverse lexicographic order (rlex)**.

A homogeneous ideal I is **saturated** if

$$(I : (x_1, \ldots, x_n)) = I.$$

The **saturation** of I is

$$I^{\text{sat}} = \cup_{k \geq 0}(I : (x_1, \ldots, x_n)^k).$$

A homogeneous ideal I is **m-saturated** if

$$I_d^{\text{sat}} = I_d \quad \text{for all } d \geq m.$$

The **satiety** of I, denoted $\operatorname{sat}(I)$, is the smallest m for which I is m-saturated.

A homogeneous ideal I is **m-regular** if in the minimal free resolution of I, for all $p \geq 0$, every p-th syzygy has degree $\leq m + p$. If I is saturated, this is equivalent to the geometric condition that the associated sheaf $\mathcal{I}$, on projective space $\mathbf{P}^{n-1}$ satisfies the condition of **Castelnuovo m-regularity**

$$H^q(\mathbf{P}^{n-1}, \mathcal{I}(m-q)) = 0 \text{ for all } q > 0.$$

The **regularity** of I, denoted $\operatorname{reg}(I)$, is the smallest m for which I is m-regular.

For a Borel-fixed monomial ideal M, it is easy to see that

$$(M : (x_1, \ldots, x_n)) = (M : x_n),$$

from which it follows:

PROPOSITION. *For a Borel-fixed monomial ideal M,*

$$\operatorname{sat}(M) = \textit{highest degree of a minimal generator involving } x_n.$$

Thus

$$M \text{ is saturated} \leftrightarrow \text{no minimal generator of } M \text{ involves } x_n.$$

From the Eliahou-Kervaire Theorem, it follows that:

PROPOSITION. *For a Borel-fixed monomial ideal M,*

$$\operatorname{reg}(M) = \text{highest degree of a minimal generator of } M.$$

The projective dimension of S/M can be read off as well:

PROPOSITION. *For a Borel-fixed monomial ideal M,*

$$\textit{proj. dim}(S/M) = r,$$

where r is the maximum j such that x_j occurs in some minimal generator of M.

One then has:[BS]

THEOREM (Bayer-Stillman). *For any homogeneous ideal I, using the reverse lexicographic order,*

$$\begin{aligned} \operatorname{sat}(I) &= \operatorname{sat}(\operatorname{gin}(I)), \\ \operatorname{reg}(I) &= \operatorname{reg}(\operatorname{gin}(I)), \end{aligned}$$

and

$$\textit{proj. dim}(S/I) = \textit{proj. dim}(S/\operatorname{gin}(I)).$$

Thus

$$\text{sat}(I) = \textit{highest degree of a minimal generator of } \text{gin}(I) \textit{ involving } x_n,$$

$$\text{reg}(I) = \textit{highest degree of a minimal generator of } \text{gin}(I),$$

and

$$\text{proj.dim}(S/I) = \textit{largest } r \textit{ such that } x_r \textit{ occurs in a minimal generator of } \text{gin}(I).$$

For many geometric applications, and also for doing inductive arguments, it is useful to know what happens when we restrict to a generic hyperplane. If h is a general linear form and H is the corresponding hyperplane, let

$$I_H = I + (h),$$

thought of as a homogeneous ideal in $n-1$ variables; of course this only makes sense modulo a linear choice of coordinates.

THEOREM. *[BS, Gr89] For a homogeneous ideal I and a generic hyperplane H, then for the rlex order,*

$$\text{gin}(I_H) = (\text{gin}(I))_{x_n};$$

more generally for $k \geq 0$,

$$\text{gin}((I : h^k)_H) = (\text{gin}(I) : x_n^k)_{x_n}.$$

This result implies, with some combinatorial work, the following analog of Macaulay's Theorem:[Gr89]

THEOREM. *If I is a homogeneous ideal and H is a generic hyperplane, and S_H is the coordinate ring of H, then if*

$$h_{S/I}(d) = \binom{k_d}{d} + \binom{k_{d-1}}{d-1} + \cdots + \binom{k_\delta}{\delta},$$

where $k_d > k_{d-1} > \cdots > k_\delta > 0$, then

$$h_{S_H/I_H}(d) \leq \binom{k_d - 1}{d} + \binom{k_{d-1} - 1}{d-1} + \cdots \binom{k_\delta - 1}{\delta},$$

where we adopt the convention $\binom{a}{b} = 0$ if $a < b$.

With these basics in place, we now set out to answer a number of natural questions about how and why gins work and how to use them.

Why does the gin exist?

An elegant way to think about this is the following: If G is any algebraic group and

$$\rho: G \to \mathrm{GL}(V)$$

is any representation of G, let us assume that V has a fixed basis $e_1, \ldots, e_n$ and that $W \subseteq V$ is a linear subspace of dimension k. If we regard $e_1 > e_2 > \cdots > e_n$, then the **initial term** $\mathrm{in}(v)$ of an element $v \in V$ is defined by

$$\mathrm{in}(\sum_{i=1}^{n} v_i e_i) = v_i e_i, \text{ where } i = \min\{j \mid v_j \neq 0\}.$$

We define

$$\mathrm{in}(W) = \mathrm{span}\{\mathrm{in}(v) \mid v \in W\}.$$

If $J = (j_1, \ldots, j_k)$ with $j_1 < j_2 < \cdots < j_k$, let

$$V_J = \mathrm{span}(e_{j_1}, \ldots, e_{j_k}).$$

Thus

$$\mathrm{in}(W) = V_{J(W)}$$

for some index set $J(W) = (j_1(W), \ldots, j_k(W))$ of size k.

Let

$$k_i(W) = \dim(W \cap \mathrm{span}(e_1, \ldots, e_i)).$$

Note

$$k_i(W) = \max\{m \mid j_m(W) \leq i\}.$$

We have the **Schubert cell decomposition** of the Grassmannian $G(k, V)$:

$$\begin{aligned} S_{a_1,\ldots,a_n} &= \{W \mid k_i(W) \geq a_i\} \\ S^o_{a_1,\ldots,a_n} &= \{W \mid k_i(W) = a_i\}. \end{aligned}$$

We note that

$$W \in S^o_{a_1,\ldots,a_n} \leftrightarrow \mathrm{card}(J(W) \cap \{1, 2, \ldots, i\}) = a_i \text{ for all } i,$$

so that knowing which open Schubert cell W lies in is equivalent to knowing $\mathrm{in}(W)$.

If we now fix $W \subseteq V$, we may define

$$\begin{aligned} G_{a_1,\ldots,a_n}(W) &= \{g \in G \mid \rho(g)(W) \in S_{a_1,\ldots,a_n}\} \\ G^o_{a_1,\ldots,a_n}(W) &= \{g \in G \mid \rho(g)(W) \in S^o_{a_1,\ldots,a_n}\}. \end{aligned}$$

The $G_{a_1,\ldots,a_n}(W)$ are algebraic subvarieties of G, defined by the vanishing of certain $k \times k$ minors of the $k \times n$ matrix of $\rho(g)(W)$ in terms of the basis $e_1, \ldots, e_n$. The union of these subvarieties is G. If we order n-tuples

$(a_1, \dots, a_n)$ lexicographically, then there is a unique largest $(a_1, \dots, a_n)$ such that

$$G = G_{a_1,\dots,a_n},$$

and then

$$\mathrm{in}(\rho(g)(W)) \text{ is constant on } G^o_{a_1,\dots,a_n}(W),$$

and $G^o_{a_1,\dots,a_n}(W)$ is a non-empty Zariski open subset of G.

If U is the vector space of linear forms with basis $x_1, \dots, x_n$, we take

$$V = S^d U$$

with fixed basis the monomials of degree d in decreasing order for a given monomial order, and

$$\rho\colon \mathrm{GL}(U) \to \mathrm{GL}(S^d U)$$

the standard representation, then in the above construction, $G^o_{a_1,\dots a_n}(I_d)$ is the Zariski open set guaranteed in degree d by Galligo's Theorem, and $\mathrm{in}(\rho(g)(I_d))$ gives $\mathrm{gin}(I_d)$ for $g \in G^o_{a_1,\dots a_n}(I_d)$.

For such a g, $\mathrm{gin}(\rho(g)(I))$ is finitely generated, so this process stops after a finite number of degrees.

Why is the gin Borel-fixed?

It is enough to consider a given degree d, and thus we are in the situation discussed in the preceding topic. A subgroup $B \subset G$ with Lie algebra $\mathcal{B}$ satisfies the **infinitesimal property** if, for all $\gamma \in \mathcal{B}$ and all $i < j$, either both $\gamma(e_i) = 0$ and $\gamma(e_j) = 0$, or $\mathrm{in}(\gamma(e_i)) > \max(e_i, \mathrm{in}(\gamma(e_j)))$. If we differentiate this, we obtain the elementary proposition that if B has the infinitesimal property, then for any W, if

$$\mathrm{gin}(W) = V_J,$$

then

$$i \in J \text{ and } \mathrm{in}(\gamma(e_i)) = e_j \to j \in J.$$

One now checks that the Borel subgroup $B \subset \mathrm{GL}(U)$ satisfies the infinitesimal property for $\rho\colon \mathrm{GL}(U) \to \mathrm{GL}(S^d U)$, and that the action of its Lie algebra is precisely the elementary moves.

Why does the rlex gin compute satiety and regularity?

The Eliahou-Kervaire resolution computes the regularity of a Borel fixed monomial ideal—it's the highest degree of a minimal generator.

The satiety of a Borel-fixed monomial ideal is also seen to be the degree of the highest minimal generator involving x_n.

In rlex, for any homogeneous ideal, we have for a generic hyperplane H,

$$\mathrm{reg}(I) = \max(\mathrm{sat}(I), \mathrm{reg}(I_H));$$

essentially by an application of the long exact sequence for Tor. By induction, the problem of showing that I and $\mathrm{gin}(I)$ have the same regularity for all I reduces to the problem of showing that I and $\mathrm{gin}(I)$ have the same satiety. In rlex, a polynomial is divisible by x_n if and only if its initial term is, and this allows us to show that I is m-saturated if and only if there are no minimal generators of $\mathrm{gin}(I)$ of of degree $> m$ involving x_n. This then implies that I and $\mathrm{gin}(I)$ have the same satiety.

How is the gin computed?

The easiest method in practice to compute the gin is to make a "random" linear change of variables, apply it to I, and then compute the initial ideal. Of course, it is possible that you will be unlucky, and the choice of random linear change of variables will not lie in the Zariski open set of elements of $\mathrm{GL}(n)$ that give $\mathrm{gin}(I)$. Executing the sequence of commands a few more times lowers the probability of getting an erroneous answer.

The following session shows how one might compute gin's using *Macaulay2* ([GS]). We compute the reverse lexicographic and lexicographic generic initial ideals for a complete intersection of type (3,3) in $\mathbf{P}^3$. The *Macaulay2* output has been modified to save space.

First we define the ring R and the ideal I. The default monomial order in *Macaulay2* is the reverse lexicographic order.

```
i1 : R = ZZ/101[a..d]

i2 : I = ideal(a^3+c^2*d, b^3-a*d^2)
```

Next, we create a random 1×4 matrix of linear forms (a random matrix $F : R^1 \longleftarrow R(-1)^4$), and then apply this random change of coordinates to I, placing the result into genericI.

```
i3 : F = random(R^1, R^{4:-1})

o3 = | -22a+43b+41c+47d -2a-37b-16c+d 35a+49b-7c-49d 38a-39b-5c-11d |

                 1      4
o3 : Matrix R  <--- R

i4 : genericI = substitute(I,F)
```

Finally, compute a Gröbner basis of genericI, and set ginI to be the matrix of lead terms (initial terms). *Macaulay2* computes automatically the Gröbner basis, using Buchberger's algorithm.

```
i5 : ginI = leadTerm genericI

o5 = | a3 a2b ab3 b5 |

             1      4
o5 : Matrix R  <--- R
```

To find the lex gin, create a ring which uses the lexicographic order.

```
i6 : S = ZZ/101[a..d,MonomialOrder=>Lex]

o6 = S

o6 : PolynomialRing
```

First move the ideal genericI to this new ring S, and then compute the lead terms as before.

```
i7 : lexginI = leadTerm substitute(genericI, S)

o7 = | a3 a2b a2c2 ab4 a2cd2 a2d4 ab3c2 ab3cd2 ab3d3 ab2c5
       ab2c4d ab2c3d2 ab2c2d4 ab2cd5 b9 ab2d7 abc8 abc7d2
       abc6d4 abc5d6 abc4d8 abc3d10 abc2d12 abcd14 abd16 ac18 |

             1      26
o7 : Matrix S  <--- S
```

The graded betti numbers of S/I and $S/\text{gin}(I)$ are given below. The pth entry of the d th row is the betti number $b_{p,p+d}(S/M)$.

```
i8 : betti resolution I
    total: 1 2 1
        0: 1 . .
        1: . . .
        2: . 2 .
        3: . . .
        4: . . 1

i9 : betti resolution cokernel ginI
    total: 1 4 3
        0: 1 . .
        1: . . .
        2: . 2 1
        3: . 1 1
        4: . 1 1
```

In Macaulay (classic!), computing the gin is done as follows: say you are in 5 variables, your ring is named r, and your ideal is named i. You might then give the commands:

```
<random_mat 1 5 r f
ev f i gi
```

This applies the random linear change of coordinates (which is now named f) to i and gives you a new ideal named gi. Execute the commands:

```
std gi gistd
in gistd gin
```

You now have a monomial ideal, which we have named gin, which is (with high probability) the generic initial ideal of i, in whatever monomial order you are using in your ring.

What is the gin used for?

(1) The gin is an incredibly useful computational tool. One can easily compute regularity and satiety of any ideal. For points in the plane, one can compute the numerical character. For space curves, one can easily read off the arithmetic genus, numerical character of the general hyperplane section, ranks in all degrees of the Rao module (this requires the trick of intersecting with a general hypersurface section of suitable degree (see [FG])).

(2) A number of geometric results have been proved using the rlex gin, including a new proof of Laudal's Lemma ([Gr96]), a generalization by M. Cook [C] of Gruson-Peskine's [GP] result on connectedness of the numerical character for space curves, and the best known bound on the degree of a smooth surface in $\mathbf{P}^4$ not of general type (see [BF, BC]). Of course, one expects more geometric results as these techniques pass into wider use. So far, progress in this direction has hinged on combining gin techniques with some explicitly geometric techniques.

(3) Gins give some of the most appealing proofs of the basic theorems on growth of ideals of Macaulay and Gotzmann, see [Gr96].

(4) Lex gins allow the computation of geometric information about multiple point loci of projections, see [Gr96].

(5) Gins are useful in determining the structure of the Hilbert schemes. For example, Reeves [Re] has shown that if H denotes the Hilbert scheme parametrizing subschemes of $\mathbf{P}^{n-1}$ with Hilbert polynomial $P(t)$ of degree d, then the radius of H (that is, the radius of the graph whose vertices are the components of H, and where two vertices are connected with an edge whenever the two components meet) is $\leq d+2$. She proves this by starting with any Borel-fixed monomial ideal, forming a fan (as Hartshorne defined them), and taking its lexicographic generic initial ideal. Now repeat the process. Reeves shows that this process yields, in at most $d+1$ steps, the lexicographic ideal.

(6) The lexicographic ideal is a very interesting Borel-fixed monomial ideal. It attains the bounds for all of the Macaulay/Kruskal-Katona type results. It has the maximum graded betti numbers of any ideal with the same Hilbert function, and it has the highest regularity of any ideal with the same Hilbert function or polynomial. One might expect from all of this that the corresponding point on the Hilbert scheme is highly singular. In fact just the opposite is true: Reeves and Stillman [RS] show that the point on the Hilbert scheme H corresponding to the lexicographic ideal is always a smooth point on H.

Why choose lex or rlex for a particular problem?

The basic rule is that rlex is well-adapted to taking hyperplane sections, and lex is well-adapted to projections. The lex gin of an ideal tends to be extremely complicated, which is bad computationally, but often is good geometrically, since it contains an enormous amount of information. For example, the lex gin of a rational normal curve of degree d contains the lex gins of the general rational curve of degree d in projective spaces of all lower dimensions.

What do the gins of some geometric objects look like?

The first remark is that if X is a projective variety in $\mathbf{P}^{n-1}$ and I_X is the homogeneous ideal of X, then since I_X is saturated,

no minimal generator of $\text{gin}(I_X)$ in rlex involves x_n.

If Λ is a general linear space of codimension k, then in rlex,

$$\text{gin}(I_{X\cap\Lambda}) = \text{gin}(I_X)|_{x_{n-k+1}\to 1,\dots,x_n\to 1}.$$

For example, if X is a set of points in $\mathbf{P}^1$ of length d, then the only possibility is

$$\text{gin}(I_X) = (x_1^d).$$

If X is a set of points in $\mathbf{P}^2$, the only possibility is that in rlex

$$\text{gin}(I_X) = (x_1^s, x_1^{s-1}x_2^{\lambda_{s-1}}, \dots x_1x_2^{\lambda_1}, x_2^{\lambda_0})$$

where

$$\lambda_0 > \lambda_1 > \cdots > \lambda_{s-1} > 0.$$

The lowest degree of a curve containing X is s;

$$\lambda_0 + \cdots + \lambda_{s-1} = \text{length}(X);$$

the invariants $\lambda_0, \dots, \lambda_{s-1}$ contain the same information as the **numerical character** of Gruson-Peskine [GP]. There are some quite subtle and beautiful relations between the geometry of X and the λ-invariants—in particular (Gruson-Peskine),

$$X \text{ in uniform position} \to \lambda_i \geq \lambda_{i-1} - 2 \text{ for } i = 1, \dots, s-1.$$

One can relate larger gaps in the λ_i to the extent to which uniform position fails.

For curves in $X \subset \mathbf{P}^3$, there is a quite beautiful story, of which the first step is that for a general hyperplane H, using the reverse lexicographic order, $\text{gin}(I_{X \cap H})$ must be the gin of a set of points in uniform position in $\mathbf{P}^2$. A number of well-known results about curves in $\mathbf{P}^3$ can be proved using gin methods, in combination with geometric arguments (see [Gr96], section 4, for a survey), and there is clearly more to do in this direction.

Lex gins of even comparatively simple geometric objects are quite interesting and revealing; e.g. codimension 2 complete intersections. For any homogeneous ideal I, the elements of $I \cap \mathbf{C}[x_2, \ldots, x_n]$ are the **elimination ideal**, i.e. the ideal of the projection from the point $(1, 0, \ldots, 0)$. The elements of

$$I \cap \oplus_{i=0}^{k} x_1^i \mathbf{C}[x_2, \ldots, x_n]$$

modulo elements of

$$I \cap \oplus_{i=0}^{k-1} x_1^i \mathbf{C}[x_2, \ldots, x_n]$$

are of the form $x_1^k p$, where $p \in \mathbf{C}[x_2, \ldots, x_n]$. The set of such polynomials is the **k'th partial elimination ideal** of I, denoted $K_k(I) \subseteq \mathbf{C}[x_2, \ldots, x_n]$. One may verify that $K_k(I)$ defines set-theoretically the image in $\mathbf{P}^{n-1}$ of the $k+1$-fold multiple points of the projection of X from $(1, 0, \ldots, 0)$. The lex gin of $K_k(I)$ is just the ideal generated by the terms of the lex gin of I which are divisible by x_1^k. See [Gr96] for more information.

Lex gins contain a lot of information. For example, the lex gin of the first partial elimination ideal $K_1(I_X)$ of a complete intersection of type (2,3) in $\mathbf{P}^n$ is the lex gin of a complete intersection of type (2,3) in $\mathbf{P}^{n-1}$; these lex gins thus increase in complexity in a nested way as the dimension goes up.

What are some interesting examples of gins?

For d generic points X in $\mathbf{P}^2$, the rlex gin in degree k contains the highest $\text{rank}(I_{X,k})$ monomials in rlex order. This is easy to prove, because for points in the plane, the gin and the Hilbert function contain equivalent information, and the latter is known for generic points. For example, the rlex gin of 8 generic points in the plane is

$$(x_1^3, x_1^2 x_2, x_1 x_2^3, x_2^4).$$

A very interesting example of an rlex gin is the gin of a **generic rational space curve** of degree d. For all known examples, the gin is described as follows: the general hyperplane section is d general points, and the rest of the rlex gin is gotten by filling in the highest unused monomials in rlex order needed to fill out the Hilbert function. For example, a generic rational space curve of degree 8 has rlex gin

$$(x_1^4, x_1^3 x_2, x_1^2 x_2^3, x_1 x_2^4, x_2^5, x_1^3 x_3^2, x_1^2 x_2 x_3^2, x_1^2 x_2^2 x_3, x_1 x_2^3 x_3, x_2^4 x_3).$$

The rlex gin of the restriction of the ideal I_X to a generic hyperplane is gotten by sending $x_4 \to 0$, i.e. it is unchanged. The saturation of this is gotten by sending $x_3 \to 1$, which then yields the ideal of 8 generic points in $\mathbf{P}^2$, as above. The maximal rank conjecture is known to hold for a generic rational space curve [BE], so proving this is probably not too hard.

A typical example of a lex gin is that of a generic complete intersection X of type $(3,3)$ in $\mathbf{P}^3$. The lex gin is:

$$x_1^3, x_1^2(x_2, x_3^2, x_3x_4^2, x_4^4), x_1(x_2^4, x_2^3x_3^2, x_2^3x_3x_4^2, x_2^3x_4^3, x_2^2x_3^5, x_2^2x_3^4x_4,$$
$$x_2^2x_3^3x_4^2, x_2^2x_3^2x_4^4, x_2^2x_3^5, x_2^2x_4^7, x_2x_3^8, x_2x_3^7x_4^2, x_2x_3^6x_4^4, x_2x_3^5x_4^6,$$
$$x_2x_3^4x_4^8, x_2x_3^3x_4^{10}, x_2x_3^2x_4^{12}, x_2x_3x_4^{14}, x_2x_4^{16}, x_3^{18}), x_2^9.$$

The x_2^9 at the end comes from the fact that the projection of the curve X to $\mathbf{P}^2$ is a plane curve of degree 9. The genus of X is 10 by the adjunctions formula, from which we conclude that the general projection of X to $\mathbf{P}^2$ has 18 double points—by the discussion of partial elimination ideals above, this is why there is an $x_1x_3^{18}$ in the lex gin. There is quite a bit more interesting geometry that can be read off of this lex gin (see [Gr96]).

How do gins work in the exterior algebra?

In the exterior algebra (see [AHH], [Gr96])

$$E = \oplus_{k=0}^n \wedge^k V,$$

much of the theory goes through, but with some important changes:
(1) The gin of a homogeneous ideal is still a Borel-fixed monomial ideal in E, but these are simpler, since repeated variables are impossible.
(2) Homogeneous ideals I in E have minimal free resolutions by twists of E, but these are usually of infinite length. Nevertheless, there exists a maximum over all p of the maximum of $a_{pi} - p$, where the p-th syzygies have degrees a_{pi}, and this is the **regularity** of I. Unlike the symmetric case, the regularity has an *a priori* upper bound of $n = \dim(V)$.
(3) The initial terms of the p-th syzygies of a Borel-fixed monomial ideal M are precisely

$$x_{i_p} \otimes x_{i_{p-1}} \otimes \cdots \otimes x_{i_1} \otimes x^J,$$

where $J = (j_1, \ldots, j_n)$, x^J a minimal generator of M, and

$$x_{i_p} \le x_{i_{p-1}} \le \cdots \le x_{i_1} \le \max\{m \mid j_m > 0\}.$$

The change from the symmetric case is that $<$ is replaced by $\le$, which allows resolutions of infinite length.
(4) The graded betti numbers of M are now

$$b_{p,d+p}(M) = \sum_{x^J} \binom{\operatorname{maxvar}(x^J) - 1 + p}{p},$$

where the sum is over the minimal generators of M of degree exactly d.
(5) The regularity of I is still the degree of the highest generator of gin(I) if we use rlex.
(6) The analog of Macaulay's bound is:

THEOREM (Kruskal-Katona). *If*

$$h_{E/I}(d) = \binom{k_d}{d} + \binom{k_{d-1}}{d-1} + \cdots + \binom{k_\delta}{\delta},$$

where

$$k_d > k_{d-1} > \cdots > k_\delta > 0,$$

then

$$h_{E/I}(d+1) \leq \binom{k_d}{d+1} + \binom{k_{d-1}}{d} + \cdots + \binom{k_\delta}{\delta+1}.$$

The analogue of Gotzmann's theorem also holds when this bound is achieved.

(7) There is also an analog for the bound for restriction to a generic hyperplane.
(8) There is a natural way to associate in a 1-1 correspondence a monomial ideal in E with a simplicial complex that is a sub-simplicial complex of the standard simplex on n vertices. This leads to a number of beautiful results about simplicial complexes which can be proved using exterior gins. See [S].

Computing exterior gins in Macaulay2 can be done in the following way. (We have abbreviated the Macaulay 2 output, to save on space).

```
i1 : R = ZZ/101[a..e,SkewCommutative=>true]

i2 : I = ideal(a*c,a*d,b*d,b*e,c*e)

i3 : F = random(R^1, R^{5:-1});

i4 : genericI = substitute(I,F);

i5 : ginI = leadTerm genericI

o5 = | ab ac bc ad bd cde |

              1      6
o5 : Matrix R  <--- R
```

The betti displays of I and gin(I) (for reverse lex order) are in accord with the fact that they have the same regularity:

```
i6 : betti res(I, LengthLimit=>8)
   total: 1 5 15 31 55 90 140 210 306
       0: 1 .  .  .  .  .   .   .   .
       1: . 5 15 30 50 75 105 140 180
```

```
        2: . . .  . 1  5 15  35  70 126

i7 : betti res(coker ginI, LengthLimit=>8)
     total: 1 6 21 50 99 175 286 441 650
         0: 1 .  .  .  .   .   .   .   .
         1: . 5 16 35 64 105 160 231 320
         2: . 1  5 15 35  70 126 210 330
```

What happens in positive characteristic?

The main problem in characteristic $p > 0$ is that Borel-fixed monomial ideals have a more complicated combinatorial form. See Pardue [Pa94] or [E] for details. For example, in characteristic p, $M = (x_1^p, x_2^p)$ is Borel-fixed. Borel ideals which satisfy the combinatorial criterion involving the elementary moves are called **standard Borel-fixed monomial ideals**. The others, such as the M above, are called non-standard. If $\text{gin}(M)$ is standard, then almost every result continues to work: One may compute regularity, satiety, projective dimension as before, and the Eliahou-Kervaire resolution still holds. For a non-standard Borel-fixed monomial ideal M, there is no known formula for the resolution of M.

Results such as Galligo's theorem, the first three statements of the Bayer-Stillman theorem, and the facts about hyperplane sections still hold in positive characteristic. Similarly the Macaulay type upper bound theorems hold in this case (see [BH]).

Pardue [Pa96] has generalized Reeve's theorem on the radius of the Hilbert scheme. The proof of the smoothness of the lexicographic point in [RS] is valid in any characteristic.

What are some interesting open problems about gins and directions for future research?

There are lots of interesting questions:

(1) Develop a good theory of gins for syzygies, and understand what the correct analogue of Borel fixed is in this case. Some work in this direction is the thesis of C. Rippel [Ri].

(2) A related problem is to understand the constraints on possible cancellations in the minimal free resolution of $\text{gin}(I)$ to produce the minimal free resolution of I; there are some obvious ones coming from the order, but it is not clear what other constraints there are.

(3) Develop a good understanding of how $\text{gin}(I_{X_t})$ is related to $\text{gin}(I_{X_0})$ for a degenerating family of varieties.

(4) Develop a better understanding of the rlex gins of geometric objects. Points in $\mathbf{P}^2$ are well-understood, but points in uniform position in $\mathbf{P}^3$ are

still a mystery. Curves in $\mathbf{P}^3$ are partially understood, but curves which are not arithmetically Cohen-Macaulay still are quite puzzling (see [Gr96].) It would be helpful to understand the rlex gin of generic complete intersections; low codimension cases are known.

(5) Develop a better understanding of lex gins of geometric objects. This is still in its infancy; even points in $\mathbf{P}^2$ are not understood. One problem is to understand the relationship between the partial elimination ideals and the various scheme structures on multiple point loci. There are a lot of potentially important consequences of doing this.

(6) Develop a technique for using rlex gins to construct algebraic varieties. Whenever one want to construct a projective variety with specified properties, especially information about the Hilbert function, a first step is "Cherchez le gin."

(7) Can one combine the combinatorial methods used to prove Macaulay's theorem with geometric information? A specific example is a special case of a conjecture of Eisenbud-Green-Harris that if a homogeneous ideal contains a regular sequence of quadrics, then its growth follows the stronger bound of the Kruskal-Katona theorem, i.e. it behaves as if it were in the exterior algebra. For a generic regular sequence of quadrics, this is known, but not for all regular sequences of quadrics. This fits into a family of larger conjectures [EGH].

(8) (Suggested by Bernd Sturmfels) A simplicial complex gives rise to a monomial ideal in the exterior algebra, not necessarily Borel-fixed. The gin of this monomial ideal gives rise to a new simplicial complex. What is the relationship between these two simplicial complexes?

(9) How does one read off the geometrically interesting information about a simplicial complex from the associated monomial ideal in the exterior algebra? For example, what about the chromatic polynomial?

References

[AHH] Aramova, A., Herzog, J., Hibi, T. (1997): Gotzmann theorems for exterior algebras and combinatorics, J. Algebra 191, no. 1, 174-211.

[BE] Ballico, E., Ellia, Ph. (1985): The maximal rank conjecture for non-special curves in $\mathbf{P}^3$, Invent. Math. 79, no. 3, 541-555.

[BS] Bayer,D., Stillman, M.(1987): A criterion for detecting m-regularity, Invent. Math. 87, no. 1, 1-11.

[BR] Behnke, K., Riemenschneider, O.(1995): Quotient surface singularities and their deformations, in Singularity Theory (Trieste, 1991), World Scientific, River Edge, N.J., 1-54.

[BC] Braun, R., Cook, M. (1997): A smooth surface in $\mathbf{P}^4$ not of general type has degree at most 66, Compositio Math. 107, no. 1, 1-9.

[BF] Braun, R., Floystad, G. (1994): A bound for the degree of smooth surfaces in $\mathbf{P}^4$ not of general type. Compositio Math. 93, no 2, 211-229.

[BH] Bruns, W., Herzog, J. (1993): Cohen-Macaulay Rings, Cambridge University Press, Cambridge.

[C] Cook, M. (1993):The connectedness of space curve invariants, Thesis (UCLA).

[E] Eisenbud, D. (1995): Commutative Algebra with a View Toward Algebraic Geometry, Springer, New York.

[EGH] Eisenbud, D., Green, M., Harris,J. (1993): Some conjectures extending Castelnuovo theory, Asterisque 218, 187-202.

[EK] Eliahou,S., Kervaire, M. (1990): Minimal resolutions of some monomial ideals, J. Algebra 129, 1-25.

[FG] Floystad, G., Green, M., The information contained in initial ideals, to appear.

[Ga] Galligo, A. (1979): Théorème de division et stabilité en géométrie analytique locale, Ann. Inst. Fourier Grenoble 29, 107-184.

[Go] Gotzmann, G. (1978): Eine Bedingung für die Flachheit und das Hilbertpolynom eines graduierten Ringes, Math. Zeitschrift 158, 61-70.

[GS] Grayson, D., Stillman, M., Macaulay2 (1997): A system for computing in algebraic geometry and commutative algebra. Available via `http://math.uiuc.edu/Macaulay2`.

[Gr89] Green, M. (1989): Restriction of linear series to hyperplanes and some results of Macaulay and Gotzmann, In Algebraic Curves and Projective Geometry, Springer LNM 1389, Berlin, pp. 76-88.

[Gr89b] Green, M. (1989): Koszul cohomology and geometry, in Lectures on Riemann Surfaces, World Scientific, Singapore, 177-200.

[Gr96] Green, M. (1996): Generic initial ideals, preprint.

[GP] Gruson, L., Peskine, C. (1978): Genre des courbes de l'espace projectif, In Algebraic Geometry: Proc. Symposium University of Tromso 1977, Springer LNM 687, Berlin, pp. 31-59.

[M] Macaulay, F. (1927): Some properties of enumeration in the theory of modular systems, Proc. London Math. Soc. 26, 531-555.

[Pa94] Pardue, K. (1994): Non-standard Borel-fixed monomial ideals, Thesis, Brandeis University.

[Pa96] Pardue, K. (1996): Deformation classes of graded modules and maximal betti numbers. Ill. J. Math. 40, no. 4, 564-585.

[Re] Reeves, A. (1995): The radius of the Hilbert scheme, J. Alg. Geom. 4, 639-657.

[RS] Reeves, A., Stillman, M. (1997): Smoothness of the lexicographic point, J. Alg Geom. 6, 235-246.

[Ri] Rippel, C. (1994): Generic initial ideal theory for coordinate rings of flag varieties, Thesis (UCLA).

[S] Stanley, R.(1996): Combinatorics and Commutative Algebra, 2nd ed., Birkhauser, Boston.

Mark L. Green, Department of Mathematics U.C.L.A., Los Angeles, CA 90095; `mlg@math.ucla.edu`

Michael Stillman, Department of Mathematics, Cornell University, Ithaca, NY 14853, USA; `mike@math.cornell.edu`

Gröbner bases and algebraic geometry

Gert-Martin Greuel and Gerhard Pfister

Universität Kaiserslautern, Fachbereich Mathematik

Contents

Preface

After the notion of Gröbner bases and an algorithm for constructing them was introduced by Buchberger [Bu1, Bu2] algebraic geometers have used Gröbner bases as the main computational tool for many years, either to prove a theorem or to disprove a conjecture or just to experiment with examples in order to obtain a feeling about the structure of an algebraic variety. Nontrivial problems coming either from logic, mathematics or applications usually lead to nontrivial Gröbner basis computations, which is the reason why several improvements have been provided by many people and have been implemented in general purpose systems like Axiom, Maple, Mathematica, Reduce, etc., and systems specialized for use in algebraic geometry and commutative algebra like CoCoA, Macaulay and Singular.

The present paper starts with an introduction to some concepts of algebraic geometry which should be understood by people with (almost) no knowledge in this field.

In the second chapter we introduce standard bases (generalization of Gröbner bases to non–well–orderings), which are needed for applications to local algebraic geometry (singularity theory), and a method for computing syzygies and free resolutions.

The last chapter describes a new algorithm for computing the normalization of a reduced affine ring and gives an elementary introduction to singularity theory. Then we describe algorithms, using standard bases, to compute infinitesimal deformations and obstructions, which are basic for the deformation theory of isolated singularities.

It is impossible to list all papers where Gröbner bases have been used in local and global algebraic geometry, and even more impossible to give an overview about these contributions. We have, therefore, included only references to papers mentioned in this tutorial paper. The interested reader will find many more in the other contributions of this volume and in the literature cited there.

1 Introduction by simple questions

The basic problem of algebraic geometry can be formulated as a very simple question: "What is the structure of the set of solutions of finitely many polynomial equations in finitely many indeterminates?"

That is, we try to understand the set of points $x = (x_1, \ldots, x_n) \in K^n$ satisfying

$$f_1(x_1, \ldots, x_n) = 0,$$
$$\vdots$$
$$f_k(x_1, \ldots, x_n) = 0,$$

where K is a field and $f_1, \ldots, f_k$ are elements of the polynomial ring $K[x] = K[x_1, \ldots, x_n]$. The solution set of $f_1, \ldots, f_k$ is called the algebraic set, or algebraic variety of $f_1, \ldots, f_k$ and is denoted by $V(f_1, \ldots, f_k)$.

Here are three simple examples, which will be used to illustrate some of our subsequent questions:

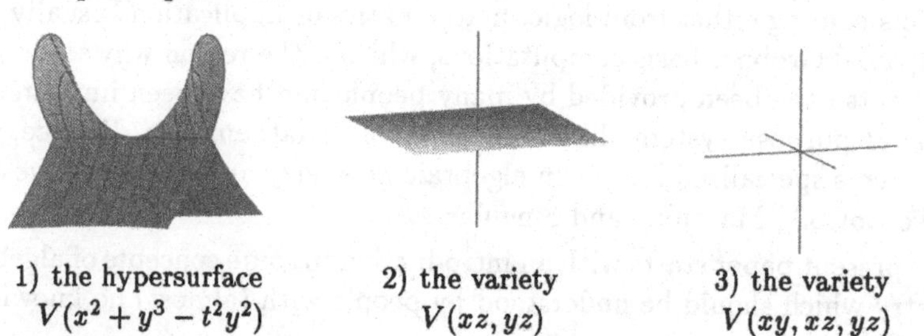

1) the hypersurface $V(x^2 + y^3 - t^2y^2)$ 2) the variety $V(xz, yz)$ 3) the variety $V(xy, xz, yz)$

The simple question, however, does not have an easy answer at all. On the contrary, the mathematical discipline algebraic geometry, which provides tools for possible answers, belongs with its long history to one of the highly developed branches of mathematics, which has created deep and quite sophisticated theories in geometry as well as in algebra. It has been estimated, as Kunz states in the introduction to his book on commutative algebra and algebraic geometry [Ku], that one can teach a course on algebraic geometry for 200 terms without repetition.

Of course, *understanding* is relative to the status of the theory but also to the cultural, economical and technical status of the society. Nowadays, faced by the technical revolution through computers, understanding requires, more and more, a computational approach to a problem, if possible. This is evident in algebraic geometry, as one can see, for instance, from recent textbooks (for example, [CLO], [St]). It is also evident that the majority of computational tools developed for algebraic geometry is based on Gröbner basis techniques.

Of course, any linear combination $f = \sum a_i f_i$, $a_i \in K[x]$, vanishes on $V = V(f_1, \dots, f_k)$ and V is equal to the solutions of all $f \in I = \langle f_1, \dots, f_k \rangle_{K[x]}$, the ideal generated by $f_1, \dots, f_k$ in $K[x]$. Even the radical of I,

$$\sqrt{I} = \{f \in K[x] \mid \exists\, d,\ f^d \in I\}$$

has the same solution set and, by the Hilbert Nullstellensatz, there is the following tight relation between ideals of $K[x]$ and algebraic sets, provided the field K is algebraically closed:

For any variety $V \subset K^n$ let $I(V) = \{f \in K[x] \mid f(x) = 0\ \forall\ x \in V\}$ the ideal of V, then

$$V = V(J) \Rightarrow I(V) = \sqrt{J}.$$

The converse is trivially true.

This theorem is the reason why the couple algebra and geometry married and produced so many wonderful theorems. Using this ideal–variety correspondence, we may formulate several geometric question and their algebraic counterparts.

One word about the role of the field K. Algebraic geometers usually draw real pictures, think about it as complex varieties and perform computations over some finite field. This attitude is justified by successful practice.

From the geometric point of view, the field K is, however, extremely important. Algebraic geometry over $\mathbb{R}$, for instance, is much more complicated and by far not as complete as over $\mathbb{C}$.

The following **questions and problems** together with those mentioned in Buchberger's article [Bu4] belong to the very basic ones in algebraic geometry. They are also quite natural and are motivated already from the above examples. Note that for these examples, the answers are more or less obvious from the figures but, nevertheless, they require a mathematical proof which is usually given by algebra. The answers given by algebra however coincide with our geometric intuition only in the case of an algebraically closed field.

- ***Is $V(I)$ irreducible or may it be decomposed into several algebraic varieties? If so, find its irreducible components. Algebraically this means to compute a primary decomposition of I or of $\sqrt{I}$, the latter means to compute the associated prime ideals of I.***

  The first example is irreducible, the second has two components (one of dimension 2 and one of dimension 1), while the third example has three components (all of dimension 1).

- ***A natural question to ask is "How independent are the generators $f_1, \ldots, f_k$ of I?" that is, we ask for all relations***

  $$(r_1, \ldots, r_k) \in K[x]^r, \text{ such that } \sum r_i f_i = 0.$$

  These relations form a submodule of $K[x]^r$, which is called the *syzygy module* of I and is denoted by $\text{syz}(I)$. It is the kernel of the $K[x]$–linear map

  $$K[x]^k \longrightarrow K[x], \; (r_1, \ldots, r_k) \mapsto \sum r_i f_i.$$

- ***More generally, we may ask for generators of the kernel of a $K[x]$–linear map $K[x]^r \to K[x]^s$, or, in other words, for solutions of a system of linear equations over $K[x]$.***

  A direct geometric interpretation of syzygies is not so clear, but there are instances where properties of syzygies have important geometric consequences (cf. [Sch2]).

  In example 1 we have $\text{syz}(I) = 0$, in example 2, $\text{syz}(I) = \langle(-y, x)\rangle \subset K[x]^2$ and in example 3, $\text{syz}(I) = \langle(-z, y-0), \; (-z, 0, x)\rangle \subset K[x]^3$.

- ***A more geometric question is the following. Let $V(I) = V(I_1) \cup V(I_2)$ be a union of not necessarily irreducible varieties and let us assume that $V(I)$ and $V(I_1)$ are known. How can we describe $V(I_2)$? Algebraically, we want to compute generators for I_2 if we know those of I and I_1. This amounts to finding generators for the ideal quotient***

  $$I : I_1 = \{f \in K[x] \mid f I_1 \subset I\}.$$

  Geometrically, $V(I : I_1)$ is the smallest variety containing $V(I) \smallsetminus V(I_1)$, which is the (Zariski) closure of $V(I) \smallsetminus V(I_1)$.

  In example 2 we have $\langle xz, yz\rangle : \langle x, y\rangle = z$ and in example 3 $\langle xy, xz, yz\rangle : \langle x, y\rangle = \langle z, xy\rangle$,which gives, in both cases, equations for the complement of the z–axis $x = y = 0$.

- ***Geometrically important is the projection of a variety $V(I) \subset K^n$ into a linear subspace K^{n-r}. Given generators $f_1, \ldots, f_k$ of I, we want to find*** *generators for the (closure of the) image of* $V(I)$ *in* $K^{n-r} = \{x | x_1 = \cdots = x_r = 0\}$. *The image is defined by the ideal* $I \cap K[x_{r+1}, \ldots, x_n]$ *and finding generators for this intersection is known as eliminating* $x_1, \ldots, x_r$ *from* $f_1, \ldots, f_k$.

Projecting the three varieties above to the (x, y) plane is, in the first two cases, surjective and in the third case it gives the two coordinate axes in the (x, y) plane. This corresponds to the fact that the intersection with $K[x, y]$ of the first two ideals is 0, while the last one is xy.

- Another problem is related to the Riemann removable theorem, which states that a function on a complex manifold, which is holomorphic and bounded outside a subvariety of codimension 1, is actually holomorphic everywhere. This is well-known for open subsets of $\mathbb{C}$, but in higher dimension there exists a second removable theorem, which states that a function, which is holomorphic outside a subvariety of codimension 2 (no assumption on boundedness), is holomorphic everywhere.

  For singular complex varieties this is not true in general, but those for which the two removable theorems hold are called *normal*. Moreover, each reduced variety has a normalization and there is a morphism with finite fibres from the normalization to the variety, which is an isomorphism outside the singular locus.

  The problem is, given a variety $V(I) \subset K^n$, *find a normal variety* $V(J) \subset K^m$ *and a polynomial map* $K^m \to K^n$ *inducing the normalization map* $V(J) \to V(I)$.

  The problem can be reduced to irreducible varieties (but need not be, as we shall see) and then the equivalent algebraic problem is to find the normalization of $K[x_1, \ldots, x_n]/I$, *that is the integral closure of* $K[x]/I$ *in the quotient field of* $K[x]/I$ *and present this ring as an affine ring* $K[x_1, \ldots, x_m]/J$ *for some* m *and* J.

  In the above examples it can be shown that the normalization of all three varieties are smooth, the last two are the disjoint union of the (smooth) components. The corresponding rings are $K[x_1, x_2]$, $K[x_1, x_2] \oplus K[x_3]$, $K[x_1] \oplus K[x_2] \oplus K[x_3]$.

- The significance of *singularities* appears not only in the normalization problem. The study of singularities is also called *local algebraic geometry* and belongs to the basic tasks of algebraic geometry. Nowadays, singularity theory is a whole subject on its own.

  A singularity of a variety is a point which has no neighbourhood in which the Jacobian matrix of the generators has constant rank.

  In the first example the whole x–axis is singular, in the two other examples only the origin.

One task is to compute generators for the ideal of the singular locus, which is itself a variety. This is just done by computing subdeterminants of the Jacobian matrix, if there are no components of different dimensions. In general, however, we need ideal quotients.

In the above examples, the singular locus is given by $\langle x, y\rangle, \langle x, y, z\rangle$ and $\langle x, y, z\rangle$, respectively.

- *Studying a variety* $V(I)$, $I = (f_1, \ldots, f_k)$, *locally at a singular point, say the origin of* K^n, *means studying the ideal* $IK[x]_{(x)}$ *generated by* I *in the local ring*

$$K[x]_{(x)} = \left\{ \frac{f}{g} \mid f, g \in K[x],\ g \notin \langle x_1, \ldots, x_n\rangle \right\}.$$

In this local ring the polynomials g *with* $g(0) \neq 0$ *are units and* $K[x]$ *is a subring of* $K[x]_{(x)}$.

Now all the problems we considered above can be formulated for ideals in $K[x]_{(x)}$ *and modules over* $K[x]_{(x)}$ *instead of* $K[x]$.

The geometric problems should be interpreted as concerning properties of the variety in a neighbourhood of the given point.

It should not be a surprise to say that all the above problems have algorithmic and computational solutions, which use, at some place, Gröbner basis methods. Moreover, algorithms for most of these have been implemented quite efficiently, in several computer algebra systems, such as CoCoA [CNR], Macaulay [GS] and SINGULAR [GPS]. The most complicated problem by far is the primary decomposition, the latest achievement is the normalization, both being implemented in SINGULAR.

At first glance, it seems that computation in the localization $K[x]_{(x)}$ requires computation with rational functions. It is an important fact that this is not necessary, but that basically the same algorithms which were developed for $K[x]$ can be used for $K[x]_{(x)}$. This is achieved by the choice of a special ordering on the monomials of $K[x]$ where, loosely speaking, the monomials of lower degree are considered to be bigger.

However, such orderings are no longer well–orderings and the classical Buchberger algorithm would not terminate. Mora discovered [Mo] that a different normal form algorithm, or, equivalently, a different division with remainders, leads to termination. Thus, Buchberger's algorithm with Mora's normal form is able to compute in $K[x]_{(x)}$ without denominators.

Several algorithms for $K[x]$ use elimination of (some auxiliary extra) variables. But variables to be eliminated have, necessarily, to be well–ordered. Hence, to be able to apply the full power of Gröbner basis methods also for the local ring $K[x]_{(x)}$, we need mixed orders, where the monomial ordering restricted to some variables is not a well–ordering, while restricted to other variables it is. In [GP] the authors described a modification of Mora's normal form, which terminates for mixed ordering and, more generally, for any monomial ordering which is compatible with the natural semigroup structure.

The corresponding modification of Buchberger's algorithm with this general normal form computes, in the case of a well–ordering (which we also call global ordering) *Gröbner bases* while, in the case of a local ordering (which was called tangent cone ordering by Mora), it computes so–called *standard bases*, which enjoy similar nice properties as Gröbner bases. We follow a suggestion by Mora and call bases computed by the general algorithm, standard bases, whilst, following the tradition of the last 33 years, reserving the name Gröbner basis for the established case of well–orderings.

2 Standard bases

Let K be a field and $K[x] = K[x_1, \dots, x_n]$ be the polynomial ring in n variables over K.

2.1 Monomial orderings and associated rings

Definition 2.1. A **monomial ordering** on $K[x]$ is a total order on the set of monomials $\{x^\alpha | \alpha \in \mathbb{N}^n\}$ satisfying

$$x^\alpha > x^\beta \Rightarrow x^{\alpha+\gamma} > x^{\beta+\gamma} \text{ for all } \alpha, \beta, \gamma \in \mathbb{N}^n.$$

We call a monomial ordering $>$ a **global** (respectively **local**, respectively **mixed**) ordering if $x_i > 1$ for all i (respectively $x_i < 1$ for all i, respectively if there exist i, j so that $x_i > 1$ and $x_j < 1$).

This notion is justified by the associated ring to be defined later. Note that $>$ is global if and only if $>$ is a well–ordering.

Definition 2.2. Any $f \in K[x] \setminus \{0\}$ can be written uniquely as

$$f = cx^\alpha + f'$$

with $c \in K \setminus \{0\}$ and $\alpha > \alpha'$ for any non–zero term $c'x^{\alpha'}$ of f'. We set

$\mathrm{lm}(f) = x^\alpha$, the **leading monomial** of f,

$\mathrm{lc}(f) = c$, the **leading coefficient** of f.

For a subset $G \subset K[x]$ we define the **leading ideal** of G as

$$L(G) = \langle\, \mathrm{lm}(g) \mid g \in G \setminus \{0\} \rangle_{K[x]},$$

the ideal generated by $\{\mathrm{lm}(g) \mid g \in G \setminus \{0\}\}$ in $K[x]$.

Typical global orderings are the **lexicographical ordering lp** ($x^\alpha >_{\mathrm{lp}} x^\beta :\Leftrightarrow$ the first non–zero entry of $\alpha - \beta$ is positive) and the **degree reverse lexicographical ordering dp** ($x^\alpha >_{\mathrm{dp}} x^\beta :\Leftrightarrow \deg x^\alpha > \deg x^\beta$ or $\deg x^\alpha = \deg x^\beta$ and the last non–zero entry of $\alpha-\beta$ is negative), typical local orderings are the **negative lexicographical ordering ls** ($x^\alpha >_{\mathrm{ls}} x^\beta :\Leftrightarrow$ the first non–zero entry of $\alpha - \beta$ is negative) and the **negative degree reverse lexicographical ordering ds** ($x^\alpha >_{\mathrm{ds}} x^\beta :\Leftrightarrow \deg x^\alpha < \deg x^\beta$ or $\deg x^\alpha = \deg x^\beta$ and the last non–zero entry of $\alpha - \beta$ is negative). In the abbreviations lp, dp, ls, ds the p refers to a polynomial ring and the s to a series ring (cf. Definition 2.3).

For practical purposes, as well as for certain theoretical arguments, it is important to extend the definitions of dp, respectively ds, to weighted degree orderings, where the variables have positive, respectively negative, weights.

Given monomial orderings $>_1$ on $K[x_1, \ldots, x_n]$ and $>_2$ on $K[y_1, \ldots, y_m]$, we define the **product ordering** or **block ordering** $> = (>_1, >_2)$ on $K[x, y]$ by $x^\alpha y^\beta > x^\gamma y^\delta :\Leftrightarrow x^\alpha >_1 x^\gamma$ or $x^\alpha = x^\gamma$ and $y^\beta >_2 y^\delta$.

Definition 2.3. For a given monomial ordering $>$ define the multiplicatively closed set

$$S_> := \{u \in K[x] \setminus \{0\} \mid \mathrm{lm}(u) = 1\}$$

and the K–algebra

$$\mathrm{Loc}\, K[x] := S_>^{-1} K[x] = \{\frac{f}{u} \mid f \in K[x], u \in S_>\},$$

the localization (ring of fractions) of $K[x]$ with respect to $S_>$.
We call $\mathrm{Loc}\, K[x]$ also the **ring associated to** $K[x]$ **and** $>$.

Remark 2.1. 1) $K[x] \subset \mathrm{Loc}\, K[x] \subset K[x]_{(x)}$ where $K[x]_{(x)}$ denotes the localization of $K[x]$ with respect to the maximal ideal $(x_1, \ldots, x_n)$. $\mathrm{Loc}\, K[x]$ is noetherian, it is $K[x]$–flat and $K[x]_{(x)}$ is $\mathrm{Loc}\, K[x]$–flat.

2) $\mathrm{Loc}\, K[x] = K[x]$ if and only if $>$ is global and $\mathrm{Loc}\, K[x] = K[x]_{(x)}$ if and only if $>$ is local (which justifies the names).

Mixed orderings occur as a product ordering of two orderings with one global and the other local. Many constructions with Gröbner bases in $K[x]$ use a set of auxiliary variables which have to be eliminated later. If one wants to perform such constructions in $K[x]_{(x)}$, the auxiliary variables must be bigger than 1, hence, mixed orderings occur naturally in this context.

3) The product ordering on $K[x, y] = K[x_1, \ldots, x_n, y_1, \ldots, y_m]$ with $>_1$ global on $K[x]$ and $>_2$ arbitrary on $K[y]$ is an **elimination ordering**

for $x_1, \ldots, x_n$ on $K[x,y]$, that is, for $g \in K[x,y]$ and $\text{lm}(g) \in K[y]$ we have $g \in K[y]$. It is easy to see that for an arbitrary monomial ordering $>$ to be an elimination ordering for $x_1, \ldots, x_r$ it is necessary that $x_i > 1$ for $i = 1, \ldots, r$. For example, let $>_1$ be global on $K[x]$ and $>_2$ local on $K[y]$, then the product ordering $> = (>_1, >_2)$ on $K[x,y]$ satisfies $S_> = K^* + (y)K[y]$, hence

$$\text{Loc}\, K[x,y] = (K[y]_{(y)})[x].$$

Note that lm and lc have natural **extensions to the localization**. For $f \in \text{Loc}\, K[x]$ there exists $u \in S_>$, $\text{lc}(u) = 1$, such that $uf \in K[x]$ and we define

$$\text{lm}(f) := \text{lm}(uf), \quad \text{lc}(f) := \text{lc}(uf).$$

Since

$$\begin{aligned} \text{lm}(fg) &= \text{lm}(f)\,\text{lm}(g) \text{ and} \\ \text{lc}(fg) &= \text{lc}(f)\text{lc}(g) \end{aligned}$$

this definition is independent of the choice of u. Moreover, for a subset $G \subset \text{Loc}\, K[x]$ set

$$L(G) = \langle \text{lm}(g) | g \in G \setminus \{0\} \rangle_{K[x]} \subset K[x]$$

and call it the **leading ideal** of G. Note also that $u \in \text{Loc}\, K[x] \setminus \{0\}$ is a unit in $\text{Loc}\, K[x]$ if and only if $\text{lm}(u) = 1$, that is, if $u \in S_>$.

For our intended applications of standard bases, but also for an elegant proof of Buchberger's standard basis criterion, we have to extend the notion of monomial orderings to the free module $K[x]^r = \sum_{i=1,\ldots,r} K[x]e_i$ where

$$e_i = (0, \ldots, 1, \ldots, 0) \in K[x]^r$$

denotes the i–th canonical basis vector of $K[x]^r$. We call

$$x^\alpha e_i = (0, \ldots, x^\alpha, \ldots, 0) \in K[x]^r$$

a **monomial (involving component i)**.

Definition 2.4. Let $>$ be a monomial ordering on $K[x]$. A **monomial ordering** or a **module ordering** on $K[x]^r$ is a total ordering $>_m$ on the set of monomials $\{x^\alpha e_i | \alpha \in \mathbb{N}^n, i = 1, \ldots, r\}$ satisfying

$$\begin{aligned} x^\alpha e_i >_m x^\beta e_j &\Rightarrow x^{\alpha+\gamma} e_i >_m x^{\beta+\gamma} e_j, \\ x^\alpha > x^\beta &\Rightarrow x^\alpha e_i >_m x^\beta e_i, \end{aligned}$$

for all $\alpha, \beta, \gamma \in \mathbb{N}^n$, $i, j = 1, \ldots, r$.
Two module orderings are of particular practical interest:

$$x^\alpha e_i > x^\beta e_j \Leftrightarrow i > j \text{ or } i = j \text{ and } x^\alpha > x^\beta,$$

giving priority to the components and

$$x^\alpha e_i > x^\beta e_j \Leftrightarrow x^\alpha > x^\beta \text{ or } x^\alpha = x^\beta \text{ and } i > j,$$

which gives priority to the monomials in $K[x]$.

Note that, by the second condition, each component of $K[x]^r$ carries the ordering of $K[x]$. Hence, $>_m$ is a well-ordering on $K[x]^r$ if and only if $>$ is a well–ordering on $K[x]$. We call $>_m$ **global** respectively **local** respectively **mixed**, if this holds for $>$ respectively.

Now we fix a module ordering $>_m$ and denote it also with $>$. Since any $f \in K[x]^r \setminus \{0\}$ can be written uniquely as

$$f = cx^\alpha e_i + f'$$

with $c \in K \setminus \{0\}$ and $x^\alpha e_i > x^{\alpha'} e_j$ for any non–zero term $c' x^{\alpha'} e_j$ of f' we can define as before

$$\begin{aligned} \operatorname{lm}(f) &= x^\alpha e_i, \\ \operatorname{lc}(f) &= c \end{aligned}$$

and call it the **leading monomial** respectively the **leading coefficient** of f. Moreover, for $G \subset K[x]^r$ we call

$$L(G) = \langle \operatorname{lm}(g) | g \in G \setminus \{0\} \rangle_{K[x]} \subset K[x]^r$$

the **leading submodule** of G.

As from $K[x]$ to $\operatorname{Loc} K[x]$ these definitions carry over naturally from $K[x]^r$ to $(\operatorname{Loc} K[x])^r$.

We say that $x^\alpha e_i$ is divisible by $x^\beta e_j$ if $i = j$ and $x^\beta | x^\alpha$. For any set of monomials $G \subset K[x]^r$ and any monomial $x^\alpha e_i$, we have

$$x^\alpha e_i \notin \langle G \rangle_{K[x]} \Leftrightarrow x^\alpha e_i \text{ is not divisible by any element of } G.$$

2.2 Standard bases and normal forms

Let $>$ be a fixed monomial ordering on $K[x]$. In order to have a short notation, we write

$$R := \operatorname{Loc} K[x] = S_>^{-1} K[x]$$

to denote the localization of $K[x]$ with respect to $>$.

We define the notion of standard basis respectively Gröbner basis and give an algorithm to compute such a basis. In the case of a well-ordering this is Buchberger's [Bu1], [Bu2], [Bu3], [Bu4] celebrated algorithm, in the general case it is a variation of Mora's tangent cone algorithm [Mo], first published in [Getal], [GP], [Gra]. We like to stress that it is important to work consequently with the ring R and not with $K[x]$, even if the input is polynomial.

Definition 2.5.

1) Let $I \subset R^r$ be a submodule. A finite set $G \subset I$ is called a **standard basis** of I if and only if $L(G) = L(I)$, that is, for any $f \in I \setminus \{0\}$ there exists a $g \in G$ satisfying $\mathrm{lm}(g) | \mathrm{lm}(f)$.

2) If the ordering is a well-ordering, then a standard basis G is called a **Gröbner basis**. In this case $R = K[x]$ and, hence, $G \subset I \subset K[x]^r$.

With the above notation, we follow the suggestion of [MPT], reserving the name Gröbner basis exclusively for well-orderings.

A set $G \subset R^r$ is called **interreduced** if $0 \notin G$ and if $\mathrm{lm}(g) \notin L(G \setminus \{g\})$.

Note that any standard basis can be made interreduced by deleting successively those g with $\mathrm{lm}(g) | \mathrm{lm}(h)$ for some $h \in G \setminus \{g\}$. An interreduced standard basis is also called **minimal**.

For $f \in K[x]^r$ and $G \subset K[x]^r$ we say that f is **reduced with respect to G** if no monomial of f is contained in $L(G)$. If $>$ is not a well-ordering, we extend this to $f \in R^r$ and $G \subset R^r$ by saying that f is (completely) reduced with respect to G if there exist $u_f \in S_>$, and for each $g \in G$, $u_g \in S_>$ such that $u_f f, u_g g \in K[x]^r$ and $u_f f$ is reduced with respect to $\{u_g g | g \in G\}$.

A set $G \subset R^r$ is called **reduced** if $0 \notin G$ and if each $g \in G$ is reduced with respect to $G \setminus \{g\}$ and if, moreover, $g - \mathrm{lc}(g)\,\mathrm{lm}(g)$ is reduced with respect to G. For $>$ a well-ordering this just means that for each $g \in G \subset K[x]^r$, $\mathrm{lm}(g)$ does not divide any monomial of any element of $G \setminus \{g\}$.

We shall see later that reduced Gröbner bases do always exist, but reduced standard bases, in general, do not.

Definition 2.6. Let $\mathcal{G}$ denote the set of all finite and ordered subsets $G \subset R^r$.

1) A map

$$\mathrm{NF} : R^r \times \mathcal{G} \to R^r, \ (f, G) \mapsto \mathrm{NF}(f|G),$$

is called a **normal form** on R^r if, for all f and G,

(i) $\mathrm{NF}(f|G) \neq 0 \Rightarrow \mathrm{lm}(\mathrm{NF}(f|G)) \notin L(G)$,

(ii) $f - \mathrm{NF}(f|G) \in \langle G \rangle_R$.

NF is called a **reduced normal form** if, moreover, $\mathrm{NF}(f, G)$ is reduced with respect to G. NF is called a **weak normal form** if, instead of (ii), only the condition (ii') holds:

(ii') for each $f \in R^r$ and each $G \in \mathcal{G}$ there exists a unit $u \in R$, so that $uf - \mathrm{NF}(f|G) \in \langle G \rangle_R$.

2) Let $G = \{g_1, \ldots, g_s\} \in \mathcal{G}$. A representation of $f \in \langle G \rangle_R$,

$$f = \sum_{i=1}^{s} a_i g_i, \quad a_i \in R,$$

satisfying $\mathrm{lm}(f) \geq \mathrm{lm}(a_i g_i)$, whenever both sides are defined, is called **a standard representation** of f (with respect to G).

Remark 2.2. The reason for introducing weak normal forms is twofold. On the one hand, they are usually more easy to compute and as good as normal forms for practical applications. On the other hand, and more seriously, normal forms may not exist, while weak normal forms do. For example, it is easy to see that $f = x \in R = K[x]_{(x)}$ (with ls) does not have a normal form with respect to $G = \{x - x^2\}$. On the other hand, since $(1 - x)f = x - x^2$ and $1 - x$ a unit in R, f is a weak normal form of itself with respect to G.
$\mathrm{NF}(f|G)$ is by no means unique. For applications (weak) normal forms are most useful if G is a standard basis of $\langle G \rangle_R$. We shall demonstrate this with a first application, which follows immediately from the definitions.

Lemma 2.1. *Let $I \subset R^r$ be a submodule, $G \subset I$ a standard basis of I and $\mathrm{NF}(-|G)$ a weak normal form on R^r with respect to G.*

1) For any $f \in R^r$ we have $f \in I \Leftrightarrow \mathrm{NF}(f|G) = 0$.

2) If $J \subset R^r$ is a submodule with $I \subset J$, then $L(I) = L(J)$ implies $I = J$.

3) $I = \langle G \rangle_R$, that is, G generates I as R-module.

4) If $\mathrm{NF}(-|G)$ is a reduced normal form, then it is unique.

For describing Buchberger's normal form algorithm, we need the notion of an s-polynomial.

Definition 2.7. Let $f, g \in R^r \setminus \{0\}$ with $\mathrm{lm}(f) = x^\alpha e_i$ and $\mathrm{lm}(f) = x^\beta e_j$, respectively. Let

$$\gamma := \mathrm{lcm}(\alpha, \beta) := (\max(\alpha_1, \beta_1), \ldots, \max(\alpha_n, \beta_n))$$

be the **least common multiple** of x^α and x^β and define the **s–polynomial** of f and g to be

$$\text{spoly}(f,g) := \begin{cases} x^{\gamma-\alpha}f - \dfrac{\text{lc}(f)}{\text{lc}(g)}x^{\gamma-\beta}g, & \text{if } i = j, \\ 0, & \text{if } i \neq j. \end{cases}$$

If $\text{lm}(g) | \text{lm}(f)$, say $\text{lm}(g) = x^\beta e_i$, $\text{lm}(f) = x^\alpha e_j$, then the s–polynomial is especially simple,

$$\text{spoly}(f,g) = f - \frac{c(f)}{c(g)}x^{\alpha-\beta}g,$$

and $\text{lm}(\text{spoly}(f,g)) < \text{lm}(g)$. For the normal form algorithm, the s–polynomial will only be used in this form, while for the standard basis algorithm we need it in the general form above. In order to be able to use the same expression in both algorithms, we prefer the definition of spoly above and not the more symmetric form $\text{lc}(g)x^{\gamma-\alpha}f - \text{lc}(f)x^{\gamma-\beta}g$. Both are, of course, equivalent, since our ground ring K is a field.

Algorithm 2.1. Assume that $>$ is a well–ordering on $K[x]^r$.
NFBuchberger($f|G$)
Input: $f \in K[x]^r$, $G \in \mathcal{G}$.
Output: $h \in K[x]^r$, a normal form of f with respect to G.

- $h = f$;
- while ($h \neq 0$ and $G_h = \{g \in G \mid \text{lm}(g) \mid \text{lm}(h)\} \neq \emptyset$)
  choose any $g \in G_h$;
  $h = \text{spoly}(h, g)$;
- return h;

For termination and correctness see [Bu4]. Note that each specific choice of “any” gives a different normal form function.

It is easy to extend NFBuchberger to a reduced normal form.

The idea of many standard basis algorithms may be formalized as follows:

Algorithm 2.2. Let $>$ be any monomial ordering on R^r and assume that a weak normal form algorithm NF on R^r is given.
Standard(G,NF)
Input: $G \in \mathcal{G}$
Output: $S \in \mathcal{G}$ such that S is a standard basis of the submodule $I = \langle G \rangle_R \subset R^r$

- $S = G$;
- $P = \{(f,g) | f,g \in S\} \subset S \times S$
- while $(P \neq \emptyset)$
  choose $(f,g) \in P$;
  $P = P \smallsetminus \{(f,g)\}$;
  $h = \mathrm{NF}(\mathrm{spoly}(f,g)|S)$;
  If $(h \neq 0)$
    $P = P \cup \{(h,f) | f \in S\}$;
    $S = S \cup h$;
- return S;

To see termination of STANDARD, note that if $h \neq 0$ then $\mathrm{lm}(h) \notin L(S)$ by property 1) of NF. Hence, we obtain a strictly increasing sequence of monomial submodules of $K[x]^r$, which becomes stationary by Dickson's lemma or by the Noether property of $K[x]$. That is, after finitely many steps, we always have $\mathrm{NF}(\mathrm{spoly}(f,g)|S)) = 0$ for $(f,g) \in P$ and, after some finite time, the pairset P will become empty.

Correctness follows from applying Buchberger's fundamental standard basis criterion below.

Theorem 2.1. *[Buchberger's criterion] Let $I \subset R^r$ be a submodule and $G = \{g_1, \ldots, g_s\}$ be a subset of I. Let NF$(-|G)$ be a weak normal form on R^r with respect to G, satisfying: for each $f \in R^r$ there exists a unit u such that $uf - \mathrm{NF}(f|G)$ has a standard representation with respect to G.*
Then the following are equivalent

1) G is a standard basis of I,

2) $\mathrm{NF}(f|G) = 0$ for all $f \in I$,

3) each $f \in I$ has a standard representation with respect to G,

4) G generates I and $\mathrm{NF}(\mathrm{spoly}(g_i, g_j)|G) = 0$ for $i,j = 1, \ldots, s$.

The implications 1) $\Rightarrow$ 2) $\Rightarrow$ 3) $\Rightarrow$ 4) are easy.

The implication 4) $\Rightarrow$ 1) is the important criterion which allows the checking and construction of standard bases in finitely many steps. The proof is most easily done by using syzygies and is, therefore, postponed to the next section (Theorem 2.2).

We present now a general normal form algorithm, which works for any monomial ordering. It is basically due to Mora [Mo], with a different notion of ecart, as given in [Getal], [GP]).

Before doing this, let us first analyze Buchberger's algorithm in the case of a local ordering.

The standard example is in one variable x, with $x < 1$, $f = x$ and $G = \{g = x - x^2\}$. We obtain

$$x - \left(\sum_{i=0}^{\infty} x^i\right)(x - x^2) = 0$$

in $K[[x]]$, which is checked to be true, since $\sum x^i = \frac{1}{1-x}$ in $K[[x]]$. However, this is not a normal form in our sense, since $\sum x^i \notin R$.

Mora's idea was to allow more elements for reduction in order to create a standard expression of the form

$$uf = \sum_{i=1}^{s} a_i g_i + \mathrm{NF}(f|G),$$

with u a unit, $a_i \in K[x]$ and $\mathrm{NF}(f|G) \in K[x]^r$ in the case when the input data f and $G = \{g_1, \ldots, g_s\}$ are *polynomial*. In the above example he arrives at an expression

$$(1 - x)x = x - x^2$$

instead of $x = (\sum_{i=0}^{\infty} x^i)(x - x^2)$.

Definition 2.8. For a monomial $x^\alpha e_i \in K[x]^r$ set $\deg x^\alpha e_i = \deg x^\alpha = \alpha_1 + \cdots + \alpha_n$. For $f \in K[x]^r \smallsetminus \{0\}$, let $\deg f$ be the maximal degree of all monomials occurring in f. We define the **ecart** of f as

$$\text{ecart }(f) = \deg f - \deg \mathrm{lm}(f).$$

For a homogeneous $f = \sum f_i e_i$ (all components f_i are homogeneous polynomials of the same degree), we have $\mathrm{ecart}(f) = 0$.

Algorithm 2.3. Let $>$ be any monomial ordering on $K[x]^r$, $R = S_>^{-1}K[x]$.
NFMORA$(f|G)$
Input: $f \in K[x]^r$, $G = \{g_1, \ldots, g_s\} \subset K[x]^r$
Output: $h \in K[x]^r$ a weak normal form of f with respect to G. Moreover, there exists a standard representation $uf - h = \sum_{i=1}^{s} a_i g_i$ with $a_i \in K[x]$, $u \in S_>$.

- $h = f$;
- $T = G$,

- while($h \neq 0$ and $T_h = \{g \in T \mid \text{lm}(g) \mid \text{lm}(h)\} \neq \emptyset$)
  choose $g \in T_h$ with ecart(g) minimal;
  if (ecart(g) > ecart(h))
    $T = T \cup \{h\}$;
  $h = \text{spoly}(h, g)$;
- return h;

If the input is homogeneous, then the ecart is always 0 and NFMORA is equal to NFBUCHBERGER. If $>$ is a well–ordering, then $\text{lm}(g) \mid \text{lm}(h)$ implies that $\text{lm}(g) \leq \text{lm}(h)$, hence $T = G$ during the algorithm. Thus, NFMora is the same as NFBuchberger, but with a special selection strategy for the elements from G.

For termination and correctness see [GP].

It is clear that, with a little extra storage, the algorithm does also return $u \in S_>$.

Algorithm 2.4. Let $>$ be any monomial ordering on $K[x]^r$, $R = S_>^{-1}K[x]$.
STANDARD BASIS(G)
Input: $G = \{g_1, \ldots, g_s\} \subset K[x]^r$
Output: $S = \{h_1, \ldots, h_t\} \subset K[x]^r$ such that S is a standard basis of $I = \langle G \rangle_R \subset R^r$.

- S = Standard(G,NFMora);
- return S;

2.3 Syzygies and free resolutions

Let K be a field and $>$ a monomial ordering on $K[x]^r$. Again R denotes the localization of $K[x]$ with respect to $S_>$.

We shall give a method, using standard bases, to compute syzygies and, more generally, free resolutions of finitely generated R–modules. Syzygies and free resolutions are very important objects and basic ingredients for many constructions in homological algebra and algebraic geometry. On the other hand, the use of syzygies gives a very elegant way to prove Buchberger's criterion for a standard basis. Moreover, a close inspection of the syzygies of the generators of an ideal allows detection of useless pairs during a computation of a standard basis (cf. [MM], [Ei]). Our presentation follows partly that of Schreyer [Schl], [Schl2], cf. also [Ei]. The generalization to arbitrary monomial orderings was first formulated and proved in [Getal] and [GP].

A **syzygy** or a **relation** between k elements $f_1, \ldots, f_k \in R^r = \bigoplus_{i=1}^{r} Re_i$ is a k–tuple $(g_1, \ldots, g_k) \in R^k$ satisfying

$$\sum_{i=1}^{k} g_i f_i = 0.$$

The set of all syzygies between $f_1, \ldots, f_k$ is a submodule of R^k. Indeed, it is the kernel of the ring homomorphism

$$\varphi_1 : F_1 := \bigoplus_{i=1}^{k} R\varepsilon_i \longrightarrow F_0 := \bigoplus_{i=1}^{r} Re_i,$$
$$\varepsilon_i \longmapsto f_i,$$

where e_i respectively ε_i denote the canonical bases of R^r respectively R^k. φ_1 surjects onto $I = \langle f_1, \ldots, f_k \rangle_R$ and

$$\operatorname{syz} I = \operatorname{Ker} \varphi_1$$

is called the **module of syzygies** of I with respect to the generators $f_1, \ldots, f_k$.

We shall now define a monomial ordering on F_1, which behaves perfectly well with respect to standard bases. This was first introduced and used by Schreyer [Schl].

Set

$$x^\alpha \varepsilon_i > x^\beta \varepsilon_j \Leftrightarrow \operatorname{lm}(x^\alpha f_i) > \operatorname{lm}(x^\beta f_j) \text{ or}$$
$$\operatorname{lm}(x^\alpha f_i) = \operatorname{lm}(x^\beta f_j) \text{ and } i < j.$$

The left–hand side $>$ is the new ordering on F_1 and the right–hand side $>$ is the ordering on F_0. In order to distinguish them, we occasionally call them $>_1$ respectively $>_0$. $>_0$ and $>_1$ induce the same ordering on R. We call the ordering $>_1$ the Schreyer ordering. Note that it depends on $f_1, \ldots, f_k$.

Now we are going to prove Buchberger's criterion, stating that $G = \{f_1, \ldots, f_k\}$ is a standard basis of I, if, for all $i < j$, $\operatorname{NF}(\operatorname{spoly}(f_i, f_j)|G) = 0$. The proof uses syzygies and is basically due to Schreyer [Schl], [Schl2], although our generalization (to general monomial orderings) seems to be simpler. It gives, at the same time, a proof of Schreyer's result that the syzygies derived from a standard representation of $\operatorname{spoly}(f_i, f_j)$ form a standard basis of $\operatorname{syz} I$ for the Schreyer ordering.

We introduce some notations. For each $i < j$ such that f_i and f_j have leading term in the same component, say $\operatorname{lm}(f_i) = x^{\alpha_i} e_\nu$, $\operatorname{lm}(f_j) = x^{\alpha_j} e_\nu$, define the monomial

$$m_{ji} := x^{\gamma - \alpha_i} \in K[x],$$

where $\gamma = \operatorname{lcm}(\alpha_i, \alpha_j)$. If $c_i = \operatorname{lc}(f_i)$ and $c_j = \operatorname{lc}(f_j)$ then

$$m_{ji} f_i - \frac{c_i}{c_j} m_{ij} f_j = \operatorname{spoly}(f_i, f_j).$$

Assume now that for $i < j$

$$\operatorname{NF}(\operatorname{spoly}(f_i, f_j)|G) = 0,$$

for some weak normal form NF on R^r.

Then we have a standard representation

$$m_{ji}f_i - \frac{c_i}{c_j}m_{ij}f_j = \sum_{\nu=1}^{k} a_\nu^{(ij)} f_\nu, \; a_\nu^{(ij)} \in R.$$

Define for $i < j$ such that $\operatorname{lm}(f_i)$ and $\operatorname{lm}(f_j)$ involve the same component

$$s_{ij} = m_{ji}\varepsilon_i - \frac{c_i}{c_j}m_{ij}\varepsilon_j - \sum_{\nu} a_\nu^{(ij)}\varepsilon_\nu.$$

Then $s_{ij} \in \operatorname{syz} I$ and it is easy to see that

Lemma 2.2. $\operatorname{lm}(s_{ij}) = m_{ji}\varepsilon_i$.

Theorem 2.2. *Let $G = \{g_1, \ldots, g_s\}$ be a set of generators of $I \subset R^r$ satisfying, for some weak normal form NF on R^r,*

$$NF(\operatorname{spoly}(g_i, g_j) \mid G) = 0, \; i < j,$$

then the following holds:

1) G is a standard basis of I.

2) $\{s_{ij}\}$ is a standard basis of syz I *with respect to the Schreyer ordering. In particular, $\{s_{ij}\}$ generates* syz I.

Proof. We give a proof of 1) and 2) at the time time.

Take any $f \in I$ and a preimage $g \in F_1$ of f,

$$g = \sum_{i=1}^{s} a_i\varepsilon_i, \; f = \varphi(g) = \sum_{i=1}^{s} a_i g_i.$$

This is possible as G generates I.

In case 1), we assume $f \neq 0$, in case 2) $f = 0$.

Consider a standard representation of $g - h$,

$$g = \sum a_{ij}s_{ij} + h, \; a_{ij} \in R,$$

where $h = \sum h_j\varepsilon_j \in F_1$ is a normal form of g with respect to $\{s_{ij}\}$ for some weak normal form on F_1 (we need only know that it exists). We have, if $h \neq 0$,

$$\operatorname{lm}(h) = \operatorname{lm}(h_\nu) \cdot \varepsilon_\nu \text{ for some } \nu$$

and $\operatorname{lm}(h) \notin L(\{s_{ij}\}) = \langle\{m_{ji}\varepsilon_i\}\rangle$ by Lemma 2.2. This shows

$$m_{j\nu} \nmid \operatorname{lm}(h_\nu) \text{ for all } j.$$

Since $g - h \in \langle\{s_{ij}\}\rangle \subset \operatorname{syz} I$, we obtain

$$f = \varphi(g) = \varphi(h) = \sum h_j g_j.$$

Assume that for some $j \neq \nu$, $\operatorname{lm}(h_j g_j) = \operatorname{lm}(h_\nu g_\nu)$. Then $\operatorname{lm}(h_\nu g_\nu)$ is divisible by $\operatorname{lm}(g_\nu)$ and by $\operatorname{lm}(g_j)$ and hence, by

$$\operatorname{lm}(g_\nu)\operatorname{lm}(g_j)/\gcd\big(\operatorname{lm}(g_\nu),\operatorname{lm}(g_j)\big) = \operatorname{lm}(g_\nu)m_{j\nu}.$$

This contradicts $m_{j\nu} \nmid \operatorname{lm}(h_\nu)$.

In case 1) we obtain $\operatorname{lm}(f) = \operatorname{lm}(h_\nu g_\nu) \in L(G)$, in case 2) it shows that $h \neq 0$ leads to a contradiction. In case 1) G is a standard basis by definition and in case 2) $\{s_{ij}\}$ is a standard basis by Theorem 2.1, 2) $\Rightarrow$ 1), which was already proved. □

We shall now see, as an application, that the Hilbert syzygy theorem holds for the rings $R = S_>^{-1}K[x]$, stating that each finitely generated R–module has a free resolution of length at most n, the number of variables.

Lemma 2.3. *Let $G = \{g_1, \ldots, g_s\}$ be a standard basis of $I \subset R^r$, ordered in such a way that the following holds: if $i < j$ and $\operatorname{lm}(g_i) = x^{\alpha_i}e_\nu$, $\operatorname{lm}(g_j) = x^{\alpha_j}e_\nu$ for some ν, then $\alpha_i \geq \alpha_j$ lexicographically. Let s_{ij} denote the syzygies defined above. Suppose that $\operatorname{lm}(g_1), \ldots, \operatorname{lm}(g_s)$ do not depend on the variables $x_1, \ldots, x_k$. Then the $\operatorname{lm}(s_{ij})$, taken with respect to the Schreyer ordering, do not depend on $x_1, \ldots, x_{k+1}$.*

Proof. Given s_{ij}, then $i < j$ and $\operatorname{lm}(g_i)$ and $\operatorname{lm}(g_j)$ involve the same component, say e_ν. By assumption $\operatorname{lm}(g_i) = x^{\alpha_i}e_\nu$, $\operatorname{lm}(g_j) = x^{\alpha_j}e_\nu$ satisfy $\alpha_i = (0, \ldots, \alpha_{i,k+1}, \ldots)$ $\alpha_j = (0, \ldots, \alpha_{j,k+1}, \ldots)$ with $\alpha_{i,k+1} \geq \alpha_{j,k+1}$. Therefore, $\operatorname{lm}(s_{ij}) = m_{ji}e_i$, $m_{ji} = x^{\operatorname{lcm}(\alpha_i,\alpha_j)-\alpha_i}$, does not depend on x_{k+1}. □

Applying the lemma successively to the higher syzygy modules, we obtain (cf. [GP] for a detailed proof):

Theorem 2.3. *Let $>$ be any monomial ordering on $K[x] = K[x_1, \ldots, x_n]$ and $R = S_>^{-1}K[x]$ be the associated ring. Then any finitely generated R–module M has a free resolution*

$$0 \longrightarrow F_m \longrightarrow F_{m-1} \longrightarrow \ldots \longrightarrow F_0 \longrightarrow M \longrightarrow 0,$$

F_i free R–modules, of length $m \leq n$. In particular, R is a regular ring.

It is clear that the methods of this section provide an algorithm to compute (non–minimal) free resolutions. This algorithm has been implemented in SINGULAR.

3 Applications

3.1 The normalization

Here we describe an algorithm which goes back to Grauert and Remmert [GR] and was proposed by T. de Jong ([J]). There are also algorithms by Gianni, Trager ([GT]) and Vasconcelos ([V]).

The algorithm is based on the following criterion for normality:

Proposition 3.1. *Let R be a noetherian reduced ring and J be a radical ideal containing a non–zero divisor such that the zero set of J, $V(J)$ contains the non– normal locus of Spec(R). Then R is normal if and only if $R = \mathrm{Hom}_R(J, J)$.*

Remark 3.1. Let J, R be as in the proposition and x a non–zero divisor, then

1) $xJ : J = x \cdot \mathrm{Hom}_R(J, J)$

and, consequently,

2) $R = \mathrm{Hom}_R(J, J)$ if and only if $xJ : J \subseteq \langle x \rangle$.

3) Let $u_0 = x, u_1, \ldots, u_s$ be generators of $xJ : J$. Because of the fact that $\mathrm{Hom}_R(J, J)$ is a ring we have $\frac{s(s+1)}{2}$ relations $\frac{u_i}{x} \cdot \frac{u_j}{x} = \sum_{k=0}^{s} \xi_k^{ij} \frac{u_k}{x}$, $s \geq i \geq j \geq 1$, $\xi_k^{ij} \in R$ in $\frac{1}{x}(xJ : J)$, which, together with the linear relations, the syzygies, between $u_0, \ldots, u_s$ define the ring structure of $\mathrm{Hom}_R(J, J)$:

$$\begin{array}{ccc} R[T_1, \ldots, T_s] & \twoheadrightarrow & \mathrm{Hom}_R(J, J) \\ T_i & \rightsquigarrow & \dfrac{u_i}{x}. \end{array}$$

The kernel of this map is the ideal generated by the $T_iT_j - \sum_{k=0}^{s} \xi_k^{ij} T_k$, where $T_0 = 1$, and $\sum_{k=0}^{s} \eta_k T_k$ such that $\sum_{k=0}^{s} \eta_k u_k = 0$.

Algorithm 3.1.
NORMAL(I)

Input: a radical ideal $I \subseteq K[x_1, \ldots, x_n]$,
Output: s polynomial rings $R_1, \ldots, R_s$ and s prime ideals $I_1 \subset R_1, \ldots, I_s \subset R_s$ and s maps $\pi_i : R \to R_j$, such that the induced map $\pi : K[x_1, \ldots, x_n]/I \to R_1/I_1 \times \cdots \times R_s/I_s$ is the normalization of $K[x_1, \ldots, x_n]/I$

- Result $= \emptyset$;

- compute idempotents of $K[x_1,\ldots,x_n]/I$;
  Assume $K[x_1,\ldots,x_n]/I = K[x_1,\ldots,x_n]/I_1 \times \cdots \times K[x_1,\ldots,x_n]/I_s$.
- For $i = 1$ to s do
  - compute $J =$ singular locus of I_i
  - choose $f \in J \setminus I_i$ and compute $I_i : f$ to check whether f is a zero divisor
  - if $I_i : f \supsetneqq I_i$ then
    Result = Result $\cup$ NORMAL($\sqrt{I_i, f}$)$\cup$ NORMAL($I_i : f$)

    else

    $J = \sqrt{I_i, f}$

    $H = fJ : J$

    if $H = \langle f \rangle$

    Result = Result $\cup\{K[x_1,\ldots,x_n], I_i, \text{id}\}$

    else

    assume $H = fJ : J = \langle f, u_1, \ldots, u_s \rangle$

    then compute an ideal $L \subseteq K[x_1,\ldots,x_n,T_1,\ldots,T_s]$ such that

    $K[x_1,\ldots,x_n,T_1,\ldots,T_s]/L \xrightarrow{\sim} \text{Hom}(J,J)$, $T_i \rightsquigarrow \frac{u_i}{f}$

    let $\varphi : K[x_1,\ldots,x_n] \to K[x_1,\ldots,x_n,T_1,\ldots,T_s]$ be the inclusion.

    $S = \text{normal}(L)$;

    compose the maps of S with φ;

    Result = Result $\cup S$;
- return Result

It remains to give an algorithm to compute the idempotents.

We shall explain this for the case when the input ideal I is (weighted) homogeneous with strictly positive weight.

An idempotent e, that is, $e^2 - e \in I$ has to be homogeneous of degree 0, will, therefore, not occur in the first step. It can, however, occur after one normalization loop in $\text{Hom}(J,J) \simeq K[x_1,\ldots,x_n,T_1,\ldots,T_s]/L$ because some of the generators may have the same degree.

Let $T \subseteq \{T_1,\ldots,T_s\}$ be the subset of variables of degree 0.

Then $L \cap K[T]$ is zero–dimensional because $T_j^2 - \sum \xi_k^{jj} T_k \in L \cap K[T]$ for all $T_j \in T$ (the weights are ≥ 0 and, therefore, $\xi_k^{jj} \in K$, $T_k \in T$).

For this situation there is an easy algorithm:

Algorithm 3.2.
IDEMPOTENTS(I)

Input: $I \subseteq K[x_1,\ldots,x_n]$ a (weighted) homogeneous radical ideal, $\deg(x_i) = 0$ for $i \leq k$, $\deg(x_i) > 0$ for $i > k$, $I \cap K[x_1,\ldots,x_k]$ 0–dimensional.

Output: ideals $I_1, \dots, I_s$ such that

$$K[x_1, \dots, x_n]/I = K[x_1, \dots, x_n]/I_1 \times \cdots \times K[x_1, \dots, x_n]/I_s$$

and $I \cap K[x_1, \dots, x_k] = \cap(I_v \cap K[x_1, \dots, x_k])$ is the prime decomposition

- Result $= \emptyset$;
- compute $J = I \cap K[x_1, \dots, x_k]$;
- compute $J = P_1 \cap \cdots \cap P_s$ the (0–dimensional) prime decomposition;
- For $i = 1$ to s do
  - choose $g_i \neq 0$ in $\bigcap\limits_{v \neq i} P_v$;
  - Result = Result $\cup\{I : g_i\}$;
- return Result;

3.2 Singularities

The basic concepts and ideas of singularity theory are best explained over the field $\mathbb{C}$ of complex numbers, although, algebraically, most invariants make sense over arbitrary fields.

Let $U \subset \mathbb{C}^n$ be an open subset in the usual euclidian topology, and $f_1, \dots, f_k$ holomorphic (complex analytic) functions on U, then we may consider

$$V = V(f_1, \dots, f_k) = \{x \in U | f_1(k) = \cdots = f_k(x) = 0\},$$

the complex analytic subvariety defined by $f_1, \dots, f_k$ in U.

In practice, $f_1, \dots, f_k$ will be polynomials, but singularity theory is interested only in the behaviour of $V(f_1, \dots, f_k)$ in an arbitrary small neighbourhood of some point $p \in V$, that is, the **germ** of V at p, which is denoted by (V, p). Algebraically, this means that we are not interested in the ideal generated by $f_1, \dots, f_k$ in the polynomial ring $\mathbb{C}[x_1, \dots, x_n]$ but in the ideal I generated by $f_1, \dots, f_k$ in the convergent power series ring $\mathbb{C}\{x_1 - p_1, \dots, x_n - p_n\} = \mathbb{C}\{x - p\}$.

For arbitrary fields K, where the notion of convergence does not make sense, we consider instead the formal power series ring $K[[x]] = K[[x_1, \dots, x_n]]$ and ideals I generated by formal power series (in practice polynomials) $f_1, \dots,$ $f_k \in K[[x]]$. In order to have a uniform notation, we write

$$K\langle x\rangle = K\langle x_1, \dots, x_n\rangle$$

to denote both $K[[x]]$ and $K\{x\} = K\{x_1, \ldots, x_n\}$ if K is a complete valued field (for example, $K = \mathbb{C}$).

The ring $\mathcal{O}_{V,p} = \mathbb{C}\{x-p\}/I$ (respectively $K\langle x\rangle/I$) is called the **analytic local ring** of the singularity (V,p).

If $f_1, \ldots, f_k$ are polynomials, we may also consider the corresponding **algebraic local ring** $K[x]_{(x-p)}/\langle f_1, \ldots, f_k\rangle$, where $K[x]_{(x-p)}$ is the localization of $K[x]$ in the maximal ideal $\langle x_1 - p_1, \ldots, x_n - p_n\rangle$. Indeed in this ring we are able to compute standard bases (cf. Section 1).

As in the affine case, we have the Hilbert Nullstellensatz (also called the Hilbert–Rückert Nullstellensatz), stating that

$$\sqrt{I} = I(V,p) := \{f \in \mathbb{C}\{x-p\} \mid f|_{(V,p)} = 0\}$$

for $I \subset \mathbb{C}\{x-p\}$ and (V,p) the complex analytic germ defined by I.

A **(complex) singularity** is, by definition, nothing but a complex analytic germ (V,p) (together with its analytic local ring $\mathbb{C}\{x-p\}/I$). (V,p) is called **non–singular** or **regular** or **smooth** if $\mathbb{C}\{x_1 - p_1, \ldots, x_n = p_n\}/I$ is isomorphic (as local ring) to a power series ring $\mathbb{C}\{y_1, \ldots, y_d\}$ (respectively $K\langle x_1, \ldots, x_n\rangle/I \cong K\langle y_1, \ldots, y_d\rangle$). By the implicit function theorem, this is equivalent to the fact that I has a system of generators $g_1, \ldots, g_{n-d}$ such that the Jacobian matrix of $g_1, \ldots, g_{n-d}$ has rank $n-d$ in some neighbourhood of p. (V,p) is called an **isolated singularity** if there is a neighbourhood $W \subset \mathbb{C}^n$ of p such that $W \cap (V \setminus \{p\})$ is regular everywhere.

Isolated Singularities

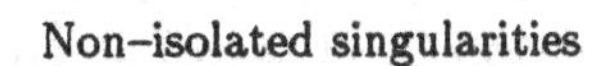

Non–isolated singularities

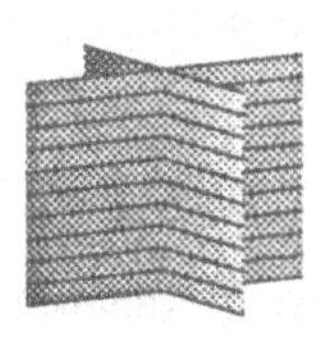

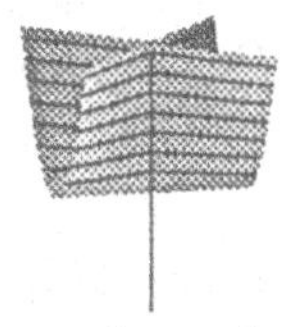

$A_1 : x^2 - y^2 + z^2 = 0$ $D_4 : z^3 - zx^2 + y^2 = 0$ $A_\infty : x^2 - y^2 = 0$ $D_\infty : y^2 - zx^2 = 0$

The **dimension** of the singularity (V,p) is, by definition, the Krull dimension of the analytic local ring $\mathcal{O}_{V,p} = K\langle x\rangle/I$, which is the same as the Krull dimension of the algebraic local ring $K[x]_{(x-p)}/I$ if $I = \langle f_1, \ldots, f_k\rangle$ is generated by polynomials, which follows easily from the theory of dimensions by Hilbert–Samuel series. Using this fact, we can compute $\dim(V,p)$ by computing a standard basis of the ideal $\langle f_1, \ldots, f_k\rangle_{\mathrm{Loc}\, K[x]}$ with respect to any *local* monomial ordering on $K[x]$. The dimension is equal to the dimension of the corresponding monomial ideal (which is a combinatorial problem).

It is important to compute a standard basis with respect to a local ordering. For example, the leading ideal of $\langle yx - y, zx - z\rangle$, with respect to *dp*, is $\langle xy, xz\rangle$ (hence of dimension 2), but, with respect to *ds*, it is $\langle y, z\rangle$ (hence of dimension 1). Geometrically, this means that the dimension of the affine variety $V = V(yx - y, zx - z)$ is 2 but the dimension of the singularity $(V, 0)$ (that is, the dimension of V at the point 0) is 1:

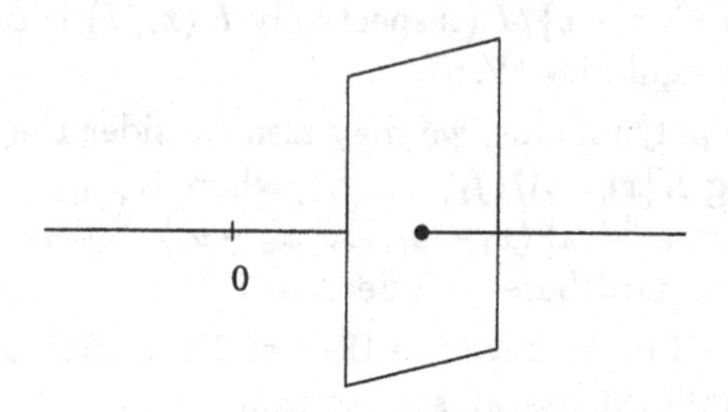

$$V : y(x-1) = z(x-1) = 0, \ \dim(V, 0) = 1$$

Another basic invariant is the **multiplicity** $\mathrm{mt}(V, p)$ of the singularity (V, p). If (V, p) is reduced, that is, $I = \sqrt{I}$, then $\mathrm{mt}(V, p)$ has a nice geometric interpretation: it is the number of intersection points of a sufficiently small representative $V \subset \mathbb{C}^n$ of (V, p) with a general $(n-d)$–dimensional plane in $\mathbb{C}^n$ close to p (but not through p). If (V, p) is a **hypersurface singularity**, that is, $I = \langle f \rangle$ is a principal ideal in $\mathbb{C}\{x - p\}$, then $\mathrm{mt}(V, p)$ is the smallest degree in the Taylor expansion of f in p.

Algebraically, $\mathrm{mt}(V, p)$ is the Hilbert–Samuel multiplicity of the ideal $\langle x_1 - p_1, \ldots, x_n - p_n \rangle$ in the analytic or in the algebraic local ring of (V, p).

As before the Hilbert-Samuel function of the ideal defining V coincides with the Hilbert-Samuel function of the leading ideal of a standard basis with respect to a local degree ordering.

One of the most important invariants of an isolated hypersurface singularity (V, p) given by $I = \langle f \rangle \subset K\langle x \rangle$ is the **Milnor number**

$$\mu(f) := \dim_K K\langle x \rangle / \langle f_{x_1}, \ldots, f_{x_n} \rangle$$

$(\mathrm{char}(K) = 0)$ where f_{x_i} denotes the partial derivative of f with respect to x_i.

For $K = \mathbb{C}$, μ is even a topological invariant and has the following topological meaning, due to Milnor [Mil]. For $f \in \mathbb{C}\{x_1, \ldots, x_n\}$ defining an isolated singularity at 0, let

$$V_t = B_\varepsilon(0) \cap f^{-1}(t),$$

$0 < |t| << 1$ and $B_\varepsilon(0)$ a small ball of radius ε around 0, then V_t (the "Milnor fibre of f") has the homotopy type of a 1–point union of $\mu(f)$ $(n-1)$–dimensional spheres. In particular, $\mu(f) = \dim_{\mathbb{C}} H_n(V_t, \mathbb{C})$.

Algorithm 3.3. (Assume $\mathrm{char}(K) = 0$).
MILNOR(f)
Input: $f \in K[x_1, \ldots, x_n]$
Output: $\mu(f)$

- compute a standard basis $\{g_1, \ldots, g_s\}$ of $\langle f_{x_1}, \ldots, f_{x_n} \rangle_{\mathrm{Loc}\, K[x]}$ with respect to any *local monomial ordering* on $K[x]$;
- the number of monomials of $K[x]$ not in $\langle \mathrm{lm}(g_1), \ldots, \mathrm{lm}(g_s) \rangle$ is equal to $\mu(f)$;

The correctness of this algorithm follows from [GP, 3.7].

Similarly, we can compute the **Tjurina number** ($\mathrm{char}(K) \geq 0$)

$$\tau(f) = \dim_K K\langle x \rangle / \langle f, f_{x_1}, \ldots, f_{x_n} \rangle.$$

This number plays an important role in the deformation theory of the singularity defined by f and will be considered in the next section.

There is an interesting conjecture, due to Zariski, stating that the multiplicity of a complex hypersurface singularity is a topological invariant. This conjecture is still open. For a formulation, using the Milnor number, and for a partial positive answer (which was prompted by computer experiments using SINGULAR with local standard bases) see [GP].

3.3 Deformations

Let $(V, 0) \subset (\mathbb{C}^n, 0)$ be a singularity given by convergent power series $f_1, \ldots, f_k$, converging in a neighbourhood U of $0 \in \mathbb{C}^n$. The idea of deformation theory is to perturb the defining functions, that is to consider functions $F_1(t, x), \ldots, F_k(t, x)$ with $F_i(0, x) = f_i(x)$, where t are small parameters of a parameter space S. For $t \in S$ the functions $f_{i,t}(x) = F_i(t, x)$ define a complex analytic set

$$V_t = V(f_{1,t}, \ldots, f_{k,t}) \subset U$$

which, for t close to 0, may be considered to be a small deformation of $V = V_0$. It may be hoped that V_t is simpler than V_0 but still contains enough information about V. For this hope to be fulfilled, it is, however, necessary to restrict the possible perturbations of the equations to flat perturbations, which are called deformations.

The formal definition is as follows: a **deformation** of the singularity $(V, 0)$ over a complex analytic germ $(S, 0)$ consists of a cartesian diagram

$$\begin{array}{ccccl} (V,0) & \stackrel{i}{\hookrightarrow} & (\mathcal{U},0) & = \{(t,x) \in S \times U \mid F_1(t,x) = \cdots = F_k(t,x) = 0\} \\ \downarrow & & \downarrow \phi & \\ \{0\} & \in & (S,0) & \end{array}$$

such that ϕ, which is the restriction of the second projection, is flat, that is, $\mathcal{O}_{\mathcal{U},0}$ is, via ϕ^*, a flat $\mathcal{O}_{S,0}$–module.

Grothendieck's criterion of flatness states that ϕ is **flat** if and only if any relation between the f_i, say $\sum r_i(x)f_i(x) = 0$, lifts to a relation $\sum R_i(t,x)F_i(t,x) = 0$, where $R_i(x,0) = r_i(x)$, between the F_i. Equivalently, for any generator $(r_1,\ldots,r_k)$ of $\mathrm{syz}(\langle f_1,\ldots,f_k\rangle)$ there exists an element $(R_1,\ldots,R_k) \in \mathrm{syz}(\langle F_1,\ldots,F_k\rangle)$ satisfying $R_i(0,x) = r_i(x)$.

The notion of flatness is not easy to explain geometrically. It has, however, important geometric consequences. For example, the fibres of a flat morphism have all the same dimension. Topologists would call a flat morphism perhaps transversal. In any case, the intuitive meaning is that the fibres of a flat morphism vary in some sense continuously with the parameter.

By a theorem of Grauert [Gr] (see also Schlessinger [Schl1] for the formal case), every isolated singularity admits a semiuniversal or miniversal deformation $\phi : (\mathcal{U},0) \longrightarrow (S,0)$ of $(V,0)$, which, in some sense, contains the information upon all possible deformations.

By a power series Ansatz it is possible to compute the miniversal deformation up to a given order. In general, the algorithm will, however, not stop. The existence of such an algorithm follows from the work of Laudal [La]. This algorithm has been implemented in SINGULAR by Martin.

We are not going to describe this algorithm here but just mention that for an isolated hypersurface singularity $f(x_1,\ldots,x_n)$ the semiuniversal deformation is given by

$$F(t,x) = f(x) + \sum_{j=1}^{\tau} t_j g_j(x),$$

where $1 = g_1, g_2, \ldots, g_\tau$ represent a basis of the Tjurina algebra

$$K\langle x\rangle/\langle f, f_{x_1}, \ldots, f_{x_n}\rangle,$$

τ being the Tjurina number.

Instead we describe algorithms to compute the modules $T^1_{V,0}$ respectively $T^2_{V,0}$ of first order deformations of $(V,0)$ respectively of obstructions, which are the first objects one likes to know about the semiuniversal deformation.

We switch now to an algebraic setting where deformations are described on the algebra level.

Since the infinitesimal deformation theory of an affine algebra and an analytic algebra is pretty much the same (cf. [Ar]), we use from now on the same notation $K\langle x_1,\ldots,x_n\rangle$ for the polynomial ring over the field K as well as for power series ring over K.

Let $I = \langle f_1,\ldots,f_k\rangle \subset K\langle x\rangle = K\langle x_1,\ldots,x_n\rangle$ be an ideal and let $R = K\langle x\rangle/I$.

An **embedded deformation of R** over an analytic algebra $A = K\langle t\rangle/J = K\langle t_1,\ldots,t_m\rangle/J$ is given by

$$F_i(t,x) = f_i(x) + \sum_{j=1}^{m} t_j g^i_j(t,x) \in A\langle x\rangle$$

satisfying that every relation (syzygy) between the f_i,

$$(r_1, \ldots, r_k) \in K\langle x\rangle^k, \quad \sum_{i=1}^{k} r_i(x) f_i(x) = 0,$$

lifts to a relation between the F_i,

$$(R_1, \ldots, R_k) \in A\langle x\rangle^k, \quad \sum R_i(t,x) F_i(t,x) = 0,$$

$R_i(0,x) = r_i(x)$.

By definition, $F_1, \ldots, F_k$ and $F'_1, \ldots, F'_k$ define the same embedded deformation, if they generate the same ideal.

Setting

$$R_A = A\langle x\rangle / \langle F_1, \ldots, F_k\rangle$$

we obtain a commutative cartesian diagramme

$$\begin{array}{ccc} R & \twoheadleftarrow & R_A \\ \uparrow & & \uparrow\varphi \\ K & \twoheadleftarrow & A \end{array}$$

with φ flat, which is called a **deformation of R over A** (and which is just the algebraic translation of the geometric definition). Two deformations $A \longrightarrow R_A$ and $A \longrightarrow R'_A$ of R over A are called **isomorphic** if there is an A–isomorphism $R_A \cong R'_A$ compatible with the given isomorphisms to the "special fibre",

$$R_A / \langle t\rangle \cong R \cong R'_A / \langle t\rangle,$$

where $\langle t\rangle = \langle t_1, \ldots, t_m\rangle$. It is not difficult to see that every deformation of R is isomorphic to an embedded deformation.

We like to stress the fact that the base algebras A for deformations have to be local analytic K–algebras with $K \xrightarrow{\cong} A/\mathfrak{m}$, $\mathfrak{m}$ the maximal ideal of A, even if R is affine.

Infinitesimal deformations

Let $K[\varepsilon] = K\langle\varepsilon\rangle / \langle\varepsilon^2\rangle$ denote the two–dimensional analytic algebra $K + K\varepsilon$, $\varepsilon^2 = 0$ (the space Spec($K[\varepsilon]$) may be considered as a "thick" point, that is, a point together with a tangent direction). An (embedded) deformation of R over $K[\varepsilon]$ is called an **infinitesimal (embedded) deformation.**

Proposition 3.2.

1) *The R–module of infinitesimal embedded deformations is isomorphic to the normal module*

$$N_R = \mathrm{Hom}_R(I/I^2, R).$$

2) *The R–module of isomorphism classes of infinitesimal deformations of R is isomorphic to T^1_R, where* $\mathbf{T^1_R}$ *is defined by the exact sequence*

$$\Theta \otimes_{K\langle x\rangle} R \xrightarrow{\alpha} N_R \longrightarrow T^1_R \longrightarrow 0.$$

Here, $\Theta = \mathrm{Der}_K K\langle x\rangle = \bigoplus_{i=1}^{n} K\langle x\rangle \frac{\partial}{\partial x_i}$ *and the map* α *sends the derivation* $\frac{\partial}{\partial x_i}$ *to the homomorphism sending* h *to* $\frac{\partial h}{\partial x_i}$.

For the proof we refer to [Schl2].

Remarks:

1) Schlessinger's theorem [Schl1] states that R admits a formal semiuniversal deformation $B \longrightarrow R_B = B\langle x\rangle/\langle F_1, \ldots, F_k\rangle$ over a complete local K–algebra B if and only if $\dim_K T^1_R < \infty$.

   This assumption is fulfilled in the affine case $R = K[x_1, \ldots, x_n]/I$ if the affine variety $V(I)$ has only isolated singularities (necessarily finitely many) or in the analytic case $R = K\langle x_1, \ldots, x_n\rangle/I$ if the singularity $(V(I), 0)$ has an isolated singularity.

2) In the complex analytic case with $R = \mathbb{C}\{x_1, \ldots, x_n\}/I$ and $\dim_{\mathbb{C}} T^1_R < \infty$, R admits even a convergent complex analytic semiuniversal deformation with base algebra $B = \mathbb{C}\{t_1, \ldots, t_m\}/J$ and total algebra $R_B = \mathbb{C}\{t_1, \ldots, t_m, x_1, \ldots, x_n\}/\langle F_1, \ldots, F_k\rangle$.

The proof of the convergence is quite difficult and was given by Grauert in 1972 [Gr] and it was in this paper that he proved the "division theorem by an ideal". In our language, he introduced the notion of standard bases and proved the existence of normal forms for complex analytic convergent power series. An equivalent theorem had already been proved before in 1964 by Hironaka in his famous resolution paper ([Hi]). It is interesting to notice that the analog of Gröbner bases in power series rings was invented for proving deep theoretical results. The proofs were, however, not constructive and did not contain Buchberger's criterion.

3) It follows from Grothendieck's definition of tangent spaces that, if a semiuniversal deformation $B \longrightarrow R_B$ of R exists, then T^1_R is isomorphic to the Zariski tangent space to Spec B at the maximal ideal of B. This shows, with $t_1, \ldots, t_m$ a K–basis of T^{1*}_R, that $B \cong K\langle T^1_R\rangle/J \cong K\langle t_1, \ldots, t_m\rangle/J$ for some ideal of J. Hence, the base algebra B of the semiuniversal deformation of R is defined by analytic relations between the elements $t_1, \ldots, t_m$ of a K–basis of the dual of T^1_R and these relations generate J.

To compute T^1_R, let $0 \longleftarrow R \longleftarrow K\langle x\rangle \longleftarrow K\langle x\rangle^k \xleftarrow{r} K\langle x\rangle^\ell$ be a representation of R, then, applying $\mathrm{Hom}_{K\langle x\rangle}(-, R)$ to the sequence

$$0 \longleftarrow R \longleftarrow K\langle x\rangle \longleftarrow K\langle x\rangle^k \xleftarrow{r} K\langle x\rangle^\ell,$$

we obtain $N_R = \ker(R^k \xrightarrow{r^t} R^\ell)$.

Now choose a resolution of N_R

$$\begin{array}{ccccc} R^k & \xleftarrow{s_1} & R^s & \xleftarrow{s_2} & R^t \\ \rotatebox{-45}{$\subseteq$} & \swarrow & & & \\ & N_R & & & \end{array}$$

The canonical map $\pi : R^n \simeq \Theta_{K\langle x\rangle} \otimes R \longrightarrow N_R$ is induced by the map $j : R^n \longrightarrow R^k$ defined by the jacobian matrix $\left(\frac{\partial f_i}{\partial x_j}\right)_{\substack{i\leq k\\ j\leq n}}$. We can lift j to a map $lj : R^n \longrightarrow R^s$ such that $s_1 \circ lj = j$ because j induces the map π.

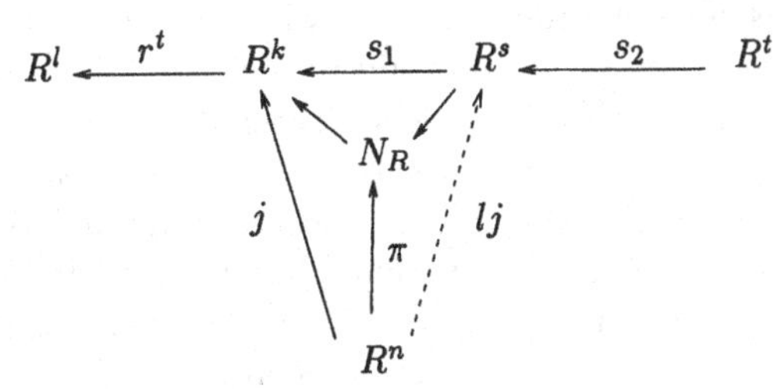

Now $T^1_R = N_R/\,\mathrm{Im}(\pi) \simeq R^s/\,\mathrm{Im}(s_2) + \mathrm{Im}(lj)$ gives the required representation

$$0 \longleftarrow T^1_R \longleftarrow R^s \xleftarrow{s_2 \oplus lj} R^t \oplus R^n.$$

Algorithm 3.4.
T1(I)
Input: an ideal $I = \langle f_1, \ldots, f_k\rangle \subseteq K\langle x\rangle$,
Output: a matrix $M \in M_{a,b}(K\langle x\rangle)$ which defines a representation

$$T^1_{K\langle x\rangle/I} \twoheadleftarrow K\langle x\rangle^a \xleftarrow{M} K\langle x\rangle^b$$

- compute $r = \mathrm{syz}(I)$ the matrix of the syzygies of $f_1, \ldots, f_k$;
- compute the jacobian matrix $j = \left(\frac{\partial f_i}{\partial x_j}\right)$;
- compute in $K\langle x\rangle/I$ a representation of the kernel of the transposed matrix r^t of r:

$$(K\langle x\rangle/I)^k \xleftarrow{s_1} (K\langle x\rangle/I)^s \xleftarrow{s_2} (K\langle x\rangle/I)^t;$$

- lift the jacobian matrix j to a $s \times n$–matrix lj such that $s_1 \cdot lj = j$;
- concatenate lj and s_2 to obtain the matrix $t1 = lj, s_2$;
- choose a matrix $M_0 \in M_{s,n+t}(K\langle x\rangle)$ such that $M_0 \mod I = t1$;
- choose a matrix $L \in M_{s,k\cdot s}$ (corresponding to $IK\langle x\rangle^s$), such that

$$0 \leftarrow (K\langle x\rangle/I)^s \twoheadleftarrow K\langle x\rangle^s \xleftarrow{L} K\langle x\rangle^{k\cdot s}$$

  is exact;
- concatenate M_0 and L to obtain $M = M_0, L$;
- return M

Obstructions

The construction of a semiuniversal deformation of R, in case T^1_R is finite dimensional, starts with the preceding remark 3): we start with the infinitesimal deformations of first order, that is, with elements of T^1_R, and try to lift these to second order. This is not always possible, there are obstructions against lifting. That is, a lifting to second order is possible if and only if the corresponding obstruction is zero. Assuming that the obstruction is zero, we choose a lifting to second order (which is not unique) and try to lift this to third order. Again there are obstructions, but if these are zero, the lifting is possible and we can continue. In any case, the obstructions yield formal power series in $K[[t_1, \ldots, f_n]]$, $t_1, \ldots, t_n$ a K–basis of T^{1*}_R, and if J denotes the ideal generated by them, $B = K[[t_1, \ldots, t_m]]/J$ will be the base algebra of the formal semiuniversal deformation of R.

The following proposition describes the module of obstructions to lift a deformation from an artinian algebra to an infinitesimally bigger one, where we may think of starting with $A = K\langle t_1, \ldots, t_n\rangle/\langle t_1, \ldots, t_n\rangle^2$.

For this, let $R = K\langle x\rangle/I$ and consider a presentation of $I = \langle f_1, \ldots, f_k\rangle$,

$$0 \longleftarrow I \xleftarrow{\alpha} K\langle x\rangle^k \xleftarrow{\beta} K\langle x\rangle^\ell$$

with $\alpha(e_i) = f_i$ and $\mathrm{syz}(I) = \ker(\alpha) = \mathrm{im}(\beta)$ is the module of relations of $f_1, \ldots, f_k$, which contains the module of **Koszul relations**,

$$\mathrm{Kos} = \langle f_i e_j - f_j e_i | 1 \leq i < j \leq j\rangle.$$

Set $\mathrm{Rel} = K\langle x\rangle^\ell / \ker(\beta)$ which is isomorphic to $\mathrm{syz}(I)$ and $\mathrm{Rel}_0 = \beta^{-1}(\mathrm{Kos})$. We define the module $\mathbf{T^2_R}$ by the exact sequence

$$\mathrm{Hom}_R(R^k, R) \xrightarrow{\beta*} \mathrm{Hom}_R(\mathrm{Rel}/\mathrm{Rel}_0, R) \longrightarrow T^2_R \longrightarrow 0.$$

Proposition 3.3.

1) *Let $A' \twoheadrightarrow A$ be a surjection of artinian local K-algebras with kernel an ideal J satisfying $J^2 = 0$. There is an obstruction map*

$$ob\colon \operatorname{Def}_R(A) \longrightarrow T^2_R \otimes_K J$$

satisfying: a deformation $A \longrightarrow R_A$ of R admits a lifting $A' \longrightarrow R_{A'}$,

$$\begin{array}{ccc} R_A & \twoheadleftarrow & R_{A'} \\ \uparrow & \square & \uparrow \\ A & \twoheadleftarrow & A', \end{array}$$

if and only if $ob([A \longrightarrow R_A]) = 0$ ($[A \longrightarrow R_A]$ denotes the deformation class of $A \longrightarrow R_A$).

2) *If T^1_R is finite dimensional over K and if $T^2_R = 0$, then the semiuniversal deformation $B \longrightarrow R_B$ of R has a smooth base space, that is B is a free analytic algebra $K\langle t_1, \ldots, t_m\rangle$ for some $m \geq 0$.*

For the computation of T^2_R we choose, as before, a representation

$$0 \longleftarrow R \longleftarrow K\langle x\rangle \longleftarrow K\langle x\rangle^k \xleftarrow{\;r\;} K\langle x\rangle^\ell \xleftarrow{\;s\;} K\langle x\rangle^t.$$

Then $\mathrm{Rel} = \mathrm{syz}(I) = \mathrm{Im}(r)$ and Rel_0 is the submodule of Rel generated by the $\binom{k}{2}$ Koszul relations Kos.

Now $\mathrm{Rel}/\mathrm{Rel}_0 \longrightarrow R^k$ is the induced map defined by the following diagram

$$\begin{array}{ccccc} IK\langle x\rangle^k & \hookrightarrow & K\langle x\rangle^k & \longrightarrow & R^k \\ \cup\mid & & \cup\mid & & \uparrow \\ \mathrm{Rel}_0 & \hookrightarrow & \mathrm{Rel} & \longrightarrow & \mathrm{Rel}/\mathrm{Rel}_0\,. \end{array}$$

To obtain a representation of $\mathrm{Rel}/\mathrm{Rel}_0$ we lift the Koszul relations to $K\langle x\rangle^\ell$:

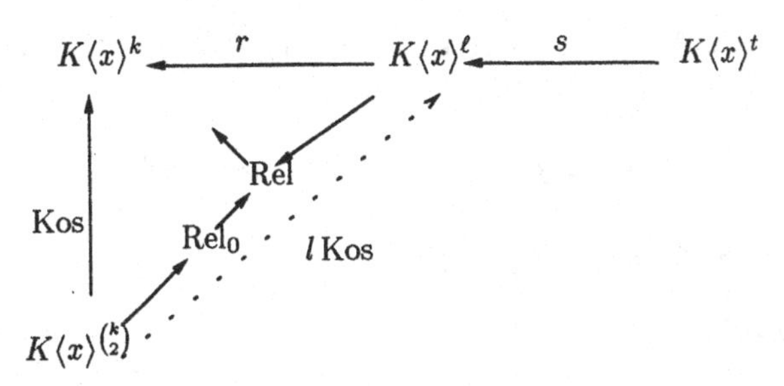

Then $\mathrm{Rel}/\mathrm{Rel}_0 \simeq K\langle x\rangle^\ell / \mathrm{Im}(s) + \mathrm{Im}(\ell\,\mathrm{Kos})$ and if we denote by $s\ell$ the $\ell \times (t + \binom{k}{2})$–matrix s, $\ell\,\mathrm{Kos}$:

$$0 \longleftarrow \mathrm{Rel}/\mathrm{Rel}_0 \longleftarrow K\langle x\rangle^\ell \xleftarrow{\;s\ell\;} K\langle x\rangle^{t_1},\ t_1 = t + \binom{k}{2},$$

is a representation of $\mathrm{Rel}/\mathrm{Rel}_0$.

Now we are interested in a representation of T^2_R which is just

$$\mathrm{Coker}\big(\mathrm{Hom}_R(R^k,R)\longrightarrow \mathrm{Hom}_R(\mathrm{Rel}/\mathrm{Rel}_0,R)\big).$$

We dualize the representation of $\mathrm{Rel}/\mathrm{Rel}_0$ and obtain

$$\begin{array}{ccccccc} 0 \to & \mathrm{Hom}_{K\langle x\rangle}(\mathrm{Rel}/\mathrm{Rel}_0,R) & \to & \mathrm{Hom}_{K\langle x\rangle}(K\langle x\rangle^{\ell},R) & \xrightarrow{s\ell^t} & \mathrm{Hom}_{K\langle x\rangle}(K\langle x\rangle^{t_1},R) \\ & \| & & \| & & \| \\ & \mathrm{Hom}_R(\mathrm{Rel}/\mathrm{Rel}_0,R) & & \mathrm{Hom}_R(R^{\ell},R) & & \mathrm{Hom}_R(R^{t_1},R), \end{array}$$

that is, $\mathrm{Hom}_R(\mathrm{Rel}/\mathrm{Rel}_0,R)=\ker(s\ell^t)$.

Now we take a representation of $\mathrm{Hom}_R(\mathrm{Rel}/\mathrm{Rel}_0,R)$

$$\begin{array}{ccccc} R^{t_1} \xleftarrow{s\ell^t} & R^{\ell} & \xleftarrow{r_1} & R^{\ell_1} & \xleftarrow{r_2} R^{\ell_2} \\ & \cup\!\!\mid & & \swarrow & \\ & & \mathrm{Hom}_R(\mathrm{Rel}/\mathrm{Rel}_0,R) & & \end{array}$$

The map $R^k \simeq \mathrm{Hom}_R(R^k,R)\longrightarrow \mathrm{Hom}_R(\mathrm{Rel}/\mathrm{Rel}_0,R)$ is defined by $r^t: R^k\longrightarrow R^{\ell}$. We can lift this map to a map $\ell rt: R^k\longrightarrow R^{\ell_1}$ such that $r_1\circ \ell rt = r^t$.

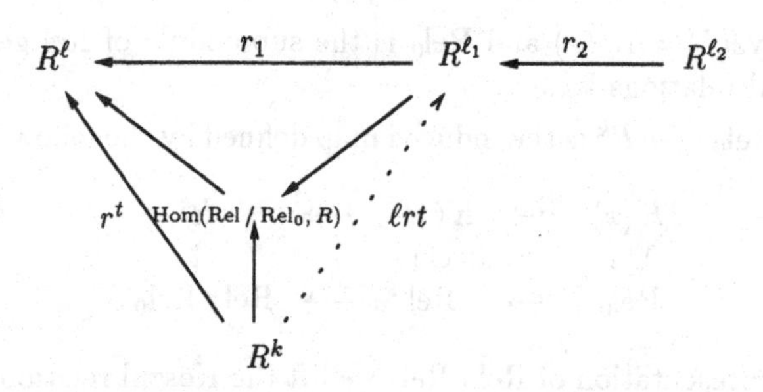

Then $T^2_R=\mathrm{Coker}\big(R^k\longrightarrow \mathrm{Hom}(\mathrm{Rel}/\mathrm{Rel}_0,R)\big)\simeq R^{\ell_1}/\mathrm{Im}(r_2)+\mathrm{Im}(\ell rt)$ gives the required representation

$$0\longleftarrow T^2_R\longleftarrow R^{\ell_1}\xleftarrow{r_2\oplus \ell rt} R^{\ell_2}\oplus R^k.$$

We obtain the following algorithm:

Algorithm 3.5.
T2(I)
Input: an ideal $I=\langle f_1,\dots,f_k\rangle\subseteq K\langle x\rangle$
Output: a matrix $M\in M_{a,b}(K\langle x\rangle)$ which defines a representation

$$T^2_{K\langle x\rangle/I}\twoheadleftarrow K\langle x\rangle^a \xrightarrow{M} K\langle x\rangle^b$$

- compute $r = \mathrm{syz}(I) \in M_{k,\ell}(K\langle x\rangle)$, the matrix of the syzygies of $f_1, \ldots, f_k$ and $s = \mathrm{syz}(\mathrm{syz}(I)) \in M_{\ell,t}(K\langle x\rangle)$ to obtain a representation of $K\langle x\rangle/I$;
- compute the matrix Kos $\in M_{k,\binom{k}{2}}$, the Koszul matrix of the relations of $f_1, \ldots, f_k$;
- lift Kos to a matrix ℓKos $\in M_{\ell,\binom{k}{2}}$ such that $r \cdot \ell\mathrm{Kos} = \mathrm{Kos}$;
- concatenate ℓKos and s to obtain the matrix $s\ell = \ell\mathrm{Kos}, s \in M_{\ell,\binom{k}{2}+t}$;
- compute in $K\langle x\rangle/I$ a representation of the kernel $\ker(s\ell^t)$ given by matrices r_1 and r_2 $(r_1 = \mathrm{syz}(s\ell^t) \in M_{\ell,\ell_1}(K\langle x\rangle/I)$, $r_2 = \mathrm{syz}(r_1) \in M_{\ell_1,\ell_2}(K\langle x\rangle/I))$;
- lift the matrix r^t to a matrix $\ell rt \in M_{\ell_1,k}(K\langle x\rangle/I)$ such that $r_1 \cdot \ell rt = r^t$;
- concatenate ℓrt and r_2 to obtain $t2 = \ell rt, r_2 \in M_{\ell_1,k+\ell_2}(K\langle x\rangle/I)$;
- choose a matrix $M_0 \in M_{\ell_1,k+\ell_2}(K\langle x\rangle)$ such that $M_0 \mod I = t2$;
- choose a matrix $L \in M_{\ell_1,k\cdot\ell_1}$ (corresponding to $IK\langle x\rangle^{\ell_1}$) such that

  $$0 \leftarrow (K\langle x\rangle/I)^{\ell_1} \twoheadleftarrow K\langle x\rangle^{\ell_1} \xleftarrow{L} K\langle x\rangle^{k\cdot\ell_1}$$

  is exact;
- concatenate M_0 and L to obtain $M = M_0, L$;
- return M.

References

[Ar] Artin, M.: Lectures on deformations of singularities. Tata Institute, Bombay (1976).

[Bu1] Buchberger, B.: Ein Algorithmus zum Auffinden der Basiselemente des Restklassenringes nach einem nulldimensionalen Polynomideal. PhD Thesis, University of Innsbruck, Austria (1965).

[Bu2] Buchberger, B.: Ein algorithmisches Kriterium für die Lösbarkeit eines algebraischen Gleichungssystems. In: Aequ. Math. **4**, 374-383 (1970).

[Bu3] Buchberger, B.: Gröbner bases: an algorithmic method in polynomial ideal theory. In: Recent trends in multidimensional system theory, N.B. Bose, ed., Reidel (1985).

[Bu4] Buchberger, B.: Introduction to Gröbner bases. This volume.

[BWe] Becker, T.; Weispfennig, V.: Gröbner Bases, A Computational Approach to commutative Algebra. Graduate Texts in Mathematics 141, Springer 1993.

[CLO] Cox, D.; Little, J.; O'Shea, D.: Ideals, Varieties and Algorithms. Springer Verlag (1992).

[CNR] Capani, A.; Niesi, G.; Robbiano, L.: Some Features of CoCoA 3. Moldova Journal of Computer Science. To appear.

[Ei] Eisenbud, D: Commutative Algebra with a view toward Algebraic Geometry. Springer 1995.

[ERS] Eisenbud, D.; Riemenschneider, O.; Schreyer, F.-O.: Resolutions of Cohen-Macaulay Algebras. Math Ann.**257** (1981).

[Getal] Grassmann, H.; Greuel, G.-M.; Martin, B.; Neumann, W.; Pfister, G.; Pohl, W.; Schönemann, H.; Siebert, T.: Standard bases, syzygies and their implementation in SINGULAR. In: Beiträge zur angewandten Analysis und Informatik, Shaker, Aachen, 69-96, 1994.

[GP] Greuel, G.-M.; Pfister,G.: Advances and improvements in the theory of standard bases and syzygies. Arch. Math. **66**, 163–1796 (1996).

[GPS] Greuel, G.-M.; Pfister, G.; Schönemann, H.: SINGULAR Reference Manual, Reports On Computer Algebra Number 12, May 1997, Centre for Computer Algebra, University of Kaiserslautern from www.mathematik.uni-kl.de/zca/Singular.

[GR] Grauert, H.; Remmert, R.: Analytische Stellenalgebren. Springer 1971.

[Gr] Grauert, H.: Über die Deformation isolierter Singularitäten analytischer Mengen. Invent. Math. **15**, 171–198 (1972).

[Gra] Gräbe, H.-G.: The tangent cone algorithm and homogeneization. J. Pure Appl. Alg. 97, 303–312 (1994).

[GS] Grayson, D.; Stillmann, M.: A computer software system designed to support research in commutative algebra and algebraic geometry. Available from math.uiuc.edu.

[GT] Gianni, P.; Trager, B.: Integral closure of noetherian rings. Preprint, to appear.

[Hi] Hironaka, H.: Resolution of singularities of an algebraic variety over a field of characteristic zero. Ann. of Math. **79**, 109–326 (1994).

[J] de Jong, T.: An algorithm for computing the integral closure. Preprint, Saarland University, Saarbrücken.

[Ku] Kunz, E.: Einführung in die kommutative Algebra und algebraische Geometrie. Vieweg (1980).

[La] Laudal, O.A.: Formal Moduli of Algebraic Structures. LNM **754**, Springer (1979).

[Mil] Milnor, J.: Singular Points of Complex Hypersurfaces. Ann. of Math. Studies 61, Princeton (1968).

[Mo] Mora, T.: An algorithm to compute the equations of tangent cones. Proc. EUROCAM 82, Lecture Notes in Comput. Sci. (1982).

[MM] Möller, H.M.; Mora, T.: Computational aspects of reduction strategies to construct resolutions of monomial ideals. Proc. AAECC 2, Lecture Notes in Comput. Sci. **228** (1986).

[MPT] Mora, T.; Pfister, G.; Traverso, C.: An introduction to the tangent cone algorithm. In: Issues in non–linear geometry and robotics, JAI Press (1992).

[Sch1] Schreyer, F.-O.: Die Berechnung von Syzygien mit dem verallgemeinerten Weierstrass'schen Divisionssatz. Diplomarbeit, Hamburg (1980).

[Sch2] Schreyer, F.-O.: A standard basis approach to syzygies of canonical curves. J. reine angew. Math. **421**, 83–123 (1991).

[Sch3] Schreyer, F.-O.: Syzygies of canonical curves and special linear series. Math. Ann. **275** (1986).

[Schl1] Schlessinger, M.: Functors of Artin rings. Trans. AMS **130**, 208-222, (1968).

[Schl2] Schlessinger, M.: On rigid singularities. Rice. Univ. Stud. **59**, 147–162 (1973).

[St] Sturmfels, B.: Algorithms in Invariant Theory. Springer Verlag (1993).

[V] Vasconcelos, W.: Computing the integral closure of an affine domain. Proc. AMS 113 (3), 633–638 (1991).

Gröbner bases and integer programming

Serkan Hoşten[1] *and Rekha Thomas*[2]

1 Introduction

This article is a brief survey of recent work on Gröbner bases (Buchberger 1965) of toric ideals and their role in integer programming. Toric varieties and ideals are crucial players in the interaction between combinatorics, discrete geometry, commutative algebra and algebraic geometry. For a detailed treatment of this topic see (Sturmfels 1995). Our survey focuses on the application of toric ideals to integer programming, a specific branch of discrete optimization, and for the sake of brevity we leave details to the references that are included.

We study a family of integer programs associated with a fixed matrix A and the corresponding toric ideal I_A. In Section 2, we show that reduced Gröbner bases of I_A are *test sets* for these integer programs. In Section 3, we define the *universal Gröbner basis* of I_A and we identify it as a subset of the *Graver basis* of A. Section 4 deals with the effect of varying the cost function while keeping the matrix A fixed; there we give a self contained construction of the *state polytope* and *Gröbner fan* of I_A. Moreover, we show that the edge directions of the state polytope are precisely the elements in the universal Gröbner basis of I_A. We conclude in Section 5 with a discussion of practical issues that arise while computing Gröbner bases of toric ideals. In particular we discuss algorithms for finding generating sets for I_A.

2 Gröbner bases as test sets in integer programming

Let $IP_{A,c}(b)$ denote the integer program

$$\text{minimize } cx : \ Ax = b, \ x \in \mathbf{N}^n$$

where the *coefficient matrix* $A \in \mathbf{Z}^{d \times n}$ has rank d, the *right hand side vector* $b \in \mathbf{Z}^d$ and the *cost vector* $c \in \mathbf{R}^n$. The program $IP_{A,c}(b)$ has a solution

[1]George Mason University, Mathematical Sciences Department, Fairfax, VA 22030, USA; **shosten@gmu.edu**

[2]Department of mathematics, Texas A&M University, College Station, TX 77843, USA; **rekha@math.tamu.edu**

if and only if b lies in the additive monoid $\mathbf{N}(A) := \{Ax : x \in \mathbf{N}^n\}$. We study $IP_{A,c}$, the family of integer programs of the form $IP_{A,c}(b)$ obtained by varying b in $\mathbf{N}(A)$, while keeping A and c fixed. We call the polyhedron $P_b^I = \mathit{convex\,hull}\{x \in \mathbf{N}^n : Ax = b\}$, the b-fiber of A and the minimum value of $c \cdot x$ over all $x \in P_b^I$, the *optimal value* of $IP_{A,c}(b)$. For simplicity, we assume that the *recession cone* of A, $\{x \in \mathbf{R}_+^n : Ax = 0\} = \{0\}$, which guarantees that P_b^I is a polytope for each $b \in \mathbf{N}(A)$. Hence every cost vector $c \in \mathbf{R}^n$ attains a bounded optimal value on each program in $IP_{A,c}$.

The *toric ideal* of A, denoted I_A, is the prime ideal in the polynomial ring $S = k[x_1, \ldots, x_n]$ generated by monomial differences (binomials) as follows:

$$I_A = \langle \mathbf{x}^\alpha - \mathbf{x}^\beta : \alpha, \beta \in \mathbf{N}^n, A(\alpha - \beta) = 0 \rangle.$$

Proposition 2.1 *Any reduced Gröbner basis of the toric ideal I_A consists of binomials.*

Proof: By Hilbert's basis theorem we can find a finite binomial generating set for I_A. If the input to the Buchberger algorithm is a set of binomials, then all intermediate polynomials created, as well as the final output, consists of binomials. □

The cost vector c typically induces a partial order on $\mathbf{N}^n$ and hence on the monomials in S via the inner product $c \cdot x$. In this situation, we pick a term order $\succ$ on $\mathbf{N}^n$ like say the lexicographic order and we replace c in $IP_{A,c}$ by the total order $\succ_c$ defined as:

$$\mathbf{x}^\alpha \succ_c \mathbf{x}^\beta \text{ if } c \cdot \alpha > c \cdot \beta, \quad \text{or if } c \cdot \alpha = c \cdot \beta, \text{ and } \mathbf{x}^\alpha \succ \mathbf{x}^\beta. \tag{2.1}$$

In general $\succ_c$ is not a term order on $\mathbf{N}^n$ but since all P_b^Is are bounded with respect to c, $\succ_c$ behaves like a term order. Our primary object of study is the reduced Gröbner basis $\mathcal{G}_{\succ_c}$ of the toric ideal I_A with respect to the total order $\succ_c$.

Definition 2.2 *A set $\mathcal{T} \subseteq \{t \in \mathbf{Z}^n : At = 0, t \succ_c 0\}$ is a* **test set** *for $IP_{A,\succ_c}$ if*

1. *given any non-optimal solution α to any program in $IP_{A,\succ_c}$, there exists $t \in \mathcal{T}$ such that $\alpha - t$ is a feasible solution to the same program, and*
2. *for the optimal solution β to a program in $IP_{A,\succ_c}$, $\beta - t$ is infeasible for every t in $\mathcal{T}$.*

Various test sets can be found in the integer programming literature (see Scarf (1986), Graver (1975), and Cook et al. (1986)). Whenever a test set $\mathcal{T}$ is known for $IP_{A,\succ_c}$, there is a natural optimization algorithm which finds the unique optimal solution to any program in $IP_{A,\succ_c}$: given a feasible solution α for $IP_{A,\succ_c}(b)$, we *test* whether $\alpha - t$ is non-negative for some $t \in \mathcal{T}$. If such a t exists we replace α with $\alpha - t$ which has a better cost value since $t \succ_c 0$ for all $t \in \mathcal{T}$. We iterate until no $t \in \mathcal{T}$ can improve the current feasible solution β. Then, since $\mathcal{T}$ is a test set for $IP_{A,\succ_c}$ we may conclude that β is the optimal solution of $IP_{A,\succ_c}(b)$. The following theorem connects reduced Gröbner bases of I_A and test sets for $IP_{A,\succ_c}$.

Theorem 2.3 *The reduced Gröbner basis $\mathcal{G}_{\succ_c}$ of I_A is a minimal test set for $IP_{A,\succ_c}$.*

Proof: Let $u \in \mathbf{N}^n$ be a non-optimal solution for the integer program $IP_{A,\succ_c}(Au)$ and let $v \in \mathbf{N}^n$ be the optimal solution for the same program. If we let $u - v = (u-v)_+ - (u-v)_-$, the binomial $\mathbf{x}^{(u-v)_+} - \mathbf{x}^{(u-v)_-}$ is in I_A and $\mathbf{x}^{(u-v)_+}$ is the initial term of this binomial with respect to $\succ_c$. This means there is a Gröbner basis element $\mathbf{x}^\alpha - \mathbf{x}^\beta$ such that the initial term $\mathbf{x}^\alpha$ divides $\mathbf{x}^{(u-v)_+}$. Therefore $u - (\alpha - \beta)$ is an improved solution for $IP_{A,\succ_c}(Au)$. By the same reasoning the optimal solution v cannot be improved by any reduced Gröbner basis element. This shows that $\mathcal{G}_{\succ_c}$ is a test set. To prove minimality, we have to show that no Gröbner basis element can be dropped from this test set. For any reduced Gröbner basis element $\mathbf{x}^\alpha - \mathbf{x}^\beta$ the non-optimal solution α for the integer program $IP_{A,\succ_c}(A\alpha)$ cannot be improved by any other element of $\mathcal{G}_{\succ_c}$ since $\mathcal{G}_{\succ_c}$ is a reduced Gröbner basis. □

The Gröbner basis algorithm for $IP_{A,\succ_c}$ was introduced by Conti and Traverso (1991) and consists of the following steps:

1. Compute the reduced Gröbner basis $\mathcal{G}_{\succ_c}$ for I_A.
2. For a solution α to the program $IP_{A,\succ_c}(b)$, compute the *normal form* $\mathbf{x}^\beta$ of the monomial $\mathbf{x}^\alpha$. Then β is the unique optimal solution of $IP_{A,\succ_c}(b)$.

Two important issues that are bypassed in the above version of the Conti-Traverso algorithm are those of 1) finding a generating set for the toric ideal I_A and 2) finding an initial solution α for $IP_{A,\succ_c}(b)$. The original algorithm of Conti and Traverso (1991) meets both of these requirements by computing the reduced Gröbner basis of a toric ideal in a larger polynomial ring with respect to an elimination order that depends on $\succ_c$. We use the above short version to highlight the main features of the procedure. The issue of finding a generating set for I_A will be discussed in Section 5.

Example 2.4

$$A = \begin{bmatrix} 2 & 4 & 3 & 1 \\ 7 & 2 & 4 & 0 \end{bmatrix}, \quad c = \begin{bmatrix} 10 & 3 & 1 & 8 \end{bmatrix}, \quad b = \begin{bmatrix} 72 \\ 79 \end{bmatrix}$$

Suppose we refine the cost vector c by the graded reverse lexicographic order to obtain $\succ_c$. Then $\mathcal{G}_{\succ_c} = \{\underline{cd^5} - b^2, \underline{a^2d^9} - bc^3, \underline{a^2bd^4} - c^4, \underline{a^2b^3} - c^5d\}$ where $S = k[a,b,c,d]$. The vector $\alpha = (3,7,11,5)$ is a feasible solution for the above program, and the normal form of $a^3b^7c^{11}d^5$ with respect to $\mathcal{G}_{\succ_c}$ is $ab^6c^{15}d$. So the optimal solution to this program is $(1,6,15,1)$. □

Corollary 2.5 *Let $\mathcal{G}_{\succ_c}$ be the reduced Gröbner basis of I_A.*

1. *The initial monomial ideal of I_A with respect to $\succ_c$, denoted $in_{\succ_c}(I_A)$, is generated by the monomials $\mathbf{x}^\alpha$ where α is a non-optimal solution for some program in $IP_{A,\succ_c}$. The minimal generators of $in_{\succ_c}(I_A)$ are the initial terms of the binomials in $\mathcal{G}_{\succ_c}$.*

2. *A vector $\beta \in \mathbf{N}^n$ is the optimal solution to the program $IP_{A,\succ_c}(A\beta)$ if and only if $\mathbf{x}^\beta$ is a standard monomial of $in_{\succ_c}(I_A)$.*

We saw earlier that an element $\mathbf{x}^\alpha - \mathbf{x}^\beta \in \mathcal{G}_{\succ_c}$ can be thought of as the vector $\alpha - \beta \in ker_{\mathbf{Z}}(A)$. Alternately we may think of $\mathbf{x}^\alpha - \mathbf{x}^\beta \in \mathcal{G}_{\succ_c}$ as the line segment $[\alpha, \beta]$ in the $A\alpha$-fiber of A directed from α to β. This interpretation of the elements of $\mathcal{G}_{\succ_c}$ says that there exists a monotone directed path in P_b^I, from *any* solution α of the program $IP_{A,\succ_c}(b)$ to the optimum, whose edges are elements of $\mathcal{G}_{\succ_c}$. This observation can be used for enumerating lattice points in polyhedra (see Chapter 5, Sturmfels 1995). It was also used in Tayur et. al (1995) to solve a special family of stochastic integer programs that arose in a manufacturing setting. In fact, the entire Buchberger algorithm in the toric situation can be reduced to a geometric process on vectors in $ker_{\mathbf{Z}}(A)$ (Thomas 1995).

3 Universal Gröbner bases of toric ideals

A subset $\mathcal{U}$ of a polynomial ideal $I \subseteq S$ is called a *universal Gröbner basis* of I (Weispfenning 1987) if it contains a Gröbner basis for I with respect to every cost vector. If we let IP_A denote all integer programs with fixed coefficient matrix A, then we call $\mathcal{U}_A \subset ker_{\mathbf{Z}}(A)$ a *universal test set* for IP_A if $\mathcal{U}_A$ contains a test set for every family $IP_{A,c}$ as c varies. Therefore a universal Gröbner basis for I_A would be a universal test set for IP_A.

The *Graver basis* of A (Graver 1975) is a universal test set for IP_A, constructed as follows. For each $\sigma \in \{+,-\}^n$, consider the semigroup $S_\sigma = ker_{\mathbf{Z}}(A) \cap \mathbf{R}^n_\sigma$, where $\mathbf{R}^n_\sigma$ is the orthant with sign pattern σ. Then $pos(S_\sigma) = \{\sum_{i=1}^p \lambda_i x_i : x_i \in S_\sigma, \lambda_i \geq 0\}$ is a pointed polyhedral $(n-d)$-dimensional cone in $\mathbf{R}^n$. The *Hilbert basis* of a polyhedral cone K in $\mathbf{R}^n$ is a minimal subset of $K \cap \mathbf{Z}^n$ such that every integral vector in K can be written as a non-negative integral combination of the elements in the basis. Further, pointed cones have unique Hilbert bases (see Chapter 16, Schrijver 1986). Let H_σ denote the unique Hilbert basis of S_σ. This is finite since $A \in \mathbf{Z}^{d\times n}$. The Graver basis of A is the set $Gr_A := \cup_\sigma H_\sigma \backslash \{0\}$ which was shown by Graver (1975) to be a universal test set for IP_A. Blair and Jeroslow (1982) and Cook et al. (1986) gave equivalent test sets (see also Section 17.4, Schrijver 1986).

Let UGB_A denote the union of all reduced Gröbner bases $\mathcal{G}_c$ as c varies over all *generic* cost vectors in $\mathbf{R}^n$. A cost vector c is generic for A if the optimal solution with respect to c in every fiber P_b^I is a unique vertex. Clearly, UGB_A is a uniquely defined universal test set for IP_A and we call it the *universal Gröbner basis* of A. The sets Gr_A and UGB_A are related as follows.

Theorem 3.1 *(cf. Thomas 1995) The Graver basis Gr_A contains the universal Gröbner basis UGB_A.*

Corollary 3.2 *There exists only finitely many distinct reduced Gröbner bases for I_A as the cost function is varied. In particular, UGB_A is finite.*

Corollary 3.2 is a well known result for polynomial ideals (Mora and Robbiano 1988, Bayer and Morrison 1988) and what we provide above is an independent proof in the case of toric ideals. Both Gr_A and UGB_A can be computed using Gröbner basis methods as described by Sturmfels and Thomas (1997). We outline below the main ingredients of these algorithms. In order to compute Gr_A we consider the *Lawrence lifting* of A which is the enlarged matrix $\Lambda(A) = \begin{bmatrix} A & 0 \\ I & I \end{bmatrix} \in \mathbf{Z}^{(n+d)\times 2n}$ where 0 is a $d \times n$ matrix of all zeros and I is the $n \times n$ identity matrix. The matrices A and $\Lambda(A)$ have isomorphic kernels: $ker_{\mathbf{Z}}(\Lambda(A)) = \{(u,-u) : u \in ker_{\mathbf{Z}}(A)\}$. The toric ideal $I_{\Lambda(A)}$ is the homogeneous prime ideal

$$I_{\Lambda(A)} = \langle \mathbf{x}^\alpha \mathbf{y}^\beta - \mathbf{x}^\beta \mathbf{y}^\alpha \; : \; \alpha, \beta \in \mathbf{N}^n, A\alpha = A\beta \rangle$$

in the polynomial ring $k[x_1, \ldots, x_n, y_1, \ldots, y_n]$.

Theorem 3.3 *For the matrix $\Lambda(A)$, the following sets coincide:*
1. the Graver basis of $\Lambda(A)$,
2. the universal Gröbner basis of $\Lambda(A)$,
3. any reduced Gröbner basis of $I_{\Lambda(A)}$, and
4. any minimal generating set of $I_{\Lambda(A)}$ (up to scalar multiples).

Algorithm 3.4 How to compute the Graver basis of A.

1. Compute the reduced Gröbner basis $\mathcal{G}$ of $I_{\Lambda(A)}$ with respect to any term order.
2. The Graver basis Gr_A consists of all elements $\alpha - \beta$ such that $\mathbf{x}^\alpha \mathbf{y}^\beta - \mathbf{x}^\beta \mathbf{y}^\alpha$

appears in $\mathcal{G}$.

Proof: By Theorem 3.3 any reduced Gröbner basis of $I_{\Lambda(A)}$ is also the Graver basis of $\Lambda(A)$. The bijection between the kernels of A and $\Lambda(A)$ implies that a reduced Gröbner basis of $I_{\Lambda(A)}$ with the variables y_j set to one, is the Graver basis of A. □

Since $UGB_A \subseteq Gr_A$, all we need now is a characterization of the elements of UGB_A so that they can be identified from among the elements of Gr_A. In the next section we will see a geometric characterization of the elements in UGB_A that can be used for this purpose. A second test can be found in Sturmfels and Thomas (1997).

Example 2.4 continued. For the matrix A in Example 2.4, $Gr_A = UGB_A = \{b^2 - cd^5, a^2d^9 - bc^3, a^2d^{14} - b^3c^2, a^2d^{19} - b^5c, a^2d^{24} - b^7, a^2bd^4 - c^4, a^2b^3 - c^5d, a^4d^{13} - c^7, a^4b^4d^3 - c^9, a^6b^7d^2 - c^{14}, a^8b^{10}d - c^{19}, a^{10}b^{13} - c^{24}\}$. □

A natural question now is to ask whether it is possible to bound the *degree* of elements in UGB_A and/or Gr_A in terms of the matrix A. By the degree of a binomial $\mathbf{x}^{v^+} - \mathbf{x}^{v^-}$ we mean $\sum_{i=1}^{n} |v_i|$, the 1-norm of the vector v. Let $D(A) = max\{|det(a_{i_1}, \ldots, a_{i_d})| : 1 \leq i_1 < i_2 < \cdots < i_d \leq n\}$. (Recall that A is assumed to have rank d.) Then the following theorem of Sturmfels (1991) gives a bound on the degree of elements in Gr_A.

Theorem 3.5 *The degree of a binomial in the Graver basis of A is at most $(d+1)(n-d)D(A)$.*

Improving degree bounds of elements in Gr_A is an important open problem in this area. It has been conjectured that the true degree bound for elements in Gr_A should not involve n. It was shown by Hoşten (1997) that if the cone $pos(S_\sigma)$ is simplicial then the maximum degree of a Graver basis element from this cone is at most $(d+1)D(A)$. A bound that is independent of n but depends exponentially on d is also given by Hoşten (1997). In the special case of $2 \times n$ matrices, this new bound can be improved further and the proof also yields an algorithm to compute Gr_A.

4 Variations of cost functions in integer programming

In this section we collect certain polyhedral results for integer programming that arise from the Gröbner basis approach to the theory. These results first appeared in (Sturmfels and Thomas 1997) and we refer the reader to that paper for proofs and details. For the basics of polyhedral theory see (Ziegler 1995).

A polytope P is a bounded polyhedron that can either be written as $P = \{x \in \mathbf{R}^n : Mx \leq m\}$ for some $M \in \mathbf{R}^{p \times n}$ and $m \in \mathbf{R}^p$, or as $P = \{\lambda_1 v_1 + \cdots + \lambda_t v_t : \sum \lambda_i = 1, \lambda_i \geq 0, \forall i\}$ where the points $v_1, \ldots, v_t$ are the vertices of P. The dimension of P, denoted $dim(P)$, is the dimension of its affine span. For a polytope $P \subset \mathbf{R}^n$ and a vector $c \in \mathbf{R}^n$, we define $face_c(P)$ to be the set of all points in P at which the linear functional $c \cdot x$ is minimized. Hence $face_c(P) = \{x \in P : c \cdot x \leq c \cdot y, \ \forall y \in P\}$. If F is any face of P, then $\mathcal{N}(F; P)$ denotes the cone of (inner) normals, called the *inner normal cone* of P at F. In symbols, $\mathcal{N}(F; P) = \{c \in \mathbf{R}^n : c \cdot x \leq c \cdot y \text{ for all } x \in F, y \in P\}$. A *polyhedral fan* in $\mathbf{R}^n$ is a collection of polyhedral cones (also called *cells* of the fan) with the property that if C is in the fan then so is every face of C, and if C_1 and C_2 are in the fan then their intersection is a common face of each. A polyhedral fan in $\mathbf{R}^n$ is said to be *complete* if the union of all elements in the fan is $\mathbf{R}^n$. It may be noted that the collection of cones $\mathcal{N}(F; P)$ as F ranges over all faces of P, denoted $\mathcal{N}(P)$, is a complete polyhedral fan in $\mathbf{R}^n$ and we call it the (inner) *normal fan* of P. We say that two polytopes are *normally equivalent* if they have the same normal fan.

Given two polytopes P and Q in $\mathbf{R}^n$, their *Minkowski sum* is the polytope $P + Q = \{p + q : p \in P, q \in Q\} \subset \mathbf{R}^n$, and P and Q are called *Minkowski summands* of $P + Q$. As in the usual extension of addition to integration, the operation of taking Minkowski sums of finitely many polytopes extends naturally to the operation of taking *Minkowski integrals* of infinitely many polytopes (see Billera and Sturmfels 1992). The common *refinement* of two fans $\mathcal{F}$ and $\mathcal{G}$ in $\mathbf{R}^n$, denoted $\mathcal{F} \cap \mathcal{G}$, is the fan of all intersections of cones from $\mathcal{F}$ and $\mathcal{G}$. We say that $\mathcal{F} \cap \mathcal{G}$ is a refinement of $\mathcal{F}$ (respectively $\mathcal{G}$). The following are two useful facts in this context: (i) for polytopes P and Q in $\mathbf{R}^n$, the fan $\mathcal{N}(P+Q) = \mathcal{N}(P) \cap \mathcal{N}(Q)$ and (ii) the fan $\mathcal{N}(P)$ is a refinement of $\mathcal{N}(Q)$ if and only if λQ is a Minkowski summand of P for some positive real number λ.

We are interested in studying the effect of varying the cost vector c on integer programs in IP_A. We start with the following definition of when two cost vectors (not just generic ones) are equivalent with respect to A.

Definition 4.1 *Two cost vectors c and c' in $\mathbf{R}^n$ are equivalent (with respect to A) if the integer programs $IP_{A,c}(b)$ and $IP_{A,c'}(b)$ have the same set of optimal solutions for all b in $\mathbf{N}(A)$.*

Our goal in this section is to provide a structure theorem (Theorem 4.8) for these equivalence classes. Using the properties of Gröbner bases of toric ideals, one can characterize generic equivalence classes as follows. The proof follows from properties of the reduced Gröbner bases of I_A.

Proposition 4.2 *Given two generic cost vectors c and c' in $\mathbf{R}^n$, the following are equivalent:*
1. For every $b \in \mathbf{N}(A)$, the programs $IP_{A,c}(b)$ and $IP_{A,c'}(b)$ have the same optimal solution.
2. The cost vectors c and c' support the same optimal vertex in each fiber P_b^I.
3. The reduced Gröbner bases $\mathcal{G}_c$ and $\mathcal{G}_{c'}$ of I_A are equal.

Let $u = u^+ - u^- \in ker_{\mathbf{Z}}(A)$. Both u^+ and u^- lie in the Au^+-fiber of A, and we may think of u as the line segment $[u^+, u^-]$ in this fiber. We shall refer to the polytope $P_{Au^+}^I$ as *the fiber of u*. By a *Gröbner fiber* of A we mean the fiber of an element $u \in UGB_A$. Let $St(A)$ denote the Minkowski sum of all Gröbner fibers. This is a well-defined polytope in $\mathbf{R}^n$ which we call the *state polytope* of A. Since the elements in $\mathcal{G}_c$ span $ker_{\mathbf{Z}}(A)$ over the integers, it follows that $St(A)$ is an $(n-d)$-dimensional polytope. The complete polyhedral fan $\mathcal{N}(St(A))$ is called the *Gröbner fan* of A. The state polytope and Gröbner fan of a graded polynomial ideal were introduced by (Bayer and Morrison, 1988) and (Mora and Robbiano, 1988) respectively. What we provide here is a self contained construction of these entities in the toric situation.

We now state a lemma that plays a crucial role in proving Theorem 4.8.

Lemma 4.3 *Every fiber P_b^I of A is a Minkowski summand of $St(A)$.*

Proof: By the facts stated at the beginning of this section, it suffices to show that $\mathcal{N}(St(A))$ is the common refinement of $\mathcal{N}(P_b^I)$ for all $b \in \mathbf{N}(A)$. Let c be a generic cost vector and $w \neq c$ belong to the interior of the cone $\mathcal{N}(face_c(St(A)); St(A))$. Then w lies in $\mathcal{N}(\beta_i; P_{A\beta_i}^I)$ for each element $\alpha_i - \beta_i$ in the reduced Gröbner basis $\mathcal{G}_c$. This implies that $w \cdot \alpha_i > w \cdot \beta_i$ for all i, and therefore, $\mathcal{G}_w = \mathcal{G}_c$.

Now consider an arbitrary $b \in N(A)$. Let u be the unique optimum of $IP_{A,c}(b)$. The equality of test sets $\mathcal{G}_w = \mathcal{G}_c$ implies that u is also the unique optimum of $IP_{A,w}(b)$. Hence w lies in the interior of $\mathcal{N}(u; P_b^I)$. Therefore, $\mathcal{N}(face_c(St(A)); St(A)) \subseteq \mathcal{N}(u; P_b^I)$, as desired. □

Proposition 4.4 *Let db denote any probability measure with support $\mathbf{N}(A)$ such that $\int_b b\,db$ is finite. Then the Minkowski integral $\int_b P_b^I db$ is a polytope normally equivalent to $St(A)$.*

Proof: The hypothesis $\int_b b\,db < \infty$ guarantees that $\int_b P_b^I db$ is bounded. By Lemma 4.3, $\int_b P_b^I db$ is a summand of $St(A)$ and is hence a polytope. However, each Gröbner fiber is a summand of $\int_b P_b^I db$ and hence $\int_b P_b^I db$ is a polytope of dimension $n-d$ in $\mathbf{R}^n$ that has the same normal fan as $St(A)$. □

From now on we shall use the term *state polytope* for any polytope normally equivalent to $\int_b P_b^I db$. We define the *Gröbner cone* associated with $\mathcal{G}_c$ to be the closed convex polyhedral cone

$$\mathcal{K}_c \quad := \quad \{\, x \in \mathbf{R}^n \,:\, g_i \cdot x \geq 0,\ g_i \in \mathcal{G}_c \,\}$$

Observation 4.5 *The Gröbner cone $\mathcal{K}_c$ is n-dimensional. Its lineality space (largest subspace contained in the cone) $\mathcal{K}_c \cap -\mathcal{K}_c$ equals rowspan(A) $\simeq \mathbf{R}^d$.*

Proposition 4.6 *The Gröbner fan of A is the collection of all Gröbner cones $\mathcal{K}_c$ together with their faces, as c varies over all generic cost vectors.*

Proof: The argument in the proof of Lemma 4.3 shows that, for c generic, the Gröbner cone $\mathcal{K}_c$ equals $\mathcal{N}(face_c(St(A)); St(A))$. □

Corollary 4.7 *The equivalence classes of cost vectors with respect to IP_A are precisely the cells of the Gröbner fan.*

Proof: Two cost vectors c and c' are equivalent if and only if they support the same optimal face in each fiber of A. Using Propositions 4.4 and 4.6, it follows that c and c' are equivalent if and only if they lie in the relative interior of the same cell in $\mathcal{N}(St(A))$. □

A cost vector w lies in the interior of a Gröbner cone $\mathcal{K}_c$ if and only if w is generic and equivalent to c. Hence the interiors of the maximal cells in the Gröbner fan are precisely the equivalence classes of generic cost vectors. The following theorem summarizes the above discussion.

Theorem 4.8 *1. There are only finitely many equivalence classes of cost vectors with respect to A.*
2. Each equivalence class is the relative interior of a convex polyhedral cone in $\mathbf{R}^n$.
3. The collection of these cones defines a complete polyhedral fan in $\mathbf{R}^n$ called the Gröbner fan of A.
4. Let db denote any probability measure with support $\mathbf{N}(A)$ such that $\int_b b\,db < \infty$. Then the Minkowski integral $St(A) = \int_b P_b^I db$ is an $(n-d)$-dimensional convex polytope, called the state polytope of A. The normal fan of St(A) equals the Gröbner fan of A.

Example 2.4 continued. The toric ideal I_A of the matrix A in Example 2.4 has fifteen distinct reduced Gröbner bases. We list the distinct initial ideals below along with a representative cost vector from each Gröbner cone. The state polytope is hence a two dimensional 15-gon in $\mathbf{R}^4$. The initial ideals are listed in the order you would encounter them as you walk along the boundary of $St(A)$.

No.	Initial ideal	Cost vector
1.	$\{a^2bd^4, cd^5, a^2d^9, a^2b^3\}$	$[66, 32, 15, 62]$
2.	$\{a^8b^{10}d, a^6b^7d^2, a^4b^4d^3, a^2bd^4, cd^5, a^2d^9, c^5d, a^{10}b^{13}\}$	$[29/2, 0, 6, 0]$
3.	$\{a^2bd^4, a^2d^9, b^2\}$	$[9/4, 1/2, 0, 0]$
4.	$\{a^2bd^4, bc^3, a^4d^{13}, b^2\}$	$[1/4, 3/2, 0, 0]$
5.	$\{a^2bd^4, bc^3, b^2, c^7\}$	$[0, 11/7, 1/7, 0]$
6.	$\{bc^3, b^2, c^4\}$	$[0, 9/7, 4/7, 0]$
7.	$\{cd^5, bc^3, b^3c^2, b^5c, b^7, c^4\}$	$[0, 1/7, 9/7, 0]$
8.	$\{cd^5, bc^3, b^3c^2, b^5c, a^2d^24, c^4\}$	$[1/2, 0, 2, 0]$
9.	$\{cd^5, bc^3, b^3c^2, a^2d^{19}, c^4\}$	$[3/2, 0, 2, 0]$
10.	$\{cd^5, bc^3, a^2d^{14}, c^4\}$	$[5/2, 0, 2, 0]$
11.	$\{cd^5, a^2d^9, c^4\}$	$[11/2, 0, 3, 0]$
12.	$\{a^2bd^4, cd^5, a^2d^9, c^5d, c^9\}$	$[13/2, 0, 3, 0]$
13.	$\{a^4b^4d^3, a^2bd^4, cd^5, a^2d^9, c^5d, c^{14}\}$	$[23/2, 0, 5, 0]$
14.	$\{a^6b^7d^2, a^4b^4d^3, a^2bd^4, cd^5, a^2d^9, c^5d, c^{19}\}$	$[33/2, 0, 7, 0]$
15.	$\{a^8b^{10}d, a^6b^7d^2, a^4b^4d^3, a^2bd^4, cd^5, a^2d^9, c^5d, c^{24}\}$	$[43/2, 0, 9, 0]$

We close this section with a geometric characterization of the elements in the universal Gröbner basis UGB_A.

Theorem 4.9 *A vector $\alpha - \beta \in ker_{\mathbf{Z}}(A)$ lies in the universal Gröbner basis UGB_A if and only if $\alpha - \beta$ is primitive and the line segment $[\alpha, \beta]$ is an edge of the $A\alpha$-fiber of A.*

For a proof of this theorem see Sturmfels and Thomas (1997). In view of Proposition 4.4, this says that the edge directions of the fibers of A are precisely the edge directions of the state polytope. If $[\alpha, \beta]$ is the primitive representative of an edge direction, then $[\alpha, \beta]$ is an edge of the $A\alpha$-fiber of A. Therefore, Theorem 4.9 is equivalent to the following assertion: *the universal Gröbner basis UGB_A consists of the edge directions of the state polytope.*

Theorem 4.9 implies several interesting corollaries.

Corollary 4.10 *For an element $\alpha - \beta$ in UGB_A, there exists two cost vectors c and c' in $\mathbf{R}^n$ such that $\alpha - \beta \in \mathcal{G}_c$ and $\beta - \alpha \in \mathcal{G}_{c'}$.*

Proof: Every element in UGB_A appears as a facet normal of some cell in the Gröbner fan. Take as $\mathcal{G}_c$ and $\mathcal{G}_{c'}$ the Gröbner bases associated with the two Gröbner cones that share this facet. □

Corollary 4.11 *For every generic cost vector $c \in \mathbf{R}^n$, the reduced Gröbner basis of $IP_{A,c}$ consists only of edges of certain fibers P_b^I.*

Theorem 4.9 implies that we can trace a monotone *edge path* from every vertex of P_b^I that is non-optimal with respect to the cost vector c to the optimal vertex, using only elements in UGB_A. Thus reduction with respect to the universal Gröbner basis can be viewed as an integer analogue to the simplex method for linear programming.

Theorem 4.9 gives rise to the following algorithm for computing the universal Gröbner basis.

Algorithm 4.12 How to compute the universal Gröbner basis UGB_A

1. Compute the Graver basis Gr_A (see Section 3).
2. For each element $\mathbf{x}^\alpha - \mathbf{x}^\beta$ of Gr_A decide whether $[\alpha, \beta]$ is an edge of its fiber.

5 Implementing toric Gröbner basis algorithms

In principle, one can compute Gröbner bases of toric ideals using available software like Macaulay2 or CoCoA. However, more careful implementations of Buchberger's algorithm along with new ideas for improving this algorithm are possible due to the nice geometric properties of toric ideals. This in turn considerably speeds up computations. Pioneering attempts in this direction are implementations like BaStaT (Pottier 1996) and GRIN (Hoşten and Sturmfels 1995). More implementations of the same kind are emerging (see Bigatti et al.). First of all, these implementations benefit from the fact that binomials require very simple data structures since they can be identified with pairs of lattice points. Moreover, S-pair computations and reduction operations in the Buchberger algorithm can be all formulated in terms of vector additions and translations (see Thomas 1995, and the discussion at the end of Section 2). Another simple observation which leads to considerable speed-up is that I_A is a prime ideal which does not contain monomials; so whenever the terms of a binomial which is created during the computation has a common monomial factor, we can simply divide by that factor.

Although the above ideas improve algorithms for computing Gröbner bases of toric ideals, it seems that the real bottleneck is computing an initial generating set for I_A. In fact, in most of the applications such a generating set is not part of the data. One is usually presented with the matrix A and a term order only. A first idea to circumvent this problem is an elimination technique as described by Conti and Traverso (1991). For this, let $A = [a_1, \ldots, a_n]$, and for simplicity assume that the columns a_i are nonnegative integral vectors.

Proposition 5.1 *Let $J = \langle x_i - \mathbf{y}^{a_i} : i = 1, \ldots, n \rangle \subset k[x_1, \ldots, n, y_1, \ldots, y_d]$. Then $I_A = J \cap k[x_1, \ldots, x_n]$ and the reduced Gröbner basis $\mathcal{G}_\succ(I_A) = k[x_1, \ldots, x_n] \cap \mathcal{G}_{\succ'}(J)$ where $\succ'$ is a term order with $y_1, \ldots, y_d \succ' x_1, \ldots, x_n$ and monomials in $k[x_1, \ldots, x_n]$ are compared by $\succ$.*

The method that this proposition suggests is in general time-consuming for computing a Gröbner basis when a generating set for I_A is absent, since it involves d extra variables and $\mathcal{G}_{\succ'}(J)$ is typically quite large compared to $\mathcal{G}_{\succ}(I_A)$. Now let $\mathcal{B} = \{u_1, \ldots, u_{n-d}\} \subset \mathbf{Z}^n$ be a lattice basis for $\ker_{\mathbf{Z}}(A)$, and let $J = \langle f_{u_1}, \ldots, f_{u_{n-d}} \rangle$ where $f_{u_i} = \mathbf{x}^{u_{i+}} - \mathbf{x}^{u_{i-}}$. The following proposition appears in (Conti and Traverso 1991), (Fischer and Shapiro 1994) and (Hoşten and Sturmfels 1995).

Proposition 5.2 *$I_A = J : (\prod_{i=1}^n x_i)^\infty$ and $I_A = (J + \langle t(\prod_{i=1}^n x_i) - 1 \rangle) \cap k[x_1, \ldots, x_n]$.*

Using the second statement of the proposition and an elimination term order where t is the most expensive variable we can compute a Gröbner basis of I_A. But although there is just one extra variable involved in this method we still get a fairly large Gröbner basis which contains the wanted Gröbner basis of I_A as a subset. The nice thing about the above ideal quotient is that it can be computed one variable at a time as follows:

$$J : (\prod_{i=1}^n x_i)^\infty = ((\cdots (J : x_1^\infty) : x_2^\infty) \cdots) : x_n^\infty).$$

Suppose that A contains a positive vector $c = (c_1, \ldots, c_n)$ in its row space, i.e. both I_A and J are homogeneous with respect to the grading $\deg(x_i) = c_i$. Now each one of the above ideal quotients can be computed as follows: first we compute a Gröbner basis where we use a c-graded reverse lexicographic order which makes the corresponding variable x_i the cheapest variable; then we divide each element of this Gröbner basis by the highest power of x_i. This allows us to compute a generating set for I_A by doing n Gröbner basis computations. These individual computations are observed to be short in practice. See the discussion in (Hoşten and Sturmfels 1995). A similar algorithm is suggested by DiBiase and Urbanke (1995) which involves at most $\lfloor n/2 \rfloor$ Gröbner basis computations. In fact, one needs to make at most $\lfloor n/2 \rfloor$ Gröbner basis computations instead of n in the above algorithm as well: Proposition 5.2 implies that I_A is an associated prime of J and any other associated prime of J contains pure variables. Now suppose M is the $(n-d) \times n$ matrix whose rows are the elements of $\mathcal{B} = \{u_1, \ldots, u_{n-d}\}$. We will call a vector *mixed* if it contains both positive and negative entries. Otherwise it will be called *unmixed.* Since we assume that $\ker(A)$ does not contain nonnegative vectors, all the rows of M are mixed. The next result is new (Hoşten and Shapiro 1997).

Theorem 5.3 *There exist variables $x_{i_1}, \ldots, x_{i_k}$ such that $I_A = J : (\prod_{j=1}^k x_{i_j})^\infty$ where $k \leq \lfloor n/2 \rfloor$.*

Proof: We construct a subset $T = \{x_{i_1}, \ldots, x_{i_k}\}$ of the variables such that for every associated prime P of J we have $P \cap T \neq \emptyset$. We start with $T = \emptyset$.

We can assume that the first k_1 entries of the first row of M are nonzero and the rest of u_1's entries are zero. Furthermore we can also assume that the first s_1 entries of u_1 are positive where $s_1 \leq \lfloor k_1/2 \rfloor$. We set $T = T \cup \{x_1, \ldots, x_{s_1}\}$. Then we delete the first row and the first k_1 columns of M. If the resulting matrix have all unmixed rows we stop, if not, we can assume that the top row is mixed. We can further assume that the first k_2 entries of this row are nonzero and the first $s_2 \leq \lfloor k_2/2 \rfloor$ of them are positive; and we set $T = T \cup \{x_{k_1+1}, \ldots, x_{k_1+s_2}\}$ and we repeat. When we are done, T will have $s_1 + s_2 + \cdots + s_p \leq \lfloor k_1/2 \rfloor + \lfloor k_2/2 \rfloor + \cdots + \lfloor k_p/2 \rfloor \leq \lfloor n/2 \rfloor$ variables. So M will be in the form

$$\left(\begin{array}{c|c} B & 0 \\ \hline C & D \end{array} \right)$$

where B is an $p \times (\sum_{j=1}^{p} k_j)$ matrix of the form

$$\begin{pmatrix} + & \cdots & + & - & \cdots & - & 0 & \cdots & 0 & 0 & \cdots & 0 & \cdots & 0 & \cdots & 0 & 0 & \cdots & 0 \\ * & \cdots & * & * & \cdots & * & + & \cdots & + & - & \cdots & - & \cdots & 0 & \cdots & 0 & 0 & \cdots & 0 \\ & \vdots & & & \vdots & & & \vdots & & & \vdots & & \ddots & & \ddots & & & \ddots & \\ * & \cdots & * & * & \cdots & * & * & \cdots & * & * & \cdots & * & \cdots & + & \cdots & + & - & \cdots & - \end{pmatrix}$$

and D is a matrix whose rows are unmixed. Now suppose x is a variable contained in an associated prime of J. If $x \in T$ then we are done. So suppose that $x = x_i \notin T$ and it corresponds to a column of B. This means that x_i corresponds to an entry in the final block of negative entries in a row u_j of B. Then P must contain yet another variable $x_{i'}$ which corresponds to one of the positive entries of row j. If $x_{i'}$ is in T, we are done. If not, since $i' < i$, we repeat this finitely many times and eventually find a variable in $T \cap P$. If x corresponds to a column of D, then we know that P must contain another variable corresponding to one of the columns of B because each row of D is unmixed. This brings us back to the above case. □

Often times one is interested in solving $IP_{A,c}(b)$ for just one fixed b. In this situation, the full Gröbner basis might be much larger than a sufficient test set for this problem. Given a fixed b, Thomas and Weismantel (1995) show how one can truncate the Buchberger algorithm using b as a *degree bound* to obtain a *truncated* Gröbner basis that will solve the particular $IP_{A,c}(b)$ as well as other integer programs whose right hand side vector are "lower" than b in a specific sense. When one is interested in $0-1$-solutions to integer programs even more can be done: Urbaniak et al. (1997) give a variant of the Buchberger algorithm in this setting.

Gröbner bases of toric ideals can be used to gain more information about the structure of integer programs. The results in Section 4 are good examples of such use. We believe the future will bring more results in this direction. Conversely, we expect that the integer programming literature and in general convex geometry will help our understanding of toric ideals, toric varieties and their Gröbner bases. For an example for this see (Hoşten 1997) and (Hoşten and Thomas 1997).

References

[1] Bayer, D., Morrison, I. (1988): Gröbner bases and geometric invariant theory I. J. Symb. Comp. 6:209–217.

[2] Bigatti, A.M., LaScala, R., Robbiano, L.: Computing toric ideals, in preparation.

[3] Billera, L.J., Sturmfels, S. (1992): Fiber polytopes. Annals of Math. 135:527–549.

[4] Blair, C.E., Jeroslow, R.G. (1982): The value function of an integer program. Mathematical Programming 23:237–273.

[5] Buchberger, B. (1965): On Finding a Vector Space Basis of the Residue Class Ring Modulo a Zero Dimensional Polynomial Ideal (German). PhD Thesis, Univ. of Innsbruck, Austria.

[6] Conti, P., Traverso, C. (1991): Gröbner bases and integer programming. In: Proceedings AAECC-9, New Orleans. Springer-Verlag, LNCS **539**, pp. 130–139.

[7] Cook, W., Gerards, A.M.H., Schrijver, A., Tardos, É. (1986): Sensitivity theorems in integer linear programming. Mathematical Programming, 34:251–264.

[8] DiBiase, F., Urbanke, R. (1995): An algorithm to calculate the kernel of certain polynomial ring homomorphisms. Experimental Mathematics, 4:227–234.

[9] Fischer, K.G., Shapiro, J. (1994): Generating prime ideals in the Minkowski ring of polytopes. In: Fischer, K.G., Loustaunau, P., Shapiro, J., Green, E.L., Farkas, D. (eds.): Computational Algebra, Marcel Dekker Inc., Lecture Notes in Pure and Applied Mathematics **151**, New York, pp. 111–130.

[10] Graver, J.E. (1975): On the foundations of linear and integer programming I. Mathematical Programming, 8:207–226.

[11] Hoşten, S. (1997): Degrees of Gröbner bases of integer programs. Ph.D thesis, Cornell University.

[12] Hoşten, S., Shapiro, J. (1997): Primary decompositions of lattice basis ideals, in preparation.

[13] Hoşten, S., Sturmfels, B. (1995): GRIN: An implementation of Gröbner bases for integer programming. In: Balas, E., Clausen, J. (eds.): Integer Programming and Combinatorial Optimization. Springer Verlag, LNCS **920**, pp. 267–276.

[14] Hoşten, S., Thomas, R.R. (1997): Arithmetic degree of initial ideals of toric ideals, in preparation.

[15] Mora, T., Robbiano, L. (1988): Gröbner fan of an ideal. J. Symb. Comp., 6:183–208.

[16] Tayur, S.R., Thomas, R.R., Natraj, N.R., (1995): An algebraic geometry algorithm for scheduling in presence of setups and correlated demands. Mathematical Programming, 69, Ser. A:396–401.

[17] Pottier, L. (1996): BaStaT (downloadable from `http://www.inria.fr/safir/WHOSWHO/Loic/Bastat/bastatdemo.html`)

[18] Scarf, H.E. (1986): Neighborhood systems for production sets with indivisibilities. Econometrica, 54:507-532.

[19] Schrijver, A. (1986): Theory of Linear and Integer Programming. Series in Discrete Mathematics. Wiley-Interscience, New York.

[20] Sturmfels, B. (1991): Gröbner bases of toric varieties. Tôhoku Math. Journal, 43:249-261.

[21] Sturmfels, B. (1995): Gröbner Bases and Convex Polytopes. AMS University Lecture Series **8**, Providence.

[22] Sturmfels, B., Thomas, R.R. (1997): Variation of cost functions in integer programming. Mathematical Programming, 77:357–387.

[23] Thomas, R.R. (1995): A geometric Buchberger algorithm for integer programming. Mathematics of Operations Research, 20:864–884.

[24] Thomas, R.R., Weismantel, R. (1995): Truncated Gröbner bases for integer programming. Preprint SC 94-29, ZIB Berlin.

[25] Urbaniak, R., Weismantel, R., Ziegler, G. (1997): A variant of Buchberger's algorithm for integer programming. SIAM J. Discrete Math., 10:96–108.

[26] Weispfenning, V. (1987): Constructing universal Gröbner bases. In: Proceedings AAEEC 5, Menorca 1987, Springer Verlag, LNCS **356**, pp. 408-417.

[27] Ziegler, G. (1995): Lectures on Polytopes. Graduate Texts in Math., Springer-Verlag, New York.

Gröbner Bases and Numerical Analysis

H. Michael Möller

Abstract

By concentrating on system solving, numerical interpolation, integration, and differentiation, we show the use of Gröbner bases in numerical analysis. The ideas of the factorizing Gröbner algorithm, of system solving by solving Eigenproblems, of computing interpolation polynomials with algorithms for computing Gröbner bases, and of constructing numerical integration and differentiation formulas by Gröbner bases are presented. A short section on Gröbner bases computation using floating point arithmetics is included.

1 Introduction

In the nineties, there is an increasing interest in combining symbolic and numerical methods. This can be seen at diverse instances. There are now international symposia supported by organizations from both sides, and the number of contributes displaying the symbolic - numerical interplay is increasing. Other examples are the facilities of using floating point arithmetics and simple numerical procedures in Computer Algebra Systems on the one hand and the (eventually partly) integration of Computer Algebra Systems into numerical software packages on the other hand. The most prominent example is here the migration of the Computer Algebra System AXIOM to NAG, the Numerical Algorithm Group.

Many interesting results have been obtained by combining symbolic and numerical methods, like in polynomial continuation the avoiding of solution paths diverging to infinity by means of concepts from toric ideals or like the numerical solving of systems of polynomial equations using resultants, see for instance Canny and Manocha (1993). On the occasion of 33 years of Gröbner bases, I will concentrate on the use of Gröbner bases in numerical analysis, stressing the word analysis, because Gröbner bases are used in numerics, theoretically and practically.

Whenever polynomials, especially multivariate ones, are considered, they can be manipulated symbolically using Gröbner bases. And if in theoretical considerations algebraic objects occur, Gröbner bases can help for studying the intrinsic structure. In this paper, we will present some details of this link from algebra to numerics. Being unable to give a complete survey on every numerical method, where Gröbner bases can be used, we concentrate on solving polynomial systems of equations, multivariate interpolation, numerical differentiation and integration, and Gröbner bases in floating point arithmetics.

In system solving by Gröbner bases, which will be presented in detail in section 2, the application of Gröbner bases comes typically first and then the numerical method. Using Gröbner bases, a polynomial system can be transformed symbolically to one or several systems, which are expected to be simpler for numerical solving, or even to an Eigenvalue problem. It depends on the users' view to say, the Gröbner bases task is a preprocessing for the numerical solving, or to call the numerical solving just a postprocessing, if the algebraic solution is found by Gröbner basis techniques.

In the numerical analysis of polynomial interpolation, numerical differentiation and integration, a polynomial ideal I and the polynomial ring $\mathbf{P} := K[x_1,\ldots,x_n]$ modulo the ideal I, $\mathbf{P}/I$ for short, are considered. Here K is a field. In numerical analysis, it is typically the field of real or complex numbers. For dealing with I and $\mathbf{P}/I$, Gröbner bases are a useful tool, as shown in the sections 3 and 4.

Since many Computer Algebra Systems offer the application of floating point arithmetics to polynomials and have integrated a variant of Buchberger's algorithm, it is seducing to run Buchberger's algorithm with floats. But because of intermediate rounding errors, the result is in almost all cases the Gröbner basis $\{ 1 \}$. A careful analysis is needed for understanding how relevant the result and the computations are for the ideal generated by the input polynomials and for the common zeros of the input polynomials. This analysis is begun in the last few years. In the last section, we give a short summary of two actual contributions of this topic.

For the notations, we follow Buchberger(1998).

2 Polynomial system solving by Gröbner bases

Let a system of polynomial equations be given,

$$f_1(x_1,\ldots,x_n) = 0,\ldots,f_s(x_1,\ldots,x_n) = 0, \tag{2.1}$$

where $f_1,\ldots,f_s$ are polynomials in $\mathbf{P} := K[x_1,\ldots,x_n]$. Then solving this system roughly means finding a description of the *variety*

$$V := \{ y \in \mathbb{C}^n \mid f_1(y) = 0,\ldots,f_s(y) = 0 \}. \tag{2.2}$$

Solving in the pure numerical sense means to find for every point of V a floating point approximation within a given precision. This notion gives only sense if the variety is a finite set. An other severe restriction is, that numerical algorithms for solving nonlinear systems of equations require $s = n$, i.e. as many equations as unknowns. This holds because underdetermined systems $s < n$ have infinitely many solutions, and for overdetermined systems $s > n$ the only recommendation found in numerical literature was to transform the problem of solving the system into the problem of minimizing $\sum_{i=1}^{s} |f_i|^2$, see for instance the recent book of Ueberhuber (1997). Unfortunately, by numerical reasons, it can not be decided whether a minimum is 0, i.e. whether a point of V is found.

If the variety (2) is finite, then the goal for an algebraic preprocessing should be to transform the system (1) into one or several systems, whose varieties cover the variety (2), but which are solvable numerically (same number of variables and equations) and if possible simpler equations, say lower degrees. If the variety (2) is infinite, then a useful information for the numerical analysist could be the size of V and some parameter depending solutions, which he can trace by assigning values to the parameters.

Example 1. The system of n equations in n unknowns

$$\begin{aligned}
&x_1 + x_2 + \ldots + x_n = 0, \\
&x_1x_2 + x_2x_3 + \ldots + x_{n-1}x_n + x_nx_1 = 0, \\
&\vdots \\
&x_1 \cdots x_{n-1} + x_2x_3 \cdots x_n + \ldots + x_nx_1 \cdots x_{n-2} = 0, \\
&x_1 \cdots x_n = 1,
\end{aligned} \tag{2.3}$$

is known under the name cyclic n-roots and serves for long as a benchmark problem in Computer Algebra. This system is completely solved for $n < 8$ by Backelin and Fröberg (1991) and for $n = 8$ by G. Björck and Fröberg (1994). There is a conjecture by Fröberg saying, that if n has a quadratic divisor, then the system has infinitely many solutions, and if the solution set is finite, then there are exactly $\binom{2n-2}{n-1}$ of them.

Example 2 Take a cyclic carbon hydrogen molecule with nodes $P_1, \ldots, P_6$ where the edges joining neighbouring points P_i, P_{i+1}, with $P_7 := P_1$, are all normalized to length 1. Then, by the carbon atom structure, the distance between P_i and P_{i+2} is $\frac{2}{3}\sqrt{2}$. The geometric configuration of the molecule is then fixed by the lengths of the three diameters $d_1 := |P_1 - P_4|$, $d_2 := |P_2 - P_5|$, $d_3 := |P_3 - P_6|$. From distance geometry, one knows, that $x_1 := d_1^2$, $x_2 := d_2^2$, $x_3 := d_3^2$ satisfy the four equations in three unknowns

$$f(x_1, x_2) = 0, \; f(x_2, x_3) = 0, \; f(x_3, x_1) = 0, \; g(x_1, x_2, x_3) = 0, \tag{2.4}$$

where f and g are the Cayley-Menger determinants

$$f(a,b) := \begin{vmatrix} 0 & 1 & 1 & 1 & 1 & 1 \\ 1 & 0 & 1 & \frac{8}{3} & a & \frac{8}{3} \\ 1 & 1 & 0 & 1 & \frac{8}{3} & b \\ 1 & \frac{8}{3} & 1 & 0 & 1 & \frac{8}{3} \\ 1 & a & \frac{8}{3} & 1 & 0 & 1 \\ 1 & \frac{8}{3} & b & \frac{8}{3} & 1 & 0 \end{vmatrix}, \quad g(a,b,c) := \begin{vmatrix} 0 & 1 & 1 & 1 & 1 & 1 & 1 \\ 1 & 0 & 1 & \frac{8}{3} & a & \frac{8}{3} & 1 \\ 1 & 1 & 0 & 1 & \frac{8}{3} & b & \frac{8}{3} \\ 1 & \frac{8}{3} & 1 & 0 & 1 & \frac{8}{3} & c \\ 1 & a & \frac{8}{3} & 1 & 0 & 1 & \frac{8}{3} \\ 1 & \frac{8}{3} & b & \frac{8}{3} & 1 & 0 & 1 \\ 1 & 1 & \frac{8}{3} & c & \frac{8}{3} & 1 & 0 \end{vmatrix}.$$

This system was solved by several authors, for instance by Melenk et al. (1989) and Gatermann (1996). In this example, only solutions with positive entries are of interest.

2.1 Solving with triangular sets

The essential observation for applying Gröbner techniques is the fact, that the variety V of (2) does not depend on the polynomials $F := \{f_1, \ldots, f_s\}$, but on the ideal I generated by F. Hence $V = V(I)$ and one may take instead of F another generating set (an ideal basis), say a Gröbner basis.

The first investigation of polynomial systems with Gröbner bases were made by Buchberger (1970). He showed, that $V(I)$ is void, if and only if the (reduced) Gröbner basis of I is $\{1\}$ independent of the term ordering, gave a criterion for deciding $V(I)$ finite, and showed that by means of a Gröbner basis polynomials from the so called *elimination ideals*

$$I \cap K[x_1],\ I \cap K[x_1, x_2], \ldots,\ I \cap K[x_1, \ldots, x_{n-1}]$$

can be computed. Trinks (1978) remarked, that having for I a Gröbner basis G with respect to a lexical term ordering with $x_1 \prec x_2 \prec \ldots \prec x_n$, then every $I \cap K[x_1, \ldots, x_i]$ has as lexical Gröbner basis $G \cap K[x_1, \ldots, x_i]$. This allows in many cases to determine successively the first, second, third etc. component of the points of $V(I)$. But in most cases a postprocessing is necessary.

Example 3. The lexical Gröbner basis for the cyclic 5-roots example consists in 11 basis elements. Let d_k denote the number of basis elements in $K[x_1, \ldots, x_k] \setminus K[x_1, \ldots, x_{k-1}]$, then $d_1 = 1$, $d_2 = 2$, $d_3 = 3$, $d_4 = 4$, $d_5 = 1$. The maximal degree in x_1 is 15, in x_2 is 7, in x_3 cubic, in x_4 quadratic, and in x_5 linear.

For the cyclohexane example, the Gröbner basis with respect to the lexical term ordering with $x_1 \prec x_2 \prec x_3$ is of type

$$\{\ x_3^2 p_2(x_1) - x_3 q_2(x_1) + r_2(x_1),\ \ x_3 x_2 - (x_3 + x_2) p_3(x_1) + q_3(x_1),\\ (x_3 + x_2) p_4(x_1) + q_4(x_1),\ \ x_2^2 p_2(x_1) - x_2 q_2(x_1) + r_2(x_1)\ \}.$$

Here $p_k(x_1)$ denotes a polynomial of degree k in x_1. Analogously $q_k(x_1)$ and $r_2(x_1)$. From the Gröbner basis one can read off, that the variety is not finite.

As the examples show, further manipulations are needed to obtain the solutions. If I is zero-dimensional, i.e. if the system (1) has only finitely many solutions, then the lexical Gröbner basis has a specific shape. This can be used to split the variety (2) into a finite number of subvarieties V_j, where every V_j is the variety of an ideal generated by polynomials $P_j := \{p_{j1}, \ldots, p_{jn}\}$ and every P_j is *triangular,* i.e. p_{jk} belongs to $K[x_1, \ldots, x_k]$ and not to $K[x_1, \ldots, x_{k-1}]$, $k = 1, \ldots, n$. The triangular systems

$$p_{j1}(x_1) = 0, \; p_{j2}(x_1, x_2) = 0, \; \ldots, \; p_{jn}(x_1, \ldots, x_n) = 0 \tag{2.5}$$

can be solved numerically either by applying a method for n-dimensional problems or by solving a series of univariate problems and substituting the solutions of $p_{j,k} = 0$ into the equations for $p_{j,\ell}$, $\ell > k$. Two different strategies for the decomposition into triangular sets are described by Lazard (1992) and Möller (1993). In Lazard's paper one can find the decomposition of the cyclic 5-roots system into one triangular system containing 20 solutions and five triangular systems containinig 10 solutions each, giving in total $\binom{8}{4} = 70$ solutions.

Thus, if the system (1) has only finitely many solutions, independently how big s is compared to n, its solution set is the union of the solutions of finitely many triangular systems (5), where the p_{jk} in (5) are obtained by symbolic manipulations using a lexical Gröbner basis of the ideal I generated by $\{f_1, \ldots, f_s\}$.

In the case the system (1) has more than finitely many solutions, then some subsets of solutions can be described using parameters. If the variety (2) is irreducible, i.e. if it can not be represented as union of two other varieties, then there is exactly one set of parameters on which the points of the variety depend. By adjunction of these parameters, meaning a transcendental field extension, the ideal becomes zero-dimensional and formally the decomposition of Lazard (1992) and Möller (1993) can be applied. In all other cases, an appropriate subdivision of the variety has to be found, such that the subvarieties are irreducible or can be treated as if (depending on just one set of parameters). For a discussion of the positive-dimensional case see Lazard (1991).

2.2 Factorizing Gröbner algorithm

The decomposition of a system into subsystems can be done earlier than at the end of the lexical Gröbner basis calculation. Many Computer Algebra Systems have a variant of Buchberger's algorithm, which splits the calculation whenever a factorizable polynomial is found, $f = f_1 \cdot f_2$, and then the calculation is continued once for each factor instead of f itself. The result is, instead of the variety V, a subvariety V_1 where the points are zeros of f_1 as well, and a subvariety V_2 on which f_2 vanishes. V_2 contains especially the points of V on which f_1 does not vanish. Hence $V = V_1 \cup V_2$. Such a procedure is sometimes called a *factorizing Gröbner algorithm* .

Example 4. As shown by Melenk et al. (1989), the factorizing Gröbner algorithm returns, if the input (4) is given, seven triangular systems. Six of them have only finitely many solutions, in fact just one each. If only solutions with positive components are considered, then three out of the six may be cancelled. Then only the points

$$(\frac{11}{3}, \frac{25}{9}, \frac{11}{3}),\ (\frac{25}{9}, \frac{11}{3}, \frac{11}{3}),\ (\frac{11}{3}, \frac{11}{3}, \frac{11}{3})$$

remain plus the points (y_1, y_2, y_3) with positive components from the seventh subvariety, where y_3 is a free parameter and $(y_1, y_2) = (a, b)$ or (b, a) with

$$a = \frac{99y_3^2 - 582y_3 - 125 + 4\sqrt{486y_3^4 - 6696y_3^3 + 30564y_3^2 - 52200y_3 + 23750}}{27y_3^2 - 198y_3 + 75},$$

$$b = \frac{99y_3^2 - 582y_3 - 125 - 4\sqrt{486y_3^4 - 6696y_3^3 + 30564y_3^2 - 52200y_3 + 23750}}{27y_3^2 - 198y_3 + 75}.$$

Here the numerical task reduces to determine numerically all values of y_3 leading to positive solutions (y_1, y_2, y_3). By chance, the decomposition found by the factorizing Gröbner algorithm is a decomposition into irreducible varieties.

A more systematic variety decomposition can be performed by using ideal quotients. Let I be an ideal, f be a polynomial. Then

$$I : f^* := \{g \in K[x_1, \ldots, x_n] \mid \exists k \in \mathbb{N} : g \cdot f^k \in I\}$$

is an ideal. It contains I and hence its variety $V(I : f^*)$ is contained in the variety $V(I)$. Using arguments from ideal theory, one can show that $V(I : f^*)$ is the least variety containing the set

$$\{y \in V \mid f(y) \neq 0\}.$$

The ideal $I + (f)$ is the least ideal containing I and f. Its variety is the set of all points in V on which f vanishes. Therefore one has the following decomposition.

Lemma 1. *Let I be an ideal, f a polynomial. Then*

$$V(I) = V(I + (f)) \cup V(I : f^*). \tag{2.6}$$

The decomposition is not trivial, if f vanishes in at least one and not in all points of $V(I)$.

Proof. $I + (f)$ and $I : f^*$ are contained in I. Hence $V(I + (f))$ and $V(I : f^*)$ are subsets of $V(I)$. Therefore $V(I+(f)) \cup V(I : f^*) \subseteq V(I)$. Let V_1 be the subset of V on which f is 0, and V_2 the subset, where f is not 0, then $V = V_1 \cup V_2$ and as shown before $V_1 \subseteq V(I+(f))$, $V_2 \subseteq V(I : f^*)$. This gives the assertion. □

This decomposition can be realized algorithmically. If the set $\{f_1, \ldots, f_s\}$, generates I, then $I+(f)$ is generated by $\{f_1, \ldots, f_s, f\}$; and $I : f^*$ is the ideal (in $K[x_1, \ldots, x_n, t]$) generated by $\{f_1, \ldots, f_s, 1-t\cdot f\}$ intersected with the ring $K[x_1, \ldots, x_n]$. The computation of the ideal intersection is done by standard Gröbner techniques, as described for instance in the book of Becker and Weispfenning (1993).

The finer the decomposition into subvarieties is, the easier is the subsequent numerical work. The finest decomposition is obtained, if one has the primary decomposition of the ideal. The computation of such decomposition is an important problem in constructive polynomial algebra. There are some investigations made on how to get the primary decomposition, see for instance Gräbe (1997) and the references quoted by him.

2.3 Solving and symmetries

If the variety (2) is invariant under a finite group Γ of mappings $\gamma : \mathbb{C}^n \mapsto \mathbb{C}^n$,

$$y \in V,\ \gamma \in \Gamma \implies \gamma(y) \in V,$$

then methods from invariant theory can be applied. Associated to every finite group Γ, there are polynomials, so called primary invariants $\theta_1, \ldots, \theta_n$ and secondary invariants $\eta_1, \ldots, \eta_t$, such that every polynomial f invariant under Γ, i.e. $f(x_1, \ldots, x_n) = f \circ \gamma(x_1, \ldots, x_n)$ for all $\gamma \in \Gamma$, has a representation

$$f(x_1, \ldots, x_n) = \sum_{i=1}^{t} \eta_i(x_1, \ldots, x_n) \cdot p_i(\theta_1(x_1, \ldots, x_n), \ldots, \theta_n(x_1, \ldots, x_n)). \tag{2.7}$$

The p_i are polynomials in $\theta_1, \ldots, \theta_n$ of degrees considerably lower than $deg(f)$. If the polynomials in (1) are invariant under the group actions, then one can use their corresponding representations (7) in order to solve a simpler problem in the unknowns $\theta_1, \ldots, \theta_n, \eta_1, \ldots, \eta_t$. This proposal was made for instance by Sturmfels (1993). Having solved this system, for each solution point $y = (y_1, \ldots, y_n, z_1, \ldots, z_t)$ there is a secondary problem of $n + t$ nonlinear equations, namely

$$\theta_i(x_1, \ldots, x_n) = y_i,\ i = 1, \ldots, n \qquad \eta_j(x_1, \ldots, x_n) = z_j,\ j = 1, \ldots, t,$$

determining the points of the variety (2). As pointed out for instance in the book of Sturmfels (1993), the primary and secondary invariants as well as the representations (7) can be found by Gröbner techniques using slack variables.

If the variety (2) is invariant under a finite group Γ, but the polynomials f_i are not, then a finer analysis is needed. Gatermann (1996) uses representation theory and so called *equivariants* for transforming the system (1) into a simpler one. This procedure also has the advantage, that the solution sets are often more transparent than those obtained by factorising Gröbner algorithms.

Example 5. The system (4) has, using factorizing techniques and using the invariance of the solution sets under the dihedral group D_3, a decomposition into two subsets. With the additional variables (invariants) $\sigma_1, \sigma_2, \sigma_3$, in fact

$$\sigma_1 = x_1 + x_2 + x_3,\ \sigma_2 = x_1^2 + x_2^2 + x_3^2,\ \sigma_3 = x_1^3 + x_2^3 + x_3^3,$$

then, as shown by Gatermann (1996), one system is

$$\begin{array}{l} 3x_1 - 11 = 0,\ \ 3x_2 - 11 = 0,\ \ 3x_3 - 11 = 0, \\ 9\sigma_3 - 1331 = 0, 3\sigma_2 - 121 = 0, \sigma_1 - 11 = 0, \end{array}$$

and the second system is

$$\begin{array}{l} -\sigma_1 + x_1 + x_2 + x_3 = 0, \\ -3\sigma_1 x_2 - 3\sigma_1 x_3 + 22\sigma_1 + 3x_2^2 + 3x_2 x_3 + 3x_3^2 - 121 = 0, \\ -27\sigma_1 x_3^2 + 198\sigma_1 x_3 - 75\sigma_1 + 27x_3^3 - 1089x_3 - 250 = 0, \\ -9\sigma_1^3 + 198\sigma_1^2 - 1164\sigma_1 + 9\sigma_3 - 250 = 0, \\ -3\sigma_1^2 + 44\sigma_1 + 3\sigma_2 - 242 = 0. \end{array}$$

2.4 The method of Stetter

In the univariate case, it is known, that the zeros of a polynomial

$$p(x) = \sum_{k=0}^{d} a_k x^k,\ a_d = 1,$$

are the Eigenvalues of the *Frobenius companion matrix*

$$\begin{pmatrix} 0 & 1 & 0 & \dots & 0 \\ & \ddots & \ddots & & \vdots \\ \vdots & & 0 & 1 & 0 \\ 0 & 0 & \dots & 0 & 1 \\ -a_0 & -a_1 & \dots & \dots & -a_{d-1} \end{pmatrix}.$$

This allows to apply software for solving Eigenvalue problems to the solving of a polynomial equation. Depending on the available software, this transformation may lead to a faster solving as demonstrated by Goedecker (1994).

The generalization to multivariate polynomial systems is attributed to H.J.Stetter, who initiated and investigated this Eigenvalue method in articles partly with different coauthors. For details and further references see the paper of Stetter (1996) and the subsequent column of Corless (1996). The idea of this approach is as follows.

Let I be an ideal of dimension 0 and G be a Gröbner basis of I with respect to an arbitrary term ordering. Then the vector space $\mathbf{P}/I$ has finite dimension. Fix a basis of it, say

$$[t_1], \ldots, [t_N],$$

where the $t_1, \ldots, t_N$ are for instance the power products, which are not divisible by a leading power product of an element of G. With this basis, the representation of an equivalence class in terms of the basis is easily found using normal forms.

Lemma 2. *Let* $\{t_1, \ldots, t_N\} := \{\, t \in \mathbf{T} \mid \nexists g \in G : \; LPP(g) | t \,\}$. *The normal form* NF *modulo* G *maps* $\mathbf{P}$ *onto* $span(\{t_1, \ldots, t_N\})$. *Then for* $f \in \mathbf{P}$ *the equivalence class* $[f]$ *satisfies*

$$[f] = \sum_{i=1}^{N} c_i [t_i], \quad \textit{if} \;\; NF(f, G) = \sum_{i=1}^{N} c_i t_i.$$

Proof. No power product of $NF(f, G)$ is a multiple of an $LPP(g)$, $g \in G$, since $NF(f, G)$ is irreducible. And $NF(t_i, G) = t_i$. Hence NF maps onto $span(\{t_1, \ldots, t_N\})$. Because of $[f] = [NF(f, G)]$ and the linearity of NF the assertion follows. □

Consider now an arbitrary polynomial $h \in \mathbf{P}$. The mapping

$$\Phi_h : \mathbf{P}/I \longrightarrow \mathbf{P}/I, \;\; \Phi_h([f]) := [h \cdot f],$$

is an endomorphism of $\mathbf{P}/I$ and has therefore Eigenvalues.

Theorem 1. *Let* A *be the matrix associated to* Φ_h *and the basis* $\{[t_1], \ldots, [t_N]\}$. *Then* A *has for every* $y \in V(I)$ *the Eigenvalue* $h(y)$ *with corresponding Eigenvector*

$$(t_1(y), \ldots, t_N(y))^T. \tag{2.8}$$

Proof. $[h \cdot t_i] = \Phi_h(t_i) = \sum_{k=1}^{N} a_{ik} [t_k]$ implies

$$h \cdot t_i - \sum_{k=1}^{N} a_{ik} t_k =: f_i \in I \text{ for } i = 1, \ldots, N.$$

Evaluating these equations at a point y, where all $f_i \in I$ are zero, gives the assertion. □

Remark. The matrix A from this theorem is sometimes called *multiplication matrix.* It was the starting point for the development of Gröbner bases because these bases are introduced first in the thesis of Buchberger (1965), when he was asked by his advisor W. Gröbner to develop methods for computing multiplication tables.

If $h(y)$ is an Eigenvalue with a onedimensional Eigenspace, then the vector (8) spans this space. By the choice of the t_i, the power products $x_1, \ldots, x_n$ are contained in $\{t_1, \ldots, t_N\}$ or are leading power products of Gröbner basis elements. Hence the components of a point $y \in V(I)$ can be read off from (8) or are easily obtained from evaluating an expression $0 = g(y) = y_k + \sum c_i t_i(y)$, $g \in G$ with leading power product x_k. This easy way of finding the components of every $y \in V$ was first stated correctly by Möller and Stetter (1995).

Denoting a polynomial h a *separating polynomial*, if all Eigenspaces of A are of dimension 1, Alonso et al. (1996) discuss how to find such polynomials and to determine the multiplicity of the zero $y \in V(I)$. Möller and Stetter (1995) gave the complete decomposition of A into a generalization of Jordan blocks.

The advantage of this Eigenmethod approach to the methods presented before is, that we do not need here an (in general) involved lexical Gröbner basis, but an arbitrary Gröbner basis. Then, usually after some trials, one has found a separating polynomial. However, the interplay of the symbolical method for computing the matrix A (using normal forms and hence Gröbner bases) and numerical software for determining Eigenvalues and corresponding Eigenvectors is not yet tested systematically. There is some hope, that we will have some positive results in the near future.

3 Polynomial interpolation

The polynomial interpolation problem can be considered as inverse to the problem of solving systems of equations. There one had the polynomials and wanted to find the zeros of the variety. Here one has the points and wants to find polynomials. In the preceding section, we did not discuss eventual multiplicities of common zeros of polynomials. Therefore we restrict first to the case of simple zeros.

3.1 The algebraic interpolation

The interpolation problem. *Given a set of M distinct points,* $\{y_1, \ldots, y_M\}$, *and M numbers* $c_1, \ldots, c_M \in K$. *Find all polynomials* p *satisfying*

$$p(y_k) = c_k \text{ for all } k = 1, \ldots, M. \tag{3.1}$$

Sometimes, at least for the univariate case $n = 1$, not all such p's, but a minimal one is sought. Having Gröbner techniques available in the multivariate case, we can make precise in what sense 'a minimal one' is to be understood.

Definition. Let an admissible term ordering $\prec$ be given. Then one defines recursively $p \in \mathbf{P}$ less than $q \in \mathbf{P}$, if either $LPP(p) \prec LPP(q)$ or $LPP(p) = LPP(q)$ and $R_{\prec}(p)$ less than $R_{\prec}(q)$. Here we wrote, following Buchberger (1998), $R_{\prec}$ for the remaining part after subtracting the leading monomial. We call a polynomial p of a polynomial set F *minimal* (with respect to F), if no other $q \in F$ is less.

Since $\prec$ is a Noetherian ordering, there is no infinite chain of polynomials f_i, where f_{i+1} is less than f_i. Hence every nonempty set F of polynomials contains a minimal one. On the other hand, there may be more than one, because the definition describes only a semiordering. Two polynomials can not be compared by this semiordering if and only if they are linear combinations (with nonzero coefficients) of the same set of power products, i.e. if they have the same support.

Theorem 2. *The solutions* $p \in \mathbf{P}$ *of* (9) *constitute an equivalence class* $[p^*]$ *in* $\mathbf{P}/I$, *where* I *denotes the ideal* $\{q \in \mathbf{P} \mid q(y_1) = \dots q(y_M) = 0\}$. *Given an arbitrary admissible term ordering* $\prec$, *the minimal element in this class with respect to* $\prec$ *is the normalform* $NF(p^*, G)$, G *denoting a Gröbner basis of* I *w.r.t.* $\prec$.

Proof. Let $p^* \in \mathbf{P}$ solve (9). Then for every polynomial p solving (9) also $p - p^* \in I$ holds. Conversely, if $p - p^* \in I$ then p solves (9). Every polynomial in $[p^*]$ reduces to the same normalform $NF(p^*, G)$, since G is a Gröbner basis. And the reduction modulo G means a replacing of a power product by a linear combination of lower order power products. Therefore $NF(p^*, G)$ is less than every other element in $[p^*]$. □

Corollary. *Let* $\prec$, I, *and* G *be as in the theorem. If* $\{t_1, \dots, t_N\}$ *is as in lemma 2, then* $N = M$ *and there exists a unique polynomial* p^* *in* $span(\{t_1, \dots, t_N\})$ *solving (9). If* f *is an other polynomial solving (9), then* p^* *is less than* f *with respect to* $\prec$.

Proof. By theorem 2, $p^* = NF(f, G)$ for every polynomial f solving (9), and, by lemma 2, NF maps onto $span(\{t_1, \dots, t_N\})$. It remains to show, that there are M basis elements, or equivalently, that $dim \mathbf{P}/I = M$. Take M polynomials p_i with $p_i(y_j) = 0$ iff $i \neq j$ (take e.g. linear polynomials ℓ_i with $\ell_i(y_j) = 0$ iff $i = j$ and define $p_i = \ell_1 \cdot \ldots \cdot \ell_M / \ell_i$, $i = 1, \dots, M$). Then the equivalence classes $[p_i]$, $i = 1, \dots, M$, constitute a basis of $\mathbf{P}/I$. □

Without using Gröbner basis terminology, de Boor and Ron (1992) discuss the construction of the space $span(\{t_1, \dots, t_N\})$ and the unique solution p^* of (9) in that space. Their approach consists in establishing a Vandermonde matrix V, where the columns are indexed by the power products t and the rows by $1, \dots, N$, and hence its entry v_{it} equals $t(y_i)$. This matrix has rank N. Assuming an ordering $\prec$ for the power products, on which de Boor and Ron impose the additional condition $deg(t_1) < deg(t_2) \Longrightarrow t_1 \prec t_2$, (this implies that the ordering is admissible,) then by Gaussian elimination the row-echelon form is obtained without any zero rows. Let in the i-th row the first nonzero entry be in the column corresponding to a power product u_i, $i = 1, \ldots, N$. Then, by construction, the first N columns of V, which are linearly independent, correspond to $\{u_1, \ldots, u_N\}$. In proposition(2.5), de Boor and Ron (1992) prove, that $span(\{u_1, \ldots, u_N\})$ is a space which contains a unique p^* solving (9). They call this space a polynomial space of smallest possible degree.

Now, $\{t_1, \ldots, t_N\} = \{u_1, \ldots, u_N\}$, because also $t_1, \ldots, t_N$ correspond to the first N linearly independent columns of V ordered with respect to the admissible term ordering. This can be seen by the following observations: Every vector with finitely many nonzero entries and orthogonal to the rows of V can be interpreted as the coefficient vector of a polynomial vanishing in $y_1, \ldots, y_N$. Hence a linear dependence relation among the columns of V corresponds to a polynomials in I. Especially, in every set of columns generating the column space of V a column corresponding to a leading power product of a polynomial in I can be replaced by columns of lower order. Doing this replacement, one ends up with N columns corresponding to those power products, which are not divisible by a power product of a Gröbner basis element, i.e. corresponding to $t_1, \ldots, t_N$.

The construction of de Boor and Ron is similar to that of Buchberger and Möller (1982), which was refined and extended to more general interpolation problems by Marinari et al. (1993). This construction can be explained also in terms of the matrix V used by de Boor and Ron. If Gaussian column operations are performed, then the new columns correspond no more to power products but to polynomials. A zero column corresponds to a polynomial in I, a unit column to a Lagrange polynomial ℓ_k with $\ell_k(y_j) = \delta_{jk}$, etc. Let an admissible term ordering $\prec$ be given. Then the columns are first ordered by the power products, later by their leading power product, since by the Gaussian column elimination one column is always transformed by preceding ones w.r.t. $\prec$. Hence the columns can be ordered by the ordering of their corresponding (leading) power products. Manipulating always the least column which is not yet cancelled and not yet transformed by Gaussian eliminations, one eliminates the column with the preceding ones. If the result is the vector 0, then it corresponds to an element of the ideal I, we store it in an initially empty set G, and we cancel all forthcoming columns which correspond to a power product divisible by the leading power product of the found ideal element. The algorithm terminates if all columns are either transformed by Gaussian elimination or cancelled. Then one has not only found N columns corresponding to polynomials q_j with $q_j(y_k) = 0$ for $k < j$ but also a Gröbner basis G of I. By construction, the supports $S(q_j)$ are $\{t_1, \ldots, t_k\}$, $k = 1, \ldots, N$. The q_j's allow an easy computation of the minimal polynomial p^* solving (9).

The difference to de Boor and Ron (1992) is that as a byproduct a Gröbner basis is obtained, allowing e.g. a description of all other polynomials solving

(9), and that by the cancellation of columns less columns than by de Boor and Ron are manipulated, if the algorithm is stopped immediately after having found the last q_N. On the other hand, de Boor and Ron say that the basis $\{t_1, \ldots, t_N\}$ is not very convenient (for numerical purposes). In fact, they show that an other space generated by homogeneous polynomials h_i with $deg(h_i) = deg(t_i), i = 1, \ldots, N$, is an interpolation space and present in their paper and in earlier ones cited there some good properties of this alternative space.

However,one can also investigate the space $span(\{t_1, \ldots, t_N\})$ using ideal theory. The Hilbert function is another feature intimately connected with Gröbner bases and already considered by Buchberger (1965). Let d_k denote the dimension of the space of all $p \in span(\{t_1, \ldots, t_N\})$ of degree k or less. Then d_k equals the Hilbert function $H(k, I)$. This connection gives immediately the numbers d_k for special settings, where the Hilbert function is known a priori. For instance let $y_1, \ldots, y_N$ be the common zeros of polynomials $p_1, \ldots, p_n$ of degrees $m_1, \ldots, m_n$ resp. Then $N \leq m_1 \cdots m_n$ and if equality holds, then Hilbert's so called Syzygienformel gives explicit expressions for d_k, esp. $max\ deg(t_i) = \sum_{k=1}^{n}(m_k - 1)$ attained by exactly one power product t_j, as shown by Möller (1976).

3.2 Hermite interpolation

If in the equations (9) also values of partial derivatives of p are prescribed, then one speaks in the univariate case $n = 1$ of Hermite-Birkhoff or of Hermite interpolation, depending on the solution set of homogeneous interpolation problem (all prescribed values $c_i = 0$). If this set is an ideal, then it is a Hermite problem, otherwise a Hermite-Birkhoff one. We use this definition also in the multivariate case.

For characterizing the cases, where the homogeneous interpolation problem is solved by the polynomials of an ideal, we need a few technical notions, see Möller and Stetter (1995) and Marinari et al. (1993). In this section, we need in addition $K = \mathbb{C}$.

Definition. For arbitrary $\alpha := (\alpha_1, \ldots, \alpha_n) \in \mathbb{N}_o^n$ and $p \in \mathbf{P}$ let

$$\partial_\alpha(p) := \frac{1}{\alpha_1! \cdots \alpha_n!} \frac{\partial^{\alpha_1 + \ldots + \alpha_n} p}{\partial x_1^{\alpha_1} \cdots \partial x_n^{\alpha_n}}$$

and $\partial_\alpha(p) := 0$ if one α_k is negative. $\mathbf{D}$ is the vector space generated by all operators ∂_α, $\alpha \in \mathbb{N}_o^n$. The shift operator σ_α, $\alpha \in \mathbb{N}_o^n$, is defined on $\mathbf{D}$ by

$$\sigma_\alpha(\sum_\beta c_\beta \partial_\beta) := \sum_\beta c_\beta \partial_{\beta - \alpha}.$$

A subspace U of $\mathbf{D}$ is *closed*, if it has a finite positive dimension and if

$$\alpha \in \mathbb{N}_o^n,\ D \in U \implies \sigma_\alpha(D) \in U.$$

In the univariate case this means, that if a differential operator of order m belongs to a closed subspace U, then applying σ_m one gets $\partial_0 = id \in U$, applying σ_{m-1} and using the linearity of U, one gets $\partial_1 \in U$, etc. Hence $U = span(\partial_0, \partial_1, \ldots, \partial_d)$ if U is $d+1$-dimensional. For $n > 1$ one can show similarly at least, that $\partial_0 = id$ is in every closed subspace.

Definition. Let U_i be a closed subspace of $\mathbf{D}$ with basis $\{D_1^{(i)}, \ldots, D_{m_i}^{(i)}\}$, $i = 1, \ldots, N$. Let $\{y_1, \ldots, y_N\} \subset \mathbb{C}^n$. Then the set of functionals

$$L_j^{(i)} : \mathbf{P} \longrightarrow \mathbb{C},\ f \mapsto D_j^{(i)}(f)(y_i),\ j = 1, \ldots, m_i,\ i = 1, \ldots, N, \qquad (3.2)$$

is called a *dual basis* generating $\mathcal{A} := \{p \in \mathbf{P} \mid L_j^{(i)}(p) = 0\ \forall j\ \forall i\}$.

As shown by Marinari et al. (1993) this $\mathcal{A}$ is an ideal with variety $\{y_1, \ldots, y_N\}$ and every ideal with a finite variety has a dual basis generating it.

The Hermite interpolation problem. Let $\{L_1, \ldots L_N\}$ be a given dual basis and $c_1, \ldots, c_N \in \mathbb{C}$. Find all polynomials p (or a minimal p resp.) satisfying

$$L_k(p) = c_k \text{ for all } k = 1, \ldots, N. \qquad (3.3)$$

Thm. 2 and its corollary can be carried over literally. The only difference is in the proof of the corollary, where the linear independence of the functionals $p \mapsto p(y_i)$ is shown by constructing suitable polynomials. Here the linear independence of the functionals (10) is proved by Marinari et al. (1993).

The construction using the matrix V works as well, if the entries of its k-th row are the numbers $L_k(t)$, t a power product. De Boor and Ron (1992) pointed out, that their construction works even with more general functionals L_k. The construction which gives also a Gröbner basis, naturally restricts to cases where the solution set of the homogeneous problem $L_k(p) = 0, k = 1, \ldots, N$, is an ideal, see Marinari et al. (1993).

For alternative constructions of solutions for (11) see for instance Sauer and Xu (1995b) and the references quoted there.

4 Numerical integration and differentiation

In numerical integration one replaces an integral

$$J(f) := \int_B f(x_1, \ldots, x_n) dB,$$

where f is an integrable function defined on $B \subseteq \mathbb{R}^n$, by a finite sum

$$C(f) := \sum_{k=1}^{N} a_k f(y_k).$$

In this context, the points y_k are called nodes and the $a_k \in \mathbb{R}$ weights. Often one restricts the consideration to functions which are defined only on B, hence admits only $y_k \in B$; but also $y_k \in \mathbb{R}^n$ are admitted, if the functions under consideration are defined on $\mathbb{R}^n$, see for instance the monograph of Sobolev and Vaskevich (1997).

C is called *a cubature formula of degree d for* J, if

$$J(p) = C(p) \text{ for all } p \in \mathbf{P} \text{ of degree} \leq d. \tag{4.1}$$

One expects that having a sequence of cubature formulas of increasing degrees, then their values $C(f)$ converge to $J(f)$ for any suitable f.

The construction of cubature formulas amounts in finding nodes and weights such that (12) is satisfied. Of course, the validity of (12) has only to be tested for a basis $\{b_1, \ldots, b_M\}$ of the space $\mathbf{P}_d$, the space of polynomials of degree $\leq d$, leading to a set of polynomial equations for the weights and the (components of the) nodes. Here Gröbner bases can be applied as pointed out in section 2.

There is another connection to Gröbner bases. We follow here the exposition of Möller (1987). For this purpose, we need only that J is a functional, hence we may also admit, that it is a partial differential operator $\sum_\alpha c_\alpha \partial_\alpha$ evaluated at a fixed point $a \in \mathbb{R}^n$ or even a linear combination of partial differential operators evaluated at different points. Then C can be interpreted as a numerical differentiation formula of degree d if (12) holds.

Using point evaluation functionals $L_k(p) := p(y_k), k = 1, \ldots, N$, (12) means that J is a linear combination of the L_k's considered as functionals defined on $\mathbf{P}_d$. Then a standard argument from functional analysis says, that (12) is equivalent to

$$p \in \mathbf{P}_d,\ L_k(p) = 0, k = 1, \ldots, N \implies J(p) = 0. \tag{4.2}$$

Theorem 3. *Let* J *be a functional defined on* $\mathbf{P}_d$ *and* $y_1, \ldots, y_N$ *be points in* $\mathbb{R}^n$. *Then there are* $a_1, \ldots, a_N \in \mathbb{R}$, *such that*

$$J(p) = \sum_{k=1}^{N} a_k p(y_k) \quad \text{for all } p \in \mathbf{P}_d,$$

iff the ideal $I := \{p \in \mathbf{P} \mid p(y_k) = 0,\ k = 1, \ldots, N\}$ *has an H-basis* $\{p_1, \ldots, p_s\}$ *with*

$$g \cdot p_k \in \mathbf{P}_d \implies J(g \cdot p_k) = 0,\ k = 1, \ldots, s.$$

(An H-basis $\{p_1, \ldots, p_s\}$ is a polynomial set with the property, that whenever a representation $f = \sum_k g_k p_k$ exists, there is also one for f, where $max_k\ deg(g_k p_k) \leq deg(f)$. Every ideal has an H-basis.)

Proof. By the equivalence of (12) and (13) it is sufficient to show $p \in I \cap \mathbf{P}_d \Longrightarrow J(p) = 0$. Using the H-representation of $p \in I \cap \mathbf{P}_d$ the assertion follows by the linearity of J. □

The theorem suggests a procedure for constructing numerical intergration and differentiation formulas.

Let H_d denote the set of polynomials p satisfying

$$g \cdot p \in \mathbf{P}_d \Longrightarrow J(g \cdot p) = 0. \tag{4.3}$$

Having a finite subset $F \subset H_d$, we start with some linear combinations $g_i := \sum_{f \in F} c_{if} \cdot f$ with undetermined coefficients c_{if}. Then we construct a Gröbner basis with respect to a term ordering $\prec$, which is compatible with the degree-semiordering,

$$deg(f) < deg(g) \Longrightarrow f \prec g.$$

Each polynomial obtained during the computations has to fulfill (14). This imposes conditions on the c_{if}. At the end, one has to test whether the ideal I generated by the Gröbner basis elements is a zero dimensional real ideal, i.e. $V \subset \mathbb{R}^n$ finite and $I = \{f \in \mathbf{P} \mid f(y) = 0 \ \forall y \in V\}$. Using the degree compatibility, one easily finds, that the Gröbner basis is as well an H-basis. Hence all assumptions of theorem 4 are satisfied and the common zeros of the Gröbner basis elements constitute the set of nodes of a numerical formula for J.

The construction has been tested for moderate degrees d and variables n, see Möller (1987). It is a challenging task, to use it for the construction of higher degree formulas or those with greater n.

5 Gröbner bases in floating point arithmetics

An infinitesimal perturbation of the coefficients of a polynomial set leads in general to a polynomial set which has no common zero although the unperturbed set may have some. This means, say by Hilbert's Nullstellensatz, that the ideal generated by the perturbed polynomials contains the constant polynomial 1, i.e. it is the whole polynomial ring, whereas the unperturbed ideal may be a nontrivial one. Hence Gröbner bases depend discontinuously on the polynomial coefficients of the ideal generating set. The computation of a Gröbner basis using a floating point arithmetic requires therefore precautions. We will present here briefly two contributions which deal with this computation.

In the univariate case $n = 1$ every ideal is a principal ideal (f), where the polynomial f is (up to normalization) the greatest common divisor, gcd for short, of all polynomials in the ideal. The set $\{f\}$ is the (reduced)

Gröbner basis of the ideal. Hence Gröbner bases can be computed using gcd computations, which can be reduced to (a sequence of) gcd computations for polynomial pairs using the recursion

$$gcd(f,g,h) = gcd(f, gcd(g,h)).$$

This way of computing a Gröbner basis can be seen as a variant of Buchberger's algorithm where for the reducing of S-polynomials $S(f,g)$ modulo a polynomial set the polynomials f and g have a higher priority than the others in the set. The gcd computations in floating point arithmetic are investigated by Corless et al. (1995).

They rely heavily on the fact, that for univariate polynomials one knows a priori a finite vector space, which contains the (reduced) Gröbner basis of an ideal (f_1, f_2). From algebra, it is known that the gcd is a linear combination of the polynomials $x^i \cdot f_1$, $0 \leq i < deg(f_2) =: d_2$, and $x^i \cdot f_2$, $0 \leq i < deg(f_1) =: d_1$. These $d_1 + d_2$ polynomials are of degree $\leq d_1 + d_2 - 1$. Arranging their coefficient vectors in a $(d_1 + d_2) \times (d_1 + d_2)$ -matrix, one gets the Sylvester matrix S_{f_1,f_2} with rank equal to $d_1 + d_2 - d$, where d denotes the degree of the gcd. This rank and hence the degree of the gcd can be computed in a numerically safe way by using the singular value decomposition. Corless et al. (1995) showed the following.

Lemma. *Let $S_{f_1,f_2} = U\Sigma V^T$ be the singular value decomposition of S_{f_1,f_2}, i.e. U and V orthogonal and Σ a diagonal matrix $diag(\sigma_1, \ldots, \sigma_r, 0, \ldots, 0)$ with $\sigma_1 \geq \sigma_2 \geq \ldots \geq \sigma_r > 0$. If k_o is the maximum k with*

$$\sigma_k > \varepsilon\sqrt{d_1 + d_2}, \ \varepsilon \geq \sigma_{k+1},$$

then there are two polynomials g_1, g_2 with $||g_i - f_i||_2 \leq \varepsilon$, $i = 1, 2$, having a gcd of degree $d_1 + d_2 - k_o$. Here $||\ .\ ||_2$ stands for the ℓ_2-norm of the corresponding coefficient vector.

This lemma is useful in the numerical standard case, when one has the polynomials f_1, f_2 only by approximate coefficient value, resulting in two (disturbed) polynomials $rd(f_1)$, $rd(f_2)$, and there is no other polynomial pair nearer with a gcd of degree greater than $deg(gcd(f_1, f_2))$. For the numerical computation of the gcd the authors give no real recommendation. They discuss briefly the usual Euclidean algorithm (or a variant of it), the computation of the row echelon form of S_{f_1,f_2}, (its last nonzero row is the coefficient vector of the gcd,) a minimization technique, and Lazard's method using u-resultants.

In the multivariate case, there is usually no finite vector space known a priori which contains the Gröbner basis elements. However, if one has a Gröbner basis of a zero dimensional ideal, then one can investigate the neighbouring ideals generated by polynomials obtained by perturbing the Gröbner basis coefficients. This can be done by using the dual basis concept, see section

3.2. For the instance that the Gröbner basis consists of n polynomials, the so called *complete intersection case*, this investigation was done by Stetter (1997).

Let $L_1, \ldots, L_N$ be a dual basis of an ideal I and let the power products $t_1, \ldots, t_N$ be chosen as in lemma 2, then the $N \times N$-Gram matrix $(L_i(t_j))$ has full rank. Stetter points out, that a nearly degenerate Gram matrix causes large Gröbner basis coefficients, if the leading coefficients are normalized to 1. Normalizing always the coefficients such that the largest one to 1, this means, that the leading coefficient of one Gröbner basis element tends to 0, the more one approaches the degenerate case. Stetter proposes a stabilized Gröbner basis computation by splitting the functionality of the leading power product. On the one hand, it is used to keep fixed the set of terms $t_1, \ldots, t_N$ which are not divisible by leading power products. On the other hand, for reducing and new S-polynomial computations, the leading power product with very small coefficient is replaced by the nearest power product with a coefficient distinguishable different from 0, where nearest means nearest to the greatest power product in the ordered set of power products of a polynomial. The practical test of this approach is under development.

References

Alonso, M.E., Becker, E., Roy, M.F., Wörmann, Th. (1996) : Zeros, multiplicities and idempotents for zerodimensional systems. In: Gonzalez-Vega, L. and Recio, T. (eds.): Algorithms in Algebraic Geometry and Applications. Birkhäuser Verlag, Basel Boston Berlin, pp. 1 - 16. (Progress in Math. 143)

Auzinger, W., Stetter, H.J. (1988): An elimination algorithm for the computation of all zeros of a system of multivariate polynomial equations. In: Proceedings of the Conference on Numerical Mathematics, Singapore 1988, Birkhäuser Verlag, Basel Boston Berlin, pp. 11 – 30. (International Series in Numerical Mathematics 86)

Backelin, J., Fröberg, R. (1991): How we proved, that there are exactly 924 cyclic 7-roots. In: Proceedings of the ACM-SIGSAM International Symposium on Symbolic and Algebraic Computation, ISSAC '91, Bonn. Association of Computing Machinery, New York, pp. 103 - 111.

Becker, Th., Weispfenning, V. (1993): Gröbner bases: A Computational Approach to Commutative Algebra. Springer, Berlin Heidelberg New York Tokyo.

Björck, G., Fröberg, R. (1994): Methods to 'divide out' certain solutions from systems of algebraic equations applied to find all cyclic 8-roots. In: Proceedings of the 21st Nordic Congr. of Math. Luleå1994, pp. 57 – 70. (Springer LN in Pure and Appl. Math. 156) de Boor, C., Ron, A. (1992): Computational aspects of polynomial interpolation in several variables. Math. Comp. 58: 705 - 727.

Buchberger, B. (1965): On finding a vector space basis of the residue class ring modulo a zero dimensional polynomial ideal (German). PhD Thesis, Univ of Innsbruck, Austria.

Buchberger, B. (1970): An algorithmical criterion for the solvability of algebraic systems of equations (German). Aequationes mathematicae 4: 374 - 383.

Buchberger, B. (1998): Introduction to Gröbner bases. This volume.

Buchberger, B., Möller, H.M. (1982): The construction of multivariate polynomials with preassigned zeros. In: Computer Algebra, EUROCAM'82, Marseille. Springer Verlag, Berlin Heidelberg New York, pp. 24 - 31. (Lecture Notes in Comp. Sci. 144)

Canny, J.F., Manocha, D. (1993): Multipolynomial resultant algorithms. J. Symb. Comp. 15: 99 - 122.

Corless, R.M. (1996): Editor's Corner: Gröbner bases and Matrix Eigenproblems. SIGSAM Bull. 30, 4: 26 - 32.

Corless, R.M., Gianni, P.M., Trager, B.M., Watt, S.M. (1995): The singular value decomposition for polynomial systems. In: Proceedings of the ACM-SIGSAM International Symposium on Symbolic and Algebraic Computation, ISSAC '95, Montreal. Association of Computing Machinery, New York, pp. 195 - 207.

Gatermann, K. (1996): Semi-invariants, equivariants, and algorithms. J. AAECC 6: 105 - 124.

Goedecker, S. (1994): Remark on algorithms to find roots of polynomials. SIAM J. Sci. Comp. 15: 1059 - 1063.

Gräbe, H.-G. (1997): Minimal primary decomposition and factorized Gröbner bases. J. AAECC 8: 265 - 278.

Lazard, D. (1991): A new method for solving algebraic systems of positive dimension. Discr. App. Math. 33: 147 - 160.

Lazard, D. (1992): Solving zero-dimensional algebraic systems. J. Symb. Comput. 13: 117 - 131.

Marinari, M.G., Möller, H.M., Mora,T. (1993): Gröbner bases of ideals defined by functionals with an application to ideals of projective points. J. AAECC 4: 103 - 145.

Melenk. H., Möller, H.M., Neun, W. (1989): Symbolic solution of large stationary chemical kinetics problems. Impact of Computing in Science and Engeneering 1: 138 - 167.

Möller, H.M. (1976): Hermite interpolation in several variables using ideal-theoretic methods. In: Schempp, W. and Zeller, K. (eds.) : Constructive Theorie of Functions of Several Variables. Springer Verlag, Berlin Heidelberg New York, pp. 155 - 163. (Lecture Notes in Mathematics 571)

Möller, H.M. (1987): On the construction of cubature formulae with few nodes using Groebner bases. In: Fairweather, G. and Keast, P. (eds.): Numerical Integration, Recent Developments, Software, and Applications. D. Reidel Publ. Comp., Dordrecht Boston Lancaster Tokyo, pp. 177 - 192. (NATO ASI Series C, Vol. 203)

Möller, H.M. (1993): On decomposing systems of polynomial equations with finitely many solutions. J. AAECC 4: 217 - 230.

Möller, H.M., Stetter, H.J. (1995):: Multivariate polynomial equations with multiple zeros solved by matrix eigenproblems. Numer. Math. 70: 311 -329.

Sauer, Th., Xu, Y. (1995a): On multivariate Lagrange interpolation. Math. Comp. 64: 1147 - 1170.

Sauer, Th., Xu, Y. (1995b): On multivariate Hermite interpolation. AICM 4: 207 - 259.

Sobolev, S.L., Vaskevitch V.L. (1997): The Theory of Cubature Formulas. Kluwer Acad. Publ., Dordrecht. (Mathematics and its applications 415)

Stetter, H.J. (1995): Matrix Eigenproblems at the heart of polynomial system solving. SIGSAM Bull. 30, 4: 22 - 25.

Stetter, H.J. (1997): Stabilization of polynomial systems solving with Groebner bases. In: Proceedings of the ACM-SIGSAM International Symposium on Symbolic and Algebraic Computation, ISSAC '97, Maui. Association of Computing Machinery, New York.

Sturmfels, B. (1993): Algorithms in Invariant Theory. Texts and Monographs in Symb. Comp., Springer Verlag, Wien New York.

Trinks, W. (1978): Über B.Buchbergers Verfahren, Systeme algebraischer Gleichungen zu lösen. J. of Number Theory 10: 475 - 488.

Ueberhuber, Chr. (1997): Numerical Computation, vol. II. Springer Verlag, Heidelberg Berlin New York.

H. Michael Möller, Universität Dortmund, Fachbereich Mathematik, Vogelpothsweg 87, D-44221 Dortmund, Germany, moeller@math.uni-dortmund.de

Gröbner Bases and Statistics

Lorenzo Robbiano [1]

Abstract

This survey describes how to use methods of Algebraic Geometry and Commutative Algebra to study some problems arising in Design of Experiments, a branch of Statistics.

Contents

1 Introduction

Whenever two branches of Science meet it is a great time for research. Sometimes such events are unexpected, sometimes they are more natural. In this survey I will try to set up a foundation for a fruitful interaction between Design of Experiments (DoE) on one side and Commutative Algebra, Algebraic Geometry and Computer Algebra on the other. DoE is a branch of Statistics, and in its history algebraic methods are not rare, so this kind of interaction falls into the category of natural events. Furthermore, it is now clear how strong was, and still is, the influence of Computer Algebra methods in Commutative Algebra and Algebraic Geometry: for instance Gröbner basis theory, which originated with the fundamental work by Buchberger (see Buchberger (1965), Buchberger (1970) and Buchberger (1985)), is nowadays recognized to be a basic tool applicable in several sectors of Science. From now on we can say that this influence extends into the realm of DoE.

Let us have a look at the main concepts in DoE by discussing an example. Suppose that a firm wants to introduce a new product, say a portable telephone, into the market. They need to obtain useful *a priori* information from the potential customers. Let's say that the strategy of the firm is to test the reaction to five features of the telephone: colour, shape, weight, material and price. Let us associate the variables C with colour, S with shape,

[1] Partially supported by the Consiglio Nazionale delle Ricerche (CNR).

W with weight, M with material, P with price and suppose that for each variable, three values are chosen, which can be coded as $\{0,1,2\}$. The interpretation is, for instance, that $C = 0$ means red, $C = 1$ means green, $C = 2$ means black; $W = 0$ means weight in between 200 and 250 grams, and so on. Therefore the particular characteristics of a portable telephone are codified by a *point* whose coordinates are five integers chosen from $\{0,1,2\}$. So the point $(0,0,0,0,0)$ could correspond to a red, rounded, light, plastic, low priced telephone, while the point $(0,1,1,0,2)$ could correspond to a red, slanted, medium weight, plastic, high priced telephone, and so on. A selected set of potential customers should be asked to rate the various possibilities, i.e. the various points, on a scale from 0 to 10, say.

Similar examples can be drawn from several fields of application, ranging from industrial design to chemistry, marketing, and finance. The method is always to draw information from experiments, and the goal can be to gain information from the market, to detect failures in a complex system, to find the correct blend of substances in a chemical product and so on (see for instance Box *et al.* (1978)). The nice feature of these examples is that all of them can be modeled in the same way, that is to say with a finite set of *experimental treatments* or *points*, which are described by their coordinates. The set of all these points is called a *Design.* A complete set where all combinations of values are allowed, such as the 3^5 points in our example, is a product set called in Statistics a *Full* (or *Full Factorial*) *Design* (see Definition 2.17).

Let us go back to our example. We call $\mathcal{D}$ the set of the $3^5 = 243$ points, and suppose for a moment that we can ask the customers to rate all of them. Having done that, to each point we associate for instance the mean value of the answers, and we get a function $y : \mathcal{D} \to \mathbb{Q}$ (the choice of the mean value as an example is motivated by the fact that we are going to treat models without replications). We observe that every function from a finite set of points to a field is a *polynomial function* (see Corollary 2.14), hence $y = f(C, S, W, M, P)$ with f a polynomial in $\mathbb{Q}[C, S, W, M, P]$. Such a polynomial function is called a *polynomial model* for our design. It encodes all the relevant information, and from it the firm can deduce fundamental knowledge for the final decision on the characteristics of the portable telephone.

However, in this case (as well as in other fundamental situations) there is an obstacle: no potential customer would be willing to rate 243 cases. The problem is to determine a suitable subset $\mathcal{F}$ of $\mathcal{D}$, called a Fraction, from which we can equally well reconstruct a good model. For instance it would be nice to be able to identify the model, simply from rating a *small* subset of $\mathcal{D}$.

Before we go on let's introduce some commutative algebra. The polynomial $X(X-1)(X-2) = X^3-3X^2+2X$ vanishes on the set $\{0,1,2\}$. This fact can be interpreted as saying that the polynomial functions X^3, X^2, X are linearly dependent over $\{0,1,2\}$. In fact the function X^3 takes the same values as $3X^2 - 2X$, which, in statistical jargon, is expressed by saying that X^3 and $3X^2 - 2X$ are *confounded* by $\{0,1,2\}$.

Also in the multivariate case a full design has *canonical confounding polynomials* (or *equations*) (see Definition 2.19) and a canonical basis of power-products (see Definition 2.20), $\mathcal{O}(\mathcal{D})$. In our example (henceforth we use X_1, X_2, X_3, X_4, X_5 instead of C, S, W, M, P) the canonical confounding polynomials are $f_i := X_i(X_i - 1)(X_i - 2)$, $i := 1, \ldots, 5$. All together they generate $I(\mathcal{D})$, the *defining ideal* of $\mathcal{D}$ (see Lemma 2.2). The canonical basis $\mathcal{O}(\mathcal{D})$ is the set of 243 power-products $\{X_1^{a_1} X_2^{a_2} \cdots X_5^{a_5} \mid a_i = 0, 1, 2;\ i := 1, \ldots, 5\}$.

Given a full design $\mathcal{D}$, a *fraction* $\mathcal{F}$ of it is simply a subset, but from the algebraic point of view, its description is no longer canonical. Of course the ideal $I(\mathcal{F})$, which defines the given fraction, contains $I(\mathcal{D})$, the defining ideal of $\mathcal{D}$, simply because every polynomial which vanishes on $\mathcal{D}$ automatically vanishes on $\mathcal{F}$. So, there are more confounding polynomials, which have to be joined to the canonical polynomials of $\mathcal{D}$ (see Definition 4.2), and the key point is to be able to draw information about a given fraction from its confounding polynomials. With the purpose of better popularizing the subject, we describe in Section 3 the best known algorithm for computing the defining ideal of any given finite set of points, the Buchberger-Möller Algorithm. Then we show in Section 4 that for every fraction there exists a unique *canonical confounding polynomial* (see Theorem-Definition 4.5), though it is often hard to extract further relevant information from it. The first attempt to solve the problem of identifying a model from a fraction can be done easily by linear algebra methods (see Theorems 4.12 and 4.14). But we want much more. For instance, if we go back to our guiding example, we may envisage the following potential situations. The first case is that the firm wants to use a particular fraction $\mathcal{F}$ of $\mathcal{D}$. The reason could be that they have already run previous experiments for similar purposes. In this case the question is: given a fraction, what are the models identifiable by it? This is our PROBLEM 1 of Section 5. The second, and most important case is that the firm already knows the *support* of the model, i.e. the set of the power-products actually occurring in the polynomial which describes the model. It is clear that such support is contained in $\mathcal{O}(\mathcal{D})$, so, to identify the model, it suffices to identify the coefficients. For instance the coefficient of X_1 quantifies the importance that the customers give to the colour of the telephone. The coefficient of X_1X_2 quantifies the importance given to the interaction between colour and shape and so on. Indeed in Statistics the indeterminates are called *main effects* and the other power-products, which involve more than one indeterminate, *interactions*. Suppose that via some previously acquired knowledge, the firm knows that only the constant and the main effects are relevant; then the model has the shape $y = a_0 + a_1X_1 + a_2X_2 + a_3X_3 + a_4X_4 + a_5X_5$. One has to identify only six coefficients. A first basic remark (see Theorem 2.6 and Corollary 2.12) shows that it suffices to consider fractions with six points, but then the question is: what are the fractions of $\mathcal{D}$, which identify the six coefficients? This is our PROBLEM 2 of Section 5, one of the most important in DOE.

At this point Gröbner bases are most welcome. It was in Pistone and Wynn (1996) that for the first time a bridge was built between the two theories, and PROBLEM 1 was introduced and partially solved. More solutions to these problems, in particular to PROBLEM 2, were given in Caboara and Robbiano (1997) and Robbiano and Rogantin (1998), where Gröbner bases play a fundamental rôle.

Section 5 is entirely devoted to the discussion of these problems, including our PROBLEM 3, which is a variant and was first tackled in Caboara *et al.* (1997). We explain how Gröbner bases can be used to clarify many aspects, and also what their limits are. Needless to say, there are many questions left open; this survey reports on some of them and includes some new results, for instance Theorem 5.16. But the main emphasis here is to clarify the importance of using algebraic methods in DoE.

I hope that this survey may be used by both statisticians and algebraists as a guide to the exploration of this beautiful field.

Here let me mention that many detailed examples are included, and it is my pleasure to say that all the computations were performed with our system CoCoA (see Capani *et al.* (1995) and Capani *et al.* (1997)); in particular a package written in CoCoAL (the high level programming language of CoCoA) by Caboara, called `statfamilies`, was intensively used to produce Example 5.5 and others. For interested researchers, the package, as well as the system CoCoA, are available at the URL `http://cocoa.dima.unige.it`.

Special thanks are due to G. Pistone, E. Riccomagno, M.P. Rogantin and H.P. Wynn, who introduced me to this beautiful subject. Finally, warm thanks are due to J. Abbott and M.P. Rogantin, who read the first draft and supplied valuable help and advice.

2 From Design of Experiments to Commutative Algebra

We start by recalling some basic facts from commutative algebra and algebraic geometry. Throughout the section we use R for $k[X_1, \dots, X_n]$. The first thing to mention is that every polynomial in R can be interpreted as a function from k^n to k. Indeed, to every element $P \in k^n$ one associates $f(P) \in k$.

Definition 2.1 *Let $\mathcal{X} \subset k^n$. Then we call*

$$I(\mathcal{X}) := \{f \in R \mid f(P) = 0 \quad \forall P \in \mathcal{X}\}$$

Lemma 2.2 *The following facts hold*

1) $I(\mathcal{X})$ is an ideal, which is called the **defining ideal** *of $\mathcal{X}$.*

2) $I(\mathcal{X})$ is radical.

PROOF. The first assertion is straightforward, while the second follows from the remark that $f^m(P) = 0$ implies $f(P) = 0$. □

Example 2.1

An ideal of this type is obtained when $\mathcal{X} := \{P\}$, and $P \in k^n$ is a rational point, i.e. a point with coordinates in k. If $P := (a_1, a_2, \ldots, a_n)$, then $I(P) := I(\{P\}) = (X_1 - a_1, X_2 - a_2, \ldots, X_n - a_n)$. And clearly $R/I(P) \cong k$.

The situation relevant to DoE is when a finite set of distinct points $P_1, P_2, \ldots, P_s \in k^n$ is given and $\mathcal{X} := \{P_1, P_2, \ldots, P_s\}$.

Lemma 2.3 *Let* $P_1, P_2, \ldots, P_s \in k^n$ *and* $\mathcal{X} := \{P_1, P_2, \ldots, P_s\}$. *Then*

$$I(\mathcal{X}) = I(P_1) \cap I(P_2) \cap \ldots \cap I(P_s)$$

PROOF. It follows immediately from the definitions. □

Although this fact suggests a way of computing such ideals, there is a much better method based on the Buchberger-Möller Algorithm, which will be sketched in Section 3. In the rest of this section we take a closer look at rational points.

Definition 2.4 *Let A be a ring, and I, J ideals in A. Then I and J are said to be* **comaximal** *if $I + J = A$.*

Lemma 2.5 (Chinese Remainder Theorem) *Let A be a ring, $I_1, I_2, \ldots, I_s$ ideals in A and consider the canonical homomorphism*
$\Phi : A/(\cap_j I_j) \to \oplus_j (A/I_j)$. *Then*

1) Φ is injective

2) If the ideals I_j are pairwise comaximal, *Φ is an isomorphism.*

PROOF. Let $J := \cap_j I_j$. Then Φ is defined by
$\Phi(a \bmod J) := (a \bmod I_1, \ldots, a \bmod I_s)$, hence it is clearly injective.

Let us prove 2). First we observe that if we call $J_i := \cap_{r \neq i} I_r$, then for every i the two ideals I_i and J_i are comaximal. This follows from the fact that if a maximal ideal contains J_i, then it must contain at least one of the I_r. Therefore, for every $i := 1, \ldots, s$ there exists $\alpha_i \in I_i$, $\beta_i \in J_i$, such that $\alpha_i + \beta_i = 1$. Now let $m := (m_1 \bmod I_1, \ldots, m_s \bmod I_s) \in \oplus_j (A/I_j)$. Then it is easy to see that $m = \Phi(\sum m_i \beta_i)$. □

Theorem 2.6 *Let $P_1, P_2, \ldots, P_s$ be distinct points in k^n and*
$\mathcal{X} := \{P_1, P_2, \ldots, P_s\}$. *Then*

$$R/I(\mathcal{X}) \cong k^s$$

PROOF. We have already said that $R/I(P) \cong k$ for every rational point P. The conclusion follows from Lemma 2.3 and Lemma 2.5, since the $I(P_i)$, being maximal ideals, are pairwise comaximal. □

We recall the following facts (see for instance Cox *et al.* (1992) from the theory of Gröbner bases.

Definition 2.7 *Let $f \in R$. Then we denote by* Supp(f), *and we call it the* **support** *of f, the set of power-products with non-zero coefficient, when f is written as a linear combination of power-products.*

Example 2.2

Let $f(X,Y,Z) := X^2 - 4XY + \frac{1}{2}Z - 3$. Then $\text{Supp}(f) = \{X^2,\ XY,\ Z,\ 1\}$

Definition 2.8 *Let $\{X_1,\ldots,X_n\}^*$ be the set of power-products in the indeterminates $X_1,\ldots,X_n$. Then a* **term-order** *σ on $\{X_1,\ldots,X_n\}^*$ is a total order (denoted by $<_\sigma$) such that*

a) $T, T', S \in \{X_1,\ldots,X_n\}^*$, $T <_\sigma T'$ imply $T \cdot S <_\sigma T' \cdot S$

b) $1 <_\sigma T$ for every $T \in \{X_1,\ldots,X_n\}^*$

Example 2.3

Examples of term-orders are Lex, DegLex, DegRevLex. A classification of term-orders is described in Robbiano (1985) and Robbiano (1986).

Definition 2.9 *Let σ be a term-order and I an ideal of R. Then we denote by* $\mathrm{Lt}_\sigma(I)$ *the ideal generated by the set* $\{\mathrm{Lt}_\sigma(f) \mid f \in I,\ f \neq 0\}$, *where* $\mathrm{Lt}_\sigma(f)$ *is the σ-leading term of f. We denote by $\mathcal{O}_\sigma(I)$ the set of power-products, which are not multiples of any of the σ-leading terms of the elements of I.*

Definition 2.10 *Let $\mathcal{O} \subseteq \{X_1,\ldots,X_n\}^*$. We say that $\mathcal{O}$ is a* **standard set of power-products** *if $T \in \mathcal{O}$ and T' divides T implies $T' \in \mathcal{O}$, i.e. all divisors of an element of $\mathcal{O}$ are also in $\mathcal{O}$.*

Proposition-Definition 2.11 *With the above assumptions we have*

1) $\mathcal{O}_\sigma(I)$ is a basis of R/I as a k-vectorspace.

2) It is a standard set of power-products.

We call it the **standard set of power-products associated to σ and I.**

Corollary 2.12 *Let $P_1, P_2, \ldots, P_s$ be distinct points in k^n and denote by $\mathcal{X} := \{P_1, P_2, \ldots, P_s\}$. Let σ be a term-order and $\mathcal{O}_\sigma(I(\mathcal{X}))$ the standard set of power-products associated to σ and $I(\mathcal{X})$. Then $\#(\mathcal{O}_\sigma(I(\mathcal{X}))) = s$ and $\mathcal{O}_\sigma(I(\mathcal{X}))$ is a basis of $R/I(\mathcal{X})$ as a k-vectorspace.*

Example 2.4

Let $n := 3$ and call the indeterminates X, Y, Z. Then let $P_1 := (\frac{1}{3}, 5, \frac{1}{2}), P_2 := (1, 2, \frac{1}{5}), P_3 := (2, 2, 3), P_4 := (2, 2, 2)$ and $\mathcal{X} := \{P_1, P_2, P_3, P_4\}$.

If $\sigma :=$ DegRevLex then the reduced σ-Gröbner Basis of $I(\mathcal{X})$ is
$\{X^2 - 3X - \frac{10}{27}Y + \frac{74}{27},$
$XY - 2X - \frac{1}{3}Y + \frac{2}{3},$
$Y^2 - 7Y + 10,$
$XZ - \frac{1}{5}X + \frac{1}{6}Y - 2Z + \frac{1}{15}$
$YZ - \frac{1}{2}Y - 2Z + 1,$
$Z^2 + \frac{126}{25}X + \frac{31}{20}Y - 5Z - \frac{359}{50}\}$
from which we may easily deduce that $\mathcal{O}_\sigma(I(\mathcal{X})) = \{1,\ X,\ Y,\ Z\}$.

In contrast, if $\sigma :=$ Lex, the reduced σ-Gröbner Basis of $I(\mathcal{X})$ is
$\{X + \frac{155}{189}Z^3 - \frac{1537}{378}Z^2 + \frac{1795}{378}Z - \frac{113}{63},$
$Y - \frac{8}{3}Z^3 + \frac{208}{15}Z^2 - \frac{56}{3}Z + \frac{6}{5},$
$Z^4 - \frac{57}{10}Z^3 + \frac{48}{5}Z^2 - \frac{47}{10}Z + \frac{3}{5}\}$
and $\mathcal{O}_\sigma(I(\mathcal{X})) = \{1,\ Z,\ Z^2,\ Z^3\}$. In both cases $\#(\mathcal{O}_\sigma(I(\mathcal{X}))) = 4$.

The elements β_i used in Lemma 2.5 behave as a canonical basis. When we deal with finite sets of points, this kind of element plays an important rôle, so we are led to the following

Definition 2.13 *Let $P_1, P_2, \ldots, P_s$ be distinct points in k^n and denote by $\mathcal{X} := \{P_1, P_2, \ldots, P_s\}$. A polynomial $s_{P_i} \in R$ is said to be a* **separator** *of P_i from $\mathcal{X}$ if $s_{P_i}(P_i) = 1$ and $s_{P_i}(P_j) = 0$ for every $j \neq i$.*

Corollary 2.14 *Let $P_1, P_2, \ldots, P_s$ be distinct points in k^n, and $b_1, \ldots, b_s \in k$. Then there exists a polynomial $f \in R$ such that $f(P_i) = b_i$ for $i := 1, \ldots, s$. In particular there exist separators.*

PROOF. It is another way of reading Theorem 2.6. □

Definition 2.15 *A polynomial $f \in R$ such that $f(P_i) = b_i$ for $i := 1, \ldots, s$ as above is called an* **interpolator** *or an* **interpolating polynomial** *of the values $b_1, \ldots, b_s$ at the points $P_1, P_2, \ldots, P_s$.*

Corollary 2.16 *Let $P_1, P_2, \ldots, P_s$ be distinct points in k^n and denote by $\mathcal{X} := \{P_1, P_2, \ldots, P_s\}$. Then let $\mathcal{T} \subseteq \mathcal{X}$, $\mathcal{Y} := \mathcal{X} \setminus \mathcal{T}$, and for every $P \in \mathcal{T}$ let s_P be a separator of P from $\mathcal{X}$. Finally let $s_\mathcal{T} := \sum_{P \in \mathcal{T}} s_P$; then $I(\mathcal{Y}) = (I(\mathcal{X}), s_\mathcal{T})$.*

PROOF. Clearly $I(\mathcal{Y}) \supseteq (I(\mathcal{X}), s_\mathcal{T})$. It is also clear that $1 - s_\mathcal{T}$ vanishes on $\mathcal{T}$. Consequently, if $f \in I(\mathcal{Y})$, then $f \cdot (1 - s_\mathcal{T})$ vanishes on $\mathcal{X}$, hence $f - f \cdot s_\mathcal{T} \in I(\mathcal{X})$, and the conclusion follows. □

Now we are ready to take the first important step in the direction of DoE.

Definition 2.17 *Let $E_1, E_2, \ldots, E_n$ be finite subsets of k. Then the cartesian product $E_1 \times E_2 \times \cdots \times E_n \subset k^n$ is a finite set of points called a* **full** *(or* **full factorial***)* **design***. If $\#(E_i) = l_i$ for $i := 1, \ldots, n$, then we say that $E_1 \times E_2 \times \cdots \times E_n$ has levels $(l_1, l_2, \ldots, l_n)$.*

An l^n-design is a full design with $l_1 = l_2 = \cdots = l_n = l$. For instance it is common to speak of 2^5, 3^4 full designs and so on.

Theorem 2.18 *Let $E_1 := \{a_{11}, \ldots a_{1l_1}\}, \ldots, E_n := \{a_{n1}, \ldots a_{nl_n}\}$ be finite subsets of k. Let $f_1 := \prod_{i=1}^{l_1}(X_1 - a_{1i}), \ldots, f_n := \prod_{i=1}^{l_n}(X_n - a_{ni})$. If we denote by $\mathcal{D}$ the full design $E_1 \times \cdots \times E_n$ then*

1) $I(\mathcal{D}) = (f_1, \ldots, f_n)$.

2) $\{f_1, \ldots, f_n\}$ is the reduced σ-Gröbner basis of $I(\mathcal{D})$ for every term-order σ.

PROOF. The proof of 1) is an easy exercise. To prove 2) we observe that the f_i are univariate polynomials, hence they are uniquely ordered independently of σ and their leading terms are pairwise coprime. □

Definition 2.19 *The polynomials $f_1, \ldots, f_n$ defined in Theorem 2.18 are called* **canonical polynomials** *(or* **equations***) of $\mathcal{D}$.*

Definition 2.20 *With the same notation as in Theorem 2.18, we denote by $\mathcal{O}(\mathcal{D})$ the set of power-products $\{X_1^{a_1} X_2^{a_2} \cdots X_n^{a_n} \mid 0 \le a_j < l_j,\ j = 1, \ldots, n\}$. From the same theorem we deduce that it is the standard set of power-products associated to $I(\mathcal{D})$ with respect to* **every** *term-order. Hence we call it* **the standard set of power-products** *of $\mathcal{D}$.*

Example 2.5

The most frequently used full design $\mathcal{D}$ with levels $(l_1, l_2, \ldots, l_n) \in \mathbb{N}^n$ is the set of points $\{0, 1, \ldots, l_1 - 1\} \times \ldots \times \{0, 1, \ldots, l_n - 1\} \subset \mathbb{N}^n$. For such a full design the canonical polynomials are $f_j := X_j\ (X_j - 1) \cdots (X_j - l_j + 1)$ for $j := 1, \ldots, n$.

Example 2.6

Let $\mathcal{D}$ be a 2^3 full design. Then $\mathcal{O}(\mathcal{D}) = \{1, X, Y, Z, XY, XZ, YZ, XYZ\}$, no matter which values of the coordinates of its points are given.

Lemma 2.21 *Let $\mathcal{X} \subset k^n$ be a finite set of points. Then there exists a unique minimal full design $\mathcal{D}_{\mathcal{X}}$ containing $\mathcal{X}$.*

PROOF. For $i := 1, \dots, n$ let $E_i := \{a_{i1}, \dots a_{il_i}\}$ be the set of the ith coordinates of the points in $\mathcal{X}$. Then clearly $\mathcal{D}_{\mathcal{X}} := E_1 \times \cdots \times E_n$ is the solution. □

Theorem 2.22 *Let s be a positive integer, $P_1, P_2, \dots, P_s$ be distinct points in k^n and $\mathcal{X} := \{P_1, P_2, \dots, P_s\}$. Then it is possible to generate $I(\mathcal{X})$ with $n+1$ polynomials.*

PROOF. Let $\mathcal{D}_{\mathcal{X}}$ be the minimal design containing $\mathcal{X}$. Then $I(\mathcal{D}_{\mathcal{X}})$ is generated by n polynomials by Theorem 2.18 and contained in $I(\mathcal{X})$. Now it suffices to apply Corollary 2.16 to the pair $(\mathcal{D}_{\mathcal{X}}, \mathcal{X})$. □

In the next section we discuss an efficient way of computing $I(\mathcal{X})$.

3 Computing Confounding Polynomials: the Buchberger-Möller Algorithm

Before we sketch the Buchberger-Möller Algorithm (BM-Algorithm), we explain the idea informally. We recall (see Definition 2.9) that given a term-order σ and an ideal I, the symbol $\mathrm{Lt}_\sigma(I)$ denotes the ideal generated by the σ-leading terms of the non-zero elements of I. Let us consider a finite set of points $\mathcal{X} := \{P_1, \dots, P_s\} \subset k^n$, let us fix a term-order σ and let $\{X_1, \dots, X_n\}^*$ be the monoid of all power-products in $X_1, \dots, X_n$. Then $\{X_1, \dots, X_n\}^* = \mathcal{O}_\sigma(I(\mathcal{X})) \cup \mathrm{Lt}_\sigma(I(\mathcal{X}))$ and the union is disjoint. On the other hand the set $\mathcal{O}_\sigma(I(\mathcal{X}))$ is finite (it has cardinality s) by Theorem 2.6 and the generators of $\mathrm{Lt}_\sigma(I(\mathcal{X}))$ delineate the border of $\mathcal{O}_\sigma(I(\mathcal{X}))$, in the sense that if T is a generator of $\mathrm{Lt}_\sigma(I(\mathcal{X}))$ and $T = X_i S$, then $S \in \mathcal{O}_\sigma(I(\mathcal{X}))$. Therefore the region to be explored in order to find $\mathcal{O}_\sigma(I(\mathcal{X}))$ and the generators of $\mathrm{Lt}_\sigma(I(\mathcal{X}))$ is a *finite region* and the idea is to explore it moving one step at a time from the σ-smallest power-product, which is 1. The exploration will eventually find the reduced σ-Gröbner basis of $I(\mathcal{X})$, the set $\mathcal{O}_\sigma(I(\mathcal{X}))$ and a set of separators from $\mathcal{X}$ for the points $P_1, \dots, P_s$.

Let us describe the BM-Algorithm in pseudocode.

BM-Algorithm

INPUT: $\mathcal{X} := \{P_1, \ldots, P_s\} \subset k^n$; σ a term-order.
OUTPUT: GB:= the reduced σ-Gröbner basis of $I(\mathcal{X})$; $\mathcal{O} := \mathcal{O}_\sigma(I(\mathcal{X}))$;
S := a list of separators of the P_i from $\mathcal{X}$.

$GB := \emptyset$; $\mathcal{O} := \emptyset$; $S := \emptyset$;
$r := 0$; $L := [1]$
WHILE $L \neq 0$ DO
  "Main Loop"
END
RETURN GB, $\mathcal{O}$, S
END

And now we describe the Main Loop.

$T := \mathrm{Min}_\sigma(L)$; $L := L \backslash \{T\}$
$f := T - \sum_{i=1}^{s} T(P_{\pi(i)}) \cdot s_i$
IF f vanishes on $\mathcal{X}$ THEN
  $GB := GB \cup \{f\}$ - updating GB
  $L := L \backslash \{\text{Multiples of } T\}$
ELSE
  $\mathcal{O} := \mathcal{O} \cup \{T\}$ - updating $\mathcal{O}$
  $r := r + 1$
  $\pi(r) := \mathrm{Min}\{i \mid f(P_i) \neq 0\}$
  $s_r := f(P_{\pi(r)})^{-1} \cdot f$
  $S := S \cup \{s_r\}$ - updating S
  FOR $i := 1, \ldots, r-1$ DO
    $s_i := s_i - s_i(P_{\pi(r)}) \cdot s_r$ - making S into a list of "partial" separators
  END
  $L := L \cup \{X_i T \mid i := 1, \ldots, n\} \backslash \mathrm{Lt}_\sigma(GB)$ - updating L

Theorem 3.1 *Given $\mathcal{X} := \{P_1, \ldots, P_s\} \subset k^n$, σ a term-order, the Buchberger-Möller algorithm terminates and computes the reduced σ-Gröbner basis of $I(\mathcal{X})$, $\mathcal{O} := \mathcal{O}_\sigma(I(\mathcal{X}))$ and a list of separators of the P_i from $\mathcal{X}$. The complexity of the algorithm is quadratic in the number of indeterminates and cubic in the number of points.*

SKETCH OF THE PROOF. The key remark is that the evaluation function separates the polynomials which are in $I(\mathcal{X})$ from the polynomials which are not in $I(\mathcal{X})$. And the algorithm keeps processing the power-products from the smallest in the list L, so that it actually separates polynomials whose leading term is a generator of $\mathrm{Lt}_\sigma(I(\mathcal{X}))$, from polynomials whose support is in $\mathcal{O}_\sigma(I(\mathcal{X}))$. If the power-product under investigation is not a generator of the leading term ideal, then it produces a separator of a point from $\mathcal{X}$ and in this case the list L can be enlarged. But only s power-products are of this type by Theorem 2.6, so that after a finite number of steps only elements in the reduced Gröbner basis can be obtained and the list eventually empties. By easy considerations on the linear algebra operations used it is not hard to check the correctness of the algorithm.

As for the complexity, see Mora and Robbiano (1993), Marinari *et al.* (1991) □

Corollary 3.2 *Using the Buchberger-Möller Algorithm it is possible to compute interpolating polynomials.*

PROOF. Once we have separators $s_{P_1}, \ldots, s_{P_s}$ and we are given s values $b_1, \ldots, b_s$, then an interpolating polynomial is $\sum_{i=1}^{s} b_i \cdot s_{P_i}$. □

Example 3.1

We use the same points as in Example 2.4 and we check that $s_{P_1} = \frac{1}{3}Y - \frac{2}{3}$ is a separator of P_1 from $\{P_2, P_3, P_4\}$, $s_{P_2} = -X - \frac{5}{9}Y + \frac{28}{9}$ is a separator of P_2 from $\{P_1, P_3, P_4\}$, $s_{P_3} = -\frac{9}{5}X - \frac{1}{2}Y + Z + \frac{13}{5}$ is a separator of P_3 from $\{P_1, P_2, P_4\}$, $s_{P_4} = \frac{14}{5}X + \frac{13}{18}Y - Z - \frac{182}{45}$ is a separator of P_4 from $\{P_1, P_2, P_3\}$. If we want to interpolate the values $\{2, 3, 0, -7\}$ at the points P_1, P_2, P_3, P_4, we find a solution by taking

$2s_{P_1} + 3s_{P_2} - 7s_{P_4} =$

$2(\frac{1}{3}Y - \frac{2}{3}) + 3(-X - \frac{5}{9}Y + \frac{28}{9}) - 7(\frac{14}{5}X + \frac{13}{18}Y - Z - \frac{182}{45}) =$

$-\frac{113}{5}X - \frac{109}{18}Y + 7Z + \frac{1634}{45}$

Many improvements can be made in an actual implementation of the BM-algorithm and many generalizations can be performed, in particular to the case of finite sets of possibly non-reduced (fat) points. On this subject it is worth mentioning the paper by Bigatti and Kreuzer (1998), which is now in preparation.

In the next section we start to explore the connection between designs and models.

4 Identifying the Models

In Section 2 we have seen that every function on a finite set of points can be represented by a polynomial function (see Corollary 2.14) and we have seen how to compute such a polynomial (see Corollary 3.2). These facts are relevant to DoE, since its main goal is to construct *good* models from a subset of a full design.

Now it is time to give the following fundamental definitions:

Definition 4.1 *Given a full design $\mathcal{D}$, a* **fractional factorial design** *is a proper subset $\mathcal{F}$ of $\mathcal{D}$. It is also called a* **fraction** *of $\mathcal{D}$. Its defining ideal $I(\mathcal{F})$ contains the ideal $I(\mathcal{D})$.*

Definition 4.2 *Any set of polynomials that, added to the canonical polynomials of a full design $\mathcal{D}$, generates the ideal of a fraction $\mathcal{F}$, is called a set of* **confounding polynomials** *of $\mathcal{F}$ in $\mathcal{D}$.*

Definition 4.3 *Let $\mathcal{D}$ be a full design and $\mathcal{F}$ a fraction of it. Then we call* **characteristic polynomial of** $\mathcal{F} \subset \mathcal{D}$ *a polynomial f such that $f(P) = 0$ for every $P \in \mathcal{F}$ and $f(P) = 1$ for every $P \in \mathcal{D} \backslash \mathcal{F}$.*

We have seen in Theorem 2.18 that $I(\mathcal{D})$ is generated by the canonical polynomials $\{f_1, \ldots, f_n\}$, which is a universal Gröbner basis, and as a consequence of Corollary 2.16 and Theorem 2.22 we have the following:

Proposition 4.4 *Let $\mathcal{D}$ be a full design, $\mathcal{F}$ a fraction of it, and $\{f_1, \ldots, f_n\}$ the canonical polynomials of $I(\mathcal{D})$. Let f be a characteristic polynomial of $\mathcal{F}$. Then $I(\mathcal{F}) = (f_1, \ldots, f_n, f)$.*

Theorem-Definition 4.5 *Let $\mathcal{D}$ be a full design, $f_1, \ldots, f_n$ the canonical polynomials (as described in Theorem 2.18), and $\mathcal{F}$ a fraction of it. Then there exists a unique characteristic polynomial f of $\mathcal{F}$ in $\mathcal{D}$ such that* $\deg_{X_i}(f) < \deg(f_i)$ *for every $i := 1, \ldots, n$. It will be called the* **canonical confounding polynomial** *of $\mathcal{F}$ in $\mathcal{D}$.*

PROOF. Let f, g be two characteristic polynomials of $\mathcal{F}$ in $\mathcal{D}$. Then, by definition, $f - g$ vanishes on $\mathcal{D}$, hence it belongs to the ideal $I(\mathcal{D}) = (f_1, \ldots, f_n)$. The set $\{f_1, \ldots, f_n\}$ is a universal Gröbner basis, hence the theory tells us that f and g have the same normal form with respect to it and such a normal form is the remainder of the division of f (or of g) with respect to $\{f_1, \ldots, f_n\}$. The conclusion follows immediately. □

Remark 4.6 *In* DoE *it is more common to use the* **indicator polynomial** *of $\mathcal{F} \subset \mathcal{D}$, which is a polynomial such that $f(P) = 1$ for every $P \in \mathcal{F}$ and $f(P) = 0$ for every $P \in \mathcal{D} \backslash \mathcal{F}$. Of course the indicator polynomial and the characteristic polynomial sum to 1 on $\mathcal{D}$, so that it is clear how to switch from one to the other.*

Example 4.1

Let $\mathcal{D}$ be a 2^3 full design as described in Example 2.5. Its canonical polynomials are

$$f_1 := X(X-1),\ f_2 := Y(Y-1),\ f_3 := Z(Z-1)$$

Let $\mathcal{F} := \{(0,0,0),\ (1,0,0),\ (1,1,0),\ (1,0,1),\ (0,1,1)\}$ be a fraction of $\mathcal{D}$. Let $P := (0,1,0)$, $Q := (0,0,1)$, $R := (1,1,1)$. Then $\mathcal{D} = \mathcal{F} \cup \{P, Q, R\}$.

Using the BM-algorithm we find the separators $s_P = Y - YZ - XY + XYZ$, $s_Q = Z - YZ - XZ + XYZ$, $s_R = XYZ$ respectively of the points P, Q, R from $\mathcal{D}$. Therefore the polynomial $s_P + s_Q + s_R = Y + Z - XY - XZ - 2YZ + 3XYZ$ is the canonical confounding polynomial of $\mathcal{F}$ in $\mathcal{D}$.

It is important to observe that all the information of $\mathcal{F} \subset \mathcal{D}$ is completely contained in its canonical confounding polynomial. In DoE this point of view has not been fully explored yet. On the other hand, the beauty of encoding the information into the single canonical confounding polynomial is somehow overshadowed by the fact that directly from it one cannot easily extract further useful information. There is a recent exception to the previous sentence, namely this approach has been used in Fontana *et al.* (1997), but only in a special case. We, however, shall change direction.

Definition 4.7 *Let $\mathcal{F}$ be a fraction of a full design $\mathcal{D}$. Let σ be a term-order and G_σ the corresponding reduced Gröbner basis of $I(\mathcal{F})$. The set of those polynomials in G_σ, which are not among the canonical polynomials of $\mathcal{D}$ is called* **the σ-canonical set of confounding polynomials** *of $\mathcal{F}$ in $\mathcal{D}$.*

Moreover we use the notation $\mathcal{O}_\sigma(\mathcal{F})$ instead of $\mathcal{O}_\sigma(I(\mathcal{F}))$.

Proposition 4.8 *Let $\mathcal{F}$ be a fraction of a full design $\mathcal{D}$. Let σ be a degree-compatible term-order. Then the BM-algorithm computes $\mathcal{O}_\sigma(\mathcal{F})$ with elements of minimal degree.*

PROOF. In fact the order which the power-products enter the computation is $-\sigma$ (i.e. σ-smallest first). $\square$

Example 4.2

Let $\mathcal{D}$ be a 2^3 full design and let $\mathcal{F}$ be the following fraction of it: $\{(1,0,0),\ (1,1,0),\ (1,0,1),\ (0,1,1)\}$.

If $\sigma :=$ DegRevLex, then the σ-canonical set of confounding polynomials of $\mathcal{F}$ in $\mathcal{D}$ is $\{XY - X - Y + 1,\ XZ - X - Z + 1,\ YZ + X - 1\}$. And $\mathcal{O}_\sigma(\mathcal{F}) = \{1,\ X,\ Y,\ Z\}$.

Alternatively, if $\sigma :=$ Lex, then the σ-canonical set of confounding polynomials of $\mathcal{F}$ in $\mathcal{D}$ is $\{X + YZ - 1\}$. And $\mathcal{O}_\sigma(\mathcal{F}) = \{1,\ Y,\ Z,\ YZ\}$.

Example 4.3

Let $\mathcal{D}$ be a 2^3 full design and let $\mathcal{F}$ be the following fraction of it (as in Example 4.1): $\{(0,0,0),\ (1,0,0),\ (1,1,0),\ (1,0,1),\ (0,1,1)\}$.

If either $\sigma :=$ DegRevLex or $\sigma :=$ Lex, then the σ-canonical set of confounding polynomials of $\mathcal{F}$ in $\mathcal{D}$ is $\{XY+YZ-Y,\ XZ+YZ-Z\}$. In both cases $\mathcal{O}_\sigma(\mathcal{F}) = \{1,\ X,\ Y,\ Z,\ YZ\}$.

Now we are ready to enter the "true" game. We have seen that over the fractions of a full design all the functions are polynomial. But what kind of polynomials are to be considered? Let us see what happens with the data of Example 4.2. For instance we know that the polynomial $X+YZ-1$ vanishes on $\mathcal{F} := \{(1,0,0),\ (1,1,0),\ (1,0,1),\ (0,1,1)\}$. This implies that the two polynomial functions X and $1-YZ$ *take the same values* on $\mathcal{F}$, i.e. they are confounded by $\mathcal{F}$. On the other hand the 4 points are not planar, hence no linear combination of $1, X, Y, Z$ except the trivial one vanishes on $\mathcal{F}$. In other words, the polynomial functions $1, X, Y, Z$ on $\mathcal{F}$ are linearly independent. Now suppose that we are looking for a linear combination of these four polynomial functions, which interpolates some given values at the four points, for instance $2, 5, -3, -6$. We have learned that the interpolation problem can be solved via separators (see Corollary 3.2), but there is another way to proceed. Namely we take a *generic* linear combination $a+bX+cY+dZ$ of these four functions, where a, b, c, d are independent parameters. Then we evaluate it at the four points, and impose the given values. We get $a+b=2$, $a+b+c=5$, $a+b+d=-3$, $a+c+d=-6$, which is a non-singular linear system. The unique solution is $a=-4$, $b=6$, $c=3$, $d=-5$. In this case, by evaluating the polynomial $a+bX+cY+dZ$ at the four points of $\mathcal{F}$, we could *identify* it, i.e. we could identify its coefficients.

Example 4.4

Let us go back to Example 4.1 . $\mathcal{O}(\mathcal{D}) = \{1,\ X,\ Y,\ Z,\ XY,\ XZ,\ YZ,\ XYZ\}$, therefore to compute the canonical confounding polynomial of $\mathcal{F}$ in $\mathcal{D}$ we can also proceed as follows. We consider the generic linear combination $a_1 + a_2X + a_3Y + a_4Z + a_5XY + a_6XZ + a_7YZ + a_8XYZ$ and we impose that its evaluation at the points of $\mathcal{F}$ is 0, and at the points of $\mathcal{D}\backslash\mathcal{F}$ is 1. We get the following linear system $a_1 = 0, a_1 + a_2 = 0, a_1 + a_2 + a_3 + a_5 = 0, a_1 + a_2 + a_4 + a_6 = 0, a_1 + a_3 + a_4 + a_7 = 0, a_1 + a_3 - 1 = 0, a_1 + a_4 - 1 = 0, a_1 + a_2 + a_3 + a_4 + a_5 + a_6 + a_7 + a_8 - 1 = 0$, from which we deduce $a_1 = 0, a_2 = 0, a_3 = 1, a_4 = 1, a_5 = -1, a_6 = -1, a_7 = -2, a_8 = 3$, hence we find $f := Y+Z-XY-XZ-2YZ+3XYZ$, in agreement with Example 4.1.

Once again in this example, by evaluating $a_1+a_2X+a_3Y+a_4Z+a_5XY+a_6XZ+a_7YZ+a_8XYZ$ at the eight points of $\mathcal{D}$, we could *identify* it, i.e. we could identify its coefficients. The considerations made before and these examples lead to the following:

Definition 4.9 *A* **polynomial model** *is a polynomial function, i.e.* $y = f(X_1, \ldots, X_n)$, *where* $f(X_1, \ldots, X_n) \in k[X_1, \ldots, X_n]$. *If* $k \subset K$ *is a field extension and* $f(X_1, \ldots, X_n) \in K[X_1, \ldots, X_n]$ *is such that its coefficients are algebraically independent over* k, *then* $y = f(X_1, \ldots, X_n)$ *is called a* **linear family of polynomial models** *or simply a* **linear polynomial model**. *The information of a linear polynomial model is encoded in the support of the polynomial, so we can call linear polynomial model the vector space generated by a set of power-products and call* **support of the model** *such a set.*

Example 4.5

Let $k := \mathbb{Q}$, $K := \mathbb{Q}(a, b, c)$, where a, b, c are transcendental over $\mathbb{Q}$. Then
$y = aX^2 - bXYZ + c$ is a linear polynomial model.
Its support is $\{X^2,\ XYZ,\ 1\}$

Definition 4.10 *Let* $y = f(X_1, \ldots, X_n)$ *be a linear polynomial model and consider a set* $\mathcal{X} := \{P_1, P_2, \ldots, P_s\}$ *of distinct points of* k^n. *We say that* y *is* **identifiable** *by* $\mathcal{X}$ *if, by evaluating it at the given points, we can uniquely determine its coefficients.*

In practice it is common to use polynomial models, whose support is a standard set of power-products. The reason is quite clear: if for a given problem it is relevant to evaluate a power-product, it seems to be even more relevant to evaluate its divisors.

Definition 4.11 *A* **complete (linear) polynomial model** *is a (linear) polynomial model, whose support is a standard set of power-products.*

Now we can rephrase some results with the new terminology, in particular we can interpret Theorem 2.6 and Corollary 2.12.

Theorem 4.12 *Let* $P_1, P_2, \ldots, P_s$ *be distinct points in* k^n *and denote by* $\mathcal{X} := \{P_1, P_2, \ldots, P_s\}$. *Then let* $y = f(X_1, \ldots, X_n)$ *be a linear polynomial model, with support* $\mathcal{S} := \{T_1, \ldots, T_r\}$ *and let* $\mathbb{X}(\mathcal{S}, \mathcal{X})$ *be the* $(s \times r)$*-matrix whose element in position* (i, j) *is* $T_j(P_i)$, *the evaluation of* T_j *at* P_i. *Then the following conditions are equivalent*

1) y *is identifiable by* $\mathcal{X}$

2) $\mathrm{rk}(\mathbb{X}(\mathcal{S}, \mathcal{X})) = r$

3) $\mathcal{S}$ *is a set of linearly independent functions on* $\mathcal{X}$.

PROOF. We know from Theorem 2.6 that the dimension of the vector-space of polynomial functions over $\mathcal{X}$ is s, therefore the equivalence is simply a statement in elementary linear algebra. □

Remark 4.13 *In* DoE *matrices such as* $\mathbb{X}(\mathcal{S},\mathcal{X})$ *above are called* **matrices of the model.**

Theorem 4.14 *Let* $P_1, P_2, \ldots, P_s$ *be distinct points in* k^n *and denote by* $\mathcal{X} := \{P_1, P_2, \ldots, P_s\}$. *Let* σ *be a term-order and* $\mathcal{O}_\sigma(I(\mathcal{X}))$ *the standard set of power-products associated to* σ *and* $I(\mathcal{X})$. *Then every polynomial model, whose support is contained in* $\mathcal{O}_\sigma(I(\mathcal{X}))$, *is identifiable by* $\mathcal{X}$.

PROOF. It follows immediately from Theorem 4.12 and Corollary 2.12. □

In the following section some important problems in DoE are discussed.

5 Some Problems

At this point many natural and fundamental questions arise. Let $\mathcal{D}$ be a full design with levels $(l_1, l_2, \ldots, l_n)$. We know that its cardinality is $\#(\mathcal{D}) = l_1 \times l_2 \times \cdots \times l_n$, which is also $\#(\mathcal{O}(\mathcal{D}))$ (see Definition 2.20 and Corollary 2.12).

PROBLEM 1: *Let* $\mathcal{F} := \{P_1, P_2, \ldots, P_s\} \subset \mathcal{D}$. *What are the complete linear polynomial models which can be identified by* $\mathcal{F}$?

PROBLEM 2: *Let* $y = f(X_1, \ldots, X_n)$ *be a complete linear polynomial model, whose support is contained in* $\mathcal{O}(\mathcal{D})$). *What are the minimal fractions, which identify* y?

PROBLEM 3: *Let* $s < \#(\mathcal{D})$. *What are the fractions of* $\mathcal{D}$ *of cardinality* s, *which identify the highest (lowest) number of complete linear polynomial models?*

5.1 Problem 1

It is important to understand the relation between PROBLEM 1 and the algebraic methods, from several points of view. As we mentioned in the introduction, this connection was the first one discovered between DoE and Gröbner bases (see Pistone and Wynn (1996)). A first step is to observe that Problem 1 can be rephrased in the following way:

PROBLEM 1′: *Let* $\mathcal{F} := \{P_1, P_2, \ldots, P_s\} \subset \mathcal{D}$. *What are the possible standard sets of power-products, which are bases of* $R/I(\mathcal{F})$ *as a* k*-vectorspace?*

Actually the exact rephrasing of PROBLEM 1 would be the following: let $\mathcal{F} := \{P_1, P_2, \ldots, P_s\} \subset \mathcal{D}$; what are the standard sets of power-products, which are k-linearly independent as sets of elements of $R/I(\mathcal{F})$? But, of course, if we are able to find the maximal ones, we get the non-maximal ones for free.

A first step can easily be made by linear algebra methods. Namely, suppose we have $\mathcal{F} := \{P_1, P_2, \ldots, P_s\} \subset \mathcal{D}$ and a standard set of power-products $\mathcal{O} \subset \mathcal{O}(\mathcal{D})$ (or even any set of power-products). If we want to check whether or not it is a basis of $R/I(\mathcal{F})$ as a k-vectorspace, it suffices to construct the matrix $\mathbb{X}(\mathcal{O}, \mathcal{F})$ (see Theorem 4.12). The answer is:

Corollary 5.1 *The following conditions are equivalent*

1) $\mathcal{O}$ is a basis of $R/I(\mathcal{F})$

2) $\mathbb{X}(\mathcal{O}, \mathcal{F})$ is invertible

PROOF. It follows immediately from Theorem 4.12. □

However, PROBLEM 1′ looks for a *complete* answer, in the sense that we want to find *all* the standard sets of power-products, which are bases of $R/I(\mathcal{F})$ as a k-vectorspace. A good approach seems to be the theory of Gröbner bases; indeed one can use Corollary 2.12 and produce many solutions to the problem. But now the question is: how many? It is well-known that, given an ideal I, for every term-order σ there is a unique standard set of power-products $\mathcal{O}_\sigma(I)$. The term-orders are infinite in number, but the standard sets of power-products associated to I are finite (see Robbiano (1985), Robbiano (1986), Mora and Robbiano (1993)) and each one is a basis of R/I as a k-vectorspace. Are they all? Let us have a look at the following important example:

Example 5.1

Let $\mathcal{D}$ be a 3^2 full design. Since X^3, Y^3 are the leading terms of the canonical polynomials, no matter which values of the coordinates are chosen, it turns out that

$$\mathcal{O}(\mathcal{D}) = \{1,\ X,\ Y,\ X^2,\ XY,\ Y^2,\ X^2Y,\ XY^2,\ X^2Y^2\}$$

Let us consider the fraction $\mathcal{F} := \{(0,0),\ (0,-1),\ (1,0),\ (1,1),\ (-1,1)\}$ of $\mathcal{D}$ and let $\mathcal{O} := \{1,\ X,\ Y,\ X^2,\ Y^2\}$, which is a standard set of power-products. We construct

$$\mathbb{X}(\mathcal{O}, \mathcal{F}) = \begin{pmatrix} 1 & 0 & 0 & 0 & 0 \\ 1 & 0 & -1 & 1 & 1 \\ 1 & 1 & 0 & 1 & 0 \\ 1 & 1 & 1 & 1 & 1 \\ 1 & -1 & 1 & 1 & 1 \end{pmatrix}$$

whose determinant is -4, hence we use Corollary 5.1 and conclude that $\mathcal{O}$ is a basis of $R/I(\mathcal{F})$. The polynomial $f := X^2 + XY - X - \frac{1}{2}Y^2 - \frac{1}{2}Y$ is in the defining ideal of $\mathcal{F}$ (this fact can be easily seen by checking that f vanishes at the 5 points of $\mathcal{F}$). Let σ be a term-order. From the definition of term-order it follows that $X^2 >_\sigma X$ and $Y^2 >_\sigma Y$. Now if $X >_\sigma Y$, then $X^2 >_\sigma XY >_\sigma Y^2$. If $Y >_\sigma X$, then $Y^2 >_\sigma XY >_\sigma X^2$. This means that there are only two possibilities for the leading term of f, either it is X^2 or Y^2. The conclusion is that $\mathcal{O} := \{1, X, Y, X^2, Y^2\}$ *cannot be* $\mathcal{O}_\sigma(\mathcal{F})$, whatever σ we choose.

The above example shows that the theory of Gröbner bases provides a good approach to the problem, but it does not always yield the full solution. More work has to be done.

5.2 Problem 2

Now we are going to discuss the more important and more complicated PROBLEM 2, which can be reformulated as:

PROBLEM 2′: *Let $\mathcal{O} \subset \mathcal{O}(\mathcal{D})$ be a standard set of power-products. What are the fractions $\mathcal{F}$ of $\mathcal{D}$, such that $\mathcal{O}$ is a basis of $R/I(\mathcal{F})$ as a k-vectorspace?*

Definition 5.2 *Let $\mathcal{O} \subset \mathcal{O}(\mathcal{D})$ be a standard set of power-products.*

If σ is a term-order, then we denote by $\mathrm{Sol}(\mathcal{O}, \sigma)$ *the set of those fractions $\mathcal{F}$, such that $\mathcal{O}_\sigma(\mathcal{F}) = \mathcal{O}$, hence $\mathcal{O}_\sigma(\mathcal{F})$ is a basis of $R/I(\mathcal{F})$ as a k-vectorspace.*

We denote by $\mathrm{Sol}(\mathcal{O}, \exists\sigma)$ *the set of those fractions $\mathcal{F}$, such that there exists a term-order σ with $\mathcal{O}_\sigma(\mathcal{F}) = \mathcal{O}$, hence $\mathcal{O}_\sigma(\mathcal{F})$ is a basis of $R/I(\mathcal{F})$ as a k-vectorspace.*

We denote by $\mathrm{Sol}(\mathcal{O})$ *the set of those fractions $\mathcal{F}$, such that $\mathcal{O}$ is a basis of $R/I(\mathcal{F})$ as a k-vectorspace, hence* $\mathrm{Sol}(\mathcal{O})$ *is the set of all the solutions to* PROBLEM 2′.

Example 5.1 shows explicitly that $\mathrm{Sol}(\mathcal{O}, \exists\sigma) \subset \mathrm{Sol}(\mathcal{O})$ may happen. This fact induces several other questions. For instance, we would like to be able to find *easy* classes of solutions. It would also be nice to have good classes of examples where Gröbner bases suffice to solve the problem completely. This investigation was carried out in Robbiano and Rogantin (1998) and in Caboara and Robbiano (1997). Here we report on some ideas explained there.

Definition 5.3 *Given n indeterminates $X_1, \ldots, X_n$, let $\mathcal{D}$ be a full design and $\mathcal{O} \subset \mathcal{O}(\mathcal{D})$ a standard set of power-products. Then there exists a unique minimal set,* $\mathrm{Min}(\mathcal{O})$, *of power-products which generates $\{X_1, \ldots, X_n\}^* \backslash \mathcal{O}$, which means that every element in $\{X_1, \ldots, X_n\}^* \backslash \mathcal{O}$ is a multiple of an element of* $\mathrm{Min}(\mathcal{O})$. *The set of power-products in* $\mathrm{Min}(\mathcal{O})$, *which are not among the leading terms of the canonical polynomials of $\mathcal{D}$, is called* $\mathrm{CutOut}(\mathcal{O})$.

The meaning is that $\mathcal{O}$ is the set of power-products not divisible by any element of the set $\text{CutOut}(\mathcal{O}) \cup \{\text{Lt}(f_1), \ldots, \text{Lt}(f_n)\}$. Let us illustrate it with an example.

Example 5.2

Let $\mathcal{D}$ be a 3^2 full design. We know that the leading terms of the two generators of $I(\mathcal{D})$ are X^3, Y^3. Moreover, we know (see Example 5.1) that $\mathcal{O}(\mathcal{D}) = \{1,\ X,\ Y,\ X^2,\ XY,\ Y^2,\ X^2Y,\ XY^2,\ X^2Y^2\}$.
If $\mathcal{O}_1 := \{1,\ X,\ Y,\ X^2,\ XY,\ Y^2\}$ then $\text{CutOut}(\mathcal{O}_1) = \{X^2Y,\ XY^2\}$.
If $\mathcal{O}_2 := \{1,\ X,\ Y,\ X^2,\ XY,\ X^2Y\}$ then $\text{CutOut}(\mathcal{O}_2) = \{Y^2\}$.

Let $\mathcal{D}$ be a full design and $\mathcal{O} \subseteq \mathcal{O}(\mathcal{D})$ a standard set of power-products. Let us denote by $f_1, \ldots, f_n$ the canonical polynomials of $\mathcal{D}$ and let $\text{CutOut}(\mathcal{O}) := \{T_1, \ldots, T_h\}$. Suppose we are able to find h more polynomials $\{g_1, \ldots, g_h\}$ and a term-order σ such that:

1) $\text{Lt}_\sigma(g_i) = T_i$, for $i := 1, \ldots, h$;

2) $\{f_1, \ldots, f_n, g_1, \ldots, g_h\}$ is a σ-Gröbner basis.

Then we denote by I the ideal generated by $\{f_1, \ldots, f_n, g_1, \ldots, g_h\}$ and we see that $I(\mathcal{D}) \subset I$, hence $I = I(\mathcal{F})$, where $\mathcal{F}$ is a fraction of $\mathcal{D}$. Moreover $\mathcal{O}_\sigma(I) = \mathcal{O}_\sigma(\mathcal{F}) = \mathcal{O}$, hence $\mathcal{O}$ is a basis of $R/I(\mathcal{F})$ by Corollary 2.12, and the fraction $\mathcal{F}$, whose cardinality is $\#(\mathcal{O})$ by Theorem 2.6, is a solution to Problem 2'.

Example 5.3

We continue with the same data as in Example 5.2. Suppose that the values of the coordinates are $\{0,\ 1,\ 2\}$ for both X and Y.
Then $f_1 = X(X-1)(X-2)$, $f_2 = Y(Y-1)(Y-2)$. If we take $g_1 := X(X-1)Y$, $g_2 := XY(Y-1)$, then one checks that $\{f_1, f_2, g_1, g_2\}$ is a σ-Gröbner basis for every σ. Therefore $I := (f_1, f_2, g_1, g_2)$ defines a fraction $\mathcal{F}$, which is easily seen to be $\{(0,0),\ (0,1),\ (0,2),\ (1,0),\ (2,0), (1,1)\}$, and $\mathcal{O}_1 = \{1,\ X,\ Y,\ X^2,\ XY,\ Y^2\}$ is a basis of $R/I(\mathcal{F})$. By the equivalence of Problem 2 and Problem 2', we conclude that any polynomial model whose support is contained in $\mathcal{O}_1$ can be identified by $\mathcal{F}$.

The example can be explained by the following facts.

Definition 5.4 *Let k be an infinite field, $T := X_1^{a_1} X_2^{a_2} \cdots X_n^{a_n}$ and let $\alpha_1, \ldots, \alpha_n$ be n sequences of elements of the base field k, where $\alpha_r = (\alpha_{r,i})_{i \in \mathbb{N}}$ for $r := 1, \ldots, n$ and $\alpha_{r,i} \neq \alpha_{r,j}$ if $i \neq j$. Then we call the polynomial*

$$D(T) := \textstyle\prod_{i=1}^{a_1} (X_1 - \alpha_{1,i}) \cdots \prod_{i=1}^{a_n} (X_n - \alpha_{n,i})$$

the **distraction** *of T with respect to $\alpha_1, \ldots, \alpha_n$.*

Proposition 5.5 *Let $T_1, \ldots, T_h$ be power-products such that T_i does not divide T_j for every $i \neq j$, let $\alpha_1, \ldots, \alpha_n$ be n sequences of elements of k and let $I := (D(T_1), \ldots, D(T_h))$ be the ideal generated by the corresponding distractions with respect to $\alpha_1, \ldots, \alpha_n$. Then $\{D(T_1), \ldots, D(T_h)\}$ is the reduced Gröbner basis of I with respect to every term-order.*

PROOF. See Carrà and Robbiano (1990), Theorem 1.4 and Corollary 2.6. □

It is clear that, in order to get a distraction of T, it suffices to have vectors of sufficiently many elements of k. So we can speak of distractions associated to vectors, not to sequences, and sometimes we can drop the assumption that k is infinite.

Theorem 5.6 *Let $\mathcal{D}$ be a full design with levels $(l_1, l_2, \ldots, l_n)$ and whose values of the coordinates are $0, 1, \ldots, l_i - 1$, for $i := 1, \ldots, n$. Let $I(\mathcal{D}) := (f_1, \ldots, f_n)$ be its defining ideal, where $f_j := X_j\ (X_j - 1) \cdots (X_j - l_j + 1)$ are the canonical polynomials, for $i = 1, \ldots, n$; let $\mathcal{O}(\mathcal{D})$ be its corresponding standard set of power-products and $\mathcal{O}$ a standard set of power-products contained in $\mathcal{O}(\mathcal{D})$. Assume that $\{T_1, \ldots, T_h\} = \mathrm{CutOut}(\mathcal{O})$. Let $\alpha_1, \alpha_2, \ldots, \alpha_n$ be permutations of $(0, 1, \ldots, l_1 - 1), \ldots, (0, 1, \ldots, l_n - 1)$ respectively and $D(T_1), \ldots D(T_h)$ the distractions of $T_1, \ldots, T_h$ with respect to $\alpha_1, \alpha_2, \ldots, \alpha_n$.*

Then, for every term-order σ, we have: $\{f_1, \ldots, f_n, D(T_1), \ldots D(T_h)\}$ is a σ-Gröbner basis and $\{D(T_1), \ldots, D(T_h)\}$ is the σ-canonical set of confounding polynomials of a fraction $\mathcal{F}$, such that $\mathcal{O}_\sigma(\mathcal{F}) = \mathcal{O}$.

PROOF. See Robbiano and Rogantin (1998), Theorem 3.4. □

Remark 5.7 *It is easy to see that the choice of coordinates can be generalized to any set. It suffices then to change the permutations $\alpha_1, \alpha_2, \ldots, \alpha_n$ accordingly.*

Definition 5.8 *Given a full design $\mathcal{D}$, a term-order σ, and a standard set of power-products $\mathcal{O} \subset \mathcal{O}(\mathcal{D})$, let $\mathcal{F}$ be a fraction of $\mathcal{D}$, whose σ-canonical set of confounding polynomials is given by distractions as indicated in Theorem 5.6. Then $\mathcal{F}$ is called a* **distracted fraction**.

Definition 5.9 *Given a full design $\mathcal{D}$ and a standard set of power-products $\mathcal{O} \subset \mathcal{O}(\mathcal{D})$, we denote by $\mathrm{Distr}(\mathcal{O})$ the set of the distracted fractions, whose associated standard set of power-products is $\mathcal{O}$ for some term-order σ. Then we denote by $\mathrm{Sol}(\mathcal{O}, \forall\sigma)$ the set of those fractions $\mathcal{F}$ such that $\mathcal{O}_\sigma(\mathcal{F}) = \mathcal{O}$ for every term-order σ.*

Corollary 5.10 *Given the hypotheses of Theorem 5.6,*

$$\mathrm{Distr}(\mathcal{O}) \subseteq \mathrm{Sol}(\mathcal{O}, \forall\sigma)$$

PROOF. It follows from Proposition 5.5. □

Remark 5.11 *In Robbiano and Rogantin (1998) some more theory is developed. For instance an explicit formula for* $\#(\mathrm{Distr}(\mathcal{O}))$ *is proved and a better insight is given to the difference between* $\mathrm{Distr}(\mathcal{O})$ *and* $\mathrm{Sol}(\mathcal{O}, \forall\sigma)$. *The following example is drawn from that paper.*

Example 5.4

Let $\mathcal{D}$ be a 2^3 full design, $\mathcal{O} := \{1, x, y, z\}$.
Let $\mathcal{F} := \{(0,0,0),\ (0,0,1),\ (1,0,0),\ (1,1,0)\}$. Then $I(\mathcal{F}) = (x^2 - x,\ y^2 - y,\ z^2 - z,\ xy - x,\ xz,\ yz)$ and $\{xy - x,\ xz,\ yz\}$ is the DegRevLex-canonical set of confounding polynomials of $\mathcal{F}$. Therefore $\mathcal{O} = \mathcal{O}_{DegRevLex}(\mathcal{F})$ and we deduce from Robbiano and Rogantin (1998), Theorem 3.13 that $\{xy - x,\ xz,\ yz\}$ is the τ-canonical set of confounding polynomials of $\mathcal{F}$ for every term-order τ, and thus that $\mathcal{F} \in \mathrm{Sol}(\mathcal{O}, \forall\sigma)$. On the other hand, $\mathcal{F} \notin \mathrm{Distr}(\mathcal{O})$. Suppose the contrary; then it follows from See Carrà and Robbiano (1990) that every pair of polynomials in $\{x^2 - x,\ y^2 - y,\ z^2 - z,\ xy - x,\ xz,\ yz\}$ should be a Gröbner basis. But clearly $\{xy - x,\ xz\}$ is not. With respect to the 2^3 full design and the standard set of power-products $\mathcal{O} := \{1,\ x,\ y,\ z\}$, we conclude that $\mathrm{Distr}(\mathcal{O}) \subset \mathrm{Sol}(\mathcal{O}, \forall\sigma)$. A suitable computation done with **`statfamilies`** shows that $\#(\mathrm{Distr}(\mathcal{O})) = 8$ and $\#(\mathrm{Sol}(\mathcal{O}, \forall\sigma)) = 32$.

We have already seen that Gröbner basis theory in general cannot solve PROBLEM $2'$ completely. But there are nice situations where it does.

We recall the following:

Theorem 5.12 *Given a full design* $\mathcal{D}$, *a term-order* σ *and a standard set of power-products* $\mathcal{O} \subset \mathcal{O}(\mathcal{D})$, *there is an algorithm which computes* $\mathrm{Sol}(\mathcal{O}, \sigma)$.

PROOF. See Caboara and Robbiano (1997), Theorem 3.3 and subsequent remarks. □

Definition 5.13 *Let* $\mathcal{O} \subset \{X_1, \ldots, X_n\}^*$ *denote a standard set of power-products,* σ *a term-order and* T *a power-product. We put* $\sigma(\mathcal{O}, T) := \{T' \in \mathcal{O} \mid T' <_\sigma T\}$.

Theorem 5.14 *Given a full design* $\mathcal{D}$ *and a standard set of power-products* $\mathcal{O} \subset \mathcal{O}(\mathcal{D})$, *we let* $\mathrm{CutOut}(\mathcal{O}) := \{T_1, \ldots, T_h\}$ *and assume that there exists a term-order* σ *such that* $\sigma(\mathcal{O}, T_i) = \mathcal{O}$ *for every* $i := 1, \ldots, h$. *Then*

1) $\mathrm{Sol}(\mathcal{O}) = \mathrm{Sol}(\mathcal{O}, \sigma)$.

2) There is an algorithm which computes them, hence it computes all the solutions to PROBLEM $2'$.

PROOF. See Caboara and Robbiano (1997), Corollary 2.14. □

Actually the proof follows from the construction of a family of ideals, which are the defining ideals of the fractions which solve the problem. The process is deeply connected with the theory of Gröbner bases as well as the theory of algebraic schemes. The details are described in Caboara and Robbiano (1997). Here we show the full computation and the solution of an instance of PROBLEM 2′.

Example 5.5

Let $\mathcal{D}$ be the 3^2 full design given by the polynomials $f_1 := X(X-1)(X-2)$, $f_2 := Y(Y-1)(Y-2)$; then $\mathcal{O}(\mathcal{D}) = \{1,\ X,\ Y,\ X^2,\ XY,\ Y^2,\ X^2Y,\ XY^2,\ X^2Y^2\}$. If $\mathcal{O} := \{1,\ X,\ Y,\ Y^2\}$, then we get $\mathrm{CutOut}(\mathcal{O}) = \{X^2,\ XY\}$. If σ is any degree-compatible term-order with $X >_\sigma Y$, then we match the assumptions of Theorem 5.14, since $\sigma(\mathcal{O}, X^2) = \sigma(\mathcal{O}, XY) = \mathcal{O}$. To solve PROBLEM 2′ completely in this specific example, it suffices to find all the σ-canonical sets of confounding polynomials, whose leading terms are X^2, XY. All the corresponding fractions $\mathcal{F}$ are such that $\mathrm{Lt}_\sigma(I(\mathcal{F}))$ is minimally generated by $\{X^2, XY, Y^3\}$.

Let $g_1 := X^2 + a_1Y^2 + a_2X + a_3Y + a_4$, $g_2 := XY + a_5Y^2 + a_6X + a_7Y + a_8$ The theory says that we have to impose that $\{f_1, f_2, g_1, g_2\}$ is a σ-Gröbner basis. This yields the following algebraic system:

$$
\begin{aligned}
&a_7^2a_8 - a_6a_7 + a_5 = 0,\\
&a_7^3 + 3a_7^2 + 2a_7 = 0,\\
&a_5a_7^2 + 3a_5a_7 + 2a_5 = 0,\\
&a_7^2a_8 - a_6a_7 + a_5 = 0,\\
&a_4a_5a_7 - a_1a_3 - a_2a_5 - 3a_1 = 0,\\
&a_5a_7a_8 + a_3a_5 - a_5a_6 - a_1a_7 = 0,\\
&a_4a_7a_8 - a_3a_4 - a_4a_6 - a_2a_8 - 3a_4a_8 - 3a_4 = 0,\\
&a_4a_7^2 - a_3^2 - a_2a_7 + a_1 - 3a_3 - 2 = 0,\\
&a_4a_6a_7 - a_2a_3 - a_4a_5 - a_2a_6 + 2a_4a_8 - 3a_2 = 0,\\
&a_6a_7^2 - a_5a_7 + 3a_6a_7 + 2a_7a_8 - 3a_5 = 0,\\
&a_7a_8^2 - a_4a_7 + a_3a_8 - 2a_6a_8 - 3a_8^2 - a_2 - 3a_4 = 0,\\
&a_6a_7a_8 + a_3a_6 - a_6^2 - a_2a_7 - a_5a_8 + 2a_8^2 - a_1 + 2a_4 = 0
\end{aligned}
$$

Solving the system yields 81 solutions, which correspond to 81 among all the $\binom{9}{4} = 126$ fractions of cardinality 4. These 81 fractions are ALL the fractions $\mathcal{F}$ of $\mathcal{D}$ such that $\{1,\ X,\ Y,\ Y^2\}$ is a basis of $R/I(\mathcal{F})$. In this case PROBLEM 2′ is solved completely.

5.3 Problem 3

We conclude the survey by saying a few words about PROBLEM 3. As before, it is easy to see that the problem can be rephrased as

PROBLEM 3′: *Let $s < \#(\mathcal{D})$. What are the fractions $\mathcal{F}$ of $\mathcal{D}$ of cardinality s, with the highest (lowest) number of standard sets of power-products, which are a basis of $R/I(\mathcal{F})$ as a k-vectorspace?*

Again Example 5.1 shows that the theory of Gröbner bases cannot give the full solution. An interesting approach is taken in Caboara *et al.* (1997), where the theory of Gröbner Fans (see Mora and Robbiano (1989)) is used. Here we show that it is possible to characterize those fractions with a *unique* standard set of power-products as a basis of $R/I(\mathcal{F})$. First, we recall from Robbiano and Rogantin (1998) the following

Theorem 5.15 *Given a design $\mathcal{D}$ and a standard set of power-products $\mathcal{O} \subset \mathcal{O}(\mathcal{D})$, let $\mathcal{F}$ be a fraction. Let σ be a term-order and $G_\sigma := \{g_1, \ldots, g_h\}$ the σ-canonical set of confounding polynomials of $\mathcal{F}$. If we write $g_i := T_i - r_i$, with $T_i := Lt_\sigma(g_i)$ for $i := 1, \ldots, h$, then the following conditions are equivalent*

1) T divides T_i for every T in the support of r_i and every $i := 1, \ldots, h$

2) G_σ is the τ-canonical set of confounding polynomials of $\mathcal{F}$ for every term-order τ

3) $\mathcal{F} \in \mathrm{Sol}(\mathcal{O}, \forall\sigma)$.

PROOF. See Robbiano and Rogantin (1998), Theorem 3.13. □

So we may conclude by proving the following

Theorem 5.16 *Given a design $\mathcal{D}$ and a standard set of power-products $\mathcal{O} \subset \mathcal{O}(\mathcal{D})$, let $\mathcal{F}$ be a fraction. Then the following conditions are equivalent*

1) $\mathcal{O}$ is the unique standard set of power-products which is a basis of $R/I(\mathcal{F})$ as a k-vectorspace

2) $\mathcal{F} \in \mathrm{Sol}(\mathcal{O}, \forall\sigma)$.

PROOF. The fact that 1) implies 2) is obvious. Let us prove that 2) implies 1). If $\mathcal{F} \in \mathrm{Sol}(\mathcal{O}, \forall\sigma)$, then we deduce from Theorem 5.15 that there exists a set of polynomials $G := \{g_1, \ldots, g_h\}$ which is the σ-canonical set of confounding polynomials, for every σ. Moreover the shape of these polynomials is $g_i := T_i - r_i$, with $T_i = \mathrm{Lt}(g_i)$ for every σ and T divides T_i for every $T \in \mathrm{Supp}(r_i)$. Let $\mathrm{Supp}(r_i) := \{T_{i1}, \ldots, T_{im_i}\}$; then $T_i, T_{i1}, \ldots, T_{im_i}$ are linearly dependent modulo $I(\mathcal{F})$. Now let $\mathcal{O}'$ be another standard set of power-products which is a basis of $R/I(\mathcal{F})$ as a k-vectorspace. The key remark is that T_i cannot be in $\mathcal{O}'$, otherwise also all the T_{ij} would be in $\mathcal{O}'$, simply because T_{ij} divides T_i. This fact implies that $\mathcal{O}' \subseteq \mathcal{O}$, hence they are equal simply because they have the same cardinality. □

Remark 5.17 *This theorem shows that not only the Distracted Fractions but also those fractions as in Example 5.4 identify a unique maximal complete linear polynomial model. In other words their* fan has only one leaf, *if we use the terminology of Caboara* et al.

6 Concluding Remarks

We conclude the survey by summarizing some of the most important facts treated above. Putting together Corollary 5.10, Example 5.1, the subsequent discussion, Theorem 5.12, Theorem 5.14 and Theorem 5.16, we conclude by saying that

$$\mathrm{Distr}(\mathcal{O}) \subseteq \mathrm{Sol}(\mathcal{O}, \forall\sigma) \subseteq \mathrm{Sol}(\mathcal{O}, \exists\sigma) \subseteq \mathrm{Sol}(\mathcal{O})$$

All the inclusions can be strict and the set $\mathrm{Sol}(\mathcal{O}, \forall\sigma)$ is precisely the set of those fractions which identify a unique maximal complete linear polynomial model. Moreover, given a term-order σ, there is an algorithm which computes all the elements in $\mathrm{Sol}(\mathcal{O}, \sigma)$, and we have established in Theorem 5.14 a set of sufficient conditions for $\mathrm{Sol}(\mathcal{O}, \sigma)$ to be equal to $\mathrm{Sol}(\mathcal{O})$.

The problem of computing $\mathrm{Sol}(\mathcal{O})$ in general, is still unsolved.

References

Adams, W. W., Loustaunau, P. (1994): *An Introduction to Gröbner Bases.* Graduate Studies in Mathematics: Amer. Math. Soc., Providence, R.I.

Bigatti, A.M., Kreuzer, M. (1998): *On the Computation of Ideals of Points.* In Preparation.

Box, G. E. P., Hunter, W. G., Hunter, J. S. (1978): *Statistics for Experimenters.* John Wiley & Sons, New York.

Buchberger, B. (1965): On Finding a Vector Space Basis of the Residue Class Ring Modulo a Zero Dimensional Polynomial Ideal (German). PhD Thesis, Univ. of Innsbruck, Austria.

Buchberger, B. (1970): An Algorithmical Criterion for the Solvability of Algebraic Systems of Equations (German). *Aequationes mathematicae* **4/3**, 374–383.

Buchberger, B. (1985): An Algorithmic Method in Polynomial Ideal Theory. Chapter 6 in *Progress, directions, and open problems in multidimensional systems theory.* N. K. Bose, Ed., D. Reidel Publishing Company, 184–232.

Buchberger, B., Möller, M. (1982): The Construction of Multivariate Polynomials with Preassigned Zeroes. *Proceedings EUROCAM 82, LNCS* **144**, 24–31.

Caboara, M., Pistone, G., Riccomagno, E. Wynn, H.P. (1997): The Fan of an Experimental Design. Preprint.

Caboara, M., Robbiano, L. (1997): Families of Ideals in Statistics. *Proceedings of the 1997 ISSAC (Maui, Hawaii, July 97)* Küchlin Ed., New York, 404–409.

Capani, A., Niesi, G., Robbiano, L. (1995): CoCoA, a system for doing Computations in Commutative Algebra.
Available via anonymous ftp from `cocoa.dima.unige.it`.

Capani, A., Niesi, G., Robbiano, L. (1997): Some Features of CoCoA 3. *Moldova Journal of Computer Science* **4/3**, 296–314.

Carrà, G., Robbiano, L. (1990): On SuperG-Bases. *J. Pure Appl. Algebra* **68**, 279–292.

Cox, D., Little, J., O'Shea, D. (1992): *Ideals, Varieties, and Algorithms.* Springer-Verlag, New York.

Dey, A. (1985): *Orthogonal Fractional Factorial Designs.* John Wiley and Sons, New York.

Diaconis, P., Sturmfels, B. (1993): Algebraic Algorithms for Sampling from Conditional Distributions. To appear in *Annals of Statistics*

Eisenbud, D. (1995): *Commutative Algebra with a View toward Algebraic Geometry.* Springer Graduate Texts in Mathematics **150**.

Fisher, R.A. (1935): *The Design of Experiments.* Oliver and Boyd, Edinburgh (1935) (1st ed.).

Fontana, M., Pistone, G., Rogantin, M.P. (1997): Classification of Two-Level Factorial Fractions. Preprint.

Geramita, A.V., Seberry, J. (1979): *Orthogonal Designs: Quadratic Forms and Hadamard Matrices* . Marcel Dekker, New York, N.Y.

Heiberger, R.M. (1989): *Computation for the Analysis of Designed Experiments.* John Wiley and Sons, New York.

Holliday, T., Riccomagno, E., Wynn, H. P., Pistone, G. (1997): The Application of Computational Algebraic Geometry to the Analysis of Designed Experiments. A Case Study. To appear in *Computational Statistics.*

Kreuzer, M, Robbiano, L. (1998): *Computations in Commutative Algebra.* In Preparation.

Marinari, M.G., Möller, M., Mora, F. (1991): Gröbner Bases of Ideals Defined by Functionals with an Application to Ideals of projective Points. *AAECC* **4**, 103–145.

Mora, F., Robbiano, L. (1989): The Gröbner Fan of an Ideal. *In* L. Robbiano, ed., *Computational Aspects of Commutative Algebra.* Academic Press, London, 183–208.

Mora, F., Robbiano, L. (1993): Points in affine and projective spaces. *In* D. Eisenbud and L. Robbiano eds., *Computational Algebraic Geometry and Commutative Algebra, Cortona-91, Symposia Mathematica Vol XXXIV* . Cambridge University Press, 106–150.

Pistone G., Wynn, H.P. (1996). Generalised Confounding and Gröbner Bases. *Biometrika* **83-3**, 653–666.

Raktoe, B.L., Hedayat, A., Federer, W.T. (1981): *Factorial Designs*. John Wiley and Sons, New York.

Robbiano, L. (1985): Term Orderings in the Polynomial Ring. *Lecture Notes in Computer Science* **203**.

Robbiano, L. (1986): On the Theory of Graded Structures. *J. Symbolic Comput.* **2** 139–170.

Robbiano, L., Rogantin, M.P. (1998): Full Factorial Designs and Distracted Fractions. This volume.

Street A.P., Street, D.J. (1987): *Combinatorics of Experimental Design*. Oxford Clarendon.

Yates, F. (1937): *Design and Analysis of Factorial Experiments*, London: Imperial Bureau of Soil Sciences.

Gröbner Bases and Coding Theory

Shojiro Sakata

The University of Electro-Communications
Department of Computer Science and Information Mathematics
Chofu-shi, Tokyo 182, JAPAN
sakata@cs.uec.ac.jp

Abstract

In this note, we present a sketch of several interplays between Gröbner bases theory and coding theory. For readers who are not so familiar to coding theory, some introductory explanations on error-correcting codes and coding theory are included. The main topics are some problems of encoding and decoding of algebraic codes which are related to Gröbner bases. Some simple examples of codes are referred to. In particular, recent developments in coding theory which have been done around multidimensional or multivariate codes have initiated and strengthened the connections, and several new problems and relevant algorithms have been explored in coding theory.

1 Introduction

In this note, we give a sketch of several interactions between coding theory (or rather algebraic coding theory) and Gröbner bases theory. Precisely, coding theory is the theory of error-correcting codes, which are used widely in digital communication and storage systems to transmit or to store digital information error-free or as correct as possible. Although error-correcting codes are probabilistic by nature in the sense that they are used to battle against random noise in transmission or storage channels, practical codes are constructed in deterministic fashions with algebraic methods, without which error control cannot be achieved with low complexity and cost. As a result, almost all of them are algebraic. Therefore we discuss nothing but the algebraic error-correcting codes.

As is well-known, Gröbner basis is one of the fundamental concepts and tools in algebraic methods. Consequently coding theory and Gröbner basis have been destined to be interrelated with each other since their births in the history of computer science. But, they encountered each other rather lately. As far as we know, Ikai et al. (1976a) were the first to treat the concept corresponding to Gröbner basis in the world of coding theory. Although the paper had been scarcely known outside Japan until it was introduced in the book by Poli and Huguet (1988), several Japanese researchers, e.g. Ikai et al. (1976b), Imai (1977), Sakata (1978), had continued their efforts to find good multidimensional or multivariate codes on the basis of that concept. Among these investigations we succeeded to give an algorithm identical with Buchberger's algorithm, which we applied to construct several two-dimensional or bivariate codes (Sakata 1981). At that time we had no knowledge of existence of Buchberger's papers, e.g. Buchberger (1965), Buchberger (1970), Buchberger (1976), although we had known the paper by Hermann (1926) which treated the same problem. Just after we had presented a paper (Sakata 1981) at the ISIT81 (the 1981 IEEE International Symposium on Information Theory) in USA, we became acquainted with Buchberger's algorithm. These happenings were the début of the Gröbner bases theory and algorithm in the history of coding theory.

Until the discovery of algebraic geometric (AG) codes by V. G. Goppa (Goppa 1981), mainly the theory of finite fields and univariate polynomial rings over finite fields had been applied to construct and use most of error-correcting codes, which belong to a wide class of so-called cyclic codes. Bose-Chaudhuri-Hocquenguem (BCH) codes and Reed-Solomon (RS) codes have been among the most well-known cyclic error-correcting codes which are used in the real world (about the details of coding theory, see e.g. Peterson 1972, Berlekamp 1968, MacWilliams and Sloane 1977). AG codes derived from algebraic curves and surfaces are a class of multidimensional codes which are expected to become practically important in future (for AG codes, see e.g. Tsfasman 1991).

To transmit or store digital information, i.e. a sequence of symbols in the form of electro-magnetic signal through a noisy channel, where some errors or exchanges of transmitted symbols can occur by the effects of noise, one preprocesses the information sequence before sending it into the channel. For the symbol alphabet A and fixed integers k, n, $0 < k < n$, the preprocessing called 'encoding' is a kind of mapping from the information space A^k (i.e. the set of information vectors $\underline{i} = (i_1, \cdots, i_k)$ of k symbols) into the signal space A^n (i.e. the set of signal vectors $\underline{c} = (c_1, \cdots, c_n)$ of n symbols). The image $\mathbf{C} := Im(A^k)$ $(\subset A^n)$ is a code, where n and k are called the code length and information dimension of the code $\mathbf{C}$, respectively, and the ratio $\frac{k}{n}$ is the rate of the code $\mathbf{C}$ which represents the degree of its 'speed' of transmission. It often is convenient to have a way of systematic encoding such that $c_j = i_j$, $1 \leq j \leq k$, and the redundant symbols c_j, $k+1 \leq j \leq n$, called the check symbols are determined from the information symbols systematically.

When a code vector $\underline{c} \in \mathbf{C}$ was sent into the channel and an erroneous signal vector $\underline{r}$ $(\in A^n)$ is received as an output from the channel, one postprocesses the received vector $\underline{r}$ to get or guess the information vector $\underline{i}$ $(\in A^k)$ or the sent code vector $\underline{c}$ $(\in \mathbf{C})$. The postprocessing called 'decoding' is done by using the gap or vacancy in the space A^n which is not covered by the code vectors $\underline{c}$ of $\mathbf{C}$, or more definitely, the minimum distance d of the code C, where the distance

$d(\underline{r}, \underline{r}')$ between any two signal vectors $\underline{r} = (r_1, \cdots, r_n), \underline{r}' = (r_1', \cdots, r_n') \in A^n$ is defined to be the number of indices i, $1 \leq i \leq n$, such that $r_i \neq r_i'$, and $d := min\{d(\underline{c}, \underline{c}') \,|\, \underline{c} \neq \underline{c}', \underline{c}, \underline{c}' \in \mathbf{C}\}$ is just the minimum of the distances between any two distinct code vectors of $\mathbf{C}$. Under the assumption that at most t $(:= \lfloor \frac{d-1}{2} \rfloor)$ errors happened, i.e. only t or less symbols among the n symbols of the received vector $\underline{r}$ $(\in A^n)$ are not equal to the corresponding symbols of the sent vector, one can find the sent vector as $\underline{c}$ $(\in \mathbf{C})$ such that $d(\underline{c}, \underline{r}) < d(\underline{c}', \underline{r})$, $^\forall \underline{c}' \in \mathbf{C}$, $\underline{c}' \neq \underline{c}$. Thus, the ratio $\frac{d}{n}$ represents the degree of exactness or correction performance of the code $\mathbf{C}$.

A good error-correcting code must confront two contradictory requirements, i.e. it must convey digital information in the highest degree of speed as well as in the highest degree of exactness. Although one might want to have good codes such that both ratios $\frac{k}{n}$ and $\frac{d}{n}$ are maximal, these values are in the relationship of trade-off. Therefore, throughout the history of coding theory, optimal codes, which have the maximum possible value $\frac{k}{n}$ together with a certain fixed value $\frac{d}{n}$ or vice versa, have been explored. In practical applications, the computational complexity of encoding and decoding is also crucial. In particular, it is NP-hard and actually intractable to find a code vector $\underline{c} \in \mathbf{C}$ which is nearest to the received vector $\underline{r}$, if we have no way different from the complete enumeration of all the code vectors of $\mathbf{C}$. What algebraic methods are suitable for devising better codes and using (i.e. encoding and decoding) such codes has been the most important issue of coding theory. Numerous approaches have been tested, and Gröbner basis was captured into the world of coding theory in that trend.

2 Code construction: Encoding

Essentially most algebraic codes are defined by three kinds of algebraic sets, i.e. the symbol alphabet A, the symbol locator space B and the function space C. Usually A is identified with a finite field F_q, i.e. the Galois field of q elements, and the signal space A^n is a linear space over the field $A = F_q$. Both encoding and decoding are executed through arithmetic operations over the field F_q or its extension field F_{q^l}.

The symbol locator space B is a Euclidean or projective space over the finite field F_{q^l} as treated in finite geometry. For a code vector $\underline{c} = (c_1, \cdots, c_n) \in \mathbf{C}$, an ordered set $P = \{P_1, \cdots, P_n\}$ $(\subset B)$ is the symbol locators with cardinality $\#P = n$. The i-th element P_i is used to label or identify the i-th symbol c_i, $1 \leq i \leq n$. This algebraic labeling will play a very important role in decoding of algebraic codes.

The function space C is composed of algebraic functions defined on the set P. In a typical case C is a coordinate ring R/I for an ideal I of the m-variate polynomial ring $R = F_{q^l}[\boldsymbol{x}] = F_{q^l}[x_1, \cdots, x_m]$ such that P $(\subset B = (F_{q^l})^m)$ is identical with the zeros $V(I) = \{\boldsymbol{x} \in (F_{q^l})^m | f(\boldsymbol{x}) = 0,\, f \in I\}$ of the ideal I.

Sometimes I is a homogeneous ideal with a projective set $P = V(I)$, e.g. the F_q-rational points of an algebraic curve $\mathcal{X} : f(\boldsymbol{x}) = 0$ in the m-dimensional (mD) projective space $\mathbf{P}^m$ for which $I =< f >$ and C is a subring of the algebraic function field on the curve $\mathcal{X}$.

Given a triplet (A, P, Q), where Q is a linear subspace of the coordinate ring $C = R/I$ (as a linear space), two kinds of codes are defined as linear subspaces of A^n (over the field $A = F_q$) as follows:

$$\mathbf{C} = \{\underline{c} = (c_j) \in A^n | \ c_j = f(P_j), 1 \le j \le n, f \in Q\}; \tag{2.1}$$

$$\mathbf{C} = \{\underline{c} = (c_j) \in A^n | \ \sum_{j=1}^{n} c_j f(P_j) = 0, f \in Q\}, \tag{2.2}$$

which we call the primal and dual linear codes, respectively. Ordinarily we assume that the embedding φ from C into A^n defined by $\varphi(f) := (f(P_1), \cdots, f(P_n))$ is injective, where the dimensions $dim_A Q$ for the primal and dual codes are assumed to equal k and $n-k$, respectively. (Remark: This embedding φ embodies the labeling of symbol locators as mentioned in the above.)

In this setting it is important to scrutinize the structure of the coordinate ring R/I. This task is just done via the Gröbner bases theory and algorithm. In case of $I \subset R = F_q[\boldsymbol{x}] = F_q[x_1, \cdots, x_m]$, one can find the Gröbner basis window or the excluded point set (in our terminology) $\Delta(I)$ in the Cartesian product Z_0^m of the set Z_0 of nonnegative integers according to a specific term order $\le_T$. Under a suitable assumption, $n = \#\Delta(I)$, and $C =< \boldsymbol{x}^{\boldsymbol{p}} : \boldsymbol{p} \in \Delta(I) >$, i.e. a linear space spanned by the monomials $\boldsymbol{x}^{\boldsymbol{p}} = x_1^{p_1} \cdots x_m^{p_m}$, $\boldsymbol{p} = (p_1, \cdots, p_m) \in \Delta(I)$. Ordinarily $Q =< \boldsymbol{x}^{\boldsymbol{p}} : \boldsymbol{x}^{\boldsymbol{p}} \le_T \boldsymbol{x}^{\boldsymbol{q}}, \boldsymbol{p} \in \Delta(I) >$ for a fixed $\boldsymbol{q} = (q_1, \cdots, q_m) \in \Delta(I)$. Although the problem which couples (B, C) or rather (P, Q) give us good codes is substantially important, the selection of the term order $\le_T$ sometimes affects that.

Example 1 (BCH codes and RS codes): *For $A = F_q$ and a primitive n-th root α ($\in F_{q^l}$) of 1, $n | q^l - 1$, a couple of $P = \{\alpha^i | 0 \le i \le n-1\} \subset B = F_{q^l}$ and $Q =< x^0, x^1, \cdots, x^{\delta-2} >$ ($\subset C = R/I$), $0 < \delta < n$, defines a BCH code as the dual linear code $\mathbf{C}$, where $R = F_{q^l}[x]$, $I =< x^n - 1 >$ ($\subset R$). The parameter δ is called its designed distance, and it is known that δ is a lower bound of the minimum distance d of $\mathbf{C}$. In the particular case of $l = 1$, i.e. $B = A$, these codes are called Reed-Solomon codes for which $d = \delta$ and $k = n - \delta + 1$. These codes are optimal in the sense that they have the best possible value of k which is less than or equal to $n - d + 1$ for any linear codes having the fixed minimum distance d.*

Example 2 (Hermitian codes): *For the Hermitian curve $\mathcal{X} : x^{a+1} - y^a - y = 0$ over $A = F_q$, where $q = a^2$, a couple of $P = \{(x, y) | x^{a+1} - y^a - y = 0\}$ ($\subset B = F_q^2$) and $Q =< \boldsymbol{x}^{\boldsymbol{p}} := x^{p_1} y^{p_2} | o(\boldsymbol{x}^{\boldsymbol{p}}) := ap_1 + (a+1)p_2 \le m, \boldsymbol{p} = (p_1, p_2) \in \Delta(I) >$ ($\subset C = R/I$), $0 < m < n$, defines a Hermitian code* *as the dual linear code, which is a typical one-point AG code. In this case,* $R = F[\boldsymbol{x}] = F_q[x,y]$, $I =< y^a + y - x^{a+1}, x^q - x >\subset R$, *and* $C = R/I$ *is identified with the ring of algebraic functions* f *having a single pole of order* $o(f)$ *at the point* P_∞ *of infinity on* $\mathcal{X}$. *Furthermore,* $\Delta(I) = \{\boldsymbol{p} = (p_1, p_2) | 0 \leq p_1 \leq q-1, 0 \leq p_2 \leq a-1\}$. *This code has the information dimension* $k = n - m + g + 1$ *and the minimum distance* $d \geq \delta_G := m - 2g + 2$ *for the genus* $g = \frac{a(a-1)}{2}$ *of the curve* $\mathcal{X}$, *where* δ_G *is called the Goppa designed distance. Another designed distance* δ_{FR} *which satisfies in general* $\delta_{FR} \geq \delta_G$ *is given by Feng and Rao. It is known that the term order corresponding to the pole number* $o(\boldsymbol{x}^{\boldsymbol{p}})$ *gives a best selection to construct a sequence of decreasing dual codes corresponding to the increasing values of the parameter* m.

Sometimes algebraic codes can be embedded into another algebraic structure. For $I =< x_1^{n_1} - 1, \cdots, x_m^{n_m} - 1 >\subset F_q[\boldsymbol{x}]$ and $P = V(I) = \{(\alpha_1^{p_1}, \cdots, \alpha_m^{p_m}) | 0 \leq p_1 < n_1, \cdots, 0 \leq p_m < n_m\} \subset (F_{q^l})^m$, where α_i is the n_i-th root of 1, $1 \leq i \leq m$, the code vectors $\underline{c} = (c_{\boldsymbol{p}})$ are represented as polynomials $c(\boldsymbol{x}) = \sum_{\boldsymbol{p} \in \Delta(I)} c_{\boldsymbol{p}} \boldsymbol{x}^{\boldsymbol{p}}$ and the code $\mathbf{C}$ can happen to be just an ideal J/I of the residue class ring R/I for an ideal J of $R = F_q[\boldsymbol{x}]$, where such a code is called an mD cyclic code (Ikai et al. 1976a, Imai 1977, Sakata 1981). In this case the Gröbner basis algorithm is invoked to find a Gröbner basis of J, which gives a systematic encoding mapping from A^k into A^n (Sakata 1981). (In this context, the introduction of the Gröbner basis algorithm and its application in coding theory emerged.)

Another application of the Gröbner basis algorithm to encoding is given for AG codes. Usually it is difficult to get a systematic encoding mapping from either of the definitions of primal and dual linear codes. Based on the automorphism group of the defining curve $\mathcal{X}$, one can get a decomposition of the code into a sum of modules using a Gröbner basis technique. This module structure gives us a nice way of encoding the code (Heegard et al. 1995).

3 Error corrrection: Decoding

To use error-correcting codes in practice, decoding or error correction procedure is indispensable. That is how to find the sent code vector $\underline{c} \in \mathbf{C}$ or equivalently the error vector $\underline{e} := \underline{r} - \underline{c}$ from a given received vector $\underline{r}$. For the dual linear codes $\mathbf{C}$, one can get some information about the error vector $\underline{e}$ as follows. In case of $Q =< f^{(1)}, \cdots, f^{(r)} >$, $r = n - k$, we have the $r \times n$ parity check matrix of the code $\mathbf{C}$

$$H := \| f^{(i)}(P_j) \|, \; 1 \leq i \leq r, \; 1 \leq j \leq n,$$

by which one can get the syndrome vector of $\underline{r}$ as $\underline{s} := H\underline{r}$ $(\in A^r)$. Because $H\underline{c} = \underline{0}$ for any $\underline{c} \in \mathbf{C}$, $\underline{s} = H\underline{e}$.

On decoding one is required to find $\underline{e} \in A^n$ such that $H\underline{e} = \underline{s}$ for the given $\underline{s}$ and the weight $w(\underline{e}) := \#\{i | e_i \neq 0, 1 \leq i \leq n\}$ is the smallest. Assuming that the error vector $\underline{e}$ has weight $w(\underline{e}) \leq t' := \lfloor \frac{d'-1}{2} \rfloor$ for some lower bound d' of the minimum distance d, we neglect the danger of losing track of $\underline{e}$ with $w(\underline{e}) > t'$ in trying to find $\underline{e}$ with $w(\underline{e}) \leq t'$, where we call such a trial a bounded distance decoding up to t' errors. In case of $d' = d$, we call it the minimum distance decoding for short. If one can find the error locators $E := \{P_j | e_j \neq 0, 1 \leq j \leq n\} \subset P$, such that $\#E \leq d - 1$ for the minimum distance d of the code $\mathbf{C}$, the system of linear equations $H\underline{e} = \underline{s}$ is equivalent to a smaller system $H'\underline{e}' = \underline{s}$, which has a unique solution $\underline{e}'$ and can be solved by Gaussian elimination, where $H' := H|_E$ and $\underline{e}' := \underline{e}|_E$ are the restrictions of H and $\underline{e}$ within the part E, respectively. Thus, the decoding problem is reduced to finding the error locators E with $\#E \leq d-1$ or rather with $\#E \leq \lfloor \frac{d-1}{2} \rfloor$.

In case of $R = F_q[\boldsymbol{x}] := F_q[x_1, \cdots, x_m]$, for $C = R/I$, we can have a natural numbering of the elements of C according to the admissible total order $\leq_T$ in $\Delta(I)$ which corresponds to the term order over R. Therefore, we can take the basis elements of the linear space C as $f^{(1)}, \cdots, f^{(n)}$ such that $f^{(1)} <_T \cdots <_T f^{(n)}$, and $f^{(i)}$ corresponds to $\boldsymbol{p}^{(i)} \in \Delta(I)$, $1 \leq i \leq n$. Putting $s_{\boldsymbol{p}^{(i)}}(\underline{e}) := \sum_{j=1}^n e_j f^{(i)}(P_j)$, we call $\underline{s}(\underline{e}) := (s_{\boldsymbol{p}^{(i)}}(\underline{e}))$, $\boldsymbol{p}^{(i)} \in \Delta(I)$, the error syndrome array, among the elements of which we know only the values $s_{\boldsymbol{p}^{(i)}}(\underline{e}) = s_{\boldsymbol{p}^{(i)}}(\underline{r}) := \sum_{j=1}^n r_j f^{(i)}(P_j)$, for $f^{(i)} \in Q$, $1 \leq i \leq r$. If we get all the values $s_{\boldsymbol{p}^{(i)}}$, $\boldsymbol{p}^{(i)} \in \Delta(I)$, by any appropriate method, we can obtain the error values e_j, $1 \leq j \leq n$, because the injective embedding φ from $C = R/I$ into A^n is surjective in view of $\#\Delta(I) = n$. This procedure is called the generalized inverse Fourier transform (Saints and Heegard 1995). Thus, the decoding can be viewed also as finding the unknown error syndrome values.

Example 3 (RS codes) *According to the natural numbering of the basis elements*
$\{x^0, x^1, \cdots, x^{n-1}\}$ *of* $R/I = F_q[x]/ < x^n - 1 >$, *we define the error syndrome sequence*

$$s_i(\underline{e}) := \sum_{j=1}^n e_j x^{i-1}(\alpha^j) = \sum_{j=1}^n e_j \alpha^{(i-1)j}, \ 1 \leq i \leq n, \tag{3.1}$$

where $s_i(\underline{e}) = s_i(\underline{r}) := \sum_{j=1}^n r_j x^{i-1}(\alpha^j)$, $1 \leq i \leq \delta - 1$, *are known from the received vector* $\underline{r} = (r_1, \cdots, r_n)$.

Example 4 (Hermitian codes) *According to the numbering corresponding to the pole numbers* $o(\cdot)$, *we have the 2D error syndrome array*

$$s_{\boldsymbol{p}}(\underline{e}) := \sum_{j=1}^n e_j \boldsymbol{x}^{\boldsymbol{p}}(P_j), \ \boldsymbol{p} \in \Delta(I) = \{\boldsymbol{p} = (p_1, p_2) | 0 \leq p_1 \leq q-1, 0 \leq p_2 \leq a-1\}, \tag{3.2}$$

where $s_{\boldsymbol{p}}(\underline{e}) = s_{\boldsymbol{p}}(\underline{r}) := \sum_{j=1}^{n} r_j \boldsymbol{x}^{\boldsymbol{p}}(P_j)$, $o(\boldsymbol{x}^{\boldsymbol{p}}) = ap_1 + (a+1)p_2 \leq m$, $\boldsymbol{p} \in \Delta(I)$, *are known from the received vector* $\underline{r} = (r_1, \cdots, r_n)$.

For the error locators E, we can define the error locator ideal

$$I(E) := \{f \in C = R/I | f(P_j) = 0, \ ^{\forall}P_j \in E\}.$$

In Section 5, we show a method of finding a Gröbner basis of the error locator ideal from a given syndrome array $\underline{s} = (s_{\boldsymbol{p}^{(i)}})$ for $\boldsymbol{p}^{(i)}$ corresponding to $f^{(i)} \in Q$. At the same time, we obtain the unknown error syndrome values. In the case of Hermitian codes, the residue class ring $C = R/I$ is isomorphic to an $F_q[x]$-module because any element f of R/I can be represented as an a-tuple $(f_1(x), \cdots, f_a(x))$ of univariate polynomials such that $f = \sum_{i=1}^{a} f_i(x) y^{i-1}$. In this terminology, we can deal with the error locator module $M(E)$ of C as the $F_q[x]$-module, where the definition of $M(E)$ is the same as $I(E)$ in form (Sakata 1997).

4 Minimum distance decoding

After publication of the paper (Buchberger 1985) in the book 'Multidimensional Systems Theory' edited by N. K. Bose, the Gröbner basis theory and algorithm became known gradually also in the world of coding theory. One of the applications of the Buchberger algorithm to decoding appeared at the beginning of 90's. First it was devised to decode binary BCH codes, i.e. BCH codes over the binary field $A = F_2$ (as the symbol alphabet), up to $\lfloor \frac{\delta-1}{2} \rfloor$ errors for the designed distance δ (Cooper 1990). Now this idea is generalized to decoding of the dual linear codes including nonbinary BCH codes, RS codes (Chen et al. 1994a, Chen et al. 1994b) and AG codes (Chen et al. 1994c) up to the minimum distance.

In case of the symbol locator space $B = \tilde{A}^m$ for some extension $\tilde{A} = F_{q^l}$ of the symbol field $A = F_q$, the syndrome equations $H\underline{e} = \underline{s}$ can be viewed as a system of algebraic equations for $(m+1)t$ unknowns. For simplicity, we change the notations as follows. Assuming that the true number τ of errors is less than or equal to $t := \lfloor \frac{d-1}{2} \rfloor$, let the error locators $E \subset P$ be such that $\#E = t$ and that E contains τ true error locators ($\tau \leq t$). Now, for $E = \{P_{i_1}, \cdots, P_{i_t}\}$, we put $P_{i_j} = (x_{j1}, \cdots, x_{jm})$, $1 \leq j \leq t$, and denote e_{i_j} as e_j, $1 \leq j \leq t$. Then, we have the algebraic system for the unknowns $e_j, x_{j1}, \cdots, x_{jm}$, $1 \leq j \leq t$,

$$\sum_{j=1}^{t} e_j f^{(i)}(x_{j1}, \cdots, x_{jm}) = s_i, \ 1 \leq i \leq r, \tag{4.1}$$

where $t - \tau$ values e_j should vanish if $\tau < t$. Herein the most important fact is that there corresponds one and only one syndrome vector $\underline{s}$ to each error vector $\underline{e}$ with $w(\underline{e}) \leq t$. It implies that the algebraic system (4.1) has a unique solution $\{(e_j, x_{j1}, \cdots, x_{jm}) | 1 \leq j \leq t\}$ if $\tau \leq t$ (as it is assumed in our context). Since $e_1, \cdots, e_t \in F_q$, $x_{11}, \cdots, x_{tm} \in F_{q^l}$, the following equations are added:

$$\begin{aligned} e_j^q - e_j = 0, \qquad & 1 \leq j \leq t, \\ x_{ja}^{q^l} - x_{ja} = 0, \quad & 1 \leq j \leq t, 1 \leq a \leq m, \end{aligned} \tag{4.2}$$

If the functions $f^{(i)}$, $1 \leq i \leq r$, are polynomials as in most cases, it is possible to solve the system of algebraic equations composed of (4.1) and (4.2) by applying the Buchberger algorithm. By treating s_i as variables y_i, $1 \leq i \leq r$, we have a set of $r + t(1+m)$ polynomials $\in F_{q^l}[e_1, \cdots, e_t, x_{11}, \cdots, x_{tm}, y_1, \cdots, y_r]$:

$$\begin{aligned} \sum_{j=1}^{t} e_j f^{(i)}(x_{j1}, \cdots, x_{jm}) - y_i, \qquad & 1 \leq i \leq r, \\ e_j^q - e_j, \qquad & 1 \leq j \leq t, \\ x_{ja}^{q^l} - x_{ja}, \quad & 1 \leq j \leq t, 1 \leq a \leq m, \end{aligned} \tag{4.3}$$

from which we obtain a triangular system of algebraic equations via the Buchberger algorithm with respect to the lexicographic term order with

$$y_1 < \cdots < y_r < e_1 < \cdots < e_t < x_{11} < \cdots < x_{tm}.$$

The advantage of this method is that one can get the error vector $\underline{e}$ easily by substituting any given syndrome values s_i for the variables y_i, $1 \leq i \leq r$, provided that one finds the triangular system peculiar to the code and a prescribed way of calculating a set of values for unknowns $e_1, \cdots, e_t, x_{11}, \cdots, x_{tm}$ in advance in the stage of preprocessing.

Example 5 (Binary BCH codes) *In the case of $A = F_2$ and $\tilde{A} = F_{2^l}$, decoding is just to find the error locators since the error values $e_j = 1$ for true error locators P_j. The syndrome equation gives the algebraic system for unknowns $x_1, \cdots, x_t$*

$$\sum_{j=1}^{t} x_j^{i-1} = s_i, 1 \leq i \leq \delta - 1, \tag{4.4}$$

where $s_i = \sum_{j=1}^{n} r_j \alpha^{(i-1)j}$ ($\in F_{2^l}$), $1 \leq i \leq \delta - 1$ for a primitive n-th root α ($\in F_{2^l}$), $n | 2^l - 1$, and only nonzero values of x_j give the true error locators. In this case, we have the simpler additional equations

$$x_j^{n+1} - x_j = 0, 1 \leq j \leq t, \tag{4.5}$$

instead of $x_j^{2^l} - x_j = 0, 1 \leq j \leq t$.

Generally, application of the Buchberger algorithm to this method has high computational complexity for large n. But, for some codes, a better way of attacking this problem with a Gröbner basis approach has been proposed recently (Loustaunau and York 1997). It is based on an efficient computational method of zero-dimensional Gröbner basis by a change of term order (Buchberger 1970, Faugère et al. 1993). Although it is very difficult to compute a Gröbner basis of an ideal generated by any given set of polynomials, it is easier to obtain a Gröbner basis with respect to a specific term order by the method if a GB with respect to another term order is given a priori. For example, in the case of binary BCH codes, the system of polynomials (from Example 5)

$$\sum_{j=1}^{t} x_j^{i-1} - y_i, \quad 1 \le i \le r,$$

$$x_j^{n+1} - x_j, \quad 1 \le j \le t,$$

is fortunately a Gröbner basis with respect to the lexicographic term order with

$$x_1 < \cdots < x_t < y_1 < \cdots < y_r.$$

Therefore, one can apply Faugère et al.'s algorithm (ibid.) to obtain the desired Gröbner basis with respect to another lexicographic term order with

$$y_1 < \cdots < y_r < x_1 < \cdots < x_t.$$

This method reduces the complexity $n^{O(t^3)}$ of the Buchberger algorithm to $n^{O(t)}$, although the exponential complexity still remains.

5 Fast decoding

One of the basic problems in the Gröbner bases theory is raised in connection with the theory of linear recurrence arrays. Let $Supp(u)$ be a subset of Z_0^m. An m-dimensional (mD) array u over a finite field F_q is a mapping from $Supp(u)$ into F_q and denoted as $u = (u_{\boldsymbol{p}})$, where $u_{\boldsymbol{p}} := u(\boldsymbol{p})$, $\boldsymbol{p} \in Supp(u)$. For an array u with $Supp(u) = Z_0^m$, called a perfect array, a linear recurrence (or a linear homogeneous partial difference equation with constant coefficients in F_q)

$$\sum_{\boldsymbol{r} \in Supp(f)} f_{\boldsymbol{r}} u_{\boldsymbol{r}+\boldsymbol{p}} = 0, \ \boldsymbol{p} \in Z_0^m, \tag{5.1}$$

can be represented by the m-variate polynomial

$$f = \sum_{\boldsymbol{r} \in Supp(f)} f_{\boldsymbol{r}} \boldsymbol{x}^{\boldsymbol{r}} \in F_q[\boldsymbol{x}], \ \boldsymbol{p} \in Z_0^m, \tag{5.2}$$

called its characteristic polynomial, where $Supp(f)$ is a finite subset of Z_0^m and $\boldsymbol{x}^{\boldsymbol{r}} := x_1^{r_1} \cdots x_m^{r_m}$. We say that f is valid for u if the identity (5.1) holds for any $\boldsymbol{p}$ for which the lefthand side of (5.1) is meaningful, i.e. it can be calculated. If we are given a set of polynomials f or linear recurrences (5.1) in advance, we can ask not only what the solutions u satisfying all the linear recurrences simultaneously are, but also how we can get the solution set. This primal problem had been attacked earlier in the investigations of systematic encoding of 2D cyclic codes so that an algorithm (Sakata 1981) identical with the Buchberger algorithm was devised as mentioned in Sections 1 and 2.

Given a perfect array u, the polynomials f that are *valid* for u compose an ideal of the m-variate polynomial ring $F_q[\boldsymbol{x}]$, which is called the characteristic ideal $I(u)$ of u. Now, one can ask the inverse problem, i.e. how to find the characteristic ideal of a given u or its Gröbner basis. The problem of finding a Gröbner basis of $I(u)$ can be solved efficiently by an iterative algorithm (Sakata 1988, Sakata 1990) called the BMS algorithm which is a generalization of the BM algorithm (Berlekamp 1968, Massey 1969) for 1D arrays $u = (u_i)$, $i \in Z_0$, to multidimensional arrays. From the standpoint of system theory, this is a kind of realization problem, i.e. one is required to synthesize an mD linear feedback shift-register capable of generating a given mD array. The structure of this switching circuit is represented by a set of polynomials called the connection polynomials, among which some consistency should be maintained. This implies the fact that it is possible to generate or determine an array consistently based on the linear recurrences characterized by the connection polynomials from a set of arbitrary initial values if and only if the connection polynomials compose a Gröbner basis (Sakata 1989). In passing, we notice that it was no accident that the Buchberger algorithm appeared in the book on system theory (Buchberger 1985), by which the Gröbner bases theory became popular in the engineering world.

In case of BCH codes and RS codes, when one applies the BM algorithm to a given finite 1D syndrome array $\underline{s} = (s_i)$, $1 \le i \le \delta - 1$, one can get a generator polynomial $f(x) \in F_q[x]$ of the characteristic ideal $I(\underline{s}) \subset F_q[x]$, where $f(x)$ is the error locator polynomial whose roots are just the error locators. In addition, the linear recurrence characterized by $f(x)$ can be used to determine the unknown error syndrome values $s_i(\underline{e})$, $\delta \le i < n$. In the case of such a Euclidean ring $F_q[x]$, we have only to treat single polynomials, neglecting notions such as ideal and Gröbner basis. On the other hand, for the m-variate polynomial ring $F_q[\boldsymbol{x}]$ and mD arrays, we must treat ideals, sets of polynomials, and Gröbner bases explicitly. To enumerate the components $u_{\boldsymbol{p}}$ of an array u, according to the term order $\le_T$ over $F_q[\boldsymbol{x}]$, we often take the corresponding total order $\le_T$ (denoted by the same symbol) over Z_0^m. In addition, we have the partial order $\le_P$ over Z_0^m such that $\boldsymbol{p} = (p_1, \cdots, p_m) \le_P \boldsymbol{q} = (q_1, \cdots, q_m)$ if and only if $p_i \le q_i$, $1 \le i \le m$. We consider a finite array u with $Supp(u) = \Sigma^{\boldsymbol{q}} := \{\boldsymbol{p} \in Z_0^m | \boldsymbol{p} <_T \boldsymbol{q}\}$ for $\boldsymbol{q} \in Z_0^m$, where we restrict ourselves to consider the total order $\le_T$ such that such a subset $\Sigma^{\boldsymbol{q}}$ is finite for $\boldsymbol{q} \in Z_0^m$, and denote such an array as $u = u^{\boldsymbol{q}}$. For $f \in F_q[\boldsymbol{x}]$ and its leading power product $LPP(f) = \boldsymbol{x}^{\boldsymbol{r}}$ with respect to $\le_T$, we call $\boldsymbol{r} (\in Z_0^m)$

the degree of f and denote it as $\boldsymbol{r} = Deg(f)$. With respect to the partial order $\leq_P$, we define a point set $\Gamma_{\boldsymbol{p}} := \{\boldsymbol{r} \in Z_0^m | \boldsymbol{r} \leq_P \boldsymbol{p}\}$ ($\subset Z_0^m$) for a fixed point $\boldsymbol{p}$.

For a finite array $u^{\boldsymbol{q}} = (u_{\boldsymbol{p}} | \boldsymbol{p} <_T \boldsymbol{q})$ and a point $\boldsymbol{p} <_T \boldsymbol{q}$, we say that a polynomial F with $Deg(f) = \boldsymbol{s}$ is valid at the point $\boldsymbol{p}$ if the following identity holds:

$$\sum_{\boldsymbol{r} \in Supp(f)} f_{\boldsymbol{r}} u_{\boldsymbol{r}+\boldsymbol{p}-\boldsymbol{s}} = 0. \tag{5.3}$$

(In the expressions (5.1) and (5.3), the meanings of the point $\boldsymbol{p}$ are different, i.e. the relative coordinates in (5.1) and the absolute coordinates in (5.3), where the former depend on the choice of f, but the latter not.) Thus, we say that f is valid for a finite array $u = u^{\boldsymbol{q}}$ if f is valid at any $\boldsymbol{p}$ such that $\boldsymbol{s} \leq_P \boldsymbol{p} <_T \boldsymbol{q}$. For a finite array u, we define a *minimal* polynomial set of u as a finite set F of polynomials which satisfies the following conditions:

(1) Any element f of F is valid for the array u;

(2) All elements $f \in F$ are *minimal* for u in the sense that, if a polynomial $g \in F_q[\boldsymbol{x}]$ is valid for u, there exists $f \in F$ such that $Deg(f) \leq_P Deg(g)$.

The condition (2) can be paraphrased in terms of the point set $\Delta(F) := \cup_{f \in F} \Gamma_{Deg(f)}$ ($\subset Z_0^m$), which is similar to a so-called Gröbner basis window, i.e. there exists no polynomial g with $Deg(g) \in \Delta(F)$ which is valid for u. We observe that F is not always unique for u, but that $\Delta(F)$ depends only on u, and we can denote it as $\Delta(u)$, called the excluded point set of u. If $\boldsymbol{p} \leq_T \boldsymbol{q}$, then $\Delta(u^{\boldsymbol{p}}) \subseteq \Delta(u^{\boldsymbol{q}})$. For a given array $u = u^{\boldsymbol{q}}$, the BMS algorithm finds a minimal polynomial set F of u iteratively. One checks each component $u_{\boldsymbol{p}}$ of the array u successively according to the total order $\leq_T$, and update a minimal polynomial set F of $u^{\boldsymbol{p}}$ to that of $u^{\boldsymbol{p} \oplus 1}$ by a set of simple calculations for the next point $\boldsymbol{p} \oplus 1$ with respect to the total order $\leq_T$. Thus, one tests whether every element f of F is valid still at $\boldsymbol{p}$ by calculating the linear recurrence (5.3). If some $f \in F$ is not valid at $\boldsymbol{p}$, one should alter F. Sometimes the excluded point set strictly increases, i.e. $\Delta(u^{\boldsymbol{p}}) \subset \Delta(u^{\boldsymbol{p} \oplus 1})$. For a perfect array u with a meaningful $I(u)$ such that $I(u) \neq \emptyset, R$, a certain finite part $u^{\boldsymbol{q}}$ has a minimal polynomial set F such that F is a Gröbner basis of $I(u)$ and $\Delta(u^{\boldsymbol{q}})$ is its Gröbner basis window.

In the context of decoding of one-point AG codes such as Hermitian codes, a minimal polynomial set of the error syndrome array $\underline{s}$ is a Gröbner basis of the error locator ideal, which can be obtained provided that the values $s_{\boldsymbol{p}}(\underline{e})$ are known at any $\boldsymbol{p} \in \Sigma^{\boldsymbol{q}}$ for a sufficiently large $\boldsymbol{q}$, where the term order $\leq_T$ is taken corresponding to the pole numbers. As a matter of fact, the part of $s_{\boldsymbol{p}}(\underline{e})$, $o(\boldsymbol{x}^{\boldsymbol{p}}) \leq m$, known from the received vector $\underline{r}$, is not enough for decoding up to $\lfloor \frac{\delta-1}{2} \rfloor$ errors, where δ is the Feng-Rao designed distance δ_{FR} of the code. However, by a modification of the BMS algorithm with a trick based on majority vote logic, one can get the unknown syndrome values $s_{\boldsymbol{p}}(\underline{e})$ for any $\boldsymbol{p} \in \Delta(I)$ (Sakata et al. 1995a, Sakata et al. 1995b,

Sakata et al. 1995c), where a minimal polynomial set of $\underline{s}(\underline{e}) = (s_{\boldsymbol{p}}(\underline{e}))$, $\boldsymbol{p} \in \Delta(I)$, coincides with a Gröbner basis of the error locator ideal $I(E)$. This method is realized also in the form of finding a Gröbner basis of the error locator module $M(E)$ (Sakata 1997). Its computational complexity is $\mathcal{O}(n^a)$ with $2 < a < 3$, which is generally less than $\mathcal{O}(n^3)$ for the Feng-Rao algorithm based on Gaussian elimination with majority vote logic (Feng and Rao 1993). Almost simultaneously, there were published several decoding algorithms (Feng and Rao 1994, Leonard 1995, O'Sullivan 1995, Kötter 1996) of one-point AG codes, which are essentially equivalent to the BMS algorithm. The first application of the BMS algorithm to decoding of AG codes appeared in Justesen et al. (1992), and that to 2D cyclic codes in Sakata (1991). Saints et al. (1993) treated a case where there is no need of majority vote logic in using the BMS algorithm straightforwardly for decoding. To get a wider view of decoding of AG codes, see Høholdt and Pellikaan (1995).

Example 6 (Hermitian codes) *First one applies the BMS algorithm to the syndrome array* $s(\underline{e}) := (s_{\boldsymbol{p}}(\underline{e}))$ *defined on the extended region* $\{\boldsymbol{p} = (p_1, p_2) | o(\boldsymbol{x}^{\boldsymbol{p}}) = ap_1 + (a+1)p_2 \leq m,\ 0 \leq p_2 \leq 2a-2\}$, *on which the values* $s_{\boldsymbol{p}}(\underline{e})$ *are obtained based on linear recurrences implied by the equation* $x^{a+1} - y^a - y = 0$ *defining the curve* $\mathcal{X}$ *from the known values. Then, successively for each* $j = 1, \cdots, 2g$, *one continues to apply the BMS algorithm with majority vote logic, by which one can get a set of correct syndrome values among the candidates values for the unknown syndromes* $s_{\boldsymbol{p}}$ *for* $o(\boldsymbol{x}^{\boldsymbol{p}}) = m + j$. *A minimal polynomial set of the obtained array is just a Gröbner basis of the error locator module, whose zeros give the error locators. The complexity of decoding Hermitian codes by the BMS algorithm is* $\mathcal{O}(n^{\frac{7}{3}})$.

The method can be extended to some more general problems of decoding (e.g. Kötter 1996, Sakata et al. 1997). Furthermore, it has been shown that the BMS algorithm for decoding of AG codes can be realized efficiently by a parallel processing hardware architecture (Sakata and Kurihara 1997).

6 Concluding remarks

In this note, we have reviewed significance and application of the Gröbner bases theory and related algorithms in coding theory, by getting the encoding and decoding processes into focus. In treating the algebraic codes, particularly multidimensional or multivariate codes including the most recent and promising AG codes, one cannot dispense with the concept of Gröbner basis and its methodology. In the other engineering applications, numerical errors accompanying finite precision computations are unavoidable. Coding theory is a rare exception where there is no need of worrying about exactness of the calculated values and coefficient explosion in exact computations free from numerical errors, because computations over finite fields prevail. This fact ensures that the Gröbner bases theory and algorithm match coding theory better than any other branches of science and engineering.

In fact, there have been more investigations in coding theory which are connected to the Gröbner bases theory and algorithm in various aspects within or beyond the present scope. From shortage of pages, we have omitted descriptions of several relevant works in this note.[1] The following are a few examples of them:

(1) implementation of the BM algorithm as converting a Gröbner basis of the solution module with respect to one term order to a Gröbner basis with respect to another term order (Fitzpatrick 1995). (2) weight enumeration of small code vectors of cyclic codes by using the Buchberger algorithm (Augot 96).

References

Augot, D. (1996): Descrption of minimum weight codewords of cyclic codes by algebraic systems. Finite Fields and Their Appl. 2: 138–152. Berlekamp, E. R. (1968): Algebraic Coding Theory. MacGraw-Hill, New York.

Buchberger, B. (1965): Ein Algorithmus zum Auffinden der Basiselemente des Restklassenringes nach einem nulldimensionalen Polynomideal. Ph. D. Dissertation University of Innsbruck, Innsbruck, Austria.

Buchberger, B. (1970): Ein algorithmisches Kriterium für die Lösbarkeit eines algebraischen Gleichungssystems. Aequ. Math. 4: 374–383.

Buchberger, B. (1976): A theoretical basis for the reduction of polynomials to canonical form. ACM SIGSAM Bullet. 10: 19–29.

Buchberger, B. (1985): Gröbner bases: An algorithmic method in polynomial ideal theory. In: N.K. Bose (ed.): Multidimensional Systems Theory. Reidel, Dordrecht, pp.184–232.

Chen. X., Reed, I. S., Helleseth, T., Truong, T. K. (1994a): Algebraic decoding of cyclic codes: A polynomial ideal point of view. Contemp. Math. 168: 15–22.

Chen. X., Reed, I. S., Helleseth, T., Truong, T. K. (1994b): Use of Gröbner bases to decode binary cyclic codes up to the true minimum distance. IEEE Trans. Inform. Theory. 40: 1654–1661.

Chen. X., Reed, I. S., Helleseth, T., Truong, T. K. (1994c): General principles

[1] I am afraid I might have ignored some other important works by limitation of knowledge. I should appreciate it much if you, kind readers, would give me any information about missing items.

for the algebraic decoding of cyclic codes. IEEE Trans. Inform. Theory. 40: 1661–1663.

Cooper, A. B. III (1990): Direct solution of BCH decoding equations. In: Arikan, E. (ed.): Communication, Control, and Signal Processing. Elsevier, Amsterdam, pp.281–286.

Faugère, J. G., Gianni, P., Lazard, D., Mora, T. (1993): Efficient computation of zero-dimensional Gröbner bases by a change of ordering. J. Symbolic Comp. 16: 329–344.

Feng, G.-L., Rao, T. R. N. (1993): Decoding of AG codes up to the designed minimum distance. IEEE Trans. Inform. Theory. 39: 37–45.

Feng, G.-L., Rao, T. R. N. (1994): Simplified understanding and efficient decoding of a class of AG codes. IEEE Trans. Inform. Theory. 40: 981–1001.

Fitzpatrick, P. (1995): On the key equation. IEEE Trans. Inform. Theory. 41: 1290–1302.

Goppa, V. G. (1981): Codes on algebraic curves. Soviet Math. Dokl. 24: 170–172.

Heegard, C., Little, J., Saints, K. (1995): Systematic encoding via Gröbner bases for a class of AG Goppa codes. IEEE Trans. Inform. Theory. 41: 1752–1761.

Hermann, G. (1926); Die Frage der endlich vielen Schritte in der Theorie Polynomideale. Math. Ann. 95: 736–788.

Høholdt, T., Pellikaan, R. (1995): On the decoding of AG codes. IEEE Trans. Inform. Theory. 41: 1589–1614.

Ikai, T., Kosako, H., Kojima, Y. (1976a): Basic theory of 2D cyclic codes: Generator polynomials and the positions of check symbols. Trans. Inst. Electr. Comm. Eng. J59-A: 33-41.

Ikai, T., Kosako, H., Kojima, Y. (1976b): Basic theory of 2D cyclic codes: Periods of ideals and fundamental theorems. Trans. Inst. Electr. Comm. Eng. J59-A: 216–223.

Imai, H. (1977): A theory of 2D cyclic codes. Inform. Contr. 34: 1-21.

Justesen, J., Larsen, K.J., Jensen, H. E., Høholdt, T. (1992): Fast decoding of codes from algebraic plane curves. IEEE Trans. Inform. Theory. 38: 111–119.

Kötter, R. (1996): Fast generalized minimum-distance decoding of AG codes and RS codes. IEEE Trans. Inform. Theory. 42: 721–737.

Leonard, D. (1995): Error-locator ideals for AG codes. IEEE Trans. Inform. Theory. 41: 819–824.

Loustaunau, P., York, E. V. (1997): On the decoding of cyclic codes using Gröbner bases. To appear in J. AAECC.

MacWilliams, F. J., Sloane, N. J. A. (1977): The Theory of Error-Correcting Codes, Part I & Part II. North-Holland Publish., Amsterdam.

Massey, J. L. (1969): Shift-register synthesis and BCH decoding. IEEE Trans. Inform. Theory. 15: 122–127.

O'Sullivan, M. E. (1995): Decoding of codes defined by a single point on a curve. IEEE Trans. Inform. Theory. 41: 1709–1719.

Peterson, W. W., Weldon, E. J. Jr. (1972): Error-Correcting Codes. 2nd Ed. MIT Press, Cambridge, Massachusetts, London.

Poli, A., Huguet, L. (1988): Codes Correcteurs: Théorie et Applications, Masson, Paris. (English ed. (1992): Error-Correcting Codes: Theory and Applications. Prentice-Hall, Hemel Hempstead.)

Saints, K., Heegard, C. (1993), On hyperbolic cascaded RS codes. In: Cohen, G., Mora., T., Moreno, O. (eds.): Applied Algebra, Algebraic Algorithms, and Error-Correcting Codes. Springer, Berlin Heidelberg New York Tokyo, pp.291–303 (Lecture Notes in Computer Science, vol.673).

Saints, K., Heegard, C. (1995): AG codes and multidimensional cyclic codes: A unified theory and algorithms for decoding using Gröbner bases. IEEE Trans. Inform. Theory. 41: 1733-1751.

Sakata, S. (1978): General theory of doubly periodic arrays over an arbitrary finite field and its applications. IEEE Trans. Inform. Theory. 24: 719–730.

Sakata, S. (1981): On determining the independent point set of doubly periodic arrays and encoding 2D cyclic codes and their duals. IEEE Trans. Inform. Theory. 27: 556-565.

Sakata, S. (1988): Finding a minimal set of linear recurring relations capable of generating a given finite 2D array. J. of Symbolic Comp. 5: 321–337.

Sakata, S. (1989): Synthesis of 2D linear feedback shift registers and Gröbner bases. In: Huguet, L., Poli, A. (eds.): Applied Algebra, Algebraic Algorithms, and Error-Correcting Codes. Springer, Berlin Heidelberg New York Tokyo, pp.394–407, (Lecture Notes in Computer Science, vol.356).

Sakata, S. (1990): Extension of the BM algorithm to N dimensions. Inform. and Comput. 84: 207-239.

Sakata, S. (1991): Decoding binary 2D cyclic codes by the 2D BM algorithm. IEEE Trans. Inform. Theory. 37: 1200-1203.

Sakata, S. (1997): A vector version of the BMS algorithm for implementing fast erasure-and-error decoding of one-point AG codes. In: Mora, T., Mattson, H. (eds): Applied Algebra, Algebraic Algorithms, and Error-Correcting Codes. Springer, Berlin Heidelberg New York Tokyo, pp.291–310 (Lecture Notes in Computer Science, vol.1255).

Sakata, S., Jensen, H. E., Høholdt, T. (1995a): Generalized Berlekamp-Massey decoding of AG codes up to half the Feng-Rao bound. IEEE Trans. Inform. Theory. 41: 1762–1768.

Sakata, S., Justesen, J., Madelung, Y., Jensen, H. E., Høholdt, T. (1995b): A fast decoding method of AG codes from Miura-Kamiya curves C_{ab} up to half the Feng-Rao bound. Finite Fields and Their Appl. 11: 83–101.

Sakata, S., Justesen, J., Madelung, Y., Jensen, H. E., Høholdt, T. (1995c): Fast decoding of AG codes up to the designed minimum distance. IEEE Trans. Inform. Theory. 41: 1672–1677.

Sakata, S., Kurihara, M. (1997): A systolic array architecture for implementing a fast parallel decoding algorithm of one-point AG codes. Submitted for IEEE Trans. Inform. Theory.

Sakata, S., Leonard, D., Jensen, H. E., Høholdt, T. (1997): Fast erasure-and-error decoding of AG codes up to the Feng-Rao bound. Submitted for IEEE Trans. Inform. Theory.

Tsfasman, M. A., Vlăduţ, S. G. (1991): Algebraic-Geometric Codes. Kluwer Academic Publish., Dordrecht.

Janet Bases for Symmetry Groups

Fritz Schwarz

1 Introduction

The subject of this article are systems of linear homogeneous partial differential equations (pde's) of various kinds. Above all such equations are characterized by the number m of dependent and the number n of independent variables. Additional quantities of interest are the number of equations, the order of the highest derivatives that may occur and the function field in which the coefficients are contained. Without further specification it is the field of rational functions in the independent variables. The basic new concept to be considered in this article is the *Janet base*. This term is chosen because the French mathematician Maurice Janet (Janet 1920) recognized its importance and described an algorithm for obtaining it. After it had been forgotten for about fifty years, it was rediscovered (Schwarz 1992) and utilized in various applications as it is described later on.

The theory of systems of linear homogeneous pde's is of interest for its own right, independent of its applications e. g. for finding symmetries and invariants of differential equations. Any such system may be written in infinitely many ways by linearly combining its members or derivatives thereof without changing its solution set. In general it is a difficult question whether there exist nontrivial solutions at all, or what the degree of arbitrariness of the general solution is. It may be a set of constants, or a set of functions depending on a differing number of arguments. This problem was the starting point for Janet. A *Janet base* is a unique representation of such systems of pde's that provides important information on its solutions similar to a Gröbner base representation of a system of algebraic equations, (Buchberger 1970) and (Buchberger 1985). Both a Gröbner base and a Janet base have one important feature in common: Except in very simple cases it is virtually impossible to calculate any of them by pencil-and-paper, i. e. an efficient computer algebra implementation is crucial for utilizing it in practical problems.

Basic concepts in Janet's theory are *terms* and *term orderings*. A term $t_{i,j}$ in a system of linear homogeneous pde's is either a dependent variable or a derivative of it. A *term ordering* is a linear order that is consistent with taking derivatives. This means a relation $t_{i,j} > t_{i',j'}$ remains valid if the same derivative ∂ is applied to either side of the order relation. Any linear homogeneous system of pde's in this article is arranged in the form

$$\begin{array}{ccccccc}
\boxed{t_{1,1}} & > & t_{1,2} & > & \ldots & > & t_{1,k_1} \\
\wedge & & & & & & \\
\boxed{t_{2,1}} & > & t_{2,2} & > & \ldots & > & t_{2,k_2} \\
\wedge & & & & & & \\
\vdots & & & & & & \\
\boxed{t_{m,1}} & > & t_{m,2} & > & \ldots & > & t_{m,k_m}
\end{array}$$

such that the ordering relations are valid as indicated for the ordering that is applied. Each line of this scheme corresponds to an equation of the system of pde's. In order to save space, sometimes several equations are arranged into a single line. In these cases, in any line the leading terms *increase* from left to right. The terms in the square boxes are the *leading terms*, i. e. the leading term is that term preceding any other term in the respective equation.

The leading terms subdivide the totality of derivatives into two subsets: Those that cannot be obtained by proper derivation of a leading term are called *parametric*, and the remaining ones, they are called *principal*. In any term ordering there is a *lowest term*, it is always one of the dependent variables itself. For a given number of dependent and independent variables there is a finite number of term orderings. Due to its importance the following definition is introduced.

Definition 1 *The pattern of leading terms is called the type of a Janet base.*

It will always be assumed in this article that the general solution of the systems of pde's considered contains a finite set of constants as arbitrary elements, i. e. its general solution is a linear combination of a fundamental system. There are two basic achievements of a Janet base representation for such a system of linear homogeneous pde's.

- ▷ The dimension of the solution space is identical to the number of parametric derivatives.
- ▷ In a Janet base the coefficients are uniquely defined and have a meaning, they may describe additional features of the solutions. This is not true for an arbitrary system of pde's where the coefficients may be changed at will.

Janet himself described all steps of an algorithm for transforming any given system of linear homogeneous pde's into a Janet base (Janet 1920). They are subsumed under the algorithm *JanetBase*. A detailed description of all steps may also be found in (Schwarz 1992).

Algorithm 1 *JanetBase(S). Given a linear homogeneous system of pde's $S = \{e_1, e_2, \ldots\}$ in a fixed term ordering, the Janet base corresponding to S is returned.*

$S1:$ *(Autoreduction) Assign $S := AutoReduce(S)$.*

$S2:$ *(Completion) Assign $S := CompleteSystem(S)$.*

$S3:$ *(Find Integrability Conditions) Find all pairs of leading terms v_i and v_j such that differentiation w.r.t. a non-multiplier x_{i_k} and multipliers $x_{j_1}, \ldots x_{j_l}$ respectively leads to*

$$\frac{\partial v_i}{\partial x_{i_k}} = \frac{\partial^{p_1+\ldots+p_l} v_j}{\partial x_{j_1}^{p_1} \ldots \partial x_{j_l}^{p_l}}$$

and determine the integrability conditions

$$c_{i,j} = Lcoef(e_j) \cdot \frac{\partial e_i}{\partial x_{i_k}} - Lcoef(e_i) \cdot \frac{\partial^{p_1+\ldots+p_l} e_j}{\partial x_{j_1}^{p_1} \ldots \partial x_{j_l}^{p_l}}$$

$S4:$ *(Reduce Integrability Conditions) For all $c_{i,j}$ assign*

$$c_{i,j} := ReduceEquation(c_{i,j}, S)$$

$S5:$ *(Termination?) If all $c_{i,j}$ are zero return S, otherwise assign $S := S \cup \{c_{i,j} | c_{i,j} \neq 0\}$, reorder S properly and go to $S1$.*

The first fundamental operation of Janet's algorithm is the *reduction*. Given any pair of equations e and f of the system, it may occur that the leading term of the latter equation f, or of a suitable derivative of it, equals some term in the former equation e. This coincidence may be applied to remove this term from e by an operation which is called a *reduction step*. Due to the properties of the term orderings described above and the genuine lowering of terms in any reduction step, it is assured that for any pair of equations this process terminates after a finite number of iterations until no further reduction steps are possible. In this case e is called *reduced* with respect to f. Similarly an equation is called reduced w.r.t. to a system of equations if it is reduced w.r.t. any equation of the system. If in a system any equation is reduced w.r.t. to the system of remaining equations, the full system is called *autoreduced*.

Consider any linear homogeneous system, choose two equations with the same dependent variable in its leading term and solve both equations with respect to its leading term. By suitable cross differentiation it is always possible to obtain two new equations such that the derivatives at the left hand sides are identical to each other. Intuitively it is expected that the same should be true for the right hand sides if all reductions w.r.t. to the remaining equations of the system are performed. A priori there are infinitely many conditions of this kind, they are called *integrability conditions*. It is one of the basic achievements of Janet to identify a *finite* number of constraints which guarantee that all integrability conditions reduce to zero if the algorithm terminates. To this end the concept of a *complete system* is introduced. For a complete system, there is a systematic procedure for obtaining all integrability conditions algorithmically as it is done in step $S3$. If an integrability condition may be reduced to zero by the equations of the system it is said to be *satisfied*, otherwise the reduced conditions are added as new equations to the system and the same procedure is applied to the extended system. Upon termination, an autoreduced system of pde's is returned for which all integrability conditions are satisfied. This property is of fundamental importance and a special term is introduced for it.

Definition 2 *(Janet Base). An autoreduced system of linear homogeneous pde's is called a Janet base for a given term ordering if all integrability conditions are reduced to zero.*

2 Classification of Janet Bases

From now on it will be assumed that $m = n = 2$ because this will be true for the applications considered later on. Furthermore it is assumed that a fixed term ordering is applied. Under these constraints a fairly explicit description of all possible Janet bases is possible. To any given dimension of the solution space there corresponds a finite number of possible types of Janet bases. They are obtained by the following considerations. In the total degree term ordering with $\eta > \xi$ and $y > x$ the derivatives up to order two are arranged as follows

$$\ldots \eta_{yy} > \eta_{yx} > \eta_{xx} > \xi_{yy} > \xi_{yx} > \xi_{xx} > \eta_y > \eta_x > \xi_y > \xi_x > \eta > \xi$$

According to the definition of a parametric derivative, for a r-dimensional solution space, r derivatives must be selected from this arrangement such that none of it may be obtained by differentiation from any of the remaining ones. For $r = 1$, 2 or 3 the possible selections are as follows.

$$\begin{aligned} r = 1: &\quad \{\eta\}, \{\xi\} \\ r = 2: &\quad \{\eta_x, \eta\}, \{\eta_y, \eta\}, \{\eta, \xi\}, \{\xi_y, \xi\}, \{\xi_x, \xi\} \\ r = 3: &\quad \{\eta_{xx}, \eta_x, \eta\}, \{\eta_y, \eta_x, \eta\}, \{\eta_{yy}, \eta_y, \eta\}, \{\eta_x, \eta, \xi\}, \{\eta_y, \eta, \xi\}, \\ &\quad \{\xi_x, \eta, f\xi\}, \{\xi_y, \eta, \xi\}, \{\xi_{yy}, \xi_y, \xi\}, \{\xi_y, \xi_x, \xi\}, \{\xi_{xx}, \xi_x, \xi\} \end{aligned}$$

In Table 1 the structure of the corresponding Janet bases is indicated and a unique notation for any of them is introduced. Each line corresponds to an equation of the Janet base, its leading term is enclosed in a box. The parametric derivatives that may occur in any equation are listed to the right of its leading term in decreasing order. In a concrete Janet base they are multiplied by a coefficient depending on the independent variables. Similar

$\mathbf{j}_{1,1}$	$\mathbf{j}_{1,2}$	$\mathbf{j}_{2,1}$	$\mathbf{j}_{2,2}$	$\mathbf{j}_{2,3}$	$\mathbf{j}_{2,4}$	$\mathbf{j}_{2,5}$
$\boxed{\xi}$	$\boxed{\eta}\ \xi$	$\boxed{\xi}$	$\boxed{\xi}$	$\boxed{\xi_x}\ \eta, \xi$	$\boxed{\eta}\ \xi$	$\boxed{\eta}\ \xi$
$\boxed{\eta_x}\ \eta$	$\boxed{\xi_x}\ \xi$	$\boxed{\eta_y}\ \eta_x, \eta$	$\boxed{\eta_x}\ \eta$	$\boxed{\xi_y}\ \eta, \xi$	$\boxed{\xi_y}\ \xi_x, f$	$\boxed{\xi_x}\ \xi$
$\boxed{\eta_y}\ \eta$	$\boxed{\xi_y}\ \xi$	$\boxed{\eta_{xx}}\ \eta_x, \eta$	$\boxed{\eta_{yy}}\ \eta_y, \eta$	$\boxed{\eta_x}\ \eta, \xi$	$\boxed{\xi_{xx}}\ \xi_x, \xi$	$\boxed{\xi_{yy}}\ \xi_y, \xi$
				$\boxed{\eta_y}\ \eta, \xi$		

$\mathbf{j}_{3,1}$	$\mathbf{j}_{3,2}$	$\mathbf{j}_{3,3}$	$\mathbf{j}_{3,4}$	$\mathbf{j}_{3,5}$
$\boxed{\xi}$	$\boxed{\xi}$	$\boxed{\xi}$	$\boxed{\xi_x}\ \eta, \xi$	$\boxed{\xi_x}\ \eta, \xi$
$\boxed{\eta_y}\ \eta_x, \eta$	$\boxed{\eta_{xx}}\ \eta_x, \eta$	$\boxed{\eta_x}\ \eta$	$\boxed{\xi_y}\ \eta, \xi$	$\boxed{\xi_y}\ \eta, \xi$
$\boxed{\eta_{xxx}}\ \eta_{xx}, \eta_x, \eta$	$\boxed{\eta_{yx}}\ \eta_y, \eta_x, \eta$	$\boxed{\eta_{yyy}}\ \eta_{yy}, \eta_y, \eta$	$\boxed{\eta_y}\ \eta_x, g, f$	$\boxed{\eta_x}\ \eta, \xi$
	$\boxed{\eta_{yy}}\ \eta_y, \eta_x, \eta$		$\boxed{\eta_{xx}}\ \eta_x, \eta, \xi$	$\boxed{\eta_{yy}}\ \eta_y, \eta, \xi$

$\mathbf{j}_{3,6}$	$\mathbf{j}_{3,7}$	$\mathbf{j}_{3,8}$	$\mathbf{j}_{3,9}$	$\mathbf{j}_{3,10}$
$\boxed{\xi_y}\ \xi_x, \eta, \xi$	$\boxed{\xi_x}\ \eta, \xi$	$\boxed{\eta}\ \xi$	$\boxed{\eta}\ \xi$	$\boxed{\eta}\ \xi$
$\boxed{\eta_x}\ \xi_x, \eta, \xi$	$\boxed{\eta_x}\ \xi_y, \eta, \xi$	$\boxed{\xi_x}\ \xi$	$\boxed{\xi_{xx}}\ \xi_y, \xi_x, \xi$	$\boxed{\xi_y}\ \xi_x, \xi$
$\boxed{\eta_y}\ \xi_x, \eta, \xi$	$\boxed{\eta_y}\ \xi_y, \eta, \xi$	$\boxed{\xi_{yyy}}\ \xi_{yy}, \xi_y, \xi$	$\boxed{\xi_{yx}}\ \xi_y, \xi_x, \xi$	$\boxed{\xi_{xxx}}\ \xi_{xx}, \xi_x, \xi$
$\boxed{\xi_{xx}}\ \xi_x, \eta, \xi$	$\boxed{\xi_{yy}}\ \xi_y, \eta, \xi$		$\boxed{\xi_{yy}}\ \xi_y, \xi_x, \xi$	

Table 1: **Table 1.** The types of Janet bases for linear homogeneous systems in $\xi(x, y)$ and $\eta(x, y)$. The dimension $r \leq 3$ of the solution space is attached as the first index to $\mathbf{j}_{r,k}$, the second index enumerates the various alternatives for each dimension. The term ordering is total-degree lexicographic with $\eta > \xi$ and $y > x$. Any type is uniquely characterized by its pattern of leading terms in the square boxes. Further details may be found in the main text.

schemes may be set up for higher dimensional solution spaces or for values of m and n different from 2. The number of alternatives increases quickly if these values increase. This classification of Janet bases provides a complete survey of all possible alternatives for $m = n = 2$ if the dimension of the solution space is not higher than three. If this is *a priori* known to be true for the problem at hand, it is guaranteed that the resulting Janet base is included in this listing.

In some applications however the classification problem occurs in another form. The type is fixed and the most general coefficients are desired such that the Janet base property is assured. This information is particularly useful if it is combined with additional coefficient constraints expressing further properties of the solutions. The desired constraints for the coefficients are essentially obtained in steps $S2$, $S3$ and $S4$ of the algorithm *JanetBase* given above. Because the type of the Janet base is fixed now, upon completion of step $S4$ the nonvanishing integrability conditions are *not* added to the system. In order to preserve the type of the Janet base, the coefficients of the various parametric derivatives have to vanish identically. The conditions obtained represent a system of pde's for the coefficient functions. If it is satisfied, the full system is guaranteed to be the most general Janet base of the specified type. This is achieved by replacing step $S5$ by step $S5a$ as follows

$S5a$: *(Separate Integrability Conditions)* Assign
$$C := \cup_{c_{i,j}} Coefficients(c_{i,j})$$
and return C.

In general the integrability conditions obtained in step $S5a$ are a nonlinear system of pde's that may comprise redundant equations. However the following important property makes it amenable for further simplification.

Lemma 1 *The integrability conditions obtained in step* $S5a$ *are quasilinear, i. e. they are linear in the highest derivatives.*

Proof: In steps $S2$ and the first part of $S3$ only differentiations of the given linear homogeneous equations are performed, i. e. the resulting system is such that the reductum of any equation is linear and homogeneous in the coefficients. The same is true for the integrability conditions $c_{i,j}$ obtained in the second part of $S3$ because the leading terms cancel each other. For any given $c_{i,j}$ two kinds of reductions in step $S4$ may occur. In the first place there are those applying an equation e_k with an index different from i and j. They can never generate a quadratic term because there is no intersection between the coefficient variables of any equation involved. On the other hand, any reduction w.r.t. equation e_i or e_j involves only derivatives that are lower than those involved in forming the integrability condition. As a consequence any derivative of maximal order can only occur linearly. This completes the proof. □

The quasilinearity of the constraints generated in step $S5a$ allows it to apply algorithm *JanetBase* to the system C with one modification: It terminates with an error message if an equation is generated that is not quasilinear. It turns out that this does not occur for the types listed in Table 1. The term ordering applied for the coefficient constraints is always total degree, then lexicographic in inverse alphabetical order, i. e. $a < b < c < \ldots$, then by ranking with $x < y$. In most cases the Janet base property is either obtained upon proper reordering or by autoreduction. A typical case is type $\mathbf{j}_{3,3}$ for which in the first place the integrability conditions

$$b_x - 3a_y = 0, \;\; a_{yy} - \frac{1}{3}c_x + \frac{2}{3}a_y b = 0, \;\; a_{yyy} + a_{yy}b - d_x + a_y c = 0$$

are obtained. Autoreduction generates the Janet base from it by replacing the third equation with

$$c_{yx} - 3d_x + \frac{1}{3}c_x b - 2b_y a_y + 3a_y c - \frac{2}{3}a_y b^2 = 0$$

The complete answer for the types listed in Table 1 is given elsewhere. Here only those types are covered that occur in the applications later in this article. For each case at first the notation for the coefficients is fixed by explicitly defining them, they are denoted by small roman letters a, b, c etc. After that the Janet base for the integrability conditions (IC's) is given.

One-dimensional solution space

$j_{1,1}$: $\{\xi, \eta_x + a\eta, \eta_y + b\eta\}$. IC: $a_y - b_x = 0$.

$j_{1,2}$: $\{\eta + a\xi, \xi_x + b\xi, \xi_y + c\xi\}$. IC: $b_y - c_x = 0$.

Two-dimensional solution space

$j_{2,3}$: $\{\xi_x + a\eta + b\xi, \xi_y + c\eta + d\xi, \eta_x + p\eta + q\xi, \eta_y + r\eta + s\xi\}$.

IC's: $c_x - a_y + ra - pc - da + cb = 0$, $d_x - b_y + sa - qc = 0$,

$r_x - p_y - sa + qc = 0$, $s_x - q_y + sp - sb - rq + qd = 0$.

Three-dimensional solution space

$j_{3,4}$: $\{\xi_x + a\eta + b\xi, \xi_y + x\eta + d\xi, \eta_y + p\eta_x + q\eta + r\xi, \eta_{xx} + u\eta_x + v\eta + w\xi\}$.

IC's: $pa + c = 0$, $c_x - a_y + qa - da + cb = 0$, $d_x - b_y + ra = 0$

$p_{xx} - u_y - u_x p + 2q_x - p_x u - ra = 0$,

$q_{xx} - v_y - v_x p - 2r_x a + q_x u - 2p_x v - a_x r - ura + rba = 0$,

$r_{xx} - w_y - w_x p + r_x u - 2r_x b - 2p_x w - b_x r - wq + wpb + wd + vr - urb + rb^2 = 0$.

$j_{3,6}$: $\{\xi_y + a\xi_x + b\eta + c\xi, \eta_x + d\xi_x + e\eta + f\xi, \eta_y + p\xi_x + q\eta + r\xi, \xi_{xx} + u\xi_x + v\eta + w\xi\}$.

IC's: $p_x - d_y + a_x d - up - uda + r - qd + pe + ha - d^2 b + dc = 0$,

$q_x - e_y + b_x d - vp - vda + hb - edb = 0$,

$r_x - h_y + c_x d - wp - wda + re - qh - hdb + hc = 0$,

$a_{xx} - u_y - u_x a - d_x b + 2c_x - 2b_x d - a_x u + vp + vda - hb + edb = 0$,

$b_{xx} - v_y - v_x a - e_x b + b_x u - 2b_x e - 2a_x v$

$+ wb + vq + vea + vdb - vc - ueb + e^2 b = 0$.

$c_{xx} - w_y - w_x a - h_x b + c_x u - 2b_x h - 2a_x w + wdb + vr + vha - uhb + heb = 0$

$j_{3,7}$: $\{\xi_x, a\eta + b\xi, \eta_x + d\xi_y + e\eta + f\xi, \eta_y + p\xi_y + q\eta + r\xi, \xi_{yy} + u\xi_y + v\eta + w\xi\}$.

IC's: $p_x - d_y + ud - qd + p^2 a - pb + pe - h = 0$, $q_x - a_y p + vd - ra + qpa - e_y = 0$,

$r_x - b_y p + wd + rpa - rb + re - qh = 0$,

$u_x + p_y a - 2b_y + 2a_y p - vd + ra - qpa = 0$,

$a_{yy} - v_x - q_y a + a_y u - 2a_y q + wa + vpa - vb + ve - uqa + q^2 a = 0$,

$b_{yy} - w_x - r_y a + b_y u - 2a_y r + wpa + vh - ura + rqa = 0$,

This explicit form provides a profound understanding of the features of a given system of linear homogeneous pde's. Of particular importance is the clear separation of all correlations for the coefficients due to the integrability conditions given in this listing, and additional properties that may be due to the structure of a specific problem. This feature is particularly striking for zero order conditions. For example, the lowest equation in any Janet base of type $j_{3,4}$ has always the form $pa + c = 0$. This constraint is *not* due to a particular feature of the problem described by the Janet base of this type. The importance of this distinction will become clear in the next section where Janet bases of these types occur as determining systems for certain symmetry groups.

3 Symmetries of Ordinary Differential Equations

This section deals with the most important means for finding closed form solutions of ordinary differential equations (ode's) that has been introduced by Lie, i. e. the *symmetries* that a given ode may admit, for details see (Schwarz 1997). Roughly speaking a symmetry is a variable transformation that leaves the ode invariant. It is obvious that the entirety of symmetry transformations of any given ode forms a group. The term *symmetry group* is applied to the *largest* group of transformations sharing this property.

The symmetry problem for differential equations comes in various versions. On the one hand, there is the classification problem. The aim is to obtain a complete survey of all possible symmetry groups for a class of ode's, e. g. for ode's of a given order. Usually this is achieved in terms of some *canonical variables.* For second order ode's the complete answer has been given by Lie as follows.

Theorem 1 *Any non-trivial symmetry group of a second order ode is similar to one given below in canonical variables u and $v(u)$.*

One-parameter group

S^2_1: $\{\partial/\partial v\}$.

Two-parameter groups

$S^2_{2,1}$: $\{\partial/\partial u, \partial/\partial v\}$. $S^2_{2,2}$: $\{\partial/\partial v, u\partial/\partial u + v\partial/\partial v\}$.

Three-parameter groups

$S^2_{3,1}$: $\{\partial/\partial u + \partial/\partial v, u\partial/\partial u + v\partial/\partial v, u^2\partial/\partial u + v^2\partial/\partial v\}$.

$S^2_{3,2}$: $\{\partial/\partial u, 2u\partial/\partial u + v\partial/\partial v, u^2\partial/\partial u + uv\partial/\partial v\}$.

$S^2_{3,3}$: $\{\partial/\partial u, \partial/\partial v, u\partial/\partial u + \alpha v\partial/\partial v\}$, $\alpha \neq 1$.

$S^2_{3,4}$: $\{\partial/\partial u, \partial/\partial v, u\partial/\partial u + (u+v)\partial/\partial v\}$.

Eight-parameter group:

S^2_8: $\{\partial/\partial u, \partial/\partial v, u\partial/\partial v, v\partial/\partial v, u\partial/\partial u, v\partial/\partial u,$
$u^2\partial/\partial u + uv\partial/\partial v, uv\partial/\partial u + v^2\partial/\partial v\}$.

This listing shows in particular that there does not exist any second order ode allowing a group of point symmetries with 4, 5 or 6 parameters.

On the other hand, any particular ode is usually given in non-canonical variables x and y, i. e. the *actual variables*, they are related to each other by

$$x = \phi(u, v), \quad y = \psi(u, v), \quad \text{and} \quad u = \sigma(x, y), \quad v = \rho(x, y) \qquad (3.1)$$

Let an ode in x and y of order two be given as

$$\omega(x, y, y', y'') = 0 \tag{3.2}$$

Any symmetry transformation of this equation with the generator

$$U = \xi(x, y)\frac{\partial}{\partial x} + \eta(x, y)\frac{\partial}{\partial y} \tag{3.3}$$

has to be determined from the condition

$$U^{(2)}\omega(x, y, y', y'') = 0 \quad for \quad \omega = 0 \tag{3.4}$$

The second prolongation $U^{(2)}$ is determined by

$$U^{(2)} = U + \zeta^{(1)}\frac{\partial}{\partial y'} + \zeta^{(2)}\frac{\partial}{\partial y''} \tag{3.5}$$

where

$$\begin{gathered}\zeta^{(1)} = \eta_x + (\eta_y - \xi_x)y' - \xi_y y'^2 \\ \zeta^{(2)} = \eta_{xx} + (2\eta_{xy} - \xi_{xx})y' + (\eta_{yy} - 2\xi_{xy})y'^2 - \xi_{yy}y'^3\end{gathered} \tag{3.6}$$

and

$$D = \frac{\partial}{\partial x} + y'\frac{\partial}{\partial y} + y''\frac{\partial}{\partial y'} \cdots \tag{3.7}$$

is the operator of total differentiation with respect to x. Separation of (3.4) w.r.t. the derivatives y' and y'' yields a system of linear homogeneous pde's for ξ and η, it is usually called the *determining system* because its general solution determines the symmetries of (3.2). It turns out that the type of the symmetry group may be obtained from a Janet base for the determining system *without* solving it. This is established by the following Theorem.

Theorem 2 *The following criteria provide a decision procedure for the type of symmetry group of a second order ode if its Janet base in a total-degree lexicographic term ordering with $\eta > \xi$, $y > x$ is given.*

One-parameter group

S_1^2: *Janet base of type* $\mathbf{j}_{1,1}$ *if* $\sigma_y = 0$, *type* $\mathbf{j}_{1,2}$ *otherwise.*

Two-parameter groups

$S_{2,1}^2$: *Janet base of type* $\mathbf{j}_{2,3}$, *it has the form*

$$\begin{gathered}\xi_x + a\eta + b\xi = 0, \ \xi_y + c\eta + d\xi = 0, \\ \eta_x + p\eta + q\xi = 0, \ \eta_y + r\eta + s\xi = 0\end{gathered} \tag{3.8}$$

with $a = d$ and $p = s$.

$S_{2,2}^2$: *Janet base of type* $\mathbf{j}_{2,3}$ *(3.8) with $a \neq d$ or $p \neq s$.*

Three-parameter groups

$S^2_{3,1}$: *Janet base of type* $\mathrm{j}_{3,6}$*, it has the form*

$$\begin{gathered}\xi_y + a\xi_x + b\eta + c\xi = 0,\ \ \eta_x + d\xi_x + e\eta + f\xi = 0\\ \eta_y + p\xi_x + q\eta + r\xi = 0,\ \ \xi_{xx} + u\xi_x + v\eta + w\xi = 0\end{gathered} \tag{3.9}$$

with $p = 1$ *and* $ad + 1 \neq 0$.

$S^2_{3,2}$: *Janet base of type* $\mathrm{j}_{3,4}$ *if* $\sigma_y = 0$, *of type* $\mathrm{j}_{3,7}$ *if* $\sigma_x = 0$, *or of type* $\mathrm{j}_{3,6}$ *(3.9) with* $p = 1$ *and* $ad + 1 = 0$ *otherwise.*

$S^2_{3,3}$: *Janet base of type* $\mathrm{j}_{3,6}$ *(3.9) with* $p \neq 1$ *and* $4ad + (p+1)^2 \neq 0$.

$S^2_{3,4}$: *Janet base of type* $\mathrm{j}_{3,6}$ *(3.9) with* $p \neq 1$ *and* $4ad + (p+1)^2 = 0$.

Eight-parameter group

S^2_8: *Janet base of type* $\{\xi_{yy}, \eta_{xx}, \eta_{xy}, \eta_{yy}, \xi_{xxx}, \xi_{xxy}\}$.

Proof: The basic idea behind the proof may be explained as follows. Take the Janet base for any group defined by Theorem 1 in the canonical variables u and v, and transform it into the actual variables via (3.1). In general this change of variables will destroy the Janet base property. In order to reestablish it, the algorithm *JanetBase* has to be applied. During this process it may occur that a leading coefficient of an equation that is applied for reduction vanishes due to a special choice of the transformation (3.1). This has the consequence that alternatives occur which may lead to different types of Janet bases. In order to obtain the complete answer, each of these alternatives has to be investigated separately. Finally all Janet bases have to be combined such that a minimal number of *generic cases* is retained, and all those that are discarded may be obtained from them by specialization. This resembles the proceeding of computing an algebraic Gröbner base with parameters in its coefficients. It turns out that all alternatives are generated by vanishing first order partial derivatives of the transformation functions σ and ρ. Taking into account the constraint $\Delta \equiv \sigma_x\rho_y - \sigma_y\rho_x \neq 0$ there are altogether seven possible combinations. There exists a partial order between these cases which is described by the following diagram. Specialization may only occur along the arrows.

According to what has been said above, in order to obtain a complete survey of all possible Janet bases that may occur for a second order ode, it is sufficient to give a complete set of generic bases and to indicate those that are obtained by specialization from them, e. g. *Case k*→Case j means that *Case j* is obtained by specialization from *Case k*. If there is only a single generic case, it must obviously be *Case 7*, all others are obtained from it. This is usually not indicated. The calculations involved cannot be given, they are performed by using a computer algebra software package. Further details may be found in (Schwarz 1997).

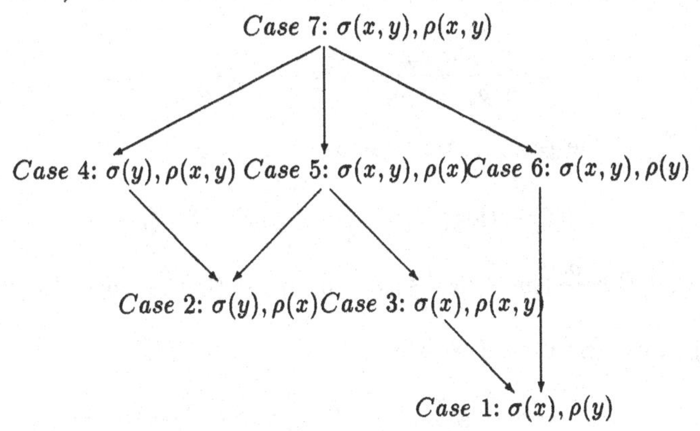

Group $\mathcal{S}_1^2$. Type $\mathbf{j}_{1,1}$, generic *Case 3→Case 1.*

$$\xi = 0,\ \eta_x + (\log \rho_y)_x \eta = 0, \quad \eta_y + (\log \rho_y)_y \eta = 0$$

Type $\mathbf{j}_{1,2}$, generic *Case 7→Case 2,4,5,6.*

$$\eta + \frac{\sigma_x}{\sigma_y}\xi = 0$$

$$\xi_x + \Big(\frac{\sigma_x\rho_{xy} - \sigma_y\rho_{xx}}{\Delta} + \frac{\rho_y}{\sigma_y}\frac{\sigma_y\sigma_{xx} - \sigma_x\sigma_{xy}}{\Delta}\Big)\xi = 0$$

$$\xi_y + \Big(\frac{\sigma_x\rho_{yy} - \sigma_y\rho_{xy}}{\Delta} - \frac{\rho_y}{\sigma_y}\frac{\sigma_x\sigma_{xy} - \sigma_y\sigma_{xy}}{\Delta}\Big)\xi = 0$$

Group $\mathcal{S}_{2,1}^2$. Type $\mathbf{j}_{2,3}$, single generic *Case 7.*

$$\begin{aligned}
&\xi_x + \frac{\sigma_y\rho_{xy} - \sigma_{xy}\rho_y}{\Delta}\eta + \frac{\sigma_y\rho_{xx} - \sigma_{xx}\rho_y}{\Delta}\xi = 0\\
&\xi_y + \frac{\sigma_y\rho_{yy} - \sigma_{yy}\rho_y}{\Delta}\eta + \frac{\sigma_y\rho_{xy} - \sigma_{xy}\rho_y}{\Delta}\xi = 0\\
&\eta_x - \frac{\sigma_x\rho_{xy} - \sigma_{xy}\rho_x}{\Delta}\eta - \frac{\sigma_x\rho_{xx} - \sigma_{xx}\rho_x}{\Delta}\xi = 0\\
&\eta_y - \frac{\sigma_x\rho_{yy} - \sigma_{yy}\rho_x}{\Delta}\eta - \frac{\sigma_x\rho_{xy} - \sigma_{xy}\rho_x}{\Delta}\xi = 0
\end{aligned} \tag{3.10}$$

Group $\mathcal{S}_{2,2}^2$. Type $\mathbf{j}_{2,3}$, single generic *Case 7.*

$$\begin{aligned}
&\xi_x - \Big(\frac{\sigma_y\rho_{xy} - \sigma_{xy}\rho_y}{\Delta} + \frac{\sigma_y}{\sigma}\Big)\eta - \Big(\frac{\sigma_y\rho_{xx} - \sigma_{xx}\rho_y}{\Delta} - \frac{\sigma_x}{\sigma}\Big)\xi = 0\\
&\xi_y - \frac{\sigma_y\rho_{yy} - \sigma_{yy}\rho_y}{\Delta}\eta - \frac{\sigma_y\rho_{xy} - \sigma_{xy}\rho_y}{\Delta}\xi = 0\\
&\eta_x + \frac{\sigma_x\rho_{xy} - \sigma_{xy}\rho_x}{\Delta}\eta + \frac{\sigma_x\rho_{xx} - \sigma_{xx}\rho_x}{\Delta}\xi = 0\\
&\eta_y + \Big(\frac{\sigma_x\rho_{yy} - \sigma_{yy}\rho_x}{\Delta} - \frac{\sigma_y}{\sigma}\Big)\eta + \Big(\frac{\sigma_x\rho_{xy} - \sigma_{xy}\rho_x}{\Delta} - \frac{\sigma_x}{\sigma}\Big)\xi = 0
\end{aligned} \tag{3.11}$$

Group $\mathcal{S}^2_{3,1}$. Type $\mathbf{j}_{3,6}$, single generic *Case 7*. The coefficients a, d and c in (3.9) are

$$a = \frac{-2\sigma_y \rho_y}{\sigma_x \rho_y + \sigma_y \rho_x}, \quad d = \frac{2\sigma_x \rho_x}{\sigma_x \rho_y + \sigma_y \rho_x}, \quad p = 1 \tag{3.12}$$

Group $\mathcal{S}^2_{3,2}$. Type $\mathbf{j}_{3,4}$, generic *Case 3*→*Case 1.*

$$\xi_x - (\log \rho^2)_y \eta + (\log \frac{\sigma_x}{\rho^2})_x \xi = 0, \quad \xi_y = 0$$

$$\eta_y + (\log \frac{\rho_y}{\rho})_y \eta + (\log \frac{\rho_y}{\rho})_x \xi = 0, \quad \eta_{xx} + (\log \frac{\rho_y^2}{\sigma_x})_x \eta_x + \Omega(\eta) = 0.$$

Type $\mathbf{j}_{3,7}$, generic *Case 4*→*Case 2.*

$$\xi_x + (\log \frac{\rho_x}{\rho})_y \eta + (\log \frac{\rho_x}{\rho})_x \xi = 0, \quad \eta_x = 0$$

Type $\mathbf{j}_{3,6}$, generic *Case 7*→*Case 5,6.* The coefficients a, d and c in (3.9) are

$$a = -\frac{1}{d}, \quad p = 1, \quad d = \frac{\sigma_x}{\sigma_y}, \quad e = \frac{\sigma_x}{\sigma_y}(\log \frac{\sigma_x}{\rho^2})_y, \quad f = \frac{\sigma_x}{\sigma_y}(\log \frac{\sigma_x}{\rho^2})_x \tag{3.13}$$

Group $\mathcal{S}^2_{3,3}$. Type $\mathbf{j}_{3,6}$, single generic *Case 7*. The coefficients a, d and c in (3.9) are

$$a = -\frac{(c-1)\sigma_y \rho_y}{\sigma_x \rho_y - \alpha \sigma_y \rho_x}, \quad d = \frac{(c-1)\sigma_x \rho_x}{\sigma_x \rho_y - c\sigma_y \rho_x}, \quad p = \frac{\alpha \sigma_x \rho_y - \sigma_y \rho_x}{\sigma_x \rho_y - \alpha \sigma_y \rho_x} \tag{3.14}$$

Group $\mathcal{S}^2_{3,4}$. Type $\mathbf{j}_{3,6}$, single generic *Case 7*. The coefficients a, d and c in (3.9) are

$$a = \frac{-\sigma_y^2}{\sigma_x \sigma_y - \Delta}, \quad d = \frac{\sigma_x^2}{\sigma_x \sigma_y - \Delta}, \quad p = \frac{\sigma_x \sigma_y + \Delta}{\sigma_x \sigma_y - \Delta} \tag{3.15}$$

Group $\mathcal{S}^2_8$. Type $\{\xi_{yy}, \eta_{xx}, \eta_{yx}, \eta_{yy}, \xi_{xxx}, \xi_{yxx}\}$, single generic *Case 7.*
From this listing the type of the symmetry group may be uniquely identified. This is obvious in all cases where a certain Janet base type occurs only for a single group. This is true for $\mathbf{j}_{1,1}$, $\mathbf{j}_{1,2}$, $\mathbf{j}_{3,4}$, $\mathbf{j}_{3,7}$ and the eight-parameter projective group.

For any two-parameter group with generators U_1 and U_2 and Janet base (3.8) the commutator U_3 has the form

$$U_3 = (\xi_1 \eta_2 - \eta_1 \xi_2)[(d-a)\frac{\partial}{\partial x} + (s-p)\frac{\partial}{\partial y}]$$

From this representation and the Janet bases (3.10) and (3.10) the distinction for type $\mathbf{j}_{2,3}$ given in the Theorem above is obvious.

Finally for Janet base type $\mathbf{j}_{3,6}$ of the form (3.9) the four three-parameter groups are allowed. A first distinctive property is the value of the coefficient

p. Its value is $p = 1$ for the former two groups and $p \neq 1$ for the latter. This is obvious from (3.14) due to $\alpha \neq 0$ and (3.15). A second distinctive feature is obtained by considering the number of systems of imprimitivity. For any group with Janet base (3.9) this number is one or two depending on whether the discriminant $4ad + (p + 1)^2$ does or does not vanish (Schwarz 1997). Therefore the following table yields a complete set of criteria for identifying the type of any three-parameter group from the Janet base type $\mathbf{j}_{3,6}$.

$\mathcal{S}^2_{3,1}$	$\mathcal{S}^2_{3,2}$	$\mathcal{S}^2_{3,3}$	$\mathcal{S}^2_{3,4}$
$p = 1$	$p = 1$	$p \neq 1$	$p \neq 1$
$ad + 1 \neq 0$	$ad + 1 = 0$	$4ad + (p+1)^2 \neq 0$	$4ad + (p+1)^2 = 0$

This completes the proof. □

4 Concluding Remarks

The discussion in the preceding section shows clearly the overwhelming importance of the coefficients of a Janet base. Of particular interest is the information on the problem at issue that is contained in these coefficients. In order to identify it, the integrability conditions that are *always* satisfied must be completely known.

The integrability conditions given in Section 2 have another important application. If the transformation functions $\sigma(x, y)$ and $\rho(x, y)$ are to be determined in terms of the known coefficients $a, b, c, \ldots$ of the various Janet bases, a Janet base for the system expressing the relation between the two sets of functions has to be generated in a lexicographic term ordering with σ and ρ higher than the $a, b, c, \ldots$. This system has to be completed by the respective integrability conditions in order to get a manageable problem. As an example, consider the group $\mathcal{S}^2_{2,1}$. The desired system is obtained by equating the coefficients in (3.10) to the respective coefficients of (3.8), adding the integrability conditions of Section 2 and the constraints given in Theorem 2. The four highest equations of a Janet base for this system with σ and ρ ordered ahead of the remaining functions are

$$\sigma_{xxx} - (\tfrac{q_x}{q} + b + p)\sigma_{xx} + (b\tfrac{q_x}{q} - b_x + bp - aq)\sigma_x = 0$$

$$\sigma_y - \frac{1}{q}\sigma_{xx} + \frac{b}{q}\sigma_x = 0$$

and an identical set of equations with σ replaced by ρ. This is the most complete information that may be obtained for the equivalence transformation to canonical coordinates *without* leaving the coefficient field of the given ode because the $a, b, c, \ldots$ are rationally known in terms of the coefficients of the latter. The power of the Janet base concept is particularly obvious from this example. A complete solution of these equivalence problems for all second order ode's with symmetries will be given elsewhere.

The software package *SPDE* that has been applied for performing all calculations presented in this article is available from the following URL: http://borneo.gmd.de/spde/

References

Buchberger, B. (1970) 'An Algorithmic Criterion for the Solvability of Algebraic Systems of Equations', Aequationes Mathematicae (4)**3**, 374-383.

Buchberger, B. (1985) 'Gröbner Bases: An Algorithmic Method in Polynomial Ideal Theory'. In: Multidimensional Systems Theory (N.K. Bose ed.), D. Reidel, Dordrecht, pp. 184-232.

Janet, M. (1920) 'Sur les systèmès d'équations aux dérivées partielles', Journal de Mathématiques', 8**3**, 65-151.

Schwarz, F. (1992) 'An Algorithm for Determining the Size of Symmetry Groups', *Computing* **49**, 95-115.

Schwarz, F. (1997) 'Algorithmic Lie Theory for Solving Ordinary Differential Equations', to appear.

Gröbner Bases in Partial Differential Equations

Daniele C. Struppa

Department of Mathematical Sciences,
George Mason University
Fairfax, VA 22030

1. In this paper, I will describe some recent (and rather unexpected) applications of the theory of Gröbner bases to the study of the structure of solutions of linear systems of constant coefficients partial differential systems. Gröbner bases first appeared in Buchbergers's Ph.D. thesis [5] (see also [6] where the main results were first published), and their theory has provided the conceptual basis for the creation of several computational algebra packages which can be utilized for the solution of polynomial problems. The approach that I have successfully applied in a series of joint papers [1], [2], [3], [4], [10], [11] is made possible by the algebrization of analysis which began in the sixties [12], [16] and was then perfected by M. Sato and his collaborators in the seventies [20], [21]. In this introductory section, I will briefly recall the foundations of this algebraic approach to partial differential equations, while in section 2, I will show a few concrete and remarkable applications of Gröbner bases to specific systems of differential equations. I would like to point out that the way in which we have been using Gröbner bases is twofold: on one hand we have used some symbolic computation packages which are based on the theory of Gröbner bases; on the other hand (see Theorem 2), we have used the theory itself to generalize results which had been computed in special cases. The reader interested in some of the basic papers on Gröbner bases should be referred to [7], [8] and [9], which is probably the first easy to read syrvey of the theory. The last section contains a series of open problems and points out a few directions for further research.

Let $R = \mathbb{C}[z_1, \ldots, z_n]$ be the ring of polynomials with complex coefficients in n variables, and let $P = [P_{ij}]$ be an $r_1 \times r_0$ matrix in R so that $P(D) = [P_{ij}(D)]$ is a matrix of linear constant coefficients differential operators, where $D = \left(-i\frac{\partial}{\partial x_1}, \ldots, -i\frac{\partial}{\partial x_n}\right)$. If $\mathcal{S}$ is a sheaf of generalized functions (in the case we are interested in, $\mathcal{S}$ will either be the sheaf $\mathcal{B}$ of hyperfunctions, or the sheaf $\mathcal{E}$ of infinitely differentiable functions), we will denote by $\mathcal{S}^P$ the kernel of the map

$$P(D) : \mathcal{S}^{r_0} \to \mathcal{S}^{r_1},$$

i.e. the sheaf of $\mathcal{S}$-solutions of the system

$$P(D)\vec{f} = \vec{0}. \tag{0.1}$$

Now the multiplication by the matrix P^t maps R^{r_1} to R^{r_0} and we know that its cokernel

$$\mathcal{M} = \frac{R^{r_0}}{P^t R^{r_1}}$$

contains all the relevant information on (1), in view of the well known isomorphism

$$\mathcal{H}om_R(\mathcal{M}, \mathcal{S}) \cong \mathcal{S}^P.$$

In particular, we know that Hilbert's syzygy theorem (proved in its most general form by Palamodov [16]) shows that there is a finite free resolution

$$0 \longleftarrow \mathcal{M} \longleftarrow R^{r_0} \xleftarrow{P^t} R^{r_1} \xleftarrow{P_2^t} R^{r_2} \longleftarrow \ldots \longleftarrow R^{r_{m-1}} \xleftarrow{P_m^t} R^{r_m} \longleftarrow 0 \tag{0.2}$$

for which $m \leq n$.

The dualized of (2) is the complex

$$0 \longrightarrow R^{r_0} \xrightarrow{P} R^{r_1} \xrightarrow{P_2} R^{r_2} \longrightarrow \ldots \longrightarrow R^{r_{m-1}} \xrightarrow{P_m} R^{r_m} \longrightarrow 0 \tag{0.3}$$

whose cohomology groups are denoted by $Ext^j(\mathcal{M}, R)$. Interestingly enough, even though (2) and (3) are not uniquely defined, the Ext groups (they are actually R-modules) are uniquely determined by $\mathcal{M}$.

Computing the resolution (2) amounts to computing the generators for the kernel of polynomial homomorphisms or, in algebraic jargon, their syzygies. This can be done explicitly (at least in principle) by using Gröbner bases techniques, and if the module $\mathcal{M}$ is graded (i.e. if the matrix has homogeneous columns) one can actually compute a minimal length free resolution.

There are several reasons why (3) is an important object, and among them the following result of Ehrenpreis et al. (see [13] for the complete references).

Theorem 1 *If Ω is a convex open (or convex compact) set in $\mathbb{R}^n$ and $\mathcal{S}$ is a sheaf on $\mathbb{R}^n$, then the sequence*

$$0 \longrightarrow \mathcal{S}^P(\Omega) \longrightarrow \mathcal{S}(\Omega)^{r_0} \xrightarrow{P(D)} \mathcal{S}(\Omega)^{r_1} \xrightarrow{P_2(D)} \dots \xrightarrow{P_m(D)} \mathcal{S}(\Omega)^{r_m} \longrightarrow 0$$

is exact, which in particular means that $P_2(D)$ is a compatibility system for $P(D)$.

On the other hand, the *Ext* modules also carry much relevant information. $Ext^0(\mathcal{M}, R) = 0$ means that P is one-to-one, i.e. the columns of P are linear independent over R (and the system corresponding to P is determined [13]). Also, it is known that $Ext^1(\mathcal{M}, R) = 0$ is equivalent to the fact that the Hartogs' phenomenon holds for $\mathcal{S}^P$ (i.e. for any compact set $K \subseteq \mathbb{R}^n$ such that $\mathbb{R}^n \backslash K$ is connected, every element in $\mathcal{S}^P(\mathbb{R}^n \backslash K)$ extends to an element in $\mathcal{S}^P(\mathbb{R}^n)$); thanks to the pioneering work of Palamodov, we also know that there are several important analytic consequences of the vanishing of the higher *Ext*-modules (still related to various results on the removability of special singularities); the reader may refer to [12] and [16] for more details. Palamodov's work also makes evident the strong geometric flavor of this entire construction; in fact, if V is the characteristic variety associated to the operator $P(D)$, i.e. if V is the set of common zeroes of the maximal size minors of $[P_{ij}]$, then $dim(V) = n - p - 1$ if and only if $Ext^j(\mathcal{M}, R) = 0$ for all $j \leq p$ and $Ext^{p+1}(\mathcal{M}, R) \neq 0$.

2. As I have shown in section 1, much analytic information is contained in the algebra of the module $\mathcal{M}$; more precisely, the information is contained in the syzygies of $\mathcal{M}$. As it is well known, the complexity of computing syzygies is daunting and one cannot expect to compute effectively the modules $Ext^j(\mathcal{M}, \mathcal{R})$, unless $\mathcal{M}$ has some nice geometrical properties. In this section, I will show that there are many important examples (mostly motivated by physical considerations) in which $\mathcal{M}$ is simple enough so that we can employ Gröbner bases arguments to compute explicitly the resolution (2) and its cohomology!

Let $f = (f_0, f_1, f_2, f_3)$ be a vector whose components are $\mathcal{C}^\infty$ functions in $4n$ real variables $(\xi_{i0}, \xi_{i1}, \xi_{i2}, \xi_{i3}; i = 1, \dots, n)$. Using a classical terminology due to Fueter, we say that f is left regular if,

$$\begin{cases} \frac{\partial f_0}{\partial \xi_{i0}} - \frac{\partial f_1}{\partial \xi_{i1}} - \frac{\partial f_2}{\partial \xi_{i2}} - \frac{\partial f_3}{\partial \xi_{i3}} = 0 \\ \\ \frac{\partial f_0}{\partial \xi_{i1}} + \frac{\partial f_1}{\partial \xi_{i0}} - \frac{\partial f_2}{\partial \xi_{i3}} + \frac{\partial f_3}{\partial \xi_{i2}} = 0 \\ \\ \frac{\partial f_0}{\partial \xi_{i2}} + \frac{\partial f_1}{\partial \xi_{i3}} + \frac{\partial f_2}{\partial \xi_{i0}} - \frac{\partial f_3}{\partial \xi_{i1}} = 0 \\ \\ \frac{\partial f_0}{\partial \xi_{i3}} - \frac{\partial f_1}{\partial \xi_{i2}} + \frac{\partial f_2}{\partial \xi_{i1}} + \frac{\partial f_3}{\partial \xi_{i0}} = 0 \end{cases} \tag{0.4}$$

for $i = 1, \dots, n$. We can view f as a function $f : \mathbb{H}^n \to \mathbb{H}$, where $\mathbb{H}$ is the space of quaternions. If we let $q = (q_1, \dots, q_n)$ be the variable in $\mathbb{H}^n$, then condition (4) is equivalent to

$$\frac{\partial f}{\partial \bar{q}_i} = 0, \quad i = 1, \dots, n.$$

By taking the Fourier transform of the matrix of differential operators associated to Equation (4), one is led to consider the matrix

$$P_n = \begin{bmatrix} U_1 & U_2 & \cdots & U_n \end{bmatrix}^t,$$

where

$$U_i = \begin{bmatrix} z_{i0} & z_{i1} & z_{i2} & z_{i3} \\ -z_{i1} & z_{i0} & z_{i3} & -z_{i2} \\ -z_{i2} & -z_{i3} & z_{i0} & z_{i1} \\ -z_{i3} & z_{i2} & -z_{i1} & z_{i0} \end{bmatrix}$$

for $i = 1, \dots, n$, and where the variables z_{ij} are the dual variables of the variables ξ_{ij}, so that now $R = \mathbb{C}[z_{i0}, z_{i1}, z_{i2}, z_{i3} | i = 1, \dots, n]$.

Based on what we have discussed in section 1, it is clear that we need to construct a minimal resolution for $M_n := \frac{R^4}{P_n^t R^{4n}}$. The first step consists in computing the projective dimension $pd(M_n)$. Let me point out that, for $n = 1$, it is straightforward to see that the syzygy module of P_1^t is zero, and so $pd(M_1) = 1$. For $n = 2$, 3, and 4, we (i.e. the authors of [1]) have been able to use CoCoA [1] and to explicitly compute a minimal resolution for M_n, which shows, in particular, that $pd(M_2) = 3$, $pd(M_3) = 5$ and $pd(M_4) = 7$. These numbers seem to suggest the following natural conjecture:

$$pd(M_n) = 2n - 1 \tag{0.5}$$

In fact, in [3], we gave a rigorous proof of this result; the proof seems difficult to achieve unless techniques from the theory of Gröbner bases are heavily used.

Let me describe here the main steps of the proof of (5), referring the reader to [3] for its complete details.

From now on we will assume that $n > 1$. We will use the Auslander-Buchsbaum formula

[1] CoCoA is a special purpose system for doing computations in commutative algebra. It is the ongoing product of a research team in Computer Algebra at the University of Genova, Italy. It is freely available, and more information can be obtained by sending an e-mail message to cocadima.unige.it.

$$pd(M_n) = \text{depth}(\wp_n, R) - \text{depth}(\wp_n, M_n),$$

where $\wp_n$ is the maximal ideal of R generated by the $4n$ variables.

We recall that, for an ideal I of R and an R-module M, the depth of I on M, denoted depth(I, M) is the length of any maximal M-regular sequence in I. The polynomials $f_1, \ldots, f_s \in I$ form an M-regular sequence if

1. f_ν is a non-zerodivisor on $M/\langle f_1, \ldots, f_{\nu-1}\rangle M$, for $\nu = 1, \ldots, s$;

2. $M \neq \langle f_1, \ldots, f_s\rangle M$.

Clearly, depth$(\wp_n, R) = 4n$, so we only need to compute depth$(\wp_n, M_n)$. To do this we will exhibit a maximal M_n-regular sequence in $\wp_n$.

This will be accomplished using the theory of Gröbner bases. Related ideas were used in [16, ch. II, Section 2.4]. We first need a Gröbner basis for $\langle P_n^t\rangle$. We use the degree reverse lexicographic (degrevlex) term ordering on R with

$$x_{10} > x_{20} > \cdots > x_{n0} > x_{11} > \cdots > x_{n1} > x_{12} > \cdots > x_{n3}, \tag{0.6}$$

and the TOP (TOP stands for term over position) ordering on R^4 with $e_1 > e_2 > e_3 > e_4$, where e_i is the ith column of the 4×4 identity matrix. That is, for monomials $X = x_{10}^{\alpha_{10}} \cdots x_{n3}^{\alpha_{n3}}$ and $Y = x_{10}^{\beta_{10}} \cdots x_{n3}^{\beta_{n3}}$, we have

$$Xe_r > Ye_s \iff \begin{cases} \deg(X) = \sum_{\substack{i=1,\ldots,n \\ j=0,1,2,3}} \alpha_{ij} > \deg(Y) = \sum_{\substack{i=1,\ldots,n \\ j=0,1,2,3}} \beta_{ij} & \textit{or} \\ \deg(X) = \deg(Y) \text{ and } \alpha_{ij} < \beta_{ij} \text{ for the index } ij, \\ \text{last with respect to (6) such that } \alpha_{ij} \neq \beta_{ij} & \textit{or} \\ X = Y \text{ and } r < s. \end{cases}$$

Lemma 1 *The reduced Gröbner basis for the R-module $\langle P_n^t\rangle$ is given by the columns of P_n^t together with the columns of the $\binom{n}{2}$ matrices $U_rU_s - U_sU_r$. Moreover the module generated by the leading terms of all the elements of P_n^t, denoted $Lt(P_n^t)$, is*

$$Lt(P_n^t) = \langle x_{i0}e_\ell, x_{r2}x_{s1}e_\ell\rangle_{\substack{i=1,\ldots,n \\ 1\leq r<s\leq n \\ \ell=1,2,3,4}}$$

This result allows us to start an M_n-regular sequence in $\wp_n$.

Corollary 1 *The variables x_{11}, x_{n2}, x_{i3}, $i = 1, \dots, n$ form an M_n-regular sequence of length $n+2$.*

To enlarge this regular sequence, we consider the module

$$\begin{aligned} M_n^* &= M_n / \langle x_{11}, x_{n2}, x_{i3}, i = 1, \dots, n \rangle M_n \\ &\simeq R^4 / \left\langle A_n + \langle x_{11}, x_{n2}, x_{i3}, i = 1, \dots, n \rangle R^4 \right\rangle \\ &= R^4 / \langle U_i, U_r U_s - U_s U_r, x_{11} e_\ell, x_{n2} e_\ell, x_{i3} e_\ell \rangle_{\substack{i=1,\dots,n \\ 1 \le r < s \le n \\ \ell = 1,2,3,4}} \end{aligned}$$

We can then prove

Proposition 1 *The polynomials $x_{21} + x_{12}, x_{31} + x_{22}, \dots, x_{n1} + x_{n-1,2}$ form a maximal M_n^*-regular sequence in $\wp_n$.*

We can now obtain the formula for the projective dimension of M_n.

Theorem 2

$$pd(M_n) = 2n - 1.$$

Proof. - By the Auslander-Buchsbaum formula we have

$$pd(M_n) = \text{depth}(\wp_n, R) - \text{depth}(\wp_n, M_n) = 4n - \text{depth}(\wp_n, M_n).$$

By Corollary 1 and Proposition 1 we have

$$\text{depth}(\wp_n, M_n) = 2n + 1.$$

□

Remark. - We now have a free resolution of M_n

$$0 \to R^{r_{2n-1}} \xrightarrow{C} R^{r_{2n-2}} \xrightarrow{B} \dots \to R^{r_2} \to R^{4n} \to R^4 \to M_n \to 0 \tag{0.7}$$

(by the well-known Quillen-Suslin Theorem, we know that every projective R-module is free). By taking the dual of Resolution (7) we obtain a complex

$$0 \to R^4 \to R^{4n} \to R^{r_2} \to \dots \to R^{r_{2n-2}} \xrightarrow{C^t} R^{r_{2n-1}} \to 0, \tag{0.8}$$

whose homology groups are, by definition, $Ext^i(M_n, R)$. We see that the last homology, $Ext^{2n-1}(M_n, R)$ in (8) is not zero. Indeed, if the map $R^{r_{2n-2}} \to$

$R^{r_{2n-1}}$ is onto, then we obtain a matrix D with D^t defining a map $R^{r_{2n-1}} \to R^{r_{2n-2}}$ such that $C^t D^t = I$, the identity. So we get that $DC = I$ as well and the map C in Resolution (7) splits, $R^{r_{2n-2}} = \mathrm{im}C \oplus \ker D$. Since $\ker D$ is free and B restricted to $\ker D$ is one to one we have obtained a shorter free resolution for M_n than (8), which violates Theorem 2.

It is actually possible to say more than this.

Proposition 2 *The characteristic variety $V(M_n)$ of M_n has dimension $2n+1$.*

Theorem 3 *Sequence (8) is exact except at the last spot, i.e.*

$$Ext^j(M_n, R) = 0, \text{ for all } j = 0, \ldots, 2n-2$$

and

$$Ext^{2n-1}(M_n, R) \neq 0.$$

Let me now denote by $\mathcal{R} = \mathcal{E}^P$ the sheaf of $\mathcal{C}^\infty$ solutions of P, i.e. the sheaf of regular functions. Note that, since P is an elliptic system, we have $\mathcal{R} = \mathcal{E}^P = \mathcal{D}'^P$, where $\mathcal{D}'$ is the sheaf of distributions. However a fundamental result of Bengel-Harvey-Komatsu (see [13]) shows that we also have $\mathcal{R} = \mathcal{B}^P$, where $\mathcal{B}$ is the sheaf of hyperfunctions. This fact immediately allows us to prove the following

Theorem 4 *The sheaf $\mathcal{R}$ has flabby dimension equal to $2n-1$.*

Remark. - Theorem 4 generalizes to the sheaf of germs of regular functions the well-known fact that fl.dim$(\mathcal{O}) = n$, where $\mathcal{O}$ is the sheaf of germs of holomorphic functions. Such a result was probably hard to imagine before our computations in [1].

As it is easy to see, all open sets U in $\mathbb{H}$ are cohomologically trivial in the sense that

$$H^p(U, \mathcal{R}) = 0 \quad p \geq 1.$$

In [1], on the other hand, we showed that this result fails for $p > 1$, since a Hartogs' phenomenon occurs. This situation clearly mirrors what happens for the sheaf $\mathcal{O}$ of holomorphic functions. In that case, the most important result, due to Malgrange, states that, for any open set $U \subseteq \mathbb{C}^n$,

$$H^p(U, \mathcal{O}) = 0 \quad p \geq n.$$

In our case, the analog of such a statement is an immediate corollary of Theorem 4.

Corollary 2 *If U is any open set in $\mathbb{H}^n$, then*

$$H^p(U, \mathcal{R}) = 0 \quad p \geq 2n - 1.$$

Once again, we believe this result to be quite unexpected. We do not know of any analytic proof for it.

Another general consequence of the ellipticity of the system is the following result (see also [3]):

Theorem 5 *Let K be a compact convex set in $\mathbb{H}^n$. Then*

$$[H^0(K, \mathcal{R})]' \cong H_K^{2n-1}(\mathbb{H}, \mathcal{B}^Q).$$

For the case $n = 1$, $\mathcal{B}^Q \cong \mathcal{R}$, and for the case $n = 2$, Q is nothing but the system associated to $\partial/\partial q_2 - \partial/\partial q_1$. The situation is more complex for $n \geq 3$.

As we pointed out earlier, the vanishing of the *Ext*-modules has relevant applications to the removability of singularities of regular functions. We have the following two results:

Theorem 6 *Let Ω be a convex connected open set in $\mathbb{H}^n$, and let K be a compact subset of Ω. Let $\Sigma_1, \ldots, \Sigma_{2n-2}$ be closed half-spaces in $\mathbb{R}^{4n}$ and set $\Sigma = \Sigma_1 \cup \ldots \cup \Sigma_{2n-2}$. Then every regular function f in $\Omega \backslash (K \cup \Sigma)$ extends to a regular function in $\Omega \backslash \Sigma$, which coincides with f in $\Omega \backslash (K' \cup \Sigma)$ for K' a compact subset of Ω.*

Theorem 7 *Let L be a subspace of $\mathbb{H}^n = \mathbb{R}^{4n}$ of dimension $2n + 2$. Then for every compact set $K \subseteq L$, and every connected open set Ω, relatively compact in K, every regular function defined in the neighborhood of $K \backslash \Omega$ can be extended to a regular function defined in a neighborhood of K.*

Partial improvements of these general results have recently been obtained in [15], where some classical ideas of Severi are used to remove compact singularities in $\mathbb{H} \times \mathbb{R}$ and $\mathbb{H} \times \mathbb{C}$; also, in [15] we have shown how some subclasses of $\mathcal{R}(\mathbb{H}^2)$ admit stronger singularity removability properties. It may be worth pointing out that Theorem 6 and 7 can be improved for a special subclass of regular functions defined in the recent [15]. These abstract methods also provide a concrete way to represent all regular functions on any open convex set $\Omega \subseteq \mathbb{H}^n$. Indeed, the Ehrenpreis-Palamodov Fundamental Principle implies the following representation:

Theorem 8 *Let $\Omega \subseteq \mathbb{H}^n$ be an open convex set. Then every regular function f in Ω can be represented as*

$$f(q_q, \cdots, q_n) = \int_{Ch(M_n)} \exp(\xi \cdot q) \cdot \bar{\xi} d\mu$$

where $\xi \in \mathcal{H}^n$, *and* $d\mu = (d\mu_1, \ldots, d\mu_n)$ *is a vector of quaternionic valued densities supported on* $Ch(M_n)$ *and satisfying, for every compact* $K \subseteq \Omega$,

$$\int_{Ch(M_n)} \exp(\max_K(\xi \cdot q))|d\mu| < +\infty.$$

Note that this representation corresponds to a particular choice for the noetherian operator associated to the module M_n. Additional representation for regular functions in several quaternionic variables are also given in the recent [17].

Let me point out that the methods which I have just described can be applied to a variety of different systems. In a forthcoming paper [18], it is shown how a substantially similar analysis can be made for the so called Moisil-Theodoresco system [9], whose solutions are functions $f : \mathbb{R}^{3n} \to \mathbb{R}^4$ which satisfy a sort of odd-dimensional version of the Cauchy-Riemann system. In [13] we prove, among other things that

Theorem 9 *The projective dimension of the module associated to the Moisil-Theodoresco system is* $2n - 1$.

Other systems are examined in [10], [11], where variations of the Dirac equations are studied.

3. In this final section, I wish to discuss some interesting open problems which, I believe, will be solved by clever applications of the theory of Gröbner bases.

Problem 1: Let us go back, for an instant, to the Cauchy- Fueter system (4); when one resolves the module M_n, in all the known cases (i.e. $n = 2, 3, 4$), one sees that the first syzygies are quadratic, while all the remaining are linear. Can we prove this in general (i.e. for $n \geq 5$)?

Problem 2: Our Theorem 5 shows that there is a very important link between the first and the last matrices in the resolution of M_n. Such a link is well understood for $n = 1, 2, 3$, where the construction can be made explicitely. But, can one construct Q directly from P, at least in the case of the Cauchy-Fueter system?

Problem 3: Our Theorem 8 is nothing but an application of the well known Ehrenpreis-Palamodov Fundamental Principle [12], [16] which states that if $I = (P_1, \ldots, P_n)$ is an ideal in R, then it is possible to find a finite number of algebraic varieties $V_j \subseteq \mathbb{C}^n$ and differential operators ∂_j with polynomial coefficients on V_j such that an active function f belongs to the ideal generated by $P_1, \ldots, P_n$ in the space of entire functions if and only if $\partial_j f_{|V_j} = 0$ for any j. The collection (V_j, ∂_j) which allows this sort of "improved Nullstellensatz" is usually referred to as the "multiplicity variety" associated to the ideal I and can also be constructed if the ideal I is replaced by a module.

In the case of the Cauchy-Fueter operator (as well as in the case of the Moisil-Theodoresco operator) we have an explicit construction for such a variety. Can we use Gröbner bases to construct the multiplicity variety associated to an arbitrary module?

Problem 4: Algebraic Analysis, as defined in [20] and in the more introductory [21], applies not just to constant coefficients p.d.e.'s but also (and more interestingly) to p.d.e.'s with variable coefficients. If one considers the case, say, of polynomial coefficients, then the role of R is taken by the Weyl algebra $A_n(\mathbb{C}) := \mathbb{C}\langle x_1, \dots, x_n, \partial_1, \dots, \partial_n \rangle$. The algebraic theory of such equations is by now quite well understood. However, can we use Gröbner bases in $A_n(\mathbb{C})$ to explicitly solve computational problems for such equations? This, I believe, would be a substantial progress which could give algebraic analysis a new, computational, flavor.

I wish to conclude by expressing my gratitude to Prof. Bruno Buchberger for inviting me to contribute this paper, and to the friends of the CoCoA team for providing me the opportunity to experiment with preliminary versions of their product.

References

[1] Adams, W.W., Berenstein, C.A., Loustaunau, P., Sabadini, I., Struppa, D.C. (1997): Regular functions of several quaternionic variables and the Cauchy-Fueter complex. In corso di stampa su J. Geom. Anal.

[2] Adams, W.W., Berenstein, C.A., Loustaunau, P., Sabadini, I., Struppa, D.C. (1996): On compact singularities for regular functions of one quaternionic variable. Compl. Var. 31: 259-270.

[3] Adams, W.W., Loustaunau, P., Palamodov, V.P., Struppa, D.C. (1997): Hartog's phenomenon and projective dimension of related modules. Ann. Inst. Fourier 47: 623-640.

[4] Berenstein, C.A., Sabadini, I., Struppa, D.C. (1996): Boundary values of regular functions of quaternionic variables. Pitman Res. Notes Math. Ser. 347: 220-232

[5] Buchberger, B. (1965): On Finding a Vector Space Basis of the Residue Class Ring Modulo a Zero Dimensional Polynomial Ideal (German). Ph.D. Thesis, Univ. of Innsbruck, Austria.

[6] Buchberger, B. (1970): An algorithmical criterion for the solvability of algebraic systems of equations (German). Aequationes Mathematicae 4/3: 374-383.

[7] Buchberger, B. (1976): A theoretical basis for the reduction of polynomials to canonical forms. ACM SIGSAM Bull. 10/3: 19-29, and 10/4: 19-24.

[8] Buchberger, B. (1979): A criterion for detecting unnecessary reductions in the construction of Groebner bases. Proc. of teh EUROSAM 79 Symp. on Symbolic and Algebraic Computation, Marseille, June 26-28, 1979. Lecture Notes in Computer Science 72, Springer, 3-21.

[9] Buchberger, B. (1985): An algorithmic method in polynomial ideal theory. Chapter 6 in: Multidimensional Systems Theory (N.K.Bose ed.), Reidel Publishing Company, Dordrecht: 184-232.

[10] Colombo, F., Loustaunau, P., Sabadini, I., Struppa, D.C. (1996): Regular functions of biquaternionic variables and Maxwell's Equations. To appear in Geometry and Physics.

[11] Colombo, F., Loustaunau, P., Sabadini, I., Struppa, D.C. (1997): Dirac Equation in the Octonion Algebra. Preprint

[12] Ehrenpreis, L. (1970): Fourier Analysis in Several Complex Variables. Wiley Interscience, New York

[13] Komatsu, H. (1973): Relative cohomology of sheaves of solutions of differential equations. Springer LNM 287: 192-261

[14] Moisil, G., Theodoresco, N. (1931): Functions holomorphes dans l'espace. Mathematica (Cluj) 5: 142-159

[15] Napoletani, D., Sabadini, I., Struppa, D.C. (1996): Variation on a theorem of Severi. To appear in the Proceedings of "Simposio Internacional de Analisis Complejo y Temas Afines", (M. Shapiro ed.), Cuernavaca

[16] Palamodov, V.P. (1970): Linear Differential Operators with Constant Coefficients. Springer, Berlin

[17] Palamodov, V.P. (1997): Holomorphic synthesis of monogenic functions of several quaternionic variables. Preprint

[18] Sabadini, I., Shapiro, M., Struppa, D.C.: Algebraic analysis of the Moisil-Theodoresco system. In preparation

[19] Sabadini, I., Struppa, D.C. (1997): Some open problems on the analysis of the Cauchy-Fueter system in several variables. RIMS Kokyuroku 1001: 1-21

[20] Sato, M., Kawai, T., Kashiwara, M. (1973): Microfunctions and Pseudo-Differential Equations. Springer Lectures Notes in Mathematics 287: 265-529

[21] Struppa, D.C., Kato, G. (1998): Fundamentals of Microlocal Algebraic Analysis. Marcel Dekker

Gröbner Bases and Hypergeometric Functions

Bernd Sturmfels[1] *and Nobuki Takayama*

1 Introduction

The purpose of this tutorial is to illustrate the use of Gröbner bases and Buchberger's algorithm in the algebraic study of linear partial differential equations. Our reference example is the following system of six differential equations for a function $\Phi(x_1, x_2, x_3, x_4)$ in four complex variables:

$$\begin{aligned}
\frac{\partial^2 \Phi}{\partial x_2 \partial x_3} &= \frac{\partial^2 \Phi}{\partial x_1 \partial x_4}, & \frac{\partial^3 \Phi}{\partial x_1^2 \partial x_3} &= \frac{\partial^3 \Phi}{\partial x_2^3}, \\
\frac{\partial^3 \Phi}{\partial x_2 \partial x_4^2} &= \frac{\partial^3 \Phi}{\partial x_3^3}, & \frac{\partial^3 \Phi}{\partial x_1 \partial x_3^2} &= \frac{\partial^3 \Phi}{\partial x_2^2 \partial x_4},
\end{aligned} \tag{1.1}$$

$$\begin{aligned}
x_1 \frac{\partial \Phi}{\partial x_1} + x_2 \frac{\partial \Phi}{\partial x_2} + x_3 \frac{\partial \Phi}{\partial x_3} + x_4 \frac{\partial \Phi}{\partial x_4} &= a \cdot \Phi, \\
x_2 \frac{\partial \Phi}{\partial x_2} + 3x_3 \frac{\partial \Phi}{\partial x_3} + 4x_4 \frac{\partial \Phi}{\partial x_4} &= b \cdot \Phi,
\end{aligned}$$

where a and b are complex parameters. Experts on hypergeometric functions will recognize these equations: this is the $\mathcal{A}$-hypergeometric system of Gel'fand, Kapranov and Zelevinsky (1989) for the particular configuration

$$\mathcal{A} = \left\{ \binom{1}{0}, \binom{1}{1}, \binom{1}{3}, \binom{1}{4} \right\}.$$

We shall use Gröbner bases to prove that, for every open ball U in $\mathbf{C}^4$, the dimension of the $\mathbf{C}$-vector space of holomorphic functions Φ on U which satisfy (1.1) is at most five. Moreover, for $(a, b) = (1, 2)$ and suitable U it is exactly five. See Propositions 2.1 and 4.1 for precise statements.

One remark for experts: our example shows that Theorem 2 in (Gel'fand et al. 1989) is incorrect without an additional Cohen-Macaulayness assumption. The statement of this important theorem was corrected in (Gel'fand et al. 1993) but no explicit counterexample was given. This tutorial shows that it is easy to construct and study counterexamples by using Gröbner bases.

In the remainder of the introduction we motivate the equations (1.1) by describing some special solutions: polynomial solutions, algebraic solutions,

[1]Partially supported by the David and Lucile Packard Foundation

rational solutions, and solutions admitting integral representations. In the sequel, any function Φ which satisfies the differential equations (1) is called $\mathcal{A}$-*hypergeometric*. The relation to classical hypergeometric functions, mentioned in the tutorial by Chyzak (1998), can be seen in Theorem 4.1.

We first ask whether the system (1) has any polynomial solutions. A necessary condition for this is that (a,b) lies in the monoid $\mathbf{N}\mathcal{A}$ spanned by the set $\mathcal{A}$ in $\mathbf{N}^2$, because the last two equations in (1.1) are equivalent to

$$\Phi(sx_1, stx_2, st^3x_3, st^4x_4) \quad = \quad s^a t^b \cdot \Phi(x_1,x_2,x_3,x_4). \tag{1.2}$$

To see this equivalence, apply the operators $\partial/\partial s|_{s=1}$ and $\partial/\partial t|_{t=1}$ to (2). Conversely, if (a,b) lies in $\mathbf{N}\mathcal{A}$ then the following set is finite and non-empty

$$\mathcal{F}_{(a,b)} = \{(u_1,u_2,u_3,u_4) \in \mathbf{N}^4 \mid u_1+u_2+u_3+u_4 = a,\ u_2+3u_3+4u_4 = b\},$$

and there is exactly one $\mathcal{A}$-hypergeometric polynomial up to scalar multiples:

$$\Phi_{(a,b)}(x) \quad = \quad \sum_{u \in \mathcal{F}_{(a,b)}} \frac{x_1^{u_1} x_2^{u_2} x_3^{u_3} x_4^{u_4}}{u_1!\, u_2!\, u_3!\, u_4!}, \tag{1.3}$$

which is a solution of (1.1).

Next, consider the zeros of the following polynomial in one variable ϕ,

$$f(\phi) \quad = \quad x_1 + x_2\phi + x_3\phi^3 + x_4\phi^4.$$

They can be written in terms of radicals using Cardano's formula. These algebraic functions are $\mathcal{A}$-hypergeometric for $(a,b) = (0,-1)$. In fact, if U is an open ball in $\mathbf{C}^4$ which does not intersect the zero set of the discriminant

$$\Delta_{\mathcal{A}} \; = \; x_1x_4\cdot(27x_1^2x_3^4 + 4x_2^3x_3^3 + 6x_4x_1x_2^2x_3^2 + 192x_4^2x_1^2x_2x_3 + 27x_4^2x_2^4 - 256x_4^3x_1^3),$$

then the four roots of $f(\phi) = 0$ are well-defined holomorphic functions on U, and they form a $\mathbf{C}$-basis for the space of $\mathcal{A}$-hypergeometric functions on U. More generally, if m is a non-zero integer and $(a,b) = (0,-m)$ then a solution basis is given by the m-th powers of the roots $\phi_1, \phi_2, \phi_3, \phi_4$ of f.

This observation leads to a nice rational $\mathcal{A}$-hypergeometric function:

$$\phi_1^m + \phi_2^m + \phi_3^m + \phi_4^m.$$

Indeed, the m-th power sum of the roots is a Laurent polynomial in $x_1, \ldots, x_4$ which satisfies (1.1) with parameters $(a,b) = (0,-m)$. For instance, for $m = 1$ we get the $\mathcal{A}$-hypergeometric function x_3/x_4, and for $m = 6$ we get

$$\phi_1^6 + \phi_2^6 + \phi_3^6 + \phi_4^6 \quad = \quad (x_3^6 - 6x_1x_3^2x_4^3 + 6x_2x_3^3x_4^2 + 3x_2^2x_4^4)\,/\,x_4^6.$$

Further algebraic solutions to (1.1) for other integral parameter values can be found by considering residue integrals such as

$$\frac{\phi_i^m}{f'(\phi_i)} \quad = \quad \frac{1}{2\pi\sqrt{-1}} \int_{C_i} \frac{z^m}{f(z)} dz,$$

where C_i is a small, positively oriented loop around the point ϕ_i in the complex plane. This is an algebraic function of x_1, x_2, x_3, x_4, and it satisfies (1.1) for $(a, b) = (-1, -1-m)$. If we take a big loop C which positively encircles all four roots, then we get the rational $\mathcal{A}$-hypergeometric function

$$\frac{\phi_1^m}{f'(\phi_1)} + \frac{\phi_2^m}{f'(\phi_2)} + \frac{\phi_3^m}{f'(\phi_3)} + \frac{\phi_4^m}{f'(\phi_4)}.$$

Finally, we mention a general integral representation for $\mathcal{A}$-hypergeometric functions. If the complex parameters a and b are sufficiently generic, then every solution to (1.1) on an open ball $U \subset \mathbf{C}^4$ can be written as

$$\Phi(x_1, x_2, x_3, x_4) \quad = \quad \int_\Gamma f(z)^a z^{-b-1} dz \tag{1.4}$$

where Γ is a segment with endpoints belonging to $D = \{0, \phi_1, \phi_2, \phi_3, \phi_4\}$. If the real parts of a and $-b-1$ are less than or equal to -1, then the integral does not converge. In this case, we take the finite part of the integral in the sense of Hadamard. How many segments are linearly independent? We assume that all the roots ϕ_i are real numbers for simplicity. Among $\binom{5}{2} = 10$ segments, only four segments are independent; when all roots are ordered as

$$\phi_0 = 0 < \phi_1 < \phi_2 < \phi_3 < \phi_4,$$

the integral (1.4) can be expressed as a $\mathbf{C}$-linear combination of the four integrals $\int_{[\phi_i, \phi_{i+1}]} f(z)^a z^{-b-1} dz$. The segments $[\phi_i, \phi_{i+1}]$ are the compact chambers of a 1-dimensional hyperplane arrangement. They form a basis of the twisted homology group $H_1^{lf}(\mathbf{C} \backslash D, \nabla)$; see page 53 in (Aomoto and Kita 1994). Hypergeometric differential equations play an important role in algebraic geometry because for certain rational parameters (a, b) the integral (1.4) can be regarded as periods of a family of curves; see (Yoshida 1997).

2 Characteristic variety and holonomicity

In this section we show how to prove that the space of $\mathcal{A}$-hypergeometric functions is finite-dimensional, how to evaluate its dimension, and how to compute the characteristic variety. We use Gröbner bases in the Weyl algebra

$$W \quad = \quad k\langle x_1, x_2, x_3, x_4, \partial_1, \partial_2, \partial_3, \partial_4 \rangle$$

over a field k, which in our case is either $\mathbf{Q}$ or $\mathbf{Q}(a, b)$. The Weyl algebra W is the free associative algebra over k modulo the commutation rules

$$x_i x_j = x_j x_i, \quad \partial_i \partial_j = \partial_j \partial_i, \quad x_i \partial_j = \partial_j x_i \text{ for } i \neq j\,, \quad \text{and } \partial_i x_i = x_i \partial_i + 1.$$

A linear system of partial differential equations with polynomial coefficients corresponds to a left ideal in W. Our system (1.1) corresponds to the left ideal $H_{\mathcal{A}}$ which is generated by

$$\partial_2\partial_3 - \partial_1\partial_4,\ \partial_1^2\partial_3 - \partial_2^3,\ \partial_2\partial_4^2 - \partial_3^3,\ \partial_1\partial_3^2 - \partial_2^2\partial_4, \tag{2.1}$$
$$x_1\partial_1 + x_2\partial_2 + x_3\partial_3 + x_4\partial_4 - a,\ x_2\partial_2 + 3x_3\partial_3 + 4x_4\partial_4 - b. \tag{2.2}$$

Here a, b are either specific numbers or indeterminate parameters. A function Φ on $U \subset \mathbf{C}^4$ is $\mathcal{A}$-hypergeometric if it is annihilated by the left W-ideal $H_{\mathcal{A}}$.

The theory (and practise) of Gröbner bases works perfectly well for left ideals in the Weyl algebra W. We quickly review the relevant basics. Every element f in W can be written uniquely as a k-linear combination of *normally ordered* monomials $x^a\partial^b = x_1^{a_1}x_2^{a_2}x_3^{a_3}x_4^{a_4}\partial_1^{b_1}\partial_2^{b_2}\partial_3^{b_3}\partial_4^{b_4}$. This representation of f is called *normally ordered representation.* For example, the monomial $\partial_1 x_1 \partial_1$ is not normally ordered. Its normally ordered representation is $x_1\partial_1^2 + \partial_1$.

Consider the commutative polynomial ring in eight variables

$$\mathrm{gr}\,(W) \quad = \quad k[x_1, x_2, x_3, x_4, \xi_1, \xi_2, \xi_3, \xi_4]$$

and the k-linear map $\mathrm{gr} : W \to \mathrm{gr}\,(W)$, $x^a\partial^b \mapsto x^a\xi^b$. Let $<$ be any term order on $\mathrm{gr}\,(W)$. This gives a total order among normally ordered monomials in W via $x^A\partial^B > x^a\partial^b \Leftrightarrow x^A\xi^B > x^a\xi^b$. For any element $f \in W$ let $in_<(f)$ denote the highest monomial $x^A\partial^B$ in the normally ordered representation of f. If I is a left ideal in W then its *initial ideal* is the ideal $\mathrm{gr}\,(in_<(I))$ in $\mathrm{gr}\,(W)$ generated by all monomials $\mathrm{gr}\,(in_<(f))$ for $f \in I$. Clearly, $\mathrm{gr}\,(in_<(I))$ is generated by finitely many monomials $x^a\xi^b$. A finite subset G of I is called *a Gröbner basis* of I with respect to the term order $<$ if $\{\mathrm{gr}\,(in_<(g))\,|\,g \in G\}$ generates $\mathrm{gr}\,(in_<(I))$. Noting that $in_<(f) \le in_<(g)$ implies $in_<(hf) \le in_<(hg)$ for all $h \in W$, one proves that the reduced Gröbner basis of I is unique and finite, and can be computed using Buchberger's algorithm. An efficient implementation is available in the system **kan** by Takayama (1991).

The definition of the initial ideal extends to partial term orders given by weight vectors. Of importance in algebraic analysis is the partial term order v defined on W by the weight vector $(0,0,0,0,1,1,1,1)$, that is, by the rule:

$$\begin{array}{ccc} x_1^{A_1}x_2^{A_2}x_3^{A_3}x_4^{A_4}\partial_1^{B_1}\partial_2^{B_2}\partial_3^{B_3}\partial_4^{B_4} & \text{if and only if} & B_1 + B_2 + B_3 + B_4 \\ \ge x_1^{a_1}x_2^{a_2}x_3^{a_3}x_4^{a_4}\partial_1^{b_1}\partial_2^{b_2}\partial_3^{b_3}\partial_4^{b_4} & & \ge b_1 + b_2 + b_3 + b_4. \end{array}$$

The v-leading form $in_v(f)$ of an element $f \in W$ is also called the *v-principal part* of f. For an m-th order operator $f = \sum_{B_1+\cdots+B_4\le m} C_{AB}x^A\partial^B$ it equals

$$in_v(f) \quad = \sum_{B_1+\cdots+B_4=m} C_{AB}x^A\partial^B.$$

Let $\mathrm{gr}\,(in_v(I))$ be the ideal in $\mathrm{gr}\,(W)$ generated by the polynomials $\mathrm{gr}\,(in_v(f))$ for all $f \in I$. We call $\mathrm{gr}\,(in_v(I))$ the *characteristic ideal* of $I \subset W$.

The algebraic subset of affine 8-space defined by gr $(in_v(I))$ is the *characteristic variety* of I. It can be computed as follows: let $<$ be any term order which refines the partial term order v and compute the reduced $<$-Gröbner basis G for I. Then gr $(in_v(I))$ is generated by the polynomials gr $(in_v(g))$ for all $g \in G$. This algorithm was introduced by Oaku (1994), (1996). For the general setting of D-modules and computation of functors for D-modules see (Borel et al. 1987), (Kashiwara 1983), (Oaku 1997) and their references.

The *Fundamental Theorem of Algebraic Analysis* (Sato et al. 1973) states that each associated prime of the characteristic ideal is at least middle-dimensional, that is, has dimension ≥ 4. If each associated prime of gr $(in_v(I))$ has dimension exactly 4 then the left ideal I is called *holonomic*. For any connected open subset U of $\mathbf{C}^4$ let $Sol(U, I)$ denote the $\mathbf{C}$-vector space of holomorphic functions on U which are annihilated by all operators in I. If I is holonomic then $dim_{\mathbf{C}}(Sol(U, I))$ is finite, and, moreover, these dimensions have a finite maximum $r(I)$ as U ranges over all *connected* open subsets of $\mathbf{C}^4$. The integer $r(I)$ is called the *rank* of the holonomic ideal I. It is known that $dim_{\mathbf{C}}(Sol(U, I)) = r(I)$ if U is sufficiently small and does not meet the *singular locus* of I. In our example $I = H_{\mathcal{A}}$ the singular locus is the discriminant hypersurface $\Delta_{\mathcal{A}} = 0$.

We shall express the rank $r(I)$ purely algebraically. Consider the ring $R = k(x_1, \ldots, x_4)\langle \partial_1, \ldots, \partial_4 \rangle$ of differential operators with rational function coefficients. Let RI be the left ideal generated by I in R. Then

$$r(I) \quad = \quad \dim_{k(x_1,\ldots,x_4)} R/RI. \tag{2.3}$$

The equivalence of this formula with the analytic definition of $r(I)$ and an algorithm to evaluate the dimension were announced in (Takayama 1987).

To evaluate the rank $r(I)$ we proceed as follows. Fix any block term order

$$x^A \partial^B > x^a \partial^b \quad \Longleftrightarrow \quad \partial^B >_1 \partial^b \text{ or } (\partial^B = \partial^b \text{ and } x^A >_2 x^a)$$

where $>_1$ is a term order in $k[\partial]$ that refines the partial term order by the weight vector $(1, 1, 1, 1)$ and $>_2$ is a term order in $k[x]$. Compute the reduced Gröbner basis G of I with respect to this block order. Let gr $(in_<(G))$ be the set of leading monomials $x^a \xi^b$ of all elements in G. This set generates the ideal gr $(in_<(I))$ in gr (W). Now substitute $x_1 = x_2 = x_3 = x_4 = 1$ in gr $(in_<(G))$. The remaining monomials generate a monomial ideal M in $k[\xi_1, \xi_2, \xi_3, \xi_4]$. Since G is also a Gröbner basis for RI in R, we have

$$r(I) \quad = \quad dim_k(k[\xi_1, \xi_2, \xi_3, \xi_4]/M).$$

Let us see how this algorithm works for our hypergeometric ideal $H_{\mathcal{A}}$. We select $<$ to be the reverse lexicographic refinement of v such that $\partial_1 < \ldots < \partial_4$ and $x_1 < \ldots < x_4$. We regard a and b as parameters, that is, we work over the rational function field $k = \mathbf{Q}(a, b)$. The reduced Gröbner basis G of $H_{\mathcal{A}}$

with respect to $<$ consists of 34 elements. We list only their leading terms, the 34 minimal generators of the initial ideal $\mathrm{gr}\,(in_<(H_\mathcal{A}))$ in $\mathrm{gr}\,(W)$:

$$\begin{gathered}\xi_2\xi_3,\xi_2^3,\ \xi_3^3,\ x_3\xi_1\xi_4^2,\ x_3\xi_2\xi_4^2,\ \xi_2^2\xi_4,\ x_4\xi_4,\ x_3\xi_3,\ \underline{x_4\xi_3^2},\ x_3\xi_1\xi_2\xi_4,\ x_3^3\xi_1^2\xi_4,\underline{x_3^2\xi_2\xi_4},\\ x_1x_4\xi_1\xi_2^2,x_1x_3\xi_1\xi_2^2,x_1x_3^2\xi_1^2\xi_4,x_2x_3^2\xi_1^2\xi_4,x_2\xi_2^2,\underline{x_1x_3x_4\xi_2^2},x_1x_2^2x_4\xi_1^2\xi_3,x_1^2x_4^2\xi_1^2\xi_3,\\ x_1x_2x_4^2\xi_1^2\xi_3,x_2^2x_4^2\xi_1^2\xi_3,x_1x_4^3\xi_1^2\xi_3,\underline{x_2x_3^3\xi_1\xi_4},x_1x_2^3x_4\xi_1^2\xi_2,x_2x_3^4\xi_1^2\xi_2,x_1x_2x_3^3\xi_1^2\xi_2,\\ x_1x_2^2x_3^2\xi_1^2\xi_2,x_2^2x_3^3\xi_1^2\xi_2;x_1x_2^2x_3x_4\xi_1^2\xi_2,\underline{x_1x_2x_4^3\xi_1\xi_3},x_1^2x_3^4\xi_1^2\xi_2,\underline{x_2^2x_3^4\xi_1\xi_2},x_1x_2^3x_3^3\xi_1^3.\end{gathered}$$

From the Gröbner basis G we can extract a subset of seven elements whose v-leading forms minimally generate the characteristic ideal $\mathrm{gr}\,(in_v(H_\mathcal{A}))$:

$$\xi_2\xi_3-\xi_1\xi_4,\quad \xi_2^3-\xi_1^2\xi_3,\quad \xi_2^2\xi_4-\xi_1\xi_3^2,\quad \xi_3^3-\xi_2\xi_4^2, \tag{2.4}$$

$$x_1\xi_1+x_2\xi_2+x_3\xi_3+x_4\xi_4,\quad x_2\xi_2+3x_3\xi_3+4x_4\xi_4, \tag{2.5}$$

$$(b-2)x_1\xi_2^2+(b-a-1)x_2\xi_1\xi_3+(b-3a+1)x_3\xi_2\xi_4+(b-4a+2)x_4\xi_3^2 \tag{2.6}$$

The polynomials in (8) and (9) obviously lie in $\mathrm{gr}\,(in_v(H_\mathcal{A}))$ since they are the v-leading forms of (5) and (6). The polynomial (10) is non-trivial and depends on the parameters (a,b). It is easy to verify that (8), (9) and (10) generate $\mathrm{gr}\,(in_v(H_\mathcal{A}))$: just compute their (commutative) Gröbner basis with respect to $<$ and check that you get the same 34 initial monomials as above.

The 34 initial monomials define an ideal of dimension 4. We conclude that the characteristic ideal has dimension 4, and therefore the system $H_\mathcal{A}$ is holonomic. To evaluate the rank of $H_\mathcal{A}$, we set $x_1=x_2=x_3=x_4=1$ in these 34 monomials. The resulting ideal M has codimension 4 in $k[\xi_1,\xi_2,\xi_3,\xi_4]$:

$$M \quad = \quad \langle \xi_1^3,\ \xi_1\xi_2,\ \xi_2^2,\ \xi_3,\ \xi_4\rangle.$$

Thus the hypergeometric ideal $H_\mathcal{A}$ is holonomic of rank 4 for (a,b) generic. By looking at the Gröbner basis G we can see that $\mathrm{gr}\,(in_<(H_\mathcal{A}))$ remains the same if a,b are specialized to numerical values, provided $(a,b)\neq(1,2)$.

For $(a,b)=(1,2)$ things are dramatically different: the polynomial (10) vanishes identically and the reduced Gröbner basis G has only 29 elements. To get the 29 monomial generators of $\mathrm{gr}\,(in_<(H_\mathcal{A}))$ erase the six underlined monomials above and replace them by $x_4\xi_1\xi_3^2$. Setting $x_i=1$ we now find

$$M \quad = \quad \langle \xi_1^3,\ \xi_1^2\xi_2,\ \xi_2^2,\ \xi_3,\ \xi_4\rangle.$$

This ideal has codimension 5 in $k[\xi_1,\xi_2,\xi_3,\xi_4]$. We summarize our results:

Proposition 2.1 *The hypergeometric ideal $H_\mathcal{A}$ is holonomic of rank* 4 *for $(a,b)\neq(1,2)$, and it is holonomic of rank* 5 *for $(a,b)=(1,2)$.*

We remark that the square of (10) lies in the ideal generated by (8) and (9). This shows that the radical of $\mathrm{gr}\,(in_v(H_\mathcal{A}))$ is independent of the parameters (a,b), and hence so is the characteristic variety. The reader will find it instructive to compute (using Gröbner bases) the radical and primary decomposition of the characteristic ideal. Hint: The characteristic variety has four irreducible components, by Theorem 4 in (Gel'fand et al. 1989).

3 Creation operators

An important issue in the study of differential equations is to find operators that create new solutions from known solutions. Many nice families of equations such as hypergeometric systems, Painlevé equations and KP equations admit such operators. In this section we discuss creation operators for the system (1.1), and we show how to compute them using Gröbner bases.

Suppose we are given any solution $\phi_{a,b}(x)$ to (1.1), for instance, one of the polynomial or algebraic functions discussed in the introduction, or one of the series solutions in Section 4. By taking partial derivatives we get new solutions for smaller parameter values indicated by subscripts:

$$\begin{aligned} \partial_1 \phi_{a,b}(x) &= \phi_{a-1,b}(x), & \partial_2 \phi_{a,b}(x) &= \phi_{a-1,b-1}(x), \\ \partial_3 \phi_{a,b}(x) &= \phi_{a-1,b-3}(x), & \partial_4 \phi_{a,b}(x) &= \phi_{a-1,b-4}(x). \end{aligned} \tag{3.1}$$

In this context ∂_i is called the i-th *annihilation operator*. Our problem is to construct an inverse operator C_i, called a *creation operator*, which takes solutions of low parameter values to solutions of higher parameter values.

Saito et al. (1997) showed that the creation operators have the form $C_i = r_i/B_i(a,b)$, where r_i is an element of $W = \mathbf{Q}\langle x_1, x_2, x_3, x_4, \partial_1, \partial_2, \partial_3, \partial_4 \rangle$ and B_i is a product of linear polynomials in a and b with integer coefficients. For example, the creation operator for $i = 1$ equals $C_1 = r_1/B_1(a,b)$ where

$$B_1(a,b) \quad = \quad (b-2)(4a-b)(4a-b-1)(4a-b-2)(4a-b-3)$$

and r_1 has the following longish expression in the Weyl algebra W:

$$\begin{aligned}
&256x_2x_1^4\partial_2\partial_1^3 + 768x_2^2x_1^3\partial_2^2\partial_1^2 + 864x_2^3x_1^2\partial_3\partial_1^3 + 768x_3x_1^4\partial_3\partial_1^3 + 864x_2^4x_1\partial_4\partial_1^3 \\
&+2560x_3x_2x_1^3\partial_4\partial_1^3 + 1024x_4x_1^4\partial_4\partial_1^3 + 81x_2^5\partial_4\partial_2\partial_1^2 + 3168x_3x_2^2x_1^2\partial_4\partial_2\partial_1^2 \\
&+3072x_4x_2x_1^3\partial_4\partial_2\partial_1^2 + 1728x_3x_2^3x_1\partial_4\partial_2^2\partial_1 + 768x_3^2x_1^3\partial_4\partial_2^2\partial_1 + 3456x_4x_2^2x_1^2\partial_4\partial_2^2\partial_1 \\
&+351x_3x_2^4\partial_4\partial_3\partial_1^2 + 1824x_3^2x_2x_1^2\partial_4\partial_3\partial_1^2 + 1728x_4x_2^3x_1\partial_4\partial_3\partial_1^2 + 1024x_4x_3x_1^3\partial_4\partial_3\partial_1^2 \\
&+1440x_3^2x_2^2x_1\partial_4^2\partial_1^2 + 234x_4x_2^4\partial_4^2\partial_1^2 + 2304x_4x_3x_2x_1^2\partial_4^2\partial_1^2 + 378x_3^2x_2^3\partial_4^2\partial_2\partial_1 \\
&+288x_3^3x_1^2\partial_4^2\partial_2\partial_1 + 1728x_4x_3x_2^2x_1\partial_4^2\partial_2\partial_1 + 448x_3^3x_2x_1\partial_4^2\partial_2^2 + 432x_4x_3x_2^3\partial_4^2\partial_2^2 \\
&+384x_4x_3^2x_1^2\partial_4^2\partial_2^2 + 174x_3^3x_2^2\partial_4^2\partial_3\partial_1 + 576x_4x_3^2x_2x_1\partial_4^2\partial_3\partial_1 + 48x_4x_3^3x_2\partial_4^2\partial_3^2 \\
&+48x_3^4x_1\partial_4^3\partial_1 + 216x_4x_3^2x_2^2\partial_4^3\partial_1 + 37x_3^4x_2\partial_4^3\partial_2 + 64x_4x_3^3x_1\partial_4^3\partial_2 + 3x_3^5\partial_4^3\partial_3 \\
&+4x_4x_3^4\partial_4^4 - 512x_1^4\partial_1^3 + 384x_2x_1^3\partial_2\partial_1^2 + 2304x_2^2x_1^2\partial_2^2\partial_1 + 1944x_2^3x_1\partial_3\partial_1^2 \\
&+3712x_3x_1^3\partial_3\partial_1^2 + 8640x_3x_2x_1^2\partial_4\partial_1^2 + 4608x_4x_1^3\partial_4\partial_1^2 + 1620x_3x_2^3\partial_4\partial_2^2 \\
&+486x_2^4\partial_4\partial_1^2 + 6552x_3x_2^2x_1\partial_4\partial_2\partial_1 + 9216x_4x_2x_1^2\partial_4\partial_2\partial_1 + 2112x_3^2x_1^2\partial_4\partial_2^2 \\
&+6048x_4x_2^2x_1\partial_4\partial_2^2 + 2952x_3^2x_2x_1\partial_4\partial_3\partial_1 + 1296x_4x_2^3\partial_4\partial_3\partial_1 + 2304x_4x_3x_1^2\partial_4\partial_3\partial_1 \\
&+204x_3^3x_2\partial_4\partial_3^2 + 288x_4x_3^2x_1\partial_4\partial_3^2 + 1008x_3^2x_2^2\partial_4^2\partial_1 + 2880x_4x_3x_2x_1\partial_4^2\partial_1 \\
&+328x_3^3x_1\partial_4^2\partial_2 + 864x_4x_3x_2^2\partial_4^2\partial_2 + 144x_4x_3^2x_2\partial_4^2\partial_3 + 10x_3^4\partial_4^3 - 4608x_1^3\partial_1^2 \\
&-1488x_2x_1^2\partial_2\partial_1 + 816x_2^2x_1\partial_2^2 + 504x_2^3\partial_3\partial_1 + 3024x_3x_1^2\partial_3\partial_1 + 576x_3^2x_1\partial_3^2 \\
&+720x_4x_2^2\partial_3^2 + 3984x_3x_2x_1\partial_4\partial_1 + 3264x_4x_1^2\partial_4\partial_1 + 1212x_3x_2^2\partial_4\partial_2 \\
&+3264x_4x_2x_1\partial_4\partial_2 + 252x_3^2x_2\partial_4\partial_3 + 384x_4x_3x_1\partial_4\partial_3 + 96x_4x_3x_2\partial_4^2 \\
&-1632x_1^2\partial_1 - 792x_2x_1\partial_2 + 168x_3x_1\partial_3 + 48x_3x_2\partial_4 + 96x_4x_1\partial_4 - 48x_1.
\end{aligned}$$

The creation operator C_1 is a left inverse to ∂_1 modulo $H_\mathcal{A}$. This implies

$$r_1\,\phi_{a,b}(x) \quad = \quad B_1(a+1,b)\cdot\phi_{a+1,b}(x).$$

The algorithm to be described below gives us similar formulas for $i = 2,3,4$:

$$\begin{aligned} r_2\,\phi_{a,b}(x) &= B_2(a+1,b+1)\cdot\phi_{a+1,b+1}(x),\\ r_3\,\phi_{a,b}(x) &= B_3(a+1,b+3)\cdot\phi_{a+1,b+3}(x),\\ r_4\,\phi_{a,b}(x) &= B_4(a+1,b+4)\cdot\phi_{a+1,b+4}(x). \end{aligned} \tag{3.2}$$

Outside the zero set of the *b-function* B_i, the creation operator $C_i = r_i/B_i$ is well-defined and gives an isomorphism of left W-modules between $\mathcal{A}$-hypergeometric systems with different parameter values. The relations (3.1) and (3.2) are also called *contiguity relations* of $\mathcal{A}$-hypergeometric functions. See (Chyzak 1998) for contiguity relations in the setting of Ore algebra.

The relations (12) can be used to generate solutions to (1) for high parameter values from solutions with small parameter values. For instance, all $\mathcal{A}$-hypergeometric polynomials (3) can be generated from the trivial $\mathcal{A}$-hypergeometric function $\Phi_{(0,0)} = 1$ by successively applying the creation operators. More precisely, there exists an identity

$$r_1^{i_1} r_2^{i_2} r_3^{i_3} r_4^{i_4}\,(1) \quad = \quad const\cdot\Phi_{(i_1+i_2+i_3+i_4,\,i_2+3i_3+4i_4)},$$

where the multiplier "*const*" is a positive integer.

We now present an algorithm from (Saito et al. 1997) for computing creation operators. It uses Buchberger's algorithm in the extended Weyl algebra

$$W[a,b] \quad = \quad \mathbf{Q}[a,b]\langle x_1,\ldots,x_4,\partial_1,\ldots,\partial_4\rangle.$$

The two new variables a and b commute with the eight old variables x_j, ∂_j. Let $>$ be any term order on $W[a,b]$ which satisfies the following rule

$$a^{C_1}b^{C_2}x_1^{A_1}x_2^{A_2}x_3^{A_3}x_4^{A_4}\partial_1^{B_1}\partial_2^{B_2}\partial_3^{B_3}\partial_4^{B_4} \;\geq\; a^{c_1}b^{c_2}x_1^{a_1}x_2^{a_2}x_3^{a_3}x_4^{a_4}\partial_1^{b_1}\partial_2^{b_2}\partial_3^{b_3}\partial_4^{b_4}$$
$$\text{whenever}\quad A_1+\cdots+A_4+B_1+\cdots+B_4 \geq a_1+\cdots+a_4+b_1+\cdots+b_4.$$

Thus $>$ is an *elimination order* with blocks $\{x_1,\ldots,x_4,\partial_1,\ldots,\partial_4\} > \{a,b\}$.

Let $H_\mathcal{A}(a,b)$ be the *parametric hypergeometric ideal*, that is, the left ideal in $W[a,b]$ generated by the six polynomials in (5) and (6). To find the i-th creation operator C_i, we consider the left ideal generated by $H_\mathcal{A}(a,b)$ together with the i-th annihilation operator ∂_i. We compute the reduced Gröbner basis G for this ideal with respect to $>$. It is proved in (Saito et al. 1994) that $H_\mathcal{A}(a,b) + W[a,b]\cdot\partial_i$ has non-zero intersection with $\mathbf{Q}[a,b]$. The elimination property of $>$ ensures that this intersection is generated by $G \cap \mathbf{Q}[a,b]$.

Choose any element $B_i(a,b)$ of $G \cap \mathbf{Q}[a,b]$. Keeping track of multipliers during the Gröbner basis computation, we find $r_i \in W[a,b]$ such that

$$B_i(a,b) - r_i\cdot\partial_i \quad\in\quad H_\mathcal{A}(s).$$

In fact, we may assume $r_i \in W$ since any occurrence of a and b in r_i can be replaced by $x_1\partial_1 + x_2\partial_2 + x_3\partial_3 + x_4\partial_4$ and $x_2\partial_2 + 3x_3\partial_3 + 4x_4\partial_4$ respectively.

Here is the punch line: any $\mathcal{A}$-hypergeometric function $\phi_{a,b}(x)$ is annihilated by $B_i(a,b) - r_i \cdot \partial_i$. Thus we can recover $\phi_{a,b}(x)$ by applying the operator $r_i / B_i(a,b)$ to the "lower" $\mathcal{A}$-hypergeometric function $\partial_i \phi_{a,b}(x)$. This is exactly what we wanted: by shifting (a,b) we get the formulas in (11).

4 Series solutions

The study of hypergeometric series is important for many applications, including type II_B string theory (Hosono et al. 1997), singularity theory (Noumi 1984), and Gauss-Schwarz theory (Yoshida 1997). Gröbner bases give a nice description of the asymptotic behavior of $\mathcal{A}$-hypergeometric functions and can be used to construct series expansions.

Let $\mathcal{A} = \begin{pmatrix} 1 & 1 & 1 & 1 \\ 0 & 1 & 3 & 4 \end{pmatrix}$ and consider the corresponding *toric ideal*

$$I_{\mathcal{A}} \quad = \quad (\partial^u - \partial^v \; : \; \mathcal{A}u = \mathcal{A}v, \; u, v \in \mathbf{N}^4) \quad \subset \quad k[\partial_1, \ldots, \partial_4].$$

This ideal is generated by the four binomials in (2.1). Fix $\omega \in \mathbf{R}^4$ which defines a term order for $I_{\mathcal{A}}$. This means that $in_\omega(I_{\mathcal{A}})$ is a monomial ideal in $k[\partial_1, \ldots, \partial_4]$. The set of *all* weight vectors which give the same initial ideal is an open convex polyhedral cone C_ω, called the *Gröbner cone* of $I_{\mathcal{A}}$ and ω:

$$C_\omega \quad = \quad \{ \omega' \in \mathbf{R}^4 \mid in_{\omega'}(I_{\mathcal{A}}) = in_\omega(I_{\mathcal{A}}) \}.$$

For background on Gröbner bases and Gröbner cones of toric ideals see (Sturmfels 1995).

The weight vector $(-\omega, \omega) \in \mathbf{R}^8$ defines a partial term order on the Weyl algebra W. The initial ideal $in_{(-\omega,\omega)}(H_{\mathcal{A}})$ of the hypergeometric ideal $H_{\mathcal{A}}$ is called the *initial system* of $H_{\mathcal{A}}$ with respect to ω. Note that the initial system contains $in_\omega(I_{\mathcal{A}})$ and the two polynomials in (6). Saito, Sturmfels and Takayama (1998) showed that $in_{(-\omega,\omega)}(H_{\mathcal{A}})$ is generated by $in_\omega(I_{\mathcal{A}})$ and (6) if the parameters (a,b) are generic. For special parameters, the initial system may be larger than the subideal generated by $in_\omega(I_{\mathcal{A}})$ and (6). It can be found easily by computing a Gröbner basis for $H_{\mathcal{A}}$ with respect to a term order refining $(-\omega, \omega)$.

Let $\phi(x_1, x_2, x_3, x_4)$ be any holomorphic solution of (1.1). Note that it can be analytically continued to any point outside the zeros of the discriminant $\Delta_{\mathcal{A}}$. We introduce a parameter $t \in \mathbf{R}$ and consider the asymptotic behavior of the function

$$\phi(t^\omega x) \quad = \quad \phi(t^{\omega_1}x_1, t^{\omega_2}x_2, t^{\omega_3}x_3, t^{\omega_4}x_4) \qquad \text{for} \quad t \longrightarrow 0.$$

Let $in_\omega(\phi)$ denote the dominant term of $\phi(t^\omega x)$ around $t = 0$. It is easy to see that $in_\omega(\phi)$ is annihilated by the initial system $in_{(-\omega,\omega)}(H_A)$.

This observation suggests that we construct series solutions to (1) whose lowest terms are monomials x^u, $u \in \mathbf{C}^4$, which are annihilated by the initial system. This makes sense because the solution space to the initial system $in_{(-\omega,\omega)}(H_A)$ is spanned by monomials if (a, b) is generic; cf. (Saito et al. 1998).

Let x^u be any monomial solution to $in_{(-\omega,\omega)}(H_A)$. Let L be the integer kernel of the matrix A. This is a rank 2 sublattice of $\mathbf{Z}^4$. Consider the monoid $M_{\omega,u} := \{ v \in L \mid u_i + v_i \in \mathbf{N} \text{ for all } i \in \{1,2,3,4\} \text{ such that } u_i \in \mathbf{N}\}$.

Theorem 4.1 *The following series is a formal solution to the equations (1):*

$$\Phi_{\omega,u}(x) \quad := \quad \sum_{v \in M_{\omega,u}} x^{u+v} / \prod_{i=1}^{4} \Gamma(u_i + v_i + 1)$$

Moreover, there exists a vector $\tau_\omega \in \mathbf{R}^4$ *such that this series converges whenever* $(-\log|x_1|, \ldots, -\log|x_4|)$ *lies in the translated Gröbner cone* $C_\omega + \tau_\omega$.

This construction of series solutions appears in the work of Hosono, Lian and Yau (1997) for explicit constructions of mirror maps. See also (Sturmfels 1992). It differs from the triangulation construction of Gel'fand, Kapranov and Zelevinsky (1989). The key point in the proof of Theorem 4.1 is that the inner product $\omega' \cdot v$ is positive for all non-zero elements v in the monoid $M_{\omega,u}$ and all vectors ω' in the Gröbner cone C_ω.

Let us see how this construction works for the weight vector $\omega = (0, 1, 2, 0)$. For this term order the four generators of I_A are already a Gröbner basis:

$$\underline{\partial_2\partial_3} - \partial_1\partial_4, \ \underline{\partial_1\partial_3^2} - \partial_2^2\partial_4, \ \underline{\partial_2^3} - \partial_1^2\partial_3, \ \underline{\partial_3^3} - \partial_2\partial_4^2.$$

The initial system $in_{(-\omega,\omega)}(H_A)$ contains the underlined leading terms and the two polynomials in (6). Suppose that $x^u = x_1^{u_1}x_2^{u_2}x_3^{u_3}x_4^{u_4}$ is annihilated by $in_{(-\omega,\omega)}(H_A)$. Then the exponents satisfy following six algebraic equations

$$u_2u_3 \ = \ u_1u_3(u_3-1) \ = \ u_2(u_2-1)(u_2-2) \ = \ u_3(u_3-1)(u_3-2) \ = \ 0$$

$$\text{and} \qquad \begin{pmatrix} 1 & 1 & 1 & 1 \\ 0 & 1 & 3 & 4 \end{pmatrix} \begin{pmatrix} u_1 \\ u_2 \\ u_3 \\ u_4 \end{pmatrix} \ = \ \begin{pmatrix} a \\ b \end{pmatrix}.$$

Suppose that a and b are generic. Then these six equations have four solutions

$$\begin{aligned} u^{(1)} &= (\ (4a-b)/4, \ 0, \ 0, \ b/4 \ ), \\ u^{(2)} &= (\ (4a-b-3)/4, \ 1, \ 0, \ (b-1)/4 \ ), \\ u^{(3)} &= (\ (4a-b-1)/4, \ 0, \ 1, \ (b-3)/4 \ ), \\ u^{(4)} &= (\ (4a-b-6)/4, \ 2, \ 0, \ (b-2)/4 \ ). \end{aligned}$$

The corresponding monomials $x^{u^{(1)}}$, $x^{u^{(2)}}$, $x^{u^{(3)}}$ and $x^{u^{(4)}}$ form a basis for the solution space of $in_{(-\omega,\omega)}(H_{\mathcal{A}})$. They give rise to $\mathcal{A}$-hypergeometric series

$$\Phi^{(i)} \quad = \quad \sum_{\substack{m,n\in\mathbf{Z}:\\ u_2^{(i)}+4m\geq 3n,\ u_3^{(i)}+n\geq 0}} c_{mn}\cdot x^{u^{(i)}+m(-3,4,0,-1)+n(2,-3,1,0)}, \qquad \text{where}$$

$$c_{mn} = 1/(\Gamma(u_1^{(i)}-3m+2n+1)\Gamma(u_2^{(i)}+4m-3n+1)\Gamma(u_3^{(i)}+n+1)\Gamma(u_4^{(i)}-m+1)).$$

By Theorem 4.1, the series $\Phi^{(1)}, \Phi^{(2)}, \Phi^{(3)}$ and $\Phi^{(4)}$ converge on an open subset U of $\mathbf{C}^4$ specified by the Gröbner cone. They are linearly independent since their ω-lowest terms $x^{u^{(i)}}$ are distinct. Hence, by Proposition 2.1, they form a basis for the space of $\mathcal{A}$-hypergeometric functions on U for generic (a, b).

Let us specialize the parameters (a, b) to $(1, 2)$. In this case we have

$$\begin{array}{lcl} u^{(1)} & = & (\ 1/2,\ 0,\ 0,\ 1/2\ ), \\ u^{(2)} & = & (\ 1/4,\ 1,\ 0,\ 1/4\ ), \\ u^{(3)} & = & (\ 1/4,\ 0,\ 1,\ -1/4\ ), \\ u^{(4)} & = & (\ -1,\ 2,\ 0,\ 0\ ). \end{array}$$

$\Phi^{(1)}, \Phi^{(2)}$ and $\Phi^{(3)}$ are convergent series solutions, but $\Phi^{(4)} \equiv 0$. For example, the series $\Phi^{(3)}$ with terms ordered by ω-weight starts like this:

$$c_{00}\frac{x_1^{1/4}x_3}{x_4^{1/4}} + c_{0,-1}\frac{x_2^3}{x_1^{7/4}x_4^{1/4}} + c_{11}\frac{x_2x_3^2}{x_1^{3/4}x_4^{5/4}} + c_{10}\frac{x_2^4x_3}{x_1^{11/4}x_4^{5/4}} + c_{1,-1}\frac{x_2^7}{x_1^{19/4}x_4^{5/4}}$$
$$+ c_{22}\frac{x_2^2x_3^3}{x_1^{7/4}x_4^{9/4}} + c_{21}\frac{x_2^5x_3^2}{x_1^{15/4}x_4^{9/4}} + c_{20}\frac{x_2^8x_3}{x_1^{23/4}x_4^{9/4}} + c_{2,-1}\frac{x_2^{11}}{x_1^{31/4}x_4^{9/4}} + c_{33}\frac{x_2^3x_3^4}{x_1^{11/4}x_4^{13/4}} + \cdots$$

By examining the initial system, we find the two additional $\mathcal{A}$-hypergeometric functions x_2^2/x_1 and x_3^2/x_4. Proposition 2.1 now implies

Proposition 4.1 *Let* $(a, b) = (1, 2)$. *The set* $\{\Phi^{(1)}, \Phi^{(2)}, \Phi^{(3)}, x_2^2/x_1,\ x_3^2/x_4\}$ *is a basis for the five-dimensional space of* $\mathcal{A}$*-hypergeometric functions on* U.

In closing we mention an important result about series solutions. Hotta (1991) showed that all $\mathcal{A}$-hypergeometric ideals are *regular holonomic.* This implies that any formal series solution of the form

$$\sum_{k\in\mathbf{N}^4} c_k(x_1-a_1)^{k_1}(x_2-a_2)^{k_2}(x_3-a_3)^{k_3}(x_4-a_4)^{k_4}, \quad a_i\in\mathbf{C}$$

converges in a neighborhood of the point (a_1, a_2, a_3, a_4).

References

[1] Aomoto, K., Kita, N. (1994): *Theory of Hypergeometric Functions*, (in Japanese) Springer-Verlag, Tokyo.

[2] Borel, A., Grivel, P.-P., Kaup, B., Haefliger, A., Malgrange, B., Ehlers, F. (1987): *Algebraic D-modules*, Academic Press, Boston.

[3] Chyzak, F. (1998): Gröbner bases, symbolic summation and symbolic integration, in this volume.

[4] Gel'fand, I.M., Zelevinskii, A.V., Kapranov, M.M. (1989): Hypergeometric functions and toral manifolds, *Functional Analysis and its Applications* 23: 94–106.

[5] Gel'fand, I.M., Zelevinskii, A.V., Kapranov, M.M. (1993): Correction to the paper: "Hypergeometric functions and toric varieties", *Functional Analysis and its Applications* 27: 4; see also Math. Reviews 95a:22010.

[6] Hosono, S., Lian, B.H., Yau, S.-T. (1997): Maximal degeneracy points of GKZ systems, *Journal of the American Math. Society* 10: 427–443.

[7] Hotta, R. (1991): Equivariant D-modules, Proceedings of ICPAM Spring School in Wuhan, ed. by P. Torasso, Travaux en cours, Hermann, Paris, to appear.

[8] Kashiwara, M. (1983): *Systems of Microdifferential Equations*, Birkhäuser, Boston.

[9] Noumi, M. (1984). Expansion of the solutions of a Gauss-Manin system at a point of infinity, *Tokyo Journal of Mathematics* 7: 1–60.

[10] Oaku, T. (1994): Computation of the characteristic variety and the singular locus of a system of differential equations with polynomial coefficients. *Japan Journal of Industrial and Applied Math.* 11: 485–497.

[11] Oaku, T. (1996): Gröbner bases for D-modules on a non-singular affine algebraic variety, *Tohoku Mathematical Journal* 48: 575-600.

[12] Oaku, T. (1997): Algorithms for b-functions, restrictions, and algebraic local cohomology groups of D-modules, *Advances in Applied Mathematics* 19: 61–105.

[13] Saito, M., Sturmfels, B., Takayama, N. (1997): Hypergeometric polynomials and integer programming, submitted.

[14] Saito, M., Sturmfels, B., Takayama, N. (1998): Initial systems of $\mathcal{A}$-hypergeometric equations, in preparation.

[15] Saito, M., Takayama, N. (1994): Restrictions of $\mathcal{A}$-hypergeometric systems and connection formulas of the $\Delta_1 \times \Delta_{n-1}$-hypergeometric function, *International Journal of Mathematics* 5: 537–560.

[16] Sato, M., Kawai, T., Kashiwara, M. (1973): Microfunctions and pseudodifferential equations, in: Springer Lecture Notes in Mathematics, 287, pp. 265–529.

[17] Sturmfels, B. (1992): Asymptotic analysis of toric ideals, *Memoirs of the Faculty of Sciences, Kyushu University, Ser. A: Math.* 46: 217–228.

[18] Sturmfels, B. (1995): *Gröbner Bases and Convex Polytopes*, American Mathematial Society, University Lecture Series 8, Providence, RI.

[19] Takayama, N. (1987): Gröbner bases in the ring of differential operators and its applications (in Japanese), *Sūrikaisekikenkyūsho Kokyūroku* 612: 67–77.

[20] Takayama, N. (1991): *Kan: A system for computation in algebraic analysis*, source code available at http://www.math.s.kobe-u.ac.jp/KAN/

[21] Yoshida, M. (1997): *Hypergeometric Function, My Love: Modular Interpretations of Configuration Spaces*, Vieweg Verlag, Wiesbaden.

Bernd Sturmfels, Research Institute for Mathematical Sciences, Kyoto University, Kyoto, 606-11, Japan and Department of Mathematics, University of California, Berkeley, CA 94720, USA; `bernd@math.berkeley.edu`

Nobuki Takayama, Department of Mathematics, Kobe University, Rokko, Kobe, 657, Japan; `takayama@math.kobe-u.ac.jp`

Introduction to Noncommutative Gröbner Bases Theory

V. Ufnarovski

Matematiska institutonen, Lunds tekniska högskola, Box 118, S-221 00, Lund, Sweden; e-mail: ufn@maths.lth.se

Abstract

The definitions and main results connected with Gröbner bases, Poincaré series, Hilbert series and Anick's resolution are formulated.

1 Introduction: Some Motivating Examples

The main difficulty in the studying the finitely presented noncommutative algebras is that in general most of the problems are unsolvable. For an illustration, consider the following example.

Example 1 *(Tsejtin, 1958) Let*

$$A = \langle a, b, c, d | ac = ca, ad = da, bc = cb, bd = db, eca = ce, edb = de, cca = ccae \rangle.$$

It is impossible to give an algorithm determining whether two given words in the alphabet a, b, c, d, e are equal in the algebra A.

Nevertheless rather often (e.g. in the commutative or finite-dimensional cases) we are able to solve the corresponding word problem. The reason for this is the regular behaviour of the Gröbner bases in those cases. The aim of this article is to describe how to construct, analyse and use Gröbner bases in noncommutative algebras.

Let $A = \langle X | R \rangle$ be a finitely presented associative algebra over the field K. If I is the ideal generated by R in the free algebra $K\langle X \rangle$ then A is isomorphic to the factor-algebra $K\langle X \rangle / I$ and every element of A can be presented as a linear combination of words in the alphabet X. Example 1 shows that one of the most important questions – which linear combinations present the same element in A – is far from trivial.

One important case in which an algorithmic decision for this problem is possible is the graded case:

Definition 1 *An algebra A (not necessarily associative), decomposable into direct sum of finite-dimensional subspaces $A = \bigoplus_0^\infty A_n$ such that $A_n A_m \subseteq A_{n+m}$ is called* graded. *The formal series $H_A = \sum_0^\infty (\dim A_n) t^n$ is called* the Hilbert series of the algebra. *In the associative case we will always assume $A_0 = K$, in the non-associative $A_0 = 0$.*

One can easily check whether an algebra is graded: every generator $x \in X$ should have a *degree* – a natural number $\deg x$ such that $x \in A_{\deg x}$ and (if the number of generators is infinite) only finitely many of them should have the same degree. Besides that the defining relations $r_i = 0$ can be chosen to be *homogeneous*, i.e. every word $f = x_{i_1} x_{i_2} \dots x_{i_n}$ in r_i has the same degree $\deg f = \sum \deg_{x_{i_k}}$. It is natural to denote this common degree as $\deg r_i$ and to call it the *degree of the relation.*

Example 2 *The algebra $A = \langle x, y | x^3 - y^2 \rangle$ is graded with* $\deg x = 2, \deg y = 3$. *Its Hilbert series is $(1 - t^2 - t^3 + t^6)^{-1}$, as we will see later in Section 6.*

Example 3 *The algebra $\langle x, y | xy - x \rangle$ is not graded:*

$$\deg xy = \deg x \Rightarrow \deg y = 0.$$

Other useful invariants are growth and Gel'fand-Kirillov dimension (briefly GK-dimension), and they are defined for non-graded algebras too. We recall the definition and some facts (see Krause and Lenagan (1985) for more details).

Definition 2 *Let V be a finite dimensional subspace of an algebra A and $d_V(n) = \dim(V + V^2 + \dots V^n)$. Then*

$$\textit{GK-Dim}(A) = \sup_V \overline{\lim_{n\to\infty}} \frac{\ln d_V(n)}{\ln n}$$

If A is generated by V, then the function $d_V(n)$ up to a certain equivalence is the *growth* of A. For example, polynomial growth means finite GK-dimension and exponential growth means $d_V(n) > a^n$ for some $a > 1$.

A useful generalisation of the GK-dimension was found by Petrogradsky (1993).

We will see later how to conclude that the algebra from Example 2 has exponential growth, but the algebra from Example 3 has polynomial growth of degree 2.

The Hilbert series of a graded algebra is one of our main objects of interest. It is a very useful invariant in the commutative case, but also in the noncommutative case it contains a lot of important information about the algebra. It plays the role of generalised dimension of an algebra. For example, it has the following trivial properties:

- $H_{A\oplus B} = H_A + H_B; H_{A\otimes B} = H_A H_B$, $\frac{1}{H_{A*B}} = \frac{1}{H_A} + \frac{1}{H_B} - 1$,
- If L is a graded Lie algebra (superalgebra), $H_L = \sum_1^\infty a_n t^n$, and $A = U(L)$ is the universal enveloping algebra, then

$$H_A = \prod_1^\infty \frac{1}{(1-t^n)^{a_n}} \ (H_A = \prod_{k=1}^\infty \frac{(1+t^{2k-1})^{a_{2k-1}}}{(1-t^{2k})^{a_{2k}}}).$$

Example 4 *Let* $\deg e_i = i$ *and*

$$A = \langle e_1, e_2, e_3, ... | [e_i, e_j] = (i-j)e_{i+j} \rangle,$$

where $[x, y] = xy - yx$. *If* L *is a corresponding Lie algebra then*

$$H_L = t + t^2 + t^3 + \cdots \Rightarrow H_A = \prod_1^\infty \frac{1}{1-t^n} = \sum p(n)t^n,$$

where $p(n)$ *is the number of partitions of* n.

Thus we have an example with a non-rational Hilbert series. Moreover, this is the shortest example of a finitely presented associative algebra with non-rational Hilbert series! Though this example does not look like finitely presented, it is evident that e_1 and e_2 generate the algebra ($e_{k+1} = \frac{[e_k, e_1]}{k-1}$ for $k \geq 2$ and we assume in this example that the characteristic is zero). Slightly less trivial is the fact that it is sufficient to have only two relations:

$$[e_3, e_2] = e_5; \ [e_4, e_3] = e_7$$

(Ufnarovskij 1980). Note that graded algebras with one relation or two quadratic relations always have rational Hilbert series (Backelin 1978), (Ufnarovskij 1984).

It is convenient to introduce two generating functions

$$H_X = \sum_{x \in X} t^{\deg x}, H_R = \sum_{r \in R} t^{\deg r}.$$

For example, one can calculate the Hilbert series of a free algebra:

$$H_{K\langle X \rangle} = (1 - H_X)^{-1}.$$

If $\deg x = 1$ for every $x \in X$ then $H_{K\langle X \rangle} = (1 - dt)^{-1}$, where d is the number of generators.

Free algebras are exactly graded algebras of global dimension 1. The following important result of D.Anick (1982) shows that the Hilbert series can be used for calculating the global dimension of non-free algebras.

Theorem 1 *Let X be a minimal generating set of a graded algebra and R be a non-empty minimal set of homogeneous relations. Then*

$$gl.dim A = 2 \Leftrightarrow H_A = (1 - H_X + H_R)^{-1}.$$

Thus the algebra in Example 2 has global dimension 2, but this is not true for the algebra in Example 4 (the Hilbert series is not equal to the series $(1 - t^2 - t^3 + t^5 + t^7)^{-1}$).

One could expect that the Hilbert series of a finitely presented graded algebra can be computed in a similar way as in commutative case. Unfortunately this is not the case.

Theorem 2 *(Anick 1985) For every system of diophantine equations $S = 0$, there exists a finitely presented algebra A (which can be constructively expressed in terms of the coefficients of S) such that A has global dimension 2 if and only if the system $S = 0$ has no solutions.*

Moreover this algebra is the universal enveloping algebra of a Lie superalgebra, defined by quadratic relations only. Together with the previous theorem this implies the following important, though negative corollary:

Corollary 1 *All of the following is impossible:*

- *To find an algorithm that calculates the global dimension of an arbitrary finitely presented algebra.*
- *To find an algorithm that takes relations as input and gives the Hilbert series as output. (Even, one cannot decide in general whether the Hilbert series of a given algebra is equal to a given series).*
- *To predict in general the behaviour of a Hilbert series, knowing only finitely many of its coefficients.*

As to the growth, theoretically there exists an algorithm that checks whether a graded finitely-presented algebra has linear or constant growth. More exactly, in the case when all generators have degree 1: for a fixed number of generators, relations and their degrees there exists a constant n such that $\dim A_n \leq n$ if and only if A has a linear growth, and $\dim A_n = 0$ if and only if A is finite dimensional as it follows from Anick's arguments (1988) and Bergman's GK-gap theorem (see Krause, Lenagan (1985) or Ufnarovski (1990)).

It would be interesting to know if there exists an algorithm to determine if A has polynomial growth.

Despite the negative conclusions above it is rather often possible to calculate the Hilbert series and even to solve the word problem. One of the best approaches is to use the Gröbner bases technique. The main idea is to replace the original algebra by the monomial algebra with the same linear basis (and Hilbert series). But before starting this we should carefully study monomial algebras themselves.

2 Graphs, Languages and Monomial Algebras

Let F be a set of words in the alphabet X. An algebra $A = \langle X|F\rangle$ is called a *monomial algebra.* In other words, the defining relations should be monomials.

Example 5 *The algebra $A = \langle x, y|x^2, xy^2\rangle$ is monomial.*

Example 6 *Let S^* be the set of all non-empty words in the alphabet $X = \{x, y\}$. The algebra $A = \langle x, y|f^3 = 0, f \in S^*\rangle$ is monomial. Is it finite-dimensional? The answer is no. Indeed, let $f_0 = x; f_{n+1} = f_n\phi(f_n)$, where ϕ is the homomorphism defined by $\phi(x) = y, \phi(y) = x$. This is an infinite sequence of distinct nonzero words and its limit is the so-called* Thue sequence

$$xyyxyxxyyxxyxyyx\ldots,$$

that has the following nice property proved by M.E.Prouhet in 1851(!).

Let $X(n)$ be the set of all integers i for which f_n has the letter x on the i–th place, and $Y(n)$ its complement in the set $\{1, 2, 3, \ldots 2^n\}$. Then

$$\sum_{j\in X(n)} j^k = \sum_{j\in Y(n)} j^k; k = 0, 1, 2\ldots n-1.$$

For example, for $n = 3$ we have:

$$1+4+6+7 = 2+3+5+8; 1^2+4^2+6^2+7^2 = 2^2+3^2+5^2+8^2.$$

We normally demand that no word from F be a subword of another word in F. The previous example was an exception from this rule.

The advantage of monomial algebras is that we can directly construct a linear basis. It consists of all the words that do not contain any word from F as a subword. We call such words *normal.* So, the problem to calculate the Hilbert series is equivalent to the problem of calculating the generating function of the number of normal words. This does not look too difficult. However, from a computational point of view, it appears to be a less trivial problem than one might expect. To verify this, the reader can try to invent an *efficient* algorithm, answering the following easy question: Is a given monomial algebra finite dimensional?

One possible way to deal with monomial algebras is to use various kinds of graphs. We start with the easiest case: finitely-presented monomial algebras, i.e. algebras of the form $A = \langle X|F\rangle$, where F is a finite set of words in the alphabet X.

Let $n+1$ be the maximal length of words from F. Let us construct a graph $\Gamma(F) = (V, E)$, where the set of vertices V consists of all normal words of length n and the edges E are described as follows: $f \longrightarrow g$ if and only if there exist generators $x, y \in X$ such that $fx = yg$ and this word is normal.

Example 7 *For $A = \langle x, y | xy \rangle$ we have $n = 1$ and $\Gamma(A)$ looks as follows:*

$$\circlearrowleft x \longrightarrow y \circlearrowleft$$

Example 8

$$A = \langle x, y | x^2, xy^2 \rangle$$

has the following graph:

$$\circlearrowleft y^2 \longrightarrow xy \rightleftarrows yx$$

The following theorem (Ufnarovskij 1982) solves the problem of growth:

Theorem 3 *Let $A = \langle X | F \rangle$ be a monomial finitely presented algebra.*

- *The growth of A is exponential if and only if there are two different cycles in the graph $\Gamma(F)$ with a common vertex.*
- *Otherwise it is polynomial of degree d, where d is the maximal possible number of different cycles through which one path can pass.*
- *There is a bijective correspondence between the set of normal words of length $\geq n$ and pathes in the graph.*
- *The Hilbert series H_A is a rational function.*
- *The algebra has a polynomial identity if and only if it has the polynomial growth (the result of Borisenko (1985)).*

For example, the growth of the algebras in the last two examples is polynomial of degree 2. Using this theorem the reader can easily check that the growth of the algebra $A = \langle x, y | x^2 \rangle$ is exponential. A rather efficient algorithm for calculating the growth, based on this theorem is presented in (Ufnarovski, 1993).

The graph $\Gamma(F)$ has found several theoretical applications. For example the Jacobson radical $J(A)$ can be calculated using this graph. Also such properties as being primary, left and right noetherian can be easily expressed and checked using this graph. See work of Gateva-Ivanova and Latyshev (1988) for more details.

Another approach is to consider the set F as a language and to use the corresponding concepts and results (here F may be infinite). Especially important are so called regular languages. Recall, that a *regular language* is a set of words, obtained from a finite number of words using a finite number of the following operations:

$$A \cup B; AB; A^* = \cup_0^\infty A^n$$

They can be described by finite state automata. One can show that F is regular if and only if the set N of normal words is regular.

Example 9

$$A = \langle x, y | xy^n x; n = 0, 1, 2, \ldots \rangle$$

The regular set F can be described as $x(y^*)x$ and the automata to recognise F and N look as follows:

The automaton for normal words can be used for calculating both the growth (with the same criteria as above) and the Hilbert series. The reader can find more details about this and other types of graphs for normal words in (Ufnarovski 1990).

We finish this section with the following description of the Jacobson radical $J(A)$ and nil-radical $Nil(A)$, obtained by Belov and Gateva-Ivanova (1996).

Definition 3 *An infinite word W is called* uniformly recurrent *if for every of its finite subwords f there exists an integer $l = l(f)$ such that f is contained in every subword of W of length l.*

An infinite word W is normal if every its finite subword is normal.

For example an infinite word $\ldots xyxyxyxy \ldots$ is a normal uniformly recurrent word for the algebra $A = \langle x, y | x^2 \rangle$.

Theorem 4 *Let A be a finitely generated monomial algebra. Then $J(A) = Nil(A)$. Furthermore, as a linear space, $J(A)$ is spanned by all words that are not subwords of any uniformly recurrent normal word.*

3 $n-$Chains and Poincaré Series

It is useful to investigate monomial algebras from the point of view of homological algebra. Let $A = \langle X | F \rangle$ be a monomial algebra. Let us consider a graph $\Gamma = (V, E)$, where the set of vertices V consist of the union of the unit 1, the alphabet X, and all proper suffices of the words from F. The edges E are defined as follows: $1 \to x$ for every $x \in X$, and in the other cases $f \to g$

if and only if $fg = 0$ in A, but all proper left segments of the word fg are normal.

In Example 8 ($F = \{x^2, xy^2\}$) the graph Γ looks as follows:

1
x y
yy

In the Example 9 ($F = \{xy^n x\}$) the vertices (except 1 and y) have the form $y^n x$ and are connected to each other (including themselves).

Definition 4 *An $n-$chain is a word that can be read in the graph Γ during a path of length $(n+1)$, starting from 1. Let C_n be the set of all $n-$chains.*

From this definition it follows that

the only $-1-$chain is 1 itself: $C_{-1} = 1$,

the only $0-$chains are the letters from the alphabet: $C_0 = X$,

the only $1-$chains are the elements of F: $C_1 = F$.

Let us enumerate the $n-$chains for $n \geq 2$ in the above examples. In Example 8,

$$C_n = \{x^{n+1}, x^n y^2\}.$$

In Example 9

$$C_n = \{xy^{k_1}xy^{k_2}x \cdots xy^{k_n}x | k_i \geq 0\}.$$

Let us recall the definition of Poincaré series of a graded algebra.

Definition 5 *The (ordinary) Poincaré series is defined as*

$$P_A(t) = \sum \dim(Tor_n^A(K,K))t^n.$$

It exists only in the case when the corresponding dimensions are finite. That is why in the graded (for example in the monomial) algebras it is more natural to use its graded version:

The double Poincaré series for an algebra A is defined as the generating function

$$P_A(s,t) = \sum \dim(Tor_{n,m}^A(K,K))t^m s^n,$$

where $Tor_n^A(K,K)$ is considered as a graded module.

Theorem 5 *The double Poincaré series for an algebra A is equal to*

$$P_A(s,t) = \sum c_{m,n} t^m s^n,$$

where $c_{m,n}$ is the number of $(n-1)-$chains of degree m. (Thus, the $n-$chains corresponds to the homology of the monomial algebra).

An ordinary Poincaré series $P_A(t)$ for a monomial algebra exists (and is equal to $P_A(1,t)$) if and only if the algebra is finitely presented.

Thus in Example 8

$$P_A(s,t) = 1 + 2ts + (t^2 + t^3)s^2 + (t^3 + t^4)s^3 + \cdots,$$

and in Example 9

$$P_A(s,t) = 1 + 2ts + \sum_{m \geq n \geq 2} \binom{m-2}{n-2} t^m s^n =$$

$$= 1 + 2ts + t^2s^2 \left(\sum_{k=0}^{\infty} (st+t)^k \right) = 1 + 2ts + \frac{t^2s^2}{1-t-st}.$$

Theorem 6

$$H_A^{-1} = P_A(-1,t).$$

For example, $H_A^{-1} = 1 - 2t + t^2$ in Example 9.

This theorem provides a rather efficient method for calculating the Hilbert series.

In Example 8 we get

$$H_A^{-1} = 1 - 2t + t^2 + t^3 - t^3 - t^4 + t^4 + t^5 - \cdots = 1 - 2t + t^2.$$

By the same method we obtain the Hilbert series in the following example:

Example 10 *Let* $\deg x = 2, \deg y = 3$ *and* $A = \langle x, y | x^3, xy^2 \rangle$. *Then*

$$C_0 = \{x, y\}; C_1 = \{x^3, xy^2\}; C_2 = \{x^4, x^3y^2\}; C_3 = \{x^6, x^4y^2\}; \ldots$$

$$P_{s,t} = 1 + (t^2 + t^3)s + (t^6 + t^8)s^2 + (t^8 + t^{12})s^3 + (t^{12} + t^{14})s^4 \cdots.$$

$$H_A^{-1} = 1 - t^2 - t^3 + t^6 + t^8 - t^8 - t^{12} + t^{12} + t^{14} - \cdots = 1 - t^2 - t^3 + t^6$$

The growth of the algebra is exponential.

4 Admissible Orderings

The next notion we need is a total ordering on the set S of all the words in the alphabet X (including the empty word which we will identify with the unity **1**.)

Definition 6 *A total ordering* $>$ *on* S *is* admissible *if*

- *it is stable under multiplication:*

$$f \geq h, g \geq k \Rightarrow fg \geq hk; gf \geq kh.$$

- *every infinite sequence*

$$f_1 \geq f_2 \geq f_2 \geq \dots$$

stabilises: $f_n = f_{n+1} = \dots$ *for some* n.

Note that the unity is the least element in S: If $1 > f$ for a non-empty word f then $f > ff > fff\dots$ is an infinite decreasing sequence. More generally, if f is a subword of g, then $f < g$.

In particular and in contrast to the commutative case, the usual lexicographical order is not admissible: If $x > y$ then the sequence $x > yx > yyx > yyyx\dots$ is infinite.

A standard example of an admissible ordering is DEGLEX:

$$f > g \Leftrightarrow \begin{cases} \deg f > \deg g \\ \text{or} \\ \deg f = \deg g \text{ and } f \text{ is greater then } g \text{ lexicographically.} \end{cases}$$

In our examples we will take $x > y > z$ and DEGLEX with $\deg x = \deg y = \deg y = 1$ (so $\deg f$ coincides with the length of f) as the default admissible ordering.

For example, we use this ordering in the following algebra:

Example 11 $A = \langle x, y | x^2 = y^2 \rangle$.

In contrast, it is natural to use $e_1 > e_2 > \dots$ and DEGLEX with $\deg e_k = k$ in Example 4.

Let Y be a non-empty subset of the generating set X. We can define $\deg_Y$ as follows:

$$\deg_Y x = \begin{cases} \deg x & \text{if } x \in Y \\ 0 & \text{otherwise.} \end{cases}$$

Then every word $f = x_{i_1} x_{i_2} \dots x_{i_n}$ has degree $\deg_Y f = \sum \deg_Y x_{i_k}$. For any admissible ordering $>$ we can define a new *eliminating ordering*

$$f >_Y g \Leftrightarrow \begin{cases} \deg_Y f > \deg_Y g \\ \text{or} \\ \deg_Y f = \deg_Y g \text{ and } f > g. \end{cases}$$

Note that according to standard commutative arguments, this ordering is admissible.

Example 12 *If* $X = \{a, b, c\}, Y = \{a, c\}, a > b > c$ *in DEGLEX then*

$$aa >_Y ac >_Y ca >_Y cc >_Y ab >_Y ba >_Y bc >_Y cb >_Y a >_Y c >_Y bb >_Y b.$$

Using several such orderings we find that the following ordering is admissible: first we compare words lexicographically as commutative words, but, if they are equal as commutative words, use DEGLEX. More ingenious orderings can be found in (Saito et al. 1992).

5 Normal Words, Obstruction Set and Gröbner Bases

Let $>$ be an admissible ordering. Every element u from the free algebra has a unique representation as a linear combination of words $u = \sum \alpha_f f$, and has a greatest word f with non-zero coefficient α_f. This word (let us denote it by $L(u)$) is usually called *the leading monomial* or leading term, and its coefficient $\alpha_{L(u)}$ is called *the leading coefficient.*

If I is an ideal in $K\langle X\rangle$, then the set S of all words in the alphabet X is divided into two classes:

- the set of all leading monomials $\{L(u)|u \in I\}$,
- the remaining set N of *normal* words.

Note that *normal* words are exactly the words that *cannot* be rewritten in $A = K\langle X\rangle/I$ as a linear combinations of smaller words, and that definition coincides with the previous definition of normal words in the case of monomial algebras. The most important facts about normal words are formulated in the following theorem:

Theorem 7 *Let I be an ideal of a free algebra $K\langle X\rangle$ and N be the set of normal words in the admissible ordering. Then we have the following properties:*

- *The free algebra can be decomposed as direct sum of two vector spaces:*
$$K\langle X\rangle = KN \oplus I$$
(here KN is a vector subspace with the set N as a basis).
- *For every element $u \in K\langle X\rangle$ there exists a* normal form, *the unique linear combination $\overline{u}$ of the normal words such that u and $\overline{u}$ represent the same element in the factor-algebra $A = K\langle X\rangle/I$.*
- *The vector space KN together with the operation $*$*
$$u * v = \overline{uv},$$
is an associative algebra, which is isomorphic to A.

Of course, the normal form is nothing else than the projection onto the first component in the decomposition of the free algebra. This theorem explains why we can work in the factor-algebra in terms of words, i.e. elements of the free algebra. Note that both the decomposition and the normal form are depending on the choice of the ordering.

In the monomial algebra the normal form of a word is either 0 or (if it is normal) the word itself. In general the situation is more complicated.

Example 13 *We consider the algebra $A = \langle x, y | x^2 - y^2 \rangle$ from Example 11 (more exactly, the ideal, generated in the free algebra by $x^2 - y^2$ and defining this algebra).*

It is clear that both the words x^2 and y^2 have the same normal form, namely the word y^2, because y^2 is the lowest word in degree 2. It is less evident that the words xy^2 and y^2x have the same normal form ($xy^2 = x^3 = y^2x$ in A). We conclude that the normal words cannot contain both x^2 and xy^2 as a subword. It requires much more effort to prove that this necessary condition is also sufficient, and therefore the set of normal words consists of the words of the form

$$y^n(xy)^k; y^n(xy)^k x.$$

One possible way to prove this is to calculate the generating function of the number of such words and to notice that it is equal to the minimal Hilbert series: $(1 - H_X + H_R)^{-1}$, which is $(1 - 2t + t^2)^{-1}$, and conclude that all those words are normal because of minimality.

However, as we will soon see, it is much easier to use the Gröbner bases approach. The shortest way to this notion is to investigate the words that are not normal (i.e. words of the form $L(u), u \in I$), and to introduce the following definition:

Definition 7 *Let I be an ideal. A word f is called an* obstruction *if it is not normal (i.e. it is a leading monomial $L(u)$ for some element $u \in I$), but each of its proper subword is normal. If F is the set of all obstructions then the set*

$$G = \{f - \overline{f} | f \in F\}$$

is called a (reduced) Gröbner basis *of the ideal I.*

For example, in the monomial algebra $A = \langle X | F \rangle$ (remember that we require F to be minimal) the set F itself is the obstruction set and the Gröbner basis. In the algebra $A = \langle x, y | x^2 = y^2 \rangle$, as we have seen in Example 13, the obstruction set consists of $F = \{x^2, xy^2\}$, and the Gröbner basis contains only two elements: $G = \{x^2 - y^2, xy^2 - y^2x\}$.

The algebra in Example 4 has the following Gröbner basis $G = \{e_ie_j - e_je_i - (i - j)e_{i+j}, i > j\}$. The shortest proof of this is to show that the remaining words $e_1^{k_1} e_2^{k_2} \dots e_n^{k_n}$ are, according to the Poincaré-Birkhoff-Witt theorem, linear independent in A and form a basis in the normal complement to I in theorem 7.

We use the word "obstruction" following Anick (1986). Another synonym is "tip" (Anick and Green 1987). More generally, a Gröbner basis can be defined as a subset $G \subseteq I$, such that the set $L(G) = \{L(g) | g \in G\}$ contains the obstruction set F. For example, the ideal I itself is its own Gröbner basis, and this shows that the general notion is not as useful as the uniquely determined reduced Gröbner basis. More useful is the notion of the *minimal Gröbner basis,* i.e. a Gröbner basis such that any proper subset is not a Gröbner basis. For example, the reduced Gröbner basis is minimal. In the sequel Gröbner basis will always mean reduced Gröbner basis.

6 Computational Approach

The purely theoretical definition of the Gröbner basis above has the advantage that one can apply it even in such complicated cases as Example 1 where one cannot construct the reduced Gröbner basis algorithmically (otherwise the word problem would be solvable). Nevertheless, as in the commutative case, it is the computational aspect that makes this notion particularly useful. The theory here is quite parallel to the commutative case (compare with the classical works of Buchberger (1965, 1970)) except the possible problems with infinite Gröbner bases) and can be considered as a special case of the term rewriting theory (see Book and Otto (1982)). Let us recall the key ideas.

If u is any element of I then we can consider its leading monomial $L(u) = f$ and form a rewriting rule:

$$f \rightarrow_u v,$$

where u is proportional to $f - v$ (normally, when the leading coefficient is equal to 1, simply equal $f - v$). The rule, applied to any element $w \in K\langle X\rangle$ means the (successive) substitution of occurrences of f as a subword by the element v. If U is a subset of I then we have such rules for every $u \in U$ and $a \Rightarrow_U b$ means that b is obtained from a by several (successive) applications of these rules.

Example 14 *If* $U = \{x^2 - 4xy, 2xy^2 - 3y^3\}$ *then*

$$x^3 \Rightarrow_U 4x^2y \Rightarrow_U 16xy^2 \Rightarrow_U 24y^3.$$

After replacing as above we always obtain lower elements. Since the ordering is admissible this process of applying rules (called *reduction*) cannot be infinite. Unfortunately the final results can be different depending on the order of applications of the rewriting rules (for example, the rewriting rule $x^2 \rightarrow y^2$ can be applied to x^3 in two different ways giving xy^2 or y^2x correspondingly). But if G is a Gröbner basis then the reduction by G of any element a is independent of the order - it is always the normal form of a :

$$a \Rightarrow_G \overline{a}.$$

But the converse is also true!

Theorem 8 *G is a Gröbner basis (not necessarily reduced) if and only if the result of reduction by G, applied to any element a is independent of order of the applications of the rewriting rules.*

Reducing leading monomials from U by the help of U itself we can always replace it by a *self-reduced* set U' i.e. such that no leading monomial from $\{L(u), u \in U'\}$ is a subword of another leading monomial. If U is finite we can do it in a finite number of steps, if U is infinite sometimes only theoretically. Nevertheless this is a way to obtain a minimal Gröbner basis from an arbitrary one (to obtain a reduced one from a minimal it is sufficient to divide every element by its leading coefficient and reduce all non-leading terms).

This gives also a method to check and construct Gröbner bases. The only thing we need after self-reduction is to solve all ambiguities. To be more precise let us introduce the notion of composition.

Definition 8 *Let $u, w \in K\langle X\rangle$, $f = L(u), g = L(w)$ be their leading monomials and $f \to v, g \to t$ be corresponding rewriting rules. A triple (a, b, c) of words (normally written as a single word abc) is a* composition *of u and w if $ab = f, bc = g$. The* result of the composition *is an element $at - vc$ (it is simply the difference of two different applications of rewriting rules to the composition).*

Example 15 *If $u = w = x^2 - y^2$ then the composition xxx gives the result $xy^2 - y^2x$. The composition $x \cdot x \cdot y^2$ of this new element with u gives the result $xy^2x - y^4$.*

Theorem 9 *(Bergman's Diamond Lemma (Bergman 1978)). Let G be a self-reduced set. G is a (minimal) Gröbner basis if and only if the results of all possible compositions between elements of G are reduced by G to zero.*

Example 16 *We can check once again that $G = \{x^2 - y^2, xy^2 - y^2x\}$ is a Gröbner basis for the algebra $A = \langle x, y | x^2 - y^2\rangle$. According to Example 15 we have only two compositions, and reductions of their results gives*

$$xy^2 - y^2x \Rightarrow_G 0; xy^2x - y^4 \Rightarrow_G y^2x^2 - y^4 \Rightarrow_G y^4 - y^4 = 0.$$

As in the commutative case (Buchberger 1979) there exists an improved version of theorem 9 that says that it is not necessary to check all the compositions but it is sufficient to check only 2—chains of the corresponding monomial algebra.

Example 17 *In the algebra $A = \langle x, y | x^3 - y^2\rangle$ from Example 2 the Gröbner basis consists of two elements $x^3 - y^2, xy^2 - y^2x$. It is not necessary to check the composition x^5. It is sufficient to check that the results of the compositions x^4 and x^3y^2 reduce to zero:*

$$xy^2 - xy^2 \Rightarrow_G 0; x^2y^2x - y^4 \Rightarrow_G xy^2x^2 - y^4 \Rightarrow_G y^2x^3 - y^4 \Rightarrow_G y^4 - y^4 = 0.$$

The Hilbert series H_A is equal to the Hilbert series H_B, where the algebra $B = \langle x, y | x^3, xy^2\rangle$ is a monomial algebra from Example 10. Thus $H_A = (1 - t^2 - t^3 + t^6)^{-1}$. Algebra A has the exponential growth, and $gl.dim A = 2$.

This theorem gives also a direct analogue of Buchberger's algorithm of constructing Gröbner bases for finitely presented algebras, known in the non-commutative theory as Mora's algorithm (Mora 1986) (the term rewriting analogue is known as the Knuth-Bendix algorithm see (Book and Otto 1982)). The algorithm can be described as successive resolving of all appearing ambiguities.

Namely, let G_0 be the set obtained by the self-reduction from the defining relations R (normally $G_0 = R$), and L be an ordered list of all compositions (or better – 2-chains). The simplest version of the algorithm takes in step number $n+1$ the first composition from the list and adds its result (if non-zero) to G_n. The finite set obtained after application of self-reduction is G_{n+1}. The list L should be changed too: the considered composition and compositions, connected with the elements, eliminated during the self-reduction, are thrown away; new compositions, connected with new elements are added at the end of the list.

The algorithm terminates if the list becomes empty. Unfortunately, as a rule it is not the case, and the typical Gröbner basis is infinite.

Example 18 $A = \langle x, y | x^2 = xy \rangle$. *The Gröbner basis here is infinite and looks as* $G = \{xy^n x - y^{n+1}; n = 0, 1, 2 \ldots\}$. *The compositions are* $xy^n xy^k x$ *and their results are reduced to zero:*

$$xy^n xy^{k+1} - xy^{n+1}y^k x = xy^n xy^{k+1} - xy^{n+k+1}x \Rightarrow_G xy^{n+1}y^{k+1} - xy^{n+k+2} = 0.$$

If we in this example select another ordering $y > x$ the Gröbner basis will be finite (and consisting of $xy - y^2$ only), but the Gröbner basis in Example 4 cannot be finite in any ordering:

Theorem 10 *An algebra A with finite Gröbner basis has either polynomial or exponential growth. If it is graded, it has rational Hilbert series and well defined ordinary Poincaré series* $P_A(t)$.

The proof follows immediately from the monomial case: if F is the obstruction set, then by the definition, the monomial algebra $B = \langle X | F \rangle$ has the same set of the normal words and the same growth and Hilbert series. As to Poincaré series, it is sufficient to show that the minimal, resolution consisting of $n-$chains, implicitly constructed for the monomial algebra can be extended to the non-monomial case (usually loosing the property of being minimal).

7 Anick's Resolution

In order to calculate the Poincaré series in the general case we construct Anick's resolution (Anick 1986):

$$\cdots \to C_n \otimes A \to C_{n-1} \otimes A \to \cdots C_{-1} \otimes A \to K \to 0$$

where C_n is the set of $n-$chains, corresponding to the obstruction set F (see Section 3). It is sufficient to define module homomorphisms $d_n : C_n \otimes A \to C_{n-1} \otimes A$ only for terms $f \otimes 1$. It is convenient to identify $C_n \otimes N$ with $C_n N$. Then the map d_n is defined by induction as

$$d_{n+1}(f) = f - i_n d_n(f)$$

and $i_n : \ker d_{n-1} \to KC_nN$ is defined recursively:

$$i_n(u) = \alpha L(u) + i_n(u - \alpha d_n(L(u))),$$

where α is the leading coefficient and $L(u)$ is the leading monomial.

Note, that:

- d_0 calculates, for every non-empty word f, its normal form $\bar{f}$, i_0 acts identically.
- d_1 calculates, for any obstruction f an element $f - \bar{f}$, i.e. it recovers the element of the Gröbner basis from its obstruction. To apply d_1 to arbitrary words of form fs one needs to be more careful: Let R_0 be a linear map that, applied to a non-empty word g does not change the first letter, but reduces the remaining part of the word to its normal form. Then $d_1(fs) = R_0(fs - \bar{f}s)$ (in the general case one needs to use the map $R_n : C_{n+1}N \to C_nN$, that leaves the $n-$chain in the beginning unchanged and reduces the remaining part to normal form, for example R_{-1} is reduction to the normal form).

In our main examples :

1. For every monomial algebra: $d_n(f) = f$ for $f \in C_n$ and $d_n(fs) = R_{n-1}(fs)$ in general.
2. in Example 13 we have using tensor language:

$$d_2 : x^3 \otimes 1 \to x^2 \otimes y - xy^2 \otimes 1,$$
$$d_2 : x^2y^2 \otimes 1 \to x^2 \otimes y^2 - xy^2 \otimes x.$$

The calculations of d_2 can replace the calculations of the results of the compositions and can therefore be used for calculating the Gröbner basis through Anick's resolution. This is also an important resource for optimisation: A lot of compositions produce results that are reduced to zero. It would be better to recognise them from the very beginning. The images of d_3 can be used for this (see (Ufnarovski 1990)). Another important idea is to use known Hilbert series in graded algebras: If we already achieved the minimal possible number of normal words for some degree, we may be sure that the results of all remaining compositions in this degree reduce to zero (compare with the arguments in the Example 13).

8 Gröbner bases in Lie algebras

The definition above of Gröbner basis in the noncommutative case can be transformed to the Lie algebra case. The main difficulty here is in the construction of the linear basis of the free Lie algebra, and in the reduction. It was Shirshov (1962) who found that also for Lie algebras it is possible to use the associative words and their compositions to construct the Gröbner bases! We give here a brief sketch of this approach.

The key notion is the *regular or Lyndon-Shirshov word.* A convenient instrument is the following partial ordering (note – not admissible!) Let g, h be words in the alphabet X, and $>$ be the lexicographical order. We define

$$g \rhd h \Leftrightarrow gh > hg.$$

Definition 9 *A non-empty word f is called* regular or Lyndon-Shirshov *if for any representation $f = gh$ we have $g \rhd h$.*

For example, the words $x, xxy, xyyxz$ are regular, but the words yx, xx, xyx, $xzxy$ are not (if $x > y > z$).

Theorem 11 *Every associative word F has a unique presentation in the form $F = f_1^{k_1} f_2^{k_2} \dots f_m^{k_m}$, where f_i are regular words and $f_1 \lhd f_2 \lhd f_3 \dots \lhd f_m$.*

Theorem 12 *Let f be a regular word.*

- *Let h be its longest regular proper right segment: $f = gh$. Then g is regular.*
- *Any regular subword of f is either a subword of g, a subword of h, or a beginning of f.*

Let f be a regular word. Using this theorem we can inductively define the *non-associative regular word* $[f]$:

$[x] = x$ for $x \in X$;

If $f = gh$ is the decomposition from theorem 12 then $[f] = [[g][h]]$.

Example 19

$$[xyyxyz] = [[xyy][xyz]] = [[[xy][y]][[x][yz]]] = [[[x\ y]y][x[y\ z]]].$$

Theorem 13 *Non-associative regular words form a linear basis of a free Lie algebra.*

Thus we can use associative words to enumerate the linear basis, and therefore we can introduce an admissible ordering (for example DEGLEX) and leading terms. If I is an ideal, then the normal words are those non-associative regular words that cannot be rewritten in the factor-algebra as a linear combination of smaller words. It is important that they are exactly those words that are not leading terms of the elements of I, so we have a decomposition similar to the decomposition in Theorem 7, and the projection gives us a normal form. The obstruction can now be defined as a non-associative word $[f]$ such that it is not normal, but for any *regular(!)* proper subword g of the associative word f the non-associative word $[g]$ is normal.

At last a Gröbner basis (or following Bokut and Malcolmson (1996), Gröbner-Shirshov) basis is the set $G = \{[f] - \overline{[f]}\}$ for all obstructions $[f]$.

The computational aspect here is a little more complicated: The main problem is the reduction. For example, xy is a subword of xyz, but $[xy]$ is not part of $[xyzz] = [x[[yz]z]]$, so how can we apply a rewriting rule $[xy] \to [xz]$? The solution is as follows.

Let a be a regular subword of a regular word f. To apply a rewriting rule $[a] \to v$ to f we can, using induction and Theorem 12, consider only the case $f = ab$. If $b = b_1b_2 \ldots b_n$ is the decomposition into regular words from Theorem 11, then the reduction is $[f] - [\ldots[[[v[b_1]][b_2]]\ldots[b_n]]$. Note that the last commutator is not regular word so we need to rewrite this using Jacoby identity. During this process $[f]$ should be cancelled.

Thus in the example above we have:

$$[x[[yz]z]] - [[[xy]z]z] = [x[[yz]z]] - [[[xz]y]z] - [[x[yz]]z] =$$

$$[x[[yz]z]] - [[[xz]z]y] - [[xz][yz]] - [[xz][yz]] - [x[[yz]z]] = -[xzzy] - 2[xzyz].$$

Thus monomial algebras are here more complicated. For example if the defining relations are regular words it does not mean that they form a Gröbner basis. But the compositions of associative words work even here. The reader can find more details (including the similar case of Lie superalgebras) in the book of Mikhalev and Zolotykh (1995). It is also useful to read the original article of Shirshov (1962).

9 More Applications

In this section we would like to mention some applications of noncommutative Gröbner bases. First of all, with help of eliminating orderings one can solve a lot of standard problems, e.g. to find the intersection of two ideals, to calculate the kernel of a homomorphism. As in commutative case the base for such calculations is

Theorem 14 *Let Y be a subset of X, and $Z = X - Y$ its complement. If G is a Gröbner basis for an ideal I in $K\langle X\rangle$ in the eliminating ordering $>_Y$, then $G \cap K\langle Z\rangle$ is a Gröbner basis for the ideal $I \cap K\langle Z\rangle$ in $K\langle Z\rangle$.*

In the article of Nordbeck (1998) in this issue the reader can find both proofs and applications.

Those, who are interested in the commutative applications should read the important article of Eisenbud, Peeva and Sturmfels (1997) where it was shown that after a general linear change of variables a commutative algebra (considered as noncommutative) has a finite Gröbner basis. In representation theory one can work in the so-called path algebras instead of free algebras. All the definitions (including the construction of Anick's resolution) can be found in (Anick and Green 1987).

Ed Green (1994) and Teo Mora (1994) have nice lecture notes on noncommutative Gröbner basis (available also by Internet). A lot of deep theoretical applications can be found in book of Bokut' and Kukin (1994). Some interesting articles of L. Gerritzen on noncommutative Gröbner basis should be published in the nearest future. Applications for the operator theory can be found in (Wavrik 1996).

Algebras which have a finite Gröbner basis are very special algebras. A lot of problems (starting from the word problem) are solvable here, so this class of algebras, containing commutative and finite dimensional algebras, is a good test for any new problem. For example, it is not known how to calculate the Jacobson radical here. A very interesting property in this class is the noetherian property (Gateva-Ivanova 1996).

Especially important is the class of algebras that are close to commutative algebras. The works of Gateva-Ivanova (1996), Kandri-Rody and Weispfenning (1987) are good starting points.

Another, very important class is described in the article of Artin and Schelter (1987). The article of J.E-Roos (1994) is a nice example of computer-aided study both commutative and noncommutative algebras.

We finish this introduction with a list of some packages for calculating the Gröbner bases.

- GRB by Ed Green – calculations in path algebras. Available by anonymous ftp from **calvin.math.vt.edu**.

- Bergman by J.Backelin – calculations in commutative and noncommutative graded algebras. Available by ftp from **ftp.matematik.su.se**

- Anick by A.Podoplelov and V.Ufnarovski – calculations of Anick resolutions and Gröbner bases in the noncommutative case. Nonhomogeneous relations and eliminating ordering are implemented. Available by anonymous ftp from **ftp.riscom.net**, pub/anick/anick.tar.gz, The alias address for ftp.riscom.net is riscom.moldnet.md.

The author is grateful to Prof. Buchberger and Gert Almkvist for their valuable remarks.

References

Anick, D. (1982): Non-commutative graded algebras and their Hilbert series. J. Algebra, 78, No 1: 120-140.

Anick, D. (1985): Diophantine equations Hilbert series and undecidable spaces. Ann. Math.,II Ser. 122: 87-112.

Anick, D. (1986): On the homology of associative algebras, Trans. Am. Math. Soc.,296, No 2: 641-659.

Anick, D. (1988): Generic algebras and CW-complexes. In proc. Conf. Alg. Topol. and K-theory in honour of John Moore, Prinston univ. 1983: 247-331.

Anick, D., Green, E.L. (1987): On the homology of quotients of path algebras. Commun. Algebra, 15, No 1,2: 309-341.

Artin M., Schelter, W. (1987): Graded algebras of Global Dimension 3, Adv. Math., 66, No 2: 171-216.

Backelin, J. (1978): La série de Poincaré-Betti d'une algébre graduée de type fini a une relation est rationnelle. C.R.Acad. Sci., Paris, Ser. A 287: 843-846.

Belov, A. Gateva-Ivanova, T. (1996): Radicals of monomial algebras. In: Fong, Y. (ed) et al. First international Tainan-Moscow algebra workshop. Proc. of the Int. Conf., Tainan, Taiwan, Republica of China, July 23 - August 22, 1994. Berlin: de Gruyter, 159-169

Bergman, G. (1978): The diamond lemma for ring theory. Adv. Math. 29, No 2: 178-218.

Bokut' L.A., Kukin G.P. (1994): Algorithmic and Combinatorial Algebra. Kluwer Academic Publisher, Dordrecht Boston London.

Bokut, L., Malcolmson P.(1996): Gröbner-Shirshov bases for quantum enveloping algebras. Isr. J.of Math. 96: 77-113.

Book, R., Otto, F. (1982): String-rewriting systems,. Springer, Berlin Heidelberg New York.

Borisenko, V. V. (1985): On matrix representation of finitely generated algebras, given by a finite number of relations. Vestn. Mosk. Univ., Ser. I, No 4: 75-77. English transl: Mosc. univ. Bull. 40, No 4, 80-83 (1985).

Buchberger, B. (1965): On Finding a Vector Space Basis of the Residue Class Ring Modulo a Zero Dimensional Polynomial Ideal (German). PhD Thesis, Univ of Innsbruck, Austria.

Buchberger, B. (1970): A Theoretical Basis for the Reduction of Polynomials to Canonical Forms. ACM SIGSAM Bull. 10/3: 19-29, and 10/4: 19-24.

Buchberger, B. (1979): A Criterion for Detecting Unnecessary Reductions in the Construction of Groebner Bases. In: Proc. of the EUROSAM 79 Symp. on Symbolic and Algebraic Computation, Marseille, June 26-28, 1979, (Lect. Notes in Comp. Sci. 72, Springer): 3-21.

Eisenbud, D., Peeva. I., Sturmfels. B. (1997): Non-commutative Gröbner bases for Commutative Algebras. to appear in Proc. AMS.

Gateva-Ivanova, T. (1996): Skew polynomial rings with binomial relations. J. Algebra 185, No.3: 710-753.

Gateva-Ivanova, T., Latyshev, V. (1988): On the recognizable properties of associative algebras. J. Symb. Comput. 6, No 2/3: 371-388.

Green, Ed. (1994): An Introduction To Noncommutative Gröbner bases. In: Fisher K.G. (ed.), Computational Algebra, Dekker, New York. (Lect. Notes Pure Appl. Math, 151): 167-190.

Kandri-Rody, A., Weispfenning, V. (1987): Non-commutative Gröbner bases in algebras of solvable type. J. Symb. Comput. 9, No 1: 1-26.

Krause, G.R., Lenagan T.H. (1985): Growth of Algebras and Gelfand-Kirillov Dimension. Pitman Publ., London.

Mikhalev, A.A., Zolotykh A.A. (1995): Combinatorial aspects of Lie Superalgebras. CRC Press, Boca Raton, New York.

Mora T. (1986): Gröbner bases for non-commutative polynomial rings. In: J.Calmet (ed.), AAECC-3, Lect. Notes Comp. Sc. 229: 353-362.

Mora T. (1994): An introduction to commutative and non-commutative Gröbner bases, Theor. Comp. Sci., 134: 131-173.

Nordbeck P. (1998): On some basic Applications of Gröbner bases in Noncommutative Polynomial Rings. In this voloum.

Petrogradsky V.M. (1993): On some types of intermediate growth in Lie algebras, Uspechi Mat.Nauk, 48, No 5: 181-182.

Roos J.-E. (1994): A computer-aided study of the graded Lie algebra of a local commutative noetherian ring, J. Pure and Appl. Algebra 91: 253-315.

Saito, T., Katsura, M., Kobayashi, Y., Kajitori, K. (1992): On total ordered free monoids. In Words, Language and Combinatorics, Word Scientific: 454-479.

Shirshov A.I. (1962): Some algorithmic problems for Lie algebras. Sib Mat. Zh. 3, No 2: 292-296.

Tsejtin, G.S. (1958) Associative computations with unsolvable equivalence problem. Tr. Mat. Inst. Steklova 52, 172-189.

Ufnarovskij, V. (1980): Poincaré series of graded algebras. Mat.Zametki 27, No 1: 21-32. English transl: Math. Notes 27 (1980), 12-18.

Ufnarovskij, V. (1984): Algebras, defined by two quadratic relations. Mat. Issled. 76: 148-171.

Ufnarovski ,V. (1990): Combinatorial and Asymptotic Methods of Algebra. Itogi Nauki Tekh., Ser. Sovrem. Probl. Mat., Fundam. Napravleniya 57, 5-177. English transl. in: Kostrikin, A.I., Shafarevich, I.R. (eds.): Algebra-VI. Springer, Berlin Heidelberg New York, (1995): pp. 5-196. (Encycl. Math. Sci., vol. 57).

Ufnarovski, V. (1993): Calculations of growth and Hilbert series by computer. In: Fisher K.G. (ed.), Computational Algebra, Dekker, New York. (Lect. Notes Pure Appl. Math, 151): 247–256. Ufnarovskij V. (1980): A growth criterion for graphs and algebras defined by words. Mat Zametki 31, No 3: 465-472. English transl: Math. Notes 31 (1982), 238-241.

Wavrik, J. (1996): Rewrite Rules and Simplification of matrix Expressions, Comp.Sc. J. of Moldova, v.4, No 3: 360-398.

Gröbner Bases Applied to Geometric Theorem Proving and Discovering

Dongming Wang

LEIBNIZ–IMAG, 46, avenue Félix Viallet, 38031 Grenoble Cedex, France

This tutorial explains how interesting geometric theorems may be proved or even discovered automatically and effectively by using Gröbner bases. Several examples including theorems named after Morley, Steiner and Poncelet are provided to illustrate the underlying ideas. Three algorithms are presented formally for proving and discovering theorems in elementary geometry.

1 Simple examples

One of the most impressive and successful applications of Gröbner bases is to automated theorem proving in elementary geometry: hundreds of non-trivial theorems have been proved by using Gröbner bases in the matter of seconds. To illustrate this remarkable achievement, let us start with the following example.

Example 1 (Gauss' line). The midpoints M_1, M_2, M_3 of the three diagonals A_1B_1, A_2B_2, A_3B_3 of any complete quadrilateral are collinear.

We take an affine coordinate system with A_1 as the origin, A_1A_2 as the x-axis and A_1B_2 as the y-axis. The coordinates of the involved points may be assigned, without loss of generality, as follows

$$A_1(0,0),\quad A_2(x_1,0),\quad A_3(x_2,0),$$
$$B_1(x_5,x_6),\quad B_2(0,x_3),\quad B_3(0,x_4),$$
$$M_1(x_7,x_8),\quad M_2(x_9,x_{10}),\quad M_3(x_{11},x_{12}).$$

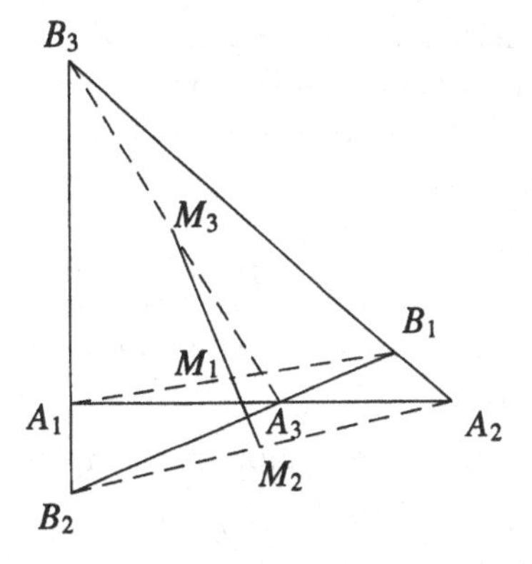

Then the geometric hypotheses of the theorem may be translated into polynomial relations:

$$\mathsf{int}(A_2,B_3,A_3,B_2,B_1) \iff \begin{cases} h_1 = x_1x_6 + x_4x_5 - x_1x_4 = 0, \\ h_2 = x_2x_6 + x_3x_5 - x_2x_3 = 0; \end{cases}$$

$$\text{midp}(A_1, B_1, M_1) \iff \begin{cases} h_3 = 2x_7 - x_5 = 0, \\ h_4 = 2x_8 - x_6 = 0; \end{cases}$$

$$\text{midp}(A_2, B_2, M_2) \iff \begin{cases} h_5 = 2x_9 - x_1 = 0, \\ h_6 = 2x_{10} - x_3 = 0; \end{cases}$$

$$\text{midp}(A_3, B_3, M_3) \iff \begin{cases} h_7 = 2x_{11} - x_2 = 0, \\ h_8 = 2x_{12} - x_4 = 0. \end{cases}$$

★ int(A, B, C, D, E) *stands for "the two lines AB and CD intersect at point E."*

★ midp(A, B, M) *stands for "M is the midpoint of the segment AB."*

The conclusion of the theorem is

$$\text{col}(M_1, M_2, M_3) \iff c = x_9x_{12} - x_7x_{12} - x_{10}x_{11} + x_8x_{11} + x_7x_{10} - x_8x_9 = 0.$$

To prove the theorem, compute a Gröbner basis $\mathbb{G}$ of the polynomial set $\mathbb{H} = \{h_1, \ldots, h_8\}$ with respect to plex determined by ω_{12}. In this special case $\mathbb{H}$ is almost triangularized. The Gröbner basis is obtained by simply adding one polynomial to $\mathbb{H}$:

$$\mathbb{G} = [x_2x_4x_5 - x_1x_3x_5 - x_1x_2x_4 + x_1x_2x_3, h_1, \ldots, h_8].$$

★ col$(A,\ B,\ C)$ *stands for "the three points A, B and C are collinear."*

★ plex *is abbreviated from "the purely lexicographical term ordering," and ω_n denotes the variable ordering $x_1 \prec \cdots \prec x_n$.*

It is easy to verify that the normal form of c modulo $\mathbb{G}$ is identically equal to 0. This means that c belongs to the polynomial ideal generated by $h_1, \ldots, h_8$, i.e., c can be expressed as a linear combination of the h_i with polynomial coefficients. Hence, $c = 0$ follows from $h_1 = 0, \ldots, h_8 = 0$. This proves the theorem.

Alternatively, one may compute a Gröbner basis $\mathbb{G}^*$ of $\mathbb{H} \cup \{cz - 1\}$, where z is a new indeterminate; the constant 1 is contained in $\mathbb{G}^*$. By Hilbert's Nullstellensatz, this implies that c belongs to the radical of the ideal generated by $h_1, \ldots, h_8$, so

$$(\forall x_1, \ldots, x_{12})[h_1 = 0 \wedge \cdots \wedge h_8 = 0 \Longrightarrow c = 0].$$

The theorem is again proved. □

The above proof is easy and straightforward; it is not always so simple. Let us consider another example.

Example 2. Let A, B, C be any three points in the plane, D the midpoint of AB and E the midpoint of CD. Extend AE to intersect BC at point F. Show that $|FB| = 2|FC|$.

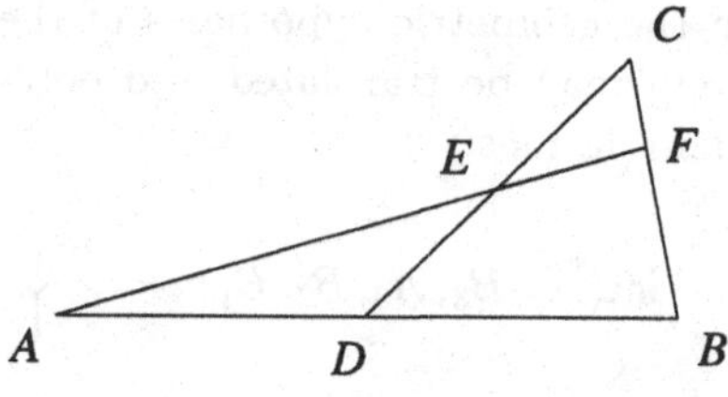

Without loss of generality, let the points be located as

$$A(-x_1,0),\ B(x_1,0),\ C(x_2,x_3),\ D(0,0),\ E(x_4,x_5),\ F(x_6,x_7).$$

This special choice of coordinates already takes into account of the hypothesis that M is the midpoint of AB. The other hypotheses of the theorem are:

$$\mathsf{midp}(C,D,E) \iff \begin{cases} h_1 = 2x_4 - x_2 = 0, \\ h_2 = 2x_5 - x_3 = 0; \end{cases}$$
$$\mathsf{int}(A,E,C,B,F) \iff \begin{cases} h_3 = (x_4+x_1)x_7 - x_5x_6 - x_1x_5 = 0, \\ h_4 = (x_2-x_1)x_7 - x_3x_6 + x_1x_3 = 0. \end{cases}$$

We want to prove that

$$|FB| = 2|FC| \iff \begin{cases} c_1 = 3x_6 - 2x_2 - x_1 = 0, \\ c_2 = 3x_7 - 2x_3 = 0. \end{cases}$$

As in the previous example, compute a plex Gröbner basis $\mathbb{G}$ of $\mathbb{H} = \{h_1,\ldots,h_4\}$ with respect to ω_7:

$$\mathbb{G} = [h_1, h_2, g_3, g_4, g_5],$$

where

$$g_3 = 3x_1x_3x_6 - 2x_1x_2x_3 - x_1^2x_3,$$
$$g_4 = 3x_1x_7 - 2x_1x_3,$$
$$g_5 = 3x_2x_7 - 3x_3x_6 + x_1x_3.$$

Now we have $\mathrm{nform}(c_1,\mathbb{G}) \not\equiv 0$ and $\mathrm{nform}(c_2,\mathbb{G}) \not\equiv 0$. So it is not known whether the theorem is true or not. Also, 1 is not contained in the Gröbner basis of $\mathbb{H} \cup \{c_iz - 1\}$ for $i = 1, 2$. Therefore, c_1 and c_2 do not belong to the radical of the ideal generated by $h_1,\ldots,h_4$. In other words, the theorem is not true logically over the complex number field.

- ⋆ *$\boldsymbol{K}$ is a geometry-associated field of characteristic* 0.
- ⋆ $\boldsymbol{x} = (x_1,\ldots,x_n)$ *are n variables.*
- ⋆ $\mathrm{nform}(f,\mathbb{G})$ *denotes the* normal form *of a polynomial f modulo a Gröbner basis $\mathbb{G}$ in $\boldsymbol{K}[\boldsymbol{x}]$ (see Buchberger 1985).*
- ⋆ *For any k $(1 \le k \le n)$ and $f \in \boldsymbol{K}[x_1,\ldots,x_k] \setminus \boldsymbol{K}[x_1,\ldots,x_{k-1}]$,* the leading variable *of f, denoted $\mathrm{lv}(f)$, is defined to be x_k.*
- ⋆ *The leading coefficient of any $f \in \boldsymbol{K}[\boldsymbol{x}]$ with respect to $\mathrm{lv}(f)$ is called the* initial *of f.*
- ⋆ *For any $g, f \in \boldsymbol{K}[\boldsymbol{x}]$ and $1 \le k \le n$, $\mathrm{prem}(g,f,x_k)$ denotes the* pseudo-remainder *of g with respect to f in x_k.*

Why? This is because most geometric theorems are true only *generically*. Assumptions about the genericness of the involved figures are not explicitly stated in the hypothesis of a theorem. For the above example, the algebraic formulation also includes the case in which A, B, C are collinear. In this case, the theorem is meaningless or false.

There are two ways to deal with this situation:

(a) Take a q-basic set $\mathbb{B} = [h_1, h_2, g_3, g_4]$ of $\mathbb{G}$ (note that $\mathbb{B} \subset \mathbb{G}$). Pseudo-dividing c_i successively by the polynomials g_4, g_3, h_2, h_1 with respect to their leading variables $x_7, \ldots, x_4$ for $i = 1, 2$, one may find that the final pseudo-remainders are both 0. This implies that the theorem is true as far as the initials of the polynomials in $\mathbb{B}$ are non-zero, viz.,

$$x_1x_3 \neq 0.$$

In other words, the theorem is proved to be true under the found *non-degeneracy condition* $x_1x_3 \neq 0$. The geometric meaning of the condition may be easily interpreted:

$$x_1x_3 \neq 0 \iff A, B, C \text{ are not collinear.}$$

⋆ *For any* $\mathbb{F} \subset \boldsymbol{K}[\boldsymbol{x}]$, *define*

$$\mathbb{F}^{(i)} = \{f \in \mathbb{F} \mid \mathrm{lv}(f) = x_i\}$$

for $1 \leq i \leq n$. *Let* $0 < p_1 < \cdots < p_r \leq n$ *be all the indices for which* $\mathbb{F}^{(p_j)} \neq \emptyset$, *and let* b_j *be a polynomial in* $\mathbb{F}^{(p_j)}$ *which has minimal degree in* x_{p_j} *for* $1 \leq j \leq r$. *The ordered set of polynomials*

$$\mathbb{B} = [b_1, \ldots, b_r]$$

is called a q-basic set *of* $\mathbb{F}$. *Clearly, for any* $1 \leq j \leq r$ *we have* $x_{p_j} = \mathrm{lv}(b_j)$.

Alternatively, the polynomial x_1x_3 can be obtained from the Gröbner bases of $\mathbb{H} \cup \{c_iz - 1\}$ for $i = 1, 2$.

(b) We predetermine the degenerate case in which A, B, C are collinear. To exclude this case, let us compute a plex Gröbner basis $\mathbb{G}^*$ of $\mathbb{H}^* = \mathbb{H} \cup \{x_1x_3z_1 - 1\}$ under $\omega_7 \prec z_1$, where z_1 is a new variable. One may find that

$$\mathbb{G}^* = [h_1, h_2, c_1, c_2, x_1x_3z_1 - 1].$$

As $c_1, c_2 \in \mathbb{G}^*$, $\mathrm{nform}(c_1, \mathbb{G}^*) = \mathrm{nform}(c_2, \mathbb{G}^*) \equiv 0$. Similarly, one can verify that 1 is contained in the Gröbner bases of $\mathbb{H}^* \cup \{c_1z - 1\}$ and $\mathbb{H}^* \cup \{c_2z - 1\}$ with $\omega_7 \prec z_1 \prec z$, where z is another new variable. In any case, the theorem is proved to be true under the *given* non-degeneracy condition $x_1x_3 \neq 0$. □

Adding non-degeneracy conditions to the hypotheses is a good heuristic for geometric theorem proving using Gröbner bases. So one should figure out such conditions in the way of formulating a geometric theorem. However, in practice it is not realistic to predetermine all the possible non-degeneracy conditions to make every geometric theorem rigorously stated; the inclusion of all the conditions also makes the hypotheses tedious and leads to high computational complexity. We look at the following theorem.

Example 3. Let the line EF joining the midpoints of the diagonals of a trapezoid $ABCD$ intersect one side of the trapezoid. Then EF bisects that side.

Refer to the figure. Let the points be assigned coordinates as follows

$$A(x_1, 0),\ D(x_2, 0),\ B(x_3, x_4),\ C(x_5, x_4),\ E(x_6, x_7),\ F(x_8, x_9),\ M(x_{10}, x_{11}).$$

The hypothesis of the theorem consists of the following relations

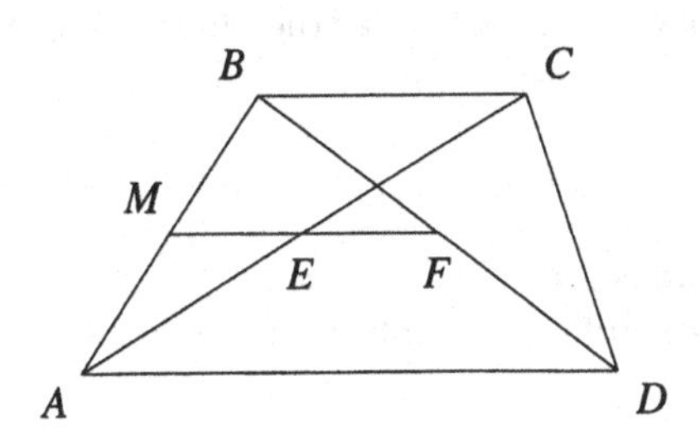

$$\mathsf{midp}(A,C,E) \iff \begin{cases} h_1 = 2x_6 - x_5 - x_1 = 0, \\ h_2 = 2x_7 - x_4 = 0; \end{cases}$$

$$\mathsf{midp}(B,D,F) \iff \begin{cases} h_3 = 2x_8 - x_3 - x_2 = 0, \\ h_4 = 2x_9 - x_4 = 0; \end{cases}$$

$$\mathsf{int}(A,B,E,F,M) \iff \begin{cases} h_5 = (x_8 - x_6)x_{11} - (x_9 - x_7)x_{10} + x_6x_9 \\ \qquad -x_7x_8 = 0, \\ h_6 = (x_3 - x_1)x_{11} - x_4(x_{10} - x_1) = 0. \end{cases}$$

We add the condition

- $x_4 \neq 0 \iff$ the lines AD and BC do not coincide

to rule out the simple degenerate case. The conclusion of the theorem to be proved is

$$\mathsf{midp}(A,B,M) \iff \begin{cases} c_1 = 2x_{10} - x_3 - x_1 = 0, \\ c_2 = 2x_{11} - x_4 = 0. \end{cases}$$

Following the preceding example, let $\mathbb{H}^* = \{h_1, \ldots, h_6, x_4z - 1\}$. With respect to the variable ordering $\omega_{11} \prec z$, a plex Gröbner basis of $\mathbb{H}^*$ is

$$\mathbb{G} = [h_1, \ldots, h_4, g_5, h_6, g_7, x_4z - 1],$$

where

$$g_5 = (x_5 - x_3 - x_2 + x_1)(2x_{10} - x_3 - x_1),$$
$$g_7 = 2(x_5 - x_2)x_{11} - x_4(2x_{10} + x_5 - x_3 - x_2 - x_1).$$

Now, neither c_1 nor c_2 has normal form 0 modulo $\mathbb{G}$. So one cannot tell whether the theorem is true or not. As in the previous example, take $\mathbb{B} = [h_1, \ldots, h_4, g_5, h_6]$, a q-basic set of $\mathbb{G}$. It may be easily verified that $\mathrm{prem}(c_i, \mathbb{B}) \equiv 0$ for both $i = 1, 2$.

⋆ *The* pseudo-remainder *of any* $g \in \boldsymbol{K}[\boldsymbol{x}]$ *with respect to a q-basic set* $\mathbb{B} = [b_1, \ldots, b_r]$ *is defined to be*

$$\mathrm{prem}(g, \mathbb{B}) = \mathrm{prem}(\cdots \mathrm{prem}(g, b_r, \mathrm{lv}(b_r)), \ldots, b_1, \mathrm{lv}(b_1)).$$

This implies that the theorem is true when the initials of the polynomials in $\mathbb{B}$ are non-zero. The non-constant initials are

$$I_5 = 2(x_5 - x_3 - x_2 + x_1),\ I_6 = x_3 - x_1.$$

Hence, the theorem is proved to be true under the found *subsidiary conditions* $I_5 I_6 \neq 0$. It is not difficult to interpret the geometric meanings of the two conditions:

- $x_5 - x_3 - x_2 + x_1 \neq 0 \iff ABCD$ is not a parallelogram;
- $x_3 - x_1 \neq 0 \Longleftarrow AB$ is not perpendicular to AD.

The first is a somewhat surprising case, in which $ABCD$ is a perfect trapezoid, but the theorem is not true because the line EF is not uniquely determined. It is not immediate to be aware of this special case *a priori*. The second condition is redundant. In other words, the theorem is true as well when $I_5 \neq 0$ and $I_6 = 0$. □

The above examples indicate that the method under discussion not only proves theorems but also finds out some degenerate or special cases in which the theorems are meaningless or even false.

2 Direct and refutational approaches

As shown in Examples 1–3, the first step of proving geometric theorems algebraically is to introduce a coordinate system and express geometric statements by means of polynomial relations. For theoretical and practical considerations, we restrict ourselves to the class of theorems whose algebraic formulations involve polynomial equations and inequations of the form

- hypothesis: $h_1(\boldsymbol{x}) = 0, \ldots, h_s(\boldsymbol{x}) = 0,\ d_1(\boldsymbol{x}) \neq 0, \ldots, d_t(\boldsymbol{x}) \neq 0;$
- conclusion: $c(\boldsymbol{x}) = 0,$

where $\boldsymbol{x} = (x_1, \ldots, x_n)$ are geometric entities such as the coordinates of points and areas of triangles; the polynomials are all in $\boldsymbol{x}$ with coefficients in the geometry-associated field $\boldsymbol{K}$. Proving a theorem amounts to deciding whether the formula

$$(\forall \boldsymbol{x})[h_1(\boldsymbol{x}) = 0 \wedge \cdots \wedge h_s(\boldsymbol{x}) = 0 \wedge d_1(\boldsymbol{x}) \neq 0 \wedge \cdots \wedge d_t(\boldsymbol{x}) \neq 0 \Longrightarrow c(\boldsymbol{x}) = 0] \tag{T}$$

is valid over $\boldsymbol{K}$ or some extension field of $\boldsymbol{K}$.

As illustrated by Examples 2 and 3, geometric theorems are true usually under subsidiary conditions. For many theorems the formula (T) cannot be proved valid logically, because some non-degeneracy conditions may be missing, or not explicitly expressed in the hypothesis.

Therefore, if (T) is not valid, one must find "appropriate" subsidiary conditions $d_1^*(\boldsymbol{x}) \neq 0, \ldots, d_r^*(\boldsymbol{x}) \neq 0$ so that the formula

$$(\forall \boldsymbol{x})[h_1(\boldsymbol{x}) = 0 \wedge \cdots \wedge h_s(\boldsymbol{x}) = 0 \wedge d_1(\boldsymbol{x}) \neq 0 \wedge \cdots \wedge d_t(\boldsymbol{x}) \neq 0 \wedge \\ d_1^*(\boldsymbol{x}) \neq 0 \wedge \cdots \wedge d_r^*(\boldsymbol{x}) \neq 0 \Longrightarrow c(\boldsymbol{x}) = 0]$$

becomes valid.

The simple algorithm below can be applied to prove a large number of theorems in elementary geometry. Its termination and correctness are quite obvious. The standard Rabinowitsch's trick is used to deal with inequations.

Algorithm ProveA. Given the algebraic form (T) of a geometric theorem $\mathbb{T}$, this algorithm either proves that $\mathbb{T}$ is true under some found subsidiary conditions, or reports that the hypothesis of $\mathbb{T}$ is self-contradictory or $\mathbb{T}$ is not confirmed.

A1. Compute a Gröbner basis $\mathbb{G}$ of $\{h_1, \ldots, h_s, d_1 z_1 - 1, \ldots, d_t z_t - 1\}$ over $\boldsymbol{K}$ with respect to plex determined by $\omega_n \prec z_1 \prec \cdots \prec z_t$, where $z_1, \ldots, z_t$ are new indeterminates. If $1 \in \mathbb{G}$ then report that "the hypothesis of $\mathbb{T}$ is self-contradictory" and the algorithm terminates.

A2. Compute $h \leftarrow \text{nform}(c, \mathbb{G})$. If $h \equiv 0$ then return "$\mathbb{T}$ is universally true" and the algorithm terminates.

A3. Take a q-basic set $\mathbb{B}$ of $\mathbb{G}$ and compute $r \leftarrow \text{prem}(h, \mathbb{B})$. If $r \equiv 0$ then let $I_1, \ldots, I_r$ be all the distinct irreducible factors of the initials of the polynomials in $\mathbb{B}$ which do not divide any d_i and return "$\mathbb{T}$ is true under the subsidiary conditions $I_1 \cdots I_r \neq 0$" else report that "$\mathbb{T}$ is not confirmed."

Of course, the found subsidiary conditions should not rule out interesting cases of the geometric theorem. In ProveA, whether the conditions $I_1 \cdots I_r \neq 0$ are appropriate in this sense is not examined; for the examination, extra computation is required.

The variables $x_1, \ldots, x_n$ usually consist of two kinds: one of parameters and the other of geometric dependents. The former are free variables, while the latter are constrained by the geometric conditions. It is not absolutely necessary to distinguish between the two kinds of variables for confirming theorems, but it is very helpful to do so for determining non-degeneracy conditions. The identification of parameters can be done quite easily when the theorem is stated constructively step by step. Let us assume that $x_1, \ldots, x_m$ $(m < n)$ are the parameters. If one is not interested in the exact non-degeneracy conditions provided as inequations in $x_1, \ldots, x_m$, the Gröbner basis in A1 of

ProveA can be computed over $\boldsymbol{K}(x_1, \ldots, x_m)$. In this case, the Gröbner basis computation is less expensive and thus is the proof algorithm more efficient.

In contrast with ProveA in which the conclusion-polynomial is directly reduced to 0 by using the Gröbner basis of the hypothesis-polynomial set, the following algorithm employs a refutational approach that verifies the inconsistence of the hypothesis-relations with the negation of the conclusion-equation.

Algorithm ProveB. The same specification as that of ProveA.

B1. Compute a Gröbner basis $\mathbb{G}$ of $\{h_1, \ldots, h_s, d_1 z_1 - 1, \ldots, d_t z_t - 1, cz - 1\}$ over $\boldsymbol{K}$ with respect to plex determined by $\omega_n \prec z_1 \prec \cdots \prec z_t \prec z$, where $z_1, \ldots, z_t, z$ are new indeterminates. If $1 \in \mathbb{G}$ then return "$\mathbb{T}$ is universally true" and the algorithm terminates.

B2. For each $g \in \mathbb{G}$ do:

If $g \in \boldsymbol{K}[x_1, \ldots, x_m]$ and $g \notin \{h_1, \ldots, h_s\}$ then:

Compute a Gröbner basis $\mathbb{G}^*$ of $\{h_1, \ldots, h_s, d_1 z_1 - 1, \ldots, d_t z_t - 1, gz - 1\}$ under any admissible term and variable ordering. If $1 \notin \mathbb{G}^*$ then return "$\mathbb{T}$ is true under the found subsidiary condition $g \neq 0$" and the algorithm terminates.

B3. Return "$\mathbb{T}$ is not confirmed."

To demonstrate the power of the algorithms described above, let us recall one of the most surprising and beautiful theorems in elementary geometry that was discovered around 1899 by F. Morley. The first automated proof of Morley's theorem in the generalized form stated below is attributed to Wu (1984), who worked out a tricky and elegant algebraic formulation. Since then, several simplified machine proofs of the theorem have been given by other researchers. Here the theorem is used to illustrate proving methods based on Gröbner bases.

Example 4 (Morley's theorem). The neighboring trisectors of the three angles of an arbitrary triangle intersect to form 27 triangles in all, of which 18 are equilateral.

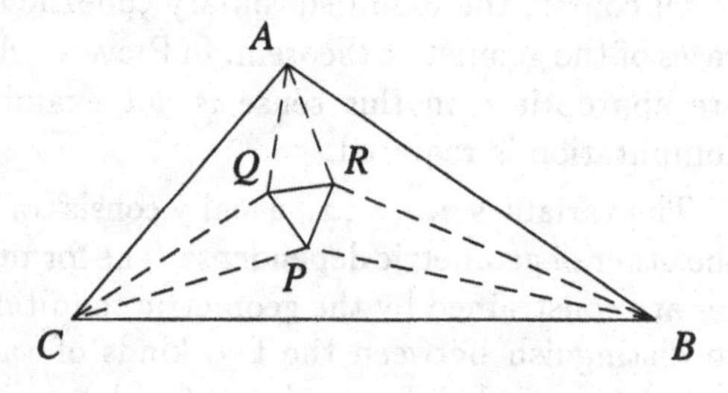

Following Wu (1984), the hypothesis of the theorem consists of

$$\angle ABC = 3\angle PBC,\ \angle ACB = 3\angle PCB,\ \tan^2\theta = 3,$$
$$\angle ABR = \angle PBC,\ \angle ACQ = \angle PCB,\ \angle BAR = \angle QAC,$$
$$\angle CBP + \angle PCB + \angle BAR \equiv \theta \bmod 2\pi,$$

and the conclusion to be proved is

$$\angle QPR = \angle RQP = \frac{\pi}{3}.$$

Let $x_6 = \tan\theta$ and take the coordinates of the points as

$$A(x_4, x_5),\ B(x_1, 0),\ C(x_2, 0),\ P(0, x_3),\ Q(x_{10}, x_9),\ R(x_8, x_7).$$

Then, by taking tangent for the equalities of angles both the hypothesis and the conclusion of Morley's theorem can be expressed as polynomial equations with index triples

⋆ *For any polynomial $f \in \boldsymbol{K}[\boldsymbol{x}]$, an index triple $[t\ x_p\ d]$ is used to characterize f, where t is the number of terms in f, $x_p = \mathrm{lv}(f)$, and d is the degree of f in x_p.*

$$[6\ x_5\ 1],\ [6\ x_5\ 1],\ [2\ x_6\ 2],\ [9\ x_8\ 1],\ [9, x_{10}\ 1],\ [41\ x_{10}\ 1],\ [40\ x_8\ 1]$$

and

$$[9\ x_{10}\ 1],\ [10\ x_{10}\ 1]$$

with respect to the variable ordering ω_{10}. A plex Gröbner basis of the hypothesis-polynomial set under $x_4 \prec \cdots \prec x_{10}$ consists of 7 polynomials with index triples

$$[7\ x_4\ 1],\ [9\ x_5\ 1],\ [2\ x_6\ 2],\ [10\ x_7\ 1],\ [13\ x_8\ 1],\ [10\ x_9\ 1],\ [13\ x_{10}\ 1].$$

The normal forms of the conclusion-polynomials modulo this Gröbner basis are 0. Therefore, the theorem is proved to be true under some possible non-degeneracy conditions which are not explicitly provided.

The above proof of Morley's theorem took about 23 CPU seconds in Maple V.3 running on a SUN SparcServer 690/51. The computing times reported elsewhere in Sects. 1–3 were obtained in Maple V.3 on the same machine; wherever timing is not given, it is less than five seconds.

Without using Wu's trick, let us consider a natural formulation of the theorem, where the hypothesis consists of

$$\begin{aligned} &\angle ABC = 3\angle PBC, \quad \angle ACB = 3\angle PCB, \quad \angle CAB = 3\angle RAB, \\ &\angle ABR = \angle PBC, \quad \angle ACQ = \angle PCB, \quad \angle BAR = \angle QAC \end{aligned}$$

and the conclusion to be proved is

$$PQ = PR,\ PQ = QR.$$

Let the coordinates of the points be chosen as

$$A(y_2, y_1),\ B(u_1, 0),\ C(u_2, 0),\ P(0, 1),\ Q(y_6, y_5),\ R(y_4, y_3).$$

The hypothesis and the conclusion can both be expressed as polynomial equations with index triples

- hypothesis: [6 y_2 1], [6 y_2 1], [191 y_4 3], [9 y_4 1], [9, y_6 1], [41 y_6 1];
- conclusion: [6 y_6 2], [6 y_6 2]

with respect to the variable ordering $y_1 \prec \cdots \prec y_6$. By means of polynomial factorization, the set $\mathbb{H}$ of hypothesis-polynomials can be decomposed over $\boldsymbol{Q}(u_1, u_2)$ into two plex Gröbner bases $\mathbb{G}_1$ and $\mathbb{G}_2$ such that

$$\mathbb{H} = 0 \iff \mathbb{G}_1 = 0 \text{ or } \mathbb{G}_2 = 0,$$

where

$$\mathbb{G}_1 = \begin{bmatrix} g_1, g_2, Iy_3 + 2u_1(u_2 - u_1)(3u_2^2 - 1), \\ Iy_4 - 3(u_1^2 - 1)u_2^3 - 2u_1u_2^2 - 7u_1^2u_2 - u_2 - 2u_1, \\ Iy_5 - 2u_2(u_2 - u_1)(3u_1^2 - 1), \\ Iy_6 - 3(u_2^2 - 1)u_1^3 - 2u_1^2u_2 - 7u_1u_2^2 - u_1 - 2u_2 \end{bmatrix},$$

$$\mathbb{G}_2 = \begin{bmatrix} g_1, g_2, Iy_3^2 - 4u_1(3u_1u_2^2 + 4u_2 - u_1)y_3 + 4u_1^2(3u_2^2 - 1), \\ 2u_1y_4 + (u_1^2 - 1)y_3 - 2u_1^2, \\ u_1(u_2^2 + 1)y_5 + (u_1^2 + 1)u_2y_3 - 2u_1u_2(u_2 + u_1), \\ 2u_1(u_2^2 + 1)y_6 - (u_1^2 + 1)(u_2^2 - 1)y_3 + 2u_1(u_1u_2^2 - 2u_2 - u_1) \end{bmatrix};$$

$$\begin{aligned} g_1 &= Iy_1 - (3u_1^2 - 1)(3u_2^2 - 1), \\ g_2 &= Iy_2 - 8u_1u_2(u_2 + u_1), \\ I &= (3u_1^2 - 1)u_2^2 + 8u_1u_2 - u_1^2 + 3. \end{aligned}$$

Note that $\boldsymbol{Q}$ denotes the field of rational numbers and the decomposition took 7.8 CPU seconds. It is easy to verify that the normal forms of the two conclusion-polynomials are both 0 with respect to $\mathbb{G}_2$, but not 0 with respect to $\mathbb{G}_1$. Therefore, under some non-degeneracy conditions the theorem is true for one component and false for the other. This reflects the fact that among the 27 triangles 18 are equilateral and not so are the other 9. □

The above example illustrates the effect of algebraic formulations in geometric theorem proving. Different formulations of the same theorem may produce quite different proofs. Although the method does not depend on how theorems are formulated algebraically, appropriate formulations may considerably reduce the computational cost and yield simple proofs for theorems that appear beyond the method's applicability. For a straightforward formulation of the following Steiner theorem using square of distance, one may encounter the reducibility problem because on which side of a line an isosceles triangle is drawn cannot be easily distinguished. Using vector rotation in which orientation is taken into account, we have a simple formulation shown below. With this formulation, the machine proof becomes quite trivial.

Example 5 (Steiner's theorem generalized). Let ABC', BCA' and CAB' be three similar isosceles triangles drawn all inward or all outward on the three sides of an arbitrary triangle ABC. Then the three lines AA', BB' and CC' are concurrent.

As $\triangle ABC'$, $\triangle BCA'$ and $\triangle CAB'$ are similar, their altitudes are proportional to the corresponding bases $|AB|$, $|BC|$ and $|CA|$. Let the ratio be α and the six points be located as

$$A(0,0),\ B(x_1,0),\ C(x_2,x_3),\ A'(x_4,x_5),\ B'(x_6,x_7),\ C'(x_8,x_9).$$

To avoid the problem of reducibility, we consider the point A' as the end of the vector starting from the midpoint of BC with length equal to $\alpha|BC|$ and the same direction as the vector obtained by rotating $\overrightarrow{BC}$ 90° anticlockwise. Similarly, the points B' and C' are so constructed. Then the hypothesis of the theorem may be expressed as

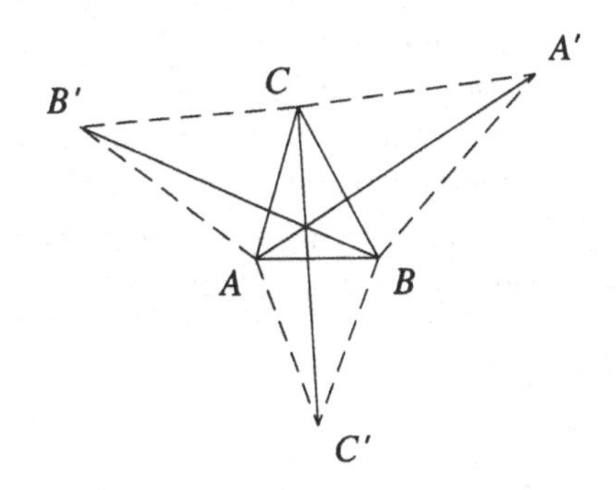

$$\begin{cases} h_1 = 2x_4 - (x_1 + x_2) + 2\alpha x_3 = 0, \\ h_2 = 2x_5 - x_3 + 2\alpha(x_1 - x_2) = 0, \\ h_3 = 2x_6 - x_2 - 2\alpha x_3 = 0, \\ h_4 = 2x_7 - x_3 + 2\alpha x_2 = 0, \\ h_5 = 2x_8 - x_1 = 0, \\ h_6 = x_9 - \alpha x_1 = 0. \end{cases}$$

The polynomial set $\mathbb{G} = [h_1, \ldots, h_6]$ is already a plex Gröbner basis with respect to $\omega_3 \prec \alpha \prec x_4 \prec \cdots \prec x_9$. The conclusion of the theorem is

$$\begin{aligned} c = {} & x_2x_4x_7x_9 - x_1x_4x_7x_9 - x_2x_5x_6x_9 + x_1x_2x_5x_9 + x_1x_5x_7x_8 - x_3x_4x_7x_8 \\ & + x_3x_5x_6x_8 - x_1x_3x_5x_8 - x_1x_2x_5x_7 + x_1x_3x_4x_7 = 0. \end{aligned}$$

It is easy to verify that $\mathrm{nform}(c, \mathbb{G}) \equiv 0$ and 1 is contained in the Gröbner basis of $\mathbb{G} \cup \{cz - 1\}$. So the theorem is proved to be true universally. □

*Example 3**. Consider the theorem proved in Example 3 by using ProveA, where a redundant subsidiary condition $x_3 - x_1 \neq 0$ was obtained. This redundant condition would not be produced when ProveB is used. With reference to the theorem in question, we explain a different yet complete procedure that can derive the *simplest* or *minimal* subsidiary conditions. Assume that the non-degeneracy condition $x_4 \neq 0$ is not provided artificially, and let $\mathbb{H} = \{h_1, \ldots, h_6\}$. Some computation verifies that the ideal $\mathfrak{I}$ generated by $\mathbb{H}$ is radical. For $i = 1, 2$, let the ideal generated by the single polynomial c_i be denoted (c_i). A finite basis $\mathbb{P}_i$ of the quotient ideal $\mathfrak{I} : (c_i)$ may be obtained as follows:

- Compute a plex Gröbner basis $\mathbb{G}_i$ of $\{(z-1)h_1, \ldots, (z-1)h_6, zc_i\}$ under $\omega_{11} \prec z$, where z is a new variable;

- Set $\mathbb{P}_i = \{g/c_i |\ g \in \mathbb{G}_i \cap \boldsymbol{Q}[x_1, \ldots, x_{11}]\}$.

9-second computation shows that

$$\mathbb{P}_1 = [x_4 I_5, h_1, \ldots, h_4, h_6, (x_5 - x_2)x_{11} - x_4(x_{10} - x_1)],$$
$$\mathbb{P}_2 = [I_5, h_1, \ldots, h_4, h_6].$$

In fact, both $\mathbb{P}_1$ and $\mathbb{P}_2$ are plex Gröbner bases with respect to ω_{11}, and the first polynomial p_i of $\mathbb{P}_i$ is not contained in $\mathfrak{J}$. For $i = 1, 2$, $p_i \neq 0$ gives the simplest subsidiary condition for the conclusion $c_i = 0$ to hold true. Hence, the simplest conditions for the theorem to be true are $x_4 I_5 \neq 0$, i.e., AD does not coincide with BC and $ABCD$ is not a parallelogram. □

Note that the translation of geometric statements into algebraic expressions can be done easily and automatically. The geometric interpretation of the algebraic subsidiary conditions can also be done automatically in most cases. From the geometric meanings of the conditions, one may see whether the excluded cases are degenerate ones.

The term ordering used for most cases in the paper is plex. For the sake of efficiency one can choose other elimination orderings instead. In some situation, the total degree term ordering is sufficient. We do not go further in this direction.

3 Discovering theorems

The method explained in the previous section can be adapted to discover "new" geometric theorems and to derive unknown geometric relations automatically. A typical example is the automated discovery of Qin-Heron's formula that represents the area of a triangle in terms of its three sides. This simple example has been repeatedly used for illustration in the recent literature. Here let us consider a more complicated example.

Example 6 (Brahmagupta's formula). Find the representation of the signed area of any oriented cyclic quadrilateral $ABCD$ in terms of its four sides.

Let the coordinates of the points be chosen as

$$A(0,0),\ B(a,0),\ C(x_1,x_2),\ D(x_3,x_4).$$

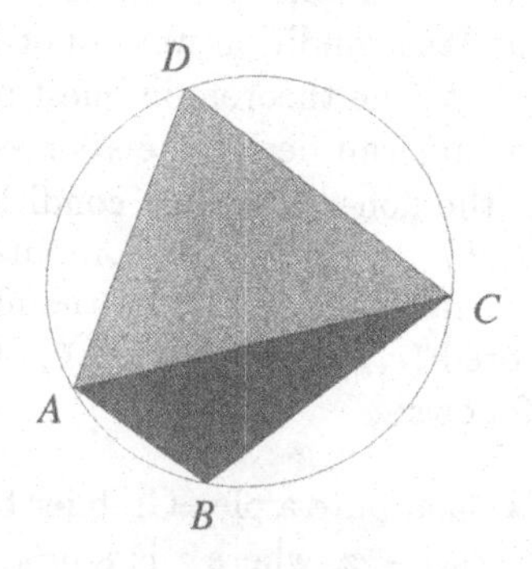

Denote the lengths of the three sides BC, CD, DA by b, c, d, respectively, and the sum of the signed areas of $\triangle ABC$ and $\triangle ACD$ by Θ. Then the conditions relating these geometric entities may be expressed as

$$\begin{cases} h_1 = x_2^2 + x_1^2 - 2ax_1 - b^2 + a^2 = 0, & \leftarrow b = |BC| \\ h_2 = x_4^2 - 2x_2x_4 + x_3^2 - 2x_1x_3 + x_2^2 & \\ \quad +x_1^2 - c^2 = 0, & \leftarrow c = |CD| \\ h_3 = x_4^2 + x_3^2 - d^2 = 0, & \leftarrow d = |DA| \\ h_4 = a[x_2x_4^2 - (x_2^2 + x_1^2 - ax_1)x_4 & \\ \quad +x_2x_3^2 - ax_2x_3] = 0, & \leftarrow A, B, C, D \text{ are concyclic} \\ h_5 = x_1x_4 - x_2x_3 + ax_2 - 2\Theta = 0. & \leftarrow \Theta = \text{signed area of } ABCD \end{cases}$$

We wish to decide if there exists some relation among $a, \ldots, d$ and Θ. To this end, set $\mathbb{H} = \{h_1, \ldots, h_5\}$ and compute a Gröbner basis $\mathbb{G}$ of $\mathbb{H}$ under $\Theta \prec \omega_4$: $\mathbb{G}$ may be found (in 128 CPU seconds) to consist of five polynomials with index triples

$$[46\ \Theta\ 4],\ [26\ x_1\ 1],\ [13\ x_2\ 1],\ [26\ x_3\ 1],\ [13\ x_4\ 1].$$

The first polynomial in $\mathbb{G}$ factors as

$$r = (r_0 + 8abcd)(r_0 - 8abcd),$$

where

$$r_0 = 16\Theta^2 + d^4 - 2(c^2 + b^2 + a^2)d^2 + c^4 - 2(b^2 + a^2)c^2 + (b^2 - a^2)^2.$$

It follows from the properties of Gröbner bases that

$$\mathbb{H} = 0 \Longrightarrow \mathbb{G} = 0 \Longrightarrow r = 0.$$

Thus, we get the algebraic relation $r = 0$ under some possible non-degeneracy conditions. In fact, $r = 0$ holds in all the degenerate cases; namely, the relation follows from the geometric hypotheses universally. This can be verified, for instance, by computing a Gröbner basis of $\mathbb{H} \cup \{rz - 1\}$ over $\boldsymbol{Q}$ with respect to the total degree term ordering; 1 is contained in the basis and the verification took 23 seconds.

Let $p = (a+b+c+d)/2$; $r = 0$ leads to either of the following two equalities

$$\Theta^2 = (p-a)(p-b)(p-c)(p-d),$$
$$\Theta^2 = p(p-a-b)(p-a-c)(p-a-d).$$

The first, which is the well-known Brahmagupta formula, gives the real result when the number k of positive variables among $a, \ldots, d$ is even; and so does the second when k is odd. □

Algorithm Derive. Given a set of geometric hypotheses expressed as a system of polynomial equations and inequations

$$h_1(\boldsymbol{u}, \boldsymbol{x}) = 0, \ldots, h_s(\boldsymbol{u}, \boldsymbol{x}) = 0,\ d_1(\boldsymbol{u}, \boldsymbol{x}) \neq 0, \ldots, d_t(\boldsymbol{u}, \boldsymbol{x}) \neq 0$$

in two sets of geometric entities $\boldsymbol{u} = (u_1, \ldots, u_m)$ and $\boldsymbol{x} = (x_1, \ldots, x_n)$ with coefficients in $\boldsymbol{K}$ and a fixed integer k, without loss of generality, say $k = 1$, this algorithm either reports that the geometric hypotheses are self-contradictory, or determines whether there exists a polynomial relation $r(\boldsymbol{u}, x_1) = 0$ between $\boldsymbol{u}$ and x_1 such that

$$(\forall \boldsymbol{u}, \boldsymbol{x})[h_1(\boldsymbol{u}, \boldsymbol{x}) = 0 \wedge \cdots \wedge h_s(\boldsymbol{u}, \boldsymbol{x}) = 0 \wedge d_1(\boldsymbol{u}, \boldsymbol{x}) \neq 0 \wedge \cdots \wedge d_t(\boldsymbol{u}, \boldsymbol{x}) \neq 0 \\ \Longrightarrow r(\boldsymbol{u}, x_1) = 0]$$

is valid; and if so, the algorithm finds the polynomial $r(\boldsymbol{u}, x_1)$.

D1. Compute a Gröbner basis $\mathbb{G}$ of $\{h_1, \ldots, h_s, d_1 z_1 - 1, \ldots, d_t z_t - 1\}$ over $\boldsymbol{K}$ with respect to plex determined by $u_1 \prec \cdots \prec u_m \prec \omega_n \prec z_1 \prec \cdots \prec z_t$, where $z_1, \ldots, z_t$ are new indeterminates. If $1 \in \mathbb{G}$ then report that "the geometric hypotheses are self-contradictory" and the algorithm terminates.

D2. Set $\mathbb{G}^{\langle 1 \rangle} \leftarrow \mathbb{G} \cap (\boldsymbol{K}[\boldsymbol{u}, x_1] \setminus \boldsymbol{K}[\boldsymbol{u}])$. If $\mathbb{G}^{\langle 1 \rangle} \neq \emptyset$ then return the polynomial $r(\boldsymbol{u}, x_1) \in \mathbb{G}^{\langle 1 \rangle}$ that has minimal degree in x_1 else report that "there is no algebraic relation between $\boldsymbol{u}$ and x_1 in general."

The above algorithm makes use of the elimination property of Gröbner bases. If $\mathbb{G} \cap \boldsymbol{K}[\boldsymbol{u}] = \emptyset$ then $\boldsymbol{u}$ are independent variables; otherwise, $\boldsymbol{u}$ are algebraically dependent. If $\boldsymbol{u}$ are known to be independent (parameters) and any inequation in $\boldsymbol{u}$ may be assumed to be a non-degeneracy condition of the geometric problem, then the Gröbner basis in D1 can be computed over $\boldsymbol{K}(\boldsymbol{u})$ (which is much easier in general). In case no relation between $\boldsymbol{u}$ and x_1 is found, one can analyze the computed Gröbner basis $\mathbb{G}$ and try to get possible relations by supplying appropriate subsidiary conditions of the form $d_i^* \neq 0$ to exclude some components.

Example 7 (Poncelet's theorem). Let R be the radius of the circumscribed circle and r the radius of the inscribed circle of an arbitrary triangle, and let d be the distance between the centers of the two circles. Determine the relation among R, r and d.

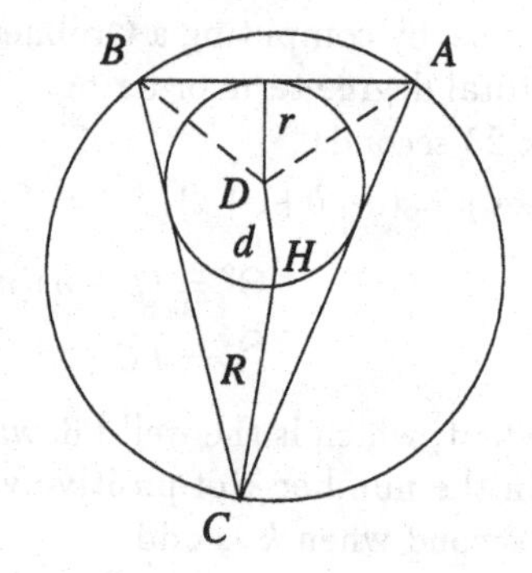

Let ABC be an arbitrary triangle, D and H be the incenter and circumcenter of $\triangle ABC$ and the coordinates be assigned as

$$A(x_1, 0),\ B(x_2, 0),\ D(0, x_3),\ C(x_4, x_5),\ H(x_6, x_7).$$

Now the geometric hypotheses are:

- C lies on the reflection line of AB with respect to AD

$$\iff h_1 = (x_2 - x_1)[(x_3^2 - x_1^2)x_5 - 2x_1x_3(x_4 - x_1)] = 0;$$

- C lies on the reflection line of BA with respect to BD

$$\iff h_2 = (x_2 - x_1)[(x_3^2 - x_2^2)x_5 - 2x_2x_3(x_4 - x_2)] = 0;$$

- H is the circumcenter of $\triangle ABC$

$$\iff \begin{cases} h_4 = (x_2 - x_1)(2x_6 - x_2 - x_1) = 0, \\ h_3 = 2x_5x_7 + 2x_4x_6 - 2x_2x_6 - x_5^2 - x_4^2 + x_2^2 = 0; \end{cases}$$

- r is the radius of the inscribed circle of $\triangle ABC \implies h_5 = r^2 - x_3^2 = 0$;
- R is the radius of the circumcircle of $\triangle ABC$

$$\implies h_6 = R^2 - x_7^2 - (x_6 - x_1)^2 = 0;$$

- d is the distance between D and $H \implies h_7 = d^2 - (x_7 - x_3)^2 - x_6^2 = 0.$

Assume that $\triangle ABC$ does not degenerate into a line, so that $(x_2 - x_1)x_5 \neq 0$.

Computing a plex Gröbner basis $\mathbb{G}$ of $\{h_1, \ldots, h_7, (x_2 - x_1)z_1 - 1, x_5z_2 - 1\}$ with respect to $d \prec x_2 \prec \cdots \prec x_7 \prec z_1 \prec z_2$, one finds (in 282 seconds) that there is one polynomial g in $\mathbb{G}$ which involves d, R, r only:

$$g = d^4 - 2d^2R^2 + R^4 - 4R^2r^2 = (d^2 - R^2 + 2Rr)(d^2 - R^2 - 2Rr).$$

Hence, the geometric hypotheses imply that $g = 0$. In the above derivation, we have not used the implicit assumption that $R > 0$ and $r > 0$. Moreover, it is obvious that $R > d$ because the inscribed circle is contained in the circumcircle of $\triangle ABC$. Therefore, we have

$$R^2 - 2Rr = d^2.$$

This is the great Poncelet theorem; it has been rediscovered automatically by using Derive. □

*Example 6**. Consider the geometric problem in Example 6. The theorem can be "discovered" in a different way as follows. Motivated by the Qin-Heron formula, we may conjecture that the Brahmagupta formula holds for an arbitrary oriented quadrilateral $ABCD$. In other words, we wish to show that

$$(\forall a, b, c, d, x_1, \ldots, x_4, \Theta)[h_1 = 0 \wedge h_2 = 0 \wedge h_3 = 0 \wedge h_5 = 0 \implies r_0 + 8abcd = 0],$$

where the polynomials are as in Example 6. The conjecture is clearly true when two of the points A, B, C, D coincide. If it is true not for arbitrary A, B, C, D, there should exist some relation which keeps the four points constrained. So we order one of the variables $a, x_1, \ldots, x_4$ at the beginning of the increasing queue, e.g.,

$$x_4 \prec \Theta \prec b \prec c \prec d.$$

With respect to this variable ordering, a plex Gröbner basis $\mathbb{G}$ of $\{h_1, h_2, h_3, h_5, r_0 + 8abcd\}$ may be computed in 9.6 seconds. One finds that $\mathbb{G}$ contains the polynomial $(h_4/a)^2$. In consequence,

$$h_1 = 0, \ h_2 = 0, \ h_3 = 0, \ h_5 = 0, \ r_0 + 8abcd = 0$$

imply that $h_4 = 0$. Hence, the conjecture holds only if $h_4 = 0$, i.e., A, B, C, D are concyclic. One may verify that the conjecture becomes true indeed when $h_4 = 0$ is added to the hypothesis. In this way, the theorem about Brahmagupta's formula is rediscovered. □

4 Coordinate-free techniques

The methods of proving and discovering geometric theorems presented in the previous sections proceed by first taking coordinates for points. The proof steps are provided as polynomial computations and reductions. These steps are not geometrically interesting and readable because the involved polynomials are usually large. It is also possible to prove geometric theorems using Gröbner bases without taking coordinates. We give two examples to show this coordinate-free technique. In the first example geometric relations are represented by means of vector sum and outer product in Grassmann algebra. The theorem is proved by computing a non-commutative Gröbner basis of the set of Grassmann algebraic expressions together with normal form reduction.

Now let a point A be considered also as a vector from the origin O to A, and $A + B$ as the vector from O to the opposite vertex of the parallelogram formed by OA and OB. Then the midpoint M of AB can be simply expressed as

$$M = \frac{A + B}{2}.$$

Let $A \wedge B$ denote the *outer product* of two vectors A and B. Geometrically, it is a bivector corresponding to the parallelogram obtained by sweeping the vector A along the vector B. The parallelogram obtained by sweeping B along A differs only in orientation from that obtained by sweeping A along B. This is simply expressed as

$$B \wedge A = -A \wedge B.$$

In other words, the outer product is anticommutative; it is associative and distributive with addition. Thus,

$$\mathsf{col}(A, B, C) \iff (C - A) \wedge (B - A) = 0.$$

*Example 1**. The same theorem in Example 1. Its hypothesis may be expressed as

$$\begin{cases} h_1 = (A_1 - A_2) \wedge (A_1 - A_3) = 0, & \leftarrow \mathsf{col}(A_1, A_2, A_3) \\ h_2 = (A_1 - B_2) \wedge (A_1 - B_3) = 0, & \leftarrow \mathsf{col}(A_1, B_2, B_3) \\ h_3 = (B_1 - A_2) \wedge (B_1 - B_3) = 0, & \leftarrow \mathsf{col}(B_1, A_2, B_3) \\ h_4 = (B_1 - B_2) \wedge (B_1 - A_3) = 0, & \leftarrow \mathsf{col}(B_1, B_2, A_3) \\ h_5 = 2M_1 - (A_1 + B_1) = 0, & \leftarrow \mathsf{midp}(A_1, B_1, M_1) \\ h_6 = 2M_2 - (A_2 + B_2) = 0, & \leftarrow \mathsf{midp}(A_2, B_2, M_2) \\ h_7 = 2M_3 - (A_3 + B_3) = 0. & \leftarrow \mathsf{midp}(A_3, B_3, M_3) \end{cases}$$

With respect to the inverse lexicographical term ordering determined by

$$A_1 \prec A_2 \prec A_3 \prec B_1 \prec B_2 \prec B_3 \prec M_1 \prec M_2 \prec M_3,$$

a two-sided non-commutative Gröbner basis $\mathbb{G}$ of $\{h_1, \ldots, h_7, A_1 \wedge A_1, \ldots, M_3 \wedge M_3\}$ consists of $A_1 \wedge A_1, A_2 \wedge A_2, A_3 \wedge A_3, B_1 \wedge B_1, B_2 \wedge B_2, B_3 \wedge B_3$ and

$$\begin{gathered} A_2 \wedge A_3 - A_1 \wedge A_3 + A_1 \wedge A_2, \; B_1 \wedge B_2 - A_3 \wedge B_2 + A_3 \wedge B_1, \\ B_1 \wedge B_3 - A_2 \wedge B_3 + A_2 \wedge B_1, \; B_2 \wedge B_3 - A_1 \wedge B_3 + A_1 \wedge B_2, \\ M_1 - \frac{1}{2}B_1 - \frac{1}{2}A_1, \; M_2 - \frac{1}{2}B_2 - \frac{1}{2}A_2, \; M_3 - \frac{1}{2}B_3 - \frac{1}{2}A_3. \end{gathered}$$

The conclusion of the theorem is

$$c = (M_1 - M_2) \wedge (M_1 - M_3) = -M_1 \wedge M_3 - M_2 \wedge M_1 + M_2 \wedge M_3 = 0.$$

As $\mathrm{nform}(c, \mathbb{G}) \equiv 0$, the theorem is proved to be true universally. □

Stifter (1993) proposed to use Gröbner bases in vector spaces. In this case, the outer product is not used in the formulation of geometric theorems; instead one introduces scale variables with existential quantifiers.

*Example 2**. The same theorem as in Example 2. Using vector formalism, the hypothesis of the theorem may be expressed as follows

$$\begin{aligned} \mathsf{midp}(A, B, D) &\iff h_1 = 2D - A - B = 0; \\ \mathsf{midp}(C, D, E) &\iff h_2 = 2E - C - D = 0; \\ \mathsf{col}(A, E, F) &\iff \exists\, a \text{ such that } h_3 = A + a(A - E) - F = 0; \\ \mathsf{col}(C, B, F) &\iff \exists\, b \text{ such that } h_4 = B + b(B - C) - F = 0. \end{aligned}$$

We need to show that

$$(\forall A, B, C, D, E, F)(\exists a, b)[h_1 = 0 \wedge \cdots \wedge h_4 = 0 \Longrightarrow c = 0],$$

where the conclusion-polynomial is

$$c = 2(F - C) - (B - F).$$

Let $\boldsymbol{Z}$ be the ring of integers; compute a Gröbner basis $\mathbb{G}$ of $\mathbb{H} = \{h_1, \ldots, h_4\}$ in the reduction ring $\boldsymbol{R} = (\boldsymbol{Z}[a,b])^2[A,B,C,D,E,F]$ using the lexicographical term ordering with $b \prec a$ and $A \prec B \prec C \prec D \prec E \prec F$. We have

$$\mathbb{G} = \begin{bmatrix} -2(a-2b)C - (a+4b+4)B + (3a+4)C, \\ (3a+4)D - (a-2b)C - 2(a+b+2)B, \\ (3a+4)E - (2a-b+2)C - (a+b+2)B, \\ F + bC = (b+1)B \end{bmatrix}.$$

The first polynomial, say g_1, in $\mathbb{G}$ only involves a, b and the free points A, B, C. $g_1 = 0$ if and only if either the coefficients of g_1 with respect to A, B, C are identically 0, or A, B, C are collinear. The latter is clearly a degenerate case, so we can assume that A, B, C are not collinear. Hence,

$$h_5 = a - 2b = 0, \;\; h_6 = a + 4b + 4 = 0, \;\; h_7 = 3a + 4 = 0.$$

Compute moreover a Gröbner basis $\mathbb{G}^*$ of $\mathbb{H} \cup \{h_5, h_6, h_7\}$ in $\boldsymbol{R}$ under the above-mentioned orderings: $\mathbb{G}^*$ comprises the following polynomials

$$3b + 2, \; 3a + 4, \; 2D - B - A, \; 4E - 2C - B - A, \; 3F - 2C - B.$$

The last polynomial is identical to c, so $\mathrm{nform}(c, \mathbb{G}^*) \equiv 0$. Therefore, the theorem is proved to be true under the non-degeneracy condition that A, B, C do not lie on the same line. □

The above coordinate-free approaches have quite limited applicability. Without further extension they can prove only a small number of geometric theorems. How to enlarge the application domain of Gröbner bases with coordinate-free techniques remains to be an interesting research problem.

5 Bibliographical notes

Proving geometric theorems automatically on modern computers is one of the classical research subjects in artificial intelligence. The early axiomatic approaches represented by the *Geometry Machine* of Gelernter (1959) and his co-workers were designed for a class of theorems in plane Euclidean geometry. Although various strategies and heuristics were subsequently adopted and implemented, the methods following Gelernter's idea could not prove geometric theorems that mathematicians find difficult to prove. It was W.-t. Wu who introduced in the later 1970s a powerful algebraic method which can prove a large number of non-trivial theorems in elementary geometry and differential geometry (see Wu 1978, 1984, 1994). The discovery and success of this method, now known as *Wu's method*, marked a breakthrough and have motivated the lasting research interest and activity on the subject of geometric theorem proving. In particular, Kapur (1986a, b), Kutzler and Stifter (1986a, b), and Chou and Schelter (1986) showed that the Gröbner bases method of Buchberger (1965, 1985) can be applied to prove the same class of elementary geometric theorems that Wu's method addresses. Extensive experiments were made subsequently on Wu's method and methods based on Gröbner bases. Empirical comparisons show that Wu's method is superior in general because of its suitability in dealing with non-degeneracy conditions. However, Gröbner bases remain to be a powerful method for geometric theorem proving. This point of view is supported partially by the examples given in the present paper. The interested reader may refer to Wang (1996b) for other historical background, the state-of-the-art and a long list of references for the subject.

The algorithm ProveA is essentially due to Kutzler and Stifter (1986a, b), and ProveB due to Kapur (1986a, b; 1988). A combination of ProveA and ProveB was proposed by Chou and Schelter (1986). Winkler (1990, 1992) devised a decision procedure that is able to find simplest subsidiary conditions as shown in Example 3*. Gröbner bases are also used to evaluate the dimension of (differential) algebraic varieties in the method introduced by Carrà Ferro and Gallo (1990, 1994). It seems that the use of q-basic set as in ProveA and factorized Gröbner bases to handle reducibility as in Example 4 has not been investigated before. Methods based Gröbner bases have been implemented by different researchers including Chou (1988, 1990), Ko (1986) and her co-workers Cyrluk et al. (1988), Kutzler and Stifter (1986c) and their co-workers Kusche et al. (1987) and this author (Wang 1996a). The approach of computing Gröbner bases in vector spaces is proposed by Stifter (1993). The use of non-commutative Gröbner bases is suggested in Wang (1996b). Following Wu (1986), the automatic derivation of geometric formulae using Gröbner bases was investigated in Chou (1987), Chou and Gao (1990), and Wang (1991). Relevant work has also been done by several other researchers: the reader may consult the following list of selected references for various discussions, refinements, and improvements on Gröbner-basis-related methods for geometric theorem proving and discovering.

Acknowledgments. This work has been supported partially by AFCRST under Project PRA M94-1 and by CEC under Reactive LTR Project 21914 – CUMULI. Part of the paper was written during the author's stay at the "Laboratoire franco-chinois de Recherche en Informatique, Automatique et Mathématiques Appliquées" in September/October 1997.

References

Buchberger, B. (1965): Ein Algorithmus zum Auffinden der Basiselemente des Restklassenringes nach einem nulldimensionalen Polynomideal. Ph.D thesis, Universität Innsbruck, Austria.

Buchberger, B. (1985): Gröbner bases: An algorithmic method in polynomial ideal theory. In: Bose, N. K. (ed.): Multidimensional systems theory. Reidel, Dordrecht, pp. 184–232.

Buchberger, B. (1987): Applications of Gröbner bases in non-linear computational geometry. In: Rice, J. R. (ed.): Mathematical aspects of scientific software. Springer, New York, pp. 59–87.

Buchberger, B., Collins, G. E., Kutzler, B. (1988): Algebraic methods for geometric reasoning. Ann. Rev. Comput. Sci. 3: 85–119.

Carrà Ferro, G. (1994): An extension of a procedure to prove statements in differential geometry. J. Automat. Reason. 12: 351–358.

Carrà Ferro, G., Gallo, G. (1990): A procedure to prove statements in differential geometry. J. Automat. Reason. 6: 203–209.

Chou, S.-C. (1987): A method for the mechanical derivation of formulas in elementary geometry. J. Automat. Reason. 3: 291–299.

Chou, S.-C. (1988): Mechanical geometry theorem proving. Reidel, Dordrecht.

Chou, S.-C. (1990): Automated reasoning in geometries using the characteristic set method and Gröbner basis method. In: Proceedings ISSAC '90, Tokyo, August 20–24, 1990, ACM Press, New York, pp. 255–260.

Chou, S.-C., Gao, X.-S. (1990): Mechanical formula derivation in elementary geometries. In: Proceedings ISSAC '90, Tokyo, August 20–24, 1990, ACM Press, New York, pp. 265–270.

Chou, S.-C., Schelter, W. F. (1986): Proving geometry theorems with rewrite rules. J. Automat. Reason. 2: 253–273.

Chou, S.-C., Schelter, W. F., Yang, J.-G. (1989): Characteristic sets and Gröbner bases in geometry theorem proving. In: Aït-Kaaci, H., Nivat, M. (eds.): Resolution of equations in algebraic structures. Academic Press, San Diego, pp. 33–92.

Cyrluk, D. A., Harris, R. M., Kapur, D. (1988): GEOMETER: A theorem prover for algebraic geometry. In: Proceedings CADE-9, Argone, May 23–26, 1988, Springer, Berlin Heidelberg New York Tokyo, pp. 770–771 (Lecture notes in computer science, vol. 310).

Gelernter, H. (1959): Realization of a geometry theorem proving machine. In: Proc. Int. Conf. Info. Process., Paris, June 15–20, 1959, Unesco, Paris, pp. 273–282.

Kapur, D. (1986a): Geometry theorem proving using Hilbert's Nullstellensatz. In: Proceedings SYMSAC '86, Waterloo, July 21–23, 1986, ACM Press, New York, pp. 202–208.

Kapur, D. (1986b): Using Gröbner bases to reason about geometry problems. J. Symb. Comput. 2: 399–408.

Kapur, D. (1988): A refutational approach to geometry theorem proving. Artif. Intell. 37: 61–93.

Ko, H.-P. (1986): ALGE-prover II: A new edition of ALGE-prover. Tech. Rep. 86CRD-081, General Electric Co., Schenectady, USA.

Kusche, K., Kutzler, B., Stifter, S. (1987): Implementation of a geometry theorem proving package in Scratchpad II. In: Proceedings EUROCAL '87, Leipzig, June 2–5, 1987, Springer, Berlin Heidelberg New York Tokyo, pp. 246–257 (Lecture notes in computer science, vol. 387).

Kutzler, B. (1988): Algebraic approaches to automated geometry theorem proving. Ph.D thesis, RISC-Linz, Johannes Kepler University, Austria.

Kutzler, B. (1989): Careful algebraic translations of geometry theorems. In: Proceedings ISSAC '89, Portland, July 17–19, 1989, ACM Press, New York, pp. 254–263.

Kutzler, B., Stifter, S. (1986a): Automated geometry theorem proving using Buchberger's algorithm. In: Proceedings SYMSAC '86, Waterloo, July 21–23, 1986, ACM Press, New York, pp. 209–214.

Kutzler, B., Stifter, S. (1986b): On the application of Buchberger's algorithm to automated geometry theorem proving. J. Symb. Comput. 2: 389–397.

Kutzler, B., Stifter, S. (1986c): A geometry theorem prover based on Buchberger's algorithm. In: Proceedings CADE-8, Oxford, July 27 – August 1, 1986, Springer, Berlin Heidelberg New York Tokyo, pp. 693–694 (Lecture notes in computer science, vol. 230).

Stifter, S. (1993): Geometry theorem proving in vector spaces by means of Gröbner bases. In: Proceedings ISSAC '93, Kiev, July 6–8, 1993, ACM Press, New York, pp. 301–310.

Wang, D. (1991): Reasoning about geometric problems using algebraic methods. In: Proceedings MEDLAR 24-Month Workshop, Grenoble, December 8–11, 1991, DOC, Imperial College, University of London, UK.

Wang, D. (1996a): GEOTHER: A geometry theorem prover. In: Proceedings CADE-13, New Brunswick, July 30 – August 3, 1996, Springer, Berlin Heidelberg New York Tokyo, pp. 166–170 (Lecture notes in artificial intelligence, vol. 1104).

Wang, D. (1996b): Geometry machines: From AI to SMC. In: Proceedings AISMC-3, Steyr, September 23–25, 1996, Springer, Berlin Heidelberg New York Tokyo, pp. 213–239 (Lecture notes in computer science, vol. 1138).

Winkler, F. (1988): A geometrical decision algorithm based on the Gröbner bases algorithm. In: Proceedings ISSAC '88, Rome, July 4–8, 1988, Springer, Berlin Heidelberg New York Tokyo, pp. 356–363 (Lecture notes in computer science, vol. 358).

Winkler, F. (1990): Gröbner bases in geometry theorem proving and simplest degeneracy conditions. Math. Pannonica 1: 15–32.

Winkler, F. (1992): Automated theorem proving in nonlinear geometry. In: Hoffmann, C. (ed.): Issues in robotics and nonlinear geometry. JAI Press, Greenwich, pp. 183–197.

Wu, W.-t. (1978): On the decision problem and the mechanization of theorem-proving in elementary geometry. Sci. Sinica 21: 159–172 [also in Automated theorem proving: After 25 years. Contemp. Math. 29: 213–234 (1984)].

Wu, W.-t. (1984): Basic principles of mechanical theorem proving in elementary geometries. J. Syst. Sci. Math. Sci. 4: 207–235 [also in J. Automat. Reason. 2: 221–252 (1986)].

Wu, W.-t. (1986): A mechanization method of geometry and its applications I. Distances, areas and volumes. J. Syst. Sci. Math. Sci. 6: 204–216.

Wu, W.-t. (1994): Mechanical theorem proving in geometries: Basic principles. Springer, Wien New York [translated from the Chinese edition — published in 1984 by Science Press, Beijing — by X. Jin and D. Wang].

Research Papers

The Fractal Walk

Beatrice Amrhein[1] *and Oliver Gloor*[2]

Abstract

The Gröbner Walk is a method which converts a Gröbner basis of an arbitrary dimensional ideal I to a Gröbner basis of I with respect to another term order. The walk follows a path of intermediate Gröbner bases according to the Gröbner fan of I. One of the open problems in the walk algorithm is path finding in the Gröbner fan. In order to avoid intersection points in the fan, paths are perturbed up to a certain degree. The *Fractal Walk* allows us to perturb the path locally in each step rather than globally. Thus, it removes the difficulty of finding the globally best perturbation degree. Our implementation shows that we even obtain speedups over the best perturbation degree because of the "tunneling" effect of the Fractal Walk. In addition, the Fractal Walk is compared to other Gröbner basis conversion methods.

1 Introduction

It is well known that the term ordering strongly determines the complexity of the Gröbner basis computation. The choice of the term ordering usually depends on the type of problem we want to solve. Elimination orders such as lexicographic, which we need for polynomial system solving, are known to be *slow* term orders, that is, they lead to particularly long computations.

A strategy to overcome this difficulty is to apply basis conversion. One approach is the well-known *FGLM* method [8], another is Hilbert-driven Gröbner basis conversion [15]. A different method, which is independent of the dimension of the ideal and does not need homogenization, is the *Gröbner Walk* developed by Collart, Kalkbrener, and Mall [4].

Practical experience with the walk [1, 2, 14] shows that lexicographic Gröbner bases can often be computed orders of magnitude faster via a total degree basis followed by a walk rather than directly with the Buchberger algorithm. As the walk is independent of the dimension of the ideal, we see a broad variety of applications, divided into two categories. The first is elimination and system solving where we compute elimination bases via "fast" ordered Gröbner bases in the way sketched above. The second are pure conversion problems like inverse kinematics in robotics or implicitization in geometric modeling [2, 7, 11], where the input is already a Gröbner basis.

[1]Software-Schule Schweiz, Bern, Switzerland; amrhein@isbe.ch.

[2]Institute for Computer Science, University of Tübingen, Germany; gloor@informatik.uni-tuebingen.de

In this paper, we address the problem of path finding by perturbation. Our examples show that, among the many different paths in a Gröbner fan, some may be much faster than others. Path perturbation is therefore an important issue in practice. A low degree of perturbation leads to large initial Gröbner bases and therefore to heavy Gröbner basis computations in the steps. A perturbation of high degree however produces very large order vectors, which slows down every comparison, and increases the number of steps.

Here, we propose a new algorithm, which is independent of the perturbation degree: the Fractal Walk. With this enhancement of the walk algorithm, we can not only find the best perturbation, we can often obtain speedups over the *fastest* path, as we adapt the perturbation according to the local circumstances.

The paper is organized as follows: Section 2 recalls some definitions about orders and weight vectors, followed by an introduction to the walk algorithm. In Section 3 we introduce the concept of path perturbation and in Section 4 we describe the Fractal Walk algorithm. Comparative timings of our implementation to the usual Gröbner walk as well as to FGLM are presented in Section 5. Finally, we give some conlusions and hints for further investigations; in particular we address the question whether Hilber-driven Gröbner basis conversion can speedup the walk.

2 The Gröbner Walk

2.1 Orders and Weight Vectors

For further theoretical background we refer to [1, 2, 4] and [16]. Throughout this paper let $R = K[x_1, \ldots, x_n]$ be a polynomial ring over an arbitrary field K, and let I be an ideal. The ideal generated by a set of polynomials $G \subseteq R$ is denoted by $\langle G \rangle$. For an admissible term ordering $\prec$, $in_\prec(f)$ is the head monomial of a polynomial f, and $in_\prec(G)$ is the set $\{in_\prec(g) \mid g \in G\}$. By $G(I, \prec)$, we denote a Gröbner basis of the ideal I with respect to $\prec$.

A rational weight vector ω is an element of

$$\{(\nu_1, \ldots, \nu_n) \in \mathbb{Q}^n \mid \nu_i \geq 0 \text{ for } i = 1, \ldots, n\}.$$

For a monomial $t = cx_1^{e_1} \cdots x_n^{e_n}$ we define its *ω-degree* by

$$deg_\omega(t) := \sum_{i=1}^{n} e_i \nu_i, \quad \text{where } \omega = (\nu_1, \ldots, \nu_n).$$

A weight vector ω is *compatible* with a term ordering $\prec$ on G, if for each polynomial $g = m_1 + \cdots + m_s \in G$ ordered in descending order with respect to $\prec$, $deg_\omega(m_1) \geq deg_\omega(m_i)$ holds for all $1 < i \leq s$. We say that the *initial form* of g

$$in_\omega(g) := m_1 + \cdots + m_i, \quad \text{where } deg_\omega(m_1) = deg_\omega(m_i) > deg_\omega(m_{i+1})$$

degenerates with respect to ω if $i > 1$ (i.e., the initial form consists of more than the leading monomial).

Following [16], a regular $n \times n$ matrix A over $\mathbb{Q}_+ = \{q \in \mathbb{Q} \mid q \geq 0\}$ determines a term ordering $\prec := \mathcal{O}(A)$ by

$$t \prec r \;:\Longleftrightarrow\; \bigvee_{i=1}^{n} \left(\bigwedge_{j=1}^{i-1} deg_{A_j}(t) = deg_{A_j}(r) \right) \wedge deg_{A_i}(t) < deg_{A_i}(r).$$

Vice versa, for each term order $\prec$ there is a matrix A that determines $\prec$.

For the walk, we assume that we are given a Gröbner basis $G(I,<)$ with respect to the start order $<$. Our aim is to compute the Gröbner basis of $G(I,\ll)$ with respect to the target order $\ll$. First, we determine the respective order matrices A and B with $\mathcal{O}(A) = <$ and $\mathcal{O}(B) = \ll$. σ and τ are the first rows of A and B and they are compatible (on R) with $<$ and $\ll$, respectively. The path of our walk is along the line segment $\overline{\sigma\tau}$, where we denote the occurring intermediate weight vectors by

$$\sigma =: \omega_1, \;\ldots,\; \omega_k, \;\ldots,\; \omega_m := \tau.$$

During the walk, we gradually change the order matrix from A to B, and hence we successively compute the (reduced) Gröbner bases of I with respect to the orderings $\prec_k$, $k = 1, \ldots m$. These are called the *intermediate* Gröbner bases $G(I, \prec_k)$, which form the stepping-stones for the walk.

2.2 Fans and Cones

Let G be a reduced Gröbner basis of I with respect to a term ordering $\prec$. By $C_\prec(I)$ we denote the set of weight vectors

$$C_\prec(I) \;:=\; \{\omega \in \mathbb{Q}_+^n \mid \omega \text{ compatible with } \prec \text{ on } G\}.$$

$C_\prec(I)$ contains those vectors ω which can be extended to orders that lead to the same reduced Gröbner basis as $\prec$. This is a convex cone in $\mathbb{Q}^n$ with nonempty interior. The set $F(I) := \{C_\prec(I) \mid \prec \text{ a term order}\}$ is called the *Gröbner Fan* of I [12]. By a result from [12], the Gröbner fan of a polynomial ideal has finite cardinality.

As multiples of ω lie in the same cone as ω itself, it suffices to consider the slice of the fan with

$$\{\omega \mid \omega = (\nu_1, \ldots, \nu_n), \sum_{i=1}^{n} \nu_i \;=\; 1\}$$

(cf. Figure 2 for a slice of a sample fan in $\mathbb{Q}^3$).

2.3 The Steps

Here we present the key idea of the original walk algorithm, describing one walking step from ω_{k-1} to ω_k. We refer to [2] for further details.

Given the Gröbner basis $G(I, \prec_{k-1})$, we first determine $in_{\omega_k}(G(I, \prec_{k-1}))$ (cf. Step ❶ in Figure 1). This is a Gröbner basis of $\langle in_{\omega_k}(I)\rangle$ with respect to $\prec_{k-1}$, as ω_k is compatible with $\prec_{k-1}$.

In Step ❷ we compute a reduced Gröbner basis $M = G(\langle in_{\omega_k}(I)\rangle, \prec_k)$ applying the Buchberger algorithm. By abuse of language, we sometimes call this basis of initials an *initial Gröbner basis.*

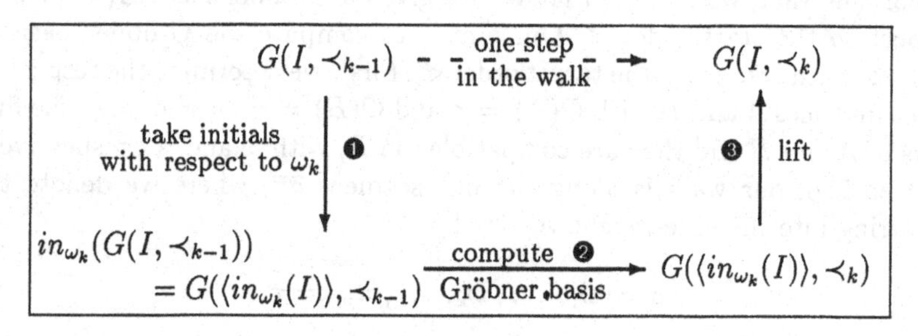

Figure 1: Step k of the Gröbner Walk

In the lifting step ❸, we construct a representation of the new intermediate Gröbner basis $G(I, \prec_k)$ by reducing each polynomial in $M = G(\langle in_{\omega_k}(I)\rangle, \prec_k)$ by the previous Gröbner basis $G(\langle in_{\omega_k}(I)\rangle, \prec_{k-1}) = in_{\omega_k}(G(I, \prec_{k-1}))$. Replacing all occurrences of polynomials $in_{\omega_k}(g_j)$ in their representation by the full polynomials $g_j \in G(I, \prec_{k-1})$, we obtain a new set of polynomials $F := \{f_1, \ldots, f_s\}$, which forms a (minimal) Gröbner basis of I with respect to $\prec_k$.

To finish a step of the walk, it remains to determine the next weight vector, that is, the point on the path where we leave the cone $C_{\prec_k}(I)$ (i.e., some other initial forms of the reduced Gröbner basis $G(I, \prec_k)$ degenerate). To detect a change in the initial forms, we determine the first weight vector $\omega(t) := \omega_k + t(\tau - \omega_k)$, $0 < t \leq 1$, on the directed path $\overline{\omega_k\tau}$ with the following property.

$$t := \min(\ \{s \mid deg_{\omega(s)}(p_1) = deg_{\omega(s)}(p_i),\ g = p_1 + \ldots + p_n,\ g \in G(I, \prec_k)\} \cap (0,1]) \tag{1}$$

In this way, we successively compute the Gröbner bases and weight vectors

$$G(I,<) = G(I, \prec_0),\ \omega_1,\ G(I, \prec_1),\ \ldots,\ \omega_m,\ G(I, \prec_m) = G(I, \ll).$$

The walk is finished as soon as $\omega(t) = \tau$ or if there is no $t \leq 1$ anymore, hence $in_{\prec_k}(G) = in_{\ll}(G)$.

3 Path Perturbation

The path of the walk starts at σ and ends at τ. More precisely, given the reduced Gröbner basis $G(I, <)$, we start at the cone $C_<(I)$ and aim for the cone $C_\ll(I)$, where σ and τ belong to $C_<(I)$ and $C_\ll(I)$, respectively. However, we can start the walk anywhere in $C_<(I)$ and even take detours through other cones before we reach τ (or any point in $C_\ll(I)$). This insight, although rather obvious, turns out to be crucial in practice. Naive paths tend to be rather slow, and the walk gains very significant speed advantages through careful path planning [1].

3.1 Path Finding

In the original walk, path finding is a very important issue, as, among the many paths from $C_<(I)$ to $C_\ll(I)$ in a Gröbner fan, some may be computationally much faster than others.

Whenever a path leaves a cone of the Gröbner fan, some of the initial monomials of the Gröbner basis degenerate with respect to the weight vector ω. At intersection points of several cones, several monomials in a polynomial have the same maximal weight and generate long initial forms, or initial forms of several polynomials degenerate.

Especially in a complicated fan, meeting-points where several cones adjoin are frequent. Moreover, if the chosen path in the walk moves along the intersection of two or even more cones, there are monomials which keep the same maximal weight, and therefore remain in the initial form of a polynomial on this line. Hence, the initial forms become unnecessarily heavy on such a path.

3.2 Global Path Perturbation

We can avoid walking through meeting-points or along arbitrary dimensional faces of cones if we slightly perturb the starting point (i.e., the starting weight vector σ) and the end-point (i.e., the target weight vector τ) of the Walk, providing we stay in their cones. Then the walk passes through a sequence of maximally adjacent cones, avoiding intersection points. However, we may have to take many more steps since we have to walk through more cones on the perturbed path.

Figure 2 shows a slice of a sample fan of an ideal in three variables (x, y, z) as intersection with the plane $x + y + z = 1$. Path segment ① goes through a common edge of three cones, path segment ② runs along a surface of two cones.

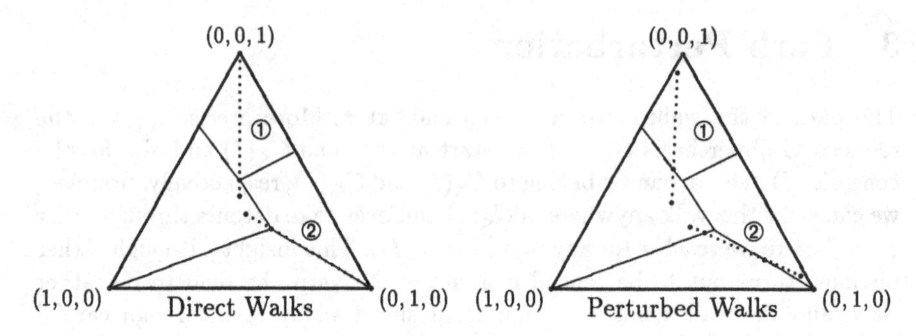

Figure 2: A Slice of a Gröbner Fan

3.3 Computation of Weight Vectors

As already pointed out, with each ordering $\prec$ we can easily associate a regular $n \times n$-matrix A over $\mathbb{Q}_+$ that determines this ordering. Obviously, we can even choose A over $\mathbb{N}$ by multiplying each row with the least common multiple of the denominators of this row. The cone $C_\prec$ of a reduced Gröbner basis G with respect to $\prec$ contains the first vector A_1 of A. Now, if we perturb this vector slightly by ϵA_2, we remain in $C_\prec$, provided that ϵ is sufficiently small, that is, $1/\epsilon > tdeg(p) \cdot \max(A_2)$ for each p in G, where

$$tdeg(p) := \max\left\{\sum_{i=1}^{n} e_i \mid x_1^{e_1} \cdots x_n^{e_n} \text{ a term of } p\right\}.$$

Then, $deg_{\epsilon A_2}(p) < 1$, which asserts that for all p in G the initial form $in_{A_1+\epsilon A_2}(p)$ is a subterm of $in_{A_1}(p)$ (i.e., no new monomials enter the initial forms). Therefore, A_1 keeps the priority in the determination of the $(A_1+\epsilon A_2)$-degree.

However, if A_1 is at a vertex of the fan's slice (i.e., an edge and hence a one-dimensional face of the cone), and ϵ is small enough, then $A_1 + \epsilon A_2$ is in a face of the cone of dimension at least two.

To reach a face of dimension at least three, we additionally perturb by the third vector, resulting in a vector $A_1 + \epsilon A_2 + \epsilon^2 A_3$. Provided $1/\epsilon > tdeg(p) \cdot (\max(A_2) + \max(A_3))$ for each p in G, we move to a location contained in a three-dimensional open set of the cone.

For n variables, we can extend this procedure up to any $k \leq n$ and obtain a perturbed vector

$$A_1 + \epsilon A_2 + \epsilon^2 A_3 + \cdots + \epsilon^{k-1} A_k$$

of *degree* k (see Figure 3). It lies in a face of the cone of dimension at least k. For $k = 1$ we obtain the unperturbed weight vector. If $k = n$, we obtain a *maximally perturbed* weight vector that lies *within* the cone (and hence belongs to only one cone).

Given a reduced Gröbner basis with respect to a start ordering, an ϵ satisfying the conditions for the perturbation of the start vector can easily be determined. However, to obtain a perturbed target vector, we have to guess such an ϵ and check the validity at the end of the Walk. Note that the previous steps of the algorithm remain correct even if the ϵ becomes invalid. In this case we have to take some further steps (closer to the unperturbed target) as we did not reach the target ordering yet.

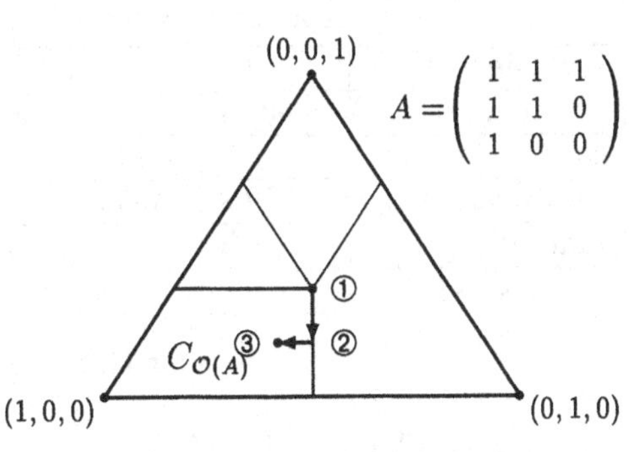

Figure 3: First, Second, and Third Degree Perturbation

3.4 Practical Influence of Perturbation Degrees

The perturbation degrees of the start and target vector are *a priori* independent and can be combined in any way. However, if we perturb either the start or the target vector by degree k, we usually[3] omit vectors on faces of dimensions $k' < k-1$. As the last step often crosses a meeting point of several cones, it is useful to perturb the target vector. In the Fractal Walk we usually perturb the target vector only.

We will speak of a *direct path/walk* to indicate no perturbation and of a *maximally perturbed path/walk* to indicate maximally perturbed start and target vectors.

Experience shows that in general a small perturbation degree leads to fewer steps with larger initial forms, whereas a large perturbation degree leads to more steps with smaller initial forms. The advantage of the maximally perturbed path is that the initial Gröbner basis computations take very little time. The main disadvantages are that very large weight vectors may occur (especially if there are many variables) and that the walk may consist of many more steps.

Table 1 shows some typical timings for walks with different perturbation degrees. We see that a maximally perturbed path is often not the best choice. A walk with a low perturbation degree usually failed (did not terminate within one hour, or aborted due to memory overflow).

In the next section, we show how the Fractal Walk removes the problem of finding the best perturbation degree.

[3]It may happen that the degree is lowered if the path direction coincides with one of the perturbation directions or a linear combination thereof.

Walk from grevlex to lexicographic term ordering

Pert. degree:	1	2	3	4	5	6	7
fateman (4 vars)	> 500 s	29 s	23 s	48 s			
cohn-2 (4 vars)	> 1 h	3098 s	11 s	16 s			
s6 (6 vars)	70 s	495 s	1.7 s	2.4 s	8.5 s	4.4 s	
jhd6 (6 vars)	> 1 h	> 1 h	> 1 h	3977 s	19 s	19 s	
s7 (7 vars)	> 1 h	> 1 h	> 1 h	> 1 h	3678 s	28 s	22 s

Table 1: Timings for Different Perturbation Degrees

4 The Fractal Walk

4.1 The Design

Let us consider a walk on a direct or only slightly perturbed path. We further assume a big step (caused by long initial forms) that would split into many small steps in the corresponding maximally perturbed path. In this case, the computation of the initial Gröbner basis ❷ consumes most of the time of this step.

As already mentioned in [4], this transformation of the initial basis $G(\langle in_\omega(I)\rangle, <)$ into the basis $G(\langle in_\omega(I)\rangle, \ll)$ (where ω is compatible with both $<$ and $\ll$) can be performed by any basis conversion algorithm, since the initials $in_\omega(G)$ form a Gröbner basis. Therefore, instead of applying Buchberger's algorithm to compute initial Gröbner bases, we may as well apply the walk.

If we stay on the direct path $\overline{\sigma\tau}$, we stumble again at ω. Taking initials with respect to ω once again does not shorten the polynomials of $in_\omega(G)$ any further. Therefore, we elude this stumbling block and perturb ω to ω' (w.r.t. $<$) in the expectation that $in_{\omega_1}(in_\omega(G))$ contains shorter polynomials than $in_\omega(G)$ (see Fig. 4). In other words, we walk locally on a perturbed path, starting with $G(\langle in_\omega(I)\rangle, <)$ at ω' and finishing with $G(\langle in_\omega(I)\rangle, \ll)$ at ω'' (where ω'' is the perturbation of ω with respect to $\ll$).

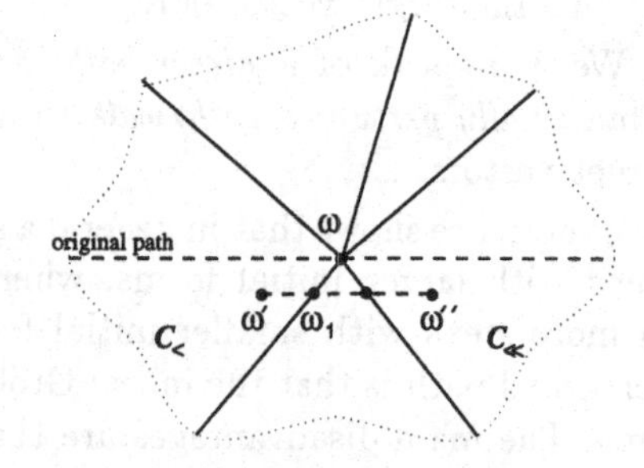

Figure 4: Local Path Perturbation for Initial Gröbner Basis Compuation

What is the difference of this approach of taking initials twice to the usual (perturbed) walk from ω' to ω''? Obviously, we take the same number of steps. However, the Fractal walk runs in $in_\omega(G)$ rather than in G. Therefore, the path is on a "lower" level (or in a "tunnel"). As the stepping-stones are now ideal bases of $\langle in_\omega(G)\rangle$, we avoid the lifting to bases of the full ideal $I = \langle G\rangle$

as long as possible (until we reach ω''). Hence, the interreductions after lifting become substantially simpler as they are on a lower level.

This method can now be applied recursively with increasing degrees of perturbation. We start with an unperturbed walk (perturbation degree 1). Each perturbation degree leads to a recursion level. At each level, we perform a walk in which each initial Gröbner basis is now converted by a recursive call instead of by Buchberger's algorithm. The deeper the recursion is, the higher the degree of perturbation and hence the shorter the initials of the Gröbner bases become. At the bottom, we perform a maximally perturbed walk.

Figure 5 shows the stepping-stones for the Fractal Walk. You can even read off the stepping-stones of the Gröbner walk for any perturbation degree.

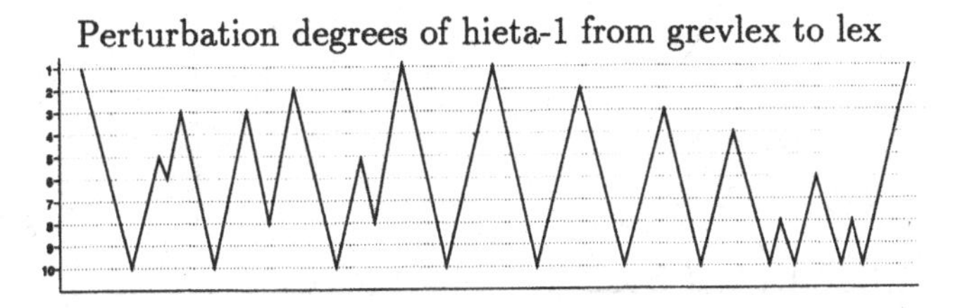

Figure 5: The Trace of a Fractal Walk

4.2 The Algorithm

The main difficulty in developing an algorithm from the design above is the appropriate perturbation of the (intermediate) weight vectors. The question is how we compute ω' and ω'' from ω. Perturbing ω in a naive way with respect to $<$ and $\ll$, respectively, can lead to perturbations which are diametrically opposed. Then, the path runs through the same edge/face of the fan and does not split into smaller steps; that is, the perturbation is useless.

The problem here is to guarantee that $\omega' \in C_<$ and $\omega'' \in C_\ll$ have a perturbation degree increased by one. This could be ensured by Gaussian elimination involving ω and the order matrices for $<$ and $\ll$. As there may be hundreds or thousands of steps (and the same number of computations of such perturbations), it is important to minimize the effort for these computations.

The following observation provides us with a simple way to maintain ω' and ω'' with the help of global perturbations.

Theorem (Collart, Mall)[6]: *Given an ideal I and a weight vector ω. The fan $F(\langle in_\omega(I)\rangle)$ of the toric degeneration $\langle\{in_\omega(g) : g \in I\}\rangle$ (of I with respect to the admissible partial order induced by ω) is the geometric localisation $Loc(F)_\omega$ of the fan $F(I)$ at ω.*

In other words, we obtain the fan of the initial ideal $\langle in_\omega(I)\rangle$ by blowing up a small neighbourhood of ω to the whole $\mathbb{Q}^n_+$. This process removes all surfaces and edges of the fan that do not contain ω (see Fig. 6). (In the special case that ω lies inside a cone, this cone is blown up to the whole fan; for $\omega = (0,\ldots,0)$ we obtain the original fan again.)

Therefore, we are free to choose ω' as a perturbation of σ, and ω'' as a perturbation of τ. Most importantly, we can determine all perturbations at the very beginning of the algorithm; they remain the same for all steps (provided the chosen ϵ is small enough—the validity of ϵ has to be checked at the end of each recursion step).

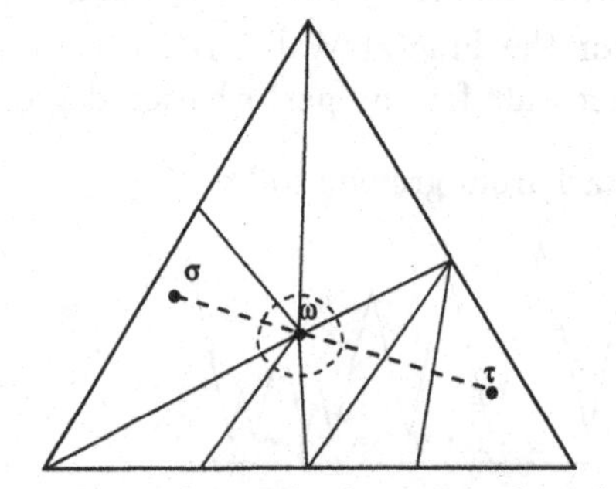

The Fan of I with the original path

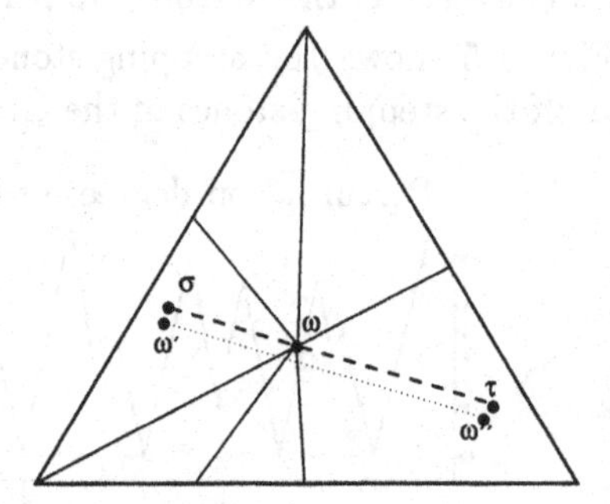

The Fan of $\langle in_\omega(I)\rangle$ and the locally perturbed path

Figure 6: Using Globally Perturbed Vectors for Local Perturbations

We first assume that we do not have to transform the Gröbner basis at the starting point[4] (we will treat the very first step at σ later). Then, the perturbation of only the target is needed (see Section 3.4). This prevents us from using diametrically opposed perturbations that cancel each other and are therefore useless. Hence, we provide a global list $(\tau = \tau^1, \ldots, \tau^n)$ of perturbations of τ with respect to $\ll$. In level p of the recursion, we assign $\omega' = \sigma$ and $\omega'' = \tau^p$.

In Figure 7 we present the principal algorithm (assuming that we do not have to perform a step at σ). The expression "$\prec$ `= order(`ω`, `B`)`" states that the ordering $\prec$ is determined first by the weight vector σ and then by the order matrix B. In ② we compute the next (first) border. t is determined according to Equation (1) in Section 2.3. This level is finished and we return G if there is no $t \leq 1$ and the chosen ϵ is still valid (that is, the precomputed perturbed target vectors $\tau^1, \ldots, \tau^p$ are still valid). Whenever we reach the maximal perturbation degree (p `==` n in ④), we apply a specialized Buchberger algorithm ⑤ (see [2]) to compute the initial Gröbner basis. Otherwise, we perform a recursive call with an increased perturbation degree ⑦.

[4]This is the case whenever σ already determines the heads of the polynomials of the starting Gröbner bases, i.e., $\langle in_\sigma(I)\rangle$ is a monomial ideal.

```
Fractal_Gröbner_Walk
Input:          Orders < = O(A) and ≪ = O(B)
                G ⊂ Q[x_1,...,x_n]: reduced Gröbner basis with resp. to <
                starting weight vector σ (first row of matrix A)
                perturbed target weight vectors τ^1,...,τ^n (comp. from B)
Function Call:  G = fractal(G, 1);
Output:         Gröbner basis G with respect to ≪

fractal(G, p):
① ω = ω' = σ;                           // Local starting weight vector
  ≺ = order(ω, B);                      // Local starting order
  ω'' = τ^p;                            // Local target weight vector
② t = determine_border(ω, ω'', G);      // Determine the next parameter
  if( undefined(t) ) then
    if( check(G,τ^1,...,τ^p) ) then     // Check validity of ε
      return(G);                        // No further conversion needed
    update( τ^1, ..., τ^n, t);          // Update perturbation vectors
  ω = (1-t)*ω + t*τ^p;                  // Determine the next border
  ≺^+ = order(ω, B);                    // Determine the next order
③ G_ω = initials(G, ω);                 // Take initials of G
④ if( p == n ) then                     // Maximal recursion depth
⑤     G_ω^+ = init_gb(G_ω, ≺^+);        // Compute interreduced GB
⑥     G_ω^+ = interreduce(G_ω^+, ≺^+);  //     of initials by Buchberger
  else                                  // Recursion step
⑦     G_ω^+ = fractal(G_ω, p+1);
⑧ G = lift(G_ω^+, ≺^+, G_ω, G, ≺);      // Lift G_ω^+ to a full Gröbner basis
                                        //   with respect to ≺^+
⑨ G = interreduce(G, ≺^+);              // Interreduce G
  ≺ = ≺^+; Goto ②;
```

Figure 7: The Fractal Walk

4.3 First Step

As already mentioned, the algorithm presented in Figure 7 covers all steps outside of σ. If we have to perform a step in σ at the top level, we start with *first step mode*, and we have to perturb the starting and target vectors. That is, we choose $\omega' = \sigma^p$ and $\omega'' = \tau^p$ in level p. We leave first step mode as soon as we reach the top level again.

4.4 Algorithm Refinements

As a result of the following algorithmic refinements, we were able to solve the original problem how to choose the perturbation degree. For a broad variety of examples, the Fractal Walk is a reliable method: it does not always run faster than the Gröbner walk with the best perturbation degree, but it is never much slower.

4.4.1 Last Step

Usually we are interested in walks from total degree Gröbner bases to elimination Gröbner bases. Let us have a look at the case of a lexicographic target. Then, we have the following order matrix and perturbations:

$$B_{lex} = \begin{pmatrix} 1 & 0 & \dots & 0 \\ 0 & 1 & & \vdots \\ \vdots & & \ddots & 0 \\ 0 & \dots & 0 & 1 \end{pmatrix} \qquad \begin{aligned} \tau^1 &= (1, 0, \dots, 0) \\ \tau^2 &= (1, \epsilon, 0, \dots, 0) \\ \tau^3 &= (1, \epsilon, \epsilon^2, 0, \dots, 0) \\ &\vdots \end{aligned} .$$

Whenever we approach a lexicographic target, one variable overrules the others and is therefore often eliminated in many polynomials of the Gröbner basis. Taking initials with respect to τ^1 does not affect those polynomials that do not contain the first variable, that is, they remain unshortened at ③. Therefore, taking initials and—much more importantly—lifting at ⑧ and interreducing at ⑨ are essentially useless, as all the real conversion effort is done in the recursive call.

We solve this problem by omitting ③ in the last step of the walk. More precisely, whenever we reach `t == 1` and hence `ω == τ` in ②, we only perturb the path to the next degree (that is, we perform a recursive call) and dispense with taking initials (and lifting/interreducing after the call).

4.4.2 Look Ahead Recursion Elimination

Up to now, we applied a kind of blind recursion. That is, regardless of the size of the initials, we performed recursive calls and applied basis conversions only at the bottom level of the recursion. However, paths already of perturbation degree two or three often use stepping-stones that connect maximally adjacent cones (i.e., the initials do not shorten further up to maximal perturbation, and there is no further simplification with recursive calls). Therefore, we better apply the Buchberger algorithm here instead of going down further levels of recursion. For that, we have to install a further criterion at ④. In our current implementation of this criterion, we already apply Buchberger's algorithm as soon as there are no initial forms that degenerate to more than two monomials. (Whenever we start a Gröbner basis computation with monomials and binomials only, the result also consists of monomials and binomials only.)

In Figure 8 we show the trace of the *Look Ahead Fractal Walk* of the same example as in Figure 5.

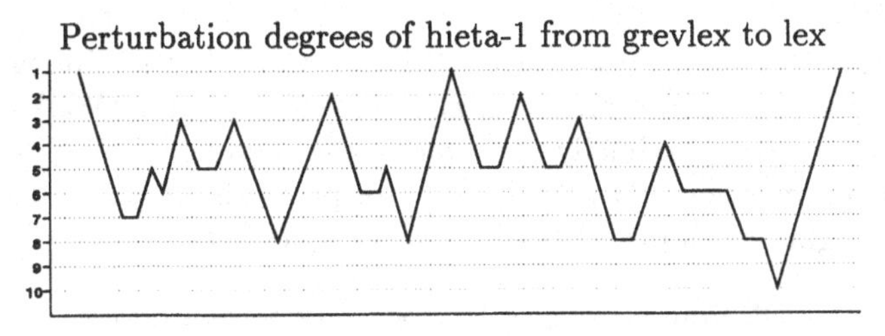

Figure 8: A Trace of a Look-Ahead Fractal Walk

4.5 Implementation

Although we compute Gröbner bases over $Q[x_1, \ldots, x_n]$, all computations were done in $Z[x_1, \ldots, x_n]$.

Like the Gröbner walk, we implemented the Fractal Walk in C. The Gröbner walk, as described in [2], was based on SACLIB [3]. We recently replaced SACLIB in our installation by a sequential variant of our own core system [9] using the GNU multi-precision package [10].

In addition to the implementation techniques described in [2], in particular for the Fractal Walk we have to use a special data structure to support the lifting step. Whenever we take initials, we also store a pointer to the full polynomial. In the lifting step (❸), we reduce each polynomial with respect to the old initial basis; in each reduction step we use the pointer to the full polynomial to add up the corresponding share of the full polynomial. In this way we easily obtain the lifted polynomial.

5 Empirical Results

The given timings of the Fractal Walk are those obtained by the implementation with all algorithm refinements. For the presented timings, we used a 140 MHz UltraSPARC[5] under Solaris 2.5 with 32MB RAM.

5.1 Comparison with the Gröbner Walk

Our results show that we succeeded in our effort to remove the need for choosing perturbation degrees in the Gröbner walk. In many cases we even gain

[5]The current compiler does not support 64-bit arithmetic.

over the best Gröbner walk timings. This can be explained by the "tunneling" effect of the Fractal Walk, as the lifting and interreduction steps become smaller. In the worst case, the Fractal Walk shows some overhead over the best walk. However, up to now we did not find any example for which the Fractal Walk is significantly slower than the Gröbner walk with the best perturbation degree.

The second column (Walk 2) of Table 2 contains the timings for the Gröbner walk with perturbation degree 2, the third column (Walk ∞) shows the timings for the maximally perturbed walk, and the fastest walk times (with the optimal perturbation degree) are presented in column four. The examples originate from the PoSSo [13] examples list.

Walk from grevlex to lexicographic term ordering

	Walk 2	Walk ∞	best Walk	(pert.deg.)	Fractal
cassou	1.8 s	0.15 s	0.13 s	(3)	0.15 s
fateman	29 s	63 s	24 s	(3)	25 s
hieta-1	3.1 s	0.62 s	0.62 s	(10)	0.82 s
jhd6	> 1 h	22 s	17 s	(5)	13 s
katsura5	> 1 h	27 s	27 s	(5)	35 s
cohn-2	3070 s	17 s	12 s	(3)	10 s

Table 2: Timings of the Fractal Walk vs. Gröbner Walk

Table 3 shows the timings of the implicitization of Bézier surfaces. The fastest Gröbner walk was always the maximally perturbed one. Walks with a perturbation degree lower than four usually did not terminate due to memory overflow. As the polynomials have five variables, the maximal perturbation degree is five. In some of these examples (listed in the appendix), the Fractal Walk is even faster than the best Walk.

	Walk 4	Walk 5	Fractal	Output Statistics		
B1	> 1 h	285 s	289 s	32	10,230	629 kB
B2	96 s	42 s	62 s	31	6,896	339 kB
B3	170 s	148 s	105 s	41	12,871	951 kB
B4	2495 s	140 s	137 s	32	8,478	448 kB
B5	153 s	48 s	55 s	34	8,421	367 kB

Table 3: Implicitization Timings

The last three columns show the number of polynomials, the number of monomials and the size (in kilobytes) of the result.

5.2 Comparison with FGLM

For zero-dimensional ideals[6] the timings of the Gröbner walk can be compared with the FGLM basis conversion method as presented in [8].

Table 4 gives a rough idea of the relative speeds of the Fractal Walk, the Gröbner walk with optimal perturbation degree, and the FGLM method. We received the timings of FGLM from an unknown referee. Although we do not know anything about how these timings were obtained, we attempt a comparison. In the second and in the last column, we give some statistics for the graded reverse lex and the lexicographic Gröbner bases of the examples: the number of polynomials, the number of monomials and the memory space (in kilobytes) needed to store the bases. For details of the examples see [1].

Walk from grevlex to lexicographic term ordering

	Input Statistics			best Walk	Fractal	FGLM	Output Statistics		
Ex1	10	133	8.2 kB	0.97 s	0.86 s	4.2 s	4	135	11.3 kB
Ex2	15	380	5.8 kB	6.5 s	5.8 s	11.8 s	5	182	38.1 kB
Ex3	15	477	7.4 kB	4.9 s	4.0 s	12.9 s	8	268	43.0 kB
s6	38	233	2.7 kB	1.7 s	1.8 s	0.54 s	17	85	1.2 kB
s7	64	622	8.2 kB	22 s	8.3 s	7.7 s	8	82	6.3 kB

Table 4: Timings of Fractal Walk vs. Gröbner Walk and FGLM

For our zero-dimensional ideals, the timings of the Walk with optimal perturbation degree, the Fractal Walk, and the FGLM basis conversion algorithm are in the same range.

6 Conclusions and Future Work

Our investigations show that the Fractal Walk is an important algorithmic enhancement of the Gröbner walk. It solves the problem of finding the obtimal path perturbation degree and often leads to additional speedups because of the "tunneling" effect.

We see three points closely related to the Fractal Walk which deserve further investigation.

- The criterion at ④ for *look ahead recursion* can certainly be improved. One example which should be covered is the case of inherent initial forms such as $x^2 + xy + y^2$ (whenever the initial form contains the first two terms, it comprises the third as well). In this case, the criterion fails although the third monomial can not be separated from the others.

[6] There are ways to lift this restriction of FGLM in special cases.

- The Fractal Walk exhibits similarities with the *Gröbner Stripping* algorithm [5]. These similarities deserve further investigation.
- Especially in the elimination case we can try to eliminate the variables one after the other, walking on an indirect path of the form $(1,1,\ldots,1)$, $(1,1,\ldots,1,0)$, $(1,1,\ldots,1,0,0)$, $\ldots$, $(1,0,\ldots,0)$, eliminating one variable at each edge of the path. Up to now, we did not find any examples where such a path is faster than the direct path. The reason could be that Gröbner bases often become very large at the border of the fan.

Poly–Algorithms

One promising approach to improve basis conversions and even general Gröbner basis computations seems to be the combined use of different algorithms.

As already mentioned in [4], we could also apply the FGLM-algorithm to convert the initial bases in the different steps of the walk. The initial ideal is zero-dimensional if and only if the original ideal is zero-dimensional (for weight vectors with nonzero components). Gröbner bases which belong to adjacent cones are usually very similar. The FGLM algorithm can not profit from this fact as much as the initial Gröbner basis computation algorithm we use for the walk. Therefore, we doubt whether it is possible to speedup the Gröbner walk by using FGLM.

A further, rather obvious refinement of the walk could be to apply the Hilbert function to predict zero reductions. To investigate whether this approach may be fruitful, we measured the time we need for the zero reductions. In Figure 9, we show an average case of our examples; as you can see, the time needed for zero reductions is almost zero in the optimally perturbed walk. (In fact, most of the time is spent for lifting and interreductions, see also [2].) Moreover, the optimal perturbation degree is independent whether we count the zero reduction time or not. Therefore, the gain we can expect by applying the Hilbert function can not exceed the loss we cause by not choosing the optimal perturbation degree.

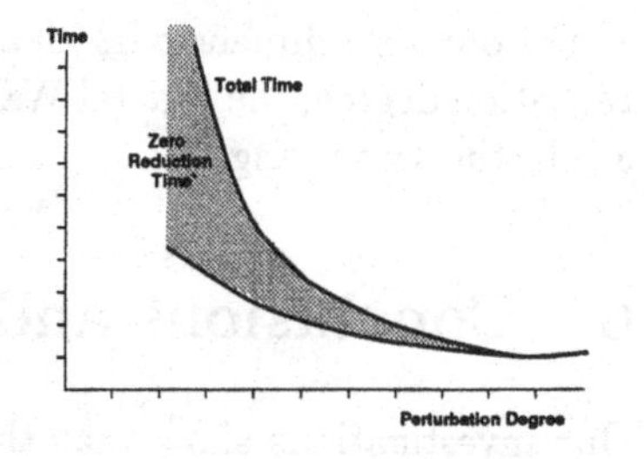

Figure 9: The Share of the Zero Reduction Time Dependent on the Perturbation Degree

However, it is still an open question, whether there are other ways to profitably apply the Hilbert function in the Gröbner walk algorithm.

Acknowledgements

We would like to thank Michael Kalbrener for his comments on an earlier version of this paper. This work is based upon research supported by grant Ku 966/2 from Deutsche Forschungsgemeinschaft (DFG).

References

[1] B. Amrhein, O. Gloor, and W. Küchlin. Walking faster. In J. Calmet and C. Limongelli, editors, *Design and Implementation of Symbolic Computation Systems*, volume 1128 of *LNCS*, pages 150–161, Karlsruhe, Germany, Sept. 1996. Springer-Verlag.

[2] B. Amrhein, O. Gloor, and W. Küchlin. On the Walk. *Theoretical Comput. Sci.*, 187, 1997. In print.

[3] Buchberger, Collins, Encarnación, Hong, Johnson, Krandick, Loos, Mandache, Neubacher, and Vielhaber. SACLIB User's Guide, 1993. On-line software documentation.

[4] S. Collart, M. Kalkbrener, and D. Mall. Converting bases with the Gröbner Walk. *J. Symbolic Computation*, 24(3–4):465–469, 1997.

[5] S. Collart and D. Mall. The ideal structure of Gröbner base computations. Preprint, 1994.

[6] S. Collart and D. Mall. Toric degenerations of polynomial ideals and geometric localization of fans. *J. Symbolic Computation*, 24(3–4):443–464, 1997.

[7] D. Cox, J. Little, and D. O'Shea. *Ideals, Varieties, and Algorithms.* Undergraduate Texts in Mathematics. Springer-Verlag, 1992.

[8] J. Faugère, P. Gianni, D. Lazard, and T. Mora. Efficient computation of zero-dimensional Gröbner Bases by change of ordering. *J. Symbolic Computation*, 16:329–344, 1993.

[9] O. Gloor and S. Müller. PARCAN—a parallel computer algebra nucleus. In preparation, 1998.

[10] T. Granlund. *The GNU MP Manual.* Free Software Foundation, Boston, 1996. Edition 2.0.2.

[11] C. Hoffmann. *Geometric and Solid Modeling: An Introduction.* Morgan Kaufmann Publishers Inc., 1989.

[12] T. Mora and L. Robbiano. The Gröbner Fan of an ideal. *J. Symbolic Computation*, 6:183–208, 1988.

[13] PoSSo. Polynomial systems library. `ftp: posso.dm.unipi.it`.

[14] A. Steel. The Magma Gröbner Walk, 1996. Preprint.

[15] C. Traverso. Hilbert functions and the Buchberger Algorithm. *J. Symbolic Computation*, 22(4):355–376, Oct. 1996.

[16] V. Weispfenning. Admissible orders and linear forms. *ACM Sigsam Bull.*, 21(2):16–18, 1987.

Appendix

	Bézier Surfaces in Parametric Form
B1	$x = -2 + 3u + 3u^2 - 3u^3 - 2u^2v + u^2v^2$ $y = -4 + 6u^2 - 3v^2 - u^2v + 2u^2v^2$ $z = 3u + 3u^2 - 6vu^3 + 3v - uv - 9u^2v - v^2$
B2	$x = 2 + 3u + 3u^2 - 6u^3 + 3v - uv - 9u^2v - v^2$ $y = -2 + 6u^2 - 3v - u^2v + 2u^2v^2$ $z = 2 - 3u - 3u^2 + 3u^3 + 2u^2v - u^2v^2$
B3	$x = 3 - 2u + 2u^2 - 2u^3 - v + uv + 2u^2v^3$ $y = 6u + 5u^2 - u^3 + v + uv + v^2$ $z = -2 + 3u - uv + 2uv^2$
B4	$x = 2u^2 - uv^2 + 6uv^3$ $y = 3u + 3uv - u^2v + 4uv^2 + v^2$ $z = u^3 + 4u^2v - 3uv + v$
B5	$x = 2u - 2u^2 + u^3 - 2uv + u^2v$ $y = u - 4u^3 + 3v - 2uv + u^2v - 6v^2 + uv^2 + 3v^3 - uv^3$ $z = 3 - 5u + u^2 - 2v + uv^2$

Gröbner Bases Property on Elimination Ideal in the Noncommutative Case

Miguel Angel Borges and Mijail Borges

Department of Mathematics, Faculty of Sciences, University of Oriente, Santiago of Cuba 1-90500, Cuba; mborges@cnm.uo.edu.cu

Abstract

Some generalizations to noncommutative algebras of the Gröbner bases property on elimination ideal are discussed here, together with their application to determine Gröbner presentations, independent polynomials modulo an ideal, algebra membership, and algebra equality. In addition, it is also shown how it may be possible to compute the normal closure of a subgroup by means of Gröbner presentations.

1 Introduction

The notion of Gröbner basis and related techniques are remarkable contributions to the solution of problems by algorithmic way in polynomial ideal theory. The greatest number of applications of the Gröbner bases method have been, for the time being, in commutative polynomial algebras. In Mora F. (1986), the concept of Gröbner basis has been generalized to noncommutative free algebras, allowing to extend the field of applications. Many Gröbner bases results, for the commutative case, are based on the well known Elimination Ideal Property of these bases; Buchberger B. (1987), Gianni P. et al. (1988), Shannon D. et al. (1988), Kalkbrener M. (1991), are some examples. In this paper we show how this property can be gene- ralized to free associative K-algebras and, by means of some applications, illustrate that this generalized property could become similarly applicable as its starting point in commutative polynomial rings.

We assume the reader to be familiar with some basic knowledge on Gröbner bases theory, which can be achieved by studying the surveys Buchberger B. (1985) and Mora T. (1994). However, trying to be somewhat self-contained, we present in section 2 a minimal background of this theory. In section 3, we exhibit the Elimination Ideal Property generalization (theorem 3.2). Theorem 3.3, similar to 3.2, deals with the reduced Gröbner basis determined by an ideal when some indeterminates are eliminated, but 3.3 does not require any additional property of the term ordering. Lastly, we devote section 4 to applications; being more precise, let us introduce some usual notation in order to give a first version of the central problems that are studied there: Let $X := \{x_1, \ldots, x_n\}$ be a finite alphabet, $\langle X \rangle$ the free monoid on X, let K be a field, $K\langle X \rangle$ the free associative K-algebra on X. On the other hand, let $K\langle X \rangle / I$ be the residue class algebra of $K\langle X \rangle$ modulo the two-sided ideal I and let $[f]_I$ be the residue class of $f \in K\langle X \rangle$ modulo I. Let $F := \{f_1, \ldots, f_m\}$ be a subset of $K\langle X \rangle$ and let $Ideal(F)$ be the two-sided ideal of $K\langle X \rangle$ generated by F. Finally, let $K\langle [F]_I \rangle := K\langle [f_1]_I, \ldots, [f_m]_I \rangle$ be the K-subalgebra of $K\langle X \rangle / I$ generated by the classes of equivalence of F. Then, the main problems are:

Problem 1:

Given $K\langle X \rangle / I$, $F := \{f_1, \ldots, f_m\} \subset K\langle X \rangle$, $f(x_1, \ldots, x_n) \in K\langle X \rangle$:

- 1.1: Find a Gröbner presentation for $K\langle [F]_I \rangle$ (4.2).
- 1.2: Decide whether:
  - 1.2.1: $f([X]_I) := f([x_1]_I, \ldots, [x_n]_I) \in K\langle [F]_I \rangle$ (4.3.i).
  - 1.2.2: $K\langle [F]_I \rangle = K\langle X \rangle / I$ (4.3.iii).

Remarks.

- We consider (as it is customary) that $K\langle X \rangle / I$ is given, from an algorithmic point of view, when a finite generating set of I is known.
- The notion of Gröbner presentation will be introduced in 4.1.

Problem 2:

Given G, a finitely presented group;
W, a finite subset of words in the generating set of G:
Find the normal closure of the subgroup of G generated by W (4.6).

2 Preliminaries

Let us consider a term ordering on $\langle X \rangle$, that is to say, a total ordering such that, for all $s, t, u \in \langle X \rangle$: i) $1 \leq s$. ii) $t < u$ implies $st < su$ and $ts < us$. Then, it is usual to define:

- For $f := \sum_1^m c_i s_i$, where $c_i \in K \setminus \{0\}$, $s_i \in \langle X \rangle$, and $s_m <, \cdots, < s_1$:
  - the maximal term of f with respect to $<$ is $T_<(f) := s_1$,
  - the leading coefficient of f w.r.t. $<$ is $C_<(f) := c_1$,
  - the rest of f w.r.t. $<$ is $Rest_<(f) := f - C_<(f)T_<(f)$.
- The set of maximal terms of F w.r.t. $<$ is:

$$T_<(F) := \{T_<(f) \mid f \in F \setminus \{0\}\}.$$

When there is no risk of confusion, the symbol $<$ is omitted from the above notation; in these cases it is only said, for example, maximal term of f, i.e. one eliminates the explicit reference to the term ordering. This convention will also be used in subsequent notations and definitions.

2.1. Theorem. Let $R(I)$ be the K-vector space whose basis is $B(I) := \langle X\rangle \setminus T(I)$. Then the following holds:

i. $K\langle X\rangle = I \oplus R(I)$ (this sum is considered as a direct sum of vector spaces).

ii. For each $f \in K\langle X\rangle$ there is a unique polynomial of $R(I)$, denoted by $Can(f, I)$, such that $f - Can(f, I) \in I$; moreover:

- $Can(f, I) = Can(g, I)$ if and only if $f - g \in I$.
- $Can(f, I) = 0$ if and only if $f \in I$.

iii. For each $f \in K\langle X\rangle$, $T(Can(f, I)) \leq T(f)$.

iv. There is a K-vector space isomorphism between $K\langle X\rangle/I$ and $R(I)$ (the isomorphism associates the class of f modulo I with $Can(f, I)$).

v. If the multiplication in $R(I)$ is defined as follows: $f \odot g := Can(fg, I)$, then the above isomorphism can be extended to an isomorphism between K-algebras.

Gröbner bases are not present in this theorem; however, they can provide, when they are finite sets, an effective procedure for computing $Can(f, I)$ (cf. theorem 1.8 in Mora T. (1994).

2.2. Definition. $G \subset I \setminus \{0\}$ is called a Gröbner basis of I if $T(G)$ generates $T(I)$, i.e., if every $t \in T(I)$ is a multiple of some $s \in T(G)$.

2.3. Definition. A Gröbner basis G of I is called reduced if, for every $g \in G$:

i. $T(g)$ is not a multiple of any $s \in T(G) \setminus \{T(g)\}$.

ii. $C(g) = 1$.

iii. $g = T(g) - Can(T(g), I)$.

Comments about reduced Gröbner bases.

- For each term ordering $<$ there is only one reduced Gröbner basis of I with respect to $<$; it will be denoted by $rGb(I, <)$, or simply $rGb(I)$. We shall also use the notation $rGb(F)$ for $rGb(Ideal(F))$.
- For many applications, it is enough to use any G-basis, in other cases the rGb is necessary. However, the latter is preferable since it is unique and also has the minimal quantity of polynomials required to be a G-basis with respect to the given term ordering.

- Moreover, as it has been already said by Buchberger (Buchberger (1987)): "In practice, however, this makes very little difference because the computation of Gröbner bases is not significantly easier if one relaxes the requirement that the Gröbner basis must be reduced".

3 Gröbner Bases Property on Elimination Ideal

Let us denote by X_k the subset of the first k elements of X, then it is known that $I_k := I \cap K\langle X_k\rangle$ is an ideal of $K\langle X_k\rangle$, which is named the k-th elimination ideal of I, also called the contraction of I to the subalgebra $K\langle X_k\rangle$. It is well known that, in commutative polynomial rings, this contraction can be easily obtained if one has a Gröbner basis of I with respect to a term ordering having the property "to separate" the variables. We are going to show here that it can be possible, in noncommutative polynomial rings, to attain a similar result; with the obstacle, of course, that the Gröbner basis of I could be infinite; in which case, in spite of the fact that the result is essentially the same, it would not be so practical as in the commutative case.

3.1. Definition. We are going to say that the term ordering $<$ has the elimination property at the position k ($k \in [1, n-1]$) if:

ε: For every $s, t \in \langle X\rangle$, $s < t$ and $t \in \langle X_k\rangle$ implies $s \in \langle X_k\rangle$.

The elimination property of a term ordering can also appear in some equivalent forms; for instance, it is easy to verify the condition ε_1 below amounts to ε; this characterization is more compatible with the property defined for a term ordering in Shannon D. et al. (1988) with the objective of solving the problems proposed there:

ε_1: For $j \in (k, n]$ and for $s \in \langle X_k\rangle$, $s < x_j$.

Mora also named in section 4.3 of Mora (1994), for commutative polynomial rings, "elimination ordering" to a term ordering with the ε_1 property.

The set of term orderings with the elimination property is not empty. The following one, due to Mora (Mora (1988)), is an example; this ordering will be denoted by $<_L$, has the elimination property at every position $k \in [1, n-1]$, and is defined as follows:

For $k = 1$, $<_L$ is the only term ordering on $\langle X_1\rangle$;

for each $k \in [2, n]$:

if $u \in \langle X_{k-1}\rangle$ and $w \in \langle X_k\rangle \setminus \langle X_{k-1}\rangle$, then $u <_L w$;

if $u, w \in \langle X_k\rangle \setminus \langle X_{k-1}\rangle$ (it can be possible to write them in the form:

$$u = t_1 x_k t_2 \cdots t_{m_1} x_k t_{m_1+1},$$

$$w = v_1 x_k v_2 \cdots v_{m_2} x_k v_{m_2+1},$$

where every t_i and v_j lie in $\langle X_{k-1}\rangle$, and x_k is the k-th element of X),

then $u <_L w$ if $m_1 < m_2$

or ($m_1 = m_2$ and $t_j <_L v_j$, where $j = \max \{i \in [1, m_1] \mid t_i \neq v_i\}$).

3.2. Theorem. Let G be a Gröbner basis of I with respect to a term ordering $<$ with the elimination property at the position k. Then, the following holds:

i. For each $f \in K\langle X_k \rangle$, $Can(f, I) \in K\langle X_k \rangle$ and $Can(f, I) = Can(f, I_k)$.

ii. If $f \in K\langle X_k \rangle$ and $T(g)$ divides $T(f)$, for some $g \in K\langle X \rangle$, then $g \in K\langle X_k \rangle$.

iii. $G_k := G \cap K\langle X_k \rangle$ is a Gröbner basis of I_k w.r.t. the restriction of $<$ onto $\langle X_k \rangle$.

iv. If $G = rGb(I)$, then $G_k = rGb(I_k)$.

Proof: i arises from 2.1.iii and the relation $\langle X \rangle \setminus T(I) \subset \langle X \rangle \setminus T(I_k)$, ii is a direct consequence of the elimination property of $<$. In regard to iii: Let $f \in I_k \setminus \{0\}$, as G is a G-basis of I, there exists $g \in G$ such that $T(g)$ divides $T(f)$, hence (see ii) $g \in G_k$; it means that G_k is a G-basis of I_k (see 2.2). Finally, on the basis of iii above, the relation $G_k \subset G$, and i (second part), the reader can easily see that G_k satisfies iv (see 2.3). □

3.3. Theorem. Let $<$ be a term ordering on $\langle X \rangle$, $x \in X \cap T(rGb(I))$, $G_{(x)} := rGb(I) \setminus \{x - Can(x, I)\}$, and $I_{(x)}$ the ideal of $K\langle X \setminus \{x\} \rangle$ given by $I \cap K\langle X \setminus \{x\} \rangle$. Then, $G_{(x)}$ is the reduced Gröbner basis of $I_{(x)}$ with respect to the restriction onto $\langle X \setminus \{x\} \rangle$ of the corresponding term ordering.

Proof: Let $f \in I_{(x)} \setminus \{0\}$; then, since $rGb(I)$ is a G-basis of I, there exists $s \in T(rGb(I))$ such that s divides $T(f)$, but $T(f)$ lies in $\langle X \setminus \{x\} \rangle$; consequently, $s \in T(G_{(x)})$. On the other hand, it is clear that $G_{(x)}$ is a subset of the ideal mentioned before; for this reason (see 2.2), $G_{(x)}$ is a Gröbner basis of this ideal. For similar reasons as the justification of 3.2.iv, $G_{(x)} = rGb(I_x)$. □

4 Applications

Warning message. In the applications that are going to be stated hereafter, it is taken for granted that one has at hand a term ordering for which rGb is finite. Hence, the proposed methods for solving the problems could fail in some cases since there are ideals that do not have finite Gröbner basis for any term ordering. In fact, it is not decidable, when given a set of polynomials, whether there exists an ordering for which rGb is finite (see Mora (1994) for a detailed and deep study about, among other topics, the finite G-basis existence problem); therefore, we are going to propose here some methods that have an implicit trial and error approach. Notwithstanding, it is also well known (see 5.1 and 5.2 in Mora T. (1987)) that for a zero dimensional ideal I (i.e. an ideal for which $K\langle X\rangle/I$ is a finite dimensional K-vector space) and for any term ordering, there always exists a finite G-basis with respect to the given term ordering; accordingly, a finite rGb.

Remark. We will say, as it is usual to speak, that a K-algebra B, generated over K by $A := \{a_1, \ldots, a_n\}$, is canonically isomorphic to $K\langle X\rangle/I$ if $Ker\ \mathsf{E} = I$, where E is the canonical projection of $K\langle X\rangle$ onto B.

4.1. Definition. Let I be an ideal of $K\langle X\rangle$ different from zero, $G \subset K\langle X\rangle$ is taken to be a Gröbner presentation for $K\langle X\rangle/I$ (or for any K-algebra canonically isomorphic to $K\langle X\rangle/I$) if there exists a term ordering $<$ on $\langle X\rangle$ for which $G = rGb(I, <)$.

The name G-presentation will also be used instead of Gröbner presentation. When we only say that G is a G-presentation, it is assumed that G is a G-presentation for $K\langle X\rangle/Ideal(G)$. Now, we are going to show the term orderings with the elimination property may be quite useful for computing Gröbner presentations.

4.2. Proposition. Let I be an ideal of $K\langle X\rangle$, $F := \{f_1, \ldots, f_m\} \subset K\langle X\rangle$, let $K\langle Y\rangle$ be the free K-algebra on the alphabet $Y := \{y_1, \ldots, y_m\}$, and let E be the homomorphism of $K\langle Y\rangle$ into $K\langle X\rangle/I$ that sends y_j into $[f_j]_I$, for $j \in [1, m]$. Then:

$$Ker\ \mathsf{E} = Ideal_{(Y,X)}(I \cup \{y_1 - f_1, \ldots, y_m - f_m\}) \cap K\langle Y\rangle.$$

Moreover, if $<$ is a term ordering on $\langle Y \cup X\rangle$ with the elimination property, at the position m, so that $Y < X$ (i.e. for each $y \in Y$ and for every $x \in X$, $y < x$), then:

$$rGb(Ker\ \mathsf{E}) = rGb_{(Y,X)}(I \cup \{y_1 - f_1, \ldots, y_m - f_m\}) \cap K\langle Y\rangle.$$

Remarks.

- Notations $Ideal_{(Y,X)}(\cdot)$, $rGb_{(Y,X)}(\cdot)$ specify that the ideal is considered to be generated in the free K-algebra on $Y \cup X$.
- Taking into consideration definition 4.1, the reader can see that $rGb(Ker\ \mathsf{E})$, when $Ker\ \mathsf{E} \neq \{0\}$, is a G-presentation for $K\langle [F]_I\rangle$, which may be then obtained if one has selected a term ordering for which $rGb_{(Y,X)}(I \cup \{y_1 - f_1, \ldots, y_m - f_m\})$ is finite.
- Proposition 4.2 is a straightforward generalization of corollary 3.2 (iii) in Gianni P. et al. (1988); therefore, we only comment some details in order to make subsequent references: The proof of the first part of 4.2 is based on the factorization of E in the form $\mathsf{E} = \mathsf{E}_2 \circ \mathsf{E}_1$, where E_1: $K\langle Y\rangle \to K\langle Y \cup X\rangle$ is the canonical injection and E_2: $K\langle Y \cup X\rangle \to K\langle X\rangle/I$ is the homomorphism that transforms y_j into $[f_j]_I$, for $j \in [1, m]$, and x_i into $[x_i]_I$, for $i \in [1, n]$. From this one infers at once that $Ker\ \mathsf{E} = Ker\ \mathsf{E}_2 \cap K\langle Y\rangle$ and $Ker\ \mathsf{E}_2 = Ideal_{(Y,X)}(I \cup \{y_1 - f_1, \ldots, y_m - f_m\})$. The rest is made for theorem 3.2.iv.

- Generalizing a similar result of Buchberger B. (1987), we can say that $f_1, \ldots, f_m$ are independent modulo I (i.e. there is no polynomial $p(y_1, \ldots, y_m)$ in $K\langle Y\rangle \setminus \{0\}$ for which $p(f_1, \ldots, f_m) \in I$) if and only if $Ker\ \mathsf{E} = \{0\}$; thus, if and only if $rGb(Ker\ \mathsf{E}_2) \cap K\langle Y\rangle = \emptyset$.

- In fact, as it has also been pointed out for the commutative case in Buchberger B. (1987), $rGb(Ker\ \mathsf{E})$ is a reduced Gröbner basis for the "ideal of algebraic relations" over F modulo I, i.e., for the set $\{g \in K\langle Y\rangle \mid g([F]_I) = 0\}$.

4.3. Proposition. Let I, F, E, and $<$ be defined as in 4.2, $f(X) \in K\langle X\rangle$. Then:

i. $f([X]_I) \in K\langle [F]_I\rangle$ if and only if there exists $q(Y) := q(y_1, \ldots, y_m) \in K\langle Y\rangle$ such that $Can(f(X), Ker\ \mathsf{E}_2) = q(Y)$; moreover, in such a case, $f([X]_I) = q([F]_I)$, where $q([F]_I) := q([f_1]_I, \ldots, [f_m]_I)$.

ii. For $x \in X$, $[x]_I \in K\langle [F]_I\rangle$ if and only if there exists $q(Y) \in K\langle Y\rangle$ such that $x - q(Y) \in rGb(Ker\ \mathsf{E}_2)$.

iii. $K\langle [F]_I\rangle = K\langle X\rangle / I$ if and only if for each $x_i \in X$ there exists $q_i(Y) \in K\langle Y\rangle$ such that $rGb(Ker\ \mathsf{E}_2) \setminus K\langle Y\rangle = \{x_1 - q_1(Y), \ldots, x_n - q_n(Y)\}$.

Proof:
i. $f([X]_I) \in K\langle [F]_I\rangle$ if and only if there exists $p(Y) \in K\langle Y\rangle$ for which $f([X]_I) = p([F]_I)$; but:

$$f([X]_I) = p([F]_I) \Longleftrightarrow f(X) - p(Y) \in Ker\ \mathsf{E}_2 \Longleftrightarrow$$
$$\Longleftrightarrow Can(f(X), Ker\ \mathsf{E}_2) = Can(p(Y), Ker\ \mathsf{E}_2) \text{ (see 2.1.ii)};$$

therefore, see 3.2.i, $f([X]_I) \in K\langle [F]_I\rangle$ implies the existence of $q(Y) \in K\langle Y\rangle$ such that $q(Y) = Can(f(X), Ker\ \mathsf{E}_2)$ ($q(Y) := Can(p(Y), Ker\ \mathsf{E}_2)$). In another direction, if $Can(f(X), Ker\ \mathsf{E}_2) \in K\langle Y\rangle$, then one can go up, through the above equivalences, until the subalgebra membership condition. Lastly, when $q(Y) = Can(f(X), Ker\ \mathsf{E}_2)$, as $f(X) - q(Y) \in Ker\ \mathsf{E}_2$, then $f([X]_I) = q([F]_I)$.
ii. $\Leftarrow$ is clear. Now let us suppose $[x]_I \in K\langle [F]_I\rangle$, then $Can(x, Ker\ \mathsf{E}_2) \in K\langle Y\rangle$ (see i); thus, remember characterization ε_1 of 3.1, $T(x - Can(x, Ker\ \mathsf{E}_2)) = x$; hence, for 2.3, the assertion is true.
iii. $K\langle [F]_I\rangle = K\langle X\rangle / I$ if and only if, for each $x_i \in X$, $[x_i]_I \in K\langle [F]_I\rangle$; so, let us apply ii and 2.3.i. □

Remarks.

- As it can be seen, the algebra membership problem (4.3.i) is reduced to computing $Can(f(X), Ker\ \mathsf{E}_2)$, which can be effectively made when $rGb(Ker\ \mathsf{E}_2)$ is finite (see warning message above).

- 4.3.i and 4.3.iii generalize respectively the membership test algorithm (of course, this generalization is not an algorithm because of the finiteness problem of reduced Gröbner bases) and theorem 10 of Shannon D. et al. (1988). Our generalizations have been made in two directions:
  - Noncommutative case.
  - We consider any two-sided ideal I (Shannon D. et al. (1988) just considers the case $I = \{0\}$.)
- As a matter of fact, many G-bases applications, for the commutative case, that are based on elimination property can be generalized to noncommutative Gröbner bases by means of considerations of section 2. Nevertheless, we prefer rather to discuss here only the above applications and show how they can be useful for solving some interesting problems than to draw up a list (probably incomplete) of these kinds of generalizations.

G-presentations for monoid and group algebras.

M_1: Let us introduce another notation: Let σ be a subset of $\langle X\rangle$ x $\langle X\rangle$ for which there exists at least a pair $(s,t) \in \sigma$ such that $s \neq t$, let $\langle\sigma\rangle$ be the congruence generated by σ, and let $P(\sigma)$ be the set $\{s - t \mid (s,t) \in \sigma\}$. Since $K\langle X\rangle/Ideal(P(\sigma))$ is (up to canonical isomorphism) the algebra over K of the monoid given by the set of defining relations σ, $rGb(P(\sigma))$ is a G-presentation for this monoid algebra (see 4.1). The computation of $rGb(P(\sigma))$ only requires, from K, the values 1 and -1 (see details in the justification of 4.4.i); hence, this rGb is the same for any K.
On the other hand, the following relation (easy to prove) allows one "to move" from congruences on $\langle X\rangle$ to ideals of $K\langle X\rangle$ generated by binomials of the form $s - t$ and vice versa: $\langle\sigma_1\rangle = \langle\sigma_2\rangle \Leftrightarrow Ideal(P(\sigma_1)) = Ideal(P(\sigma_2))$.
M_2: An ideal generated by binomials having the form $s-t$ is a particular case of a binomial ideal; by that denomination we mean a two-sided ideal of $K\langle X\rangle$ generated by binomials; i.e., polynomials of the form $cs + dt$, where $c, d \in K$ and $s, t \in \langle X\rangle$. This type of ideals, for commutative polynomial rings, has been studied in Eisenbud D. et al. (1996); now we are going to generalize, to the noncommutative case, proposition 1.1 of this paper.
4.4. Proposition. Let I be a binomial ideal, let $<$ be a term ordering on $\langle X\rangle$. Then:

i. $rGb(I,<)$ consists of binomials. Moreover, if the initial generating set of I is $P(\sigma)$, for some σ, then there exists $\beta \subset \langle X\rangle$ x $\langle X\rangle$ such that $P(\beta) = rGb(I,<)$.

ii. For $u \in \langle X\rangle$, $Can_<(u,I)$ is a monomial (i.e. a polynomial of the form cs, where $c \in K$ and $s \in \langle X\rangle$). Moreover, if the initial generating set of I is $P(\sigma)$, for some σ, then $Can(u,I) \in \langle X\rangle$.

iii. For $k \in [1, n-1]$, I_k is a binomial ideal.

Proof:

i. The extension of Buchberger's algorithm to associative free algebras (see Mora F. (1986), Mora T. (1987), Mora T. (1994)) gets a Gröbner basis, probably infinite, of I. If the initial generating set of I is formed by binomials, then it is clear that the Buchberger-Mora's procedure will only build binomials; and afterward, starting from the G-basis thus obtained, one can attain the reduced Gröbner basis just by eliminating the redundant maximal terms and taking into account 2.3.ii and 2.3.iii (of course, we are not referring to the computation of the rGb, that could not be possible, but to the structure of this G-basis). Lastly, if the polynomials of the initial generating set have the form $s - t$, then the procedure mentioned above will only generate polynomials having the same form or zero.
ii. It is immediately seen if one tries to find $Can_<(u, I)$ by using $rGb(I, <)$; which, for i above, consists of binomials.
iii. Let us choose a term ordering with the elimination property at every position and apply 3.2.iv. □
M_3: M_1 and M_2 above lead us to the following considerations:

- Given a finite set σ, one can compute $rGb(P(\sigma))$ (without forgetting the warning message) and then translate the binomials of this Gröbner basis back into relations that define the same monoid as σ (see 4.4.i and M_1).

- It may be possible to solve basic problems on monoids such as: Given a monoid M by means of σ (finite), $W := \{w_1, \ldots, w_m\} \subset \langle X \rangle$, $v \in \langle X \rangle$:

  1. Find a G-presentation for the algebra over K of the submomoid N of M generated by $[W]_\sigma := \{[w_1]_\sigma, \ldots, [w_m]_\sigma\}$ (where, as logi- cal, $[w]_\sigma$ denotes the class of congruence of the word w modulo σ).
     Solution method: Let in 4.2, $I := Ideal(P(\sigma))$, $F := W$.

  2. Decide whether $[v]_\sigma$ lies in N.
     Solution method: Let in 4.3.i, beside the above asignments, $f := v$. (Note that $q(Y)$ of 4.3.i, in this case, lies in $\langle Y \rangle$ (see 4.4.ii); i.e., $[v]_\sigma \in N$ if and only if $Can(v, Ker\ \mathsf{E}_2)$ is a word in the elements of Y.)

M_4: Let us consider a group given as a quotient of the free group on X; which means, in the first place, that a set X^{-1} (with the same cardinal as X) is added and, in the second place, that every presentation of the given group ($\sigma \subseteq \langle X \cup X^{-1} \rangle$ x $\langle X \cup X^{-1} \rangle$) contains by default, for each $x \in X$, the relations $xx^{-1} = x^{-1}x = 1$. We are going to show that it may be possible to simplify, in many cases that occur frequently, the completion procedure for getting $rGb(P(\sigma))$.

4.5. Proposition. Let us suppose that $x^{-1} - w \in Ideal(P(\sigma))$ (where $x \in X$ and $w \in \langle X \cup X^{-1} \setminus \{x^{-1}\}\rangle$, let $<$ be a term ordering on $\langle X \cup X^{-1}\rangle$ for which $w < x^{-1}$, and let β be obtained from σ by the replacement for w whenever x^{-1} appears. Then:

$$rGb(P(\beta)) = rGb(P(\sigma)) \setminus \{x^{-1} - Can(w, Ideal(P(\sigma)))\},$$

where $rGb(P(\beta))$ stands for the reduced Gröbner basis of the ideal of $K\langle X \cup X^{-1} \setminus \{x^{-1}\}\rangle$ generated by $P(\beta)$.

Proof: It is easy to see that $Ideal(P(\sigma)) \cap K\langle X \cup X^{-1} \setminus \{x^{-1}\}\rangle$ is generated, as an ideal of $K\langle X \cup X^{-1} \setminus \{x^{-1}\}\rangle$, by $P(\beta)$; thus, the reader is able to apply 3.3. □

Consequently, if the relation $x^{-1} - w \in Ideal(P(\sigma))$ is known in advance, one can compute $rGb(P(\beta))$ and, when the calculus comes to a close, add $x^{-1} - Can(w, Ideal(P(\sigma)))$ to this Gröbner basis in order to have a G-presentation of the given group algebra. See an example below.

4.6. Computing normal closure.

Specification of problem number 2 (see Introduction):

Given σ (finite), presentation of a group G;

$W := \{w_1, \ldots, w_m\} \subset \langle X \cup X^{-1}\rangle$:

Find the normal closure of the subgroup H of G generated by $[W]_\sigma$ (i.e. the smallest normal subgroup of G which contains H).

Solution method: An algorithm has already been devised so as to deal with this problem, cf. Butler G. et al. (1982); it consists, after logical adaptation to our case, on extending $[W]_\sigma$ by sufficient conjugates ($[x^{-1}]_\sigma[w]_\sigma[x]_\sigma$, where $x \in X$ and $w \in W$) until the subgroup generated by $[W]_\sigma$ no longer changes. The crucial point of this method is the subgroup membership test, which is solved in Butler G. et al. (1982) by computing, each time a new conjugate is introduced, a base and strong generating set (BSGS) for the subgroup generated by the new $[W]_\sigma$. As it is well known, for groups acting faithfully on finite sets (such as permutation groups), BSGS can be used to solve the subgroup membership problem in an efficient way. We propose here to attack this problem by computing in each step a G-presentation for the algebra over K generated by the new $[W]_\sigma$, which also allows to decide about subgroup membership (see problem 2 in M_3). The computation of such a G-presentation has been dealt in problem number 1 of M_3, thus rGb finiteness problem is moving around again; furthermore, there is also another finiteness problem, namely, we can not assure the completion procedure of $[W]_\sigma$, until getting a normal subgroup, always terminates in a finite number of steps. However, if termination is attained, we will have a G-presentation for the group algebra of the normal closure of H and so, taking into account the first comment of M_3, a presentation for this normal closure.

Remarks.

- For the wide and important class of finite groups, the above normal closure procedure always terminates, which is a consequence of the following facts:

  1. Group algebras of finite groups are finite dimensional (see warning message above).
  2. Since the group is finite, the completion procedure of $[W]_\sigma$ can be carried out successfully.

  (We highlight that it may be possible to decide whether a group given by a finite presentation is finite (see Gilman R.H. (1979) and, if so, then we are able to apply the above method in order to compute the normal closure to any subgroup of the given group. It would be interesting to find out another classes of groups for which the above procedure could be applied with success.)

- In contrast to Butler G. et al. (1982) that takes as input-output a group given by a concrete representation (permutations, matrices over a finite field), the method proposed here works with groups given by an abstract definition.

- To find normal closure is a basic task for solving another important problems such as: computing commutator subgroups of normal subgroups, derived and lower central series; deciding solvability and nilpotency (cf. Butler G. et al. (1982) and Bosma W. et al. (1992)).

Now, we are going to exhibit, by means of a simple but illustrative example, the main ideas of the normal closure procedure discussed above.

Example:

Let D_4 be the dihedral group or order 8, which is given as a polyhedral group by the following presentation (cf. Chenadec P. Le. (1986)):

$$\langle a, b, c \mid a^4 = b^2 = c^2 = abc = 1 \rangle.$$

Let us show how to compute the normal closure of the subgroup $\mathrm{H} \subset D_4$ that is generated by $W := \{a^2 b\}$:

Let $X := \{a, b, c\}$. For the sake of simplicity in the notation, we will not make differences between words in $\langle X \cup X^{-1} \rangle$ and words in D_4 (i.e. between w and $[w]_\sigma$); it will be clear from the context the precise role they are playing in each case. The set σ of M_4 is, in this occasion, equal to:

$$\sigma := \{a^4 = b^2 = c^2 = abc = aa^{-1} = a^{-1}a = bb^{-1} = b^{-1}b = cc^{-1} = c^{-1}c = 1\};$$

we are keeping on with the same group if σ is redefined as follows:

$$\sigma := \sigma \cup \{y_1 = a^2 b,\ y_1 y_1^{-1} = y_1^{-1} y_1 = 1\}.$$

Since all the inverses are words in the monoid generated by X (which always holds when the group is finite), it can be possible to build, in order to apply 4.5, the set β that is obtained from σ by substituting a^{-1}, b^{-1}, c^{-1} respectively by a^3, b, c; hence:

$$\beta := \{a^4 = b^2 = c^2 = abc = y_1y_1^{-1} = y_1^{-1}y_1 = 1,\ y_1 = a^2b\}.$$

Now, if we take the term ordering $<_L$ of section 3 in such a fashion that

$$y_1 <_L y_1^{-1} <_L a <_L b <_L c <_L a^{-1} <_L b^{-1} <_L c^{-1},$$

then we can compute $rGb(P(\sigma))$ by computing $rGb(P(\beta))$ and afterward, adding the relations of the inverses (see proposition 4.5); thus:

$$rGb(P(\beta)) = \{y_1^2 - 1,\ y_1^{-1} - y_1,\ ay_1a - y_1,\ a^2y_1 - y_1a^2,\ a^3 - y_1ay_1,$$
$$b - y_1a^2,\ c - y_1a\};$$

therefore,

$$rGb(P(\sigma)) = rGb(P(\beta)) \cup \{a^{-1} - y_1ay_1,\ b^{-1} - y_1a^2,\ c^{-1} - y_1a\}.$$

As $rGb(P(\sigma))$ reveals, b and c are superfluous generating elements; hence, we can do without the polynomials associated with them and their inverses. Furthermore, since $y_1^{-1} - y_1 \in rGb(P(\sigma))$, the condition

$$Can(x^{-1}y_1x, Ideal(P(\sigma))) \in \langle y_1 \rangle \text{ implies}$$
$$Can(x^{-1}y_1^{-1}x, Ideal(P(\sigma))) \in \langle y_1 \rangle;$$

therefore, y_1^{-1} is not really necessary. On the other hand, we are able to store the information $a^{-1} = y_1ay_1$ in another place different from the rGb and then, all things considered, we can eliminate polynomials of $rGb(P(\sigma))$ in order to get a more simplified set for a new Gröbner basis computation, i.e., from now on, we will work with

$$G := \{y_1^2 - 1,\ ay_1a - y_1,\ a^2y_1 - y_1a^2,\ a^3 - y_1ay_1\}.$$

Taking into consideration the above arguments, there is only one product $x^{-1}wx$ that needs to be analyzed, that is $a^{-1}y_1a$, thus:

$$Can(y_1ay_1y_1a, Ideal(G)) = y_1a^2;$$

accordingly, we have to introduce y_2 (which is associated to y_1a^2), redefine $G := G \cup \{y_2 - y_1a^2\}$ and compute $rGb(G)$, so:

$$rGb(G) = \{y_1^2 - 1,\ y_2y_1 - y_1y_2,\ y_2^2 - 1,\ ay_1 - y_2a,\ ay_2 - y_1a,\ a^2 - y_1y_2\}.$$

Now, the next product $x^{-1}wx$ is equal to $y_1ay_1y_2a$, whose canonical form is y_1. Hence, we can conclude that the normal closure of the subgroup of D_4 generated by a^2b has the following presentation:

$$\langle y_1, y_2 \mid y_1^2 = 1,\ y_2y_1 = y_1y_2,\ y_2^2 = 1,\ y_1^{-1} = y_1,\ y_2^{-1} = y_2\rangle.$$

For this reason, this normal closure is the abelian group Z_2 x Z_2.

Final remark.

As far as we know, Gröbner bases applications on monoids and groups have appeared in the following references: Sims C. C. (1990), where an algorithm is described for the computation of a Gröbner basis for a submodulo of a finitely generated free modulo over the ring of polynomials in finitely many indeterminates with integer coefficients; the connection of this algorithm with the problem of computing metabelian polycyclic quotient groups is shown in that paper. In Madlener K. et al. (1993), following Buchberger's approach to computing Gröbner bases, a completion procedure for finitely generated right ideals of a monoid ring is given. Termination of this procedure is assured in several classes of monoids. An application to the subgroup problem is discussed. In Rosenmann A. (1993), an algorithm is presented to computing Gröbner bases for finitely generated one-sided ideals in free group algebras. An application of this algorithm to solving the membership problem, for one-sided ideals, and to constructing free bases for subgroups of free groups is exhibited in that work. Madlener K. et al. (1995.a), Madlener K. et al. (1997), Madlener K. et al. (to appear), and Reinert B. (1996), are devoted to build Gröbner bases theory for some specific classes of groups. In Reinert B. (1995) and Madlener K. et al. (1995.b), the techniques for presenting monoids or groups by string rewriting systems are used to define several types of reduction in monoid and group rings and then, Gröbner bases are presented as a natural generalization of the corresponding known notions in the commutative and some noncommutative cases. Several results on the connection of the word problem and the congruence problem are proven. The concepts of saturation and completion are introduced for monoid rings having a finite convergent presentation by a semi-Thue system; the objective of these concepts is to repair some defects that reduction has when one tries to define it in the enviroment mentioned before. Reinert B. (1995) summarizes Gröbner bases results on monoid and group algebras and is a good source for further investigation along this line. As a general characteristic in the above references, Gröbner bases tools are redefined in order to take advantage of particular properties of the algebraic structures that are studied.
Our approach is based on the following consideration: Every monoid or group algebra can be seen as a quotient of the corresponding free associative algebra, thus, one can apply "classical" Gröbner bases theory on the two-sided ideal, of the free algebra, that determines the given monoid algebra. Of course, this way implies to live together with the finiteness problem of noncommutative

Gröbner bases (which is also present in some of the refe- rences listed before); but, on the other side, it enables us to build general methods for every monoid algebra having finite rGb, to use directly some Gröbner bases results and interpret them in the particular setting of monoid algebras. Furthermore, from a more general point of view, our main intention has been to illustrate how Gröbner bases property on elimination ideal can be used in a context different from commutative algebra.

Acknowledgements:

We owe a great deal of motivation on Gröbner Bases to Buchberger's papers. First author of this paper was invited by Bruno Buchberger to visit RISC some time ago and received his masterly education. Our gratitude also to Teo Mora, whose significant influence is also present in our work, for his very valuable suggestions in early versions of this paper. Thanks are also due to Klaus Madlener and Birgit Reinert for their helpful answer to our request of information about their results.

References

Bosma W., Cannon J. (1992) 'Structural Computation in Finite Permutation Groups.' CWI Quarterly, 5, No. 2, pp. 127-160.

Buchberger B. (1985) 'Gröbner Bases: An Algorithmic Method in Polynomial Ideal Theory.' Chapter 6 in Bose N.K. (ed.): Recent Trends in Multidimensional Systems Theory. D. Reidel. Publishing Company, Dordrecht-Boston-Lancaster, pp. 184-232.

Buchberger B. (1987) 'Applications of Gröbner Bases in Non-linear Computational Geometry.' Proc. of the Workshop on Scientific Software. IMA Volumes in Math. and its Applications. Springer. V. 14. USA, pp. 59-88.

Butler G., Cannon J. (1982) 'Computing in Permutation and Matrix Groups I: Normal Closure, Commutator Subgroups, Series.' Math. of Comp., V. 39, N. 160, pp. 663-670.

Eisenbud D., Sturmfels B. (1996) 'Binomial Ideals.' Duke Mathematical Journal, V. 84, N. 1, pp. 1-45.

Gianni P., Trager B., Zacharias G. (1988) 'Gröbner Bases and Primary Decomposition of Polynomial Ideals.' J. Symb. Comp., 6, pp. 149-167.

Gilman R.H. (1979) 'Presentations of Groups and Monoids.' J. Algebra, 57, pp. 544-554.

Kalkbrener M. (1991) 'Three Contributions to Elimination Theory.' Ph. D. Thesis. RISC, University of Linz, Austria.

Chenadec P. Le. (1986) 'A Catalogue of Complete Group Presentations.' J. Symb. Comp., 2, pp. 363-381.

Madlener K., Reinert B. (1993) 'Computing Gröbner Bases in Monoid and Group Rings.' Proc. ISSAC'93, pp. 254-263.

Madlener K., Reinert B. (1995.a) 'On Gröbner Bases for Two-Sided Ideals in Nilpotent Group Rings.'
SEKI Report SR-95-01. Universität Kaiserslautern.

Madlener K., Reinert B. (1995.b) 'String Rewriting and Gröbner Bases. A General Approach to Monoid and Group Rings.' Proc. of the Workshop on Symbolic Rewriting Techniques. Monte Verita. To appear.

Madlener K., Reinert B. (1997) 'A Generalization of Gröbner Bases Algorithms to Nilpotent Group Rings.' Applicable Algebra in Engineering, Communication and Computing, V. 8, No. 2, pp. 103-123.

Madlener K., Reinert B. (to appear) 'A Generalization of Gröbner Basis Algorithms to Polycyclic Group Rings.' J. Symb. Comp.

Mora F. (1986) 'Gröbner Bases for Non-commutative Polynomial Rings.' Proc. of AAECC 3, pp. 353-362 (Lecture Notes in Computer Science 229).

Mora T. (1987) 'Gröbner Bases and the Word Problem.' Preprint. University of Genova.

Mora T. (1988) 'Gröbner Bases for Non-commutative Algebras.' Proc. of the Joint Conf. ISSAC'88 and AAECC 6, pp. 150-161 (Lecture Notes in Computer Science 358).

Mora T. (1994) 'An Introduction to Commutative and Noncommutative Gröbner Bases.' Theoretical Computer Science 134, pp. 131-173.

Reinert B. (1995) 'On Gröbner Bases in Monoid and Group Rings.'
Ph. D. Thesis. Universität Kaiserslautern.

Reinert B. (1996) 'Introducing Reduction to Polycyclic Group Rings. A Comparison of Methods.' Reports on Computer Algebra No 9. Centre of Computer Algebra. Universität Kaiserslautern.

Rosenmann A. (1993) 'An Algorithm for Constructing Gröbner and Free Schreier Bases in Free Group Algebras.' J. Symb. Comp., 16, pp. 523-549.

Shannon D., Sweedler M. (1988) 'Using Gröbner Bases to Determine Algebra Membership, Split Surjective Algebra Homomorphisms, Determine Birational Equivalence.' J. Symb. Comp., 6, pp. 267-273.

Sims C. C. (1990) 'Implementing the Baumslag-Cannonito-Miller Polycyclic Quotient Algorithm.' J. Symb. Comp., 9, pp. 707-723.

The CoCoA 3 Framework for a Family of Buchberger-like Algorithms

Antonio Capani [1] *and Gianfranco Niesi* [2]

Abstract

In this paper we report how we have implemented in release 3.3 of the system CoCoA algorithms for computing Gröbner bases, syzygies, and minimal free resolution of submodules of finitely generated free modules over polynomial rings. In particular we describe the environment we have built where these algorithms are obtained as special instances of a single parametrized algorithm. The outcome is a framework which offers not only efficiency in computing, but also efficiency in designing new algorithms; moreover, it allows the user to interactively execute the above computations, and easily customize the way they are carried out.

1 Introduction

Computing Gröbner bases, syzygies, and finite free resolutions of ideals and of submodules of finitely generated free modules over polynomial rings is a fundamental task in Computational Commutative Algebra. Nowadays there are several specialized computer algebra systems, developed by researchers in the field, which efficiently perform such computations, for instance Asir (Noro and Takeshima 1992), CoCoA (Capani et al. 1996), Macaulay (Bayer and Stillman 1992), Macaulay 2 (Grayson and Stillman 1996), Singular (Greuel et al. 1996).

The cornerstone in this area has been the (now) classical algorithm due to Buchberger (1965) which produces a Gröbner basis of an ideal in a polynomial ring R over a field, starting from any of its possible finite sets of generators. This algorithm can be suitably modified (Schreyer 1980, Spear 1977) in such a way that it also computes the syzygy module of a list of polynomials. These procedures can be extended to submodules of the free R-module R^n and lists of vectors of polynomials (see for instance Möller and Mora 1986).

A significant situation arises from the fundamental problems of Projective Algebraic Geometry and occurs when R is a graded ring and one works

[1]Dip. di Informatica, Univ. di Genova. e-mail: capani@disi.unige.it
[2]Dip. di Matematica, Univ. di Genova. e-mail: niesi@dima.unige.it
Partially supported by the Consiglio Nazionale delle Ricerche (CNR)

with homogeneous polynomials. In this context appropriate techniques and strategies can produce remarkable improvements of the algorithms quoted above.

Since its first release (Giovini and Niesi 1990) the system CoCoA has offered an efficient implementation of algorithms for computing Gröbner bases, syzygies, and free resolutions, embodying theoretical results of several authors (Buchberger 1985, Cabŏara et al. 1996, Gebauer and Möller 1988, Giovini et al. 1991, Traverso 1996) to speed up such computations.

The system was originally designed to work with general polynomials but more recently we have enriched it with special algorithms for the homogeneous case. In particular, in a joint work with G. De Dominicis and L. Robbiano (Capani et al. 1997), we developed new algorithms for computing: minimal systems of generators, Gröbner bases, syzygy modules, and minimal free resolutions of a homogeneous submodule of a finitely generated graded free module. Variants of these algorithms use the knowledge of the Hilbert function of the submodule to speed up the computation. We discuss later in this paper our implementation of these algorithms.

It is worth mentioning other recent contributions to this subject: R. La Scala (La Scala 1996) has proposed an algorithm for computing a minimal free resolution which has been implemented in Macaulay 2. Another approach, due to M. Caboara (1997), uses an embedding of the module in a suitable polynomial ring; moreover they show how to perform various module operations and present an algorithm to compute a minimal free resolution of a homogeneous module.

The implementation of the algorithms discussed in (Capani et al. 1997) was guided by the followgin key observation: these algorithms share a common pattern and hence they can be obtained as specialization of an abstract Buchberger-like algorithm which depends on some function parameters, i.e. each of these algorithms can be obtained from the generic one by assigning to each parameter an appropriate function called *key function*. Different algorithms can share some key functions; on the other hand the same computation can be done using different settings of key functions depending on the strategy to be used and on some other options (e.g. if we want some statistics or not). We have supplied several key function sets for the most common algorithms and for some of their variations and we have designed the system in such a way that users may assign their own key function set and easily customize the behaviour of a Buchberger-like algorithm.

Another important consideration is the following: often a complete computation takes a very long time; in most cases the user wishes to execute a computation in an interactive way: that is to start with a part of the computation, explore the partial result, and, if necessary, perform further pieces of computation until a certain result is obtained.

For these reasons we have completely redesigned and reimplemented all the "Gröbner stuff" as a family of algorithms obtained from a unique generic algorithm which, after an initialization stage, can perform a single "step" of computation. We started our work by specifying all the considered algorithms in CoCoAL, the CoCoA 3 programming language and, then, we implemented them in the CoCoA kernel, which is written in the language C.

The key observation here is that CoCoA is a system for doing research, hence we concentrated not only on efficiency in computing but also on efficiency in designing new algorithms. The key functions trace a well defined and well behaved path along which the user can do good prototyping by choosing among the predefined sets of key functions. Moreover it is easier for us to extend the C library by implementing new key functions, if necessary. In both cases the overhead is greatly reduced, and hence also the possibility of introducing errors.

In this paper we present some aspects of our implementation by using the CoCoAL language. In Section 2 we revise the classical Buchberger algorithm for a module. Then we present in Section 3 the abstract version of the algorithm, and we discuss how various concrete algorithms can be obtained by specifying suitable sets of key functions. In Section 4 we present the algorithms of the homogeneous environment. Finally we conclude the paper with some remarks (see Sect. 5).

2 A review of the Buchberger algorithm

In this section we recall the classical Buchberger algorithm for a module, and present a short description of it in CoCoAL.

Let $R := k[X_1, \dots, X_n]$ be a polynomial ring over a field k and $n > 0$. Throughout this paper, for the sake of brevity, we use the word *module* to denote a submodule of the free module R^n. An element of a module is called a *vector*. Let $\{e_1, \dots, e_n\}$ be the canonical basis of R^n. The terms of R^n are the elements of the form $t\, e_i$ where t is a term of the ring R. We assume that a term-ordering σ is given on the set of the terms of R^n. If $V \in R^n$, then we denote by $\text{LT}(V)$ and $\text{LC}(V)$ the leading term and the leading coefficient of V respectively. Basic facts about polynomial rings, modules, term-orderings, and Gröbner bases can be found in (Adams and Loustaunau 1994, Möller and Mora 1986).

The Buchberger algorithm takes as input a list of vectors, L, and returns a list of vectors, *GBasis*, which is a Gröbner basis of the module generated by L. We recall that two vectors V and W in R^n give rise to a critical pair if $\text{LT}(V) = s\, e_i$ and $\text{LT}(W) = t\, e_i$ with the same i; the S-polynomial of a critical pair is the vector defined by

$$\text{LC}(W)\, \frac{\text{lcm}(s,t)}{s}\, V - \text{LC}(V)\, \frac{\text{lcm}(s,t)}{t}\, W\ .$$

The first phase of the algorithm initializes the list *IPs* of the *incoming pairs* to the list of the critical pairs of elements in L and initializes *GBasis* to L. Then the algorithm executes a loop: each incoming pair P is processed as follows. First the S-polynomial of the pair is computed; then it is normalized with respect to the list *GBasis*; if the resulting normal form N is different from 0 then the set of pairs is updated with the new critical pairs obtained pairing N with all the elements of the current Gröbner basis; furthermore the element N is added to the Gröbner basis. The algorithm terminates when the list *IPs* is empty (Buchberger 1965).

The performance of the algorithm depends heavily on the selection strategy (i.e. how to choose the pair to be processed), on the simplification strategy (i.e. how to choose the simplifier during the normalization of a vector when many choices are possible), and on the application of criteria to detect a priori useless pairs (i.e. pairs whose S-polynomial reduces to 0).

The following CoCoAL function implements the above algorithm. It uses some auxiliary functions whose behaviour is assumed to be the one described in Remark 1.

```
Define Buchberger_Classic(L)
  IPs := MakeAllPairs(L);
  GBasis := L;
  While IPs <> 0 Do
    P := GetFirst(IPs);
    V := SPoly(P);
    N := NF(V,GBasis);
    If N <> 0 Then
      IPs := UpdatePairs(GBasis,IPs,N);
      Append(GBasis,N);
    End;
  End;
  Return GBasis;
End;
```

Remark 1 The auxiliary functions used by *Buchberger_Classic(L)* behave as follows.

The function *MakeAllPairs(L)* returns the list of the critical pairs of elements in L.

The function *UpdatePairs(GBasis,IPs,N)* updates the list of pairs *IPs* by adding the new pairs which involve the vector N and elements of *GBasis*.

The function *NF(V,GBasis)* returns the normal form of V with respect to the rules *GBasis*.

The function *SPoly(P)* returns the S-polynomial of the pair P.

The function *GetFirst(L)* returns the first element of the list L and discards it from L.

The function *Append(L,X)* appends to the list L the element X.

Remark 2 The functions *MakeAllPairs(L)* and *UpdatePairs(GBasis,L,N)* may also take care of applying the criteria for detecting useless pairs (Buchberger 1985, Gebauer and Möller 1988).

A selection strategy can be implemented by keeping the list *IPs* sorted with respect to some fixed ordering (Giovini et al. 1991).

The use of the function *Append(L,X)* implies the simplification strategy that chooses the "oldest" simplifier.

3 The implementation of the Buchberger algorithm in the system CoCoA

As we said earlier, we wish to have an implementation of the Buchberger algorithm which allows us to do the computations "step by step" and allows us to implement other Buchberger-like algorithms as simple variations of the basic algorithm; computations made using such algorithms will be called *Gröbner computations.*

First of all we need to have convenient data structures. In particular we defined a structure, called "module", which contains all the relevant information about the mathematical module itself (for example one of its systems of generators and, when they are known, its Gröbner basis, its Poincaré series, etc.), but also it is the container of a Gröbner computation in progress. This means that all the information (e.g. critical pairs or reductors) is stored in the module. Hence we think of a module M as a record (in the computer science sense) and we denote respectively by *M.Gens*, *M.GBasis*, *M.IPs* the generators, the Gröbner basis, and the list of critical pairs of M. We organized the data structures with two more lists *M.Rules* and *M.IVs*. The first is the set of the rules (reductors) used to perform the normalization and is in general a superset of *M.GBasis* (we prefer to keep the two lists separated); the latter is the list of the *incoming vectors* which is initialized with the generators of the module; this list is particularly useful in the homogeneous context (see Sect. 4).

With this setting, the algorithm *Buchberger_Classic* of Section 2 is an instantiation of the following parametric algorithm *Buchberger_Abstract*, when an appropriate set of key functions is assigned. This presentation of the algorithm highlights that there is an initialization phase followed by a loop in which the computation is performed one step at a time. The boolean function *OneStep* returns TRUE if and only if some computation has been performed, otherwise it returns FALSE and the loop terminates.

```
Define Buchberger_Abstract(M)
  Initialize(M);
  Repeat
    B := OneStep(M)
  Until Not B;
End;
```

The algorithm puts the result of the computation in the module *M* passed as input. For example the Gröbner basis is stored in *M.GBasis*.

We say that this algorithm is abstract in the sense that it depends on the "parameter functions" Initialize and OneStep whose actual values are assigned at run time depending on the kind of Gröbner computation requested. Once the user has assigned to the module the setting of key functions, the computation will use these functions without having to examine any option. Thus to execute *Buchberger_Classic*, the functions Initialize and OneStep are "instantiated" (i.e. substituted) with the following two functions.

```
Define Initialize_GB(M)
  M.IVs := M.Gens;
  M.IPs := [];
  M.Rules := M.Gens;
  M.GBasis := M.Gens;
End;

Define OneStep_GB(M)
  If M.IVs <> [] Then
    V := GetFirst(M.IVs);
    UpdateBasisAndPairs(M,V);
    Return TRUE
  End;
  If M.IPs = [] Then Return FALSE End;
  P := GetFirst(M.IPs);
  V := SPoly(P);
  ProcessVector(M,V);
  Return TRUE;
End;

Define ProcessVector(M,V)
  W := NF(V,M.Rules);
  If W <> 0 Then
    IPs := UpdateBasisAndPairs(M,W);
  Else
    CaseZero(M,W);
  End
End;
```

The mechanism of "instantiation" works as follows. The user calls a function named *Start(M,KFL)* passing to it a module and a list of key functions; more precisely each key function is identified by a positive integer, and so *KFL* is a list of positive integers. This list is stored in the module and then every time we call a parameter function *KeyFun* it means that we are calling the key function assigned to *KeyFun*.

After the key functions of the module have been assigned with the function *Start*, the user can execute the complete computation by calling the function *Complete*. Since all the information about the computation in progress (i.e. *M.IVs, M.IPs, M.Rules* and *M.GBasis*) is kept in the module it is also possible to execute the computation "interactively" by repeatedly calling the function *OneStep*. So the user can examine the partial results.

Let's examine the key function *OneStep_GB*. First all the incoming vectors (which were initialized to the generators of the module) are removed one by one and the Gröbner basis, the pairs and the rules are updated with them (by using the key function *UpdateBasisAndPairs*). Each successive step takes the first incoming pair, compute its S-polynomial, and then processes it by calling the auxiliary function *ProcessVector*.

We are using the parameter functions *NF*, *SPoly*, *UpdateBasisAndPairs* and *CaseZero* because we want to be ready to adapt the algorithm to compute the syzygy module of *M.Gens*.

If we want to retrieve the algorithm *Buchberger_Classic* then the key functions assigned to the parameter functions *NF* and *SPoly* have to behave as explained in Remarks 1 and 2. The function *UpdateBasisAndPairs*(M,V) updates the list of the incoming pairs (as *UpdatePairs*) and adds the new element *V* to *M.GBasis* and to *M.Rules*.

Remark 3 The key function assigned to *UpdateBasisAndPairs* in its turn uses some parameter functions; among these there are *Add_To_Rules* used to add new vectors to the list *M.Rules* and *Pair_Prec* used to keep the list of the incoming pairs sorted with respect to a fixed ordering. By using different key functions for the parameter function *Add_To_Rules* we can apply different simplification strategies; the function *Pair_Prec* is used to keep the list of the incoming pairs sorted; different key functions for *Pair_Prec* allow us to implement different orderings and hence different selection strategies.

Let's now introduce another set of key functions which specializes the abstract algorithm to one which computes the syzygy module of the module generated by a given list of vectors.

If V is a vector belonging to the module generated by $L := \{V_1, \ldots, V_r\}$, then there exists a tuple $Rel := (P_1, \ldots, P_r)$ of polynomials such that $V = \sum_i P_i V_i$. We say that *Rel* is a "relation" of V in terms of L.

It is well known (Schreyer 1980, Spear 1977) that one can use the Buchberger algorithm also for the computation of the syzygies: it suffices to keep track of at least one relation (in terms of the given generators) for each rule, S-polynomial and vector in reduction.

So the key functions (say *SPoly_Syz* and *NF_Syz*) corresponding to the parameter functions SPoly and NF must keep the above relations updated. More precisely each arithmetical operation involving a vector V must be performed also on the corresponding relation.

When a vector reduces to 0, then we have a relation for 0 in terms of the generators L, and hence we have found a syzygy of the elements of L. Collecting all these syzygies, we get a system of generators for the syzygy module of L. So the key function for CaseZero will have to add the found syzygy to the generators of the module *M.Syz*.

```
Define CaseZero_Syz(M,N)
  If N.Res <> 0 Then Append(M.Syz.Gens,N.Rel) End;
End;
```

Note that only the key functions *SPoly_Syz, NF_Syz, CaseZero_Syz* have to be implemented to achieve the computation of the syzygy module.

4 The homogeneous algorithms

In this section we assume that the ring $R := k[X_1, \ldots, X_n]$ is graded over $\mathbb{N}$ by $\deg(X_i) > 0$ for $i = 1, \ldots, n$. Let M be an $\mathbb{N}$-graded submodule of a finitely generated graded free R-module given by a homogeneous system of generators. We consider the following problems for M: a) to find a minimal system of generators; b) to compute a Gröbner basis of M; c) to find the syzygy module of a minimal system of generators or of the given generators; d) to compute the minimal free resolution.

Algorithmic answers to these problems have been presented in (Capani et al. 1997). Here we show how we implemented these algorithms in CoCoA 3 (in the framework of the abstract Buchberger algorithm) by means of suitable key functions which take into account the homogeneity of the module.

The general idea is to sort the list of incoming vectors *M.IVs* by increasing degree and construct the Gröbner basis in the same order. Processing an incoming vector V means to normalize it with respect to *M.GBasis* and then to do some action depending on its nullity. Processing a critical pair is essentially the same as in the non-homogeneous case, but we can use the extra knowledge that the degree does not change during reduction.

In this case a degree-compatible strategy is used; this means that the elements of both the incoming queues are processed by increasing degree, and, inside each degree, first all the incoming pairs are processed and then the incoming vectors. If a degree-compatible strategy is used to compute a

Gröbner basis of M, then, after passing a certain degree d, a d-truncated Gröbner basis of M is already computed.

The pair of key functions *Initialize_Hom* and *OneStep_HGB* is used when we want to compute a Gröbner basis of the homogeneous module M using a degree-compatible strategy.

```
Define Initialize_Hom(M)
  M.IVs := M.Gens;
  M.IPs := [];
  M.Rules := [];
  M.GBasis := [];
End;

Define OneStep_HGB(M)
  If M.IPs = [] And M.IVs = [] Then Return FALSE End;
  V := GetFirstToProcess(M);
  ProcessVector(M,V);
  Return TRUE;
End;
```

The function *GetFirstToProcess* selects an element of the queues *M.IPs* and *M.IVs* using the degree-compatible strategy: if it is a pair, then the function returns the corresponding S-polynomial.

Capani et al. (1997) showed how to obtain a *minimal* set of critical pair which have to be processed by the Buchberger algorithm in the homogeneous case. This leads to an improved version of the criteria for detecting useless pairs described in (Gebauer and Möller 1988). Since we want to use different criteria in the homogeneous and non-homogeneous case, the part of *Update-BasisAndPairs* related to the criteria has been parametrized.

If we want to compute just a minimal system of generators of M, then, by Proposition 4 of (Capani et al. 1997), it suffices to assign the key function *Initialize_Hom* to <u>Initialize</u> and the following key function to <u>OneStep</u>:

```
Define OneStep_MinGens(M)
  If M.IVs = [] Then Return FALSE End;
  V := GetFirstToProcess(M);
  ProcessVector(M,V);
  Return TRUE;
End;
```

We conclude this section by sketching the implementation in our environment of the basic algorithm for computing minimal free resolutions proposed by Capani et al. (1997). This algorithm takes as input a homogeneous submodule M of a finitely generated graded free module, and gives as output the resolution as a chain of graded modules $M_0, M_1, \ldots, M_r$ such that $M_0 = M$, and for each $i \geq 1$, M_i is the ith-order syzygy module of M i.e. the syzygy module of a minimal system of generators of M_{i-1}. Since each module, as a CoCoA data type, can store inside itself the syzygy module of one of its minimal systems of generators, after the computation the whole resolution can be retrieved in M.

We look at a resolution as a growing bidimensional table having a column for each module M_i and a row for each degree. The rightmost column corresponds to M_0.

So the cell (i, d) of the table (denoted by *cell*(i, d)) represents the degree d part of the module M_i and it contains the incoming pairs and the incoming vectors of degree d of M_i. The algorithm follows a "reading pattern strategy", i.e. the cells are processed left to right and up to down.

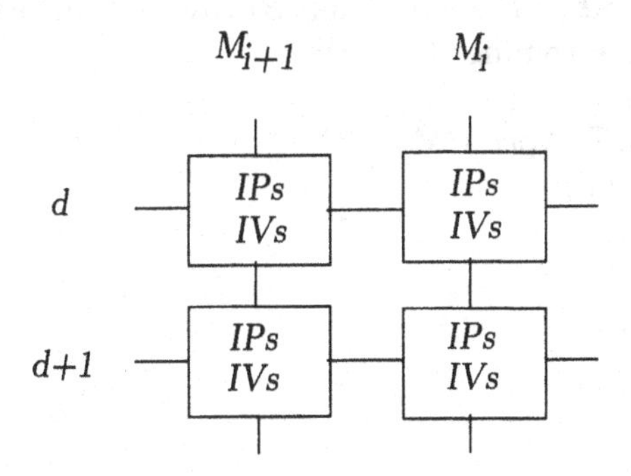

The computation matrix

The algorithm is informally described as follows (Capani et al. 1997).

> Suppose we have completed all the top part of *cell*$(i + 1, d)$ i.e. we have processed all the critical pairs related to it. Then we enter *cell*(i, d) where we find critical pairs in the top part and vectors in the bottom part. Some of the critical pairs produce new elements in the Gröbner Basis of M_i. The others produce syzygies, i.e. incoming vectors in the *cell*$(i+1, d)$. These vectors may reduce to 0 or produce new minimal generators of M_{i+1}. In the last case they possibly produce new critical pairs in the cells *cell*$(i + 1, \delta)$, with $\delta > d$. Then *cell*$(i + 1, d)$ is completely done and we enter *cell*$(i - 1, d)$.

Looking at the processing of an element (pair or vector) of *cell*(i, d) as a single step of computation, we can obtain the above algorithm from *Buchberger_Abstract* by assigning to the parameter function <u>*OneStep*</u> the following key function *OneStep_Res* which operates on the rows of the table by increasing degree, calling the auxiliary function *Do_OneStep_Row*.

```
Define OneStep_Res(M)
  D = Min_Deg_Incomings_Res(M);
  If D = NULL Then Return FALSE End;
  Return Do_OneStep_Row(M,D);
End;

Define Do_OneStep_Row(M, D)
  If M.Syz <> NULL Then
    B := Do_OneStep_Row(M.Syz,D);
    If B Then Return TRUE End;
  End;
  If M.IPs = [] And M.IVs = [] Then Return FALSE End;
  DF := Min_Deg_Incomings_Res(M);
  If DF = D Then
    V := GetFirstToProcess(M);
    ProcessVector(V);
    Return TRUE
  Else
    Return FALSE
  End
End;
```

The function *Do_OneStep_Row* reduces an element of the minimum degree *D* of the highest order syzygy module (the leftmost cell of the first row to be processed) by recursively calling *Do_OneStep_Row* on the syzygy before trying to reduce an element of the "current" module. The function *Min_Deg_Incomings_Res* returns the minimum degree of the incoming elements.

5 Final remarks

In the previous sections we have outlined our framework for the Gröbner computations. The actual implementation (available with the release 3.3 of CoCoA) uses more parameter functions for still greater flexibility. For example we have implemented a refined notion of "step" by using a parameter function Reduce (instead of NF) which allows us to consider the application of a single rule as a single step of computation. Moreover we have implemented a set of key functions that inspect some special variables stored in the module called "Truncations" and stop the execution when the truncation is reached. For example it is possible to specify a "Degree Truncation" in the variable

M.DegTrunc: the execution stops when all the incoming pairs or vectors have degree greater that M.DegTrunc. It is also possible to specify the maximum length of the resolution in the variable *M.ResTrunc* and the Castelnuovo regularity in the variable *M.RegTrunc*.

The minimal interface of CoCoA with the Gröbner framework is given by the functions *Start*, *OneStep* and *Complete*. The use of these functions together with the large number of key functions available allow the experienced user to customize the Gröbner environment.

However the system also offers to the user several predefined settings of key functions for the most common usage. For example the user may compute a Gröbner basis of a module M by typing the CoCoA command `GBasis(M);` this command selects the appropriate setting of key functions taking into account all the information already present in M: for example it check if it is homogeneous, if there are truncations, if its Poincaré Series is present, and so on.

Furthermore the system offers a CoCoAL package containing several high level functions which allow the user:

- to execute a given number of steps of computation;
- to compute for a given time;
- to obtain summary reports about the current status of the computation (how many IVs, IPs, GBasis elements and so on)

We are planning to extend the family of algorithms managed in our framework with the tangent cone algorithm (Mora et al. 1992) and an interactive elimination algorithm driven by the Hilbert function.

References

Adams, W., Loustaunau, P. (1994) *An Introduction to Gröbner Bases.* Amer. Math. Soc., Providence.

Bayer, D., Stillman, M. (1992) *Macaulay: a system for computation in Algebraic Geometry and Commutative Algebra.* Available via anonymous ftp from `math.harvard.edu`.

Buchberger, B. (1965) *An algorithm for finding a basis of the residue class ring of a zero-dimensional polinomial ideal.* PhD thesis, Math. Inst. Univ. Innsbruck.

Buchberger, B. (1985) *Gröbner bases: an algorithmic method in polynomial ideal theory.* In: Bose, N. K. (ed.): *Recent Trends in Multidimensional System Theory.* D. Reidel Publ. Co., Dordrecht, pp. 184–232.

Caboara, M. (1997) *A modified Buchberger algorithm for resolutions.* Preprint.

Caboara, M., Dominicis, G. D., Robbiano, L. (1996) *Multigraded Hilbert functions and Buchberger algorithm.* In: Proc. of ISSAC '96, ACM Press, New York, pp. 72–78.

Capani, A., Dominicis, G. D., Niesi, G., Robbiano, L. (1997) *Computing minimal finite free resolution.* J. Pure Appl. Algebra 117-118: 105–117.

Capani, A., Niesi, G., Robbiano, L. (1996) *CoCoA, a system for doing Computations in Commutative Algebra.* Available via anonymous ftp from `cocoa.dima.unige.it`.

Gebauer, R., Möller, H. M. (1988) *On an installation of Buchberger's algorithm.* J. Symb. Comput. 6: 275–286.

Giovini, A., Mora, T., Niesi, G., Robbiano, L., Traverso, C. (1991) *"One sugar cube, please" or selection strategies in the Buchberger algorithm.* In: Proc. of ISSAC '91, ACM Press, New York, pp. 49–54.

Giovini, A., Niesi, G. (1990) *CoCoA: A user-friendly system for commutative algebra.* In: Miola, A. (ed.): *Design and Implementation of Symbolic Computation Systems.* Springer-Verlag, Berlin Heidelberg New York, pp. 20–29 (Lecture Notes in Computer Science, vol. 429).

Grayson, D., Stillman, M. (1996) *Macaulay 2, a system for computations in Commutative Algebra and Algebraic Geometry.* Available via anonymous ftp from `math.uiuc.edu`.

Greuel, G. M., Pfister, G., Schönemann, H. (1996) *Singular: A system for computation in Algebraic Geometry and Singular Theory.* Available via anonymous ftp from `helios.mathematik.uni-kl.de`.

La Scala, R. (1996) *Un approccio computazionale alle risoluzioni libere minimali.* PhD thesis, Quaderno 1/96, University of Bari.

Möller, M., Mora, T. (1986) *New constructive methods in classical ideal theory.* J. of Alg. 100: 138–178.

Mora, T., Pfister, G., Traverso, C. (1992) *An introduction to the tangent cone algorithm.* Advanced in Computing Research (Issue in Robotics and non Linear Geometry) 6: 199–270.

Noro, M., Takeshima, T. (1992) *Risa/Asir – a computer algebra system.* In: Proc. of ISSAC '92, ACM Press, New York, pp. 387–396.

Schreyer, F. (1980) *Die Berechnung von Syzygien mit dem verallgemeinerten Weierstrass'schen Divisionssatz.* PhD thesis, Diplomarbeit Hamburg.

Spear, D. (1977) *A constructive approach to commutative ring theory.* In: Proc. of MACSYMA Users' Conference. NASA CP-2012, pp. 369–376.

Traverso, C. (1996) *Hilbert functions and the Buchberger algorithm.* J. Symb. Comput. 22: 355–376.

Newton Identities in the multivariate case: Pham Systems

María–José González–López and Laureano González–Vega

Dpt. Matemáticas, Estadística y Computación
Universidad de Cantabria
39071 Santander, Spain

glopez@matesco.unican.es
gvega@matesco.unican.es

Introduction

This paper is devoted to show how the classical Newton Identities relating the Newton Sums together with the elementary symmetric functions can be generalized to the multivariate case. If $\mathbb{K}$ is a field of characteristic 0, $\mathbb{L}$ an algebraically closed field with $\mathbb{K} \subseteq \mathbb{L}$ and $f(T)$ is a polynomial in $\mathbb{K}[T]$ such that:

$$f(T) = a_0 T^n + a_1 T^{n-1} + \ldots + a_{n-1} T + a_n = a_0 (T - \sigma_1) \cdot \ldots \cdot (T - \sigma_n) \qquad \sigma_i \in \mathbb{L},$$

denoting by $\mathbf{S}_k$ the sum of k–powers of the roots of $f(T)$:

$$\mathbf{S}_k = \sum_{i=1}^{n} \sigma_i^k$$

then, for every $j \geq 1$, the following equalities (*the Newton Identities*) hold:

$$j a_j = -\mathbf{S}_j a_0 - \mathbf{S}_{j-1} a_1 - \ldots - \mathbf{S}_1 a_{j-1}$$

(remark that for $j > n$ these identities involve coefficients a_i with $i > n$, which are implicitly defined as 0). In this way Newton Identities give the exact relation between coefficients and roots of an univariate polynomial through the Newton Sums (modulo the leading coefficient a_0): once the roots are known the coefficients are uniquely determined and reciprocally.

In [1] the same kind of relations is proven for a Pham system in $\mathbb{K}[\underline{X}] = \mathbb{K}[X_1, \ldots, X_n]$

$$F_i = X_i^{d_i} + Q_i(\underline{X}); \qquad \deg(Q_i(\underline{X})) < d_i, \qquad i \in \{1, \ldots, n\},$$

over the complex numbers by using the multidimensional logarithmic residue. These "Newton Identities" give the relation between the coefficients of the polynomials F_i and the roots of the system through, in this case, the trace of the different monomials: the addition of the evaluation of the considered monomial over the roots of the system. In this paper these generalized Newton Identities are proven in a completely algebraic way and thus the formulae obtained are also true over any characteristic zero field.

This paper is divided into four different parts. In the first one the algebraic technics to be used are introduced, mainly, some basic results about duality theory over finite dimensional algebras. The second section is devoted to generalize properly to the multivariate setting the notion of Newton Sum by allowing non positive exponents in the considered monomials. The third section presents the "Newton Identities" for Pham Systems together with the announced algebraic proof and finally the last section presents two immediate applications of these identities closely related to polynomial system solving: the parametrization of the Rational Univariate Representation (or Generalized Shape Lemma) and the Quantifier Elimination Problem over Pham Systems. Remark finally that these generalized Newton Identities have already been successfully used to deal with the implicitation problem of parametric surfaces (see [10] or [14]).

1 Algebraic Preliminaries

Let $\mathbb{K}$ be a field of characteristic 0 and $\mathbb{L}$ an algebraically closed field with $\mathbb{K} \subseteq \mathbb{L}$. Let $F_1, \ldots, F_n$ be polynomials in $\mathbb{K}[\underline{X}] = \mathbb{K}[X_1, \ldots, X_n]$ such that the ideal generated by them, J, is zero dimensional, ie the set of common solutions to $F_1, \ldots, F_n$ in $\mathbb{L}^n$ is finite. This set will be denoted by $\{\Delta_1, \ldots, \Delta_r\}$ and μ_i will denote the multiplicity of the solution Δ_i. This section is devoted to show the algebraic properties of the quotient ring $\mathbb{B} = \mathbb{K}[\underline{X}]/J$ (which is a finite–dimensional $\mathbb{K}$-algebra) to be used in the sequel. The terminology in [3] will be followed closely.

Since J is zero dimensional and a complete intersection (same number of equations as unknowns) then $\mathbb{B}$ is a Frobenius Algebra (or Gorenstein ring): there exists a linear form

$$\ell_{[F_1,\ldots,F_n]}: \mathbb{B} \longrightarrow \mathbb{K}$$

such that the symmetric $\mathbb{K}$-bilinear form

$$\begin{array}{rccc} \Phi_\ell: & \mathbb{B} \times \mathbb{B} & \longrightarrow & \mathbb{K} \\ & (a,b) & \longmapsto & \ell_{[F_1,\ldots,F_n]}(ab) \end{array}$$

is non degenerate. Such a linear form $\ell_{[F_1,\ldots,F_n]}$ is called *dualizing linear form* or *global residue operator*, and it can be computed explicitly (see [3]) by means of Gröbner basis computations.

When J is a radical ideal ($\mathbb{B}$ a reduced algebra or every μ_i equal to 1, or the jacobian does not vanish on any solution of J), $\ell_{[F_1,\ldots,F_n]}$ verifies:

$$\ell_{[F_1,\ldots,F_n]}(g) = \sum_{i=1}^{r} \frac{g(\Delta_i)}{\mathbf{Jac}(\Delta_i)}$$

where **Jac** denotes the Jacobian determinant of the polynomials $F_1, \ldots, F_n$. Without assumptions on J, $\ell_{[F_1,\ldots,F_n]}$ is related with the trace by the following equality:

$$\ell_{[F_1,\ldots,F_n]}(g \cdot \mathbf{Jac}) = \sum_{i=1}^{r} \mu_i g(\Delta_i) \stackrel{\text{def}}{=} \text{Trace}(g).$$

A very useful tool to deal with this dualizing form is the *Transformation Law* which is presented in the following proposition.

Proposition 1.1 *Let $G_1, \ldots, G_n$ be polynomials in $\mathbb{K}[\underline{X}]$ with a finite number of solutions in $\mathbb{L}^n$ and such that*

$$G_i = \sum_{j=1}^{n} A_{ij} F_j; \qquad A_{ij} \in \mathbb{K}[\underline{X}], \qquad i \in \{1, \ldots, n\}.$$

If $\mathcal{A}$ is the determinant of the matrix $(A_{ij})_{ij}$ then for any $h \in \mathbb{K}[\underline{X}]$,

$$\ell_{[F_1,\ldots,F_n]}(h) = \ell_{[G_1,\ldots,G_n]}(h \cdot \mathcal{A}).$$

If the polynomials $F_1, \ldots, F_n$ have the following structure (*a Pham System*),

$$F_i = X_i^{d_i} + Q_i(\underline{X}); \qquad \deg(Q_i(\underline{X})) < d_i, \qquad i \in \{1, \ldots, n\}$$

then the dualizing form $\ell_{[F_1,\ldots,F_n]}$ on the monomials defining a basis of $\mathbb{B}$, ie the monomials $\mathbf{X}^\alpha$ whose exponent vector $\alpha = (\alpha_1, \ldots, \alpha_n)$ is such that $\alpha_i < d_i$ for all i, verifies:

$$\ell_{[F_1,\ldots,F_n]}(\mathbf{X}^\alpha) = \begin{cases} 1 & \text{if } \alpha = (d_1 - 1, \ldots, d_n - 1) \\ 0 & \text{if } \alpha \neq (d_1 - 1, \ldots, d_n - 1) \end{cases} \tag{1.1}$$

The last result in this section is the classical Euler–Jacobi Theorem:

Theorem 1.1 *If the polynomials $F_1, \ldots, F_n$ have no solutions at infinity then for any polynomial $h(\underline{X})$ such that:*

$$\deg(h) < \deg(F_1) + \deg(F_2) + \ldots + \deg(F_n) - n$$

we have that $\ell_{[F_1,\ldots,F_n]}(h) = 0$.

2 Generalizing Newton Sums to the multivariate case

The first definition generalizes in the natural way the notion of Newton Sum coming from the univariate case.

Definition 2.1 *Let α be an element of $\mathbb{N}^n$ (a multi–index) and $\mathbf{X}^\alpha$ the monomial whose exponent vector is α. The Newton Sum associated to α is defined as:*

$$\mathbf{S}_\alpha = \sum_{i=1}^{r} \mu_i \mathbf{X}^\alpha(\Delta_i)$$

ie, the Trace of the monomial $\mathbf{X}_\alpha$.

Using the terminology presented in the previous section it is clear that:

$$\mathbf{S}_\alpha = \ell_{[F_1,\ldots,F_n]}(\mathbf{X}^\alpha \cdot \mathbf{Jac})$$

Remark that in the particular case of $\alpha = (0,\ldots,0)$ we have that $\mathbf{S}_\alpha = \ell_{[F_1,\ldots,F_n]}(\mathbf{Jac}) = \text{Trace}(1)$, and thus $\mathbf{S}_\alpha$ is equal to $\dim_{\mathbb{K}} \mathbb{B}$.

Now the definition of $\mathbf{S}_\alpha$ it is extended to the case of multi–indices with negative entries and Pham Systems:

$$F_i = X_i^{d_i} + Q_i(\underline{X}); \qquad \deg(Q_i(\underline{X})) < d_i$$

In the rest of this section we will consider that $\{F_1,\ldots,F_n\}$ have this Pham structure.

Definition 2.2 *Let $\alpha = (\alpha_1,\ldots,\alpha_n)$ be an element of $\mathbb{Z}^n$ (a multi–index). The Newton Sum associated to α is defined as:*

$$\mathbf{S}_\alpha = \ell_{[G_1,\ldots,G_n]}(\mathbf{X}^{\alpha^+} \cdot \mathbf{Jac})$$

where **Jac** *is the jacobian determinant of $F_1,\ldots,F_n$,*

$$\alpha^+ = (\alpha_1^+,\ldots,\alpha_n^+), \qquad \alpha_i^+ = \begin{cases} \alpha_i & \text{if } \alpha_i \geq 0 \\ 0 & \text{if } \alpha_i < 0 \end{cases}$$

and $G_1 = X_1^{\alpha_1^-} F_1,\ldots,G_n = X_n^{\alpha_n^-} F_n$, where

$$\alpha^- = (\alpha_1^-,\ldots,\alpha_n^-), \qquad \alpha_i^- = \begin{cases} 0 & \text{if } \alpha_i \geq 0 \\ -\alpha_i & \text{if } \alpha_i < 0 \end{cases}$$

The conditions imposed on the structure of the polynomials allow to conclude that the new system $G_1,\ldots,G_n$ arising when considering a multi–index with negative entries is also a *Pham System* and therefore zero dimensional. It is clear that both definitions for $\mathbf{S}_\alpha$ agree when considering a multi–index with only positive entries. The next propositions show some particular properties of these Newton Sums.

Proposition 2.1 *If $\alpha = (\alpha_1,\ldots,\alpha_n)$ is a multi–index with norm $\|\alpha\| := \alpha_1 + \ldots + \alpha_n$ strictly smaller than 0 then $\mathbf{S}_\alpha = 0$.*

Proof:
We have $\mathbf{S}_\alpha = \ell_{[G_1,\dots,G_n]}(\mathbf{X}^{\alpha^+}\cdot\mathbf{Jac})$ where $\deg(\mathbf{Jac}) = \sum_{i=1}^n d_i - n$. Besides:

$$\|\alpha\| = \sum_{i=1}^n \alpha_i = \sum_{i=1}^n (\alpha_i^+ - \alpha_i^-) < 0 \Rightarrow \sum_{i=1}^n \alpha_i^+ < \sum_{i=1}^n \alpha_i^- \Rightarrow$$

$$\Rightarrow \sum_{i=1}^n \alpha_i^+ \sum_{i=1}^n d_i - n < \sum_{i=1}^n (d_i + \alpha_i^-) - n \Rightarrow \deg(\mathbf{X}^{\alpha^+}\cdot\mathbf{Jac}) < \sum_{i=1}^n \deg(G_i) - n.$$

Applying the Euler–Jacobi Theorem we conclude that $\ell_{[G_1,\dots,G_n]}(\mathbf{X}^{\alpha^+}\cdot\mathbf{Jac}) = 0$ and thus $\mathbf{S}_\alpha = 0$. ∎

Proposition 2.2 *If α is a multi–index whose norm is 0 then*

$$\mathbf{S}_\alpha = \begin{cases} d_1\cdot\ldots\cdot d_n & \text{if } \alpha = (0,\dots,0) \\ 0 & \text{otherwise.} \end{cases}$$

Proof:
The case $\alpha = (0,\dots,0)$ is a consequence of the previous remark: $\mathbf{S}_\alpha = \mathrm{Trace}(1) = \dim_{\mathbb{K}} \mathbb{B}$. Let α be a multi–index different from $(0,\dots,0)$ such that $\|\alpha\| = 0$. Then:

$$\mathbf{S}_\alpha = \ell_{[X_1^{\alpha_1^-}F_1,\dots,X_n^{\alpha_n^-}F_n]}(\mathbf{X}^{\alpha^+}\cdot\mathbf{Jac}).$$

The special structure of the polynomials F_i allows to describe the jacobian in the following terms:

$$\mathbf{Jac} = \Big(\prod_{i=1}^n d_i\Big)\Big(\prod_{i=1}^n X_i^{d_i-1}\Big) + R(\underline{X}), \qquad \text{where } \deg(R(\underline{X})) < \sum_{i=1}^n d_i - n.$$

Since ℓ is a $\mathbb{K}$-linear form:

$$\mathbf{S}_\alpha = \Big(\prod_{i=1}^n d_i\Big)\ell_{[X_1^{\alpha_1^-}F_1,\dots,X_n^{\alpha_n^-}F_n]}\Big(\prod_{i=1}^n X_i^{d_i+\alpha_i^+-1}\Big) + \ell_{[X_1^{\alpha_1^-}F_1,\dots,X_n^{\alpha_n^-}F_n]}(\mathbf{X}^{\alpha^+}\cdot R(\underline{X})).$$

Due to the explicit behaviour of ℓ for a Pham System (see the equality (1.1) in Section 1) and since α is different from $(0,\dots,0)$, we have that the first summand is equal to 0. For the second summand we have that the degree of the polynomial whose image by $\ell_{[X_1^{\alpha_1^-}F_1,\dots,X_n^{\alpha_n^-}F_n]}$ is to be computed is strictly smaller than

$$\sum_{i=1}^n (d_i + \alpha_i^+) - n$$

and this number is equal to the sum of the degrees of the polynomials $X_i^{\alpha_i^-}F_i$ minus n, ($\|\alpha\| = 0 \implies \sum_{i=1}^n \alpha_i^- = \sum_{i=1}^n \alpha_i^+$). Thus, by Euler–Jacobi Theorem, the second summand is also equal to 0 and it is concluded that if $\|\alpha\| = 0$ and $\alpha \neq (0,\dots,0)$ then $\mathbf{S}_\alpha = 0$. ∎

After these propositions it is clear that the only interesting Generalized Newton Sums are those $\mathbf{S}_\alpha$ with $\alpha \in \mathbb{N}^n$ and $\mathbf{S}_\alpha$ with $\alpha \in \mathbb{Z}^n$ and strictly positive norm (and some negative coordinate).

3 Newton identities for Pham Systems

This section is devoted to generalize to the multivariate setting the Newton Identities when the involved polynomials $\{F_i\}_{i=1}^n$ form a Pham system. More precisely, a family of formulae relating the $\mathbf{S}_\alpha$ with the coefficients of the polynomials F_i will be derived.

Adopting the previous notation for Pham systems, the coefficients of each polynomial $Q_j(\underline{X})$ will be denoted in the following way:

$$Q_j(\underline{X}) = \sum_{\|\alpha\|<d_j} a_\alpha^{(j)} \mathbf{X}^\alpha.$$

This notation is extended to the whole polynomial F_j:

$$F_j(\underline{X}) = \sum_{\|\alpha\|\leq d_j} a_\alpha^{(j)} \mathbf{X}^\alpha$$

by imposing the conditions:

$$\|\alpha\| = d_j \Longrightarrow a_\alpha^{(j)} = \begin{cases} 1 & \text{if } \alpha = (0,\ldots,0,d_j,0,\ldots,0) \\ 0 & \text{otherwise} \end{cases}$$

The next theorem presents the first step to relate the Generalized Newton Sums with the $a_\alpha^{(j)}$'s. This will be done by computing explicitly the image by ℓ of a well precised polynomial with respect to a zero–dimensional ideal.

Theorem 3.1 *Let j be an element of $\{0,\ldots,n\}$, $\delta^{[j]} = (d_1,\ldots,d_j,0,\ldots,0)$, $\beta \in \mathbb{Z}^n$ a multi-index such that $\|\beta\| < \|\delta^{[j]}\|$ and $\gamma = -\beta$. Then:*

$$\mathbf{S}_{\delta^{[j]}-\beta} + \sum_{\|\beta\|<\|\alpha^1+\ldots+\alpha^j \ Vert<\|\delta^{[j]}\|} a_{\alpha^1}^{(1)} \cdot \ldots \cdot a_{\alpha^j}^{(j)} \cdot \mathbf{S}_{\alpha^1+\ldots+\alpha^j-\beta} +$$

$$+\prod_{i=1}^n d_i \cdot \sum_{\alpha^1+\ldots+\alpha^j=\beta} a_{\alpha^1}^{(1)} \cdot \ldots \cdot a_{\alpha^j}^{(j)} = \ell_{[X_1^{\gamma_1^-}F_1,\ldots,X_n^{\gamma_n^-}F_n]}(\mathbf{X}^{\gamma^+} \cdot \mathbf{Jac} \cdot F_1 \cdot \ldots \cdot F_j)$$

where the sums are made for the α^i's running in the set of multi-indices appearing as exponents in the corresponding polinomial F_i.

Proof:
The key point of the proof is the linearity of the form ℓ:

$$\ell_{[X_1^{\gamma_1^-}F_1,\dots,X_n^{\gamma_n^-}F_n]}(\mathbf{X}^{\gamma^+}\cdot\mathbf{Jac}\cdot F_1\cdot\ldots\cdot F_j)=$$

$$=\ell_{[X_1^{\gamma_1^-}F_1,\dots,X_n^{\gamma_n^-}F_n]}\left(\mathbf{Jac}\cdot\sum_{\alpha^1,\dots,\alpha^j}a^{(1)}_{\alpha^1}\cdot\ldots\cdot a^{(j)}_{\alpha^j}\cdot\mathbf{X}^{\alpha^1+\ldots+\alpha^j+\gamma^+}\right)=$$

$$=\sum_{\alpha^1,\dots,\alpha^j}a^{(1)}_{\alpha^1}\cdot\ldots\cdot a^{(j)}_{\alpha^j}\cdot\ell_{[X_1^{\gamma_1^-}F_1,\dots,X_n^{\gamma_n^-}F_n]}(\mathbf{X}^{\alpha^1+\ldots+\alpha^j+\gamma^+}\cdot\mathbf{Jac}).$$

For simplicity in the notation of this proof we will write $\alpha^{1\dots j}$ to denote the sum $\alpha^1+\ldots+\alpha^j$, and $a_{\alpha^{1\dots j}}$ to denote the product $a^{(1)}_{\alpha^1}\cdot\ldots\cdot a^{(j)}_{\alpha^j}$. Remark now that for $\omega=\alpha^{1\dots j}+\gamma$, we can write the following equalities, by using twice the Transformation Law (Proposition 1.1):

$$\ell_{[X_1^{\gamma_1^-}F_1,\dots,X_n^{\gamma_n^-}F_n]}(\mathbf{X}^{\alpha^{1\dots j}+\gamma^+}\cdot\mathbf{Jac})=$$

$$=\ell_{[X_1^{\omega_1^-+\gamma_1^-}F_1,\dots,X_n^{\omega_n^-+\gamma_n^-}F_n]}(\mathbf{X}^{\alpha^{1\dots j}+\gamma^++\omega^-}\cdot\mathbf{Jac})=$$

$$=\ell_{[X_1^{\omega_1^-+\gamma_1^-}F_1,\dots,X_n^{\omega_n^-+\gamma_n^-}F_n]}(\mathbf{X}^{\alpha^{1\dots j}+\gamma^+-\gamma^-+\omega^-}\cdot\mathbf{X}^{\gamma^-}\cdot\mathbf{Jac})=$$

$$=\ell_{[X_1^{\omega_1^-}F_1,\dots,X_n^{\omega_n^-}F_n]}(\mathbf{X}^{\omega+\omega^-}\cdot\mathbf{Jac})=\ell_{[X_1^{\omega_1^-}F_1,\dots,X_n^{\omega_n^-}F_n]}(\mathbf{X}^{\omega^+}\cdot\mathbf{Jac})=\mathbf{S}_\omega$$

Therefore, we can continue the previous chain of equalities:

$$\sum_{\alpha^1,\dots,\alpha^j}a_{\alpha^{1\dots j}}\cdot\ell_{[X_1^{\gamma_1^-}F_1,\dots,X_n^{\gamma_n^-}F_n]}(\mathbf{X}^{\alpha^{1\dots j}+\gamma^+}\cdot\mathbf{Jac})=\sum_{\alpha^1,\dots,\alpha^j}a_{\alpha^{1\dots j}}\cdot\mathbf{S}_{\alpha^{1\dots j}+\gamma}=$$

$$=\sum_{\|\alpha^{1\dots j}-\beta\|\geq 0}a_{\alpha^{1\dots j}}\cdot\mathbf{S}_{\alpha^{1\dots j}-\beta}=\sum_{\alpha^{1\dots j}=\beta}a_{\alpha^{1\dots j}}\cdot\mathbf{S}_{(0,\dots,0)}+$$

$$+\sum_{\|\beta\|<\|\alpha^{1\dots j}\|<\|\delta^{[j]}\|}a_{\alpha^{1\dots j}}\cdot\mathbf{S}_{\alpha^{1\dots j}-\beta}+\sum_{\|\alpha^{1\dots j}\|=\|\delta^{[j]}\|}a_{\alpha^{1\dots j}}\cdot\mathbf{S}_{\alpha^{1\dots j}-\beta}=$$

$$=\prod_{i=1}^n d_i\cdot\sum_{\alpha^{1\dots j}=\beta}a_{\alpha^{1\dots j}}+\sum_{\|\beta\|<\|\alpha^{1\dots j}\|<\|\delta^{[j]}\|}a_{\alpha^{1\dots j}}\cdot\mathbf{S}_{\alpha^{1\dots j}-\beta}+\mathbf{S}_{\delta^{[j]}-\beta}$$

since by Proposition 2.2 we have that:

$$\sum_{\|\alpha^{1\dots j}-\beta\|=0}a_{\alpha^{1\dots j}}\cdot\mathbf{S}_{\alpha^{1\dots j}-\beta}=\sum_{\alpha^{1\dots j}=\beta}a_{\alpha^{1\dots j}}\cdot\mathbf{S}_{(0,\dots,0)},$$

and the equality $\|\alpha^{1\dots j}\|=\|\delta^{[j]}\|$ is equivalent to $\alpha^i=(0,\dots,d_i,\dots,0)$ for every i. ∎

In order to derive a pure relation (without using the form ℓ) between the Generalized Newton Sums and the coefficients of the polynomials F_i's, first a technical lemma will be showed, which provides a very useful way of representing the jacobian of the polynomials $F_1, \ldots, F_n$.

Lemma 3.1 *For any family $\alpha^1, \ldots, \alpha^n$ of multi–indices in $\mathbb{Z}^n$ let us denote by $\Gamma_{[\alpha^1,\ldots,\alpha^n]}$ the determinant:*

$$\Gamma_{[\alpha^1,\ldots,\alpha^n]} = \begin{vmatrix} \alpha_1^1 & \ldots & \alpha_n^1 \\ \vdots & & \vdots \\ \alpha_1^n & \ldots & \alpha_n^n \end{vmatrix}.$$

Then the following equality holds:

$$X_1 \cdot \ldots \cdot X_n \cdot \mathbf{Jac} = \sum_{\alpha^1,\ldots,\alpha^n} a^{(1)}_{\alpha^1} \cdot \ldots \cdot a^{(n)}_{\alpha^n} \cdot \Gamma_{[\alpha^1,\ldots,\alpha^n]} \cdot \mathbf{X}^{\alpha^1} \cdot \ldots \cdot \mathbf{X}^{\alpha^n}$$

with every α^i running in the set of multi–indices appearing as exponents in the corresponding polinomial F_i.

Proof:
The previous way of representing the polynomials F_j allows to write:

$$X_i \cdot \frac{\partial F_j}{\partial X_i} = X_i \cdot \Big(\sum_{\|\alpha\| \le d_j} a^{(j)}_\alpha \frac{\partial \mathbf{X}^\alpha}{\partial X_i} \Big) = \sum_{\|\alpha\| \le d_j} a^{(j)}_\alpha \alpha_i \mathbf{X}^\alpha \qquad (2)$$

Therefore we can write:

$$X_1 \cdot \ldots \cdot X_n \cdot \mathbf{Jac} = \begin{vmatrix} X_1 \cdot \dfrac{\partial F_1}{\partial X_1} & \ldots & X_n \cdot \dfrac{\partial F_1}{\partial X_n} \\ \vdots & & \vdots \\ X_1 \cdot \dfrac{\partial F_n}{\partial X_1} & \ldots & X_n \cdot \dfrac{\partial F_n}{\partial X_n} \end{vmatrix}$$

and by replacing here the equalities in (2), the desired result is obtained by developping the determinant. ■

The previous lemma and theorem allow to introduce the Newton identities for a *Pham System*: a set of inductive relations connecting explicitly the Newton Sums with the coefficients of the considered *Pham System*.

Theorem 3.2 *(Newton Identities for Pham Systems) Let δ be the multi–index $(d_1, \ldots, d_n)$ and $\beta \in \mathbb{Z}^n$ a multi–index such that $\|\beta\| < \|\delta\|$. Then:*

$$\mathbf{S}_{\delta-\beta} + \sum_{\|\beta\| < \|\alpha^1+\ldots+\alpha^n\| < \|\delta\|} a^{(1)}_{\alpha^1} \cdot \ldots \cdot a^{(n)}_{\alpha^n} \cdot \mathbf{S}_{\alpha^1+\ldots+\alpha^n-\beta}$$

$$= \sum_{\alpha^1+\ldots+\alpha^n=\beta} (\Gamma_{[\alpha^1,\ldots,\alpha^n]} - \prod_{i=1}^n d_i) \cdot a^{(1)}_{\alpha^1} \cdot \ldots \cdot a^{(n)}_{\alpha^n}$$

Proof:
By theorem 3.1 (specialized to the case $j = n$), the proof is reduced to compute the value of

$$\ell_{[X_1^{\gamma_1^-} F_1,\ldots,X_n^{\gamma_n^-} F_n]}(\mathbf{X}^{\gamma^+} \cdot \mathbf{Jac} \cdot F_1 \cdot \ldots \cdot F_n)$$

for $\gamma = -\beta$. By using the Transformation Law (Proposition 1.1), the expression of the Jacobian for $F_1, \ldots, F_n$ presented in Lemma 3.1 and the equality (1.1) in Section 1, we obtain:

$$\ell_{[X_1^{\gamma_1^-} F_1,\ldots,X_n^{\gamma_n^-} F_n]}(\mathbf{X}^{\gamma^+} \cdot \mathbf{Jac} \cdot F_1 \cdot \ldots \cdot F_n) = \ell_{[X_1^{\gamma_1^-},\ldots,X_n^{\gamma_n^-}]}(\mathbf{X}^{\gamma^+} \cdot \mathbf{Jac}) =$$

$$= \sum_{\alpha^1+\ldots+\alpha^n=\beta} \Gamma_{[\alpha^1,\ldots,\alpha^n]} \cdot a^{(1)}_{\alpha^1} \cdot \ldots \cdot a^{(n)}_{\alpha^n}$$

and thus the proof of the theorem. ∎

The next corollary, which is no more than a rewriting of the previous theorem, gives some insight about how the Newton Sums

$$\mathbf{S}_{(\gamma_1,\ldots,\gamma_n)}; \qquad \gamma_i < d_i, \qquad i \in \{1,\ldots,n\}$$

can be derived explicitly from the coefficients of the polynomials F_i. In what follows, it will be assumed that these coefficients are variables which will be specialized to its corresponding value; in particular, the leading coefficient of F_i, $a^{(i)}_{(0,\ldots,d_i,\ldots,0)}$ will be specialized to 1. The set of the coefficients of the polynomials F_i will be denoted by $\underline{a}$:

$$\underline{a} = ((\ldots, a^{(1)}_{\alpha^1}, \ldots), \ldots, (\ldots, a^{(i)}_{\alpha^i}, \ldots), \ldots, (\ldots, a^{(n)}_{\alpha^n}, \ldots))$$

where $(\ldots, a^{(i)}_{\alpha^i}, \ldots)$ represents the list of coefficients of F_i.

Corollary 3.1 *Let $\gamma = (\gamma_1, \ldots, \gamma_n)$ be a multi-index with strictly positive norm. Then*

$$\mathbf{S}_\gamma = H_\gamma(\underline{a}) + \sum_{0 \leq \|\tau\| < \|\gamma\|} h_\tau(\underline{a}) \cdot \mathbf{S}_\tau$$

where:

- *every $h_\tau(\underline{a})$ is a homogeneous polynomial in $\mathbb{Z}[\underline{a}]$ of total degree n and each monomial in $h_\tau(\underline{a})$ has exactly one coefficient from each F_i,*
- *the polynomial $H_\gamma(\underline{a})$ can be described in the following way:*

$$H_\gamma(\underline{a}) = \sum_{\alpha^1+\ldots+\alpha^n=\delta-\gamma} \Gamma_{[\alpha^1,\ldots,\alpha^n]} \cdot a^{(1)}_{\alpha^1} \cdot \ldots \cdot a^{(n)}_{\alpha^n}$$

and it has the previously stated properties for the polynomials $h_\tau(\underline{a})$.

Thus $\mathbf{S}_\gamma$ is a polynomial in $\mathbb{Z}[\underline{a}]$ of total degree equal to $\|\gamma\|\cdot n$, with its lowest homogeneous part of degree bigger or equal to n.

For example if $\|\gamma\| = 1$ then

$$\mathbf{S}_\gamma = \sum_{\alpha^1+\ldots+\alpha^n=\delta-\gamma} (\Gamma_{[\alpha^1,\ldots,\alpha^n]} - \prod_{k=1}^{n} d_k)\cdot a^{(1)}_{\alpha^1}\cdot\ldots\cdot a^{(n)}_{\alpha^n}$$

and, in particular, when $\gamma = p_i = (0,\ldots,1,\ldots,0)$, the trace of the variable X_i is explicitly described:

$$\mathbf{S}_{(0,\ldots,1,\ldots,0)} = \sum_{\alpha^1+\ldots+\alpha^n=\delta-p_i} (\Gamma_{[\alpha^1,\ldots,\alpha^n]} - \prod_{k=1}^{n} d_k)\cdot a^{(1)}_{\alpha^1}\cdot\ldots\cdot a^{(n)}_{\alpha^n} =$$

$$= \Big((d_i-1)\prod_{k\neq i} d_k - \prod_{k=1}^{n} d_k\Big)\cdot a^{(1)}_{d_1p_1}\cdot\ldots\cdot a^{(i)}_{(d_i-1)p_i}\cdot\ldots\cdot a^{(n)}_{d_np_n} = -a^{(i)}_{(0,\ldots,d_i-1,\ldots,0)}\cdot\prod_{k\neq i} d_k$$

since $\alpha^1+\ldots+\alpha^n = \delta - p_i$ is equivalent to $\alpha^j = d_jp_j$ for $j\neq i$ and $\alpha^i = (d_i-1)p_i$.

This last particular case motivates the following computation which shows how the formulae in theorem 3.2 generalizes the Newton Identities of the one variable case. If $n = 1$ and

$$F_1 = x^d + a_1x^{d-1} + \ldots + a_d$$

then the formula in theorem 3.2 applied to $\delta = (d)$ and $\beta = (k)$ with $k < d$ gives

$$\mathbf{S}_{d-k} + \sum_{k<j<d} a_j\mathbf{S}_{j-k} = (k-d)a_k$$

which is another way of rewriting the classical Newton Identities.

This section is finished by showing how these Newton Identities work on a particular example.

Example 3.1
Let n be equal to 2 and $\delta = (d_1,d_2) = (2,2)$. In this case our Pham System has the following structure:

$$F_1 = X_1^2 + u_1X_1 + v_1X_2 + w_1,$$
$$F_2 = X_2^2 + u_2X_1 + v_2X_2 + w_2.$$

The attention will be focused in the computation of those $\mathbf{S}_\gamma$ such that every component of γ is non–negative, ie, following the notation of Theorem 3.2, we are interested in computing $\mathbf{S}_{\delta-\beta}$ such that $\|\beta\| < 4$, $\beta_1 \leq 2$ and $\beta_2 \leq 2$. Remark that in some cases these computations can imply to compute some $\mathbf{S}_\alpha$

such that some component of α is strictly negative. We start the computations with those β of norm 3: $(2,1)$ and $(1,2)$. The following results are obtained:

$$\mathbf{S}_{(0,1)} = \mathbf{S}_{(2,2)-(2,1)} = -2v_2, \qquad \mathbf{S}_{(1,0)} = \mathbf{S}_{(2,2)-(1,2)} = -2u_1.$$

Next, the case of norm 2 is considered: $(2,0)$, $(1,1)$ and $(0,2)$. In this case the corresponding Newton Sums are:

$$\begin{aligned}
\mathbf{S}_{(2,0)} = \mathbf{S}_{(2,2)-(0,2)} &= 2u_1^2 + 2v_1v_2 - 4w_1, \\
\mathbf{S}_{(1,1)} = \mathbf{S}_{(2,2)-(1,1)} &= 3v_1u_2 + v_2u_1, \\
\mathbf{S}_{(0,2)} = \mathbf{S}_{(2,2)-(2,0)} &= 2u_2u_1 + 2v_2^2 - 4w_2.
\end{aligned}$$

Remark that to determine the values $\mathbf{S}_{(2,0)}$ and $\mathbf{S}_{(0,2)}$ it was necessary to compute the following values:

$$\begin{aligned}
&\mathbf{S}_{(3,-2)} = \mathbf{S}_{(2,2)-(-1,4)} = 0, &\qquad &\mathbf{S}_{(2,-1)} = \mathbf{S}_{(2,2)-(0,3)} = -4v_1, \\
&\mathbf{S}_{(-1,2)} = \mathbf{S}_{(2,2)-(3,0)} = -4u_2, &\qquad &\mathbf{S}_{(-2,3)} = \mathbf{S}_{(2,2)-(4,-1)} = 0.
\end{aligned}$$

The computation follow the same lines for other values of β.

Other approach that we can follow to compute the Newtom Sums in this particular case is to note that once that $\mathbf{S}_{(0,0)}$, $\mathbf{S}_{(1,0)}$, $\mathbf{S}_{(0,1)}$ and $\mathbf{S}_{(1,1)}$ have been computed, we can determine the other Newton Sums for α's with positive components, as any monomial $X_1^{s_1}X_2^{s_2}$ can be written modulo F_1 and F_2 in an unique way as

$$X_1^{s_1}X_2^{s_2} = A + BX_1 + CX_2 + DX_1X_2$$

and thus:

$$\mathbf{S}_{(s_1,s_2)} = A\mathbf{S}_{(0,0)} + B\mathbf{S}_{(1,0)} + C\mathbf{S}_{(0,1)} + D\mathbf{S}_{(1,1)}.$$

Therefore, as $\mathbf{S}_{(0,0)} = 4$, the formulae in theorem 3.2 would just be used to determine $\mathbf{S}_{(1,0)}, \mathbf{S}_{(0,1)}$ and $\mathbf{S}_{(1,1)}$.

4 Applications

This section is devoted to present two applications of the Newton Identities for Pham Systems introduced in the previous section. These two applications are the parametrization of the Shape Lemma (or the Generalized one, also called Rational Univariate Representation) and the parametrization of the Trace Matrix for *Pham Systems.* They are very closely related to Polynomial System Solving. Since every Pham System is already a Gröbner Basis (see [6]) the formulae shown in the previous section can be regarded as the way of connecting the polynomials in a Gröbner Basis with the solutions of the considered ideal.

The Shape Lemma is a powerful tool in Computer Algebra since it allows to reduce many questions about the solution set of a zero dimensional ideal to univariate computations. It was introduced in Computer Algebra in [8] but its history can be traced back to Cartan or Narasimhan as a natural generalization of the Primitive Element Theorem. It can be stated in the following way (see [4], [5], [7], [8], [9], [15], [16] or [18] for several proofs).

Theorem 4.1 *(Shape Lemma) Let J be a zero–dimensional ideal in $\mathbb{K}[\underline{X}]$ which is in general position with respect to X_1, i.e. the projection of $\mathcal{V}_{\mathbf{L}}(J)$ onto the 1-st coordinate is injective. Then $\sqrt{J}$ has a lexicographical Gröbner base with respect to $X_n > X_{n-1} > \ldots > X_1$ of the form:*

$$\sqrt{J} = \langle f(X_1), X_2 - g_2(X_1), \ldots, X_{n-1} - g_{n-1}(X_1), X_n - g_n(X_1) \rangle$$

where f is a square–free polynomial and the degree of every g_i does not exceed the degree of f.

Although the Shape Lemma has a practical interest, it is difficult to apply in many practical situations due to the growing of the coefficients of the g_i's (when dealing with integer coefficients) which makes it very hard to perform the computations or to use the output. A very detailed study with bounds (on the degrees or the heights of the integers involved in the Shape Lemma) can be found in [9] or [15].

One direction trying to avoid this drawback is the Generalized Shape Lemma or Rational Univariate Representation (see [2], [13] or [18]) by allowing a more complicated structure in the base of $\sqrt{J}$ but with a better computational behaviour showed on some particular examples where the use of the Shape Lemma is complicated (see [18]).

Theorem 4.2 *(Generalized Shape Lemma) Let J be a zero–dimensional ideal in $\mathbb{K}[\underline{X}]$ which is in general position with respect to X_1. Then there exists an algorithm computing a description of $\sqrt{J}$ with the following shape:*

$$\sqrt{J} = \langle f(X_1), f'(X_1)X_2 - h_2(X_1), \ldots, f'(X_1)X_n - h_n(X) \rangle$$

where f is a square–free polynomial and the degree of every h_i does not exceed the degree of f.

The Shape Lemma is an easy corollary of this last theorem since $f(x_n)$ and $f'(x_n)$ are coprime. In [12], [13] and [18] it was proved that the computation of the Generalized Shape Lemma reduces to the determination of the Traces of a well precised set of polynomials as it is showed in the next proposition.

Proposition 4.1 *The coefficients of the polynomials in the Generalized Shape Lemma are fully determined once the following elements of $\mathbb{K}$ are computed:*

- Trace(X_1^m), *for $m \in \{1, \ldots, D\}$.*

- $\text{Trace}(X_1^i X_k)$, *for* $k \in \{2,\ldots,n\}$, $i \in \{0,\ldots,D-1\}$.

where D *is the dimension of* $\mathbb{B}$.

When dealing with a *Pham System* of type $(d_1,\ldots,d_n)$, the value of D is equal to the product of the d_i's. Assuming that X_1 is a separating element, the computation of the univariate polynomial $f(X_1)$ is performed by determining the square–free part of a polynomial $g(X_1) \in \mathbb{K}[X_1]$ of degree D. The coefficients of

$$g(X_1) = a_0 X_1^D + a_1 X_1^{D-1} + \ldots + a_D$$

are computed by using the classical Newton Identities of the univariate case on the Newton Sums:

$$ja_j = -\mathbf{S}_j a_0 - \mathbf{S}_{j-1} a_1 - \ldots - \mathbf{S}_1 a_{j-1}$$

but remark that each one of these Sums $\mathbf{S}_m$ can be determined by the formulae in theorem 3.2, as we have:

$$\mathbf{S}_m = \text{Trace}(X_1^m) = \mathbf{S}_{(m,0,\ldots,0)}; \qquad m \in \{1,\ldots,D\}.$$

The coefficients of the other polynomials in the Generalized Shape Lemma are obtained as Newton Sums

$$\mathbf{S}_{(i,0,\ldots,0,1,0,\ldots,0)} = \text{Trace}(X_1^i X_k)$$

In fact, the coefficient of X_1^{D-i} $(i \in \{1,\ldots,D\})$ in $h_k(X_1)$ is equal to

$$\text{Trace}(X_k(a_0 X_1^{i-1} + \ldots + a_{i-2}X_1 + a_{i-1})) =$$

$$= a_0 \mathbf{S}_{(i-1,0,\ldots,0,1,0,\ldots,0)} + \ldots + a_{i-2}\mathbf{S}_{(1,0,\ldots,0,1,0,\ldots,0)} + a_{i-1}\mathbf{S}_{(0,\ldots,0,1,0,\ldots,0)}.$$

The next proposition shows the degree behaviour of the coefficients of the polynomials in this construction. Its proof is an easy consequence of the degree considerations in corollary 3.1.

Proposition 4.2 *The coefficient of* X_1^{D-i} *in* $g(X_1)$ *and in* $h_k(X_1)$ *has total degree, as polynomial in* $\mathbb{Z}[\underline{a}]$, *bounded by* $i \cdot n$.

The last part of this section is devoted to show how the Generalized Newton Identities can be used to deal with the Quantifier Elimination problem for formulae involving Pham Systems. Let $F_1,\ldots,F_n$ be a Pham System of type $(d_1,\ldots,d_n)$ with coefficients in $\mathbb{Z}[t_1,\ldots,t_m]$. Since the monomials $X_1^{\alpha_1} \cdot \ldots \cdot X_n^{\alpha_n}$ with $0 \le \alpha_i < d_i$ $(1 \le i \le n)$ are a basis $\{\omega_1,\ldots,\omega_D\}$ of $\mathbb{B} = \mathbb{K}[\underline{X}]/\langle F_1,\ldots,F_n\rangle$ for any value of the parameters in any ordered field $\mathbb{K}$, then, according to [11] or [17]:

$$\exists X_1 \in \mathbb{F} \ldots \exists X_n \in \mathbb{F} \qquad F_1(\underline{X}) = 0 \wedge \ldots \wedge F_n(\underline{X}) \quad \Longleftrightarrow$$

$$\iff \quad \text{signature}\begin{pmatrix} \text{Trace}(\omega_1\omega_1) & \dots & \text{Trace}(\omega_1\omega_D) \\ \vdots & & \vdots \\ \text{Trace}(\omega_D\omega_1) & \dots & \text{Trace}(\omega_D\omega_D) \end{pmatrix} > 0$$

where $\mathbb{F}$ is a real closed field containing $\mathbb{K}$. The use of Newton Identities in theorem 3.2 allows to compute $\mathbf{S}_\alpha$ for $\alpha = (\alpha_1, \dots, \alpha_n)$ with $0 \leq \alpha_i < 2d_i$. These $\mathbf{S}_\alpha$ are exactly the entries of the previous matrix and thus the problem of determining the quantifier free formula is reduced to parametrize the signature of a symmetric matrix (see [11] or [19]); in particular, V. Weispfenning in [19] produces a symmetric matrix using Comprehensive Gröbner Basis, and then, uses an approach based on the Descartes' Rule to parametrize its signature.

As a final example, our strategy applied to the Pham system of the example 3.1 reduces any Quantifier Elimination problem on the polynomial system of equations $\{F_1 = 0, F_2 = 0\}$ to compute the signature of the symmetric matrix:

$$\begin{pmatrix} \mathbf{S}_{(0,0)} & \mathbf{S}_{(1,0)} & \mathbf{S}_{(0,1)} & \mathbf{S}_{(1,1)} \\ \mathbf{S}_{(1,0)} & \mathbf{S}_{(2,0)} & \mathbf{S}_{(1,1)} & \mathbf{S}_{(2,1)} \\ \mathbf{S}_{(0,1)} & \mathbf{S}_{(1,1)} & \mathbf{S}_{(0,2)} & \mathbf{S}_{(1,2)} \\ \mathbf{S}_{(1,1)} & \mathbf{S}_{(2,1)} & \mathbf{S}_{(1,2)} & \mathbf{S}_{(2,2)} \end{pmatrix}.$$

By applying the Newton Identities to compute these $\mathbf{S}_{(\alpha_1,\alpha_2)}$, the following results are obtained:

$$\begin{aligned}
\mathbf{S}_{(0,0)} &= 4 \\
\mathbf{S}_{(0,1)} &= -2v_2 \\
\mathbf{S}_{(1,0)} &= -2u_1 \\
\mathbf{S}_{(2,0)} &= 2u_1^2 + 2v_1v_2 - 4w_1 \\
\mathbf{S}_{(1,1)} &= 3v_1u_2 + v_2u_1 \\
\mathbf{S}_{(0,2)} &= 2u_2u_1 + 2v_2^2 - 4w_2 \\
\mathbf{S}_{(1,2)} &= -2u_2u_1^2 - 5u_2v_1v_2 + 4u_2w_1 - v_2^2u_1 + 2w_2u_1 \\
\mathbf{S}_{(2,1)} &= -5u_1v_1u_2 - v_2u_1^2 - 2v_1v_2^2 + 4v_1w_2 + 2v_2w_1 \\
\mathbf{S}_{(2,2)} &= 8u_2u_1v_1v_2 + u_1^2v_2^2 - 2u_1^2w_2 + 3v_1^2u_2^2 - 6v_1v_2w_2 \\
&\quad -6u_2u_1w_1 - 2w_1v_2^2 + 4w_1w_2 + 2u_2u_1^3 + 2v_1v_2^3
\end{aligned}$$

Acknowledgements

The authors have been partially supported by FRISCO (European Union, LTR 21.024) and DGICYT PB 95/0563-A (Sistemas de Ecuaciones Algebraicas: Resolución y Aplicaciones).

References

[1] L. A. Aizenberg and A. M. Kytmanov: *Multidimensional Analogues of Newton's Formulas for systems of nonlinear algebraic equations and some of their applications.* Siberian Mathematical Journal 22, 180–189 (1981).

[2] M.-E. Alonso, E. Becker, M.-F. Roy and T. Wörmann: *Zeros, Multiplicities and Idempotents for Zerodimensional Systems.* Algorithms in Algebraic Geometry and Applications, Progress in Mathematics 143, 1–16, Birkhaüser (1996).

[3] E. Becker, J.-P. Cardinal, M.-F. Roy and Z. Szafraniec: *Multivariate Bezoutians, Kronecker Symbol and Eisenbud–Levine Formula.* Algorithms in Algebraic Geometry and Applications, Progress in Mathematics 143, 79–104, Birkhaüser (1996).

[4] E. Becker, M. G. Marinari, T. Mora and C. Traverso: *The shape of the Shape Lemma.* Proceedings of ISSAC–94, 129–133, ACM Press (1993).

[5] E. Becker and T. Wörmann: *On the trace formula for quadratic forms and some applications.* Recent Advances in Real Algebraic Geometry and Quadratic Forms. Contemporary Mathematics **155**, 271–291, AMS Publications (1993).

[6] B. Buchberger: *Gröbner Bases: An algorithmic method in polynomial ideal theory.* Chapter 6 in: Multidimensional Systems Theory (N. K. Bose Ed.), 184–232, Reidel Publishing Company, Dodrecht, (1985).

[7] J. F. Canny: *The complexity of robot motion planning.* ACM Doctoral Dissertation Series, MIT Press, Cambridge Mass. (1988).

[8] P. Gianni and T. Mora: *Algebraic solution of polynomial equations using Gröbner bases.* Proceedings AAECC–5. Lectures Notes in Computer Science 359, 247-257, Springer-Verlag (1989).

[9] M. Giusti and J. Heintz: *La determination des points isoles et de la dimension d'une variete algebrique peut se faire en temps polynomial.* Computational Algebraic Geometry and Commutative Algebra, Symposia Mathematica, vol. XXXIV, 216–256, Cambridge University Press (1993).

[10] L. Gonzalez–Vega: *Implicitization of parametric curves and surfaces by using multidimensional Newton Formulae.* Journal of Symbolic Computation 23, 137–151 (1997).

[11] L. Gonzalez–Vega: *A special Quantifier Elimination algorithm for Pham Systems.* Submitted to the RAGOS Proceedings (1997).

[12] L. Gonzalez–Vega, F. Rouillier and M.–F. Roy: *Symbolic Recipes for Polynomial System Solving.* To appear in the book Some “Tapas” of Computer Algebra to be published by Springer–Verlag in the series “Algorithms and Computations in Mathematics” (1997).

[13] L. Gonzalez–Vega and G. Trujillo: *Using Symmetric Functions to describe the Solution Set of a Zero Dimensional Ideal.* Proceedings of AAECC 11, Lecture Notes in Computer Science 948, 232–247, Springer–Verlag (1995).

[14] L. Gonzalez–Vega and G. Trujillo: *Implicitization of parametric curves and surfaces by using symmetric functions.* Proceedings of ISSAC–95, 180–186, ACM Press (1995).

[15] T. Krick and L. M. Pardo: *A Computational Method for Diophantine Approximation.* Algorithms in Algebraic Geometry and Applications, Progress in Mathematics 143, 193–254, Birkhaüser (1996).

[16] Y. N. Lakshman and D. Lazard: *On the Complexity of Zero–dimensional Algebraic Systems.* Effective Methods in Algebraic Geometry. Progress in Mathematics 94, 217–225, Birkhaüser (1991).

[17] P. Pedersen, M.-F. Roy and A. Szpirglas: *Counting Real Zeros in the multivariate case.* Computational Algebraic Geometry. Progress in Mathematics 109, 203–224, Birkhaüser (1993).

[18] F. Rouillier: *Algorithmes efficaces pour l'étude des zéros réels des systèmes polynomiaux.* Doctoral Thesis, Université de Rennes I (1996).

[19] V. Weispfenning: *A new approach to quantifier elimination for real algebra.* To appear in the special volume of the series Texts and Monographs in Symbolic Computation (Springer-Verlag) entitled *25 years of Quantifier Elimination and Cylindrical Algebraic Decomposition* (1995).

Gröbner Bases in Rings of Differential Operators

Mariano Insa and Franz Pauer

Institut für Mathematik, Universität Innsbruck,
Technikerstrasse 25, A-6020 Innsbruck, Austria.
e-mail: Mariano.Insa@uibk.ac.at & Franz.Pauer@uibk.ac.at

1 Introduction

Let $\mathcal{R}$ be the ring of all complex rational functions without poles in a given real interval. The work of U. Oberst and S. Fröhler ([7],[8],[14]) on systems of differential equations with time-varying coefficients raised several questions for modules over the ring $\mathcal{R}[D]$ of linear differential operators with coefficients in $\mathcal{R}$.

There are a number of results ([2],[5],[6],[9],[11],[12],[13],[17],...) on Gröbner bases in rings of differential operators, but the coefficient rings are fields (of rational functions), rings of power series, or rings of polynomials over a field. In the latter case every differential operator is a K-linear combination of "terms" x^iD^j, $(i,j) \in \mathbf{N}^n \times \mathbf{N}^n$. Thus Gröbner bases are defined with respect to a term order on $\mathbf{N}^n \times \mathbf{N}^n$, the coefficients are elements of a field and commute with the terms. This approach cannot be used for other coefficient rings (like $\mathcal{R}$, for example).

The results of B. Buchberger ([3],[4]) on Gröbner bases in polynomial rings have been generalized by several authors (see for example [9]) to polynomial rings with coefficients in commutative rings. In analogy to this extension we present a basic theory of Gröbner bases for differential operators with coefficients in a commutative ring. The coefficient ring $\mathcal{R}$ is a noetherian K-subalgebra of the field of n-variate rational functions, which is stable under partial derivatives. The ring of differential operators with coefficients in $\mathcal{R}$ is a free $\mathcal{R}$-module with basis $\{D^i \mid i \in \mathbf{N}^n\}$ (the set of "terms"). We define Gröbner bases with respect to a term order on $\mathbf{N}^n$. Note that in our situation the terms do not commute with the coefficients. In the last section we give some examples and applications.

For another constructive approach to differential operators see [16].

We would like to thank U. Oberst for stimulating questions and discussions.

2 Division algorithm and Gröbner bases

Let K be a field of characteristic zero, n a positive integer, and $K[x] := K[x_1, \ldots, x_n]$ (resp. $K(x) := K(x_1, \ldots, x_n)$) the ring of polynomials (resp. the field of rational functions) in n variables over K. Let $\frac{\partial}{\partial x_i} : K(x) \to K(x)$ be the partial derivative by x_i, $1 \leq i \leq n$.

Let $\mathcal{R}$ be a noetherian K-subalgebra of $K(x_1, \ldots, x_n)$ which is stable by $\frac{\partial}{\partial x_i}$, $1 \leq i \leq n$ (i.e. $\frac{\partial}{\partial x_i}(r) \in \mathcal{R}$, for all $r \in \mathcal{R}, 1 \leq i \leq n$). Important examples for $\mathcal{R}$ are $K[x]$, $K(x)$ and $K[x]_M := \{\frac{f}{g} \in K(x) \mid f \in K[x], g \in M\}$ where M is a subset of $K[x] \setminus \{0\}$ closed under multiplication.

We assume that we can solve linear equations over $\mathcal{R}$, i.e.:

- for all $r \in \mathcal{R}$ and for all finite subsets $S \subseteq \mathcal{R}$, we can decide, if r is an element of ${}_{\mathcal{R}}\langle S \rangle$, and if yes, we can compute a family $(d_s)_{s \in S}$ in $\mathcal{R}$ such that $r = \sum_{s \in S} d_s s$.
- For all finite subsets $S \subseteq \mathcal{R}$, a finite system of generators of the $\mathcal{R}$-module
$$\left\{ (c_s)_{s \in S} \in \mathcal{R}^S \mid \sum_{s \in S} c_s s = 0 \right\}$$
can be computed.

We denote by D_i the restriction of $\frac{\partial}{\partial x_i}$ to $\mathcal{R}$, $1 \leq i \leq n$. Then $D_1, \ldots, D_n \in End_K(\mathcal{R})$. Let $\mathcal{A} := \mathcal{R}[D] := \mathcal{R}[D_1, \ldots, D_n]$ be the $\mathcal{R}$-subalgebra of $End_K(\mathcal{R})$ generated by $id_{\mathcal{R}} =: 1$ and $D_1, \ldots, D_n$.

It is well-known that $\mathcal{A}$ is a left-noetherian associative $\mathcal{R}$-algebra. By an "ideal in $\mathcal{A}$" we always mean a left-ideal of $\mathcal{A}$ (i.e. an $\mathcal{A}$-submodule of ${}_{\mathcal{A}}\mathcal{A}$). If $\mathcal{R} = K[x]$, then $\mathcal{A}$ is the Weyl-Algebra. For $r \in \mathcal{R}$ we have: $D_i r = D_i(r) + r D_i$. $\mathcal{A}$ is a free $\mathcal{R}$-module with $\mathcal{R}$-basis $\{D^i := D_1{}^{i_1} \cdots D_n{}^{i_n} \mid i \in \mathbf{N}^n\}$. The elements of $\mathcal{A}$ are called "differential operators (with coefficients in $\mathcal{R}$)", they can be written uniquely as finite sums

$$\sum_{i \in \mathbf{N}^n} r_i D^i, \qquad \text{where} \quad r_i \in \mathcal{R} .$$

Let $<$ be a term order on $\mathbf{N}^n$ (i.e. a total order on $\mathbf{N}^n$ such that $0 \in \mathbf{N}^n$ is the smallest element and $i < j$ implies $i + k < j + k$, for all $i, j, k \in \mathbf{N}^n$). For a differential operator $0 \neq f = \sum_{i \in \mathbf{N}^n} r_i D^i$, we define

- $deg(f) := \max_{<}\{i \mid r_i \neq 0\} \in \mathbf{N}^n$ ("degree of f"),
- $lc(f) := r_{deg(f)}$ ("leading coefficient of f"),
- $in(f) := lc(f) D^{deg(f)}$ (the "initial form of f").

If F is a subset of $\mathcal{A}$ we define

- $deg(F) := \{deg(f) \mid f \in F,\ f \neq 0\}$ and
- $in(F) := \{in(f) \mid f \in F,\ f \neq 0\}$.

Remark 1 *The notations $deg(f)$, $lc(f)$, $in(f)$ for $f \in \mathcal{R}[D]$ are defined formally in the same way as $deg(g)$, $lc(g)$, $in(g)$ for $g \in \mathcal{R}[y_1, \ldots, y_n]$, the commutative ring of polynomials.*

Nevertheless, we have to take into account important differences for these two situations:

- *In $\mathcal{R}[y]$ we have $y^i in(g) = in(y^i g)$, but in $\mathcal{R}[D]$ in general this is not true:*
$$D^i in(f) \neq in(D^i f)\ .$$
(For example $D_1(x_1 D_1) = D_1 + x_1 D_1{}^2 \neq in(D_1(x_1 D_1)) = x_1 D_1{}^2$). However we have
$$in(D^i in(f)) = in(D^i f) \quad \text{and} \quad deg(D^i f) = i + deg(f)\ .$$
More generally, for $f, h \in \mathcal{A} \setminus \{0\}$ we have $deg(f \cdot h) = deg(f) + deg(h)$.
- *In $\mathcal{R}[y]$ monomial ideals are "easy" (e.g. ideal membership can easily be verified), in $\mathcal{R}[D]$ this is not true (see Example 3).*

Proposition 1 (**Division in** $\mathcal{A}$)*: Let F be a finite subset of $\mathcal{A} \setminus \{0\}$ and let $g \in \mathcal{A}$. Then there is a differential operator $r \in \mathcal{A}$ and there is a family $(h_f)_{f \in F}$ in $\mathcal{A}$ such that*

- $g = \sum_{f \in F} h_f f + r$ *(r is "a remainder of g after division by F"),*
- *for all $f \in F$, $h_f = 0$* or *$deg(h_f f) \leq deg(g)$,*
- *$r = 0$ or $lc(r) \notin {}_{\mathcal{R}}\langle lc(f);\ deg(r) \in deg(f) + \mathbf{N}^n \rangle$.*

The differential operators r, $h_f (f \in F)$ can be computed as follows:
First set: $r := g$ and $h_f = 0$ $(f \in F)$.
While $r \neq 0$ and $lc(r) \in {}_{\mathcal{R}}\langle lc(f);\ deg(r) \in deg(f) + \mathbf{N}^n \rangle$ do the following:
let $F' := \{f \in F \mid deg(r) \in deg(f) + \mathbf{N}^n\}$, compute a family $(c_f)_{f \in F'}$ in $\mathcal{R}$ such that
$$\sum_{f \in F'} c_f lc(f) = lc(r)\ .$$
Replace
$$r \ \text{ by } \ r - \sum_{f \in F'} c_f D^{deg(r) - deg(f)} f$$
and
$$h_f \ \text{ by } \ h_f + c_f D^{deg(r) - deg(f)} f, \quad f \in F'.$$

Proof: Since $deg(r - \sum_{f \in F'} c_f D^{deg(r)-deg(f)} f) < deg(r)$, the algorithm terminates after finitely many steps. $\surd$

Remark 2 *We can solve linear equations over $\mathcal{R}$. Hence we are able to decide whether the condition*

$$lc(r) \notin \langle lc(f);\ deg(r) \in deg(f) + \mathbf{N}^n \rangle$$

is fulfilled or not. It would not be reasonable to replace this condition by

$$in(r) \notin_{\mathcal{A}} \langle in(F) \rangle \quad or \quad in(r) \notin in({}_{\mathcal{A}}\langle in(F) \rangle),$$

since up to now we do not have any means to decide if these conditions are fulfilled or not (see Example 3).

Example 1 *Let* $\mathcal{R} := \left\{ \frac{f}{g} \mid f, g \in \mathbf{Q}[x_1, x_2],\ g(0,0) \neq 0 \right\}$ *and let*

$$f_1 := x_2 D_1 + 1,\ f_2 := x_1 D_2,\ \text{and } g := (x_1 + x_2) D_1 D_2 + x_1 x_2 D_2$$

be differential operators in $\mathcal{A} = \mathcal{R}[D_1, D_2]$*. Then division of g by $\{f_1, f_2\}$ yields*

$$r := g - (D_2 f_1 + D_1 f_2) = (x_1 + x_2) D_1 D_2 + x_1 x_2 D_2 - (x_1 + x_2) D_1 D_2 - D_1 - 2D_2 =$$

$$= x_1 x_2 D_2 - D_1 - 2D_2 = (x_1 x_2 - 2) D_2 - D_1$$

and

$$h_1 := D_2, \qquad h_2 := D_1 .$$

Definition 1 *Let $\mathcal{J}$ be an ideal in $\mathcal{A}$ and let G be a finite subset of $\mathcal{J} \setminus \{0\}$. For $i \in \mathbf{N}^n$ let*

$$lc(i, \mathcal{J}) := {}_{\mathcal{R}}\langle lc(f);\ f \in \mathcal{J},\ deg(f) = i \rangle .$$

Then G is a Gröbner basis of $\mathcal{J}$ (with respect to $<$) iff for all $i \in \mathbf{N}^n$ the ideal $lc(i, \mathcal{J})$ is generated by

$$\{lc(g);\ g \in G,\ i \in deg(g) + \mathbf{N}^n\}.$$

Example 2 *If $\mathcal{J}$ is generated by one differential operator f, then any finite subset of $\mathcal{J} \setminus \{0\}$ containing f is a Gröbner basis of $\mathcal{J}$.*

Example 3 *Let $\mathcal{R} \subseteq K(x_1, x_2)$ be such that x_1 and x_2 are not invertible in $\mathcal{R}$ (e.g. $\mathcal{R} = K[x_1, x_2]$ or $\mathcal{R} = \left\{\frac{f}{g} \in K(x_1, x_2) \mid f, g \in K[x_1, x_2],\ g(0,0) \neq 0\right\}$). Let $<$ be a term order on $\mathbf{N}^2$ such that $(1,0) < (0,1)$. Then $\{x_1 D_2, x_2 D_1\}$ is not a Gröbner basis of $\mathcal{J} :=_{\mathcal{A}} \langle x_1 D_2, x_2 D_1 \rangle$.*

For: $\mathcal{J}$ contains $(x_2 D_1) x_1 D_2 - (x_1 D_2) x_2 D_1 = x_2 D_2 - x_1 D_1$ and $deg(x_2 D_2 - x_1 D_1) = (0,1)$. Hence $\langle x_2 \rangle \subseteq lc((0,1), \mathcal{J})$, and $lc((0,1), \mathcal{J})$ is not generated by $x_1 = lc(x_1 D_2)$.

Proposition 2 *Let $\mathcal{I}$ be an ideal in $\mathcal{A}$, let G be a Gröbner basis of $\mathcal{J}$ and let $f \in \mathcal{A}$. Then $f \in \mathcal{I}$ iff a remainder of f after division by G is zero.*

Proof: Follows from Proposition 1. $\surd$

3 Buchberger's algorithm.

Definition 2

- *Let E be a finite subset of $\mathcal{A} \setminus \{0\}$. Then*

$$m(E) := (\max_{e \in E} deg(e)_1, \ldots, \max_{e \in E} deg(e)_n) \in \mathbf{N}^n.$$

- *Let $\mathcal{R}$ be a principal ideal domain (e.g. a field) and let $f, g \in \mathcal{A} \setminus \{0\}$. Choose $c, d \in \mathcal{R}$ such that*

$$c \cdot lc(f) = d \cdot lc(g) = lcm(lc(f), lc(g)) \in \mathcal{R}\,.$$

Then

$$S(f,g) := cD^{m(\{f,g\}) - deg(f)} f - dD^{m(\{f,g\}) - deg(g)} g\,.$$

Example 4 *We consider the set $\mathcal{R} := \left\{\frac{f}{g} \mid f, g \in K[x],\ g(0) \neq 0\right\}$ and let $f := xD + 1$, $g := \frac{1}{x+1} D^2$ be differential operators in $\mathcal{A} = \mathcal{R}[D]$. Then*

$$c := \frac{1}{x+1}, \quad d := x$$

and

$$S(f,g) = \frac{1}{x+1} Df - xg = \frac{2}{x+1} D\,.$$

Proposition 3 *Let G be a finite subset of $\mathcal{A} \setminus \{0\}$ and let $\mathcal{J}$ be the ideal in $\mathcal{A}$ generated by G. For $E \subseteq G$ let S_E be a finite set of generators of the $\mathcal{R}$-module*

$$\left\{(c_e)_{e \in E} \mid \sum_{e \in E} c_e lc(e) = 0\right\} \leq_{\mathcal{R}} (\mathcal{R}^E)$$

Then the following assertions are equivalent:

(i) G is a Gröbner basis of $\mathcal{J}$.

(ii) For all $E \subseteq G$ and for all $(c_e)_{e\in E} \in S_E$ a remainder of

$$\sum_{e\in E} c_e D^{m(E)-deg(e)} e$$

after division by G is zero.

If $\mathcal{R}$ is a principal ideal domain (e.g. a field), (i) and (ii) are equivalent to

(iii) For all $f, g \in G$ a remainder of $S(f,g)$ after division by G is zero.

Proof:

(i) $\Rightarrow$ (ii): follows from Proposition 2.

(ii) $\Rightarrow$ (i): Let $h \in \mathcal{J}$. We have to show:

$$lc(h) \in {}_{\mathcal{R}}\langle lc(g);\ g \in G,\ deg(h) \in deg(g) + \mathbf{N}^n\rangle .$$

For a family $(f_g)_{g\in G}$ in $\mathcal{A}$ we define

$$\delta((f_g)_{g\in G}) := \max_{<}\{deg(f_g g);\ g \in G\})$$

Since $h \in \mathcal{J}$, there is a family $(h_g)_{g\in G}$ in $\mathcal{A}$ such that $h = \sum_{g\in G} h_g g$. We may choose $(h_g)_{g\in G}$ such that

$$\delta := \delta((h_g)_{g\in G}) \quad \text{is minimal}$$

(i.e. if $(h'_g)_{g\in G}$ is such that $h = \sum_{g\in G} h'_g g$, then $\delta((h'_g)_{g\in G}) \geq \delta$). Let $E := \{g \in G \mid deg(h_g g) = \delta\} \subseteq G$.

- Case 1: $deg(h) = \delta$. Then

$$in(h) = \sum_{g\in E} in(h_g g) \quad \text{and} \quad lc(h) = \sum_{g\in E} lc(h_g) lc(g) \in_{\mathcal{R}} \langle lc(g);\ g \in E\rangle.$$

If $g \in E$, then $deg(h) = deg(h_g g) = deg(h_g) + deg(g)$, hence $deg(h) \in deg(g) + \mathbf{N}^n$. Therefore $lc(h) \in {}_{\mathcal{R}}\langle lc(g);\ g \in G,\ deg(h) \in deg(g) + \mathbf{N}^n\rangle$.

- Case 2: $deg(h) < \delta$. Then $\sum_{g\in E} lc(h_g) lc(g) = 0$, hence

$$(lc(h_g))_{g\in E} \in \left\{(c_g)_{g\in E} \mid \sum_{g\in E} c_g lc(g) = 0\right\}.$$

Thus there are $r_c \in \mathcal{R}$ such that $(lc(h_g))_{g\in E} = \sum_{c\in S_E} r_c c$, i.e.:

$$lc(h_g) = \sum_{c\in S_E} r_c c_g, \quad \forall g \in G.$$

Now

$$h = \sum_{g\in G} h_g g = \sum_{g\in E} h_g g + \sum_{g\in G\setminus E} h_g g =$$

$$= \sum_{g\in E}(h_g - \sum_{c\in S_E} r_c c_g D^{deg(h_g)})g + \sum_{g\in E}\sum_{c\in S_E} r_c c_g D^{deg(h_g)} g + \sum_{g\in G\setminus E} h_g g\ .$$

For all $g \in E$ we have $deg(h_g) + deg(g) = \delta$, hence there is an element $u \in \mathbf{N}^n$ such that $\delta = m(E) + u$. Thus

$$\sum_{g\in E}\sum_{c\in S_E} r_c c_g D^{deg(h_g)} g = \sum_{c\in S_E} r_c D^u (\sum_{g\in E} c_g D^{m(E)-deg(g)} g) +$$

$$+ \sum_{c\in S_E} r_c (\sum_{g\in E} c_g D^{deg(h_g)} - D^u c_g D^{m(E)-deg(g)}) g\ .$$

By (ii) there are families $(h_g(c))_{g\in G}$ in $\mathcal{A}$, for all $c \in S_E$, such that

$$\sum_{g\in E} c_g D^{m(E)-deg(g)} g = \sum_{g\in G} h_g(c) g$$

and $deg(h_g(c)g) < \delta - u$, for all g $\in$ G (by division algorithm). Therefore there is a family $(h''_g)_{g\in G}$ such that $\delta((h''_g))_{g\in G} < \delta$ and

$$\sum_{c\in S_E} r_c D^u (\sum_{g\in E} c_g D^{m(E)-deg(g)} g) = \sum_{g\in G} h''_g g$$

Let

$$h'_g := (h_g - \sum_{c\in S_E} r_c c_g D^{deg(h_g)}) + h''_g +$$

$$+ \sum_{c\in S_E} r_c (c_g D^{deg(h_g)} - D^u c_g D^{m(E)-deg(g)}), \quad \text{if } g \in E$$

and let $h'_g := h_g$, if $g \in G\setminus E$. Then it is easy to verify that $h = \sum_{g\in G} h'_g g$ and $\delta((h'_g)_{g\in G}) < \delta$, which is a contradiction to the minimality of δ. Hence case 2 cannot occur.

(ii) $\Leftrightarrow$ *(iii)* For $f, g \in G$ choose $c, d \in \mathcal{R}$ such that

$$c \cdot lc(f) = d \cdot lc(g) = lcm(lc(f), lc(g))\ .$$

Let

$$s^e_{f,g} := \begin{cases} c & \text{if } e = f \\ -d & \text{if } e = g \\ 0 & \text{otherwise} \end{cases}\ .$$

If $\mathcal{R}$ is a principal ideal domain, it is well known (see for example [14], Lemma 3.4) that we can choose $S_E := \{(s_{f,g}{}^e)_{e\in E} |\ f,g \in E\}$. $\surd$

Analogous to the case of ideals in commutative polynomial rings (with coefficients in $\mathcal{R}$) the proposition above implies an algorithm to compute Gröbner bases.

Proposition 4 *Let $\mathcal{J}$ be an ideal in $\mathcal{A}$ given by a finite set G of generators. In the following way we compute in finitely many steps a Gröbner basis of $\mathcal{J}$:*

While there are a subset $E \subseteq G$ and a family $(c_e)_{e\in E} \in S_E$ such that the remainder r of

$$\sum_{e\in E} c_g D^{m(E)-deg(e)} e$$

after division by G is not zero, replace G by $G \cup \{r\}$.

Example 5 *Let $\mathcal{R}$, $\mathcal{J}$ and $\mathcal{A} = \mathcal{R}[D_1, D_2]$ be as in Example 3. Let*

$$f_1 := x_1 D_2 \quad and \quad f_2 := x_2 D_1 ,$$

then $\mathcal{J} = \langle f_1, f_2 \rangle$.

Let $<$ be the graded lexicographic order with $(0,1) > (1,0)$. Then

$$x_2 D_1 f_1 - x_1 D_2 f_2 = x_2(x_1 D_2 D_1 + D_2) - x_1(x_2 D_2 D_1 + D_1) =$$

$$= x_2 D_2 - x_1 D_1 =: f_3$$

$$x_2 f_1 - x_1 f_3 = x_1^2 D_1 =: f_4$$

$$D_2 f_2 - D_1 f_3 = x_2 D_2 D_1 + D_1 - (x_2 D_2 D_1 - x_1 D_1^2 - D_1) =$$
$$= x_1 D_1^2 + 2 D_1 =: f_5 .$$

Since $S(f_1, f_5) = S(f_2, f_4) = S(f_4, f_5) = 0$,

$S(f_1, f_4) = x_1 D_1 f_1 - D_2 f_4 = x_1 D_2 = f_1$,

$S(f_2, f_5) = x_1 D_1 f_2 - x_2 f_5 = -2 f_2$,

$S(f_3, f_4) = {x_1}^2 D_1 f_3 - x_2 D_2 f_4 = -{x_1}^3 {D_1}^2 - {x_1}^2 D_1 =$

$= -{x_1}^2 f_5 - f_4$ *and*

$S(f_3, f_5) = x_1 {D_1}^2 f_3 - x_2 D_2 f_5 = -2 D_1 f_3 - (x_1 D_1 + 1) f_5$,

$\{f_1, f_2, f_3, f_4, f_5\}$ is a Gröbner basis of $\langle f_1, f_2 \rangle$.

Example 6 *Let* $\mathcal{R} := \left\{\frac{f}{g} \in \mathbf{C}(x) \mid f, g \in \mathbf{C}[x],\ g \text{ has no zeros in } (-1,1)\right\}$ *and let* $\mathcal{J}$ *be the ideal generated by* $f_1 := 2xD^2$ *and* $f_2 := D^3 + x^2D - x$. *Then*

$$S(f_1, f_2) = \frac{1}{2}Df_1 - xf_2 = D^2 - x^3D + x^2 =: f_3 \,.$$

Since $f_2 = (D - x^3)f_3 + x^2(3 + x^4)D - x(3 + x^4)$, *we replace* f_2 *by* $x^2D - x$ *(*$3 + x^4$ *is invertible).*

Now $f_3 = -xf_2 + D^2$, *hence we replace* f_3 *by* D^2 *and eliminate* f_1 *(*$f_1 = 2xf_3$*).*

$$S(f_2, f_3) = Df_2 - x^2f_3 = x^2D^2 + 2xD - xD - 1 - x^2D^2 = xD - 1 =: f_4$$

and we can eliminate f_2 *(*$f_2 = xf_4$*). Since* $S(f_3, f_4) = 0$, $\{f_3, f_4\}$ *is a Gröbner basis of* $\langle f_1, f_2\rangle$.

Example 7 *Let* $\mathcal{R} := \left\{\frac{f}{g} \mid f, g \in \mathbf{C}[x],\ g(-1) \neq 0\right\}$ *and let* $\mathcal{J} := \langle f_1, f_2\rangle$, *where* $f_1 := xD^2 + xD + 1$ *and* $f_2 := D^3 + x^2D^2 + D$.

$$S(f_1, f_2) = Df_1 - xf_2 = (1 + x - x^3)D^2 + (2 - x)D =: f_3$$

$$S(f_1, f_3) = (1 + x - x^3)f_1 - xf_3 = (-x^4 + 2x^2 - x)D + (1 + x - x^3) =: f_4$$

$$S(f_1, f_4) = (-x^4 + 5x^3 + 2x^2 - 6x)D + (-x^3 + 3x^2 + 2x - 2) =: f_5$$

$$S(f_5, f_4) = -2x^6 + x^5 + x^4 + 6x^3 + 2x^2 - 8x =: f_6$$

which is invertible in $\mathcal{R}$. *Hence* $\{1\}$ *is a Gröbner basis of* $< f_1, f_2 >$.

Remark 3 *Let* $\mathcal{V}$ *be a finite-dimensional free (left-)* $\mathcal{A}$*-module and let* B *be an* $\mathcal{A}$*-basis of* $\mathcal{V}$. *For example:* $\mathcal{V} = \mathcal{A}^k$ *and* $B = \{e_1, \ldots, e_k\}$ *is the standard basis.*

We can easily extend the notions of Gröbner basis for ideals in $\mathcal{A}$ *to* $\mathcal{A}-$ *submodules of* $\mathcal{V}$. *We indicate here how the basic definitions can be generalized to this case. Then the extension of the propositions and their proofs is straightforward.*

The set $\{D^ib \mid i \in \mathbf{N}^n,\ b \in B\}$ *is a* $\mathcal{R}$*-basis of* $\mathcal{V}$. *A term order on* $\mathbf{N}^n \times B$ *is a total order* $<$ *on* $\mathbf{N}^n \times B$ *such that*

- $(0, b)$ *is the smallest element in* $\mathbf{N}^n \times \{b\}$, *for all* $b \in B$,
- $(i, b) < (j, c)$ *implies* $(i + k, b) < (j + k, c)$, *for all* $i, j, k \in \mathbf{N}^n$ *and for all* $b, c \in B$.

Let $<$ *be a term order on* $\mathbf{N}^n \times B$. *For a vector*

$$0 \neq f = \sum_{i \in \mathbf{N}^n,\ b \in B} r_{i,b} D^i b$$

we define

$$deg(f) := (deg_D(f), deg_B(f)) := max_<\{(i, b) |\ r_{i,b} \neq 0\} \text{ and } \ lc(f) := r_{deg(f)}.$$

Division of a vector $g \in \mathcal{V}$ *by a finite subset* $F \subseteq \mathcal{V} \setminus \{0\}$ *yields a vector* $r \in \mathcal{V}$ *("a remainder") and a family* $(h_f)_{f \in F}$ *in* $\mathcal{A}$ *such that*

- $g = \sum_{f \in F} h_f f + r$,
- *for all* $f \in F$, $h_f = 0$ or $deg(h_f f) \leq deg(g)$,
- $r = 0$ *or* $lc(r) \notin_{\mathcal{R}} \langle lc(f);\ f \in F,\ deg_B(r) = deg_B(f),\ deg(r)_D \in deg(f)_D + \mathbf{N}^n \rangle$.

Let $\mathcal{U}$ *be a (left-)* $\mathcal{A}$*-submodule of* $\mathcal{V}$. *A finite subset* $G \subseteq \mathcal{U} \setminus \{0\}$ *is a Gröbner basis of* $\mathcal{U}$ *iff for all* $i \in \mathbf{N}^n, b \in B$ *the ideal*

$$lc((i, b), \mathcal{U}) := {}_{\mathcal{R}}\langle lc(f);\ f \in \mathcal{U},\ f \neq 0,\ deg(f) = (i, b)\rangle$$

is generated by the set

$$\{lc(g);\ g \in G,\ deg_B(g) = b,\ i \in deg_D(g) + \mathbf{N}^n\}.$$

The analogue of Proposition 3 is: Let G *be a finite subset of* $\mathcal{U} \setminus \{0\}$. *For* $E \subseteq G$ *let* S_E *be a finite set of generators of the* $\mathcal{R}$*-module*

$$\left\{(c_e)_{e \in E} \mid \sum_{e \in E} c_e lc(e) = 0\right\}.$$

Then G *is a Gröbner basis of* $\mathcal{U}$ *iff for all* $E \subseteq G$ *such that*

$$|\{deg_B(e);\ e \in E\}| = 1$$

and all $c \in S_E$ *a remainder of*

$$\sum_{e \in E} c_e D^{m(\{deg_D(e) |\ e \in E\}) - deg_D(e)} e$$

after division by G *is zero.*

Example 8 *Let*

$$\mathcal{R} := \left\{ \frac{f}{g} \in \mathbf{C}(x) \mid f, g \in \mathbf{C}[x],\ g \text{ has no zeros in } (-1,1) \right\},$$

$\mathcal{V} = (\mathcal{R}[D])^2$, $B := \{e_1 := (1,0),\ e_2 := (0,1)\} \subseteq \mathcal{V}$ *and let* $\mathcal{U}$ *be the submodule of* $\mathcal{V}$ *generated by*

$$f_1 := (xD^2, D^2) \quad \text{and} \quad f_2 := (x^2D, xD).$$

Let $<$ *be the order defined by*

$$(i, e_j) < (k, e_l) :\Longleftrightarrow j > l \quad \text{or} \quad [j = l \text{ and } i < k].$$

Then the remainder of

$$S(f_1, f_2) = xf_1 - Df_2 = x(xD^2, D^2) - (x^2D^2 + 2xD, xD^2 + D) =$$
$$= (-2xD, -D) =: f_3$$

after division by $\{f_1, f_2\}$ *is* f_3.

$$S(f_1, f_3) = f_1 + \frac{1}{2}Df_3 = (xD^2, D^2) + \frac{1}{2}(-2xD^2 - 2D, -D^2) =$$
$$= (-D, \frac{1}{2}D^2) =: f_4$$

$$S(f_2, f_3) = f_2 + \frac{1}{2}xf_3 = (x^2D, xD) + (-x^2D, -\frac{1}{2}xD) =$$
$$= (0, \frac{1}{2}xD) =: f_5 .$$

We have $f_3 = 2xf_4 - 2Df_5$, $f_2 = -x^2f_4 + (xD+1)f_5$ *and hence we can eliminate* f_2, f_3. *Now it is easy to verify that* $\{f_1, f_4, f_5\}$ *is a Gröbner basis of the ideal* $\mathcal{J}$.

4 Applications

Let $\mathcal{V}$ a free $\mathcal{A}$-module with a finite $\mathcal{A}$-basis B and let $\mathcal{U}$ be an $\mathcal{A}$-submodule of $\mathcal{V}$. Then $\mathcal{V}/\mathcal{U}$ is a torsion-module iff for every element $v \in \mathcal{V}$ there is an $a \in \mathcal{A} \setminus \{0\}$ such that $a \cdot v \in \mathcal{U}$. If $n = 1$ and $K = \mathbf{C}$ then a module of differential operators is holonomic iff it is a torsion-module (cf.[1] or [6], sect. 5, Theorem 1).

The following proposition shows how to check whether $\mathcal{V}/\mathcal{U}$ is a torsion module or not.

Proposition 5

1. *$\mathcal{V}/\mathcal{U}$ is a torsion-module iff $\mathcal{A}b \cap \mathcal{U} \neq \{0\}$, for all $b \in B$.*

2. *Let $b \in \mathcal{B}$. Choose a term order $<_D$ on $\mathbf{N}^n$ and a total order $<_B$ on B such that b is the smallest element. Then the order on $\mathbf{N}^n \times B$ defined by*
$$(i,c) < (j,d) :\Longleftrightarrow c <_B d \quad \text{or} \quad [c = d \text{ and } i <_D j].$$
is a term order. If G is a Gröbner basis of $\mathcal{U}$ with respect to $<$ then $G \cap (\mathcal{A}b)$ is a Gröbner basis of $(\mathcal{A}b) \cap \mathcal{U}$.

Proof: 1. If $\mathcal{V}/\mathcal{U}$ is a torsion-module, then for all $b \in B$ there are $a_b \in \mathcal{A} \setminus \{0\}$ such that $a_b \cdot b \in \mathcal{U}$.

Now suppose that $(\mathcal{A}b) \cap \mathcal{U} \neq \{0\}$, for all $b \in B$. It is well-known that $\mathcal{A}b/(\mathcal{A}b \cap \mathcal{U})$ is a torsion-module. If M and N are torsion-modules over $\mathcal{A}$ and $x \in M$, $y \in N$, then there are $c, d \in \mathcal{A}$ such that $c \cdot x = 0$ and $d \cdot (c \cdot y) = 0$. Thus $d \cdot c \cdot (x, y) = (0,0)$, and $M \times N$ is a torsion-module, too. Hence $\prod_{b \in B} \mathcal{A}b/(\mathcal{A}b \cap \mathcal{U})$ is a torsion-module, the map

$$\prod_{b \in B} \mathcal{A}b/(\mathcal{A}b \cap \mathcal{U}) \to \mathcal{V}/\mathcal{U}$$

$$\overline{(c_b \cdot b)_{b \in B}} \longmapsto \overline{\sum_{b \in B} c_b \cdot b}$$

is well-defined, $\mathcal{A}$-linear and surjective. This implies the assertion.

2. Follows from 2 by standard arguments. $\surd$

Example 9 *Let $\mathcal{R}$, $\mathcal{V}$, B and $\mathcal{U}$ be as in Example 8. A Gröbner basis of $\mathcal{U}$ is $\{(xD^2, D^2), (-D, \frac{1}{2}D^2), (0, xD)\}$.*

Due to the choice of the term order we know that $\{(0, xD)\}$ is a Gröbner basis of $\mathcal{A}e_2 \cap \mathcal{U} \neq \{0\}$. Moreover, $f_1 - 2f_4 = (xD^2 + 2D, 0) \in \mathcal{A}e_1 \cap \mathcal{U}$, hence $\mathcal{V}/\mathcal{U}$ is a torsion module.

Let $\Omega = (a, b)$ be an open interval in $\mathbf{R}$ and let

$$\mathcal{R} := \left\{ \frac{f}{g} \mid f, g \in \mathbf{C}[x_1],\ g \text{ has no zeros in } (a,b) \right\}.$$

$\mathcal{R}$ is a principal ideal domain. Using Sturm sequences, we always can determine the number of zeros in Ω of a polynomial in $\mathbf{C}[x_1]$. This implies that we can solve linear equations over $\mathcal{R}$ (we tacitly assume that we can solve linear equations over $\mathbf{C}$; otherwise we take the algebraic closure of $\mathbf{Q}$ instead of $\mathbf{C}$). An ideal $\mathcal{J}$ in $\mathcal{A} = \mathcal{R}[D_1]$ induces the sequence

$$lc(\mathcal{J}, 0) \subseteq lc(\mathcal{J}, 1) \subseteq \ldots$$

of ideals in $\mathcal{R}$. Let r_i be a generator of the ideal $lc(\mathcal{J}, i)$ and let

$$d := \min\{j \in \mathbf{N} \mid lc(j, \mathcal{J}) \neq 0\},$$

$$q := \max\{j \in \mathbf{N} \mid lc(j-1, \mathcal{J}) \neq lc(j, \mathcal{J})\}.$$

Let m be the number of zeros of r_q (counted with multiplicities) in Ω. Let S be the set of all hyperfunctions on Ω, which are solutions of the system of differential equations defined by $\mathcal{J}$.

By ([6], Sect. 4, Theorem 1), the dimension of the C-vector space S is $d + m$. Once we have computed a Gröbner basis of $\mathcal{J}$ we easily obtain the numbers d and q and the rational function r_q. Using Sturm sequences we compute m.

Example 10 *Let $\mathcal{R}$ and $\mathcal{J}$ be as in Example 6. Then $\{xD - 1, D^2\}$ is a Gröbner basis of $\mathcal{J}$.*

We have $d = 1$ and $q = 2$. A generator of $lc(2, \mathcal{J})$ is $r_2 := 1$, hence $m = 0$. Thus the dimension of the C*-vector space S is* 1.

References

[1] Björk, J. (1979): Rings of differential operators. North-Holland, Amsterdam.

[2] Briançon, J., Maisonobe, P. (1984): Idéaux de germes d'opérateurs différentiels à une variable. L'Ens. Math. 30: 7-38.

[3] Buchberger, B. (1970): Ein algorithmisches Kriterium für die Lösbarkeit eines algebraischen Gleichungssystems. Aequationes Math. 4: 374-383.

[4] Buchberger, B. (1985): Gröbner bases: An Algorithmic Method in Polynomial Ideal Theory. In: Bose, N. (ed.): Multidimensional Systems Theory. Reidel Publishing Company, Dordrecht, pp. 184-232.

[5] Castro, F. (1986): Calcul de la dimension et des multiplicités d'un D-module monogène. C.R. Acad. Sci. Paris, t.302, Série I, 14: 487-490.

[6] Castro, F. (1987): Calculs effectifs pour les ideaux d'opérateurs differentiels. In: Aroca, J.M., Sánchez Giralda and Vicente, J.L (eds.): Géométrie algébrique et applications. Vol. III, Hermann, Paris, pp. 1-20.

[7] Fröhler, S. (1997): Linear differential systems with variable coefficients - a duality theorem. Dissertation, Universität Innsbruck.

[8] Fröhler, S., Oberst, U. (1997): Time-varying linear systems. In preparation, Innsbruck.

[9] Galligo, A. (1985): Some algorithmic questions on ideals of differential operators. Lect. Notes Comp. Sci., 204: 413-421.

[10] Jacobson, C., Löfwall, C. (1991): Standard bases for general coefficient rings and a new constructive proof of Hilbert's basis theorem. J. Symbolic Comp.12: 337-371.

[11] Kandri-Rody, A., Weispfenning, V. (1990): Non-commutative Gröbner bases in algebras of solvable type. J. Symb. Comp. 9: 1-26.

[12] Mora, T. (1986): Gröbner bases for non-commutative polynomial rings. Lect. Notes Comp. Sci. 229: 353-362.

[13] Oaku, T., Shimoyama, T. (1994): A Gröbner basis method for modules over rings of differential operators. J. Symb. Comp. 18: 223-248.

[14] Oberst, U. (1990): Multidimensional constant linear systems. Acta Applic. Math. 20: 1-175.

[15] Pauer, F., Pfeifhofer, M. (1988): The theory of Gröbner bases. L'Ens. Math., 34: 215-232.

[16] Pommaret, J.F. (1994): Partial Differential Equations and Group Theory. Kluwer Academic Publishers, Dordrecht.

[17] Takayama, N. (1989): Gröbner bases and the problem of contiguous relations. Japan J. Appl. Math. 6: 147-160.

Canonical Curves and the Petri Scheme

John B. Little[1]

1 The Ideal of A Canonical Curve

Gröbner basis methods have proved quite useful in studying problems from classical algebraic geometry. In this paper we will present several results about parameter spaces for canonically embedded (but possibly singular) algebraic curves, using Schreyer's Gröbner basis reinterpretation of Petri's analysis of the homogeneous ideals of such curves.

Let C be a complete, smooth algebraic curve over an algebraically closed field k. Among the many ways to map C into projective spaces, the *canonical mappings* are especially interesting. For each basis $\{\omega_1, \ldots, \omega_g\}$ for $H^0(C, \omega_C)$ (the regular differentials on C) there is a corresponding mapping

$$\varphi : C \to \mathbf{P}^{g-1}$$

defined by $\varphi(p) = (\omega_1(p), \ldots, \omega_g(p))$. The image is called a *canonical curve*. By the Riemann-Roch theorem, if C is not hyperelliptic (that is, if there is *no* 2-1 morphism $C \to \mathbf{P}^1$), then φ is an embedding. The image $\varphi(C)$ is a smooth algebraic curve of degree $2g - 2$ contained in no linear subspace of $\mathbf{P}^{g-1}$, whose extrinsic geometry reflects intrinsic properties of the abstract curves C. We will consider only non-hyperelliptic curves C, and sometimes identify C with its canonical image.

Theorem 1 (Enriques, Noether, Petri) *Let C be a nonsingular canonical curve of genus g in $\mathbf{P}^{g-1}$. Then C is projectively normal. The dimension of the vector space of quadrics containing C is $\binom{g-2}{2}$, and the ideal of C is generated by quadrics unless either C is trigonal (that is, if there is a 3-1 morphism $C \to \mathbf{P}^1$), or C is isomorphic to a smooth plane quintic curve ($g = 6$).*

If C is trigonal, then the fibers of the 3-1 cover give a 1-parameter family of trisecant lines to C, which sweep out a ruled surface in $\mathbf{P}^{g-1}$. Any quadric that contains C contains all of these lines as well, so the variety defined by the quadrics in the ideal of C contains (in fact equals) the ruled surface.

[1]Dept. of Mathematics, College of Holy Cross, Worcester, MA 01610, USA; `little@math.holycross.edu`

Petri (1923) completed the determination of the cases where the quadrics do not generate the ideal of C by constructing a remarkable explicit set of generators for the ideal $I(C)$ as follows. (This construction is reproduced by Saint Donat (1973), Mumford (1975), and Arbarello et. al. (1985).)

Given g general points $p_i \in C$ $(i = 1, \ldots, g)$, choose coordinates in $\mathbf{P}^{g-1}$ to make $p_i = (0, \ldots, 1, \ldots, 0)$ (the 1 in the ith coordinate). By the Riemann-Roch theorem, if ω_C denotes the canonical sheaf on C, $h^0(\omega_C^2(-p_1 - \cdots - p_{g-2})) = 2g - 1$. Identifying sections of ω_C^2 with homogeneous quadratic polynomials (restricted to the canonical image), there are $2g-1$ linearly independent quadrics vanishing at all of the points p_i, $i = 1, \ldots, g-2$:

$$\begin{array}{ll} x_0 x_s; & 1 \leq s \leq g-2 \\ x_{g-1} x_s; & 1 \leq s \leq g-2 \\ x_0^2, x_0 x_{g-1}, x_{g-1}^2. & \end{array}$$

The products $x_i x_j$, $1 \leq i < j \leq g-2$ also give elements of this space. Because there must be linear dependences, there are quadrics in the ideal of C of the form

$$f_{ij} = x_i x_j - \sum_{s=1}^{g-2} a_{sij}(x_0, x_{g-1}) x_s - q_{ij}(x_0, x_{g-1}), \; 1 \leq i < j \leq g-2, \quad (1.1)$$

where the a_{sij} are linear forms and the q_{ij} are quadratic forms, both in the variables x_0 and x_{g-1}. From the leading terms, the f_{ij} are guaranteed to be linearly independent. Hence they form a basis for the degree 2 homogeneous part $I(C)_2$ of the ideal.

If $s \neq i$ and $s \neq j$, then from the form of f_{ij} in (1.1), it can be seen that the linear form a_{sij} must vanish at least to second order at p_s. Hence for s different from i, j, we have $a_{sij} = \rho_{sij} \alpha_s$ where α_s is any one particular linear form in x_0, x_{g-1} vanishing to second order at least at p_s. The scalars ρ_{ijk} are symmetric under permutations of the indices.

Continuing on to degree 3 in a similar fashion, Petri finds cubic polynomials in $I(C)_3$:

$$G_{jk} = \alpha_j x_j^2 - \alpha_k x_k^2 + \text{other terms},$$

where $j \neq k$ in $1, \ldots, g-2$ and $c \in k$. Next, Petri shows that

$$x_j f_{ik} - x_k f_{ij} + \sum_{s=1, s\neq j}^{g-2} a_{sik} f_{sj} - \sum_{s=1, s\neq k}^{g-2} a_{sij} f_{sk} - \rho_{ijk} G_{jk} = 0 \quad (1.2)$$

The ideal $I(C)$ is *always* generated by the f_{ij} and the G_{jk}. From the "Petri syzygies" (1.2), with a little work, it follows that $I(C)$ is generated by the quadrics f_{ij} unless *all* $\rho_{ijk} = 0$.

As will be clear to Gröbner basis people, the equations (1.2) above are essentially part of a *Gröbner basis calculation* – computing the remainder on division of $S(f_{ik}, f_{ij})$ by the $f's$. This point of view has been taken up by F.O. Schreyer (1991). Specifically, consider the graded reverse lex order on monomials, with the variables arranged as follows: $x_1 > x_2 > \cdots > x_{g-1} > x_0$. Schreyer shows that if all the $\rho_{ijk} = 0$, the f_{ij} form a Gröbner basis, but not for the ideal of C. They generate the ideal of a surface of degree $g-2$ in $\mathbf{P}^{g-1}$ containing C. The cubics $G_{j,g-2}$ will always be present in a Gröbner basis for $I(C)$, and there is always one additional quartic element as well, so the full Gröbner basis for $I(C)$ for this special choice of coordinates looks like

$$
\begin{aligned}
f_{ij} &= x_i x_j - \sum_{s=1}^{g-2} a_{sij}(x_0, x_{g-1}) x_s - q_{ij}(x_0, x_{g-1}); \qquad 1 \le i < j \le g-2 \\
a_{sij} &= \rho_{sij} \alpha_s; \qquad s \neq i, s \neq j \\
G_{j,g-2} &= \alpha_j x_j^2 - \alpha_{g-2} x_{g-2}^2 + \text{other terms}; \qquad 1 \le j \le g-3 \qquad (1.3) \\
H &= \alpha_{g-2} x_{g-2}^3 + \cdots .
\end{aligned}
$$

Schreyer also establishes that all of the above remains true for *singular* canonically embedded curves C, if it is possible to find some $g-2$ points p_i which span a linear space $L \cong \mathbf{P}^{g-3}$ that intersects C transversely at the p_i, and only at those points (a "simple $(g-2)$-secant"), and if C is "non-strange" – something which is only an issue in characteristic $p > 0$. For singular curves, ω_C is the *dualizing sheaf*, and Catanese (1982) gives a detailed study of canonical mappings of curves with Gorenstein singularities. We will call any connected, pure 1-dimensional subscheme of $\mathbf{P}^{g-1}$ of arithmetic genus g, for which $\mathcal{O}_C(1) \cong \omega_C$, a *canonical curve.*

Theorem 2 (Schreyer) *Let C be a non-strange canonical curve of arithmetic genus g with a simple $(g-2)$-secant. Then after a change of coordinates (and with respect to the ordering on monomials described before) $I(C)$ has Gröbner basis of the Petri form (1.3). Conversely, any collection of polynomials of the form (1.3) which form a Gröbner basis defines a canonical curve of arithmetic genus g with $V(x_0, x_{g-1})$ as a simple $(g-2)$-secant.*

This theorem of Schreyer gives a way to construct a parameter space for ideals of possibly singular canonical curves. Namely, take polynomials in Petri's form (1.3), viewing the coefficients of the non-leading terms as *indeterminates*. The equations obtained by setting all S-polynomial remainders equal to zero (Buchberger's criterion for Gröbner bases) define a "Petri scheme" $\mathcal{P}_g$ for each genus. (Naturally, this involves a large number of variables, and an even larger number of equations, so writing them out explicitly isn't terribly illuminating. But it can be done without too much difficulty for $g \le 7$.)

$\mathcal{P}_g$ is closely related to a quotient of a subset the Hilbert scheme of curves of degree $2g-2$ and arithmetic genus g in $\mathbf{P}^{g-1}$, in which the natural $PGL(g)$-action is used to normalize the simple $(g-2)$-secant to the fixed linear subspace $V(x_0, x_{g-1})$. However, the equations for $\mathcal{P}_g$ also involve the Petri coefficients ρ_{ijk}. If not all of the ρ_{ijk} are zero, the equations of $\mathcal{P}_g$ do not determine their values uniquely, and distinct points of $\mathcal{P}_g$ can yield precisely the same ideal.

In any case, we can try to use the Petri scheme to understand the structure of the collection of canonically embedded curves. For instance, some natural questions are: Is $\mathcal{P}_g$ irreducible? If not, how many components are there? What are their dimensions? What do the curves corresponding to a general point of each look like? Is there more than one component whose general point corresponds to an irreducible curve?

It is known that the *smooth* canonical curves all correspond to points on one irreducible component P of $\mathcal{P}_g$, since the moduli space of curves of genus g is irreducible of dimension $3g-3$, and the canonical mapping depends only on a choice of basis for $H^0(\omega_C)$ as in Petri's construction.

2 The Petri scheme in genus 6

In this section, we will begin by proving the following theorem about the Petri scheme of (possibly singular) canonically-embedded curves of genus 6.

Theorem 3 *Let $\mathcal{P}_6'$ be the subset of the Petri scheme of curves of arithmetic genus 6 and degree 10 in $\mathbf{P}^5$ parametrizing curves C for which $I(C)$ is generated by quadrics. Then $\mathcal{P}_6'$ is contained in one irreducible component of $\mathcal{P}_6$.*

Before sketching the proof we want to describe the main idea and illustrate it with an example that led us to the general statement. By a classical result (Arbarello and Harris (1984) give a modern treatment), a smooth curve of genus 6, neither trigonal nor isomorphic to a smooth plane quintic, lies on a surface of degree 5 in $\mathbf{P}^5$ (a possibly degenerate quintic Del Pezzo surface, or a cone over a elliptic normal curve in $\mathbf{P}^4$). The curve is then the complete intersection of the surface and *one further quadric.* For instance, projecting a general smooth genus 6 canonical curve C from a general 4-secant 2-plane (spanned by the points of one divisor in one of the g_4^1's on C) yields a plane curve of degree 6 with four double points, birational to C. The canonical divisors on the plane model are cut by cubics passing through the four double points. This linear system of cubics maps $\mathbf{P}^2$ to a quintic Del Pezzo surface containing the original canonical curve C. What we will show is that even for singular curves, we still have a surface of degree 5 containing C that plays the same role. To prove Theorem 3 we will use the fact that the family of such surfaces is irreducible. Since the canonical curves on any one such surface are simply cut by the linear system of quadrics from $\mathbf{P}^5$, we will get the irreducibility statement to be proved.

The main step of the proof will be to isolate the quadrics defining the surface of degree 5. The whole space of quadrics $I(C)_2$ is 6-dimensional, and as before we can take a basis in Petri form (1.1) (with $g = 6$). We will see shortly that if $I(C)$ is generated by quadrics, then at least two of the ρ_{ijk} are different from zero. By renumbering the variables x_1, x_2, x_3, x_4 if necessary, we may assume ρ_{123}, ρ_{124} are non-zero. If this is the case, then we will show that the (5-dimensional) subspace of $I(C)_2$ spanned by

$$\begin{aligned} F_1 &= \rho_{134}f_{12} - \rho_{123}f_{14} \\ F_2 &= \rho_{134}f_{23} - \rho_{123}f_{34} \\ F_3 &= \rho_{124}f_{13} - \rho_{123}f_{14} \\ F_4 &= \rho_{124}f_{23} - \rho_{123}f_{24} \\ F_5 &= \rho_{234}f_{12} - \rho_{123}f_{24} \\ F_6 &= \rho_{234}f_{13} - \rho_{123}f_{34} \end{aligned} \tag{2.1}$$

generates the ideal of a surface of degree 5 and sectional genus 1.

Readers of Schreyer's paper (1991) will note a clear parallel between 2.1 and the quadrics in Claim 1 in the proof of Theorem 4.1 of that paper. However, this initial $g = 6$ case seems to have some different features from the $g \geq 7$ cases treated there, and we were not able to deduce our desired result directly from Schreyer's techniques.

Here is an amusing example of a singular C that gives one possible type of (singular) surface that appears in this context. Consider the ideal I generated by the Petri-form quadrics:

$$f_{ij} = x_i x_j - (x_k + x_l)(x_0 + x_5) - x_0 x_5,$$

where $1 \leq i < j \leq 4$, and for each pair i, j, the remaining indices k, l are chosen so that $\{i, j, k, l\} = \{1, 2, 3, 4\}$. For this example, $\rho_{ijk} = 1$, all i, j, k, and $\alpha_k = x_0 + x_5$, all k. I has a Gröbner basis of the form described in Theorem 2, so $C = V(I)$ is a canonical curve of arithmetic genus 6. It is not difficult to see that C is a union of five smooth conics C_i in planes P_i defined by

$$\begin{aligned} P_1 &= V(x_2 + x_5 + x_0, x_3 + x_5 + x_0, x_4 + x_5 + x_0) \\ P_2 &= V(x_1 + x_5 + x_0, x_3 + x_5 + x_0, x_4 + x_5 + x_0) \\ P_3 &= V(x_1 + x_5 + x_0, x_2 + x_5 + x_0, x_4 + x_5 + x_0) \\ P_4 &= V(x_1 + x_5 + x_0, x_2 + x_5 + x_0, x_3 + x_5 + x_0) \\ P_5 &= V(x_1 - x_4, x_2 - x_4, x_3 - x_4) \end{aligned}$$

The five planes all contain the line

$$L = V(x_1 + x_5 + x_0, x_2 + x_5 + x_0, x_3 + x_5 + x_0, x_4 + x_5 + x_0)$$

and the conics C_i all meet L in the same two points p, q. (The tangents to the C_i at p all lie in a hyperplane, so the singularity has δ-invariant 5; the situation at q is the same.) In this case the surface of degree 5 defined by the combinations F_k of the f_{ij} given in (2.1) above is the union of the five planes $S = P_1 \cup \cdots \cup P_5$. Note that the general hyperplane section $S \cap H$ is a union of 5 concurrent lines spanning $H \cong \mathbf{P}^4$, a curve of arithmetic genus 1. Intersecting S with any quadric in $I(C)$ not in the span of the F_k gives C.

Proof of the Theorem. Let C be a Petri-general canonically-embedded curve of arithmetic genus 6 in $\mathbf{P}^5$, whose ideal is generated by quadrics. We will begin by proving the assertion above that at least two of the Petri coefficients ρ_{ijk} must be non-zero. Let $\mathcal{T}$ be the graph with vertices $\mathcal{V} = \{1,2,3,4\}$, and an edge (i,j) if and only if there is some k such that $\rho_{ijk} \neq 0$. By Proposition 3.2 of Schreyer (1991), in the minimal free resolution of the homogeneous coordinate ring of C, the graded Betti number β_{13} (giving the number of cubics in a minimal basis for $I(C)$) satisfies

$$\beta_{13} = \#\text{ connected components of } \mathcal{T} - 1.$$

By our assumption, $\beta_{13} = 0$, so $\mathcal{T}$ must be connected. By the definition and the symmetry of the ρ_{ijk} in the indices, this implies that there is at most one edge of the complete graph on $\mathcal{V}$ that is not contained in $\mathcal{T}$. After renumbering if necessary, the potentially missing edge can be taken to be $(3,4)$, and hence $\rho_{123} \neq 0$ and $\rho_{124} \neq 0$. Under this assumption, the quadrics F_k given in (2.1) above always generate a 5-dimensional vector subspace of $I(C)_2$; there is exactly one linear dependence between them:

$$\rho_{234}(F_1 - F_3) - \rho_{124}(F_2 - F_6) + \rho_{134}(F_4 - F_5) = 0\ .$$

Step 1. Let $J = \langle F_1, \ldots, F_6\rangle$. As explained above, the first step in the proof will be to show that $V(J)$ is a surface of degree 5 in $\mathbf{P}^5$. To do this, we will analyze the form of the unique reduced Gröbner basis for J with respect to the graded reverse lexicographic order with the variables ordered $x_1 > x_2 > x_3 > x_4 > x_5 > x_0$.

The results depend on whether ρ_{134} and ρ_{234} are zero; we will consider the case where all $\rho_{ijk} \neq 0$ first. (Note that this will be the case, for example, for a generic choice of $(g-2)$-secant in Petri's construction if C is irreducible.)

For simplicity, we begin by taking linear combinations of the F_k to eliminate common terms and to isolate the possible leading monomials of quadrics in the ideal. The resulting linear basis for the degree 2 part of J is:

$$\begin{aligned}
F_1' &= \rho_{234}\rho_{134}f_{12} - \rho_{124}\rho_{123}f_{34}\\
F_2' &= \rho_{234}f_{13} - \rho_{123}f_{34}\\
F_3' &= \rho_{234}f_{14} - \rho_{124}f_{34}\\
F_4' &= \rho_{134}f_{23} - \rho_{123}f_{34}\\
F_5' &= \rho_{134}f_{24} - \rho_{124}f_{34}
\end{aligned}$$

For instance, $F_1' = \rho_{234}(F_1 - F_3) + \rho_{124}F_6$. We omit the rest of the details in this calculation. For future reference, however, we note the following observation.

Lemma 1 *All quadrics of the forms $\rho_{ijk}f_{lj} - \rho_{ljk}f_{ij}$ and $\rho_{ikl}\rho_{jkl}f_{ij} - \rho_{ijk}\rho_{ijl}f_{kl}$, (where $\{i,j,k,l\} = \{1,2,3,4\}$ in each case) belong to J.*

This may be seen directly by forming linear combinations as above. Now, applying Buchberger's algorithm, we begin the Gröbner basis computation on the F_i'. At the first step, the S-polynomial $S(F_1', F_2')$ yields

$$x_3F_1' - \rho_{134}x_2F_2' \equiv \rho_{123}(\rho_{234}x_2x_3x_4 - \rho_{124}x_3^2x_4) \bmod \langle x_5, x_0\rangle \ .$$

Replacing this last polynomial with its remainder on division by the F_i' and adjusting constants, we obtain a new Gröbner basis element

$$G \equiv \rho_{123}x_3^2x_4 - \rho_{124}x_3x_4^2 \bmod \langle x_5, x_0\rangle \ .$$

We claim that $\mathcal{G} = \{F_1', \cdots, F_5', G\}$ is the reduced Gröbner basis for J. Indeed, working modulo $\langle x_5, x_0\rangle$, it is easy to see that all further S-pairs reduce to zero, modulo $\langle x_5, x_0\rangle$. Hence to prove the claim, it suffices to show that the resulting syzygies modulo $\langle x_5, x_0\rangle$ all lift to syzygies on $\mathcal{G}$. Using the following lemma, we will show that this is a consequence of the Petri syzygies (1.2) on the generators of $I(C)$.

Lemma 2 *Let f_{ij} be a basis for the quadrics in the ideal of a canonical curve of genus $g \geq 6$ in Petri's form, and let $\{i,j,k,l\}$ be any four distinct indices in $\{1, 2, \ldots, g-2\}$. Then each syzygy on the leading terms of the f_{ij} of the form:*

$$\begin{aligned} \rho_{iln}x_k(\rho_{ikn}x_ix_l - \rho_{ikl}x_ix_n) \ &+ \ \rho_{ikn}x_l(\rho_{ikl}x_ix_n - \rho_{iln}x_ix_k) + \\ \rho_{ikl}x_n(\rho_{iln}x_ix_k \ &- \ \rho_{ikn}x_ix_l) = 0 \end{aligned}$$

lifts to a syzygy on the quadrics of the form $\rho_{\alpha\beta\gamma}f_{\delta\epsilon} - \rho_{\alpha\beta\delta}f_{\gamma\epsilon}$ in $I(C)$.

Writing S_{ijk} for the left-hand side of the Petri syzygy (1.2), the lemma can be proved by rearranging the terms in the linear combination

$$\rho_{ikn}\rho_{iln}S_{ikl} + \rho_{ikn}\rho_{ikl}S_{iln} + \rho_{iln}\rho_{ikl}S_{ink}$$

suitably and using the relation $G_{kl} + G_{ln} + G_{nk} = 0$. We omit the details here.

The two lemmas together show that any syzygy on the elements of $\mathcal{G}$ modulo $\langle x_5, x_0\rangle$ that can be expressed in terms of the syzygies from Lemma 2 can be lifted to a syzygy on $\mathcal{G}$, and hence that no new elements of the Gröbner basis will be produced in those cases. In fact, *all* syzygies on $\mathcal{G}$ modulo $\langle x_5, x_0\rangle$

can be expressed in terms of syzygies as in Lemma 1, so no new Gröbner basis elements at all are introduced after G. For example, reducing $S(F_1', F_3')$ yields

$$x_4F_1' - \rho_{134}x_2F_3' - \rho_{124}x_4F_4' \equiv 0 \bmod \langle x_5, x_0\rangle,$$

or combining the two terms with x_4,

$$\rho_{134}(x_4(\rho_{234}f_{12} - \rho_{124}f_{23}) - x_2(\rho_{234}f_{14} - \rho_{124}f_{34})) \equiv 0 \bmod \langle x_5, x_0\rangle.$$

This relation is apparently of a different form than the ones in the Lemma 2, but the fact that it too lifts it can be deduced from Lemma 2.

Our conclusion is that in the case that all $\rho_{ijk} \neq 0$, $\mathcal{G}$ is the reduced Gröbner basis of J. Computing the Hilbert function of J from this information, we see that $S = V(J)$ has degree 5 and codimension 3. Indeed, $V(J) \cap V(x_0, x_5)$ consists of the five points with homogeneous coordinates

$$(1,0,0,0,0,0),\ (0,1,0,0,0,0),\ (0,0,1,0,0,0),\ (0,0,0,1,0,0),$$

and

$$\left(\frac{1}{\rho_{234}}, \frac{1}{\rho_{134}}, \frac{1}{\rho_{124}}, \frac{1}{\rho_{123}}, 0, 0\right)$$

so S is reduced.

The remaining cases to consider are those where one or both of ρ_{134}, ρ_{234} are zero. The arguments in those cases are basically similar to the ones given here, so we will omit most of the details and give only the form of the corresponding Gröbner basis in each case. If *both* $\rho_{134} = \rho_{234} = 0$, then the quadrics in (2.1) reduce to $\{f_{13}, f_{14}, f_{23}, f_{24}, f_{34}\}$. Using the vanishing of the Petri coefficients, we see that the leading term ideal of J has the form

$$M_2 = \langle x_1x_3, x_1x_4, x_2x_3, x_2x_4, x_3x_4, x_1^2x_5\rangle$$

in this case. Computing the Hilbert function gives degree 5 and codimension 3.

Finally if just one of the Petri coefficients, say ρ_{234}, is zero, then from (2.1), we see that J is generated by

$$\rho_{134}f_{12} - \rho_{123}f_{14},\ \rho_{124}f_{13} - \rho_{123}f_{14},\ f_{23},\ f_{24},\ f_{34}.$$

The leading term ideal of J has the form

$$M_1 = \langle x_1x_2, x_1x_3, x_2x_3, x_2x_4, x_3x_4, x_1x_4^2\rangle$$

and the Hilbert function is the same as in the other cases.

Step 2. We now want to show that the family $\mathcal{S}$ of all surfaces S of degree 5 obtained in step 1 is irreducible. Looking at the minimal free resolution of the coordinate ring of S in each case, we have that J is a Gorenstein ideal of codimension three. Writing $R = k[x_1, \ldots, x_5, x_0]$, the minimal resolution has the form

$$0 \to R(-5) \to R(-3)^5 \xrightarrow{A} R(-2)^5 \to R \to R/J \to 0 \tag{2.2}$$

We can use Buchsbaum and Eisenbud's (1977) structure theorem for Gorenstein ideals of codimension 3 to give a uniform description of the ideal J valid in all cases. Namely, every ideal J that appears here is generated by the 4×4 Pfaffians of a 5×5 skew-symmetric rank-4 matrix of linear forms $A = (a_{ij})$ (the "middle matrix" of the resolution (2.2) under a suitable choice of basis for J). Apart from the requirement that $\operatorname{rank}(A) = 4$, the entries in A are arbitrary. It follows that the surfaces S obtained in Step 1 form one irreducible family.

Step 3. A general C is contained in exactly one surface S of the type described above. To complete the proof, we complete the basis (2.1) of J to a basis of $I(C)$. Recall that we are assuming that $I(C)$ is generated by quadrics, so any one further quadric in $I(C)$ not in J (such as f_{34} in the case that all $\rho_{ijk} \neq 0$) will do the job. Hence C is the complete intersection of S and a quadric hypersurface. (Conversely, given a surface S of degree 5 of the form above and a general quadric Q – one not containing any component of S – then $Q \cap S$ will be a non-degenerate Gorenstein curve C of degree 10 and arithmetic genus 6 in $\mathbf{P}^5$.) The family of surfaces S is irreducible by Step 2, and given S, to obtain $C \subset S$, the additional quadric Q can be chosen generically in $I(C)_2 \setminus J$. □

The case where the ideal of C is not generated by quadrics can be studied by the same methods.

3 Canonical cones

In this section, we will show that for large g, $\mathcal{P}_g$ has other components besides P (the component whose general point parametrizes the ideal of a smooth canonical curve). For simplicity, we will assume the characteristic of k is not 2.

Consider the ideal I generated by the Petri form quadrics

$$f_{ij} = x_i x_j - \sum_{s=1, s\neq i,j}^{g-2} (x_0 + x_{g-1}) x_s,$$

where $1 \leq i < j \leq g-2$. A calculation shows that the Gröbner basis of I has the form given in Theorem 2, so the corresponding variety is a canonical curve C. A close look at the equations shows that C is a cone over a finite set (a union of $2g-2$ concurrent lines in $\mathbf{P}^{g-1}$). More generally, we have the following statement. To prepare, we need to introduce the following terminology of

Geramita and Orecchia (1981). A collection of s points in the projective space $\mathbf{P}^n$ is said to be in *generic s-position* if the images of the points under the degree-d Veronese mappings are in linear general position for all $d \geq 1$. The points are said to be in generic t-position ($t \leq s$) if every t-element subcollection is in generic t-position.

Theorem 4 *Let $p_1, \ldots, p_{2g-2}$ be any collection of points in the hyperplane $H = V(x_0)$ in $\mathbf{P}^{g-1}$ which are in in generic $(2g-3)$-position, but not in generic $(2g-2)$-position because they impose only $2g-3$ conditions on quadrics in H. Let p be any point not on H. Then the cone C over $\{p_1, \ldots, p_{2g-2}\}$ with vertex at p is a canonically embedded curve of arithmetic genus g.*

Proof. Geramita and Orecchia (1981) show the singularity of C at p is Gorenstein, and C has arithmetic genus g. If L_i is the line spanned by p_i and p, and Y_i is the union of the other components of C, then Y_i intersects L_i with multiplicity 3 at p since every quadric containing the p_j for $j \neq i$ also contains p_i. Hence, C is canonically embedded, since the sections of the dualizing sheaf embed each irreducible component as a line. □

Theorem 5 *For all g sufficiently large, there exists an irreducible component of the Petri scheme $\mathcal{P}_g$ which is distinct from the irreducible component whose general point corresponds to a smooth Petri-general canonical curve.*

Proof. We will establish this result by a dimension count. First, the dimension of the family of smooth canonical curves of genus g with a simple $(g-2)$-secant normalized as in Petri's construction and a $(g-1)$st point normalized to $(0, \ldots, 0, 1, 0)$ is equal to the dimension of the moduli space of curves of genus g, plus the dimension of the subspace of $PGL(g)$ corresponding to choices of basis in $H^0(\omega_C)$ placing some $g-1$ points of C at $(x_1, \ldots, x_{g-1}, x_0) = (1, 0, \ldots, 0), \ldots, (0, 0, \ldots, 0, 1, 0)$, or

$$3g - 3 + (g^2 - 1) - (g-1)(g-2) = 6g - 6.$$

On the other hand, canonical cones C as in Theorem 4, in Petri form, are uniquely determined by the choice of a general hyperplane section $H \cap C$ (for instance taking $H = V(x_0)$) and a vertex p. We will show how to construct a large-dimensional family of these cones. First pick any nonsingular $(g-3)$-dimensional quadric Q in $H \cong \mathbf{P}^{g-2}$ and any collection of points $\{p_1, \ldots, p_{g-1}\}$ forming the vertices of a *self-polar simplex* with respect to Q. (That is, the polar of each p_i with respect to Q is the linear space spanned by the p_j for $j \neq i$.) By a linear change of coordinates, we can place $p_1, \ldots, p_{g-1}$ at the vertices of the standard coordinate simplex in H: $(x_1, \ldots, x_g, x_0) = (1, 0, \ldots, 0), \ldots, (0, 0, \ldots, 0, 1, 0)$ as above, and with respect to this new system of coordinates, the equation of Q is taken to the diagonal form:

$$\sum_{i=1}^{g-1} \lambda_i (p_{i1} x_1 + \cdots + p_{i,g-1} x_{g-1})^2 = 0$$

where $\lambda_i \neq 0$ for all i.

We claim that if $\{p_g, \ldots, p_{2g-2}\}$ are the vertices of *any second* self-polar simplex with respect to the same Q, then $\{p_1, \ldots, p_{2g-2}\}$ gives a set of points that impose only $2g-3$ conditions on quadrics in H – in other words, that every quadric containing all but one of the p_i must contain p_i as well (Hodge and Pedoe 1952, v. 2, pp 214-5): Following the same reasoning as for the first self-polar simplex, the equation of Q can also be written as

$$\sum_{i=g}^{2g-2} \mu_i (p_{i1}x_1 + \cdots + p_{i,g-1}x_{g-1})^2 = 0$$

for some nonzero μ_i. Multiplying this second expression by a non-zero constant ρ, writing $\lambda_i = -\rho\mu_i$ for $i = g, \ldots, 2g-2$, and adding it to the first, we get

$$\sum_{i=1}^{2g-2} \lambda_i (p_{i1}x_1 + \cdots + p_{i,g-1}x_{g-1})^2 \equiv 0$$

(i.e. equals the zero polynomial). Multiplying this out and setting the coefficient of each monomial $x_j x_k$ equal to zero yields equations

$$\sum_{i=1}^{2g-2} \lambda_i p_{ij} p_{ik} = 0, \;\; 1 < j,k < g-1. \tag{3.1}$$

If $\sum_{j,k} b_{jk} x_j x_k$ is a quadric in H vanishing at all p_i except possibly p_ℓ, then we have

$$\sum_{i=1}^{2g-2} \lambda_i \left(\sum_{j,k} b_{jk} p_{ij} p_{ik} \right) = \lambda_\ell \sum_{j,k} b_{jk} p_{\ell j} p_{\ell k}.$$

But rearranging the sum on the left, we have

$$\sum_{i=1}^{2g-2} \lambda_i \left(\sum_{j,k} b_{jk} p_{ij} p_{ik} \right) = \sum_{j,k} b_{jk} \left(\sum_{i=1}^{2g-2} \lambda_i p_{ij} p_{ik} \right) = 0,$$

by (3.1). Since $\lambda_\ell \neq 0$, we must have $\sum_{j,k} b_{jk} p_{\ell j} p_{\ell k} = 0$ as well, so the quadric also vanishes at p_ℓ.

To get a second self-polar simplex with respect to Q, p_g can be chosen arbitrarily in H (dimension $g-2$), p_{g+1} can be chosen to be any point in the polar of H with respect to p_g (a linear subspace of dimension $g-3$), the third point p_{g+2} can be chosen to be any point in the intersection of the polars of p_g and p_{g+1} (dimension $g-4$), and so forth. Hence given the quadric Q, there is a $\binom{g-1}{2}$-dimensional family of self-polar simplices with respect to Q. If $p_g, \ldots, p_{2g-2}$ are chosen sufficiently generally, $\{p_1, \ldots, p_{2g-2}\}$ will be in generic $(2g-3)$-position, so Theorem 4 applies, and any cone over this hyperplane section is canonically embedded. With the choice of the vertex, we get an irreducible family of cones of dimension $\binom{g-1}{2} + g - 1$. For all g

sufficiently large ($g > 12$) this family of cones has larger dimension than the family of smooth canonical curves above, hence some of the cones must lie on a different component of the Petri scheme. □

To conclude, we mention that a similar phenomenon has been noted by Walter (1995) for arithmetically Cohen-Macaulay curves of degree d whose hyperplane section is in *generic d-position*.

References

[1] Arbarello, E., Cornalba, M., Griffiths, P., Harris, J. (1985): Geometry of Algebraic Curves, Vol. 1, Springer Verlag, New York

[2] Arbarello, E., Harris, J. (1981): Canonical curves and quadrics of rank 4. Compositio Math. 43: 145-179

[3] Buchsbaum, D., Eisenbud, D. (1977): Algebra structures for finite free resolutions and some structure theorems for ideals of codimension 3. Amer. J. of Math. 99: 447-485

[4] Catanese, F. (1982): Pluricanonical Gorenstein curves. In: LeBarz, P., Hervier, Y. (eds.) Enumerative Geometry and Classical Algebraic Geometry. Birkhäuser, Boston, pp. 51-96 (Progress in Mathematics, vol. 24)

[5] Geramita, A., Orecchia, F. (1981): On the Cohen-Macaulay Type of s-Lines in $\mathbf{A}^{n+1}$. J. Algebra 70: 116-140

[6] Hodge, W., Pedoe, D. (1952): Methods of Algebraic Geometry. Cambridge U. Press, Cambridge

[7] Mumford, D. (1975): Curves and their Jacobians. University of Michigan Press, Ann Arbor

[8] Petri, K. (1923): Über die invariante Darstellung algebraischer Funktionen einer Veränderlichen. Math. Ann. 88: 242-289

[9] Saint-Donat, B. (1973): On Petri's analysis of the linear system of quadrics through a canonical curve. Math. Ann. 206: 157-175

[10] Schreyer, F.-O. (1991): A standard basis approach to syzygies of canonical curves. J.f.d.reine u. angew. Math. 421: 83-123

[11] Walter, C. (1995): Hyperplane sections of Arithmetically Cohen-Macaulay curves. Proc. Amer. Math. Soc. 123: 2651-2656

The Buchberger Algorithm as a Tool for Ideal Theory of Polynomial Rings in Constructive Mathematics

Henri Lombardi and Hervé Perdry

Laboratoire de Mathématiques de Besançon, UMR CNRS 6623,
UFR des Sciences et Techniques, Université de Franch-Comté;
{lombardi, perdry}@math.univ-fcomte.fr

Introduction

One of the aims of Constructive Mathematics is to provide effective methods (algorithms) to compute objects whose existence is asserted by Classical Mathematics. Moreover, all proofs should be constructive, *i.e.*, have an underlying effective content. *E.g.* the classical proof of the correctness of Buchberger algorithm, based on noetherianity, is non constructive : the closest consequence is that we know that the algorithm ends, but we don't know when.

In this paper we explain how the Buchberger algorithm can be used in order to give a constructive approach to the Hilbert basis theorem and more generally to the constructive content of ideal theory in polynomial rings over "discrete" fields.

Mines, Richman and Ruitenburg in 1988 ([5]) (following Richman [6] and Seidenberg [7]) attained this aim without using Buchberger algorithm and Gröbner bases, through a general theory of "coherent noetherian rings" with a constructive meaning of these words (see [5], chap. VIII, th. I.5). Moreover, the results in [5] are more general than in our paper and the Seidenberg version gives a slightly different result. Here, we get the Richman version when dealing with a discrete field as coefficient ring ("discrete" means the equality is decidable in k).

In classical texts (*cf.* Cox, Little and O'Shea [2]) about Gröbner bases, the correctness of the Buchberger algorithm and the Hilbert basis theorem are both proved by using a non constructive version of Dixon's Lemma. So, *from a constructive point of view*, the classical approach gives a constructive tool with a gap in the proof. *E.g.*, it is impossible to give bounds for the Buchberger algorithm by a detailed inspection of the classical proof. Moreover, the classical formulation of the Hilbert basis theorem is nonconstructive. Here we give a constructive version of Dixon's Lemma, we deduce constructively the correctness of Buchberger algorithm and from this result we get the Hilbert basis theorem in a constructive form.

In our opinion Gröbner bases are a very good tool, the more natural one in the present time, for understanding the constructive content of ideal theory in polynomial rings over a discrete field.

1 A constructive Dixon's lemma

1.1 Posets and chain conditions

Definition : A poset (partially ordered set) $(E, \leq)$ is said to satisfy the *descending chain condition* (*DCC* for short) if for every nonincreasing sequence $(u_n)_{n\in\mathbf{N}}$ in E there exists $n \in \mathbf{N}$ such that $u_n = u_{n+1}$. A poset $(E, \leq)$ is said to satisfy the *ascending chain condition* (*ACC* for short) if for every nondecreasing sequence $(u_n)_{n\in\mathbf{N}}$ in E there exists $n \in \mathbf{N}$ such that $u_n = u_{n+1}$.

Examples :

- The poset $\mathbf{N}$ with the usual order satifies DCC.
- If $(E, \preceq)$ is a poset satisfying ACC, then $E' = E \cup \{-\infty\}$, ordered with the order of E extended by $-\infty \preceq x$ for all $x \in E'$ is a poset satisfying ACC.

Remark : The above definitions of conditions DCC and ACC are equivalent (from a classical point of view) to the classical ones, but they are adapted to the constructive point of view.
In fact, even $\mathbf{N}$ fails to verify constructively the classical form of DCC : when one has a nonincreasing sequence $(u_n)_{n\in\mathbf{N}}$ in $\mathbf{N}$ without more information, it is *a priori* impossible to know when the limit of the sequence is attained. *E.g.*, call Pr_{nisi} the set of primite recursive procedures $u : n \mapsto u_n$ that produce nonincreasing sequences of integers. This is an enumerable set (in the classical meaning as well as in the constructive meaning). It is well known that there exists no recursive procedure $\Phi : Pr_{nisi} \to \mathbf{N}$ that computes the limit of a sequence $(u_n)_{n\in\mathbf{N}}$ from the primitive recursive procedure producing $(u_n)_{n\in\mathbf{N}}$. If such a Φ exist it could be used to solve recursively the Halting Problem. Dealing with more intuitive arguments, one could just observe that, given a nonincreasing sequence of integer, the only general method to compute its limit is obviously to test *infinitely many* terms of this sequence, which is impossible. On the other hand, the constructive definition of DCC is easily realized by an Oracle Turing Machine working with any sequence $(u_n)_{n\in\mathbf{N}}$ given by an oracle.

Let $(E, \leq)$ be a poset. We will denote by $\leq_d$ the order on E^d defined by $(x_1, \ldots, x_d) \leq_d (y_1, \ldots, y_d)$ if and only if $x_i \leq y_i$ for all $i \in \{1, \ldots, d\}$. We shall write $\leq$ instead of $\leq_d$ when no confusion can arise.

Lemma 1.1 *If the poset $(E, \leq)$ satisfies DCC, then so does $(E^d, \leq_d)$. More generally, the finite product of posets verifying DCC satisfies DCC.*

Proof : We first give the proof for the case $d = 2$. Let $(u_n, v_n)_{n \in \mathbf{N}}$ be a nonincreasing sequence. Since the sequence $(u_n)_{n \in \mathbf{N}}$ is nonincreasing, one can find $n_1 < n_2 < \ldots$ such that $u_{n_i} = u_{n_i+1}$ for all $i \in \mathbf{N}$. The sequence $(v_{n_i})_{i \in \mathbf{N}}$ is nonincreasing ; hence, there exists $j \in \mathbf{N}$ such that $v_{n_j} = v_{n_{j+1}}$. But $v_{n_j} \geq v_{n_j+1} \geq v_{n_{j+1}}$, thus $v_{n_j} = v_{n_j+1}$, and $(u_{n_j}, v_{n_j}) = (u_{n_j+1}, v_{n_j+1})$. The same argument can be used to prove the general case by induction. □
Note that the same lemma remains true when replacing DCC by ACC.

Remark : From this lemma we can easily deduce that if a poset satisfies DCC then for any nonincreasing sequence $(u_n)_{n \in \mathbf{N}}$ and any $k \in \mathbf{N}$ there exist infinitely many $m \in \mathbf{N}$ such that $u_m = u_{m+1} = \ldots = u_{m+k}$.

1.2 Dixon's lemma for finitely generated submodules of $\mathbf{N}^d$

We will consider $\mathbf{N}^d$ as an $\mathbf{N}^d$–module with the following law : $x \star y = x + y$. Let $\mathbf{M}_d$ be the set of finitely generated $\mathbf{N}^d$–submodules of $\mathbf{N}^d$. We denote $\mathcal{M}^+(x^1, \ldots, x^n)$ the $\mathbf{N}^d$–module generated by $\{x^1, \ldots, x^n\}$, and we let $\overline{x} := \mathcal{M}^+(x) = x + \mathbf{N}^d$. We remark that

$$\mathcal{M}^+(x^1, \ldots, x^n) = \overline{x^1} \cup \ldots \cup \overline{x^n} = \{x \in \mathbf{N}^d : x \geq_d x^1 \vee \cdots \vee x \geq_d x^n\}$$

Given any poset $(E, \leq_E)$ a *final subset of finite type* of E (generated by $x^1, \ldots, x^n$) is a set

$$\mathcal{M}^+_E(x^1, \ldots, x^n) := \overline{x^1} \cup \ldots \cup \overline{x^n} = \{x \in E : x \geq_E x^1 \vee \cdots \vee x \geq_E x^n\}$$

and $\mathbf{F(E)}$ will denote the set of final subsets of finite type of E, including the empty subset considered as generated by the empty family. So we have $\mathbf{F(N^d)} = \mathbf{M_d} \cup \{\emptyset\}$.

Proposition 1.2

(i) *Every $A \in \mathbf{M}_d$ is generated by a unique minimal family (for $\subseteq$). This family can be obtained by taking the minimal elements (for $\leq_d$) of any family of generators of A.*

(ii) *Given A, B in $\mathbf{M}_d$, one can decide whether $A \subseteq B$ or not.*

(iii) *The ordered set $(\mathbf{M}_d, \subseteq)$ satisfies ACC.*

Proof : Remark that for any $a, x^1, \ldots, x^n$ in $\mathbf{N}^d$ we have

$$\overline{a} \subseteq \overline{x^1} \cup \cdots \cup \overline{x^n} \Leftrightarrow a \in \overline{x^1} \cup \cdots \cup \overline{x^n} \Leftrightarrow x^1 \leq_d a \text{ or } \ldots \text{ or } \mathbf{x^n} \leq_\mathbf{d} \mathbf{a}$$

So, a given family $x^1, \ldots, x^n$ of generators of A is minimal (for $\subseteq$) if and only if neither $x^i \leq_d x^j$, nor $x^j \leq_d x^i$, for any $i < j \in \{1, \ldots, n\}$.

Hence we can extract a minimal family of any given family $x^1, \ldots, x^n$ of generators, keeping only the minimal (for $\leq_d$) elements $x^{k_1}, \ldots, x^{k_m}$: this gives the existence part of (i).
If $x^1, \ldots, x^n$ and $y^1, \ldots, y^m$ are minimal families of generators of A, $y^i \in A = \mathcal{M}^+(x^1, \ldots, x^n)$ for all $i \in \{1, \ldots, m\}$, hence there exists $j \in \{1, \ldots, n\}$ such that $x^j \leq_d y^i$. Applying this argument again for a given x^j, we show that for all $j \in \{1, \ldots, n\}$, there exists $k \in \{1, \ldots, m\}$ such that $y^k \leq_d x^j$.
Then for all $i \in \{1, \ldots, m\}$, there exists $j \in \{1, \ldots, n\}$ and $k \in \{1, \ldots, m\}$ such that $y^k \leq_d x^j \leq_d y^i$. The family $y^1, \ldots, y^m$ being minimal, using the above remark, we deduce that $k = i$. So for all $i \in \{1, \ldots, m\}$, there exists a unique $j \in \{1, \ldots, n\}$ such that $y^i = x^j$. The converse is also true, so we conclude that the two families are equal : we have shown the uniqueness part of (i).

The proof of (ii) is easy and left to the reader.

We prove (iii) by induction on d. The case $d = 1$ is clear. Let $d \geq 2$, let $(A^m)_{m \in \mathbf{N}}$ be a nondecreasing sequence in $\mathbf{M}_d$. Let $a = (a_1, \ldots, a_d) \in A^0$ (an element of the family of generators of A^0, for instance)
For all $i \in \{1, \ldots, d\}$ and $r \in \mathbf{N}$, let

$$H^r_{i,d} := \{(x_1, \ldots, x_d) \ : x_i = r\} \subset \mathbf{N}^d$$

There is an order isomorphism between $(H^r_{i,d}, \leq_d)$ and $(\mathbf{N}^{d-1}, \leq_{d-1})$, given by $(x_1, \ldots, x_d) \mapsto (x_1, \ldots, x_{i-1}, x_{i+1}, \ldots, x_d)$. So $\mathbf{F}(\mathbf{H}^r_{i,d}, \subseteq)$ satisfies ACC by induction hypothesis (it is isomorphic to $\mathbf{M}_{d-1} \cup \{-\infty\}$).
Now remark that $\mathbf{N}^d \setminus \overline{a}$ is a finite union of $H^r_{i,d}$'s :

$$\mathbf{N}^d \setminus \overline{a} = \bigcup_{i=1}^{d} \bigcup_{r<a_i} H^r_{i,d}.$$

Rename these sets $\mathcal{H}_1, \ldots, \mathcal{H}_k$: $\mathbf{N}^d \setminus \overline{a} = \cup_{j=1}^k \mathcal{H}_j$, each $\mathcal{H}_j$ being one of the $H^r_{i,d}$ in the above formula.
Given $A = \mathcal{M}^+(x^1, \ldots, x^n) \in \mathbf{M}_d$ and $\mathcal{H} = H^r_{i,d}$, we see easily that $\mathcal{H} \cap A$ is an explicit element of $\mathbf{F}(\mathcal{H})$:

first, for any $y \in \mathbf{N}^d$,

$$\overline{y} \cap H^r_{i,d} = \begin{cases} \emptyset & \text{if } r < y_i \\ (y_1, \ldots, y_{i-1}, r, y_{i+1}, \ldots, y_d) + \mathbf{N}^{d-1} & \text{if } y_i \leq r \end{cases}$$

then $\mathcal{H} \cap A = \cup_{i=1,\ldots,n}(\overline{x^i} \cap \mathcal{H})$.

Now for each $j \in \{1, \ldots, k\}$ we consider the sequence $m \mapsto A^m \cap \mathcal{H}_j$. This is a nondecreasing sequence in $\mathbf{F}(\mathcal{H}_j)$. Each $\mathbf{F}(\mathcal{H}_j)$ satisfies ACC, so by the lemma 1.1 there exists i such that $A^i \cap \mathcal{H}_j = A^{i+1} \cap \mathcal{H}_j$ for all $j \in \{1, \ldots, k\}$.
Since for each m

$$A^m = \overline{a} \cup \Big(\bigcup_{j=1,\ldots,k} (A^m \cap \mathcal{H}_j) \Big)$$

clearly we have $A^i = A^{i+1}$. This ends the proof. □

Let k be a field, $k[x_1, \dots, x_d]$ be the associated polynomial ring. If $\alpha = (\alpha_1, \dots, \alpha_d) \in \mathbf{N}^d$, x^α denotes the monomial $x_1^{\alpha_1} \dots x_d^{\alpha_d}$. A *monomial ideal* is an ideal generated by a monomial family $(x^\alpha)_{\alpha \in A}$, where $A \subseteq \mathbf{N}^d$. Clearly, two monomial ideals are equal if and only if they contain the same monomials, and the set of finitely generated monomial ideals is in one-to-one correspondance with $\mathbf{M}_d$. Then the third assertion of our previous proposition is equivalent to the following result.

Proposition 1.3 (Dixon's lemma, constructive version) *The set of finitely generated monomial ideals of $k[x_1, \dots, x_d]$, ordered with $\subseteq$, satisfies ACC.*

Remark : the classical proof of Dixon's lemma deals with the classical Ascending Chain Condition, and is obviously non constructive : the arguments given to show that $\mathbf{N}$ fails to verify classical DCC can be used again.

2 Acceptable orders and division algorithm

Let k be a discrete field ("discrete" means the equality is decidable in k). We are dealing with finitely generated ideals $I = \mathcal{I}(f_1, \dots, f_s)$ of $k[x_1, \dots, x_d]$. For $d = 1$, the test $f \in I$? can be made using the Euclidean division of polynomials. The Buchberger algorithm (*cf.* [1]) is based on a generalisation of this division. It is necessary to have a good total order on monomials. In fact, in sections 2 and 3, we follow the method of Galligo [3] and Cox, Little, O'Shea [2].

2.1 Acceptable orders on $\mathbf{N}^d$

A total order $\preceq$ on $\mathbf{N}^d$ is *acceptable* if :

(i) $0 \preceq \alpha$ for all $\alpha \in \mathbf{N}^d$.

(ii) If $\alpha \leq_d \beta$ (*i.e.*, $\alpha_1 \leq \beta_1, \dots, \alpha_d \leq \beta_d$) then $\alpha \preceq \beta$.

(iii) If $\alpha \preceq \beta$ then $\alpha + \gamma \preceq \beta + \gamma$.

(iv) $\alpha \preceq \beta$ is decidable.

Example: The lexicographic order is an acceptable order on $\mathbf{N}^d$.

Remark: Conditions (i) and (iii) imply condition (ii), condition (ii) implies condition (i). Our definition comes from [3]. We have added the condition (iv) (which is easily verified in all ususal cases) in order to get constructive theorems. In [2], the definition of *monomial orderings* p. 54 is different from our definition of acceptable orders, but the corollary 6 p. 71 in [2] shows that the two definitions are in fact equivalent.

Lemma 2.1 *If $\preceq$ is an acceptable order on $\mathbf{N}^d$, then $(\mathbf{N}^d, \preceq)$ satisfies DCC.*

Proof. The proof is by induction on d. It is obvious for $d = 1$. Suppose now that $d > 1$, and that the result is true for $d - 1$.
With the notations of the previous proof, the bijection $\varphi_{i,d}^r : H_{i,d}^r \to \mathbf{N}^{d-1}$,

$$(x_1, \ldots, x_{i-1}, r, x_{i+1}, \ldots, x_d) \mapsto (x_1, \ldots, x_{i-1}, x_{i+1}, \ldots, x_d)$$

induces an acceptable order on $\mathbf{N}^{d-1}$: hence, by induction, $H_{i,d}^r$ (with the order inducted by $\preceq$) satisfies DCC, and so does $H_{i,d}^r \cup \{\infty\}$.
Let $(x^m)_{m \in \mathbf{N}}$ be a nonincreasing sequence of $(\mathbf{N}^d, \preceq)$. W.l.o.g. we assume $x^1 \neq x^0 = a$. All the $(x^i)_{i>0}$ are in $\mathbf{N}^d \setminus \overline{a}$. Recall that

$$\mathbf{N}^d \setminus \overline{a} = \bigcup_{i=1}^{d} \bigcup_{r<a_i} H_{i,d}^r.$$

In each $H_{i,d}^r \cup \{\infty\}$ in the above formula, we define a nonincreasing sequence $(y_{i,d,r}^m)_{m \in \mathbf{N}}$ by $y_{i,d,r}^0 = \infty$ and $y_{i,d,r}^{m+1} = \begin{cases} y_{i,d,r}^m & \text{if } x^{m+1} \notin H_{i,d}^r \\ \varphi_{i,d}^r(x^{m+1}) & \text{if } x^{m+1} \in H_{i,d}^r \end{cases}$
Then by lemma 1.1, there exists $j > 0$ such that $y_{i,d,r}^{j+1} = y_{i,d,r}^j$ for each i, r. Remark that for all $m > 0$, x^m is the minimum of $\{(\varphi_{i,d}^r)^{-1}(y_{i,d,r}^m)\ ;\ 1 \leq i \leq d,\ 0 \leq r < a_i\}$ for $\preceq$, hence $x^j = x^{j+1}$. □

2.2 Division algorithm

An acceptable order $\preceq$ on $\mathbf{N}^d$ induces an order on the monomials of $k[x_1, \ldots, x_d]$: we will write $x^\alpha \preceq x^\beta$ instead of $\alpha \preceq \beta$ (recall that x^α means $x_1^{\alpha_1} \ldots x_d^{\alpha_d}$ for all $\alpha \in \mathbf{N}^d$).
Let $f = \sum_{\alpha \in \mathbf{N}^d} a_\alpha x^\alpha$ be a nonzero polynomial of $k[x_1, \ldots, x_d]$. The *multidegree of f* is

$$\mathrm{multideg} f = \max_{\preceq}\{\alpha \in \mathrm{N^d} : \mathrm{a}_\alpha \neq 0\}.$$

If $\alpha = \mathrm{multideg} f$, the *leading term of f*, the *leading coefficient of f*, the *leading monomial of f* are

$$\mathrm{LT}(f) = a_\alpha x^\alpha, \quad \mathrm{LC}(f) = a_\alpha, \quad \mathrm{LM}(f) = x^\alpha.$$

Now we recall a generalization of the euclidean division of polynomials.

Proposition 2.2 *Let $\mathcal{F}$ be an s-tuple $(f_1, \ldots, f_s)$ in $k[x_1, \ldots, x_d]$. Every $f \in k[x_1, \ldots, x_d]$ can be written*

$$f = a_1 f_1 + \ldots a_s f_s + r$$

where $a_1, \ldots, a_s, r \in k[x_1, \ldots, x_d]$, $\mathrm{multideg}(a_i f_i) \preceq \mathrm{multideg}(f)$, *and either $r = 0$, or $r = \sum_{\alpha \in \mathbf{N}^d} a_\alpha x^\alpha$, with, for each α such that $a_\alpha \neq 0$, x^α not divisible by any* $\mathrm{LM}(f_i)$.
By definition r is called a remainder *of the division of f by $\mathcal{F}$; it will be denoted by* $r = \overline{f}^{\mathcal{F}}$.

Proof. The following algorithm computes $a_1, \ldots, a_s$ and r.

Input : $f_1, \ldots, f_s, f$.
Output : $a_1, \ldots, a_s, r$.

$a_1 := 0, \ldots, a_s := 0,\ r := 0,\ p := f$
While p $\neq$ 0 do
$i := 1$
$div :=$ **False**
While (i $\leq$ s and div = False) do
If $\mathrm{LT}(f_i) \mid \mathrm{LT}(p)$ **then** $a_i := a_i + \frac{\mathrm{LT}(p)}{\mathrm{LT}(f_i)}$
$p := p - \frac{\mathrm{LT}(p)}{\mathrm{LT}(f_i)} f_i$
$div :=$ **True**
else i := i + 1
If div = False then r := $r + \mathrm{LT}(p)$, $p := p - \mathrm{LT}(p)$

If the algorithm stops, it is easily seen that it gives a correct result (see e.g. [2]). The fact that the algorithm stops is constructively proved by lemma 2.1 since LM (p) decreases strictly until $p = 0$. □

Remark : There is *a priori* no uniqueness result ! if we change the order of the $f_1, \ldots, f_s$, we change the computed polynomials.

3 Gröbner bases and Buchberger's algorithm for ideals

We call *Gröbner basis* of a given ideal $I \subseteq k[x_1, \ldots, x_d]$ any family $g_1, \ldots, g_s$ of polynomials in I such that, for all $f \in I$, the division of f by $g_1, \ldots, g_s$ using the previous algorithm leads to a null remainder.

Lemma 3.1

- *A given family $g_1, \ldots, g_s$ of polynomials in I is a Gröbner basis of I if and only if for all $f \in I$, $\mathrm{LT}(f) \in \mathcal{I}(\mathrm{LT}(g_1), \ldots, \mathrm{LT}(g_s))$.*
- *For a given Gröbner basis $\mathcal{G} = (g_1, \ldots, g_s)$ of I, the remainder of the division of any polynomial f by $\mathcal{G}$ is unique, irrespective of the order of $g_1, \ldots, g_s$.*

Proof. The first result is a clear consequence of the division algorithm. To show the second result, if $f = h_1 + r_1$ and $f = h_2 + r_2$, with $h_1, h_2 \in I$, then we have $(r_1 - r_2) \in I$, hence $\mathrm{LT}(r_1 - r_2) \in \mathcal{I}(\mathrm{LT}(g_1), \ldots, \mathrm{LT}(g_s))$. This leads to $r_1 - r_2 = 0$, since no monomial of r_1, r_2 is divisible by any of the $\mathrm{LT}(\mathrm{g}_1), \ldots, \mathrm{LT}(\mathrm{g}_s)$. □

Remark : We don't know yet (from a constructive point of view) if for any given finitely generated ideal I, there exists a Gröbner basis. ¿From the previous lemma, we deduce that the existence of a Gröbner basis of a finitely generated ideal I is equivalent to the existence of a (finite) basis of the monomial ideal $\mathcal{I}(\mathrm{LT}(\mathrm{I})) = \mathcal{I}(\mathrm{LT}(\mathrm{f}) \; : \mathrm{f} \in \mathrm{I})$.

3.1 Buchberger's algorithm

Now we will show that for a given ideal $I = \mathcal{I}(f_1, \ldots, f_r)$, one can always find a Gröbner basis $\mathcal{G} = (g_1, \ldots, g_s)$ of I.

If $f, g \in k[x_1, \ldots, x_d] \setminus \{0\}$, with $\mathrm{multideg}\,\mathrm{f} = \alpha$ and $\mathrm{multideg}\,\mathrm{g} = \beta$, let $\gamma = (\gamma_1, \ldots, \gamma_d)$ where $\gamma_i = \max(\alpha_i, \beta_i)$; we define $\mathrm{lcm}(\alpha, \beta) = \gamma$, $\mathrm{lcm}(\mathrm{LT}(\mathrm{f}), \mathrm{LT}(\mathrm{g})) = \mathrm{x}^\gamma$.

We define the *S-polynomial of* f, g by $\mathrm{S}[\mathrm{f}, \mathrm{g}] = \frac{\mathrm{x}^\gamma}{\mathrm{LT}(\mathrm{f})}\mathrm{f} - \frac{\mathrm{x}^\gamma}{\mathrm{LT}(\mathrm{g})}\mathrm{g}$.

We will recall without proof classical results whose proof is everywhere constructive (*e.g.*, in [2]). We insist only on the constructive proof of the correctness of Buchberger algorithm.

Lemma 3.2 *Let* $g_1, \ldots, g_s \in k[x_1, \ldots, x_d]$. *If we have some* $c_i \in k$ *and* $\alpha(i) \in \mathbf{N}^d$ *(for* $i = 1, \ldots, s$*), such that* $\alpha(i) + \mathrm{multideg}(\mathrm{g}_i) = \delta$ *for all* $i \in \{1, \ldots, s\}$ *such that* $c_i \neq 0$, *and* $\mathrm{multideg}(\sum_{i=1}^s \mathrm{c}_i \mathrm{x}^{\alpha(i)} \mathrm{g}_i) \prec \delta$ *; then there exists* $c_{j,k} \in k$ *such that :*

$$\sum_{i=1}^{s} c_i x^{\alpha(i)} g_i = \sum_{j,k} c_{j,k} x^{\delta - \gamma_{j,k}} \mathrm{S}[\mathrm{g}_j, \mathrm{g}_k]$$

where $x^\gamma_{j,k} = \mathrm{lcm}(\mathrm{LT}(\mathrm{g}_j), \mathrm{LT}(\mathrm{g}_k))$.
Furthermore, for each j, k, $\mathrm{multideg}(\mathrm{x}^{\delta - \gamma_{j,k}} \mathrm{S}[\mathrm{g}_j, \mathrm{g}_k]) \prec \delta$.

Now we give a characterization of Gröbner bases.

Proposition 3.3 *Let* I *be a polynomial ideal of* $k[x_1, \ldots, x_d]$. *A given basis* $\mathcal{G} = (g_1, \ldots, g_s)$ *of* I *is a Gröbner basis if and only if for all* $i \neq j$, *the remainder* $\overline{\mathrm{S}[\mathrm{g}_i, \mathrm{g}_j]}^{\mathcal{G}}$ *of the division of* $\mathrm{S}[\mathrm{g}_i, \mathrm{g}_j]$ *by* $\mathcal{G}$ *is zero.*

This proposition gives an algorithm which checks whether a given family $\mathcal{G}$ is a Gröbner basis or not. The idea is now, if $\mathcal{G}$ is not a Gröbner basis, to add to $\mathcal{G}$ the nonzero remainders $\overline{\mathrm{S}[\mathrm{g}_i, \mathrm{g}_j]}^{\mathcal{G}}$, and to iterate this operation until a Gröbner basis is computed. The Dixon's lemma (proposition 1.3) legitimates this method.

Theorem 1 *Let* $I = \mathcal{I}(f_1, \ldots, f_s)$ *a nonzero ideal of* $k[x_1, \ldots, x_d]$. *A Gröbner basis of* I *can be obtained by a finite number of iterations of the following algorithm :*

Input : $\mathcal{F}$ a basis of I.
Output : $\mathcal{G}$ a Gröbner basis of I.

$\mathcal{G} := \mathcal{F}$
Repeat
$\mathcal{H} := \mathcal{G}$
For all p $\leq$ q in $\mathcal{H}$ do
If $\overline{\mathrm{S[p,q]}}^{\mathcal{H}} \neq 0$ **then** $\mathcal{G} := \mathcal{G} \cup \{\overline{\mathrm{S[p,q]}}^{\mathcal{H}}\}$
until $\mathcal{H} = \mathcal{G}$

Proof : This algorithm computes a nondecreasing sequence $\mathcal{G}_1 \subseteq \mathcal{G}_2 \subseteq \ldots$
First $\mathcal{G}_0 = \mathcal{F}$ is a family of elements of I and if $\mathcal{G}_i$ is in I, then for $p, q \in \mathcal{G}_i$, we have $\mathrm{S[p,q]} \in \mathrm{I}$, hence $\overline{\mathrm{S[p,q]}}^{\mathcal{G}_i} \in I$, hence $\mathcal{G}_{i+1}$ is in I. By induction, $\mathcal{G}_m$ is in I for all m.
If the algorithm ends, proposition 3.3 says that the computed family $\mathcal{G}$ is a Gröbner basis.
Hence we just need to prove that the algorithm ends. For each i, we denote by $\mathcal{I}(\mathrm{LT}\,(\mathcal{G}_\mathrm{i}))$ the monomial ideal generated by the leading terms of the elements of $\mathcal{G}_i$. Since $\mathcal{G}_i \subseteq \mathcal{G}_{i+1}$, we have $\mathcal{I}(\mathrm{LT}\,(\mathcal{G}_\mathrm{i})) \subseteq \mathcal{I}(\mathrm{LT}\,(\mathcal{G}_{\mathrm{i+1}}))$.
But if $\mathcal{G}_i \subset \mathcal{G}_{i+1}$, then there exists $p, q \in \mathcal{G}_i$ such that $\overline{\mathrm{S[p,q]}}^{\mathcal{G}_i} \neq 0$, hence $\overline{\mathrm{S[p,q]}}^{\mathcal{G}_i} \notin \mathcal{I}(\mathrm{LT}\,(\mathcal{G}_\mathrm{i}))$; since $\overline{\mathrm{S[p,q]}}^{\mathcal{G}_i} \in \mathcal{I}(\mathrm{LT}\,(\mathcal{G}_{\mathrm{i+1}}))$, we have $\mathcal{I}(\mathrm{LT}\,(\mathcal{G}\mathrm{i})) \subset \mathcal{I}(\mathrm{LT}\,(\mathcal{G}\mathrm{i}+1))$.
Then, by Dixon's lemma (proposition 1.3), the sequence of monomial ideals $\mathcal{I}(\mathrm{LT}\,(\mathcal{G}_\mathrm{i}))$ being nondecreasing, there exits i such that $\mathcal{I}(\mathrm{LT}\,(\mathcal{G}_\mathrm{i})) = \mathcal{I}(\mathrm{LT}\,(\mathcal{G}_{\mathrm{i+1}}))$, hence $\mathcal{G}_i = \mathcal{G}_{i+1}$: which completes the proof. □
Remark : Reading carefully the proofs leading to this result, one could compute a majoration of the size of a Gröbner basis of $I = \mathcal{I}(f_1, \ldots, f_s)$ depending only on d (the number of variables) and on the degrees of the polynomials. In fact, assume that the monomial ordering, d and the degrees of $f_1, \ldots, f_s$ are fixed, and consider all the coefficients $(c_j)_{1 \leq j \leq q}$ of the f_i's as indeterminates. Then we get a "universal algorithm" that computes a Gröbner basis of I in any situation. A "situation" is specified by the answers to some tests

$$h_\ell((c_j)_{1 \leq j \leq q}) = 0 \ ?$$

for a given family $(h_\ell)_{1 \leq \ell \leq r}$ in $\mathbf{Z}[(c_j)_{1 \leq j \leq q}]$. So the "coefficient space" is decomposed into cells C_v that are $\mathbf{Z}$-Zariski constructible. In any cell C_v, the Buchberger algorithm works in a completely uniform way, and all coefficients of the polynomials in the computed Gröbner basis are given by rational functions in the c_j's with denominators nowhere vanishing on C_v.

From theorem 1, we get immediately the following important corollaries.

Theorem 2 *Finitely generated ideals in a polynomial ring over a discrete field are detachable : i.e. for any system* $(f_1, \ldots, f_s, g)$ *of polynomials in* $k[x_1, \ldots, x_d]$ *we can decide if* $g \in \mathcal{I}(f_1, \ldots, f_s)$ *or not.*

With the terminology of [5], we shall say that the ring $k[x_1, \ldots, x_d]$ *has detachable ideals*, with the meaning that finitely generated ideals are detachable.

Corollary 3.4 *Inclusion between finitely generated ideals in a polynomial ring over a discrete field is decidable.*

Theorem 3 *Let I be a finitely generated ideal in a polynomial ring over a discrete field. Then the monomial ideal $\mathcal{I}(\mathrm{LT}(\mathrm{I}))$ generated by the leading monomials of the elements of I is finitely generated*

Keeping in mind lemma 3.1 that characterizes Gröbner bases we see that the last theorem is nothing but an abstract form (*i.e.*, without specifying the algorithm which is implicit in the statement) of theorem 1.

4 A few constructions relatives to polynomial ideals

4.1 Hilbert's basis theorem

Lemma 4.1 *Let I, J be finitely generated ideals of $k[x_1, \ldots, x_d]$, with $I \subseteq J$. If $\mathcal{I}(\mathrm{LT}(\mathrm{I})) = \mathcal{I}(\mathrm{LT}(\mathrm{J}))$, then $I = J$.*

Proof : Let $p \in J$. Then $\mathrm{LT}(\mathrm{p}) \in \mathcal{I}(\mathrm{LT}(\mathrm{J})) = \mathcal{I}(\mathrm{LT}(\mathrm{I}))$. Hence we can find $f \in I$ such that $\mathrm{LT}(\mathrm{f}) = \mathrm{LT}(\mathrm{p})$. The polynomial $p' = p - f$ in in J, and $\mathrm{multideg}(\mathrm{p'}) \preceq \mathrm{multideg}(\mathrm{p})$, with $\mathrm{multideg}(\mathrm{p'}) = \mathrm{multideg}(\mathrm{p})$ if and only if $p' = p = 0$. Using this argument recursively, and using lemma 2.1 we show that $p \in I$. □

Theorem 4 (A constructive version of Hilbert's basis theorem) *Let k be a discrete field. The poset of finitely generated ideals of $k[x_1, \ldots, x_d]$ with $\subseteq$ verifies ACC.*

Proof : Let $(I_n)_{n \in \mathbf{N}}$ be a nondecreasing sequence of finitely generated ideals. Then $(\mathcal{I}(\mathrm{LT}(\mathrm{I_n})))_{n \in \mathbf{N}}$ is a non decreasing sequence of finitely generated monomials ideals. We conclude using Dixon's lemma and the previous lemma. □

4.2 Polynomial rings over discrete fields are coherent

A ring (resp. a module) is said to be coherent if every finitely generated ideal (resp. submodule) is finitely presented. In classical mathematics any noetherian ring is coherent. In constructive mathematics, a good notion replacing the classical notion of noetherian ring is the notion of coherent noetherian ring, where the constructive meaning for noetheriannity is that the set of *finitely generated* ideals in the ring R satisfies constructive ACC.

The theory of coherent rings and modules is naturally constructive (there are no shortcuts by classical arguments). In [5], this theory is explained very efficiently. A good classical reference for coherent rings and modules is [4]. Let us recall the main results (restricting ourselves to the commutative case). In a coherent ring, the intersection of two finitely generated ideals and the anihilator of any element are also finitely generated ideals. Conversely these conditions imply that the ring is coherent. Over a coherent ring R, any finitely presented module M is coherent. Moreover, if R has detachable ideals, then M has detachable submodules.

In this section we show constructively that a polynomial ring over a discrete field is coherent.

Lemma 4.2 *Let* $\alpha_1, \ldots, \alpha_s \in \mathbf{N}^d$. *Let* $\gamma_{i,j} = \sup_{\leq_d}(\alpha_i, \alpha_j)$ *(so* $x^{\gamma_{i,j}} = \mathrm{lcm}(x^{\alpha_i}, x^{\alpha_j})$*). Let* $R_{i,j} \in k[x_1, \ldots, x_d]^s$ *be the vector of polynomials corresponding to the relation*

$$x^{\gamma_{i,j}-\alpha_i}.x^{\alpha_i} - x^{\gamma_{i,j}-\alpha_j}.x^{\alpha_j} = 0,$$

i.e., $R_{i,j} = (0, \ldots, 0, x^{\gamma_{i,j}-\alpha_i}, 0, \ldots, 0, -x^{\gamma_{i,j}-\alpha_j}, 0, \ldots, 0)$, *with the nonzero terms in* i *and* j*. Then for any relation*

$$p_1 x^{\alpha_1} + \ldots + p_s x^{\alpha_s} = 0,$$

there exists polynomials $q_{i,j}$ *for* $i < j$, $i, j \in \{1, \ldots, s\}$ *such that*

$$(p_1, \ldots, p_s) = \textstyle\sum_{i<j} q_{i,j} R_{i,j}.$$

Furthermore, if $\mathrm{multideg}(p_k x^{\alpha_k}) \preceq \delta$ *for each* k, *then* $\mathrm{multideg}(q_{i,j}) \preceq \delta - \gamma_{i,j}$. *This means that the module of the relations between the monomials* $x^{\alpha_1}, \ldots, x^{\alpha_s}$ *is generated by the relations* $\mathrm{S}[x^{\alpha_i}, x^{\alpha_j}] = 0$.

Proof : The proof is by finite descending induction on s. We write the division of p_s by the family $x^{\gamma_{k,s}-\alpha_s}$, for $k = 1, \ldots, s-1$: $p_s = \sum_{k=1}^{s-1} q_k x^{\gamma_{k,s}-\alpha_s} + r_s$, with $\mathrm{multideg}(q_k x^{\gamma_{k,s}-\alpha_s}) \preceq \mathrm{multideg}(p_s)$ (i.e $\mathrm{multideg}(q_k) \preceq \delta - \gamma_{k,s}$), and no monomial of r_s divisible by one of the $x^{\gamma_{k,s}-\alpha_s}$. Hence we have

$$p_1 x^{\alpha_1} + \ldots + p_{s-1} x^{\alpha_{s-1}} + \sum_{k=1}^{s-1} q_k x^{\gamma_{k,s}-\alpha_s} x^{\alpha_s} = -r_s x^{\alpha_s}.$$

If $r_s \neq 0$, $\mathrm{multideg}(r_s x^{\alpha_s})$ is equal to the multidegree of a term of the sum at the left part. This term cannot be one of the $q_k x^{\gamma_{k,s}-\alpha_s} x^{\alpha_s}$: we would have $x^{\gamma_{k,s}-\alpha_s}$ divides $\mathrm{LT}(r_s)$. It can neither be $p_i x^{\alpha_i}$ (for an $i \in \{1, \ldots, s-1\}$) : we would have x^{α_i} divides $\mathrm{LT}(r_s) x^{\alpha_s}$, hence $x^{\gamma_{i,s}-\alpha_s}$ divides $\mathrm{LT}(r_s)$. Then $r_s = 0$. Hence $(p_1, \ldots, p_s) = (p'_1, \ldots, p'_{s-1}, 0) - \sum_{k=1}^{s-1} q_k R_{k,s}$, with $p'_i = p_i + q_i x^{\gamma_{i,s}-\alpha_s}$. Remark that $\mathrm{multideg}(p') \preceq \delta$. Repeating this operation s times, we obtain the desired result. □

Theorem 5 *Let $g_1, \ldots, g_s$ be a Gröbner basis of an ideal I of $k[x_1, \ldots, x_d]$. The module $\{(p_1, \ldots, p_s) : p_1 g_1 + \ldots + p_s g_s = 0\}$ of relations is finitely generated by the relations expressing that the remainder of the division of $S[g_i, g_j]$ by $(g_1, \ldots, g_s)$ is zero.*

Proof : Let $\alpha_i = \text{multideg}(g_i)$. We use the notations of the previous lemma. The generating relations are $x^{\gamma_{i,j}-\alpha_i} g_i - x^{\gamma_{i,j}-\alpha_j} g_j = \sum_{k=1}^{s} p_k^{i,j} g_k$, with $\text{multideg}(p_k^{i,j} g_k) \preceq \text{multideg}(S[g_i, g_j]) \prec \gamma_{i,j}$. We denote by $T_{i,j} = R_{i,j} - (p_1^{i,j}, \ldots, p_s^{i,j})$ the associated vector.
Let $(p_1, \ldots, p_s)$ such that $p_1 g_1 + \ldots + p_s g_s = 0$. Let $\delta_i = \text{multideg}(p_i g_i)$, and $\delta = \max(\delta_1, \ldots, \delta_s)$. We have $\sum_{\delta_k=\delta} \text{LT}(p_k) x^{\alpha_k} = 0$; using the previous lemma, we deduce that $(\text{LT}(p_k))_{\delta_k=\delta} = \sum_{i,j} q_{i,j} R_{i,j}$. Hence

$$\sum_{\delta_k=\delta} \text{LT}(p_k) g_k = \sum_{i<j} q_{i,j}(x^{\gamma_{i,j}-\alpha_i} g_i - x^{\gamma_{i,j}-\alpha_j} g_j)$$

Using the relations $T_{i,j}$, we have a new relation

$$\sum_{\delta_k=\delta}(p_k - \text{LT}(p_k)) g_k + \sum_{i<j} q_{i,j} \sum_{k=1}^{s} p_k^{i,j} g_k + \sum_{\delta_k<\delta} p_k g_k = 0$$

We have $\text{multideg}(q_{i,j} p_k^{i,j} g_k) \prec \delta$, then in this new relation $p'_1 g_1 + \ldots + p'_s g_s$, we have $\text{multideg}(p'_i g_i) \prec \delta$. But we have $(p'_1, \ldots, p'_s) = (p_1, \ldots, p_s) - \sum_{i<j} q_{i,j} T_{i,j}$. Repeating this operation, we obtain a sequence of vectors, with the maximum degree of the components nonincreasing, and decreasing while the vector is nonzero. After a finite number of iterations, we obtain that $(p_1, \ldots, p_s)$ is in the module generated by the $T_{i,j}$'s. □

4.3 Some classical constructions

In this section, we recall some basic constructions with finitely generated ideals in polynomial rings and we see that they are constructively proved.

Elimination ideal

Let $I = \mathcal{I}(f_1, \ldots, f_s)$ be a finitely generated ideal in $k[x_1, \ldots, x_d]$ and $1 \leq r < d$. The r-th elimination ideal of I is by definition the ideal $I_r = I \cap k[x_{r+1}, \ldots, x_d]$ of $k[x_{r+1}, \ldots, x_d]$. It is obtained by choosing the lexicographical order (with $x_1 > \cdots > x_d$) for the monomials and computing a Gröbner basis $\mathcal{G}$ of I for this order. Then $\mathcal{G}_r = \mathcal{G} \cap k[x_{r+1}, \ldots, x_d]$ is a Gröbner basis of $\mathcal{I}_r$. The proof is straightforward (see *e.g.* [2]).

Intersection of two finitely generated ideals

We recall here two usual ways to compute a finite basis for the intersection of two finitely generated ideals. The second one is similar to the construction given in [5].

Let $I = \mathcal{I}(f_1, \ldots, f_s)$ and $J = \mathcal{I}(g_1, \ldots, g_t)$ be two finitely generated ideals in $R = k[x_1, \ldots, x_d]$.

The first construction is the following trick. Consider a new variable y, consider the ideals $yI = \mathcal{I}(yf_1, \ldots, yf_s) \subset k[x_1, \ldots, x_d, y]$ and $(1-y)J = \mathcal{I}((1-y)g_1, \ldots, (1-y)g_t) \subset k[x_1, \ldots, x_d, y]$. Then $I \cap J = (yI + (1-y)J) \cap k[x_1, \ldots, x_d]$.

The second construction is given by a duality idea. To give an element $a_1f_1 + \cdots + a_sf_s = b_1g_1 + \cdots + b_tg_t$ of $I \cap J$ is the same thing as giving the relation vector $(a_1, \ldots, a_s, b_1, \ldots, b_t)$ for the polynomial system $(f_1, \ldots, f_s, -g_1, \ldots, -g_t)$. So compute a finite basis of the module of relations for this polynomial system.

Quotient of two finitely generated ideals

The quotient $I : J$ of two ideals is defined as $\{f \ : fJ \subseteq I\}$. If $I = \mathcal{I}(f_1, \ldots, f_s)$ and $J = \mathcal{I}(g_1, \ldots, g_t)$ in $R = k[x_1, \ldots, x_d]$, then obviously

$$f \in (I : J) \quad \Leftrightarrow \quad (\ fg_1 \in I \ \wedge \ \cdots \ \wedge \ fg_t \in I\ )$$

So we have to compute $I : gR$ for an arbitrary g. Compute of a finite basis $h_1, \ldots, h_u$ of $I \cap gR$: $h_1/g, \ldots, h_u/g$ is a finite basis of $I : gR$.

5 Finitely generated submodules of a free module

Here we explain how the Gröbner basis technique can be extended in order to deal with finitely generated submodules of $k[x_1, \ldots, x_d]^p$. We follow [3].

5.1 Acceptable order, Gröbner bases, Dixon's lemma, Buchberger's algorithm

If $a = (\alpha, k) \in \mathbf{N}^d \times \{1, \ldots, p\}$, we denote by x^a the vector $(0, \ldots, 0, x^\alpha, 0, \ldots, 0)$ $\in k[x_1, \ldots, x_d]^p$, with x^α at the k-th position. A vector F of $k[x_1, \ldots, x_d]^p$ is a *linear combination* of such "monomials" x^a.

Given an acceptable order $\preceq$ on $\mathbf{N}^d$, writing $(\alpha, k) \prec (\beta, \ell)$ if $\alpha \prec \beta$, and $(\alpha, k) \prec (\alpha, \ell)$ if $k < \ell$, we define a total order on $\mathbf{N}^d \times \{1, \ldots, p\}$, *i.e.* on the monomials $x^{(\alpha,k)}$, satisfying *DCC*, compatible with the multiplication by a monomial of $k[x_1, \ldots, x_d]$.

For $F \in k[x_1, \ldots, x_d]^p$, we define $\mathrm{LT}\,(\mathrm{F}), \mathrm{LC}\,(\mathrm{F}), \mathrm{LM}\,(\mathrm{F})$ as we did for polynomials. Let $a = (\alpha, k)$ and $b = (\beta, \ell)$; a and b are said to be *compatible* if $k = \ell$.

If $F, G \in k[x_1, \ldots, x_d]^p$, the *S-vector* of F, G is defined if and only if $\mathrm{LM}\,(\mathrm{F})$ and $\mathrm{LM}\,(\mathrm{G})$ are compatible, by $\mathrm{S[F,G]} = \frac{\mathrm{x}^{\gamma-\alpha}}{\mathrm{LC(F)}}\mathrm{F} - \frac{\mathrm{x}^{\gamma-\beta}}{\mathrm{LC}}\mathrm{G}$, where $\mathrm{LM}\,(\mathrm{F}) = (\alpha, \mathrm{k})$, $\mathrm{LM}\,(\mathrm{G}) = (\beta, \mathrm{k})$, $\gamma = \mathrm{lcm}\,(\alpha, \beta)$. If $\mathrm{LM}\,(\mathrm{F})$ and $\mathrm{LM}\,(\mathrm{G})$ are not compatible, it will be nice to write $\mathrm{S[F,G]} = 0$.

Proposition 5.1 *Let $\mathcal{G}$ be an s-uple $(G_1, \ldots, G_s)$ of $k[x_1, \ldots, x_d]^p$. Every $F \in k[x_1, \ldots, x_d]^p$ can be written*

$$F = a_1G_1 + \cdots + a_sG_s + R$$

where $a_1, \ldots, a_s \in k[x_1, \ldots, x_d]$, $R \in k[x_1, \ldots, x_d]^p$, multideg$(a_iG_i) \preceq$ multideg(*and either $R = 0$, or $R = \sum_{a \in \mathbf{N}^d \times \{1,\ldots,p\}} c_a x^a$, with, for each a such that $c_a \neq 0$, x^a not divisible by any* LT(G_i).
The remainder *of the division of F by $\mathcal{G}$ is, by definition, R ; it will be denoted by $R = \overline{F}^{\mathcal{G}}$.*

Proof : The algorithm used to divide polynomials can be used again. □
We call *Gröbner basis* of a finitely generated submodule M of $k[x_1, \ldots, x_d]^p$ a basis (*i.e.* a generating family) $\mathcal{G} = (G_1, \ldots, G_s)$ of M such that, for all $F \in M$, the division of F by $\mathcal{G}$ using the classical algorithm, leads to a null remainder.

Proposition 5.2 *Let M be a finitely generated submodule of $k[x_1, \ldots, x_d]^p$.*

- *A given basis $\mathcal{G} = (G_1, \ldots, G_s)$ of M is a Gröbner basis if and only if for all $F \in M$,* LT(F) *is in the submodule generated by* LT$(G_1), \ldots,$ LT(G_s).
- *A given basis $\mathcal{G} = (G_1, \ldots, G_s)$ of M is a Gröbner basis if and only if for all $i \neq j$ such that* LM(G_i) *and* LM(G_j) *are compatible, the remainder* $\overline{S[G_i, G_j]}^{\mathcal{G}}$ *of the division of* $S[G_i, G_j]$ *by $\mathcal{G}$ is zero.*

Proof : One can verify that all the proofs written for ideals are still working. □

We can define monomial submodules in the same way than monomial ideals, and prove an other Dixon's lemma : *the poset of finitely generated monomial submodules of $k[x_1, \ldots, x_d]^p$, ordered with $\subseteq$, satisfies ACC.*
This gives as in section 3.1 a constructive proof for the fact that *the Buchberger algorithm can be used to compute a Gröbner basis of any finitely generated submodule.*
This implies that the monomial module of leading terms of a finitely generated submodule M of $k[x_1, \ldots, x_d]^p$ is also finitely generated. This gives also the detachability of finitely generated submodules of $k[x_1, \ldots, x_d]^p$.

5.2 Constructive noetherianity and coherence

The monomial module $\mathcal{M}$(LT(M)) generated by the elements of a finitely generated submodule M with Gröbner basis $G_1, \ldots, G_s$ is equal to the submodule generated by LT$(G_1), \ldots,$ LT(G_s). We can prove, as for ideals, that if $M \subseteq M'$ and $\mathcal{M}$(LT(M)) $= \mathcal{M}$(LT(M')), then $M = M'$.
Hence, as for polynomial ideals, we have the following "Hilbert basis theorem".

Theorem 6 *The poset of finitely generated submodules of* $k[x_1, \ldots, x_d]^p$*, ordered with* $\subseteq$*, satisfies ACC.*

The proof of coherence we wrote for ideals is still good for submodules.

Theorem 7 *Let* $G_1, \ldots, G_s$ *be a Gröbner basis of a finitely generated submodule* M *of* $k[x_1, \ldots, x_d]^p$*. The module* $\{(p_1, \ldots, p_s) : p_1G_1 + \cdots + p_sG_s = 0\} \subseteq k[x_1, \ldots, x_d]^s$ *of relations is finitely generated by the relations expressing that, for all* $i \neq j$ *such that* $\mathrm{LM}(G_i)$ *and* $\mathrm{LM}(G_j)$ *are compatible, the remainder of the division of* $[G_i, G_j]$ *by* $(G_1, \ldots, G_s)$ *is zero.*

Finally, the computation of a finite basis for the intersection of two finitely generated submodules can be made following the same lines.

Acknowledgments : Many thanks to André Galligo, Loic Pottier and Marie-Françoise Roy for usefull discussions and encouragements.

References

[1] Buchberger: *An algorithmic method in polynomial ideal theory.* in Multidimensional systems theory, ed. by N.K. Bose. D Reidel Publishing Company, Dordrecht, (1985), 184–232.

[2] Cox Q., Little J, O'Shea D. (1992) : *Ideals, Varieties, and Algorithms,* Springer Verlag UTM.

[3] Galligo A. (1983) : *Ideals, Vaieties, and Algorithms.* Technical Report. Université de Nice.

[4] Glaz S. (1989) : *Commutative Coherent Rings.* (Springer Verlag, LNM n°1371).

[5] Mines R., Richman F., Ruitenburg W. (1988) : *A Course in Constructive Algebra* . Springer-Verlag. Universitext.

[6] Richman F. (1974) : *Constructive aspects of Noetherian rings.* In : Proc. Amer. Mat. Soc. 44 pp. 436–441.

[7] Seidenberg A. (1974) : *What is Noetherian ?* In : Rend. Sem. Mat. e Fis. di Milano 44 pp. 55–61

Gröbner Bases in Non-Commutative Reduction Rings

Klaus Madlener and Birgit Reinert[1]

Fachbereich Informatik, Universität Kaiserslautern
67663 Kaiserslautern, Germany
{madlener,reinert}@informatik.uni-kl.de

Abstract

Gröbner bases of ideals in polynomial rings can be characterized by properties of reduction relations associated with ideal bases. Hence reduction rings can be seen as rings with reduction relations associated to subsets of the ring such that every finitely generated ideal has a finite Gröbner basis. This paper gives an axiomatic framework for studying reduction rings including non-commutative rings and explores when and how the property of being a reduction rings is preserved by standard ring constructions such as quotients and sums of reduction rings, and polynomial and monoid rings over reduction rings.

1 Introduction

Reasoning and computing in finitely presented algebraic structures is widespread in many fields in mathematics, physics and computer science. Reduction in the sense of simplification combined with appropriate completion methods is one general technique which is often successfully applied in this context, e.g. to solve the word problem and hence to compute effectively in the structure.

One fundamental application of this technique to polynomial rings was provided by B. Buchberger (1965) in his uniform effective solution of the ideal membership problem establishing the theory of Gröbner bases. These bases can be characterized in various manners, e.g. by properties of a reduction relation associated with polynomials (confluence or all elements in the ideal reduce to zero) or by special representations for the ideal elements with respect to a Gröbner basis (standard representations). Since Gröbner bases can be applied to solve many problems related to ideals and varieties in polynomial rings, generalizations to other structures followed (for an overview see e.g. Becker and Weispfenning (1992) or Madlener and Reinert (1995)).

[1]The author was supported by the Deutsche Forschungsgemeinschaft (DFG).

In this context it is interesting to find sufficient conditions, which allow to define a reduction relation in a ring in such a way that every finitely generated ideal has a finite Gröbner basis. Such rings will be called reduction rings. Naturally the question arises, when and how the property of being a reduction ring is preserved under various ring constructions. This can be studied from an existential or constructive point of view. Often additional conditions can be given to ensure effectivity for the ring operations, the reduction relation and the computation of the Gröbner bases, i.e. the ring is then an effective reduction ring. The main goal is to provide universal methods for constructing new reduction rings without having to generalize the whole setting individually for each new structure: e.g. knowing that the integers $\mathbb{Z}$ are a reduction ring and that the property lifts to polynomials in one variable, we find that $\mathbb{Z}[X]$ is again a reduction ring and we can immediately conclude that also $\mathbb{Z}[X_1, \ldots, X_n]$ is a reduction ring. Similarly, as sums of reduction rings are again reduction rings, we can directly conclude that $\mathbb{Z}^k[X_1, \ldots, X_n]$ or even $(\mathbb{Z}[Y_1, \ldots, Y_m])^k[X_1, \ldots, X_n]$ are reduction rings. Moreover, since $\mathbb{Z}$ is an effective reduction ring it can be shown that these new reduction rings again are effective. Commutative effective reduction rings have been studied by Buchberger (1984), Madlener (1986) and Stifter (1987).

On the other hand, many rings of interest are non-commutative, e.g. rings of matrices, the ring of quaternions, Bezout rings and monoid rings, and since they can be regarded in many cases as reduction rings, they are again candidates for applying ring constructions. More interesting examples of non-commutative reduction rings have been studied by Pesch (1997).

Hence in this paper we present a general framework for reduction rings and ring constructions including the non-commutative case. Since, in a first step, we are not interested in effectivity, this can be done by specifying three simple and natural axioms for the reduction relation and requiring the existence of finite Gröbner bases. For different ring constructions we define natural reduction relations fulfilling the axioms and we additionally show when the property of being a reduction ring is preserved. Moreover, in case the reduction ring is effective the resulting constructions as quotients and sums again are effective reduction rings. For the special case of monoid rings (including polynomial rings) we provide characterizations which enable to test the property of being a Gröbner basis by checking certain test sets which are finite provided the effective reduction ring fulfills additional properties. Such test sets are essential and have been used in critical-pair completion procedures as introduced by D. Knuth and P. Bendix or B. Buchberger for computing equivalent confluent reduction relations.

Let us close this section by summarizing some important notations and definitions of reduction relations (more details are provided by Book and Otto (1993)): Let $\mathcal{E}$ be a set of elements and $\longrightarrow$ a binary relation on $\mathcal{E}$ called **reduction**. For $a, b \in \mathcal{E}$ we will write $a \longrightarrow b$ in case $(a, b) \in \;\longrightarrow$. A pair $(\mathcal{E}, \longrightarrow)$ will be called a **reduction system**. Obviously the reflexive symmetric transitive closure $\stackrel{*}{\longleftrightarrow}$ is an equivalence relation on $\mathcal{E}$. The **word problem** for $(\mathcal{E}, \longrightarrow)$ is to decide for $a, b \in \mathcal{E}$, whether $a \stackrel{*}{\longleftrightarrow} b$ holds. An element $a \in \mathcal{E}$ is said to be **reducible** (with respect to $\longrightarrow$) if there exists an element $b \in \mathcal{E}$ such that $a \longrightarrow b$. If there is no such b, a is called **irreducible** denoted by $a \not\longrightarrow$. In case $a \stackrel{*}{\longrightarrow} b$ and b is irreducible, b is called a **normal form** of a. $(\mathcal{E}, \longrightarrow)$ is said to be **Noetherian** (or **terminating**) in case there are no infinitely descending reduction chains $a_0 \longrightarrow a_1 \longrightarrow \ldots$, with $a_i \in \mathcal{E}$, $i \in \mathbb{N}$. It is called **confluent**, if for all $a, a_1, a_2 \in \mathcal{E}$, $a \stackrel{*}{\longrightarrow} a_1$ and $a \stackrel{*}{\longrightarrow} a_2$ implies the existence of $a_3 \in \mathcal{E}$ such that $a_1 \stackrel{*}{\longrightarrow} a_3$ and $a_2 \stackrel{*}{\longrightarrow} a_3$. We can combine these two properties to give sufficient conditions for the existence of unique normal forms: $(\mathcal{E}, \longrightarrow)$ is said to be **complete** or **convergent** in case it is both, Noetherian and confluent.

2 Reduction rings

Let R be a ring with unit 1 and a (not necessarily effective) reduction relation $\Longrightarrow_B$ associated with subsets $B \subseteq \mathsf{R}$ satisfying the following axioms:

(A1) $\Longrightarrow_B = \bigcup_{\beta \in B} \Longrightarrow_\beta$, $\Longrightarrow_B$ is terminating for all finite subsets $B \subseteq \mathsf{R}$.

(A2) $\alpha \Longrightarrow_\beta \gamma$ implies $\alpha - \gamma \in \mathsf{ideal}^{\mathsf{R}}(\beta)$.

(A3) $\alpha \Longrightarrow_\alpha 0$ for all $\alpha \in \mathsf{R} \backslash \{0\}$.

Note that (A1) does not imply termination of reduction with respect to arbitrary sets: Just assume for example the ring $\mathsf{R} = \mathbb{Q}[\{X_i \mid i \in \mathbb{N}\}]$, i.e. a polynomial ring with infinitely many indeterminates, and reduction defined as usual with respect to the length-lexicographical ordering induced by $X_1 \succ X_2 \succ \ldots$. Then although reduction with respect to finite sets is terminating, this is no longer true for infinite sets, e.g. the set $\{X_i - X_{i+1} \mid i \in \mathbb{N}\}$ gives rise to an infinite reduction sequence $X_1 \Longrightarrow_{X_1 - X_2} X_2 \Longrightarrow_{X_2 - X_3} X_3 \ldots$.

Notice that in case R is commutative (A2) implies $\gamma = \alpha - \beta \cdot \rho$ for some $\rho \in \mathsf{R}$. In the non-commutative case using a single element β for reduction in general we get $\gamma = \alpha - \sum_{i=1}^{k} \rho_{i1} \cdot \beta \cdot \rho_{i2}$ for some $\rho_{i1}, \rho_{i2} \in \mathsf{R}$, $1 \leq i \leq k$, hence involving β more than once with different multipliers. One could also define a more restricted form of reduction by demanding e.g. $\gamma = \alpha - \rho_1 \cdot \beta \cdot \rho_2$ for some $\rho_1, \rho_2 \in \mathsf{R}$.

Similarly, we can define **one-sided** (right or left) reduction in rings by refining (A2): $\alpha \Longrightarrow_\beta \gamma$ implies $\alpha - \gamma \in \mathsf{ideal}_r^{\mathsf{R}}(\beta)$ respectively $\alpha - \gamma \in \mathsf{ideal}_l^{\mathsf{R}}(\beta)$. In this case we always get $\gamma = \alpha - \beta \cdot \rho$ respectively $\gamma = \alpha - \rho \cdot \beta$ for some $\rho \in \mathsf{R}$.

Let $\mathfrak{i} = \mathsf{ideal}(B)$ and $\equiv_{\mathfrak{i}}$ denote the congruence generated by $\mathfrak{i}$. Then (A1) and (A2) immediately imply $\stackrel{*}{\Longleftrightarrow}_B \subseteq \equiv_{\mathfrak{i}}$. Hence, one method for deciding the membership problem for $\mathfrak{i}$ in case the reduction relation is effective is to transform B into a finite set B' such that B' still generates $\mathfrak{i}$ and $\Longrightarrow_{B'}$ is confluent on $\mathfrak{i}$. Notice that 0 has to be irreducible for all $\Longrightarrow_\alpha$, $\alpha \in \mathsf{R}$[2]. Therefore, 0 will be chosen as the normal form of the ideal elements. Hence the goal is to achieve $\alpha \in \mathfrak{i}$ if and only if $\alpha \stackrel{*}{\Longrightarrow}_{B'} 0$. In particular $\mathfrak{i}$ is one equivalence class of $\stackrel{*}{\Longleftrightarrow}_{B'}$. The different definitions of reductions in rings existing in literature show that for deciding the membership problem it is not necessary to enforce $\stackrel{*}{\Longleftrightarrow}_{B'} \; = \; \equiv_\mathfrak{i}$. E.g. the so-called D-reduction notion given by Pan (1985) does not have this property but still is sufficient to decide $\equiv_\mathfrak{i}$-equivalence of two elements because $\alpha \equiv_\mathfrak{i} \beta$ if and only if $\alpha - \beta \in \mathfrak{i}$. It may even happen that D-reduction is not only confluent on $\mathfrak{i}$ but confluent everywhere and still $\alpha \equiv_\mathfrak{i} \beta$ does not imply that the normal forms with respect to D-reduction are the same.

Example 2.1 *Let us illustrate different ways of introducing reduction to the ring of integers* $\mathbb{Z}$. *For* $\alpha, \beta, \gamma \in \mathbb{Z}$ *we can define:*

- $\alpha \Longrightarrow^1_\beta \gamma$ *if and only if* $\alpha = \kappa \cdot |\beta| + \gamma$ *where* $0 \leq \gamma < |\beta|$ *and* $\kappa \in \mathbb{Z}$,
- $\alpha \Longrightarrow^2_\beta 0$ *if and only if* $\alpha = \kappa \cdot \beta$, *i.e.* β *is a proper divisor of* α.

For example we have $5 \Longrightarrow^1_4 1$ *but* $5 \not\Longrightarrow^2_4$.
It is easy to show that both reductions satisfy (A1) – (A3). Moreover the elements in $\mathbb{Z}$ *have unique normal forms and belong to* $\mathsf{ideal}(4)$ *if and only if they are reducible to zero using* 4. *For* $\Longrightarrow^1$*-reduction the normal forms are unique representatives of the quotient* $\mathbb{Z}/\mathsf{ideal}(4)$. *This is no longer true for* $\Longrightarrow^2$*-reduction, e.g.* $3 \equiv_{\mathsf{ideal}(4)} 7$ *since* $7 = 3 + 4$, *but both are* $\Longrightarrow^2$*-irreducible.*

Definition 2.2 *A subset* B *of* R *is called a* **Gröbner basis** *of an ideal* $\mathfrak{i}$, *if* $\stackrel{*}{\Longleftrightarrow}_B \; = \; \equiv_\mathfrak{i}$ *and* $\Longrightarrow_B$ *is convergent.*

Definition 2.3 *A ring* $(\mathsf{R}, \Longrightarrow)$ *satisfying (A1) – (A3) is called a* **reduction ring** *if every finitely generated ideal has a finite Gröbner basis.*

The notion of **one-sided reduction rings** can be defined similarly.

Effective or **computable reduction rings** can be defined similar to Buchberger's commutative reduction rings (see Buchberger (1984) or Stifter (1987)), in our case by demanding that the ring operations are computable, reduction is effective and Gröbner bases can be computed. Procedures which compute Gröbner bases are normally completion procedures based on effective tests to decide whether a finite set is a Gröbner basis.

[2] 0 cannot be reducible by itself since this would contradict the termination property in (A1). Similarly, $0 \Longrightarrow_\beta 0$ and $0 \Longrightarrow_\beta \gamma$, both β and γ not equal 0, give rise to an infinite reduction sequence again contradicting (A1).

Remark 2.4 *If B is a finite Gröbner basis of $\mathfrak{i}$ and $\alpha \in \mathfrak{i}$, then $B' = B \cup \{\alpha\}$ is again a Gröbner basis of $\mathfrak{i}$. This follows from $\overset{*}{\Longleftrightarrow}_B \subseteq \overset{*}{\Longleftrightarrow}_{B'} \subseteq \equiv_\mathfrak{i} = \overset{*}{\Longleftrightarrow}_B$ and the fact that $\beta \Longrightarrow_\alpha \gamma$ implies $\beta \equiv_\mathfrak{i} \gamma$ and hence β and γ have the same normal form with respect to $\Longrightarrow_B$ implying that $\Longrightarrow_{B'}$ inherits its confluence from $\Longrightarrow_B$.*

Hence, if B is a Gröbner basis of an ideal $\mathfrak{i}$ and $\beta \in B$ is reducible by $B \backslash \{\beta\}$ to α, then $B \cup \{\alpha\}$ is again a Gröbner basis of $\mathfrak{i}$. In order to remove β from $B \cup \{\alpha\}$ without losing the Gröbner basis property it is important for $\Longrightarrow$ to satisfy an additional axiom:

(A4) $\alpha \Longrightarrow_\beta$ and $\beta \Longrightarrow_\gamma \delta$ imply $\alpha \Longrightarrow_\gamma$ or $\alpha \Longrightarrow_\delta$.

Lemma 2.5 *Let $(\mathsf{R}, \Longrightarrow)$ be a reduction ring satisfying (A4). Further let $B \subseteq \mathsf{R}$ be a Gröbner basis and $B' \subseteq B$ such that for all $\beta \in B$, $\beta \overset{*}{\Longrightarrow}_{B'} 0$ holds. Then B' is a Gröbner basis of $\mathsf{ideal}^\mathsf{R}(B)$. In particular, for all $\alpha \in \mathsf{R}$, $\alpha \overset{*}{\Longrightarrow}_B 0$ implies $\alpha \overset{*}{\Longrightarrow}_{B'} 0$.*

Remark 2.4 and Lemma 2.5 are related to interreduction and so-called reduced Gröbner bases.

Now the question arises which ring constructions, as e.g. extensions, products, sums or quotients, preserve the property of being a reduction ring.

To simplify notations sometimes we will identify $(\mathsf{R}, \Longrightarrow)$ with R in case $\Longrightarrow$ is known or irrelevant.

3 Quotients of reduction rings

Let $(\mathsf{R}, \Longrightarrow)$ be a reduction ring and $\mathfrak{i}$ a finitely generated ideal in R with a (finite) Gröbner basis B. Then every element $\alpha \in \mathsf{R}$ has a unique normal form $\alpha\Downarrow_B$ with respect to $\Longrightarrow_B$. We choose the set of $\Longrightarrow_B$-irreducible elements of R as representatives for the elements in $\mathsf{R}/\mathfrak{i}$. Addition is defined by $\alpha + \beta := (\alpha + \beta)\Downarrow_B$ and multiplication by $\alpha \cdot \beta := (\alpha \cdot \beta)\Downarrow_B$. Then a natural reduction can be defined on the quotient $\mathsf{R}/\mathfrak{i}$ as follows:

Definition 3.1 *Let $\alpha, \beta, \gamma \in \mathsf{R}/\mathfrak{i}$. We say β* **reduces** *α to γ in one step, denoted by $\alpha \longrightarrow_\beta \gamma$, if there exists $\gamma' \in \mathsf{R}$ such that $\alpha \Longrightarrow_\beta \gamma'$ and $(\gamma')\Downarrow_B = \gamma$.*

The properties (A1) – (A3) hold for reduction in $\mathsf{R}/\mathfrak{i}$: $\longrightarrow_S = \bigcup_{s \in S} \longrightarrow_s$ is terminating for all finite $S \subseteq \mathsf{R}/\mathfrak{i}$ since otherwise $\Longrightarrow_{B \cup S}$ would not be terminating in R although $B \cup S$ is finite. If $\alpha \longrightarrow_\beta \gamma$ for some $\alpha, \beta, \gamma \in \mathsf{R}/\mathfrak{i}$ we know $\alpha \Longrightarrow_\beta \gamma' \overset{*}{\Longrightarrow}_B \gamma$, i.e. $\alpha - \gamma \in \mathsf{ideal}^\mathsf{R}(\{\beta\} \cup B)$, and hence $\alpha - \gamma \in \mathsf{ideal}^{\mathsf{R}/\mathfrak{i}}(\beta)$. Finally $\alpha \Longrightarrow_\alpha 0$ for all $\alpha \in \mathsf{R} \backslash \{0\}$ implies $\alpha \longrightarrow_\alpha 0$.

Moreover, in case (A4) holds in R this is also true for $\mathsf{R}/\mathfrak{i}$: For $\alpha, \beta, \gamma, \delta \in \mathsf{R}/\mathfrak{i}$ we have that $\alpha \longrightarrow_\beta$ and $\beta \longrightarrow_\gamma \delta$ imply $\alpha \Longrightarrow_\beta$ and $\beta \Longrightarrow_\gamma \delta' \overset{*}{\Longrightarrow}_B \delta$ and since α is $\Longrightarrow_B$-irreducible this implies $\alpha \Longrightarrow_{\{\gamma,\delta\}}$ and hence $\alpha \longrightarrow_{\{\gamma,\delta\}}$.

Theorem 3.2 *If* $(\mathsf{R}, \Longrightarrow)$ *is a reduction ring with (A4), then for every finitely generated ideal* $\mathfrak{i}$ *the quotient* $(\mathsf{R}/\mathfrak{i}, \longrightarrow)$ *again is a reduction ring with (A4).*

In Example 2.1 we have seen how the integers can be turned into reduction rings and this theorem now states that for every $m \in \mathbb{Z}$ the quotient $\mathbb{Z}/\mathsf{ideal}(m)$ again is a reduction ring. In particular reduction rings with zero divisors can be constructed in this way.

If R is an effective reduction ring, so is $\mathsf{R}/\mathfrak{i}$.

Theorem 3.2 extends to the case of one-sided reduction rings with (A4) provided that the two-sided ideal has a finite right respectively left Gröbner basis.

4 Sums of reduction rings

Let $(\mathsf{R}_1, \Longrightarrow^1), (\mathsf{R}_2, \Longrightarrow^2)$ be reduction rings. Then $\mathsf{R} = \mathsf{R}_1 \times \mathsf{R}_2 = \{(\alpha_1, \alpha_2) \mid \alpha_1 \in \mathsf{R}_1, \alpha_2 \in \mathsf{R}_2\}$ is called the direct sum of R_1 and R_2. Addition and multiplication are defined componentwise, the unit is $(1_1, 1_2)$ where 1_i is the respective unit in R_i. A natural reduction can be defined on R as followings:

Definition 4.1 *Let* $\alpha = (\alpha_1, \alpha_2)$, $\beta = (\beta_1, \beta_2)$, $\gamma = (\gamma_1, \gamma_2) \in \mathsf{R}$. *We say that* β **reduces** α *to* γ *in one step, denoted by* $\alpha \longrightarrow_\beta \gamma$, *if either* $(\alpha_1 \Longrightarrow^1_{\beta_1} \gamma_1$ *and* $\alpha_2 = \gamma_2)$ *or* $(\alpha_1 = \gamma_1$ *and* $\alpha_2 \Longrightarrow^2_{\beta_2} \gamma_2)$ *or* $(\alpha_1 \Longrightarrow^1_{\beta_1} \gamma_1$ *and* $\alpha_2 \Longrightarrow^2_{\beta_2} \gamma_2)$.

The properties (A1) – (A3) for reduction in R hold: $\longrightarrow_B = \bigcup_{\beta \in B} \longrightarrow_\beta$ is terminating for finite $B \subseteq \mathsf{R}$ since this property is inherited from the termination of the respective reductions in R_i. $\alpha \longrightarrow_\beta \gamma$ implies $\alpha - \gamma \in \mathsf{ideal}^{\mathsf{R}}(\beta)$, and $\alpha \longrightarrow_\alpha (0_1, 0_2)$ holds for all $\alpha \in \mathsf{R}\backslash\{(0_1, 0_2)\}$. It is easy to see that if condition (A4) holds for $\Longrightarrow^1$ and $\Longrightarrow^2$ then this is inherited by $\longrightarrow$.

Theorem 4.2 *If* $(\mathsf{R}_1, \Longrightarrow^1), (\mathsf{R}_2, \Longrightarrow^2)$ *are reduction rings, then* $(\mathsf{R} = \mathsf{R}_1 \times \mathsf{R}_2, \longrightarrow)$ *is again a reduction ring.*

Special regular rings as introduced by Weispfenning (1987) provide examples of such sums of reduction rings.

If R_1 and R_2 are effective reduction rings, so is R. Due to the "simple" multiplication used when defining the structure, Theorem 4.2 extends directly to one-sided reduction rings. More complicated multiplications are possible and have to be treated individually.

5 Polynomial rings over reduction rings

For a reduction ring $(\mathsf{R}, \Longrightarrow)$ we adopt the usual notations in $\mathsf{R}[X]$ with multiplication $*$. Notice that we assume $\alpha \cdot X = X \cdot \alpha$ for $\alpha \in \mathsf{R}$ (see Pesch

(1997) for other possibilities). We specify an ordering on the set of terms in one variable by defining that if X^i divides X^j, i.e. $0 \leq i \leq j$, then $X^i \preceq X^j$. Using this ordering, head term $\mathsf{HT}(p)$, head monomial $\mathsf{HM}(p)$ and head coefficient $\mathsf{HC}(p)$ of a polynomial $p \in \mathsf{R}[X]$ are defined as usual, and $\mathsf{RED}(p) = p - \mathsf{HM}(p)$. We extend the function HT to sets of polynomials $F \subseteq \mathsf{R}[X]$ by $\mathsf{HT}(F) = \{\mathsf{HT}(f) \mid f \in F\}$.

Let $\mathfrak{i} \subseteq \mathsf{R}[X]$ be a finitely generated ideal in $\mathsf{R}[X]$. It is easy to see that given a term t the set $C(t, \mathfrak{i}) = \{\mathsf{HC}(f) \mid f \in \mathfrak{i}, \mathsf{HT}(f) = t\} \cup \{0\}$ is an ideal in R. In order to guarantee that these ideals are also finitely generated we will assume that R is a Noetherian ring. Further, for any two terms t and s such that t divides s we have $C(t, \mathfrak{i}) \subseteq C(s, \mathfrak{i})$.

We additionally define a (not necessarily Noetherian) partial ordering on R by setting for $\alpha, \beta \in \mathsf{R}$, $\alpha >_{\mathsf{R}} \beta$ if and only if there exists a finite set $B \subseteq \mathsf{R}$ such that $\alpha \overset{+}{\Longrightarrow}_B \beta$. Then we can define an ordering on $\mathsf{R}[X]$ as follows: For $f, g \in \mathsf{R}[X]$, $f > g$ if and only if either $\mathsf{HT}(f) \succ \mathsf{HT}(g)$ or ($\mathsf{HT}(f) = \mathsf{HT}(g)$ and $\mathsf{HC}(f) >_{\mathsf{R}} \mathsf{HC}(g)$) or ($\mathsf{HM}(f) = \mathsf{HM}(g)$ and $\mathsf{RED}(f) > \mathsf{RED}(g)$). Notice that this ordering in general is neither total nor Noetherian on $\mathsf{R}[X]$.

Definition 5.1 *Let p, f be two non-zero polynomials in $\mathsf{R}[X]$. We say f* **reduces** *p to q at a monomial $\alpha \cdot X^i$ in p in one step, denoted by $p \longrightarrow_f q$, if*

(a) $\mathsf{HT}(f)$ divides X^i, i.e. $\mathsf{HT}(f)X^j = X^i$ for some term X^j,

(b) $\alpha \Longrightarrow_{\mathsf{HC}(f)} \beta$, with $\alpha = \beta + \sum_{i=1}^{k} \gamma_i \cdot \mathsf{HC}(f) \cdot \delta_i$ for some $\beta, \gamma_i, \delta_i \in \mathsf{R}$, $1 \leq i \leq k$ and

(c) $q = p - \sum_{i=1}^{k} (\gamma_i \cdot f \cdot \delta_i) * X^j$.

Notice that in case f reduces p to q at a monomial $\alpha \cdot t$ the term t can still occur in the resulting polynomial q. But when using a *finite* set of polynomials for reduction we know by (A1) that reducing α in R with respect to the head coefficients of the applicable polynomials must terminate and then either the monomial containing the term t disappears or is irreducible. Hence the above defined reduction is Noetherian when using a *finite* set of polynomials and (A1) holds. It is easy to see that (A2) and (A3) are also true and if the reduction ring satisfies (A4) this is inherited by $\mathsf{R}[X]$.

Theorem 5.2 *If $(\mathsf{R}, \Longrightarrow)$ is a Noetherian reduction ring, then $(\mathsf{R}[X], \longrightarrow)$ is a Noetherian reduction ring.*

Corollary 5.3 *If $(\mathsf{R}, \Longrightarrow)$ is a Noetherian reduction ring, then $\mathsf{R}[X_1, \ldots, X_n]$ is a Noetherian reduction ring with the respective lifted reduction.*

Notice that other definitions of reduction in $\mathsf{R}[X_1,\ldots,X_n]$ are known in the literature using admissible term orderings[3].

In order to provide completion procedures to compute Gröbner bases, various characterizations of Gröbner bases by finite test sets of polynomials (called s-, t-, m- or g-polynomials) in certain commutative reduction rings (e.g. the integers and Euclidean domains) can be found in the literature (see e.g. Kapur and Narendran (1985), Kandri-Rody and Kapur (1985) and Möller (1988)). A general approach to characterize commutative reduction rings allowing the computation of Gröbner bases using Buchberger's approach was presented by Stifter (1987).

Let us close this section by providing similar characterizations for polynomial rings over non-commutative reduction rings and outlining the arising problems: Given a basis $F \subseteq \mathsf{R}[X]$ the key idea is to distinguish special elements of $\mathsf{ideal}(F)$ which have representations $\sum_{i=1}^{n} g_i * f_i * h_i$, $g_i, h_i \in \mathsf{R}[X]$, $f_i \in F$ such that the head terms $\mathsf{HT}(g_i * f_i * h_i)$ are all the same within the representation. Then on one hand the respective $\mathsf{HC}(g_i * f_i * h_i)$ can add up to zero which means that the sum of the head coefficients is in an appropriate module generated by the $\mathsf{HC}(f_i)$ – m-polynomials[4] are related to these situations. If the result is not zero the sum of the $\mathsf{HC}(g_i * f_i * h_i)$ can be described in terms of a Gröbner basis of the $\mathsf{HC}(f_i)$ – g-polynomials are related to these situations. Zero divisors in the reduction ring occur as a special instance of m-polynomials where $F = \{f\}$ and $\alpha * f * \beta$, $\alpha, \beta \in \mathsf{R}$ are considered.

In case R is a commutative or one-sided reduction ring the first problem is related to solving linear homogeneous equations in R and to the existence of finite bases of the respective modules. In case we want effectiveness, we have to require that these bases are computable. This becomes more complicated for non-commutative two-sided reduction rings, as the equations are no longer linear and we have to distinguish right and left multipliers simultaneously. In some cases the problem for two-sided ideals can be translated into the one-sided case and hence solved via one-sided reduction techniques or also by weakening the reduction relation (Kandri-Rody and Weispfenning (1990)).

The g-polynomials can successfully be treated since finite Gröbner bases exist. Here, in case we want effectiveness, we have to require that they as well as representations for their elements in terms of the generating set are computable.

Using m- and g-polynomials, Gröbner bases can be characterized similar to the characterizations in terms of syzygies (a direct generalization of the approaches by Kapur and Narendran (1985) respectively Möller (1988)): In case the respective terms $\mathsf{HT}(g_i * f_i * h_i)$ only give rise to finitely many m- and g-polynomials, these situations can be localized to finitely many terms – to the least common multiples of the $\mathsf{HT}(f_i)$, i.e. the maximal term since

[3]For any term $t \neq 1$ we have $t \succ 1$ and $t_1 \succ t_2$ implies $t_1 \circ t \succ t_2 \circ t$.

[4]Explicit definitions of m- and g-polynomials will be provided in the next section.

$f_i \in \mathsf{R}[X]$ – and we can provide a completion procedure based on this characterization which will indeed compute a finite Gröbner basis if R is Noetherian. In principal ideal rings, where the function **gcd** is defined it is sufficient to consider sets F of size 2.

We will give the details of this approach for right reduction rings and the more general case of monoid rings in the next section.

6 Monoid rings over reduction rings

While polynomial rings over Noetherian reduction rings are again reduction rings, this cannot be achieved for the more general case of monoid rings. Already "non-commutative polynomial rings" over fields presented by Mora (1985), which are in fact free monoid rings, give us negative results concerning the existence of finite Gröbner bases for two-sided ideals due to the fact that they are closely related to the word problem for monoids (Kandri-Rody and Weispfenning (1990), Reinert (1995) and Madlener and Reinert (1997a)). However, when restricting the focus to one-sided ideals in this special setting, the existence of finite one-sided Gröbner bases can be shown (Mora (1985)).

Hence, we will restrict our attention to monoid rings over a right reduction ring $(\mathsf{R}, \Longrightarrow)$ with (A4) and provide a characterization of right Gröbner bases for finitely generated right ideals in this setting – the case of left ideals in monoid rings over left reduction rings with (A4) being similar.

Given a cancellative[5] monoid $\mathcal{M}$ with multiplication $\circ$, we call $\mathsf{R}[\mathcal{M}]$ the monoid ring over R with elements presented as "polynomials" $f = \sum_{t \in \mathcal{M}} \alpha_t \cdot t$ where only finitely many coefficients are non-zero. The elements $\alpha_t \cdot t$ are called monomials consisting of a coefficient $\alpha_t \in \mathsf{R}$ and a term $t \in \mathcal{M}$. Addition and multiplication for two polynomials $f = \sum_{t \in \mathcal{M}} \alpha_t \cdot t$ and $h = \sum_{t \in \mathcal{M}} \beta_t \cdot t$ is defined as $f + h = \sum_{t \in \mathcal{M}} (\alpha_t + \beta_t) \cdot t$ and $f * h = \sum_{t \in \mathcal{M}} \gamma_t \cdot t$ with $\gamma_t = \sum_{x \circ y = t} \alpha_x \cdot \beta_y$. Assuming a total well-founded ordering $\succ$ on $\mathcal{M}$, the usual notions as $\mathsf{HT}(p)$, $\mathsf{HC}(p)$, and $\mathsf{HM}(p)$ are defined for $p \in \mathsf{R}[\mathcal{M}] \backslash \{0\}$. For a subset F of $\mathsf{R}[\mathcal{M}]$ we call the set $\mathsf{ideal}_r(F) = \{\sum_{i=1}^{n} f_i * (\alpha_i \cdot w_i) \mid n \in \mathbb{N}, \alpha_i \in \mathsf{R}, f_i \in F, w_i \in \mathcal{M}\}$ the **right ideal** generated by F in $\mathsf{R}[\mathcal{M}]$.

As before, we define a partial ordering on R by setting for $\alpha, \beta \in \mathsf{R}$, $\alpha >_{\mathsf{R}} \beta$ if and only if there exists a finite set $B \subseteq \mathsf{R}$ such that $\alpha \overset{+}{\Longrightarrow}_B \beta$. This ordering can be extended to an ordering on $\mathsf{R}[\mathcal{M}]$ as follows: For $f, g \in \mathsf{R}[\mathcal{M}]$, $f > g$ if and only if either $\mathsf{HT}(f) \succ \mathsf{HT}(g)$ or $(\mathsf{HT}(f) = \mathsf{HT}(g)$ and $\mathsf{HC}(f) >_{\mathsf{R}} \mathsf{HC}(g))$ or $(\mathsf{HM}(f) = \mathsf{HM}(g)$ and $\mathsf{RED}(f) > \mathsf{RED}(g))$. Notice that the ordering in general is neither total nor Noetherian on $\mathsf{R}[\mathcal{M}]$.

[5] In case we allow arbitrary monoids we have to be more careful in defining right reduction and critical situations corresponding to it.

Definition 6.1 *Let p, f be two non-zero polynomials in $\mathsf{R}[\mathcal{M}]$. We say f* **right reduces** *p to q at a monomial $\alpha \cdot t$ in p in one step, denoted by $p \longrightarrow^{\mathsf{r}}_{f} q$, if*

(a) $\mathsf{HT}(f * w) = \mathsf{HT}(f) \circ w = t$ *for some* $w \in \mathcal{M}$,

(b) $\alpha \Longrightarrow_{\mathsf{HC}(f)} \beta$, *i.e.* $\alpha = \beta + \mathsf{HC}(f) \cdot \gamma$, $\gamma \in \mathsf{R}$ *and*

(c) $q = p - f * (\gamma \cdot w)$.

While reduction need no longer eliminate the occurrence of a term, it is Noetherian when using a fixed *finite* set of polynomials due to (A1) for $\Longrightarrow$ (compare Section 5). It is easy to see that (A2) - (A4) also hold and we can define Gröbner bases as usual:

Definition 6.2 *A set $G \subseteq \mathsf{R}[\mathcal{M}]$ is called a* **right Gröbner basis** *of $\mathsf{ideal}_{\mathsf{r}}(G)$, if $\xleftrightarrow{*}^{\mathsf{r}}_{G} = \equiv_{\mathsf{ideal}_{\mathsf{r}}(G)}$, and $\longrightarrow^{\mathsf{r}}_{G}$ is convergent.*

Notice that $p*(\alpha \cdot w) \longrightarrow^{\mathsf{r}}_{p} 0$ need not hold since the ordering on $\mathcal{M}$ need not be compatible with the multiplication in $\mathcal{M}$, i.e. in general $\mathsf{HT}(p*w) \neq \mathsf{HT}(p) \circ w$. To repair this we introduce the concept of saturation.

Definition 6.3 *A set of polynomials $F \subseteq \{p * (\alpha \cdot w) \mid \alpha \in \mathsf{R}^*, w \in \mathcal{M}\}$ is called a (right)* **saturating set** *for a polynomial $p \in \mathsf{R}[\mathcal{M}]$, if for all $\alpha \in \mathsf{R}^*$, $w \in \mathcal{M}$, $p*(\alpha \cdot w) \longrightarrow^{\mathsf{r}}_{F} 0$ holds in case $p*(\alpha \cdot w) \neq 0$. A set F of polynomials in $\mathsf{R}[\mathcal{M}]$ is called (right)* **saturated**, *if $f * (\alpha \cdot w) \longrightarrow^{\mathsf{r}}_{F} 0$ holds for all $f \in F$, $\alpha \in \mathsf{R}^*$, $w \in \mathcal{M}$ in case $f * (\alpha \cdot w) \neq 0$.*

We do not go into the details when finite saturated sets exist (see e.g. Reinert (1995) or Madlener and Reinert (1995)). In order to characterize right Gröbner bases we now introduce special polynomials.

Definition 6.4 *Let $P = \{p_1, \ldots, p_k\}$ be a set of polynomials in $\mathsf{R}[\mathcal{M}]$ and t an element in $\mathcal{M}$ such that there are $w_1, \ldots, w_k \in \mathcal{M}$ with $\mathsf{HT}(p_i * w_i) = \mathsf{HT}(p_i) \circ w_i = t$, for all $1 \leq i \leq k$. Further let $\gamma_i = \mathsf{HC}(p_i)$ for $1 \leq i \leq k$*[6]*. Let $\{\alpha_1, \ldots, \alpha_n\}$ be a right Gröbner basis of $\{\gamma_1, \ldots, \gamma_k\}$ and*

$$\alpha_i = \gamma_1 \cdot \beta_{i,1} + \ldots + \gamma_k \cdot \beta_{i,k}$$

for $\beta_{i,j} \in \mathsf{R}$, $1 \leq i \leq n$, and $1 \leq j \leq k$. Notice that the α_i respectively the $\beta_{i,j}$ do not depend on t. Then we define the **g-polynomials (Gröbner polynomials)** *corresponding to P and t by setting*

$$g_i = \sum_{j=1}^{k} p_j * (\beta_{i,j} \cdot w_j) \text{ for each } 1 \leq i \leq k.$$

[6]Note that this definition would have to be modified for non-cancellative monoids, as then $\mathsf{HT}(p * w) = \mathsf{HT}(p) \circ w$ no longer implies $\mathsf{HC}(p * w) = \mathsf{HC}(p)$.

Notice that $\mathsf{HM}(g_i) = \alpha_i \cdot t$.
For the right module $M = \{(\delta_1, \ldots, \delta_k) \mid \sum_{i=1}^{k} \gamma_i \cdot \delta_i = 0\}$, *let the set* $\{A_i \mid i \in I \subseteq \mathbb{N}\}$ *be a basis with* $A_i = (\alpha_{i,1}, \ldots, \alpha_{i,k})$ *for* $\alpha_{i,j} \in \mathsf{R}$, $i \in I$, *and* $1 \leq j \leq k$. *Notice that the* A_i *do not depend on* t. *Then we define the* **m-polynomials (module polynomials)** *corresponding to* P *and* t *by setting*

$$m_i = \sum_{j=1}^{k} p_j * (\alpha_{i,j} \cdot w_j) \text{ for each } i \in I.$$

Notice that $\mathsf{HT}(m_i) \prec t$.

Since R is a right reduction ring the number of g-polynomials related to P and t is finite while there can be infinitely many m-polynomials. The latter can be restricted by demanding that in R every right module of solutions to linear homogeneous equations is finitely generated.

Given $F \subseteq \mathsf{R}[\mathcal{M}]$, the set of corresponding g- and m-polynomials contains those which are specified by Definition 6.4 for each finite subset $P \subseteq F$ and each term $t \in \mathcal{M}$ fulfilling the respective conditions. For a set consisting of one polynomial the corresponding m-polynomials reflect the multiplication of the polynomial with zero-divisors of the head coefficient, i.e., by a basis of the annihilator of the head coefficient.

We can use g- and m-polynomials to characterize special bases in monoid rings over a reduction ring in case they are additionally saturated.

Theorem 6.5 *For a finite saturated subset* F *of* $\mathsf{R}[\mathcal{M}]$ *the following statements are equivalent:*

1. *For all polynomials* $g \in \mathsf{ideal}_r(F)$ *we have* $g \xrightarrow{*}{}^{\mathsf{r}}_F 0$.
2. F *is a right Gröbner basis of* $\mathsf{ideal}_r(F)$.
3. *All g-polynomials and all m-polynomials corresponding to* F *right reduce to zero using* F.

We have already stated when for a finite set of polynomials and a term the respective sets of g- and m-polynomials are finite. In order to use the characterization provided in Theorem 6.5 for a test, still infinitely many terms t have to be considered. A localization of such a test to finitely many terms similar to the polynomial ring case can in many cases be obtained by weakening the reduction: Remember that for two reductions $\longrightarrow^1 \subseteq \longrightarrow^2$ with $\xleftrightarrow{*}^1 = \xleftrightarrow{*}^2$ the confluence of $\longrightarrow^1$ will imply the confluence of $\longrightarrow^2$.

Assuming that $\mathcal{M}$ is presented by a finite convergent semi-Thue system such that the ordering on $\mathcal{M}$ is compatible with the completion ordering of the semi-Thue system, we can give the following syntactical weakening of right reduction:

Definition 6.6 *Let p, f be two non-zero polynomials in* $\mathsf{R}[\mathcal{M}]$. *We say f* **prefix reduces** *p to q at a monomial $\alpha \cdot t$ of p in one step, denoted by* $p \longrightarrow^{\mathsf{p}}_{f} q$, *if*

(a) $\mathsf{HT}(f)w \equiv t$ *for some $w \in \mathcal{M}$, i.e.,* $\mathsf{HT}(f)$ *is a prefix of t as a word in the generators,*

(b) $\alpha \Longrightarrow_{\mathsf{HC}(f)} \beta$, *i.e.* $\alpha = \beta + \mathsf{HC}(f) \cdot \gamma$, $\gamma \in \mathsf{R}$, *and*

(c) $q = p - f * (\gamma \cdot w)$.

Then overlaps correspond to common prefixes and in defining g- and m-polynomials it is even possible to restrict the attention to minimal such situations giving rise only to finitely many such candidates. In substituting prefix saturation for saturation and prefix reduction for right reduction, Theorem 6.5 can be specialized to characterize prefix Gröbner bases which are of course right Gröbner bases of the same right ideal.

The existence of finite prefix Gröbner bases can be shown for the classes of finite respectively free monoids and the classes of finite, free, plain respectively context-free groups. Using different syntactical weakenings of right reduction also finitary results are gained for the class of polycyclic groups (which includes the Abelian and nilpotent groups). The details on these approaches are presented in Reinert (1995) and Madlener and Reinert (1993,1997b). Hence in fact all these monoid rings are reduction rings.

It remains to state when in fact these monoid rings are effective right reduction rings: We need that right Gröbner bases and representations of their elements in terms of the elements of the generating set are computable and that it is possible to compute finite bases for right modules of solutions to linear homogeneous equations over R. With these assumptions the bases are computable in the above mentioned cases.

References

[1] Becker, T. and Weispfenning, V. (1992): Gröbner Bases - A Computational Approach to Commutative Algebra. Springer Verlag.

[2] Buchberger, B. (1965): Ein Algorithmus zum Auffinden der Basiselemente des Restklassenrings nach einem nulldimensionalen Polynomideal. Doctoral Dissertation, Universität Innsbruck.

[3] Buchberger, B. (1984): A Critical-Pair/Completion Algorithm for Finitely Generated Ideals in Rings. In: Symposium "Rekursive Kombinatorik", Münster (FRG), May 1983. Springer LNCS 171. pp 137-161.

[4] Book, R. and Otto, F.(1993): String-Rewriting Systems. Springer Verlag.

[5] Kandri-Rody, A. and Kapur, D. (1988): Computing a Gröbner Basis of a Polynomial Ideal over a Euclidean Domain. Journal of Symbolic Computation 6. pp 37-57.

[6] Kapur, D. and Narendran, P. (1985): Constructing a Gröbner Basis for a Polynomial Ring. In: Avenhaus, J. , Madlener, K. (eds): Summary in Proceedings of Combinatorial Algorithms in Algebraic Structures, Otzenhausen. Universität Kaiserslautern.

[7] Kandri-Rody, A. and Weispfenning, V.(1990): Non-Commutative Gröbner Bases in Algebras of Solvable Type. Journal of Symbolic Computation 9. pp 1-26.

[8] Madlener, K. (1986): Existence and Construction of Gröbner Bases for Ideals in Reduction Rings. Working paper. Universität Kaiserslautern.

[9] Madlener, K. and Reinert, B. (1993): Computing Gröbner Bases in Monoid and Group Rings. In: Proc. ISSAC'93. pp 254-263.

[10] Madlener, K. and Reinert, B. (1995): String Rewriting and Gröbner Bases – A General Approach to Monoid and Group Rings. In: Proceedings of the Workshop on Symbolic Rewriting Techniques, Monte Verita 1995 pp 127-180.

[11] Madlener, K. and Reinert, B. (1997a): Relating Rewriting Techniques on Monoids and Rings: Congruences in Monoids and Ideals in Monoid Rings. Theoretical Computer Science. To appear

[12] Madlener, K. and Reinert, B. (1997b): A Generalization of Gröbner Basis Algorithms to Polycyclic Group Rings. Journal of Symbolic Computation. To appear.

[13] Möller, H. M. (1988): On the Construction of Gröbner Bases Using Syzygies. Journal of Symbolic Computation 6. pp 345-359.

[14] Mora, F. (1985) Gröbner Bases for Non-Commutative Polynomial Rings. In: Proc. AAECC-3. Springer LNCS 229. pp 353-362

[15] Pan, L. (1985): On the Gröbner Bases of Ideals in Polynomial Rings over a Principal Ideal Domain. University of California. Santa Barbara. Department of Mathematics. Internal Manuscript.

[16] Pesch, M. (1997): Gröbner Bases in Skew Polynomial Rings. Doctoral Dissertation, Universität Passau.

[17] Reinert, B. (1995): On Gröbner Bases in Monoid and Group Rings. Doctoral Dissertation, Universität Kaiserslautern.

[18] Stifter, S. (1987): A Generalization of Reduction Rings. Journal of Symbolic Computation 4. pp 351-364.

[19] Weispfenning, V. (1987): Gröbner Basis for Polynomial Ideals over Commutative Regular Rings. In: Proc. EUROCAL'87. Springer LNCS 378. pp 336-347.

Effective Algorithms for Intrinsically Computing SAGBI-Gröbner Bases in a Polynomial Ring over a Field

J. Lyn Miller

Department of Mathematics, Western Kentucky University, Bowling Green, KY 42101 USA; miller@pulsar.cs.wku.edu

Abstract

Gröbner bases have been an important tool for performing computations in polynomial rings since Buchberger first presented an algorithm to construct them in 1965. A generalized type of Gröbner basis was developed in the context of a valuation ring by Sweedler in 1988. The purpose of this paper is to concretize Sweedler's theory to the context of a k-subalgebra of a polynomial ring $k[x_1, \ldots, x_n]$, where k is a field and to present effective additional algorithms necessary for intrinsically constructing objects that we shall call SAGBI-Gröbner bases.

1 Introduction

SAGBI-Gröbner bases are simply the counterparts in ideals of subalgebras of polynomial rings to Gröbner bases of ideals of the full polynomial ring. Our intrinsic development of SAGBI-Gröbner theory will mimic that of ordinary Gröbner bases, an approach that has a certain aesthetic appeal. In addition, this parallel development provides another point of view from which to study the mechanics of Gröbner basis theory, which was first presented in an algorithmic fashion by Buchberger (1965) (see also Buchberger (1985)).

We assume little experience with Gröbner bases on the part of the reader; an overview of the subject may be found in the texts of Adams and Loustaunau (1994), Becker and Weispfennig (1993), and Cox, Little, and O'Shea (1992). The related theory of SAGBI (or canonical) bases for subalgebras of polynomial rings over a field first appeared in the independent works of Robbiano and Sweedler (1988) and Kapur and Madlener (1989) and is also summarized in Ollivier (1990) and Miller (1994). (SAGBI bases for polynomial rings over less restrictive commutative rings are addressed in Dachsel (1990) and Miller (1994, 1996).) Finally, Sweedler (1988) introduced Gröbner bases for valuation rings, of which SAGBI-Gröbner bases are a special case; Ollivier

(1990) also addressed them briefly. Our goal here is to complete Sweedler's theory in such a way as to actually compute SAGBI-Gröbner bases for ideals of subalgebras of polynomial rings over fields. Again, we emphasize that this is to be done in as intrinsic a manner as possible, avoiding the tranformation to Gröbner basis computations in a polynomial overring of the subalgebra under consideration. (On the other hand, the theory of SAGBI-Gröbner bases over commutative non-fields as presented in Miller (1994, 1996) relies on such transformations.)

2 Notation and Definitions

We will work in the polynomial ring $k[x_1, \ldots, x_n]$, abbreviated $k[X]$, over a field k. Given $F \subseteq k[X]$, the notation $k[F]$ stands for the k-subalgebra generated by F, that is, the subset of $k[X]$ of all polynomials in the elements of F; note that $k \subseteq k[F]$. For a subset G of a subring $A \subseteq k[X]$, $\langle G \rangle_A$ indicates the ideal of A generated by G; we will use this same notation to denote monoid-ideals of monoids. The set of non-negative integers will be denoted by $\mathbf{N}$; $\mathbf{T}$ represents the set of all power products $\Pi x_i^{e_i}$, $e_i \in \mathbf{N}$, of the variables $x_1, \ldots, x_n$.

Definition 2.1 Let $S \subseteq k[X]$. An S-power product is a finite product of the form $s_1^{e_1} \cdots s_m^{e_m}$ where $s_i \in S$ and $e_i \in \mathbf{N}$ for $1 \leq i \leq m$; we abbreviate this as $S^{\vec{e}}$, where $\vec{e} \in \oplus_S \mathbf{N}$ is the vector having all coordinates equal to 0 except for $e_1, \ldots, e_m$ in the positions corresponding to $s_1, \ldots, s_m$.

Definition 2.2 Given a term order on $k[X]$, for each $p \in k[X]$, $\mathrm{lp}(p)$ will denote the leading X-power product of p, $\mathrm{lc}(p)$ the leading coefficient of p, and $\mathrm{lt}(p) = \mathrm{lc}(p)\mathrm{lp}(p)$ the leading term of p. (By convention, $\mathrm{lp}(0)$ is undefined while $\mathrm{lc}(0)$ and $\mathrm{lt}(0)$ are 0.) Given $S \subseteq k[X]$, $\mathrm{Lp}S$ denotes $\{\mathrm{lp}(s) : s \in S\}$.

Throughout this paper, $A \subseteq k[X]$ will be a k-subalgebra. Note that $\mathrm{Lp}A$ is a monoid under the usual multiplication. Accordingly, Robbiano and Sweedler (1988) defined an analog to Gröbner bases for k-subalgebras as follows:

Definition 2.3 A SAGBI basis for A is a subset $F \subseteq A$ such that $\mathrm{Lp}(F)$ generates the monoid $\mathrm{Lp}(A)$. (Their acronym SAGBI stands for "Subalgebra Analog to Gröbner Bases for Ideals.")

While analogous, there is a vital contrast between SAGBI and Gröbner bases: the Hilbert Basis Theorem (which implies the ascending chain condition) assures us that any ideal of $k[X]$ has a finite Gröbner basis, but this is not necessarily true of SAGBI bases, even for finitely-generated k-subalgebras of $k[X]$. In fact, determining a criterion for recognizing whether a k-subalgebra has a finite SAGBI basis is still an open question. (See Robbiano and Sweedler (1988) and Ollivier (1990).) We now define our primary object of study:

Definition 2.4 Let I be an ideal of A. A subset $G \subseteq I$ is called a SAGBI-Gröbner basis for I, or simply an SG-basis for I, if $\mathrm{Lp}G$ generates the monoid-ideal $\mathrm{Lp}I$ of the multiplicative monoid $\mathrm{Lp}A$.

This definition agrees with that of a Gröbner basis when $A = k[X]$, as the reader can easily confirm. This comparison prompts a question: must ideals of k-subalgebras of $k[X]$ have finite SG-bases? In general, the answer is no, for if $\mathrm{Lp}A$ is not finitely generated, it may contain non-finitely-generated monoid-ideals of the form $\mathrm{Lp}I$, I an ideal of A, whence I has no finite SG-basis. (The reader should explore $A = k[x, xy - y^2, xy^2] \subseteq k[x, y]$ and $I = \langle xy - y^2, xy^2 \rangle_A$ under a term order in which $\mathrm{lp}(xy - y^2) = xy$.) However, if A has a finite SAGBI basis, — that is, if $\mathrm{Lp}A$ is finitely generated, — then it is true that every ideal of A has a finite SG-basis (see Sweedler (1988) or Miller (1994)).

3 "SI"-reduction

Polynomial reduction is a key element in Buchberger's (1965, 1985) construction theory for Gröbner bases; so too our work with SG-bases requires a parallel concept. We adapt Sweedler's definition (a generalization of Buchberger's) to our setting.

Definition 3.1 Let $G \subseteq A$. A polynomial $h \in A$ SI-reduces in one step(via G) to $h' \in A$ if there is a term $cX^{\vec{\alpha}}$ of h for which there exist $g \in G$ and $a \in A$ such that $\mathrm{lt}(ag) = cX^{\vec{\alpha}}$ and $h' = h - ag$. If there is a chain of one-step reductions from h to h'' via G, then we say simply that h SI-reduces to h'', written $h \xrightarrow{G}_{\mathrm{SI}} h''$. We call h'' a final SI-reductum (also called a normal form) of h if it cannot be further SI-reduced.

The "SI-" prefix indicates both the subalgebra and ideal components of this operation: we SI-reduce h by subtracting from it an element of the ideal of A generated by G. To perform SI-reduction, we select a term $cX^{\vec{\alpha}}$ of h and attempt to solve $X^{\vec{\alpha}} = \mathrm{lp}(g)\mathrm{lp}(a)$ for both a and g; since $k \subseteq A$, we can suitably adjust the coefficient of a later.

Let F be a SAGBI basis for A. Then, since for every $a \in A$ there exists $\vec{\eta}_a \in \oplus_F \mathbf{N}$ such that $\mathrm{lp}(a) = \mathrm{lp}(F^{\vec{\eta}_a})$, we may instead solve $X^{\vec{\alpha}} = \mathrm{lp}(g)\mathrm{lp}(F^{\vec{\eta}_a})$ for g and $\vec{\eta}_a$. The reader should confirm that we can without loss of generality replace G by a subset $\widetilde{G}$ that is maximal with respect to the properties

$$\forall g \in \widetilde{G}, \mathrm{lp}(g) | X^{\vec{\alpha}} \text{ and } \forall g \neq g' \in G, \mathrm{lp}(g) \neq \mathrm{lp}(g')$$

and replace F by a similarly defined subset $\widetilde{F}$. (Clearly, neither of the sets $\widetilde{G}$ and $\widetilde{F}$ is uniquely determined. However, both are finite due to well-ordering of the divisibility order on $\mathbf{T}$.) Thus, given $\widetilde{F} = \{f_1, \ldots, f_m\}$, our task becomes that of successively testing elements $g \in \tilde{G}$ until we are able to solve $X^{\vec{\alpha}} = \text{lp}(g)[\text{lp}(f_1)]^{\eta_1} \cdots [\text{lp}(f_m)]^{\eta_m}$ for $\eta_1, \ldots, \eta_m \in \mathbf{N}$.

The strategy is standard: write $\text{lp}(g) = x_1^{\gamma_1} \cdots x_n^{\gamma_n}$, $X^{\vec{\alpha}} = x_1^{\alpha_1} \cdots x_n^{\alpha_n}$, and each $\text{lp}(f_j) = x_1^{e_{j,1}} \cdots x_n^{e_{j,n}}$, then transform the given equation into the inhomogeneous linear diophantine system

$$\alpha_i - \gamma_i = e_{1,i}\eta_1 + \cdots + e_{m,i}\eta_m, \quad i = 1, \ldots, n \tag{3.1}$$

We can solve for $\vec{\eta} = (\eta_1, \ldots, \eta_m) \in \mathbf{N}^m$ using the subroutine Dragon of the software package Pegasus designed by Dachsel (1990). (All diophantine computations for this paper were performed using this subroutine.) Once we have any solution $\vec{\eta}_0$ for some $g_0 \in \tilde{G}$, we take $c(\text{lc}(g\tilde{F}^{\vec{\eta}_0}))^{-1}\tilde{F}^{\vec{\eta}_0}$ as the corresponding a in the definition of SI-reduction. Repeating the process until it terminates produces a final SI-reductum, which must occur due to well-ordering of $\mathbf{T}$. Note, however, that the final reductum need not be unique.

This method for computing a final SI-reductum is summarized as Algorithm 1 below. Its framework is that of the standard reduction algorithm of Gröbner basis theory, which is validated in numerous sources (see Adams and Loustaunau (1994), for example); the strategy of the inner loop and use of $\tilde{G}, \tilde{F}$ were justified above.

```
Input: h ∈ A, G ⊆ A, a SAGBI basis F for A
Output: a final SI-reductum r of h via G
Initialization: r := 0, Term := lt(h)
While Term ≠ 0
    Choose G̃ ⊆ G, F̃ ⊆ F maximal with respect to:
        lp(p) ≠ lp(p')|Term  ∀p ≠ p' ∈ G̃( resp. F̃)
  While G̃ ≠ ∅ Choose g ∈ G̃.
      If System 3.1 has solutions in ⊕_F̃ N
      Then Choose a solution η⃗.
          h := h − (lc(h)/lc(gF̃^η⃗)) gF̃^η⃗, Term := 0, G̃ := ∅.
      Else G̃ := G̃ − {g}
  r := r + Term, h := h − Term, Term := lt(h)
```

Algorithm 1 : SI-Reduction Algorithm

Example 3.2 Let $A = \mathbf{Q}[f_1, f_2, f_3] = \mathbf{Q}[x, xy, y^2] \subseteq \mathbf{Q}[x, y]$. Its generators constitute a SAGBI basis F, for they are monomials. Let $G = \{x^2y^2, x^3y^3 + x, x^5y\} \subseteq A$, $h = x^4y^4 + 2x^3y^3 + 2x$.

Initially, $\text{lt}(h) = x^4y^4$, so choose $\tilde{G} = \{x^2y^2, x^3y^3 + x\}$, $\tilde{F} = F$. Next, select x^2y^2 from $\tilde{G}$; the resulting equation $x^4y^4 = x^2y^2\text{lp}(f_1)^{\eta_1}\text{lp}(f_2)^{\eta_2}\text{lp}(f_3)^{\eta_3}$ has two solutions: $(\eta_1, \eta_2, \eta_3) = (2, 0, 1)$ or $(0, 2, 0)$. (Both coincidentally produce the same polynomial $f_1^2 f_3 = f_2^2 = x^2y^2$.) Thus, using $g = a = x^2y^2$ in Definition 3.1, we have $h \xrightarrow{G}_{\text{SI}} h - (x^2y^2)(x^2y^2) = 2x^3y^3 + 2x$. We redefine $h = 2x^3y^3 + 2x$, leave $r = 0$, and conduct a second pass of the outermost loop.

With $\text{lt}(h) = 2x^3y^3$, we choose the same $\widetilde{G}$, $\widetilde{F}$ as before, again use $x^2y^2 \in \widetilde{G}$ to create System 3.1, and obtain the unique solution $\vec{\eta} = (0, 1, 0)$. Thus, $h \xrightarrow{G}_{\text{SI}} h - 2f_2(x^2y^2) = 2x$. This becomes our newest h, r remains equal to zero, and we conduct a third pass of the loop. Now, however, $\widetilde{G} = \emptyset$ for $\text{lt}(h) = 2x$, so that r becomes $2x$, h becomes zero, and the algorithm terminates. Thus, $2x$ is a final SI-reductum for h via G. $\triangle$

We conclude this section by stating three equivalent definitions of an SG-basis. The omitted proofs are very similar to corresponding ones for Gröbner bases and may be found in Miller (1994).

Proposition 3.3 *1. G is an SG-basis for $I \subseteq A \iff \forall h \in I$, $h \xrightarrow{G}_{\text{SI}} 0$. Thus, an SG-basis for I generates I as an ideal.*

2. G is an SG-basis for $\langle G \rangle_A \iff$ every element of A has a unique final SI-reductum via G. (This shows that the set G of Example 3.2 is not an SG-basis for $\langle G \rangle_A$, for 0 is another final SI-reductum for h via G, as the reader can confirm.)

3. G is an SG-basis for $\langle G \rangle_A \iff$ every element $h \in \langle G \rangle_A$ has an SG-representation with respect to G; that is, a finite representation $h = \sum_i a_i g_i$ with $a_i \in A, g_i \in G$ such that $\max_i\{\text{lp}(a_i g_i)\} = \text{lp}(h)$.

4 Syzygy Families

The next object plays a crucial role in SG-basis construction, for it takes the place of the S-polynomial of Buchberger's method. Let $g, h \in A$, and let $T_{g,h} \subseteq \mathbf{T}$ be a generating set for the monoid-ideal $\langle \text{lp}(g) \rangle \cap \langle \text{lp}(h) \rangle$ in LpA. Then for each $t \in T_{g,h}$, there exist (not necessarily unique) polynomials $a_t, b_t \in A$ such that $\text{lt}(a_t g) = \text{lt}(b_t h) = c_t t$ for some $c_t \in k$. (While we could choose the coefficients of a_t, b_t so that $\text{lc}(a_t g) = \text{lc}(b_t h) = 1$, this would be somewhat disadvantageous in our later computations.) The pair $(\text{lt}(a_t), -\text{lt}(b_t))$ is therefore a syzygy of $(\text{lt}(g), \text{lt}(h))$, meaning that $\text{lt}(a_t g)\text{lt}(g) - \text{lt}(b_t)\text{lt}(h) = 0$. This explains the name Sweedler chose for the following:

Definition 4.1 Given $g, h \in A$ and a generating set $T_{g,h}$ in LpA for $\langle \text{lp}(g) \rangle \cap \langle \text{lp}(h) \rangle$, a syzygy family for g and h is a set that contains, for each $t \in T_{g,h}$, a polynomial of the form $a_t g - b_t h$ with $\text{lt}(a_t g) = \text{lt}(b_t h) = c_t t$ for some $c_t \in k$. We denote such a set by $\text{SyzFam}(g, h)$.
Note that for each $a_t g - b_t h \in \text{SyzFam}(g, h)$, $\text{lp}(a_t g - b_t h) < \text{lp}(a_t g) = \text{lp}(b_t h)$.

Definition 4.2 For $p \in k[X]$, the multi-degree of p, denoted $\mathrm{mdeg}(p)$, is defined to be the exponent vector of $\mathrm{lp}(p)$; that is, $X^{\mathrm{mdeg}(p)} = \mathrm{lp}(p)$. For a set $S \subseteq k[X]$, $\mathrm{Mdeg}S = \{\mathrm{mdeg}(s) : s \in S\}$.

The multi-degree determines a monoid isomorphism from $\mathbf{T} \to \mathbf{N}^n$, so that computing a generating set for $\langle \mathrm{lp}(g) \rangle \cap \langle \mathrm{lp}(h) \rangle$ in $\mathrm{Lp}A$ is equivalent to computing a generating set for $\langle \mathrm{mdeg}(g) \rangle \cap \langle \mathrm{mdeg}(h) \rangle$ in the additive submonoid $\mathrm{Mdeg}A$ of $\mathbf{N}^n$. We explore this connection in greater detail:

Fix $g, h \in A$. If $t \in \langle \mathrm{lp}(g) \rangle \cap \langle \mathrm{lp}(h) \rangle$, then there exist $a_t, b_t \in A$ such that $\mathrm{mdeg}(g) + \mathrm{mdeg}(a_t) = \mathrm{mdeg}(t) = \mathrm{mdeg}(h) + \mathrm{mdeg}(b_t)$. Now assume that $F = \{f_1, \ldots, f_m\}$ is a finite SAGBI basis for A (the finiteness is algorithmically but not theoretically necessary). By definition, there exist F-power products $F^{\vec{e}_t}$, $F^{\vec{d}_t}$ such that $\mathrm{lp}(a_t) = \mathrm{lp}(F^{\vec{e}_t})$ and $\mathrm{lp}(b_t) = \mathrm{lp}(F^{\vec{d}_t})$. We shall write

$$\mathrm{mdeg}(a_t) = \mathrm{mdeg}(F^{\vec{e}_t}) = \sum_{\nu=1}^{m} e_{t,\nu} \cdot \mathrm{mdeg}(f_\nu)$$

where $e_{t,\nu}$ is the ν-th coordinate of $\vec{e}_t \in \mathbf{N}^m$. A similar statement can be written for $\mathrm{mdeg}(b_t) = \mathrm{mdeg}(F^{\vec{d}_t})$, using $d_{t,\nu}$ for the ν-th coordinate of $\vec{d}_t \in \mathbf{N}^m$. Thus, any element $t \in \langle \mathrm{lp}(g) \rangle \cap \langle \mathrm{lp}(h) \rangle$ corresponds to an expression

$$\mathrm{mdeg}(g) + \sum_{\nu=1}^{m} e_{t,\nu} \cdot \mathrm{mdeg}(f_\nu) = \mathrm{mdeg}(h) + \sum_{\nu=1}^{m} d_{t,\nu} \cdot \mathrm{mdeg}(f_\nu).$$

Viewed another way, t corresponds to a solution $(1, \vec{e}_t, \vec{d}_t, 1) \in \mathbf{N}^{2m+2}$ of the linear diophantine system in the variables ψ_i, ν_i below:

$$\psi_0 \mathrm{mdeg}(g) + \sum_{\nu=1}^{m} \psi_\nu \mathrm{mdeg}(f_\nu) = \sum_{\nu=1}^{m} \eta_\nu \mathrm{mdeg}(f_\nu) + \eta_0 \mathrm{mdeg}(h). \qquad (4.1)$$

While it is tempting to convert this system to inhomogeneous form (so that the number of solutions would be finite), doing so requires us to solve over $\mathbf{Z}$, not $\mathbf{N}$, in order not to omit solutions corresponding to "small" generators of $\langle \mathrm{mdeg}(g) \rangle \cap \langle \mathrm{mdeg}(h) \rangle$. This use of $\mathbf{Z}$ then risks losing other generators when the terms $\mathrm{lp}(g)$ and $\mathrm{lp}(h)$ have common factors of the form $\mathrm{lp}(F^{\vec{e}}) = \mathrm{lp}(F^{\vec{d}})$ with $\vec{e} \neq \vec{d} \in \mathbf{N}$. It may be that lost generators can be reclaimed by solving $\mathrm{lp}(F^{\vec{e}}) = \mathrm{lp}(F^{\vec{d}})$ and then forming sums of these solutions and those of the inhomogeneous system; however, this would again require the use of a homogeneous system, in only two fewer variables than the original. Fortunately, there is an effective, algorithmic way to extract finitely many useful solutions to the original homogeneous system. One algorithm is due to Clausen and Fortenbacher (1989), as referenced by Dachsel (1990); finiteness is assured by Lankford (1989) in the following:

Theorem 4.3 *Let $\mathcal{S}$ be a homogeneous linear diophantine system of n equations in ℓ unknowns. The set $\mathcal{L}$ of solutions to $\mathcal{S}$ which belong to $\mathbf{N}^\ell$ is a finitely generated submonoid of $\mathbf{N}^\ell$.*

Generators for this submonoid may be computed using Dragon. Furthermore, since we keep System 4.1 in homogeneous form, these solutions are easily broken into the *non-negative* exponent vectors $\vec{e}_t$ and $\vec{d}_t$ needed above. We create the following notation for this purpose:

Definition 4.4 Let $\mathcal{L}$ denote the submonoid of solution vectors to System 4.1 all of whose coordinates are non-negative integers. For each vector $\vec{v} = (c_0, c_1, \ldots, c_m, d_1, \ldots, d_m, d_0) \in \mathcal{L}$, define

$$\vec{v}^\ell = (c_0, c_1, \ldots, c_m) \text{ and } \vec{v}^r = (d_1, \ldots, d_m, d_0).$$

The original vector $\vec{v}$ will be called a parent vector of $\vec{v}^\ell$ and $\vec{v}^r$.

Proposition 4.5 *Let $g, h \in A$, and let $\mathcal{V} = \{\vec{v}_0 = \vec{0}, \vec{v}_1, \ldots, \vec{v}_M\}$ be a finite generating set for the solution submonoid $\mathcal{L}$. Then the monoid-ideal $\langle \mathrm{mdeg}(g)\rangle \cap \langle \mathrm{mdeg}(h)\rangle \subseteq \mathrm{Mdeg}A$ is generated by all vector dot-products of the form $(\mathrm{mdeg}(g), \mathrm{mdeg}(f_1), \ldots, \mathrm{mdeg}(f_m)) \cdot \vec{v}^\ell$ where the parent vector $\vec{v}$ of each $\vec{v}^\ell$ is a sum $\vec{v}_i + \vec{v}_j$ satisfying either*

$$\begin{cases} & c_{i,0} = d_{i,0} = 1 \quad \textit{and} \quad \vec{v}_j = \vec{0} \\ \textit{OR} & c_{i,0} = 1, d_{i,0} = 0 \quad \textit{and} \quad c_{j,0} = 0, d_{j,0} = 1. \end{cases}$$

(In other words, each parent vector $\vec{v}$ is either a single vector in $\mathcal{V}$ which has first and last coordinates equal to 1 or else is a sum of two vectors in $\mathcal{V}$ of the form $(1, \ldots, 0)$ and $(0, \ldots, 1)$, respectively.)

Proof. It is straight-forward to confirm that the monoid-ideal generated by vector dot-products of the form described above is indeed contained in $\langle \mathrm{mdeg}(g)\rangle \cap \langle \mathrm{mdeg}(h)\rangle$. We prove inclusion in the other direction. Let $z \in \langle \mathrm{mdeg}(g)\rangle \cap \langle \mathrm{mdeg}(h)\rangle$. Then since $\mathrm{Mdeg}F$ generates $\mathrm{Mdeg}A$, we can write

$$\mathrm{mdeg}(g) + \sum_{\nu=1}^{m} \zeta_\nu \mathrm{mdeg}(f_\nu) = z = \mathrm{mdeg}(h) + \sum_{\nu=1}^{m} \eta_\nu \mathrm{mdeg}(f_\nu)$$

for some $\zeta_\nu, \eta_\nu \in \mathbf{N}$. Thus, $(1, \zeta_1, \ldots, \zeta_m, \eta_1, \ldots, \eta_m, 1) \in \mathcal{L}$ and can be written as an $\mathbf{N}$-linear combination of its generators: without loss of generality, let $\alpha_1, \ldots, \alpha_{M_0} \neq 0 \in \mathbf{N}$ be such that

$$(1, \zeta_1, \ldots, \zeta_m, \eta_1, \ldots, \eta_m, 1) = \sum_{k=1}^{M_0} \alpha_k \vec{v}_k.$$

Equating first coordinates, we have $1 = \sum_{k=1}^{M_0} \alpha_k c_{k,0}$. Since each $\alpha_k, c_{k,0} \in \mathbf{N}$ and every $\alpha_k \geq 1$, it must be that every $c_{k,0} = 0$ except one, which must equal 1; let it be $c_{1,0}$. This also forces $\alpha_1 = 1$. We get a similar condition on the $d_{k,0}$'s: all are 0 except one, say $d_{K,0}$, which must equal 1, as does α_K.

Now if $K = 1$, then $c_{1,0} = d_{1,0} = 1$, and $\vec{v} = \vec{v}_1$ is then a parent vector of the desired form. Because $\alpha_1 = 1$ also, we easily observe that

$$\begin{aligned} z &= (\text{mdeg}(g), \text{mdeg}(f_1), \ldots, \text{mdeg}(f_m)) \cdot (1, \zeta_1, \ldots, \zeta_m) \\ &= (\text{mdeg}(g), \text{mdeg}(f_1), \ldots, \text{mdeg}(f_m)) \cdot \left(\vec{v}_1^{\ell} + \sum_{k=2}^{M_0} \alpha_k \vec{v}_k^{\ell} \right) \\ &\in \langle (\text{mdeg}(g), \text{mdeg}(f_1), \ldots, \text{mdeg}(f_m)) \cdot \vec{v}^{\ell} \rangle. \end{aligned}$$

If $K \neq 1$, then $c_{1,0} = 1$, $d_{1,0} = 0$ while $c_{K,0} = 0$, $d_{K,0} = 1$. Thus, the parent vector $\vec{v} = \vec{v}_1 + \vec{v}_K$ is of the second acceptable type, and $\alpha_1 = \alpha_K = 1$ implies that

$$\begin{aligned} z &= (\text{mdeg}(g), \text{mdeg}(f_1), \ldots, \text{mdeg}(f_m)) \cdot (1, \zeta_1, \ldots, \zeta_m) \\ &= (\text{mdeg}(g), \text{mdeg}(f_1), \ldots, \text{mdeg}(f_m)) \cdot \left(\vec{v}_1^{\ell} + \vec{v}_K^{\ell} + \sum_{k \neq 1,K} \alpha_k \vec{v}_k^{\ell} \right) \\ &\in \langle (\text{mdeg}(g), \text{mdeg}(f_1), \ldots, \text{mdeg}(f_m)) \cdot \vec{v}^{\ell} \rangle, \end{aligned}$$

again as desired. Thus, $\langle \text{mdeg}(g) \rangle \cap \langle \text{mdeg}(h) \rangle$ indeed equals the monoid-ideal generated by elements of the specified form. □

Corollary 4.6 *Let $G = \{g, f_1, \ldots, f_m\}$ and $H = \{f_1, \ldots, f_m, h\}$, and let $\mathcal{V} = \{\vec{v}_0 = \vec{0}, \vec{v}_1, \ldots, \vec{v}_M\}$ be a finite generating set for $\mathcal{L}$ as above. Then the set $\mathcal{S}$ of all polynomials of the form $s_{\vec{v}} = \text{lc}(H^{\vec{v}^r}) \cdot G^{\vec{v}^{\ell}} - \text{lc}(G^{\vec{v}^{\ell}}) \cdot H^{\vec{v}^r}$ where the parent vector $\vec{v}$ of $\vec{v}^{\ell}$ and $\vec{v}^r$ is of the form described in Proposition 4.5 is a syzygy family for g and h.*

Proof. (Built-in to the definition of the $S^{\vec{e}}$ notation is actually an innate order on the listing of the elements in S. It is crucial here that g be associated with the first coordinate of each $\vec{v}^{\ell}$ and h with the last one of each $\vec{v}^r$.) Recall that "mdeg" is a monoid isomorphism from $\mathbf{T}$ to $\mathbf{N}^n$ or, in this case, from LpA to Mdeg$A$. By Proposition 4.5, $\langle \text{mdeg}(g) \rangle \cap \langle \text{mdeg}(h) \rangle$ is generated by the elements $(\text{mdeg}(g), \text{mdeg}(f_1), \ldots, \text{mdeg}(f_m)) \cdot \vec{v}^{\ell}$, whence its pre-image $\langle \text{lp}(g) \rangle \cap \langle \text{lp}(h) \rangle$ is generated by the various pre-images $\text{lp}(G^{\vec{v}^{\ell}})$. Note that each $\text{lt}(\text{lc}(H^{\vec{v}^r}) \cdot G^{\vec{v}^{\ell}}) = \text{lt}(\text{lc}(G^{\vec{v}^{\ell}}) \cdot H^{\vec{v}^r})$ because $\vec{v}$ is a solution to System 4.1. Thus, the set described does constitute a syzygy family for g and h. □

Corollary 4.6 asserts that the coefficient polynomials a_t and b_t for each member of SyzFam(g, h) are, up to k-coefficients, the F-power products represented in each $G^{\vec{v}^{\ell}}$ and $H^{\vec{v}^r}$, respectively. Algorithm 2 below summarizes the technique for computing a syzygy family. Since $\mathcal{V}$ is finite, it terminates, and Corollary 4.6 has confirmed that the result is a syzygy family for g and h.

> Input: $g, h \in A$, a finite SAGBI basis F for A
> Output: A syzygy family $\mathrm{SyzFam}(g,h)$ for g and h
> Initialization: $\mathrm{SyzFam}(g,h) := \emptyset, V_1 = V_2 = \mathcal{PV} := \emptyset$
> Compute a generating set $\mathcal{V}$ for the set $\mathcal{L}$ of System 4.1.
> $V_1 := \{\vec{v} \in \mathcal{V} : c_0 = 1, d_0 = 0\}$
> $V_2 := \{\vec{v} \in \mathcal{V} : c_0 = 0, d_0 = 1\}$
> $\mathcal{PV} := \{\vec{v} \in \mathcal{V} : c_0 = d_0 = 1\} \cup \{\vec{v}_i + \vec{v}_j : \vec{v}_i \in V_1, \vec{v}_j \in V_2\}$
> For Each $\vec{v} \in \mathcal{PV}$:
> $\quad s_{\vec{v}} := \mathrm{lc}(H^{\vec{v}^r}) \cdot G^{\vec{v}^\ell} - \mathrm{lc}(G^{\vec{v}^\ell}) \cdot H^{\vec{v}^r}$
> $\mathrm{SyzFam}(g,h) := \cup_{\vec{v} \in \mathcal{PV}} \{s_{\vec{v}}\}$

Algorithm 2 : Syzygy Family Construction Algorithm

Example 4.7 Let $A = \mathbf{Q}[x^2, xy] \subseteq \mathbf{Q}[x,y]$, and use the term ordering degree lex with $x > y$. The generators of A are a SAGBI basis since they are monomials. We take $g = x^3y + x^2$ and $h = x^4 + x^2y^2$ in A. The corresponding System 4.1 is $\psi_0(3,1) + \psi_1(2,0) + \psi_2(1,1) = \eta_1(2,0) + \eta_2(1,1) + \eta_0(4,0)$, for which Dragon produces 6 non-trivial elements of $\mathcal{V}$:

$$\vec{v}_1 = (0,0,1,0,1,0), \quad \vec{v}_2 = (0,1,0,1,0,0), \quad \vec{v}_3 = (0,2,0,0,0,1),$$
$$\vec{v}_4 = (1,0,0,1,1,0), \quad \vec{v}_5 = (1,1,0,0,1,1), \quad \vec{v}_6 = (2,0,0,0,2,1)$$

Thus $V_1 = \{\vec{v}_4\}$, $V_2 = \{\vec{v}_3\}$ and $\mathcal{PV} = \{\vec{v}_5 = (1,1,0,0,1,1), \vec{v}_3 + \vec{v}_4 = (1,2,0,1,1,1)\}$. We form the polynomials $s_{\vec{v}_3+\vec{v}_4} = G^{(1,2,0)} - H^{(1,1,1)} = -x^5y^3 + x^6$ and $s_{\vec{v}_5} = G^{(1,1,0)} - H^{(0,1,1)} = -x^3y^3 + x^4$. Hence, $\mathrm{SyzFam}(g,h) = \{-x^5y^3 + x^6, -x^3y^3 + x^4\}$. $\triangle$

5 Constructing SG-bases

The generalized context of Sweedler's work left some computational technicalities open, such as how to effectively perform SI-reduction and compute syzygy families. Our algorithms have filled these gaps in the current context; the final algorithm, presented in this section, incorporates them in a concrete procedure for computing an SG-basis. First, however, we lay the foundation for this construction algorithm.

Theorem 5.1 *[Sweedler] Let $G \subseteq A$; choose a syzygy family* SyzFam(g, h) *for each pair of distinct elements of G. Let $\mathcal{S} = \cup_{g \neq h \in G}$*SyzFam$(g, h)$. *Then G is an SG-basis for $\langle G \rangle_A \iff$ every $s \in \mathcal{S}$ SI-reduces to 0 via G.*

Proof. We merely outline Sweedler's proof; a detailed version may be found in Sweedler (1988) or Miller (1994). The forward implication is a trivial corollary of Proposition 3.3, so we discuss the converse direction.

Let $h \neq 0 \in \langle G \rangle_A$ and assume that $h = \sum_{i=1}^{N} a_i g_i$ has $m = \max_i\{\text{lp}(a_i g_i)\}$ minimal for all such representations of h. Assume further that this representation also has a minimal number N' of products $a_i g_i$ for which $m = \text{lp}(a_i g_i)$. If $N' = 1$, then there is precisely one product having $\text{lp}(h) = \text{lp}(a_{i_0} g_{i_0}) = \text{lp}(a_{i_0})\text{lp}(g_{i_0}) \in \langle \text{Lp}G \rangle_{\text{Lp}A}$. By definition, then, G is an SG-basis for $\langle G \rangle_A$.

Suppose that $N' > 1$; without loss of generality let $m = \text{lp}(a_1 g_1) = \text{lp}(a_2 g_2)$. Sweedler's proof involves finding $u, p, q \in A$ such that:

1. $\text{lp}[(a_1 - up)g_1] < m$ (or $(a_1 - up)g_1 = 0$).
2. $\text{lp}[(a_2 + uq)g_2] \leq m$ (or $(a_2 + uq)g_2 = 0$).
3. $u(pg_1 - qg_2)$ can be expressed as an A-linear combination of the elements of G such that every summand in the expression has leading X-power product strictly less than m (or $u(pg_1 - qg_2) = 0$).

Note that $(a_1 - up)g_1 + (a_2 + uq)g_2 + u(pg_1 - qg_2) = a_1 g_1 + a_2 g_2$. Satisfying the three conditions above allows us to replace $a_1 g_1 + a_2 g_2$ in the representation for h by this new combination of g_1 and g_2, resulting in an expression which either has a lower maximum leading X-power product than m (this happens if $N' = 2$ and equality with 0 holds in both Conditions 1 and 2 above) or has the same maximum leading power product m but strictly fewer summands contributing to m. This contradicts the "double minimality" assumption on the representation for h. Hence, $N' = 1$; the result is proved. □

The previous theorem confirms that syzygy families indeed play the same role in the algorithm for computing SG-bases as S-polynomials do for Gröbner bases. Algorithm 3 is Sweedler's construction algorithm, adjusted to our context. Verification of its validity follows directly from Theorem 5.1.

We conclude with an effective computation of an SG-basis for an ideal of a k-subalgebra of $k[X]$. Various details from our SI-reduction and syzygy family algorithms are also included to provide further illustration of those methods.

Example 5.2 We again work in $\mathbf{Q}[x, y]$, using degree lex with $x > y$. Let $A = \mathbf{Q}[x^2, xy]$, $I = \langle x^3y + x^2, x^4 + x^2y^2 \rangle_A$, with $f_1 = x^2$, $f_2 = xy$, $g_1 = x^3y + x^2$, and $g_2 = x^4 + x^2y^2$. Recall that $F = \{f_1, f_2\}$ is a SAGBI basis for A.

In the initial pass of the outer loop, $\mathcal{P}$ consists of the single pair (g_1, g_2), whose syzygy family we have already computed in Example 4.7. Thus, $\mathcal{S} =$

Input: $H \subseteq A$, F a SAGBI basis for A
Output: An SG-basis G for $\langle H \rangle_A$
Initialization: $G := H, oldG := \emptyset$
While $G \neq oldG$
    $P := \{(g,h) : g \neq h \in G\}$
    For Each $(g,h) \in P$
        Choose a syzygy family SyzFam(g,h).
    $\mathcal{S} := \cup_{(g,h)\in P}\mathrm{SyzFam}(g,h)$
    $redS := \{$final SI-reducta of each $s \in \mathcal{S}$ via $G\} - \{0\}$
    $oldG := G$
    $G := G \cup redS$

Algorithm 3 : SG-Basis Construction Algorithm [Sweedler]

$\{s_1, s_2\}$, where $s_1 = -x^5y^3 + x^6$ and $s_2 = -x^3y^3 + x^4$. Applying Algorithm 1, we see that $s_1 \xrightarrow{G}_{\mathrm{SI}} 0$ and $s_2 \xrightarrow{G}_{\mathrm{SI}} -x^3y^3 - x^2y^2$, which cannot be SI-reduced via G. Thus, G gains a new element $g_3 = -x^3y^3 - x^2y^2$ and we perform a second pass of the loop.

We already have SyzFam(g_1, g_2), whose members will now conveniently SI-reduce to 0 via G. Applying Algorithm 2 to the pair (g_1, g_3), we create polynomials $G^{(1,0,3)} - H^{(1,1,1)} = 0$ and $G^{(1,0,2)} - H^{(1,0,1)} = 0$, whence SyzFam$(g_1, g_3) = \{0\}$. Finally, applying Algorithm 2 to (g_2, g_3) leads to a single polynomial, namely, $G^{(1,0,3)} - H^{(2,0,1)} = x^5y^5 - x^6y^2$. As this polynomial, too, SI-reduces to 0, we obtain $redS = \emptyset$. Therefore G will equal $oldG$ in the next pass of the loop, terminating the algorithm and yielding an SG-basis $G = \{x^3y + x^2, x^4 + x^2y^2, -x^3y^3 - x^2y^2\}$ for I. $\triangle$

Sweedler's algorithm is theoretically more efficient if each syzygy family chosen is minimal in the sense that its corresponding generating set for $\langle \mathrm{lp}(g) \rangle \cap \langle \mathrm{lp}(h) \rangle$ is minimal. Whether this is true from an implementational viewpoint as well is unclear at this time, for the process of recognizing and discarding a superfluous generator of $\langle \mathrm{mdeg}(g) \rangle \cap \langle \mathrm{mdeg}(h) \rangle$ requires solving a single diophantine system in $|T_{g,h}| - 1$ variables (which may be as many as $|G|(|G| - 1)/2$ variables), whereas SI-reducing a superfluous member of SyzFam(g,h) often requires solving multiple diophantine systems, albeit in only $|G|$ variables. In situations where $|T_{g,h}|$ is considerably larger than $|G|$ and extra members of SyzFam(g,h) SI-reduce quickly, it could actually be more efficient to deal with extra polynomials via SI-reduction instead of via minimizing $T_{g,h}$.

6 Conclusion

An intrinsic approach to calculating SG-bases for ideals of subalgebras of $k[X]$ requires careful attention to polynomial arithmetic, for the results of any such operations must always be elements of the original subalgebra. This is hardly a concern in ordinary Gröbner basis theory, yet leads to the more elaborate algorithms of SI-reduction and syzygy family computation in SG-basis theory. Both these additional algorithms as presented here rely heavily on solving diophantine systems over $\mathbf{N}$, not $\mathbf{Z}$, although it is worth exploring whether there are methods over $\mathbf{Z}$ that could efficiently produce generators for the intersections of our monoid-ideals. Applications of SG-bases also mimic those of Gröbner bases, subject to the expected modifications and increased difficulty that accompany our restriction to subalgebras (see Miller (1994) for applications in $k[X]$ and Miller (1996) for the computation of syzygy modules in subalgebras of polynomial rings over more general commutative rings).

References

[1] Adams, W., Loustaunau, P. (1994): *An Introduction to Gröbner Bases.* American Mathematical Society, Providence, RI.

[2] Becker, T., Weispfennig, V. (1993): *Gröbner Bases: A Computational Approach to Commutative Algebra.* Springer-Verlag, New York.

[3] Buchberger, B. (1965): *Ein Algorithmus zum Auffinden der Basiselemente des Restklassenrings nach einem nulldimensionalen Polynomideal.* PhD. Thesis. Inst. University of Innsbruck, Innsbruck, Austria.

[4] Buchberger, B. (1985): Gröbner Bases: An algorithmic method in polynomial ideal theory. In: Bose, N.K. (ed.): *Multidimensional Systems Theory.* Reidel, Dordrecht, pp. 184-232.

[5] Clausen, M., Fortenbacher, A. (1989): Efficient Solution of Linear Diophantine Equations. *J. Symb. Comp.* **8**: 201-216.

[6] Cox, D., Little, J., O'Shea, D. (1992): *Ideals, Varieties, and Algorithms: An Introduction to Computational Algebraic Geometry and Commutative Algebra.* Springer-Verlag, New York.

[7] Dachsel, T. (1990): *Ein Vervollständigungsverfahren zur Berechnung kanonischer Basen für unitäre Unterringe von Polynomringen.* Diplomarbeit. Universität Kaiserslautern.

[8] Kapur, D., Madlener, K. (1989): A Completion Procedure for Computing a Canonical Basis for a k-Subalgebra. In: *Computers and Mathematics.* Springer, New York, pp. 1-11.

[9] Lankford, D. (1989): Non-negative Integer Basis Algorithms for Linear Equations with Integer Coefficients. *J. Auto. Reas.* **5**: 25-35.

[10] Miller, J.L. (1994): *Algorithms for Computing in Subalgebras of Polynomial Algebras over a Ring.* Ph.D. dissertation. University of Maryland, College Park, MD.

[11] Miller, J.L. (1996): Analogs of Gröbner Bases in Polynomial Rings over a Ring. *J. Symb. Comp.* **21**: 139-153.

[12] Ollivier, F. (1990): Canonical bases: relations with standard bases, finiteness conditions and application to tame automorphisms. In: *à paraître dans les actes de MEGA '90.* Castiglioncello, Birkhauser.

[13] Robbiano, L., Sweedler, M. (1988): Subalgebra Bases. In: Burns, W., Simis, A. (eds.): *Proc. Commutative Algebra Salvador.* Springer LNM **1430**, pp. 61-87.

[14] Sweedler, M. (1988): Ideal bases and valuation rings. Manuscript. Cornell University, Ithaca, NY.

De Nugis Groebnerialium 1: Eagon, Northcott, Gröbner

Ferdinando Mora [1]

> En el nombre del Padre que fizo toda cosa.
> Et de Don Jhesucristo, fijo de la Gloriosa.
> Et del Spiritu Sancto, que egual dellos prosa,
> De un confesor sancto quiero fer una prosa.
> Quiero fer una prosa en roman paladino,
> En qual suele el pueblo fablar a su vecino,
> Ga non se tan letrado por fer otro latino,
> Bien valdra, como credo, un vaso de bon vino.
>
> Gonzalo de Berceo

Remembrance

It was Autumn 1983, when the researchers on Gröbner could have been counted on the fingers of two hands. Michael and me were completing our algorithm to compute resolutions (Mora, Möller 1986a, 1986b) and I was invited in Naples to give an introductory tutorial on Gröbner bases.

I had plenty of free time and, since somebody had just quoted me the Eagon-Northcott formula expressing the resolution of the ideals generated by the majors of a matrix whose entries are independent variables (Eagon, Northcott 1962)), I decided to try to see whether our tools allowed me to tackle the 5×3 case.

I was really surprised when not only I got the resolution but I realized that it was sufficient to give a look to the solution to devise the complete formula (Th. 1.1) and that proving it required only to generalize the computation I did[2]: it was the first time that I realized the amazing power of Buchberger's tool.

My notes ended in a pile of other computations, and probably would have died there... until I thought it could have been curious to present here this

[1]DISI, Univ. Genova, Viale Dodecaneso 35, 16146 Genova, theomora@dima.unige.it

[2]Honestly I must confess that I needed to do a few computations over the 7×4 case to fix a bug in my guess and complete the proof.

"archaeological" result to show a piece of research in those times when the researchers on Gröbner could have been counted on the fingers of two hands ...

Up to a small polishing, removing useless remarks, adding a pair of footnotes and a chapter (§ 2) aimed to summarize what I knew at that time, the note here is nothing more than my original one, including the hand-computed example[3].

Acknowledgements

I thanks C. Ciliberto who invited me in Naples where I did the computation presented here and J. Cannon who invited me in Sydney where I polished those notes.

Mainly I thanks Michael for the wonderful research together.

1 Notation

Let $m, n \in \mathbb{N}$ s.t. $m < n$.

For each t, $0 \le t \le n - m$, we denote

$$\mathcal{C}_t := \{(\gamma_1, \ldots, \gamma_{m+t}) : 1 \le \gamma_1 < \cdots < \gamma_{m+t} \le n\}.$$

If $\mathsf{c} := (\gamma_1, \ldots, \gamma_{m+t}) \in \mathcal{C}_t$, $t > 0$, we denote for each $i, 1 \le i \le m+t$,

$$\mathsf{c}(i) := (\gamma_1, \ldots, \gamma_{i-1}, \gamma_{i+1}, \ldots, \gamma_{m+t}) \in \mathcal{C}_{t-1},$$

and, if $t > 1$, we denote also for each $i, l, 1 \le i < l \le m+t$,

$$\mathsf{c}(i,l) := (\gamma_1, \ldots, \gamma_{i-1}, \gamma_{i+1}, \ldots, \gamma_{l-1}, \gamma_{l+1}, \ldots, \gamma_{m+t}) \in \mathcal{C}_{t-2}.$$

For each t, $0 \le t \le n - m$, we denote

$$\mathcal{R}_t := \{(\rho_1, \ldots, \rho_m) : \sum_{j=1}^{m} \rho_j = m + t, 1 \le \rho_j, \forall j\}.$$

If $\mathsf{r} := (\rho_1, \ldots, \rho_m) \in \mathcal{R}_t$, $t > 0$, we denote $I_\mathsf{r} := \{j : \rho_j > 1\}$ and, $\forall j \in I_\mathsf{r}$,

$$\mathsf{r}(j) := (\rho_1, \ldots, \rho_{j-1}, \rho_j - 1, \rho_{j+1}, \ldots, \rho_m) \in \mathcal{R}_{t-1}.$$

For $t > 1$ we denote

$$I_\mathsf{r}^{(2)} := \{(j,k) : j \ne k, \rho_j > 1, \rho_k > 1\} \cup \{(k,k) : \rho_k > 2\}$$

and, remarking that

[3]Within the proof of the Eagon-Northcott formula, I remarked (Lemma 3.5.1.) that the set of the majors of a generic matrix are a Gröbner basis.

In (Narasimhan 1986), (Caniglia et. al. 1990) and (Sturmfels 1990) it is proved that the minors of any given order of a generic matrix are a Gröbner basis with respect to a diagonal term order . More strong statements of this kind can be found in (Conca 1994), (Conca 1995), (Bruns, Conca 1996).

- $(j,k) \in I_r^{(2)} \iff (k,j) \in I_r^{(2)}$,
- $I_r^{(2)} = \bigcup_{j \in I_r} \{(j,k) : k \in I_{r(j)}\}$,

we denote

$$\begin{aligned} r(j,k) &:= r(j)(k) = r(k)(j) = \\ &= \begin{cases} (\ldots,\rho_{j-1},\rho_j - 1,\rho_{j+1},\ldots,\rho_{k-1},\rho_k - 1,\rho_{k+1},\ldots) & j<k \\ (\ldots,\rho_{j-1},\rho_j - 2,\rho_{j+1},\ldots) & j=k \\ (\ldots,\rho_{k-1},\rho_k - 1,\rho_{k+1},\ldots,\rho_{j-1},\rho_j - 1,\rho_{j+1},\ldots) & j>k. \end{cases} \end{aligned}$$

For each $r := (\rho_1,\ldots,\rho_m) \in \mathcal{R}_t$, let us define $d_i, 1 \le i \le m+t$, as $d_i := j$ where j is the unique integer s.t. $\sum_{k<j} \rho_k < i \le \sum_{k \le j}$ and we will set $d(r) := (d_1,\ldots,d_{m+t})$ [4].

For each t, $0 \le t \le n-m$, we denote

$$\mathcal{S}_t := \mathcal{R}_t \times \mathcal{C}_t.$$

Let us now consider the polynomial ring

$$\mathfrak{P} := \mathbb{Q}[X_{ij} : 1 \le i \le m, 1 \le j \le n]$$

and, for each t, the free module $\mathfrak{P}^{\mathcal{S}_t}$ generated by the canonical basis

$$\{E_s : s \in \mathcal{S}_t\}.$$

Remark that, since $\mathrm{card}(\mathcal{R}_0) = 1$, it holds $\mathcal{S}_0 \cong \mathcal{C}_0$.

Let $\mathcal{M}$ be the $m \times n$ matrix $\mathcal{M} := (X_{ij})_{ij}$ and for each

$$c = (\gamma_1,\ldots,\gamma_m) \in \mathcal{C}_0 \cong \mathcal{S}_0$$

let M_c be the (determinant of the) major[5] of $\mathcal{M}$ on the m γ_j^{th} columns and let

$$\mathfrak{I} := (M_c : c \in \mathcal{C}_0) \subset \mathfrak{P}.$$

Finally denote

$$\delta_0 : \mathfrak{P}^{\mathcal{S}_0} \mapsto \mathfrak{P}$$

to be the map s.t.

$$\delta_0(E_c) = M_c$$

and, for every $t > 0$,

$$\delta_t : \mathfrak{P}^{\mathcal{S}_t} \mapsto \mathfrak{P}^{\mathcal{S}_{t-1}}$$

to be the map s.t., for each $s := (r,c) \in \mathcal{S}_t$ with $c = (\gamma_1,\ldots,\gamma_{m+t})$, it holds

$$\delta_t(E_s) := \sum_{i=1}^{m+t} \sum_{j \in I_r} (-1)^i X_{j\gamma_i} E_{(r(j),c(i))}.$$

Theorem 1.1 *The minimal free resolution of $\mathfrak{P}/\mathfrak{I}$ is*

$$0 \to \mathfrak{P}^{\mathcal{S}_{n-m}} \xrightarrow{\delta_{n-m}} \mathfrak{P}^{\mathcal{S}_{n-m-1}} \xrightarrow{\delta_{n-m-1}} \cdots \mathfrak{P}^{\mathcal{S}_t} \xrightarrow{\delta_t} \mathfrak{P}^{\mathcal{S}_{t-1}} \cdots \xrightarrow{\delta_1} \mathfrak{P}^{\mathcal{S}_0} \xrightarrow{\delta_0} \mathfrak{P}/\mathfrak{I} \to 0.$$

[4] in other words, the vector d consists of ρ_1 1's, ρ_2 2's, ρ_3 3's, ...
[5] maximal minor.

2 Recall

This section is essentially a fast resume of the results in Mora, Möller (1986)[6] which will be applied to prove the claim above.

Once we are given a well-ordering $<$ on the set T of the terms in $\mathfrak{P}$, we can use $<$ to impose an ordering $<_t$ on the set

$$\mathsf{T}_t := \{mE_{\mathsf{s}} : m \in \mathsf{T}, \mathsf{s} \in \mathcal{S}_t\}$$

of the terms in $\mathfrak{P}^{\mathcal{S}_t}$ which is "compatible" with $<$ in the sense that

$$\forall m_1, m_2 \in \mathsf{T}, \forall \mu_1, \mu_2 \in \mathsf{T}_t, \mu_1 \leq_t \mu_2, m_1 \leq m_2 \Longrightarrow m_1\mu_1 \leq_t m_2\mu_2.$$

The order $<_t$ is defined by fixing

- monomials m_{s} for each $\mathsf{s} \in \mathcal{S}_t$ and
- an ordering $\prec$ on $\mathcal{S}_t$

and setting

$$mE_{\mathsf{s}} <_t m'E_{\mathsf{s}'} \iff \begin{array}{l} \mu := mm_{\mathsf{s}} < m'm'_{\mathsf{s}} =: \mu' \\ \text{or } \mu = \mu', \mathsf{s} \prec \mathsf{s}'. \end{array}$$

For each term $\mu := mE_{\mathsf{s}} \in \mathsf{T}_t$ we denote

$$Tdeg(\mu) := mm_{\mathsf{s}};$$

for each module element $f := \sum_{\mu \in \mathsf{T}_t} c_\mu \mu \in \mathfrak{P}^{\mathcal{S}_t}$ we denote

$$Tdeg(f) := \max_{<}\{Tdeg(\mu) : c_\mu \neq 0\},$$

$$Hterm(f) := \max_{<_t}\{\mu : c_\mu \neq 0\};$$

for each submodulo $V \subset \mathfrak{P}^{\mathcal{S}_t}$ we consider a finite subset $G = \{g_1, \dots g_h\} \subset V$ generating V; G is a *Gröbner basis* of V if

$$\forall f \in V, \exists g \in G : Hterm(g) \text{ divides } Hterm(f).$$

If G is a Gröbner basis of V, for each $f \in V$ there is a *G-representation*

$$f = \sum_i h_i g_i : Hterm(h_i g_i) \leq Hterm(f), \forall i.$$

[6]Of course I must also quote Bayer (1982) whose algorithm I used to compute the resolution.

Let $G := \{g_1, \ldots g_h\} \subset V$ be a basis of V and for each pair $g', g'' \in G$, with $Hterm(g') =: m' E_{s'}$, $Hterm(g'') =: m'' E_{s''}$, let us define

$$\mathrm{lcm}(Hterm(g'), Hterm(g'')) := \\ := \begin{cases} \mathrm{lcm}(Hterm(m'), Hterm(m''))E_{s'} & \text{if } s' = s'' \\ 0 & \text{otherwise.} \end{cases}$$

If $\mathrm{lcm}(Hterm(g'), Hterm(g'')) \neq 0$, then there are $t', t'' \in \mathsf{T}$ s.t.

$$t' Hterm(g') = \mathrm{lcm}(Hterm(g'), Hterm(g'')) = t'' Hterm(g''),$$

in which case we define

$$S(g', g'') := t'g' - t''g''$$

and we remark that

$$Hterm(S(g', g'')) < \mathrm{lcm}(Hterm(g'), Hterm(g'')).$$

Denote

$$\mathfrak{S}(G) := \{(g_i, g_j) \in G \times G : i < j, \mathrm{lcm}(Hterm(g_i), Hterm(g_j)) \neq 0\}$$

and let $\mathfrak{S}_u(G) \subseteq \mathfrak{S}(G)$ be s.t. for all $(g', g'') \in \mathfrak{S}(G) \setminus \mathfrak{S}_u(G)$, there is $g''' \in G$ s.t. $\mathrm{lcm}(Hterm(g'), Hterm(g'''))$ and $\mathrm{lcm}(Hterm(g''), Hterm(g'''))$ divide properly $\mathrm{lcm}(Hterm(g'), Hterm(g''))$; then the following conditions are equivalent:

- G is a Gröbner basis of V;
- each $(g', g'') \in \mathfrak{S}_u(G)$ has a G-representation.

In case the conditions above are satisfied, consider the module $\mathfrak{P}^G$ generated by the canonical basis $\{E_g : g \in G\}$ and the map $\Delta : \mathfrak{P}^G \mapsto \mathfrak{P}$ defined by

$$\Delta(E_g) := g, \forall g \in G.$$

For each $(g', g'') \in \mathfrak{S}_u(G)$, let $\sum_{g \in G} h_g g$ be a G-representation of $S(g', g'')$, so that

$$\sum_{g \in G} h_g g = S(g', g'') = t'g' - t''g'';$$

let us define

$$\Sigma(g', g'') := \sum_{g \in G} h_g E_g - t' E_{g'} + t'' E_{g''}.$$

With this notation it holds

- $Im(\Delta) = V$;
- $\{\Sigma(g', g'') : (g', g'') \in \mathfrak{S}_u(G)\}$ generates $Ker(\Delta)$.

3 (Re)solution

Lemma 3.1 $\delta_0\delta_1 = 0$.

Proof: For any $(\mathsf{r},\mathsf{c}) \in \mathcal{S}_1$ with $\mathsf{c} = \{(\gamma_1,\ldots,\gamma_{m+1})$, denoting

$$\mathfrak{M}_j := \begin{pmatrix} X_{j\gamma_1} & \ldots & X_{j\gamma_i} & \ldots & X_{j\gamma_{m+1}} \\ X_{1\gamma_1} & \ldots & X_{1\gamma_i} & \ldots & X_{1\gamma_{m+1}} \\ \vdots & \ddots & \vdots & \ddots & \vdots \\ X_{j\gamma_1} & \ldots & X_{j\gamma_i} & \ldots & X_{j\gamma_{m+1}} \\ \vdots & \ddots & \vdots & \ddots & \vdots \\ X_{m\gamma_1} & \ldots & X_{m\gamma_i} & \ldots & X_{m\gamma_{m+1}} \end{pmatrix},$$

it holds

$$\begin{aligned} \delta_0\delta_1(E_{(\mathsf{r},\mathsf{c})}) &= \sum_{i=1}^{m+1}\sum_{j\in I_\mathsf{r}}(-1)^i X_{j\gamma_i} M_{\mathsf{c}(i)} = \\ &= \sum_{j\in I_\mathsf{r}}\sum_{i=1}^{m+1}(-1)^i X_{j\gamma_i} M_{\mathsf{c}(i)} = \\ &= \sum_{j\in I_\mathsf{r}} -\det(\mathfrak{M}_j) \\ &= 0. \quad \square \end{aligned}$$

Lemma 3.2 *If* $t > 1$, $\delta_{t-1}\delta_t = 0$.

Proof: For any $(\mathsf{r},\mathsf{c}) \in \mathcal{S}_t$ with $\mathsf{c} = \{(\gamma_1,\ldots,\gamma_{m+t})$, it holds

$$\begin{aligned} &\delta_{t-1}\delta_t(E_{(\mathsf{r},\mathsf{c})}) = \\ &= \sum_{i=1}^{m+t}\sum_{j\in I_\mathsf{r}}(-1)^i X_{j\gamma_i}\delta_{t-1}\left(E_{(\mathsf{r}(j),\mathsf{c}(i))}\right) = \\ &= \sum_{i=1}^{m+t}\sum_{j\in I_\mathsf{r}}(-1)^i X_{j\gamma_i}\sum_{l=1}^{i-1}\sum_{k\in I_{\mathsf{r}(j)}}(-1)^l X_{k\gamma_l} E_{(\mathsf{r}(k,j),\mathsf{c}(l,i))} + \\ &\quad + \sum_{i=1}^{m+t}\sum_{j\in I_\mathsf{r}}(-1)^i X_{j\gamma_i}\sum_{l=i+1}^{m+t}\sum_{k\in I_{\mathsf{r}(j)}}(-1)^{l-1} X_{k\gamma_l} E_{(\mathsf{r}(k,j),\mathsf{c}(i,l))} = \\ &= \sum_{i=1}^{m+t}\sum_{l=1}^{i-1}(-1)^{i+l}\sum_{j\in I_\mathsf{r}}\sum_{k\in I_{\mathsf{r}(j)}} X_{j\gamma_i}X_{k\gamma_l} E_{(\mathsf{r}(k,j),\mathsf{c}(l,i))} + \\ &\quad + \sum_{i=1}^{m+t}\sum_{l=i+1}^{m+t}(-1)^{i+l-1}\sum_{j\in I_\mathsf{r}}\sum_{k\in I_{\mathsf{r}(j)}} X_{j\gamma_i}X_{k\gamma_l} E_{(\mathsf{r}(k,j),\mathsf{c}(i,l))} = \\ &= \sum_{i=1}^{m+t}\sum_{l=1}^{i-1}(-1)^{i+l}\sum_{(k,j)\in I_\mathsf{r}^{(2)}} X_{j\gamma_i}X_{k\gamma_l} E_{(\mathsf{r}(k,j),\mathsf{c}(l,i))} + \end{aligned}$$

$$+\sum_{i=1}^{m+t}\sum_{l=i+1}^{m+t}(-1)^{i+l-1}\sum_{(k,j)\in I_r^{(2)}} X_{j\gamma_i}X_{k\gamma_l}E_{(r(k,j),c(i,l))} =$$

$$= \sum_{(k,j)\in I_r^{(2)}}\sum_{i=1}^{m+t}\sum_{l=1}^{i-1}\left((-1)^{i+l}-(-1)^{i+l}\right)X_{j\gamma_i}X_{k\gamma_l}E_{(r(k,j),c(l,i))} =$$

$$= 0. \quad \square$$

Let us now fix the well-ordering $<$ on the set T by choosing the total degree ordering [7] induceded by

$$X_{1n} < \cdots X_{mn} < X_{1n-1} < \cdots < X_{m2} < \cdots < X_{11} < \cdots X_{m1}.$$

With this choice it holds

Lemma 3.3 *For each* $\mathsf{c} := (\gamma_1, \ldots, \gamma_m) \in \mathcal{S}_0$, *we have*

$$Hterm(M_{\mathsf{c}}) = \prod_{i=1}^{m} X_{i\gamma_i}.^{8}$$

Proof: M_{c} is just the combination of all the possible terms $\prod_{i=1}^{m} X_{i\sigma_i}$ where $(\sigma_1, \ldots, \sigma_m)$ runs among the permutations of $(\gamma_1, \ldots, \gamma_m)$. Therefore each term of M_{c} is divisible by exactly one among the variables $X_{i\gamma_m}$ and, by definition, the greatest ones are those divisible by $X_{m\gamma_m}$. The thesis then follows by induction. $\square$

We now define the monomials m_{s} for each $\mathsf{s} \in \mathcal{S}_t$ and for each t as follows: let $\mathsf{s} := (\mathsf{c}, \mathsf{r})$, with $\mathsf{c} := (\gamma_1, \ldots, \gamma_{m+t})$, and $d(\mathsf{r}) =: (d_1, \ldots, d_{m+t})$; then

$$m_{\mathsf{s}} = Tdeg(E_{\mathsf{s}}) := \prod_{i=1}^{m+t} X_{d_i\gamma_i}{}^{9}.$$

In particular we have

Corollary 3.4 $\forall \mathsf{c} := (\gamma_1, \ldots, \gamma_m) \in \mathcal{S}_0, m_{\mathsf{c}} = Hterm(M_{\mathsf{c}}) = \prod_{i=1}^{m} X_{i\gamma_i}$.

---

[7] The total degree ordering $<$ induced by $X_1 < X_2 < \ldots X_n$ is the one defined by $X_1^{a_1}\cdots X_n^{a_n} =: m_1 < m_2 := X_1^{b_1}\cdots X_n^{b_n}$ iff

$$\deg(m_1) < \deg(m_2) \text{ or } \deg(m_1) = \deg(m_2) \text{ and } \exists j : a_j > b_j, a_i = b_i \forall i < j.$$

[8] In other words the maximal term $Hterm(M_{\mathsf{c}})$ of the determinant of M_{c} is its diagonal. This choice induces the following generalized definition of $Hterm(m_{\mathsf{s}})$ and justify the introduction of the vector $d(\mathsf{r})$.

[9] You can interpret this formula in this way: build an $(m+n)$ square matrix by writing ρ_1 copies of the 1st row of $\mathcal{M}$, ρ_2 copies of the 2nd row of $\mathcal{M}, \ldots$, and then cancelling the columns which are not indexed by the elements γ_i of c; then $Tdeg(E_{\mathsf{s}})$ is the diagonal of this matrix.

Proof: In fact we have $\mathcal{R} = \{(1,1,\ldots,1)\}$ so that $d_i = i, \forall i$. $\square$

Lemma 3.5 *It holds*

1. $\{d_0(E_{\mathsf{c}}), \mathsf{c} \in \mathcal{S}_0\}$ *is the Gröbner basis of* $Im(d_0)$;
2. $\{d_1(E_{\mathsf{s}}), \mathsf{s} \in \mathcal{S}_1\}$ *is a basis of* $Ker(d_0)$;
3. $Tdeg(d_1(E_{\mathsf{s}})) = Tdeg(E_{\mathsf{s}}), \forall \mathsf{s} \in \mathcal{S}_1$;
4. $\forall \mathsf{s} := (\mathsf{c}, \mathsf{r}) \in \mathcal{S}_1$, *with* $\mathsf{c} := (\gamma_1, \ldots, \gamma_{m+1})$ *and* $\mathsf{r} = (\rho_1, \ldots, \rho_m)$, *it holds*

$$Hterm(d_1(E_{\mathsf{s}}) = X_{\nu\gamma_\nu} E_{\mathsf{c}(\nu)},$$

where ν *is the single index s.t.* $\rho_\nu = 2$.

Proof: The argument follows by considering all possible S-pairs between the elements in $G := \{d_0(E_{\mathsf{c}}), \mathsf{c} \in \mathcal{S}_0\}$; so let

$$\mathsf{c}^{(1)} = (\gamma_1^{(1)}, \ldots, \gamma_m^{(1)}) \text{ and } \mathsf{c}^{(2)} = (\gamma_1^{(2)}, \ldots, \gamma_m^{(2)}).$$

There are two cases to be considered:

- $\mathsf{c}^{(1)}$ and $\mathsf{c}^{(2)}$ differ in more than one position; in this case, let ν be the higher index s.t. $\gamma_\nu^{(1)} \neq \gamma_\nu^{(2)}$ and w.l.o.g. $\gamma_\nu^{(1)} < \gamma_\nu^{(2)}$. Define then

  $$\mathsf{c}^{(3)} = (\gamma_1^{(1)}, \ldots, \gamma_{\nu-1}^{(1)}, \gamma_\nu^{(2)}, \ldots, \gamma_m^{(2)}).$$

  Since there is another index $\mu < \nu$ s.t. $\gamma_\mu^{(1)} \neq \gamma_\mu^{(2)}$, we know that both $\text{lcm}(Tdeg(E_{\mathsf{c}^{(1)}}), Tdeg(E_{\mathsf{c}^{(3)}}))$ and $\text{lcm}(Tdeg(E_{\mathsf{c}^{(2)}}), Tdeg(E_{\mathsf{c}^{(3)}}))$ divide and have less degree than $\text{lcm}(Tdeg(E_{\mathsf{c}^{(1)}}), Tdeg(E_{\mathsf{c}^{(2)}}))$. This guarantees that $S(E_{\mathsf{c}^{(1)}}, E_{\mathsf{c}^{(2)}})$ is an element in $\mathfrak{S}(G) \setminus \mathfrak{S}_u(G)$.

- If $\mathsf{c}^{(1)}$ and $\mathsf{c}^{(2)}$ differ exactly in a single position, say the νth one, supposing w.l.o.g. $\gamma_\nu^{(1)} < \gamma_\nu^{(2)}$, we define $\mathsf{c} := (\gamma_1, \ldots, \gamma_{m+1})$ by

  $$\gamma_i := \begin{cases} \gamma_i^{(1)} = \gamma_i^{(2)} & i < \nu \\ \gamma_\nu^{(1)} & i = \nu \\ \gamma_\nu^{(2)} & i = \nu + 1 \\ \gamma_{i-1}^{(1)} = \gamma_{i-1}^{(2)} & i > \nu + 1. \end{cases}$$

  We define then $\mathsf{r} := (\rho_1, \ldots, \rho_m)$ by $\rho_i = \begin{cases} 1 & i \neq \nu \\ 2 & i = \nu \end{cases}$ and $\mathsf{s} := (\mathsf{c}, \mathsf{r})$.

  By Lemma 3.1 we know that

  $$0 = \delta_0 \delta_1 = \sum_{i=1}^{m+1} (-1)^i X_{\nu\gamma_i} M_{\mathsf{c}(i)};$$

by Lemma 3.3 that

$$Tdeg(X_{\nu\gamma_i} M_{c(i)}) = \left(\prod_{j=1}^{i-1} X_{j\gamma_j}\right) X_{\nu\gamma_i} \left(\prod_{j=i+1}^{m+1} X_{j-1\ \gamma_j}\right),$$

so that

$$Tdeg(X_{\nu\gamma_i} M_{c(i)}) \leq Tdeg(E_s) = \left(\prod_{j=1}^{\nu} X_{j\gamma_j}\right) \left(\prod_{j=\nu+1}^{m+1} X_{j-1\ \gamma_j}\right)$$

and the equality holds if $i \in \{\nu, \nu+1\}$. As a consequence

$$\begin{aligned} \pm S(E_{c^{(1)}}, E_{c^{(2)}}) &= -(-1)^{\nu} X_{\nu\gamma_\nu} M_{c(\nu)} - (-1)^{\nu+1} X_{\nu\gamma_\nu} M_{c(\nu+1)} = \\ &= \sum_{i \notin \{\nu,\nu+1\}} (-1)^i X_{\nu\gamma_i} M_{c(i)} \end{aligned}$$

is a G-representation.

Therefore the existence of a G-representation for any element in $\mathfrak{S}_u(G)$ proves that G is a Gröbner basis (proving 1.) and that the elements

$$\delta_1(E_s) = \sum_{i=1}^{m+1} (-1)^i X_{\nu\gamma_i} E_c(i)$$

form a basis of $Ker(\delta_0)$ (proving 2.). 3. and 4. follow from the computation above. □

On the basis of Lemma 3.5 we introduce the following properties for $t \geq 1$:

P1(t) $\{d_t(E_s), s \in \mathcal{S}_t\}$ is a basis of $Ker(d_{t-1})$;

P2(t) $\{d_t(E_s), s \in \mathcal{S}_t\}$ is the Gröbner basis of $Im(d_t)$;

P3(t) $Tdeg(d_t(E_s)) = Tdeg(E_s), \forall s \in \mathcal{S}_t$;

P4(t) for each $s := (c, r) \in \mathcal{S}_1$, with $c := (\gamma_1, \ldots, \gamma_{m+1})$, let $j := \min(I_r)$; then $Tdeg(d_t(E_s)) = X_{j\gamma_j} E_{(c(j),r(j))}$.

Lemma 3.6 *If* $t \geq 1$,

$$\mathbf{P1}(t), \mathbf{P3}(t), \mathbf{P4}(t) \Longrightarrow \mathbf{P1}(t+1), \mathbf{P2}(t), \mathbf{P3}(t+1), \mathbf{P4}(t+1).$$

Proof: As we did in Lemma 3.5, we have to consider all the S-pairs of the elements in $\{d_t(E_s), s \in \mathcal{S}_t\}$: let $s^{(1)}, s^{(2)} \in \mathcal{S}_t$ be s.t.

$$\text{lcm}(Hterm(\delta_t(E_{s^{(1)}})), Hterm(\delta_t(E_{s^{(2)}}))) = mE_{s^{(3)}}, s^{(3)} \in \mathcal{S}_{t-1}.$$

For $i = 1..3$ we denote

$$\mathsf{s}^{(i)} := (\mathsf{c}^{(i)}, \mathsf{r}^{(i)}), \mathsf{c}^{(i)} = (\gamma_1^{(i)}, \gamma_2^{(i)}, \ldots), \mathsf{r}^{(i)} = (\rho_1^{(i)}, \ldots \rho_m^{(i)}),$$

and we set $j := min(I_{\mathsf{r}^{(1)}})$ and $k := min(I_{\mathsf{r}^{(2)}})$ so that

$$\mathsf{r}^{(3)} = \mathsf{r}^{(1)}(j) = \mathsf{r}^{(2)}(k).$$

Assuming w.l.o.g. $j \leq k$ we have then

$$\rho_i^{(3)} = \rho_i^{(1)} = \rho_i^{(2)} = 1, \forall i < j, \tag{3.1}$$

and either $j = k$ or

- $j < k$,
- $\rho_j^{(3)} = \rho_j^{(2)} = 1, \rho_j^{(1)} = 2;$
- $\rho_i^{(1)} = \rho_i^{(3)} = \rho_i^{(2)} = 1, \forall j < i < k;$
- $\rho_k^{(1)} = \rho_k^{(3)} = 1, \rho_k^{(2)} = 2;$
- $\rho_i^{(1)} = \rho_i^{(3)} = \rho_i^{(2)}, \forall i > k.$

We then define $\mathsf{r} := \mathsf{r}^{(4)} = (\rho_1^{(4)}, \ldots \rho_m^{(4)})$ in the case $j = k$ as

$$\rho_i^{(4)} := \begin{cases} \rho_i^{(3)} = \rho_i^{(2)} = \rho_i^{(1)} & i \neq j = k \\ \rho_i^{(3)} + 2 = \rho_i^{(2)} + 1 = \rho_i^{(1)} + 1 & i = j = k, \end{cases}$$

while in case $j < k$ we define it as

$$\rho_i^{(4)} := \begin{cases} \rho_i^{(3)} = \rho_i^{(2)} = \rho_i^{(1)} & j \neq i \neq k \\ 2 & i = j \\ \rho_i^{(2)} = \rho_i^{(1)} + 1 = \rho_i^{(3)} + 1 & i = k, \end{cases}$$

so that, in both cases

$$\mathsf{r}^{(1)} = \mathsf{r}^{(4)}(k), \mathsf{r}^{(2)} = \mathsf{r}^{(4)}(j).$$

Since $\mathsf{c}^{(3)} = \mathsf{c}^{(1)}(j) = \mathsf{c}^{(2)}(k)$, we then define $\mathsf{c} := \mathsf{c}^{(4)} = (\gamma_1^{(4)}, \ldots \gamma_{m+t+1}^{(4)})$ as follows:

- if $j = k$, supposing w.l.o.g. $\gamma_j^{(1)} < \gamma_j^{(2)}$, we set

$$\gamma_i^{(4)} := \begin{cases} \gamma_i^{(3)} = \gamma_i^{(2)} = \gamma_i^{(1)} & i < j \\ \gamma_i^{(1)} & i = j \\ \gamma_{i-1}^{(2)} & i = j + 1 \\ \gamma_{i-2}^{(3)} = \gamma_{i-1}^{(2)} = \gamma_{i-1}^{(1)} & i > j + 1; \end{cases}$$

- if $j < k$ we set

$$\gamma_i^{(4)} := \begin{cases} \gamma_i^{(3)} = \gamma_i^{(2)} = \gamma_i^{(1)} & i < j \\ \gamma_i^{(1)} & i = j \\ \gamma_{i-1}^{(3)} = \gamma_{i-1}^{(2)} = \gamma_i^{(1)} & j < i \leq k \\ \gamma_{i-1}^{(2)} & i = k+1 \\ \gamma_{i-2}^{(3)} = \gamma_{i-1}^{(2)} = \gamma_{i-1}^{(1)} & i > k+1. \end{cases}$$

In both cases

$$\mathsf{c}^{(1)} = \mathsf{c}^{(4)}(k+1), \mathsf{c}^{(2)} = \mathsf{c}^{(4)}(j).$$

Then we can define $\mathsf{s}^{(4)} := \mathsf{s} := (\mathsf{r}, \mathsf{c}) \in \mathcal{S}_{t+1}$, so that

$$d_{t+1}(E_\mathsf{s}) = \sum_{\mu=1}^{m+t+1} \sum_{\nu \in I_\mathsf{r}} (-1)^\mu X_{\nu\mu} E_{(\mathsf{r}(\nu), \mathsf{c}(\mu))}.$$

We have then to verify that it holds

$$Tdeg(X_{\nu\mu} E_{(\mathsf{r}(\nu), \mathsf{c}(\mu))}) \leq Tdeg(d_{t+1}(E_\mathsf{s})),$$

and that the equality is satisfied only in the cases $\nu = j = \mu$, $\nu = k = \mu - 1$, so that

$$Tdeg(d_{t+1}(E_\mathsf{s})) = Tdeg(X_{j\gamma_j} E_{(\mathsf{r}(j), \mathsf{c}(j))}) = Tdeg(E_\mathsf{s}). \tag{3.2}$$

Since this analysis holds for all possible pairs in $\mathfrak{S}_u$, we can conclude **P1**(t+1) and **P2**(t); **P3**(t+1) follows from Eq. 3.2 and **P4**(t+1) from Eq. 3.1[10]. □

Obviously this series of lemmata, implies a proof of Theorem.

4 Example

We can present here the computation I did with $m = 3, n = 5$ in which I denote $E_{(\mathsf{r},\mathsf{c})}$, with $\mathsf{c} := (\gamma_1, \ldots, \gamma_\mu)$ and $\mathsf{r} := (\rho_1, \ldots, \rho_\nu)$ as $E_{\rho_1,\ldots,\rho_\nu;\gamma_1,\ldots,\gamma_\mu}$.

$$\begin{aligned} E_{111;123} := \; & X_{11}X_{22}X_{33} - X_{11}X_{32}X_{23} - X_{21}X_{12}X_{33} + X_{21}X_{32}X_{13} + \\ & + X_{31}X_{12}X_{23} - X_{31}X_{22}X_{13} \end{aligned}$$

[10] To be more precise we must remark that

$$Tdeg(d_t(E_\mathsf{s})) = X_{j\gamma_h} Tdeg(E_{(\mathsf{c}(h), \mathsf{r}(j))}), j \leq h \leq j + \rho_j - 1.$$

To guarantee the claimed result, one must apply the ordering $<_t$ on T_t; such ordering depends on an ordering $\prec$ on $\mathcal{S}_t$.

My note suggested to use the ordering $(\mathsf{r}, \mathsf{c}) \prec (\mathsf{r}', \mathsf{c}')$ iff c "is lexicographically less than" c' but, unfortunately it is not clear to me of which ordering I was thinking.

$$\begin{aligned}
E_{111;124} &:= X_{11}X_{22}X_{43} - X_{11}X_{42}X_{23} - X_{21}X_{12}X_{43} + X_{21}X_{42}X_{13} + \\
&\quad +X_{41}X_{12}X_{23} - X_{41}X_{22}X_{13} \\
E_{111;125} &:= X_{11}X_{22}X_{53} - X_{11}X_{52}X_{23} - X_{21}X_{12}X_{53} + X_{21}X_{52}X_{13} + \\
&\quad +X_{51}X_{12}X_{23} - X_{51}X_{22}X_{13} \\
E_{111;134} &:= X_{11}X_{32}X_{43} - X_{11}X_{42}X_{33} - X_{31}X_{12}X_{43} + X_{31}X_{42}X_{13} + \\
&\quad +X_{41}X_{12}X_{33} - X_{41}X_{32}X_{13} \\
E_{111;135} &:= X_{11}X_{32}X_{53} - X_{11}X_{52}X_{33} - X_{31}X_{12}X_{53} + X_{31}X_{52}X_{13} + \\
&\quad +X_{51}X_{12}X_{33} - X_{51}X_{32}X_{13} \\
E_{111;145} &:= X_{11}X_{42}X_{53} - X_{11}X_{52}X_{43} - X_{41}X_{12}X_{53} + X_{41}X_{52}X_{13} + \\
&\quad +X_{51}X_{12}X_{43} - X_{51}X_{42}X_{13} \\
E_{111;234} &:= X_{21}X_{32}X_{43} - X_{21}X_{42}X_{33} - X_{31}X_{22}X_{43} + X_{31}X_{42}X_{23} + \\
&\quad +X_{41}X_{22}X_{33} - X_{41}X_{32}X_{23} \\
E_{111;235} &:= X_{21}X_{32}X_{53} - X_{21}X_{52}X_{33} - X_{31}X_{22}X_{53} + X_{31}X_{52}X_{23} + \\
&\quad +X_{51}X_{22}X_{33} - X_{51}X_{32}X_{23} \\
E_{111;245} &:= X_{21}X_{42}X_{53} - X_{21}X_{52}X_{43} - X_{41}X_{22}X_{53} + X_{41}X_{52}X_{23} + \\
&\quad +X_{51}X_{22}X_{43} - X_{51}X_{42}X_{23} \\
E_{111;345} &:= X_{31}X_{42}X_{53} - X_{31}X_{52}X_{43} - X_{41}X_{32}X_{53} + X_{41}X_{52}X_{33} + \\
&\quad +X_{51}X_{32}X_{43} - X_{51}X_{42}X_{33}
\end{aligned}$$

$$\begin{aligned}
E_{211;2345} &:= -X_{15}E_{111;234} + X_{14}E_{111;235} - X_{13}E_{111;245} + X_{12}E_{111;345} \\
E_{211;1345} &:= -X_{15}E_{111;134} + X_{14}E_{111;135} - X_{13}E_{111;145} + X_{11}E_{111;345} \\
E_{211;1245} &:= -X_{15}E_{111;124} + X_{14}E_{111;125} - X_{12}E_{111;145} + X_{11}E_{111;245} \\
E_{211;1235} &:= -X_{15}E_{111;123} + X_{13}E_{111;125} - X_{12}E_{111;135} + X_{11}E_{111;235} \\
E_{211;1234} &:= -X_{14}E_{111;123} + X_{13}E_{111;124} - X_{12}E_{111;134} + X_{11}E_{111;234} \\
E_{121;2345} &:= -X_{25}E_{111;234} + X_{24}E_{111;235} - X_{23}E_{111;245} + X_{22}E_{111;345} \\
E_{121;1345} &:= -X_{25}E_{111;134} + X_{24}E_{111;135} - X_{23}E_{111;145} + X_{21}E_{111;345} \\
E_{121;1245} &:= -X_{25}E_{111;124} + X_{24}E_{111;125} - X_{22}E_{111;145} + X_{21}E_{111;245} \\
E_{121;1235} &:= -X_{25}E_{111;123} + X_{23}E_{111;125} - X_{22}E_{111;135} + X_{21}E_{111;235} \\
E_{121;1234} &:= -X_{14}E_{111;123} + X_{23}E_{111;124} - X_{22}E_{111;245} + X_{21}E_{111;234} \\
E_{112;2345} &:= -X_{35}E_{111;234} + X_{34}E_{111;235} - X_{33}E_{111;245} + X_{32}E_{111;345} \\
E_{112;1345} &:= -X_{35}E_{111;134} + X_{34}E_{111;135} - X_{33}E_{111;145} + X_{31}E_{111;345} \\
E_{112;1245} &:= -X_{35}E_{111;124} + X_{34}E_{111;125} - X_{32}E_{111;145} + X_{21}E_{111;245} \\
E_{112;1235} &:= -X_{35}E_{111;123} + X_{33}E_{111;125} - X_{32}E_{111;135} + X_{31}E_{111;235} \\
E_{112;1234} &:= -X_{34}E_{111;123} + X_{33}E_{111;124} - X_{32}E_{111;245} + X_{31}E_{111;234}
\end{aligned}$$

$$
\begin{aligned}
E_{113;12345} &:= X_{31}E_{112;2345} - X_{32}E_{112;1345} + X_{33}E_{112;1245} - \\
&\quad -X_{34}E_{112;1235} + X_{35}E_{112;1234} \\
E_{131;12345} &:= X_{21}E_{121;2345} - X_{22}E_{121;1345} + X_{23}E_{121;1245} - \\
&\quad -X_{24}E_{121;1235} + X_{25}E_{121;1234} \\
E_{311;12345} &:= X_{11}E_{211;2345} - X_{12}E_{211;1345} + X_{13}E_{211;1245} - \\
&\quad -X_{14}E_{211;1235} + X_{15}E_{211;1234} \\
E_{122;12345} &:= X_{31}E_{121;2345} + X_{21}E_{112;2345} - X_{32}E_{121;1345} - \\
&\quad -X_{22}E_{112;1345} + X_{33}E_{121;1245} + X_{23}E_{112;1245} - \\
&\quad -X_{34}E_{121;1235} - X_{24}E_{112;1235} + X_{35}E_{121;1234} + \\
&\quad +X_{25}E_{112;1234} \\
E_{212;12345} &:= X_{31}E_{211;2345} + X_{11}E_{112;2345} - X_{32}E_{211;1345} - \\
&\quad -X_{12}E_{112;1345} + X_{33}E_{211;1245} + X_{13}E_{112;1245} - \\
&\quad -X_{34}E_{211;1235} - X_{14}E_{112;1235} + X_{35}E_{211;1234} + \\
&\quad +X_{15}E_{112;1234} \\
E_{221;12345} &:= X_{21}E_{211;2345} + X_{11}E_{121;2345} - X_{22}E_{211;1345} - \\
&\quad -X_{12}E_{121;1345} + X_{23}E_{211;1245} + X_{13}E_{121;1245} - \\
&\quad -X_{24}E_{211;1235} - X_{14}E_{121;1235} + X_{25}E_{211;1234} + \\
&\quad +X_{15}E_{121;1234}
\end{aligned}
$$

References

Bayer, D. (1982) 'The Division Algorithm and the Hilbert Scheme', Ph.D. Thesis, Harvard Univ., Cambridge, Mass.

Bruns, W., Conca, A. (1996) 'KRS and powers of determinantal ideals', Comp. Math., to appear.

Caniglia, L., Guccione, J. A., Guccione, J. J. (1990): 'Ideals of generic minors', Comm. Algebra **18**: 2633-2640.

Conca, A. (1994) 'Gröbner bases of ideals of minors of a symmetric matrix', J. of Algebra **166**: 406-421

Conca, A. (1995) 'Gröbner bases of powers of ideals of maximal minors', J. of Pure and Appl. Alg., to appear.

Eagon, J., Northcott, D.G. (1962) 'Ideals defined by matrices, and a certain complex associated to that', Proc. Royal Soc. **269**: 188–204

Mora, F., Möller, H.M. (1986) 'New constructive methods in classical ideal theory', J.Alg. **100**: 138–178

Mora, F., Möller, H.M. (1986) 'Computational aspects of reduction strategies to construct resolutions of monomial ideals,' Lect. Notes Comp. Sci **228**: 182-197

Narasimhan, H. (1986): The irreducibility of ladder determinantal varieties. J. Algebra **102**: 162–185.

Sturmfels, B. (1990): Gröbner bases and Stanley decompositions of determinantal rings. Math. Z. **205**: 137–144

An application of Gröbner bases to the decomposition of rational mappings

Jörn Müller-Quade, Rainer Steinwandt, and Thomas Beth

IAKS, Fakultät für Informatik, Universität Karlsruhe, D-76128 Karlsruhe

Abstract

Let $k(X) := \text{Quot}(k[x_1, \ldots, x_n]/\text{I}(X))$ be the field of rational functions of an irreducible affine variety X. The decomposition $\mathbf{f} = \mathbf{g} \circ \mathbf{h}$ of a rational mapping $\mathbf{f} \in k(X)^m$ is defined in terms of subfields of $k(X)$. Taking a decomposition $\mathbf{g} \circ \mathbf{h}$ of $\mathbf{f}$ for a strategy to solve equations of the form $\mathbf{f}(\mathbf{x}) = \boldsymbol{\alpha}$ in several steps the definition is extended to the "inverse" mappings $\mathbf{f}^{-1} : \boldsymbol{\alpha} \mapsto \mathbf{f}^{-1}(\boldsymbol{\alpha})$.
To compute the "inner part" $\mathbf{h}$ of a decomposition $\mathbf{g} \circ \mathbf{h}$ intermediate fields of $k(\mathbf{f})$ and $k(X)$ have to be found; e.g., for $k(X)/k(\mathbf{f})$ separable algebraic the coefficients of suitable Gröbner bases can be used for performing this task. To compute the "outer part" $\mathbf{g}$ a solution to the field membership problem for subfields of $k(X)$ by means of Gröbner basis techniques is suggested which does not make use of tag variables. Finally "monomial decompositions" are considered; in this case the required Gröbner basis computations can be performed efficiently.

1 Introduction

The problem of decomposing a function f into functions g and h such that $f = g \circ h$ is the functional composition hereof has been studied by various authors, as Dickerson (1989) for the case when f is a multivariate polynomial over a field, Zippel (1991) for f a univariate rational function or Kozen et al. (1996) for f a univariate algebraic function. A functional decomposition can be useful for solving equations of the form $f(x) = \alpha$, α a specified value, for instance; through a decomposition of f the task of solving this equation naturally splits into several (potentially easier) steps. A classical example of this technique is given by the 45th degree polynomial of Adrianus Romanus which decomposes into a quintic and two cubics (Viète 1594).

Denoting by $k(X)$ the field of rational functions of an irreducible variety $X \subseteq \mathbb{A}^n$ the issue of the present paper can be described as the decomposition of rational mappings $\mathbf{f} \in k(X)^m$ and their "inverses" which map a specified $\boldsymbol{\alpha}$ to its preimage $\mathbf{f}^{-1}(\boldsymbol{\alpha})$: In the first part rational decompositions $\mathbf{f} = \mathbf{g} \circ \mathbf{h}$

are related to chains of fields $k(\mathbf{f}) \leq k(\mathbf{h}) \leq k(X)$, thereby resuming an idea from Zippel (1991). Motivated by the "inversion" of equations of the form $\mathbf{f}(\mathbf{x}) = \boldsymbol{\alpha}$ then the decomposition of "inverses" of rational functions is characterized by means of rational decompositions; in particular this yields another approach to the problem of decomposing algebraic functions. Using the coefficients of suitable Gröbner bases to characterize certain subfields of $k(X)$ — and thereby also the "inner part" $\mathbf{h}$ of a decomposition $\mathbf{g} \circ \mathbf{h}$ — an answer to the problem of effectively determining a decomposition is obtained, e.g., for $k(X)/k(\mathbf{f})$ separable algebraic. To determine the "outer part" $\mathbf{g}$ a solution to the field membership problem in $k(X)$ is suggested where the required Gröbner basis does not depend on the use of a particular term order or tag variables.

For $k(X) = k(x_1, \ldots, x_n)$ a rational function field, $\mathbf{h}$ a tuple of Laurent monomials inverting $\mathbf{h}$ is computationally particularly simple. This kind of decomposition occurs in the design of diffractive optical systems (Aagedal et al. 1996), for example. In the last section therefore an algorithm to find a tuple $\mathbf{h}$ of Laurent monomials is given such that $k(\mathbf{h}) \leq k(X)$ is the smallest subfield of $k(x_1, \ldots, x_n)$ which is generated by Laurent monomials and contains $k(\mathbf{f})$. In this case the Gröbner basis computations for determining a suitable "outer part" $\mathbf{g}$ can be performed efficiently.

2 Rational mappings

Let $n \in \mathbb{N}$, $k(X) := \mathrm{Quot}(k[x_1, \ldots, x_n]/\mathrm{I}(X))$ be the field of rational functions of an irreducible variety $X \subseteq \mathbb{A}^n$ over an algebraically closed field k (in particular for $X = \mathbb{A}^n$ we have $k(X) = k(x_1, \ldots, x_n)$ with $x_1, \ldots, x_n$ purely transcendental over k). As usual a rational mapping $\mathbf{f} : X \to Z$ with $Z \subseteq \mathbb{A}^m$ (Zariski-)closed is a tuple $(f_1, \ldots, f_m) \in k(X)^m, m \in \mathbb{N}$, such that for $\boldsymbol{\alpha} \in X$ with $\mathbf{f}$ regular at $\boldsymbol{\alpha}$ the image $\mathbf{f}(\boldsymbol{\alpha})$ is contained in Z; the image of X under $\mathbf{f}$ is denoted by $\mathbf{f}(X) := \{\mathbf{f}(\boldsymbol{\alpha}) : \boldsymbol{\alpha} \in X, \text{ and } \mathbf{f} \text{ is regular at } \boldsymbol{\alpha}\}$, $k(\mathbf{f})$ denotes the field $k(f_1, \ldots, f_m) \leq k(X)$, and $\mathbf{1}_{k(X)}$ represents the tuple $(x_1, \ldots, x_n)$.

Definition 1 *Let* $\mathbf{f} : X \to Z$ *be a rational mapping,* $r \in \mathbb{N}$. *A* (rational) decomposition (of arity r) of $\mathbf{f} \in k(X)^m$ *is a pair* $(\mathbf{g}, \mathbf{h})$ *of rational mappings* $\mathbf{g} : Y \to Z, \mathbf{h} : X \to Y$, $Y \subseteq \mathbb{A}^r$ *irreducible, such that* $\mathbf{h}(X)$ *is dense in* Y *and* $\mathbf{f} = \mathbf{g} \circ \mathbf{h}$ *(where defined).*

It is convenient to identify "equivalent" decompositions by introducing an equivalence relation on the set of decompositions of a fixed $\mathbf{f}$; interpreting $\mathbf{h}$ as an intermediate step of the evaluation of $\mathbf{f}$ it is natural to regard decompositions $(\mathbf{g}, \mathbf{h})$ and $(\mathbf{g}', \mathbf{h}')$ as equivalent if the "width of the intermediate step", i.e., the arity, coincides, and if it is possible to "derive easily" both $\mathbf{h}$ from $\mathbf{h}'$ and $\mathbf{h}'$ from $\mathbf{h}$. This motivates

Definition 2 *Let* $\mathbf{f} : X \to Z$ *be a rational mapping,* $(\mathbf{g}, \mathbf{h})$ *and* $(\mathbf{g}', \mathbf{h}')$ *decompositions of arity* r *resp.* r' *of* $\mathbf{f}$*. Then* $(\mathbf{g}, \mathbf{h})$ *and* $(\mathbf{g}', \mathbf{h}')$ *are* equivalent *iff* $r = r'$ *and* $\mathbf{h} \in k(\mathbf{h}')^{r'}$ *and* $\mathbf{h}' \in k(\mathbf{h})^r$*. If these conditions are met we write* $\mathbf{g} \circ \mathbf{h} \simeq \mathbf{g}' \circ \mathbf{h}'$*. Furthermore* $(\mathbf{g}, \mathbf{h})$ *is called a* trivial decomposition of $\mathbf{f}$ *iff* $\mathbf{g} \circ \mathbf{h} \simeq \mathbf{f} \circ \mathbf{1}_{k(X)}$ *or* $\mathbf{g} \circ \mathbf{h} \simeq \mathbf{1}_{k(Y)} \circ \mathbf{f}$.

Let $\mathbf{g} \circ \mathbf{h} \simeq \mathbf{g}' \circ \mathbf{h}'$ be equivalent decompositions of some $\mathbf{f}$. Then the definition of $\simeq$ in particular guarantees the equality $k(\mathbf{h}) = k(\mathbf{h}')$. Letting $k(\mathbf{f})$ denote the field generated by the components of $\mathbf{f}$ we have $k(\mathbf{f}) \leq k(\mathbf{h}) \leq k(X)$. On the other hand any intermediate field k' of $k(\mathbf{f})$ and $k(X)$ yields a — in case of $k(\mathbf{f}) \neq k' \neq k(X)$ non-trivial — decomposition of $\mathbf{f}$ by expressing $\mathbf{f}$ in terms of a fixed finite generating set of k'. More precisely we have

Lemma 1 *Let* $\mathbf{f} : X \to Z$ *be a rational mapping,*

- $\mathcal{K}_\mathbf{f} := \{k' : k(\mathbf{f}) \leq k' \leq k(X)\}$,
- $\mathcal{K}'_\mathbf{f} := \{k' : k(\mathbf{f}) \leq k' \leq k(X)$ *and* k'/k *is purely transcendental*$\}$,
- $\mathcal{D}_\mathbf{f} := \{(\mathbf{g}, \mathbf{h}) : (\mathbf{g}, \mathbf{h})$ *is a decomposition of* $\mathbf{f}\}$, *and*
- $\mathcal{D}'_\mathbf{f} := \{(\mathbf{g}, \mathbf{h}) : (\mathbf{g}, \mathbf{h})$ *is a decomposition of* $\mathbf{f}$*, and the set of components of* $\mathbf{h}$ *is algebraically independent over* $k\}$.

Then there is a surjection $\phi : \mathcal{D}_\mathbf{f}/\simeq \to \mathcal{K}_\mathbf{f}, [(\mathbf{g}, \mathbf{h})] \mapsto k(\mathbf{h})$ *and a bijection* $\varphi : \mathcal{D}'_\mathbf{f}/\simeq \to \mathcal{K}'_\mathbf{f}, [(\mathbf{g}, \mathbf{h})] \mapsto k(\mathbf{h})$*. Moreover, for fixed arity and* $k(X)/k(\mathbf{f})$ *algebraic and simple, e. g.,* $\mathrm{char}(k) = 0$ *and* $k(X)/k(\mathbf{f})$ *algebraic, the number of non-equivalent decompositions of* $\mathbf{f}$ *is finite.*

Proof: The definition of $\simeq$ guarantees that both mappings are well-defined. To verify surjectivity of ϕ let $k(\mathbf{h}) \in \mathcal{K}_\mathbf{f}$ be generated by $\mathbf{h} \in k(X)^r$. As $k(\mathbf{f}) \leq k(\mathbf{h})$ there are $r \in \mathbb{N}$ and $\mathbf{g} \in k(y_1, \ldots, y_r)^m$ satisfying $\mathbf{f} = \mathbf{g} \circ \mathbf{h}$; restricting $\mathbf{g}$ to the Zariski closure of $\mathbf{h}(X)$ we obtain a decomposition of $\mathbf{f}$ as required. Surjectivity of φ is verified in the same way by choosing an algebraically independent generating set $\mathbf{h}$ of k'. To check injectivity of φ assume $\varphi([(\mathbf{g}, \mathbf{h})]) = \varphi([(\mathbf{g}', \mathbf{h}')])$, i. e., $k(\mathbf{h}) = k(\mathbf{h}')$. Since both the entries of $\mathbf{h}$ and $\mathbf{h}'$ are algebraically independent $(\mathbf{g}, \mathbf{h})$ and $(\mathbf{g}', \mathbf{h}')$ must be of equal arity, and hence $(\mathbf{g}, \mathbf{h}) \simeq (\mathbf{g}', \mathbf{h}')$.
The last statement is a trivial consequence of $\mathcal{K}_\mathbf{f}$ being finite for $k(X)/k(\mathbf{f})$ algebraic and simple. □

3 "Inverses" of rational mappings

In a sense a rational decomposition of a mapping $\mathbf{f} : X \to Z$ depicts a strategy for solving a "generic"[1] equation of the form $f(\mathbf{x}) = \boldsymbol{\alpha}$, $\boldsymbol{\alpha} \in Z$:

[1]In degenerate cases the decomposition may "lose" solutions: Setting $x^4 = 0$ we obtain $x = 0$ while solving $1/y^2 = 0$ for y followed by solving $1/x^2 = y$ yields no solution, although we have the decomposition $x^4 = (1/y^2) \circ (1/x^2)$.

Having in mind Lemma 1 we may essentially think of a rational decomposition $\mathbf{f} = \mathbf{g} \circ \mathbf{h} : X \to Z$ as being characterized by a chain of fields of the form $k(\mathbf{f}) \leq k(\mathbf{h}) \leq k(X)$. Reading this chain from right to left we obtain a strategy for solving a "generic" equation of the form $f(\mathbf{x}) = \boldsymbol{\alpha}$, $\boldsymbol{\alpha} \in Z$: (1) Solve $\mathbf{g}(\mathbf{y}) = \boldsymbol{\alpha}$ yielding a set of solutions B. (2) For $\boldsymbol{\beta} \in B$ solve $\mathbf{h}(\mathbf{x}) = \boldsymbol{\beta}$.
Solving an equation $\mathbf{f}(\mathbf{x}) = \boldsymbol{\alpha}$ may be regarded as evaluating the "inverse" $\mathbf{f}^{-1}$ of $\mathbf{f}$ at $\mathbf{z} = \boldsymbol{\alpha}$ where $\mathbf{f}^{-1} : Z \to 2^X, \mathbf{z} \mapsto \mathbf{f}^{-1}(\mathbf{z}) = \{\mathbf{x} \in X : \mathbf{f}(\mathbf{x}) = \mathbf{z}\}$. In order to be able to form compositions it is convenient to extend $\mathbf{f}^{-1}$'s domain of definition from Z to 2^Z in the obvious way by setting $\mathbf{f}^{-1} : 2^Z \to 2^X, Z' \mapsto \mathbf{f}^{-1}(Z') = \bigcup_{\mathbf{z} \in Z'} \mathbf{f}^{-1}(\mathbf{z})$. Having in mind the solution of equations of the form $\mathbf{f}(\mathbf{x}) = \boldsymbol{\alpha}$ we set $\mathbf{h}^{-1} \circ \mathbf{g}^{-1} := (\mathbf{g} \circ \mathbf{h})^{-1}$; then the following terminology seems appropriate:

Definition 3 *Let* $\mathbf{f} : X \to Z$ *be a rational mapping,* $\mathbf{f}^{-1} : 2^Z \to 2^X$ *its "inverse". A*(n inverse rational) decomposition of $\mathbf{f}^{-1}$ (of arity r) *is a pair* $(\mathbf{h}^{-1}, \mathbf{g}^{-1})$ *such that* $(\mathbf{g}, \mathbf{h})$ *is a (rational) decomposition of* $\mathbf{f}$ *(of arity* r*), and* $\mathbf{g}^{-1}$ *resp.* $\mathbf{h}^{-1}$ *is the "inverse" of* $\mathbf{g}$ *resp.* $\mathbf{h}$.

From this point of view an algebraic function $k^n \to 2^k$ given by an irreducible polynomial $m \in k[x_1, \ldots, x_n][Z]$ via

$$(x_1, \ldots, x_n) \mapsto \{z \in k : m(x_1, \ldots, x_n, z) = 0\}$$

can be represented as the "inverse" of

$$\mathbf{f} : \mathrm{V}(m) \to k^n, (x_1, \ldots, x_n, z) \mapsto (x_1, \ldots, x_n).$$

Decomposing an algebraic function hence may be regarded as a special instance of decomposing an "inverse" of a rational mapping (an approach to the decomposition of algebraic functions using resultants was discussed by Kozen et al. (1996)).
In case of $k(X)/k(\mathbf{f})$ being transcendental it is sensible to introduce the notion of a *strong decomposition*: $k(X)/k(\mathbf{f})$ being transcendental may be interpreted as indicating that there are "degrees of freedom" in the solution of the system; adjoining (over $k(\mathbf{f})$) purely transcendental elements to $k(\mathbf{f})$ may be understood as diminishing degrees of freedom resp. selecting a transcendence basis of $k(X)/k(\mathbf{f})$ as removing all degrees of freedom. Informally a strong decomposition corresponds to a somewhat optimal fixing of free parameters. Before giving a precise definition we give a very simple example to illustrate the underlying idea: Assume we are interested in solving the equation $x^2 + y^2 = \alpha$ over $\mathbb{C}$. One possibility is to fix x, thus obtaining a quadratic equation for y. This "strategy" corresponds to the chain of fields $\mathbb{C}(x^2 + y^2) \leq \mathbb{C}(x^2 + y^2, x) \leq \mathbb{C}(x, y)$. A better choice is to fix $x + i \cdot y$, however; this avoids the need for solving a quadratic equation $((x + i \cdot y) := \beta \Rightarrow (x - i \cdot y) = \alpha/\beta \Rightarrow x = (\alpha + \beta^2)/(2\beta), y = i \cdot x - i\beta)$. In terms of fields this means that $\mathbb{C}(x^2 + y^2, x)$ is a proper subfield of a purely transcendental extension of $\mathbb{C}(x^2 + y^2)$, namely $\mathbb{C}(x^2 + y^2, x + i \cdot y)$. This phenomenon is not to occur in case of a strong decomposition:

Definition 4 *Let* $(\mathbf{g}, \mathbf{h})$ *(resp.* $(\mathbf{h}^{-1}, \mathbf{g}^{-1})$*) be a non-trivial decomposition of arity* r *of* $\mathbf{f} : X \to Z$ *(resp.* $\mathbf{f}^{-1} : 2^Z \to 2^X$*). Then* $(\mathbf{g}, \mathbf{h})$ *(resp.* $(\mathbf{h}^{-1}, \mathbf{g}^{-1})$*) is called a* strong ((inverse) rational) decomposition of $\mathbf{f}$ (resp. $\mathbf{f}^{-1}$) (of arity r) *iff there is no* $k' \le k(X)$ *such that both* $k(\mathbf{h}) < k' \le k(X)$ *and* $k'/k(\mathbf{f})$ *is purely transcendental.*

Note that any decomposition $(\mathbf{g}, \mathbf{h})$ (resp. $(\mathbf{h}^{-1}, \mathbf{g}^{-1})$) with $k(\mathbf{h})/k(\mathbf{f})$ algebraic is strong. Moreover, in case of $k(X)/k(\mathbf{f})$ being purely transcendental there is no strong decomposition of $\mathbf{f}$. We are not aware of an algorithm for deciding in general whether a given decomposition is strong; however, for practical purposes — say solving a system of equations — identifying a decomposition as strong may be less important than being able to solve the given system by means of the known decomposition.

4 Computing a decomposition

According to the above the task of finding a (non-trivial) decomposition splits into three steps:

1. Finding a(n) (proper) intermediate field k' of $k(X)/k(\mathbf{f})$.
2. Fixing generators $h_1, \ldots, h_r$ of k'.
3. Determining a representation of $\mathbf{f}$ in terms of $h_1, \ldots, h_r$.

Algorithms for performing the third step are known: The algorithms of Kemper (1993) and Sweedler (1993) make use of a Gröbner basis computation in $m + n + 1$ variables (where $\mathbf{f} = (f_1(x_1, \ldots, x_n), \ldots, f_m(x_1, \ldots, x_n))$) over k w.r.t. to a block order; Müller-Quade and Steinwandt (1997) proposed an algorithm for $k(X) = k(x_1, \ldots, x_n)$ a rational function field requiring a Gröbner basis computation over $k(X)$ in n variables w.r.t. to an arbitrary term order. Assuming computations in $k(X)$ to be effective we give a generalization of the latter approach to arbitrary $k(X)$, thereby also obtaining in certain cases a possibility to determine subfields of $k(X)$; for instance in case of $k(X)/k(\mathbf{f})$ separable algebraic all intermediate fields of $k(\mathbf{f})$ and $k(X)$ can be found in this way. After having identified an intermediate field $k(\mathbf{h})$ of $k(\mathbf{f})$ and $k(X)$ (first step) one may in principle use any tuple of generators of $k(\mathbf{h})$ to form a decomposition of $\mathbf{f}$. For practical purposes, however, one would usually like their number or "degrees" (e.g., maximum of degree of numerator and denominator after reducing to lowest terms) to be minimal and/or the corresponding "outer part" $\mathbf{g}$ to be sparse. We do not know the best strategy to follow here; an example where a minimal generating set can be constructed efficiently occurs in the case of monomial decompositions (cf. Section 5).

To construct the ideals which can be used to identify intermediate fields and to tackle the field membership problem we need the somewhat technical

Proposition 1 *Let* $h_1 = \frac{n_1(\mathbf{x})}{d_1(\mathbf{x})}, \ldots, h_r = \frac{n_r(\mathbf{x})}{d_r(\mathbf{x})} \in k(X)$, $D := \{p \in k[\mathbf{Z}] : p$ *prime and* $p \mid d_i(\mathbf{Z})$ *for some* $i \in \{1, \ldots, r\}\}$, $d_{\boldsymbol{\nu}} := \prod_{p \in D} p^{\nu_p}$ *with* $\nu_p \in \mathbb{N}$ *arbitrary,* $I := \langle\{p(\mathbf{Z}) : p \in I(X)\}\rangle + \langle n_1(\mathbf{Z}) - h_1 \cdot d_1(\mathbf{Z}), \ldots, n_r(\mathbf{Z}) - h_r \cdot d_r(\mathbf{Z})\rangle$, *and* $J := I : d_{\boldsymbol{\nu}}^{\infty} \trianglelefteq k(\mathbf{h})[\mathbf{Z}]$. *Then for* $f \in k(\mathbf{h})[\mathbf{Z}]$ *we have*

$$f(Z_1, \ldots, Z_n) \in J \iff f(x_1, \ldots, x_n) = 0 \, (\in k(X)).$$

In particular $J = \langle Z_1 - x_1, \ldots, Z_n - x_n\rangle \cap k(\mathbf{h})[\mathbf{Z}]$.

Proof: Note that the definitions of I and $d_{\boldsymbol{\nu}}$ depend on the representatives n_i/d_i of $h_i, i = 1, \ldots, r$ chosen, and subsequent calculations are to be understood using this fixed set of representatives. Since the characterization $J = \langle Z_1 - x_1, \ldots, Z_n - x_n\rangle \cap k(\mathbf{h})[\mathbf{Z}]$ turns out to be valid for any set of representatives we obtain in particular that the definition of J is independent of the representatives chosen. We remark that the proof is essentially an adaption of the proof of Lemma 1.3 by Müller-Quade and Steinwandt (1997) where the case $k(X) = k(x_1, \ldots, x_n)$, i.e., $k(X)/k$ purely transcendental, is considered. Throughout the proof the ideal $\langle\{p(\mathbf{Z}) : p \in \mathrm{I}(X)\}\rangle \trianglelefteq k[\mathbf{Z}]$ is denoted by $\mathrm{I}_{\mathbf{Z}}(X)$.

$\Longrightarrow$: Let $f \in J$. Hence there is $\mu \in \mathbb{N}$ with $d_{\boldsymbol{\nu}}{}^{\mu} f \in I$, i.e., there are polynomials $q \in \mathrm{I}_{\mathbf{Z}}(X) \cdot k(\mathbf{h})[\mathbf{Z}]$, $u_1, \ldots, u_r \in k(\mathbf{h})[\mathbf{Z}]$ such that $d_{\boldsymbol{\nu}}{}^{\mu} f = q + \sum_{i=1}^{r} u_i \cdot (n_i(\mathbf{Z}) - h_i \cdot d_i(\mathbf{Z}))$, and we have $(d_{\boldsymbol{\nu}}{}^{\mu} f)(x_1, \ldots, x_n) = 0$. Since clearly $d_{\boldsymbol{\nu}}(x_1, \ldots, x_n) \notin \mathrm{I}(X)$ we have $f(x_1, \ldots, x_n) = 0$ as required.

$\Longleftarrow$: Let $f(Z_1, \ldots, Z_n) \in k(\mathbf{h})[\mathbf{Z}]$ with $f(x_1, \ldots, x_n) = 0$. By multiplying f with a suitable $c \in k[\mathbf{h}]$ we get $\widetilde{f} = c \cdot f$ with $\widetilde{f} \in k[\mathbf{h}][\mathbf{Z}]$; let

$$\widetilde{f} = \sum_{\underline{\mu}, \underline{\lambda}} \alpha_{\underline{\mu}, \underline{\lambda}} \cdot \underline{Z}^{\underline{\mu}} \prod_{i=1}^{r} (h_i(x_1, \ldots, x_n))^{\lambda_i}$$

where $\alpha_{\underline{\mu}, \underline{\lambda}} \in k$. Regarding $\widetilde{f}$ as a polynomial in $r + n$ variables we get $\widetilde{f}(h_1(\mathbf{x}), \ldots, h_r(\mathbf{x}), \mathbf{x}) = 0$, and hence

$$\widetilde{f}(h_1(\mathbf{Z}), \ldots, h_r(\mathbf{Z}), \mathbf{Z}) = 0 \in \mathrm{Quot}\,(k[\mathbf{Z}]/\mathrm{I}_{\mathbf{Z}}(X)),$$

i.e., there are $u \in \mathrm{I}_{\mathbf{Z}}(X)$, $v \in k[\mathbf{Z}] \backslash \mathrm{I}_{\mathbf{Z}}(X)$ with $v \cdot \widetilde{f}(h_1(\mathbf{Z}), \ldots, h_r(\mathbf{Z}), \mathbf{Z}) = u$. Since $\mathrm{I}_{\mathbf{Z}}(X)$ is prime we can assume $v = 1$, and we obtain

$$\begin{aligned} u &= \widetilde{f}(h_1(\mathbf{x}) + (h_1(\mathbf{Z}) - h_1(\mathbf{x})), \ldots, h_r(\mathbf{x}) + (h_r(\mathbf{Z}) - h_r(\mathbf{x})), Z_1, \ldots, Z_n) \\ &= \textstyle\sum_{\underline{\mu}, \underline{\lambda}} \alpha_{\underline{\mu}, \underline{\lambda}} \cdot \underline{Z}^{\underline{\mu}} \prod_{i=1}^{r} (h_i(\mathbf{x}) + (h_i(\mathbf{Z}) - h_i(\mathbf{x})))^{\lambda_i}. \end{aligned}$$

Expanding the latter product yields only one term not involving a factor of the form $(h_i(\mathbf{Z}) - h_i(\mathbf{x}))$; namely we have

$$u = \sum_{\underline{\mu}, \underline{\lambda}} \alpha_{\underline{\mu}, \underline{\lambda}} \cdot \underline{Z}^{\underline{\mu}} \left(\prod_{i=1}^{r} (h_i(\mathbf{x}))^{\lambda_i} + \sum_{j=1}^{r} u_{\underline{\lambda}, j} \cdot (h_j(\mathbf{Z}) - h_j(\mathbf{x})) \right)$$

with $u_{\underline{\lambda}, j} \in k[h_1(\mathbf{x}), \ldots, h_r(\mathbf{x}), h_1(\mathbf{Z}), \ldots, h_r(\mathbf{Z}), \mathbf{Z}]$.

The last equation may be written as $u = \widetilde{f} + \sum_{i=1}^{r} \widetilde{u}_i \cdot (h_i(\mathbf{Z}) - h_i(\mathbf{x}))$ with $\widetilde{u}_i \in k[h_1(\mathbf{x}), \ldots, h_r(\mathbf{x}), h_1(\mathbf{Z}), \ldots, h_r(\mathbf{Z}), \mathbf{Z}]$. By multiplying with a suitable power product of the $d_i(\mathbf{Z})$ we may remove the denominators of the $h_i(\mathbf{Z})$, and for an appropriate $\eta \in \mathbb{N}$ we have

$$d_{\nu}{}^{\eta} \cdot u = d_{\nu}{}^{\eta} \cdot \widetilde{f} + \sum_{i=1}^{r} \widehat{u}_i \cdot (n_i(\mathbf{Z}) - h_i(\mathbf{x})d_i(\mathbf{Z}))$$

with $\widehat{u}_i \in k[h_1(\mathbf{x}), \ldots, h_r(\mathbf{x}), \mathbf{Z}]$, i. e., $\widetilde{f}$ is contained in
$(\langle n_1(\mathbf{Z}) - h_1 \cdot d_1(\mathbf{Z}), \ldots, n_r(\mathbf{Z}) - h_r \cdot d_r(\mathbf{Z})\rangle + \mathrm{I}_{\mathbf{Z}}(X) \cdot k[\mathbf{h}][\mathbf{Z}]) : d_{\nu}^{\infty} \trianglelefteq k[\mathbf{h}][\mathbf{Z}]$ resp. $f = c^{-1} \cdot \widetilde{f} \in J$.
The equality $J = \langle Z_1 - x_1, \ldots, Z_n - x_n \rangle \cap k(\mathbf{h})[\mathbf{Z}]$ follows immediately from the above. □

Using a straightforward modification of the proof of Lemma 3.1 by Müller-Quade and Steinwandt (1997) we get the following characterization of J:

Lemma 2 *Let J, $k(\mathbf{h})$ be as in Prop. 1. Then $k(X) \simeq$* Quot $(k(\mathbf{h})[\mathbf{Z}]/J)$, *and the transcendence degree of $k(X)$ over $k(\mathbf{h})$ is equal to* $\dim(J)$.

Sketch of proof: The homomorphism of fields $\varphi : k(X) \to \mathrm{Quot}\,(k(\mathbf{h})[\mathbf{Z}]/J)$, $x_i \mapsto \frac{\overline{Z_i}}{1}$ is bijective. As J is prime every maximally independent set S modulo J has $\dim(J)$ elements and the residue classes of the elements of S form a transcendence basis of $\mathrm{Quot}(k(\mathbf{h})[\mathbf{Z}]/J)$ over $k(\mathbf{h})$ (Becker et al. 1993, Prop. 7.26 & Lemma 7.25). □

To be able to give the desired algorithm for determining a representation of $\mathbf{f}$ we restate Remark 1.5 by Müller-Quade and Steinwandt (1997):

Remark 1 *Let $\Lambda_1, \ldots, \Lambda_s$ be parameters, $p_i \in k[\mathbf{Z}]$, $u_i = n_i/d_i \in k(\mathbf{\Lambda}), i = 1, \ldots, t$, G a Gröbner basis of $\langle G \rangle \trianglelefteq k[\mathbf{Z}]$ w. r. t. an arbitrary term order, and $\alpha_1, \ldots, \alpha_s \in k$ such that $\prod_{i=1}^{t} d_i(\alpha_1, \ldots, \alpha_s) \neq 0$. Then specializing $\Lambda_i \mapsto \alpha_i, i = 1, \ldots, s$ in the normal form of $\sum_{i=1}^{t} u_i(\mathbf{\Lambda})p_i$ modulo G yields the normal form of $\sum_{i=1}^{t} u_i(\boldsymbol{\alpha})p_i$.*

Choosing $u_1, \ldots, u_t$ to be linear polynomials in $\mathbf{\Lambda}$ ensures the resulting normal form to be a polynomial of degree ≤ 1 in $\mathbf{\Lambda}$, of course; having in mind that a Gröbner basis of a saturation $I : d^{\infty} \trianglelefteq k[\mathbf{Z}]$ can be computed effectively, e. g., via Prop. 6.37 of Becker et al. (1993), we may determine the representations of $f_1, \ldots, f_m$ by means of the next lemma (cf. Lemma 2.1 by Müller-Quade and Steinwandt (1997)).

Lemma 3 *Let J, $k(\mathbf{h})$ be as in Prop. 1, G a Gröbner basis of J w. r. t. an arbitrary term order, $n, d \in k[\mathbf{Z}]$, $d(\mathbf{x}) \notin \mathrm{I}(X)$, Λ a formal parameter, $N(\Lambda)$ the normal form of $n - \Lambda \cdot d$ modulo G. Then we have*
$\frac{n(\mathbf{x})}{d(\mathbf{x})} \in k(\mathbf{h}) \iff$ *The linear equation $N(\Lambda) = 0$ has a solution in $k(X)$.*
If there is a solution to $N(\Lambda) = 0$ in $k(X)$ then it is contained in $k(\mathbf{h})$.

Proof: $\Longrightarrow$: As $n(\mathbf{x})/d(\mathbf{x}) \in k(\mathbf{h})$ we have $n(\mathbf{Z}) - \frac{n(\mathbf{x})}{d(\mathbf{x})}d(\mathbf{Z}) \in J$ according to Prop. 1. Thus the claim follows immediately from Remark 1.

$\Longleftarrow$: The solution is unique, since otherwise Remark 1 would imply that $\forall a \in k(\mathbf{h}) : n - a \cdot d \in J$ — a contradiction to Prop. 1. Moreover, a solution in $k(X)$ must be a solution in $k(\mathbf{h})$, because all coefficients involved are contained in $k(\mathbf{h})$. Thus the claim is an immediate consequence of Prop. 1 and Remark 1. □

Note that Lemma 3 implies that the coefficients of a Gröbner basis of J form a generating set of $k(\mathbf{h})$ over k. Fixing a generating set $x_1, \ldots, x_n$ of $k(X)$ and a term order on $\mathrm{T}(Z_1, \ldots, Z_n)$ we therefore obtain a "canonical" generating set of $k(\mathbf{h})$ over k by using a reduced Gröbner basis.[2]

Hence — performing the required computations by means of "tag parameters" analogous to the case $k(X) = k(x_1, \ldots, x_n)$ considered by Müller-Quade and Steinwandt (1997) — the representations of $f_1, \ldots, f_m$ essentially can be found by computing appropriate normal forms modulo a Gröbner basis G of J; namely the steps to follow are:

1. Compute a Gröbner basis G of J as in Prop. 1 w. r. t. an arbitrary term order (e. g., by determining a reduced Gröbner basis of $J \cdot k(X)[\mathbf{Z}]$). At this keep track of how the coefficients of the polynomials are computed from the coefficients of the original polynomials (cf. the use of "tag parameters" as discussed by Müller-Quade and Steinwandt (1997)).

2. For each $f_i = n_{f_i}/d_{f_i}$ compute the normal form $N_i - \Lambda_i \cdot D_i$ of $n_{f_i} - \Lambda_i \cdot d_{f_i}$ modulo G (with $N_i, D_i \in k(\mathbf{h})[\mathbf{Z}]$) — again keeping track of how the coefficients are computed.

3. If the linear equation $N_i - \Lambda_i \cdot D_i = 0$ cannot be solved in $k(X)$ then $f_i \notin k(\mathbf{h})$. Otherwise a solution of this equation yields a representation of f_i in terms of the coefficients of the Gröbner basis and hence also in terms of $\mathbf{h}$.

In particular for $k(X)/k(\mathbf{f})$ separable algebraic information about intermediate fields of $k(\mathbf{f})$ and $k(X)$ can be obtained from the ideal J by means of a generalization of Lemma 5.1 by Müller-Quade and Steinwandt (1997) (for the problem of determining intermediate fields of finite algebraic extensions cf. also the work of Lazard and Valibouze (1993)):

Lemma 4 *Let* $k(\mathbf{f}) \leq k(X)$,

(1). J the ideal corresponding to $k(\mathbf{f})$ according to Prop. 1,

(2). $J \cdot k(X)[\mathbf{Z}] = \bigcap_{i=1}^{l} Q_i$ a minimal primary decomposition,

(3). P_i the associated prime of Q_i, $i = 1, \ldots, l$,

(4). k' an intermediate field of $k(\mathbf{f})$ and $k(X)$, and

[2] Of course, the "canonical" generating set could in principal be made independent of the term order chosen, e. g., by considering the (finite) union of all reduced Gröbner bases of J.

(5). $k(\mathbf{f})_{\text{alg}}$ the algebraic closure of $k(\mathbf{f})$ in $k(X)$.

Then for $k(X)/k(\mathbf{f})$ algebraic the following statements hold:

(i). If $k(X)/k'$ is separable then there is $\Lambda \subseteq \{1,\ldots,l\}$: The coefficients of a reduced Gröbner basis of $\bigcap_{\lambda\in\Lambda} P_\lambda$ form a generating set of k' over k.

(ii). If $k'/k(\mathbf{f})$ is separable then there is $\Lambda \subseteq \{1,\ldots,l\}$: The coefficients of a reduced Gröbner basis of $\bigcap_{\lambda\in\Lambda} Q_\lambda$ form a generating set of k' over k.

(iii). Up to permutation $Q_1,\ldots,Q_l$ are uniquely determined.

(iv). If $k(X)/k(\mathbf{f})$ is separably generated then $\forall i = 1,\ldots,l: \ Q_i = P_i$.

For $k(X)/k(\mathbf{f})_{\text{alg}}$ separable (not necessarily algebraic) there is a $\lambda \in \{1,\ldots,l\}$: The coefficients of a reduced Gröbner basis of P_λ form a generating set of $k(\mathbf{f})_{\text{alg}}$ over k.

Sketch of proof: Since the actual proof is a straightforward modification of the proof of Lemma 5.1 by Müller-Quade and Steinwandt (1997) where $k(X)$ is assumed to be a rational function field $k(x_1,\ldots,x_n)$, i. e., $k(X)/k$ purely transcendental, the details are elided here. The basic ideas of the proof can be sketched as follows: For $k(X)/k(\mathbf{f})$ algebraic the ideal corresponding to $k(\mathbf{f})$ according to Prop. 1 is of dimension zero. For an intermediate field k' of $k(\mathbf{f})$ and $k(X)$ with $k'/k(\mathbf{f})$ separable algebraic $J\cdot k'[\mathbf{Z}]$ is radical, and the ideal $J_{k'}$ corresponding to k' is an associated prime hereof; in particular for $k(X)/k(\mathbf{f})$ separable algebraic this argument can be used to prove (iv). Over $k(\mathbf{f})_{\text{sep}}$ — the separable closure of $k(\mathbf{f})$ in $k(X)$ — the ideal $J_{k'}\cdot k(\mathbf{f})_{\text{sep}}[\mathbf{Z}]$ splits into an intersection of maximal ideals. By the purely inseparable algebraic extension $k(X)/k(\mathbf{f})_{\text{sep}}$ the latter can become primary in $k(X)[\mathbf{Z}]$. For $k(X)/k'$ separable algebraic no purely inseparable extension of the ground field takes place, and the radical ideal $J_{k'}\cdot k(X)[\mathbf{Z}]$ can be reconstructed as an intersection of primes associated to $J\cdot k(X)[\mathbf{Z}]$. Uniqueness of the primary decomposition follows from J being of dimension zero. The characterization of the algebraic closure is based on $k(\mathbf{f})_{\text{alg}}$ being algebraically closed in $k(X)$; in this situation $k(X)/k(\mathbf{f})_{\text{alg}}$ separable guarantees that a prime ideal in $k(\mathbf{f})_{\text{alg}}[\mathbf{Z}]$ remains prime in $k(X)[\mathbf{Z}]$, and by means of the equality $\dim J = \dim J_{\text{alg}}$ (with J_{alg} the ideal corresponding to $k(\mathbf{f})_{\text{alg}}$) the ideal $J_{\text{alg}}\cdot k(X)[\mathbf{Z}]$ can be identified as a minimal prime over $J\cdot k(X)[\mathbf{Z}]$ and hence as an associated prime hereof.□

By means of the leading terms of a Gröbner basis of J we can easily check whether $\dim(J) = 0$ (Buchberger 1985, Method 6.9), i. e., if $k(X)/k(\mathbf{f})$ is algebraic (Lemma 2). In the affirmative case we can apply Method 6.11 by Buchberger (1985) to find the polynomial of minimal degree in $J \cap k(\mathbf{f})[Z_i]$, $i = 1,\ldots,n$ — and thereby the minimal polynomials of $x_1,\ldots,x_n$ over $k(\mathbf{f})$ (Prop. 1). Then $k(X)/k(\mathbf{f})$ is separable algebraic iff these polynomials do not have multiple zeros (in $\text{char}(k) = 0$ the latter test is unnecessary, of course).

5 Monomial decompositions

In this section $k(X)/k$ is assumed to be purely transcendental, i. e., $k(X) = k(\mathbf{x}) := k(x_1, \ldots, x_n)$ is a rational function field. For $\mathbf{f} \in k(\mathbf{x})^m$ we aim at finding a non-trivial decomposition $\mathbf{g} \circ \mathbf{h}$ with $\mathbf{h}$ consisting of Laurent monomials. The interest in this kind of decomposition which occurs, e. g., in the design of diffractive optical systems (Aagedal et al. 1996), is motivated by the fact that the "inversion" of Laurent monomial mappings — rational mappings involving only multiplication and division — can be done far more efficient than the inversion of arbitrary rational mappings. Inversion of Laurent monomial mappings reduces to solving systems of equations of the form $\gamma_1 \mathbf{x}^{\boldsymbol{\mu}_1} - \alpha_1 = 0, \ldots, \gamma_s \mathbf{x}^{\boldsymbol{\mu}_s} - \alpha_s = 0$, $s \in \mathbb{N}$ where the $\gamma_i \mathbf{x}^{\boldsymbol{\mu}_i}$ denote Laurent monomials, i. e., $\boldsymbol{\mu}_i \in \mathbb{Z}^n$, and the α_i are constants resulting from the evaluation of that Laurent monomial mapping. If $\alpha_i \neq 0$ for $i = 1, \ldots, s$ these equations can be solved efficiently (cf. pp. 10 in the work of Eisenbud and Sturmfels (1996)).

Lemma 5 *For Laurent monomials $\gamma_1 \mathbf{x}^{\boldsymbol{\mu}_1}, \ldots, \gamma_s \mathbf{x}^{\boldsymbol{\mu}_s} \in k[\mathbf{x}, \mathbf{x}^{-1}]$ we have:*

(i). Systems of equations of the form $\gamma_1 \mathbf{x}^{\boldsymbol{\mu}_1} - \alpha_1 = 0, \ldots, \gamma_s \mathbf{x}^{\boldsymbol{\mu}_s} - \alpha_s = 0$ with $\alpha_i \neq 0$ for $i = 1, \ldots, s$ can be triangularized efficiently.

(ii). $k(\mathbf{x}^{\boldsymbol{\mu}_1}, \ldots, \mathbf{x}^{\boldsymbol{\mu}_s})/k$ is purely transcendental, and a minimal generating set of this extension can be computed efficiently.

Proof: (i): (Cf. the work of Eisenbud and Sturmfels (1996) for a (slightly different) thorough treatment.) Dividing each equation by α_i yields $\frac{\gamma_1}{\alpha_1} \mathbf{x}^{\boldsymbol{\mu}_1} = 1, \ldots, \frac{\gamma_s}{\alpha_s} \mathbf{x}^{\boldsymbol{\mu}_s} = 1$. The Laurent monomials on the left hand sides generate a multiplicative abelian group M where $\gamma \mathbf{x}^{\boldsymbol{\mu}} \in M$ is equivalent to $\gamma \mathbf{x}^{\boldsymbol{\mu}} - 1 \in \langle \frac{\gamma_1}{\alpha_1} \mathbf{x}^{\boldsymbol{\mu}_1} - 1, \ldots, \frac{\gamma_s}{\alpha_s} \mathbf{x}^{\boldsymbol{\mu}_s} - 1 \rangle \trianglelefteq k[\mathbf{x}, \mathbf{x}^{-1}]$. Consider a matrix over $\mathbb{Z}$ where the rows consist of the exponent vectors $\boldsymbol{\mu}_1, \ldots, \boldsymbol{\mu}_s$ (in any order). Using row operations of the elementary divisor algorithm we can transform this matrix into a matrix in upper triangular form with rows $\widetilde{\boldsymbol{\mu}}_1, \ldots, \widetilde{\boldsymbol{\mu}}_{\tilde{s}}$ which generate the same $\mathbb{Z}$-module as $\boldsymbol{\mu}_1, \ldots, \boldsymbol{\mu}_s$. Following these instructions, but calculating in M we can derive a basis $\frac{\tilde{\gamma}_1}{\tilde{\alpha}_1} \mathbf{x}^{\widetilde{\boldsymbol{\mu}}} - 1, \ldots, \frac{\tilde{\gamma}_{\tilde{s}}}{\tilde{\alpha}_{\tilde{s}}} \mathbf{x}^{\widetilde{\boldsymbol{\mu}}} - 1$ of the above ideal which is in triangular form. After clearing denominators we get a basis of binomials in $k[\mathbf{x}]$ in triangular form.

(ii): The Laurent monomials in $k(\mathbf{x}^{\boldsymbol{\mu}_1}, \ldots, \mathbf{x}^{\boldsymbol{\mu}_s}) \leq k(\mathbf{x})$ form an abelian group, and a generating set of this group also forms a generating set of $k(\mathbf{x}^{\boldsymbol{\mu}_1}, \ldots, \mathbf{x}^{\boldsymbol{\mu}_s})$ as a field extension of k. Again using row operations of the elementary divisor algorithm on the exponent vectors of the given Laurent monomials we get a basis in triangular form of this abelian group, and the according Laurent monomials therefore must be algebraically independent. $\square$

As will be seen in Lemma 7 the triangular form obtained here has a close connection to Gröbner bases. For the sequel we need the following result about the behaviour of the greatest common divisor of polynomials under composition:

Lemma 6 *Let* $\mathbf{h} : \mathbb{A}^n \to \mathbb{A}^m$ *be a polynomial mapping with its image being dense in* $\mathbb{A}^m$. *Then there exists* $r \in k[x_1, \ldots, x_n]$: *For* $p, q \in k[x_1, \ldots, x_m]$ *the implication* $\gcd(p, q) = g \Longrightarrow \mathrm{V}(\gcd(p \circ \mathbf{h}, q \circ \mathbf{h})) \subseteq \mathrm{V}(r \cdot (g \circ \mathbf{h}))$ *holds.*

Proof: Let W be the union of all irreducible varieties $V \subseteq \mathbb{A}^n$ of dimension $n-1$ for which $\mathbf{h}(V)$ is not dense in a hypersurface of $\mathbb{A}^m$. It follows from § 6, Satz 7 by Schafarewitsch (1972) that the Zariski closure of $\{\boldsymbol{\alpha} \in k^n : \dim(\mathbf{h}^{-1}(\mathbf{h}(\boldsymbol{\alpha}))) \neq n-m\}$ is a variety $\neq \mathbb{A}^n$. Since for each irreducible hypersurface $V \subseteq W$ the set $\{\boldsymbol{\alpha} \in V : \dim(\mathbf{h}^{-1}(\mathbf{h}(\boldsymbol{\alpha}))) = \dim(V) - \dim(\mathbf{h}(V))\}$ is dense in V and $\dim(V) - \dim(\mathbf{h}(V)) \neq (n-1) - (m-1) = n-m$ we have: V and hence W is a subset of the Zariski closure of $\{\boldsymbol{\alpha} \in k^n : \dim(\mathbf{h}^{-1}(\mathbf{h}(\boldsymbol{\alpha}))) \neq n-m\}$. As W is a union of varieties of dimension $n-1$ and W is contained in a variety $\neq \mathbb{A}^n$ the set W must itself be a variety of dimension $n-1$.
Let r be the defining polynomial of W. To prove the lemma it is sufficient to verify the implication $\gcd(p, q) = 1 \Rightarrow \mathrm{V}(\gcd(p \circ \mathbf{h}, q \circ \mathbf{h})) \subseteq \mathrm{V}(r) = W$. To check the latter we can make use of $\gcd(p, q) = 1 \iff \dim(\mathrm{V}(p, q)) < m-1$: Suppose $\mathrm{V}(\gcd(p \circ \mathbf{h}, q \circ \mathbf{h})) \not\subseteq W$, i.e., $\mathrm{V}(\langle \gcd(p \circ \mathbf{h}, q \circ \mathbf{h}) \rangle : r^\infty) =: M$ is a hypersurface but not a subset of W. Thus $\mathbf{h}(M)$ is dense in a hypersurface in $\mathbb{A}^m$ and p and q vanish there which contradicts $\gcd(p, q) = 1$. □

In the context of monomial decomposition we need

Corollary 1 *Let* $\mathbf{h} : \mathbb{A}^n \setminus \mathrm{V}(Z_1 \cdot \ldots \cdot Z_n) \to \mathbb{A}^m \setminus \mathrm{V}(Z_1 \cdot \ldots \cdot Z_m)$ *be a Laurent monomial mapping with its image being dense in* $\mathbb{A}^m$, *and* $p, q \in k[Z_1, \ldots, Z_m]$ *with* $\gcd(p, q) = 1$. *Then* $\gcd(p \circ \mathbf{h}, q \circ \mathbf{h})$ *is a Laurent monomial.*

Proof: As a first case we assume each h_i to be homogeneous, i.e., equal degrees in numerator and denominator. Then the preconditions for § 6, Satz 7 by Schafarewitsch (1972) are met, and we can conclude that for every point $\boldsymbol{\alpha}$ outside $\mathrm{V}(\prod_{i=1}^n Z_i)$ the ideal

$$I := \langle \text{clear\_denominators}(\mathbf{h}(\mathbf{Z}) - \boldsymbol{\alpha}) \rangle : (Z_1 \cdot \ldots \cdot Z_n)^\infty \trianglelefteq k[\mathbf{Z}]$$

describing the fiber of $\boldsymbol{\alpha}$ is at most of dimension $n-m$. Following Lemma 5 we get a basis of I in triangular form with $n-m$ free variables implying $\dim(I) \geq n-m$ — it cannot happen that $I = \langle 1 \rangle$ since this implies the existence of a syzygy among the $h_1, \ldots, h_m$ as the $\alpha_1, \ldots, \alpha_n$ may even be chosen to be new transcendental elements and after triangulation specialized in $k^{\times n}$. Thus all $\boldsymbol{\alpha} \notin \mathrm{V}(\prod_{i=1}^n Z_i)$ have fibers of the same dimension and the set $W \setminus \mathrm{V}(\prod_{i=1}^n Z_i)$ (notation as in Lemma 6) is empty. Hence for every irreducible hypersurface V with $\mathrm{I}(V \setminus \mathrm{V}(\prod_{i=1}^n Z_i)) = \mathrm{I}(V)$ the set $\mathbf{h}(V)$ is a dense subset of a hypersurface in $\mathbb{A}^m$. If $\gcd(p \circ \mathbf{h}, q \circ \mathbf{h}) = r$ is no monomial this implies that $p \circ \mathbf{h}$ and $q \circ \mathbf{h}$ vanish on a variety V with $\mathrm{I}(V \setminus \mathrm{V}(\prod_{i=1}^n Z_i)) = \mathrm{I}(V)$ which contradicts $\gcd(p, q) = 1$ since p and q then vanish on a hypersurface containing $\mathbf{h}(V)$. Now assume $\mathbf{h}$ to be inhomogeneous and V to be an irreducible hypersurface with $\mathrm{I}(V \setminus \mathrm{V}(\prod_{i=1}^n Z_i)) = \mathrm{I}(V)$ which is not mapped to a dense subset of a hypersurface in $\mathbb{A}^m$. We introduce a new variable Z_{n+1} and homogenize $\mathbf{h}$ to $\widetilde{\mathbf{h}}$. Then $\widetilde{\mathbf{h}}(V \times \{1\}) = \mathbf{h}(V)$ and not a hypersurface which is impossible since $\mathrm{I}(V \times \{1\} \setminus \mathrm{V}(\prod_{i=1}^{n+1} Z_i)) = \mathrm{I}(V \times \{1\})$. Thus also non-homogeneous Laurent monomial mappings map V to a dense subset of a hypersurface in $\mathbb{A}^m$, and the above proof also holds. □

Now consider the following simple *Monomial Field Algorithm* which computes from a set $\mathcal{F}$ of generators of $k(\mathbf{f})$ over k with $0 \notin \mathcal{F}$ a set $\mathcal{L}$ of Laurent monomials generating a field $k(\mathcal{L}) \geq k(\mathbf{f})$:

$\mathcal{L} \leftarrow \emptyset$
`for` $f \in \mathcal{F}$ `do`
  choose $p, q \in k[\mathbf{x}] : \gcd(p,q) = 1$ and $f = p/q$
  choose $\gamma \cdot \mathbf{x}^{\mu}$ from monomials$(p) \cup$ monomials(q)
  $\mathcal{L} \leftarrow \mathcal{L} \cup$ Laurent_monomials$(\mathbf{x}^{-\mu} \cdot p) \cup$ Laurent_monomials$(\mathbf{x}^{-\mu} \cdot q)$
`od`

Note that while the output $\mathcal{L}$ is not uniquely determined $k(\mathcal{L})$ is, e. g., for $\mathcal{F} = \{y/(x^2y + x)\}$ we obtain $k(\mathcal{L}) = k(x/y, xy)$ — in accordance with the equations $\frac{y}{x^2y+x} = \frac{1}{x^2+(x/y)} = \frac{1/x^2}{1+1/xy} = \frac{y/x}{xy+1}$.

Proposition 2 *The Monomial Field Algorithm computes, given a set $\mathcal{F}$ of generators of $k(\mathbf{f})$ over k with $0 \notin \mathcal{F}$, a generating set $\mathcal{L}$ for the unique smallest field $k(\mathcal{L}) \geq k(\mathbf{f})$ which can be generated by Laurent monomials.*

Proof: To show the existence of such a unique field $k(\mathcal{L})$ we assume f to be any rational function which is contained in two fields $k(\mathcal{M})$ and $k(\widetilde{\mathcal{M}})$ both generated by Laurent monomials. The function f then has two representations $p(\mathbf{x}^{\mu})/q(\mathbf{x}^{\mu})$ and $\widetilde{p}(\mathbf{x}^{\widetilde{\mu}})/\widetilde{q}(\mathbf{x}^{\widetilde{\mu}})$ in algebraically independent generators of $k(\mathcal{M})$ and $k(\widetilde{\mathcal{M}})$ respectively (to shorten notation $\mathbf{x}^{\mu}$, $\mathbf{x}^{\widetilde{\mu}}$, and $\mathbf{x}^{\mu_i}$ can denote single Laurent monomials or a tuple of Laurent monomials). Following Corollary 1 about gcd under composition these two representations (if cancelled out correctly) vary from the representation in $k(\mathbf{x})$ only by extension with a Laurent monomial. Applying the Monomial Field Algorithm to $\{f\}$ we get a set of Laurent monomials in which f can be expressed in a way that either the numerator or the denominator contains a constant $\in k$. All Laurent monomials appearing in this representation must be contained in both fields $k(\mathcal{M})$ and $k(\widetilde{\mathcal{M}})$ since extension with any Laurent monomial m to obtain $p(\mathbf{x}^{\mu})/q(\mathbf{x}^{\mu})$ or $\widetilde{p}(\mathbf{x}^{\widetilde{\mu}})/\widetilde{q}(\mathbf{x}^{\widetilde{\mu}})$ leads to m appearing in the representation. Thus every rational function f having a representation in both fields can be represented in Laurent monomials lying in both fields, and the intersection of fields generated by Laurent monomials can again be generated by Laurent monomials.

To prove the correctness of the Monomial Field Algorithm we consider the `for` loop. If $k(\mathbf{x}^{\mu_i})$ denotes the smallest field generated by Laurent monomials lying over $k(f_i)$ then one easily verifies the smallest field generated by Laurent monomials lying over $k(f_1,\ldots,f_s)$ is $k(\mathbf{x}^{\mu_1},\ldots,\mathbf{x}^{\mu_s})$ as the field in question must contain all Laurent monomials $\mathbf{x}^{\mu_1},\ldots,\mathbf{x}^{\mu_s}$. This justifies the `for` loop. We turn to the set Laurent_monomials($\mathbf{x}^{-\mu}p$)∪Laurent_monomials($\mathbf{x}^{-\mu}q$) to show that it generates the smallest field $k(\mathcal{L})$ generated by Laurent monomials lying over $k(f)$. The rational function f has a representation $p(\mathbf{x}^{\mu})/q(\mathbf{x}^{\mu})$ in algebraically independent generators of $k(\mathcal{L})$. Again following Corollary 1 this representation (if cancelled out correctly) varies from the representation in $k(\mathbf{x})$ only by extension with a Laurent monomial.

The set Laurent_monomials($\mathbf{x}^{-\mu}p$)∪Laurent_monomials($\mathbf{x}^{-\mu}q$) in which f can be represented is independent of this Laurent monomial factor and thus is contained in $k(\mathcal{L})$ and generates $k(\mathcal{L})$ since $k(\mathcal{L})$ is the smallest field over $k(f)$ generated by Laurent monomials. □

In order to find the "outer part" $\mathbf{g}$ of a monomial decomposition we have to express the given rational function $\mathbf{f}$ in terms of the Laurent monomials $\mathbf{x}^{\mu_1},\ldots,\mathbf{x}^{\mu_s}$ generating the intermediate field $k(\mathbf{h})$. For this we have a closer look at the connection between Gröbner bases and the triangular form considered in the proof of Lemma 5: Given a matrix over $\mathbb{Z}$ as used in the proof of Lemma 5, i. e., in upper triangular form with rows $\widetilde{\mu}_1,\ldots,\widetilde{\mu}_{\widetilde{s}}$ being exponent vectors of the Laurent monomials $\neq 1$ of

$$\frac{\gamma_1}{\alpha_1}\mathbf{x}^{\widetilde{\mu}_1}-1,\ldots,\frac{\gamma_{\widetilde{s}}}{\alpha_{\widetilde{s}}}\mathbf{x}^{\widetilde{\mu}_{\widetilde{s}}}-1$$

— in particular the ith entry in a row denotes the exponent of x_i — using row operations respecting the $\mathbb{Z}$-module this upper triangular form can be modified to a matrix

$$\begin{pmatrix} d_1 & b_{1_1} & \cdots & b_{1_i} & b_{1_{i+1}} & \cdots & \cdots \\ & & & d_2 & b_{2_1} & \cdots & \cdots \\ & & & & \ddots & \ddots & \vdots \\ & & & & & d_{\widetilde{s}} & \cdots \\ & & & & & & 0 \end{pmatrix}$$

which meets the conditions (1) $d_i > 0$, $i = 1,\ldots,\widetilde{s}$, (2) every entry $b_{i_j} > 0$ of this matrix which has in its column only zeros below it has only entries less or equal to zero above it, and (3) every entry $b_{i_j} < 0$ with only zeros below has only positive or zero entries above. If these conditions are met we denote the set of corresponding (Laurent) binomials by Δ. To make the lowermost nonzero entry in every column positive we can use a change of basis $h : k[Z_1,\ldots,Z_n] \to k[\mathbf{Z},\mathbf{Z}^{-1}]$ which maps the variable Z_i to $1/Z_i$ if the lowermost nonzero entry in the ith column is negative and to Z_i otherwise; for $\boldsymbol{\alpha} \in k^{\times^n}$ we write $h(\boldsymbol{\alpha})$ for $(h(Z_1),\ldots,h(Z_n))(\boldsymbol{\alpha})$. On $k[\mathbf{Z}]$ a mapping $\eta : k[\mathbf{Z}] \to k[\mathbf{Z}]$ with $\eta : p(\mathbf{Z}) \mapsto$ clear_denominators($p(h(\mathbf{Z}))$) is induced by h — note that η in general is not a homomorphism of rings.

Lemma 7 *Let $k(\mathbf{x}^{\mu_1}, \ldots, \mathbf{x}^{\mu_s})$ be a subfield of $k(\mathbf{x})$ generated by Laurent monomials with corresponding ideal J and $I \trianglelefteq k[\mathbf{Z}]$ a binomial ideal with a generating set Δ in triangular form as above. Then*

(i). J is a binomial ideal with $\mathrm{I}(\mathrm{V}(J) \setminus \mathrm{V}(Z_1 \cdot \ldots \cdot Z_n)) = \mathrm{I}(\mathrm{V}(J))$.

(ii). $p \in J \iff \eta(p) \in \eta(J)$.

(iii). $\eta(\Delta) := \{\eta(p) : p \in \Delta\}$ *is a Gröbner basis of* $\eta(I) := \langle \eta(\Delta) \rangle$ *w. r. t. the lexicographic term order* $Z_1 > \ldots > Z_n$.

Proof: (i): The ideal J does not contain a monomial as it is prime and would have to contain single variables then (in contradiction to J vanishing at $\mathbf{x}$). Since J is prime it contains a polynomial p iff it contains a monomial multiple of p. Thus $J = J : (\prod_{i=1}^{n} Z_i)^{\infty}$ which is equivalent to the property stated.
(ii): $p \in J \iff p$ vanishes on $\mathrm{V}(J) \setminus \mathrm{V}(\prod_{i=1}^{n} Z_i) \iff p(h)$ vanishes on $h^{-1}(\mathrm{V}(J) \setminus \mathrm{V}(\prod_{i=1}^{n} Z_i))$. Furthermore $h^{-1}(\mathrm{V}(J) \setminus \mathrm{V}(\prod_{i=1}^{n} Z_i)) = \{\boldsymbol{\alpha} : \boldsymbol{\alpha} \notin \mathrm{V}(\prod_{i=1}^{n} Z_i) \text{ and } h(\boldsymbol{\alpha}) \in \mathrm{V}(J)\} = \{\boldsymbol{\alpha} : \boldsymbol{\alpha} \notin \mathrm{V}(\prod_{i=1}^{n} Z_i) \text{ and } p(h(\boldsymbol{\alpha})) = 0\} = \{\boldsymbol{\alpha} : \boldsymbol{\alpha} \notin \mathrm{V}(\prod_{i=1}^{n} Z_i) \text{ and clear\_denominators}(p(h(\boldsymbol{\alpha}))) = 0\}$ since the denominator vanishes only in $\mathrm{V}(\prod_{i=1}^{n} Z_i)$. Thus $\eta(p)$ vanishes on $\mathrm{V}(\eta(J))$ iff $p(h)$ vanishes on $h^{-1}(\mathrm{V}(J) \setminus \mathrm{V}(\prod_{i=1}^{n} Z_i))$.
(iii): It is easy to see that the head terms of $\eta(\Delta)$ w. r. t. the lexicographic term order $Z_1 > \ldots > Z_n$ are disjoint. So we can conclude that every S-polynomial reduces to zero. Eisenbud and Sturmfels (1996) describe a different change of basis which — following all the operations of the elementary divisor algorithm — results in a Gröbner basis consisting of univariate polynomials which therefore is a Gröbner basis w. r. t. any term order. □

By means of Lemma 7 the representation problem can be solved analogously to Lemma 3. The linear equation $N(\Lambda)$ used there is now derived by changing the variables according to η, reducing modulo a Gröbner basis, and then changing back to the original variables.

References

Aagedal, H., Beth, Th., Müller-Quade, J., Schmid, M. (1996): Algorithmic design of diffractive optical systems for information processing. In: Toffoli, T., Biafore M., Leão J. (ed.): Proceedings of the Fourth Workshop on Physics and Computation PhysComp96, Boston, pp. 1–6.

Becker, Th., Weispfenning, V., Kredel, H. (1993): Gröbner bases: a computational approach to commutative algebra. Graduate texts in mathematics. Springer, New York.

Buchberger, B. (1965): Ein Algorithmus zum Auffinden der Basiselemente des Restklassenringes nach einem nulldimensionalen Polynomideal (An algorithm for finding a basis for the residue class ring of a zero-dimensional polynomial ideal). Dissertation Math. Inst. Universität Innsbruck.

Buchberger, B. (1985): Gröbner bases: An algorithmic method in polynomial ideal theory. In: Bose, N. K. (ed.): Multidimensional systems theory. D. Reidel, Dordrecht, pp. 184–232.

Dickerson, M. T. (1989): The functional decomposition of polynomials. Tech. Rep. CORNELLCS//TR89-1023, Computer Science Dep., Cornell University, Ithaca.

Eisenbud, D. (1995): Commutative algebra with a view toward algebraic geometry. Graduate texts in mathematics. Springer, New York.

Eisenbud, D., Sturmfels, B. (1996): Binomial ideals. Duke Math. J. 84(1): 1–45.

Kemper, G. (1993): An algorithm to determine properties of field extensions lying over a ground field. IWR-Preprint 93-58, Heidelberg.

Kozen, D., Landau, S., Zippel, R. (1996): Decomposition of algebraic functions. J. Symb. Comp. 22(3): 235–246.

Lazard, D., Valibouze, A. (1993): Computing subfields: reverse of the primitive element problem. In: Eyssette, F., Galligo, A. (ed.): Progress in mathematics. Birkhäuser, Boston, pp. 163–176 (Progress in mathematics, vol. 109).

Müller-Quade, J., Steinwandt, R. (1997): Basic algorithms for rational function fields. E.I.S.S.-Report, E.I.S.S, Universität Karlsruhe.

Schafarewitsch, I. R. (1972): Grundzüge der algebraischen Geometrie. Friedr. Vieweg + Sohn, Braunschweig.

Sweedler, M. (1993): Using Groebner bases to determine the algebraic and transcendental nature of field extensions: return of the killer tag variables. In: Cohen, G., Mora, T., Moreno, O. (ed.): Applied Algebra, Algebraic Algorithms and Error-Correcting Codes 10th International Symposium, AAECC-10, San Juan de Puerto Rico. Springer, Berlin Heidelberg, pp. 66–75 (Lecture notes in computer science, vol. 673).

Viète, F. (1594): Problema, quod omnibus mathematicis totius orbis construendum proposuit Adrianus Romanus. Reprinted in (Viète 1970).

Viète, F. (1970): Opera mathematica. Georg Olms, Hildesheim New York.

Zippel, R. (1991): Rational function decomposition. In: Watt, S. (ed.): Proceedings of the 1991 International Symposium on Symbolic and Algebraic Computation, ISSAC '91, Bonn. ACM Press, Baltimore, pp. 1–6.

On some Basic Applications of Gröbner Bases in Non-commutative Polynomial Rings

Patrik Nordbeck

Lund Institute of Technology, Box 118, S-221 00 Lund, Sweden;
nordbeck@maths.lth.se

Abstract

In this paper we generalize some basic applications of Gröbner bases in commutative polynomial rings to the non-commutative case. We define a non-commutative elimination order. Methods of finding the intersection of two ideals are given. If both the ideals are monomial we deduce a finitely written basis for their intersection. We find the kernel of a homomorphism, and decide membership of the image. Finally we show how to obtain a Gröbner basis for an ideal by considering a related homogeneous ideal.

The method of Gröbner bases, introduced by Bruno Buchberger in his thesis (1965), have become a powerful tool for constructive problems in polynomial ideal theory and related domains. Generalizations of the basic ideas to the non-commutative setting was done, as an theoretical instrument, by Bokut (1976) and Bergman (1978). From the constructive point of view, the non-commutative version of *Buchberger's algorithm* was presented by Mora (1986). For some special classes of non-commutative rings, Gröbner bases has been studied in more detail, e.g. solvable algebras by Kandri-Rody and Weispfenning (1990).

As the title indicates, we will here consider Gröbner bases in non-commutative polynomial rings, i.e. free associative algebras (over some field). Most of the results are just easy generalizations of the theory of Gröbner basis in commutative polynomial rings, which can be found e.g. in the textbook by Adams and Loustaunau (1994), or in the original paper by Buchberger (1985).

In section 2 we define an order allowing us to use elimination techniques. Section 3 concerns intersection of ideals. We find a Gröbner basis for the intersection of two ideals. In the special case when both the ideals are monomial, we show how automata can be used to obtain a finitely written basis. The material in section 4, where we find a Gröbner basis for the kernel of a homomorphism, and decide membership of the image, is nearly identical with the commutative case as presented in the textbook mentioned above.

As an application we mention the study of subalgebras. Since homogeneous ideals are more comfortable to work with, we will in the last section show how to obtain a Gröbner basis for an ideal by considering a related ideal with homogeneous generators.

All of the Gröbner basis calculations in the examples are performed with the program GROEBNER by Feustel and Green (1992).

The author expresses his thanks to V.A. Ufnarovski for helpful discussions.

1 Basic definitions and notations

Let $X = \{x_1, x_2, \ldots, x_n\}$ be a finite alphabet, and let $K\langle X\rangle$ denote the free associative algebra over the arbitrary field K. Denote by S the set of all words in X, including the empty word **1**. We will assume that S is given an *admissible order*, i.e. an well-order preserving multiplication: $f < g$ implies $hfk < hgk$ for all $f, g, h, k \in S$, such that the smallest word always is the unity **1**. As an example we mention the following order called *deglex* (degree lexicographical): If $|u|$ denotes the length of $u \in S$, we let $u > v$ if either $|u| > |v|$ or $|u| = |v|$ but u is larger than v lexicographically.

When we have chosen an admissible order we can with every non-zero element $f \in K\langle X\rangle$ associate its leading word $\hat{f} \in S$.

If $u, v \in S$, and u is a (not necessarily proper) subword of v, we write $u \mid v$. A word $u \in S$ is called *normal* modulo an ideal $I \subset K\langle X\rangle$ (always two-sided) if for every $f \in I$, $\hat{f} \nmid u$.

If N denotes the linear hull of the set of normal words (mod I), we have $K\langle X\rangle = N \oplus I$ as direct sum of vector spaces (see Ufnarovski 1995).

For every $f \in K\langle X\rangle$, its image by the projection $K\langle X\rangle \longrightarrow N$ will be called its *normal form*, and be denoted $\bar{f}$. Clearly $\bar{f} = 0$ if and only if $f \in I$.

Definition 1 *A subset G of the ideal I is called a* Gröbner basis *if for every $f \in I$ there is $g \in G$ such that $\hat{g} \mid \hat{f}$.*

A Gröbner basis is *minimal* if no proper subset of it is a Gröbner basis. If, moreover, every element $g \in G$ has the form $\hat{g} - \bar{\hat{g}}$, then the basis is called *reduced.* When an admissible order is selected, every ideal has an unique reduced Gröbner basis.

For the construction of Gröbner bases we refer to (Mora 1986). We point out however that we now, compared to the commutative case, have a great disadvantage caused by non-noetherianity: Even a finitely generated ideal may have no finite Gröbner basis.

However, if we are given a Gröbner basis (possibly infinite) with its leading words sorted by length, then it is not hard to see that the process of *reduction* $f \to \bar{f}$ described above can be performed in a finite number of steps. This rests on the fact that an admissible order on S by definition is well-founded, i.e. every infinite sequence $u_1 \geq u_2 \geq \ldots u_i \geq \ldots$ in S stabilizes.

The standard method of constructing a Gröbner basis, using deglex, will for a homogeneous ideal (see section 5) automatically yield a basis sorted as above. Thus the reduction is algorithmic in this case.

2 Elimination

For $u \in S$, let $\deg_{x_i} u$ denote the number of different occurrences of x_i in u. We will frequently use the following admissible order suggested by Edward Green[1]:

First use the commutative lexicographic order, i.e. if $x_1 < x_2 < \ldots < x_n$ we let $u > v$ $(u, v \in S)$ if $\deg_{x_i} u > \deg_{x_i} v$, where i is the highest number such that $\deg_{x_i} u \neq \deg_{x_i} v$. If $\deg_{x_i} u = \deg_{x_i} v$ for all i we use deglex.

We thus have for example $x_3 > x_1 x_2^2$, and more generally, if a word $u \in S$ does not contain any $x_{m+1}, \ldots, x_n$, $m < n$, and if $u > v \in S$ in the order above, then v does not contain any $x_{m+1}, \ldots, x_n$ either. We will call this order the *elimination order* as motivated by the following theorem.

Theorem 1 *Let G be a Gröbner basis for the ideal I according to the elimination order with $x_1 < x_2 < \ldots < x_n$. Then, for $m \leq n$, $G_m = G \cap K\langle x_1, x_2, \ldots, x_m\rangle$ is a Gröbner basis for the ideal $I_m = I \cap K\langle x_1, x_2, \ldots, x_m\rangle$ in $K\langle x_1, x_2, \ldots, x_m\rangle$.*

Proof. If $f \in I_m \subset I$ there is $g \in G$ such that $\hat{g}|\hat{f}$. Then $\hat{g}$ involves only $x_1, \ldots, x_m$, so by the remark above every word in g involves only $x_1, \ldots, x_m$. Thus there is, for every $f \in I_m$, an element $g \in G_m$ such that $\hat{g}|\hat{f}$. □

More generally, we could in our definition of the elimination order above first used any commutative elimination order (see e.g. Adams and Loustaunau 1994), and then any (non-commutative) admissible order to break ties.

3 Intersection

For two ideals $I = (f_1, \ldots f_k), J = (g_1, \ldots, g_l) \in K\langle X\rangle$, consider the ideal

$$H = (tf_i, (1-t)g_j, tx_m - x_m t | 1 \leq i \leq k, 1 \leq j \leq l, 1 \leq m \leq n) \in K\langle X, t\rangle. \tag{3.1}$$

Theorem 2 $I \cap J = H \cap K\langle X\rangle$.

[1] Edward L. Green, Virginia Polytechnic Institute and State University, Blacksburg, Virginia USA.

Proof. If $F \in I \cap J$, then $F = \sum_{i=1}^{k} p_{i_L} f_i p_{i_R} = \sum_{j=1}^{l} q_{j_L} g_j q_{j_R}$ for some polynomials $p_{i_L}, p_{i_R}, q_{j_L}, q_{j_R} \in K\langle X\rangle$. In $K\langle X, t\rangle$ we can write

$$F = tF + (1-t)F = \sum_{i=1}^{k} p_{i_L} t f_i p_{i_R} + \sum_{j=1}^{l} q_{j_L}(1-t) g_j q_{j_R} + \sum_{m=1}^{n} r_{m_L}(tx_m - x_m t) r_{m_R}$$

with $r_{m_L}, r_{m_R} \in K\langle X, t\rangle$. Here the last sum on the right hand side appears when moving t (and $1-t$) to the generators f_i (g_j). We conclude that $F \in H \cap K\langle X\rangle$.

Conversely, assume $F \in H \cap K\langle X\rangle$. $F \in H$ gives

$$F = \sum_{i=1}^{k} p_{i_L} t f_i p_{i_R} + \sum_{j=1}^{l} q_{j_L}(1-t) g_j q_{j_R} + \sum_{m=1}^{n} r_{m_L}(tx_m - x_m t) r_{m_R},$$

where $p_{i_L}, p_{i_R}, q_{j_L}, q_{j_R}, r_{m_L}, r_{m_R} \in K\langle X, t\rangle$. Since also $F \in K\langle X\rangle$, F is independent of the value of t. Substituting 1 for t we get $F = \sum p'_{i_L} f_i p'_{i_R} \in I$, where e.g. $p'_{i_L}(X) = p_{i_L}(X, 1) \in K\langle X\rangle$. Similarly $t = 0$ gives $F \in J$, so $F \in I \cap J$ as required. □

As an immediate consequence we get, using Theorem 1:

Corollary 1 *Let G be a Gröbner basis for H according to the elimination order in $K\langle X, t\rangle$ with $t > X$. Then $G \cap K\langle X\rangle$ is a Gröbner basis for $I \cap J$.*

Even if I and J are given with a finite number of generators, we can not hope to get a finite set of generators for the intersection. As an easy example consider the ideals $I = (x^2, xy^2)$, $J = (xyx) \in K\langle x, y\rangle$. Beginning to calculate a Gröbner basis for $H = (tx^2, txy^2, (1-t)xyx, tx - xt, ty - yt)$ according to deglex with $t > x > y$, we get, considering words up to length 8 and omitting elements containing t, the set $\{xyx^2, x^2yx, xyxy^2, xy^2xyx, xy^3xyx, xy^4xyx\}$. We might guess, despite we are using just deglex and not the order in the Corollary, that

$$\{xyx^2, x^2yx, xyxy^2, xy^ixyx | i = 2, 3, \ldots\} \tag{3.2}$$

is a Gröbner basis for $I \cap J$. The problem of predicting an infinite basis, knowing a finite subset, is discussed at some extent in (Ufnarovski 1994). We will later in this section see that (3.2) indeed is a basis for $I \cap J$.

We will call an ideal *monomial* if it is generated by finite set of words. It is clear from the definition that any set of generators consisting of words is a Gröbner basis. Obviously a word is in a monomial ideal if and only if it contains one of its generators as a subword. Thus an element is in the intersection of two monomial ideals if and only if it has one generator from each ideal as subwords.

If (as in the example) both I and J are monomial, we can find a finite representation (such as (3.2)) for their intersection using automata. It is easy to see that an element $f \in K\langle X\rangle$ is in a monomial ideal I if and only if all of the words in f are in I. Since the same must be true for the intersection of two monomial ideals, it is sufficient to find all words in $I \cap J$.

We begin with constructing, for each ideal, its *ideal automaton.* This is a finite state automaton[2] recognizing the words in the ideal. The construction is similar to the one for the *automata of normal words*, which can be found in (Ufnarovski 1994). An automaton recognizing the words in the intersection

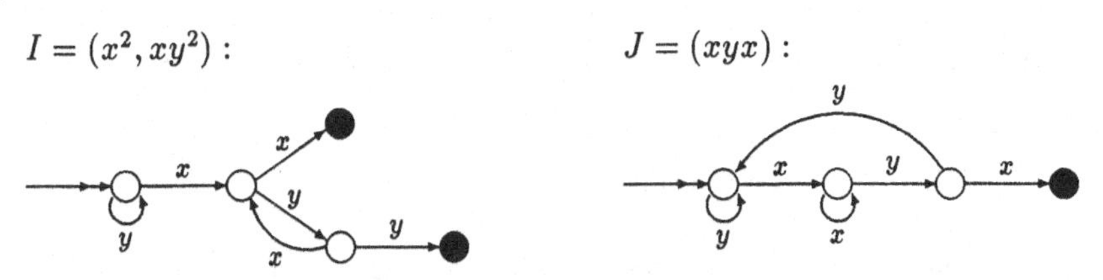

Figure 1: *Ideal automata for* $I = (x^2, xy^2)$ *and* $J = (xyx)$. *Paths not leading to accepting states corresponds to the normal words.*

of I and J can now be obtained by simultaneously tracing the two ideal automata (In the language of automata this is called the *direct product* of the automata). A word will then be in the intersection if and only if both of the automata are in accepting states.

To pick out a set of generators for $I \cap J$ we can proceed as follows: For each generator of I, find its state in the ideal automaton of J. Beginning with these states we now traverse the automaton of J, i.e. we look for all possible paths ending with an accepting state. We here have to detect the loops (cycles) in the automaton to get a finite procedure. If we use the same method again, now starting with all the generators of J and traversing the automaton of I, we have found all possible words beginning with a generator from one of the ideals, and ending with a generator from the other ideal (Here the generators may of course intersect). Since all words in $I \cap J$ must contain one such word as a subword, we have found a set of generators.

An implementation of the ideas above gives as output, when given $I = (x^2, xy^2)$ and $J = (xyx)$ as in our example, the following result:

i) Starting with xx : $xxx^*y(yy^*xx^*y)^*x$
ii) Starting with xyy : $xyyy^*xx^*y(yy^*xx^*y)^*x$
iii) Starting with xyx : $xyx(yx)^*x$ and
iiii) $xyxy(xy)^*y$

[2]For the theory of automata, see e.g. (Howie 1991).

Here * denotes the usual star operation, $a^* = \{\mathbf{1}, a, a^2, \ldots\}$, and corresponds to the loops the program has detected. The reader can check the result using the automata in Figure 1.

It is easy to see that every element in (3.2) can be obtained by choosing suitable loops in some of the *i*) – *iiii*) above. Moreover, however *iii*) is evaluated, i.e. no matter how many times we run the loop, it will always contain $xyxx$ from (3.2) as a subword. In the same way, *iiii*) will always contain $xyxyy$, and (although more difficult to see) *i*) and *ii*) will contain either $xxyx$ or some xy^ixyy (depending on how the loops are chosen). We conclude that (3.2) and the (infinite) set of elements in *i*) – *iiii*) generates the same ideal, so our guess above, that (3.2) was a Gröbner basis for $I \cap J$, was correct.

4 Homomorphisms

Let $Y = \{y_1, \ldots, y_m\}$ be another alphabet, and let $\varphi : K\langle Y\rangle \longrightarrow K\langle X\rangle$ be the K-algebra homomorphism defined by $\varphi(y_i) = h_i(X)$, $1 \le i \le m$. Consider the ideal

$$H = (y_i - h_i | 1 \le i \le m) \in K\langle X, Y\rangle.$$

We will, for convenience, in the following proofs write φ also for its natural extension $\tilde{\varphi} : K\langle X, Y\rangle \longrightarrow K\langle X\rangle$.

Theorem 3 $\ker(\varphi) = H \cap K\langle Y\rangle$.

Proof. If $f \in H \cap K\langle Y\rangle$, then $f(y_1, \ldots, y_m) = \sum_{i=1}^m p_i(y_i - h_i)q_i$ where $p_i, q_i \in K\langle X, Y\rangle$. It follows that $\varphi(f) = f(h_1, \ldots, h_m) = 0$.

Conversely, assume $f \in \ker(\varphi)$. Using that $py_iq = ph_iq + p(y_i - h_i)q$ for all $p, q \in K\langle X, Y\rangle$, we can successively replace all occurrences of the y_i in f by h_i and obtain

$$f(y_1, \ldots, y_m) = f(h_1, \ldots, h_m) + \sum_{i=1}^m p_i(y_i - h_i)q_i, \tag{4.1}$$

for some $p_i, q_i \in K\langle X, Y\rangle$. Since $f(h_1, \ldots, h_m) = \varphi(f) = 0$ we conclude that $f \in H$. □

As in the previous section we now use Theorem 1:

Corollary 2 *Let G be a Gröbner basis for H according to the elimination order in $K\langle X, Y\rangle$ with $X > Y$. Then $G \cap K\langle Y\rangle$ is a Gröbner basis for* $\ker(\varphi)$.

As an example consider the homomorphism $\varphi : K\langle y_1, y_2, y_3\rangle \longrightarrow K\langle x_1, x_2\rangle$ defined by $\varphi(y_1) = x_1 + x_2$, $\varphi(y_2) = x_1^2 + x_1x_2$ and $\varphi(y_3) = x_2^2 + x_2x_1$. Calculating the reduced Gröbner basis for

$$H = (x_1 + x_2 - y_1, x_1^2 + x_1x_2 - y_2, x_2^2 + x_2x_1 - y_3)$$

according to the elimination order ($x_2 > x_1 > y_3 > y_2 > y_1$) we obtain

$$G = \{x_1y_1 - y_2, y_3 + y_2 - y_1^2, x_2 + x_1 - y_1\}. \tag{4.2}$$

The Corollary now shows that $\ker(\varphi)$ is generated by $y_3 + y_2 - y_1^2$.

We will in the remaining of this section use, as in the Corollary above, the elimination order (in $K\langle X, Y\rangle$) with $X > Y$, so all Gröbner bases and reductions (normal forms) will be with respect to this order.

Theorem 4 *An element $f \in K\langle X\rangle$ is in $\operatorname{im}(\varphi)$ if and only if $\bar{f} \in K\langle Y\rangle$, where $\bar{f}$ is the normal form of f modulo the ideal H considered above.*

Proof. If $f \in \operatorname{im}(\varphi)$, then $f(x_1,\dots,x_n) = g(h_1,\dots,h_m)$ for some $g \in K\langle Y\rangle$. Using (4.1) with f replaced by g we see that $f(x_1,\dots,x_n) = g(y_1,\dots,y_m) + h, h \in H$. Since h reduces to zero, it is clear that the normal form of $f(x_1,\dots,x_n)$ is equal to the normal form of $g(y_1,\dots,y_m)$. But since $Y < X$, the reduction of an element in $K\langle Y\rangle$ yields another element in $K\langle Y\rangle$. Thus $\bar{f} = \bar{g} \in K\langle Y\rangle$.

Conversely, if $\bar{f} \in K\langle Y\rangle$, then, since $f - \bar{f} \in H$,

$$f(x_1,\dots,x_n) = \bar{f}(y_1,\dots,y_m) + \sum_{i=1}^{m} p_i(y_i - h_i)q_i,$$

$p_i, q_i \in K\langle X, Y\rangle$. Substituting h_i for y_i we get $f = \bar{f}(h_1,\dots,h_m) = \varphi(\bar{f})$. □

From the proof we also see that if $f \in \operatorname{im}(\varphi)$, then $f = \varphi(\bar{f})$.

For example, if we normalize $f = x_1^3 + x_1^2x_2 - x_2^3 - x_2^2x_1$ with respect to the basis in (4.2) we get $\bar{f} = y_1y_2 + y_2y_1 - y_1^3$, so the Theorem shows that f is in the image of the homomorphism defined in our example above. The reader can check that $f = \varphi(\bar{f})$.

Corollary 3 *With the notations above, let G be the reduced Gröbner basis for H. Then φ is an epimorphism if and only if there exists $f_i = x_i - g_i \in G$, $1 \le i \le n$, with $g_i \in K\langle Y\rangle$.*

Proof. Clearly φ is an epimorphism if and only if $x_i \in \operatorname{im}(\varphi)$ for all i.

Assume $x_i \in \operatorname{im}(\varphi)$. Since $f_i = x_i - \overline{x_i} \in H$ (by the theorem this is not zero), there is a element in $p \in G$ with $\hat{p} | \widehat{f_i} = x_i$, i.e. $\hat{p} = x_i$. Since G is the reduced basis we have $p = x_i - \overline{x_i}$, and $\overline{x_i} \in K\langle Y\rangle$ by the theorem.

Conversely, if $x_i - g_i \in G$, $g_i \in K\langle Y\rangle$, then, again since G is reduced, $\overline{x_i} = g_i \in K\langle Y\rangle$, so by the theorem we have $x_i \in \operatorname{im}(\varphi)$. □

We now see from (4.2) that the homomorphism in the example above is not an epimorphism.

All of the results in this section can be generalized to a homomorphism $\varphi : K\langle Y\rangle/J \longrightarrow K\langle X\rangle/I$. If φ is defined as above, and if $I = (f_1,\dots,f_s)$, we then have to work with the ideal $H = (y_i - h_i, f_j | 1 \leq i \leq m, 1 \leq j \leq s)$, and consider all the elements in $K\langle Y\rangle$ modulo J.

To conclude this section, we note that the results above can be used to study subalgebras of $K\langle X\rangle$ (or $K\langle X\rangle/I$). Because if A is the subalgebra of $K\langle X\rangle$ generated by $f_1, \ldots, f_l$, then A can be considered as the image of the homomorphism $\varphi : K\langle y_1, \ldots, y_l\rangle \longrightarrow K\langle X\rangle$ defined by $\varphi(y_i) = f_i$, $1 \leq i \leq l$.

5 Homogenization

Most computer programs devoted to non-commutative Gröbner bases works only with homogeneous ideals, i.e. ideals given with a set of homogeneous generators. It is not difficult to see that the elements of the reduced Gröbner basis then also will be homogeneous.

We will here show how to use homogenization techniques to obtain a Gröbner basis for an arbitrary ideal by considering its homogenized counterpart.

The approach below is now implemented in the program ANICK by Podoplelov and Ufnarovski (1997).

For $f \in K\langle X\rangle$, we let $f^* \in K\langle X, t\rangle$ denote the following homogenization of f: If $f = f_{(d)} + f_{(d-1)} + \ldots + f_{(0)}$, where $f_{(j)}$ is the homogeneous component of degree j, $f_{(d)} \neq 0$, then $f^* = f_{(d)} + t f_{(d-1)} + \ldots + t^d f_{(0)}$. Also, for $g \in K\langle X, t\rangle$, let $g_* = g(X, 1) \in K\langle X\rangle$. We note that $(f^*)_* = f$. Moreover, if g is homogeneous, and if all words in g has all occurrences of t at the beginning, i.e. is of form $t^k w$ for some $k \in \mathbb{N}$, $w \in K\langle X\rangle$, then $g = t^m (g_*)^*$ for some $m \in \mathbb{N}$ (the highest power of t dividing g).

For an ideal $I = (f_1, \ldots, f_s) \subset K\langle X\rangle$, let us consider the ideal

$$J = (f_1^*, \ldots, f_s^*, x_i t - t x_i \mid 1 \leq i \leq n) \subset K\langle X, t\rangle, \tag{5.1}$$

which clearly is homogeneous.

Theorem 5 *Let G be the reduced Gröbner basis for J according to deglex with $X > t$. Then $G_* = \{g_* \mid g \in G\}$ is a Gröbner basis for I (also w.r.t. deglex).*

For the proof we need a Lemma:

Lemma 1 *Let w_i and w_j be words in $K\langle X\rangle$, and let $m_i, m_j \in \mathbb{N}$ be such that $|t^{m_i} w_i| = |t^{m_j} w_j|$. Then $w_i > w_j$ w.r.t. deglex in $K\langle X\rangle$ if and only if $t^{m_i} w_i > t^{m_j} w_j$ w.r.t. deglex ($X > t$) in $K\langle X, t\rangle$.*

Proof. First assume $|w_i| = |w_j|$. Then $m_i = m_j$, so $w_i > w_j$ lexicographically is equivalent to $t^{m_i} w_i > t^{m_j} w_j$ lexicographically.

If $|w_i| \neq |w_j|$, then $|w_i| > |w_j|$ if and only if $m_i < m_j$, and this is the case if and only if $t^{m_i}w_i$ is larger than $t^{m_j}w_j$ lexicographically. $\square$

Proof. (of Theorem) Since $(f_i^*)_* = f_i$ and $(x_it - tx_i)_* = 0$ we have $G_* \subset I$, so it suffices to find, for arbitrary $f \in I$, an element $g \in G_*$ with $\hat{g} \mid \hat{f}$.

Let $f = \sum_{i=1}^{s} h_{i_L} f_i h_{i_R} \in I$. Since $h_{i_L}^* f_i^* h_{i_R}^*$ is homogeneous for all i, we can find $n_i \in \mathbb{N}$, $1 \leq i \leq s$, such that $F = \sum_{i=1}^{s} t^{n_i} h_{i_L}^* f_i^* h_{i_R}^* \in J$ is homogeneous.

After expanding F as a sum of words, we let F' denote the element obtained from F by replacing every word by the corresponding word having all occurrences of t (if any) at the beginning, e.g. $F = x_1tx_2t + x_1tx_2x_1 + x_2x_1$ gives $F' = t^2x_1x_2 + tx_1x_2x_1 + x_2x_1$. Since $x_it - tx_i \in J$ for all i, we also have $F' \in J$. Writing $f = \sum_{l=1}^{k} c_l w_l$, where the w_l are words in $K\langle X\rangle$, $w_i \neq w_j$ if $i \neq j$, and the $c_l \in K$, we have $F' = \sum_{l=1}^{k} c_l t^{m_l} w_l$ for some $m_l \in \mathbb{N}$. Because if some words cancels in f, then, since F' is homogeneous, the corresponding words also cancels in F'. Moreover, from the Lemma it follows that if $\hat{f} = w_i$, then $\widehat{F'} = t^{m_i}w_i$, i.e. the leading word of f corresponds to the leading word of F'.

There is an element $g \in G$ such that $\hat{g}|\widehat{F'}$. We claim that $\widehat{g_*}|\hat{f}$, which will prove the theorem.

As member of the reduced basis, g is homogeneous and of form $\hat{g} - \bar{\hat{g}}$. Clearly $\hat{g}$ has all occurrences of t at the beginning ($\hat{g}|\widehat{F'} = t^{m_i}w_i$). The same is true for the words in $\bar{\hat{g}}$, because they are normal modulo J and thus not divisible by any x_it (the leading word of $x_it - tx_i \in J$). It follows that $g_* \neq 0$ (since $g_* = 0$ implies $g = t^m(g_*)^* = 0$), and by the Lemma we see that $\widehat{g_*} = (\hat{g})_*$. Since clearly $\hat{f} = (\widehat{F'})_*$, it is now evident that $\hat{g}|\widehat{F'}$ implies $\widehat{g_*}|\hat{f}$. $\square$

The result above can also be proved for other admissible orders, the essential is that t is less than all the x_i. The proof of Theorem 5 need just be modified slightly if we replace deglex with e.g. the elimination order. We can thus find e.g. the intersection of the ideals $I = (f_1, \ldots f_k)$ and $J = (g_1, \ldots, g_l)$, calculating the reduced Gröbner basis for

$$H' = (sf_i, (t-s)g_j, sx_m - x_ms, st - ts, x_mt - tx_m | 1 \leq i \leq k, 1 \leq j \leq l, 1 \leq m \leq n)$$

(compare (3.1) and (5.1)) according to the elimination order with $s > X > t$. Combining Corollary 1 and Theorem 5 we see that it is sufficient to pick out the elements in H'_* (dehomogenized w.r.t. t) not containing s.

The basis G_* will not in general be minimal. As an example consider the ideal $I = (x^2 - yx, xy - y) \in K\langle x, y\rangle$. Calculating the reduced Gröbner basis for $J = (x^2 - yx, xy - ty, xt - tx, yt - ty)$ according to deglex with $x > y > t$, we get $G = \{y^2x - tyx, ty^2 - t^2y, x^2 - yx, xy - ty, xt - tx, yt - ty\}$. By the theorem $G_* = \{y^2x - yx, y^2 - y, x^2 - yx, xy - y\}$ is a Gröbner basis for I, and here the first element is clearly superfluous. The last three elements constitutes the reduced Gröbner basis.

Moreover, G_* can be infinite even if the ideal admits a finite Gröbner basis. An example is the ideal $I = (xy - z, yz - x, zx - y) \in K\langle x, y, z\rangle$ (deglex with $x > y > z$).

References

Adams, W.W., Loustaunau, P. (1994): *An Introduction to Gröbner Bases*, American Mathematical Society.

Bergman, G. (1978): *The Diamond Lemma for Ring Theory*, Advances of Mathematics 29, 178-218.

Bokut, L.A. (1976): *Embeddings in Simple Associative Algebras*, Algebra Logika 15, No. 2, 117-142. English translation: Algebra Logic 15, 73-90 (1977).

Buchberger, B. (1965): *On finding a Vector Space Basis of the Residue Class Ring Modulo a Zero Dimensional Polynomial Ideal* (German), PhD Thesis, Univ. of Innsbruck, Austria.

Buchberger, B. (1985): *Gröbner Bases: An Algorithmic Method in Polynomial Ideal Theory*, Chapter 6 in: Multidimensional Systems Theory (N.K. Bose ed.), 184-232, Reidel Publishing Company, Dodrecht.

Feustel, C.D., Green, E.L. (1992): GROEBNER, C code, available by anonymous ftp from math.vt.edu in the directory: /pub/green.

Howie, J.M. (1991): *Automata and Languages*, Oxford Science Publications.

Kandri-Rody, A., Weispfenning, V. (1990): *Non-commutative Gröbner Bases in Algebras of Solvable Type*, Journal of Symbolic Computation 9, 1-26.

Mora, F. (1986): *Gröbner Bases for Non-commutative Polynomial Rings*, Proc. AAECC-3, Lecture Notes in Computer Science, Vol. 229, Springer, 353-362.

Podoplelov, A., Ufnarovski, V.A. (1997): ANICK, C code, available by anonymous ftp from ftp.riscom.net in the directory: /pub/anick.

Ufnarovski , V.A. (1994): *Calculations of Growth and Hilbert Series by Computer*, Lecture Notes in Pure and Applied Mathematics, Vol. 151, Decker, 247-256.

Ufnarovski, V.A. (1995): *Combinatorial and Asymptotic Methods of Algebra* in "Algebra-VI" (A.I. Kostrikin and I.R.Shafarevich, Eds), Encyclopedia of Mathematical Sciences, Vol. 57, Springer, 5-196.

Full Factorial Designs and Distracted Fractions

Lorenzo Robbiano, Maria Piera Rogantin [1]

Abstract

Design of Experiments is an important branch of Statistics. One of its key problems is to find minimal Fractions of a Full Factorial Design, which identify a Complete Polynomial Model. This paper shows how to use Computer Algebra and Commutative Algebra techniques and results to produce good classes of solutions to the problem. It is known that most of them can be obtained by means of Gröbner bases, hence they generally depend on the term-order chosen; here we show how to use the Distracted Fractions to yield solutions independent of the term-order.

1 Introduction

Design of Experiments (DoE) is a branch of Statistics, which has a long tradition in the use of algebraic methods (see for example Box *et al.* 1978 and Collombier 1996). In general all these methods were developed in the case of binary experiments, with coding levels either $\{0, 1\}$ or $\{-1, 1\}$ and some generalizations to the non-binary case were also developed (see Collombier 1996).

More recently some connections were discovered between classical problems in Statistics and the methods of Computer Algebra. For instance in their recent work Pistone and Wynn (1996) address the problem of identifying polynomial models in general designs. In particular they point out the connection between DoE and Gröbner bases.

In this volume there is a survey paper (see Robbiano 1998), which gives a full account of the subject.

Let us now specify the statistical problem; the key goal of DoE is to study an input/output relation $y = f(X_1, \ldots, X_n)$ by fitting a suitable model with a finite set of experimental points. Namely, a *design* is a finite set of

[1] 1991 *Mathematics Subject Classification.* Primary 13P10; Secondary 62K15.
Key words and Phrases: Full Factorial Designs, Distractions, Gröbner Bases.
Partially supported by the Consiglio Nazionale delle Ricerche (CNR).

input values, called *treatments*, which can be described by their coordinates in $\mathbb{Z}$, hence they can be viewed as elements of $\mathbb{Z}^n$, and hence of $\mathbb{Q}^n$. In this frame they are called *experimental points* or simply *points*. A *linear model* is a vector space of functions; in particular we may have polynomial models, trigonometric models, and so on. Given a model, two functions in the model are *confounded* by a given design if both take the same values at each point of the design. A model is *identifiable* with the design if there is no confounding of different functions in the model. So the choice of the model will depend on the physical meaning of the functions involved and on the identifiability of the candidate physical models with respect to the available designs. A *full factorial design* is a design in $\mathbb{Q}^n$, which is a product of n sets of elements in $\mathbb{Q}$, called *factors*.

A *Fraction* of a full factorial design is simply a subset of it. The theory of *Gröbner bases* yields a solution to the problem of finding a *monomial basis* of the vector space of the polynomial functions on the fraction. This problem is known in Statistics as the problem of finding a full set of *estimable terms*, and it was studied by Pistone and Wynn (1996).

In a recent paper an even more important problem is studied by Caboara and Robbiano (1997). More specifically, given a full design $\mathcal{D}$, a polynomial $f(X_1, \ldots, X_n)$ and the corresponding polynomial model $y = f(X_1, \ldots, X_n)$, we want to find minimal fractions $\mathcal{F}$ of $\mathcal{D}$, such that the power products in the support of $f(X_1, \ldots, X_n)$ are not confounded by $\mathcal{F}$. As we said, Caboara and Robbiano (1997) give a full solution to a very important subproblem; namely, classes of solutions are obtained, which are related to Gröbner bases, and hence depend on the term-order chosen. But there is a drawback, since as soon as the design grows, the set of all the solutions becomes huge and practically infeasible to compute. On the other hand there is a subclass which is *easy* to compute, the subclass of the *distracted fractions*, which belongs to the wider class of solutions *independent of the term-order* chosen. This is the subject of this paper. More precisely in Section 2 we recall some basic notation and some fundamental definitions, while in Section 3 we present the main results. In particular Theorem 3.4 shows a solution to the problem by means of distracted fractions, whose number is then determined exactly (see Theorem 3.10). Finally we characterize the set $\mathrm{Sol}(\mathcal{O}, \forall\sigma)$ of the solutions which are independent of the term-order, and then we produce an example (see Example 6) which shows that these solutions are far more numerous than the distracted fractions. This fact leaves open the problem of computing the cardinality of $\mathrm{Sol}(\mathcal{O}, \forall\sigma)$.

We are pleased to thank Caboara, Pistone, Riccomagno and Wynn for some helpful conversations on the subject.

The computation of the examples was carried on with the software CoCoA (see Capani *et al.* 1996). In particular we used a CoCoA-package written by Caboara and named `statfamilies`.

2 Notation and basic definitions

Definition 2.1 *A* **full** *(or* **full factorial***)* **design** $\mathcal{D}$ *with levels* $(l_1, l_2, \ldots, l_n)$ *is the set of points* $\{0, 1, \ldots, l_1 - 1\} \times \ldots \times \{0, 1, \ldots, l_n - 1\} \subset \mathbb{N}^n$. *An* l^n*-design is a full design with* $l_1 = l_2 = \cdots = l_n = l$.

For instance it is common to speak of 2^5, 3^4 full designs. Sometimes other sets of coordinates are used, for instance the 2^2 full design with two factors at levels $-1, 1$ is the set $\{(-1,-1),\ (-1,1),\ (1,-1),\ (1,1)\}$.

Definition 2.2 *Given a full design* $\mathcal{D}$ *with levels* $(l_1, l_2, \ldots, l_n)$, *we denote by* $I(\mathcal{D})$ *its defining ideal in* $\mathbb{Q}[X_1, \ldots, X_n]$, *i.e. the ideal generated by the* n *polynomials* $f_j := X_j\ (X_j - 1) \cdots (X_j - l_j + 1)$, $j := 1, \ldots, n$.

Definition 2.3 *A* **fractional factorial design** *is a subset* $\mathcal{F}$ *of a full design* $\mathcal{D}$. *It is also called a* **fraction** *of* $\mathcal{D}$. *Its defining ideal* $I(\mathcal{F})$ *contains* $I(\mathcal{D})$. *We denote by* $R(\mathcal{F})$ *the ring* $\mathbb{Q}[X_1, \ldots, X_n]/I(\mathcal{F})$.

Definition 2.4 *Any set of polynomials that, added to the ideal* $I(\mathcal{D})$ *of a full design* $\mathcal{D}$, *generates the ideal of a fraction* $\mathcal{F}$, *is called a set of* **confounding polynomials** *of* $\mathcal{F}$ *(in* $\mathcal{D}$*).*

Definition 2.5 *A* **standard set of power-products** *is a set* E *of power products, with the following property: if* $T \in E$ *and* T' *divides* T, *then* $T' \in E$.

Definition 2.6 *Let* σ *be a term-order and* I *an ideal of* R. *Then we denote by* $\mathrm{Lt}_\sigma(I)$ *the ideal generated by the set* $\{\mathrm{Lt}_\sigma(f) \mid f \in I,\ f \neq 0\}$, *where* $\mathrm{Lt}_\sigma(f)$ *is the* σ*-leading term of* f. *We denote by* $\mathcal{O}_\sigma(I)$ *the set of power-products, which are not multiples of any of the* σ*-leading terms of the elements of* I.

A *polynomial model* is a polynomial function $y = f(X_1, \ldots, X_n)$, where the coefficients of the polynomial $f(X_1, \ldots, X_n)$ belong to a suitable extension of the base field, which contains the parameters. If the coefficients are algebraically independent parameters, then the polynomial model is *linear.* When we deal with a linear polynomial model, the whole information is coded in the support of $f(X_1, \ldots, X_n)$.

Definition 2.7 *A* **Complete (linear) Polynomial Model** *is a (linear) polynomial model, whose* **support** *is a standard set of power-products.*

Henceforth we restrict our attention to complete polynomial models and we use Gröbner bases.

Definition 2.8 *Given a σ-Gröbner basis G of the ideal $I(\mathcal{F})$ with respect to a given term-order σ, we denote by $\mathcal{O}_\sigma(I(\mathcal{F}))$, or shortly by $\mathcal{O}_\sigma(\mathcal{F})$, the corresponding* standard set of power-products, *i.e. the set of all the monomials which are not divided by any of the leading term of the elements of G.*

Different Gröbner bases are associated to different *monomial orders* and in this way we may compute many different $\mathbb{Q}$-bases of $R(\mathcal{F})$. Notwithstanding the fact that there are infinite *term-orders* (see Robbiano 1985 and Robbiano 1986), there exists only a finite number of distinct Gröbner bases (see Mora and Robbiano 1989). In particular the leading terms of the polynomials which define a full design are pure powers, hence they are pairwise coprime and consequently they are a Gröbner basis with respect to *every* term-order; it follows that the corresponding standard set of power-products $\mathcal{O}(\mathcal{D})$ is *unique.*

This suggests the following definitions.

Definition 2.9 *Given a full design $\mathcal{D}$ with levels $(l_1, l_2, \ldots, l_n) \in \mathbb{N}^n$, we denote by $\mathcal{O}(\mathcal{D})$ the unique standard set of power-products associated with it i.e. the set of the power-products $X_1^{a_1} \cdots X_n^{a_n}$, such that $a_i < l_i$ for $i := 1, \ldots, n$.*

Definition 2.10 *Let $\mathcal{F} \subset \mathcal{D}$ be a fraction of a full design. Let σ be a term-order and G_σ the corresponding reduced Gröbner basis of $I(\mathcal{F})$. The set of polynomials in G_σ, which are not among the canonical polynomials of $\mathcal{D}$ is called* **the σ-canonical set of confounding polynomials.**

3 Distracted Fractions and Complete Polynomial Models

In this section we show how to use the theory of the *SuperG-Bases* (see Carrà and Robbiano 1990) to produce a solution to the problem of finding minimal fractions which identify a complete polynomial model (see Definition 2.7). Now we state the problem, which is PROBLEM 2 in Robbiano 1998.

PROBLEM

Let $\mathcal{D}$ be a full design and let be given a Complete Polynomial Model whose support $\mathcal{O}$ is contained in $\mathcal{O}(\mathcal{D})$. What are the fractions $\mathcal{F}$ of $\mathcal{D}$, which minimally identify the model, i.e. such that $\mathcal{O}$ is a basis of $R(\mathcal{F})$ as a vector space?

Definition 3.1 *Let $X_1, X_2, \ldots, X_n$ be n independent indeterminates; we denote by $\{X_1, \ldots, X_n\}^*$ the set of power products in $\{X_1, X_2, \ldots, X_n\}$. Let $\mathcal{D}$ be a full design and $\mathcal{O} \subset \mathcal{O}(\mathcal{D})$ be a standard set of power-products. Then there exists a unique minimal set* $\mathrm{Min}(\mathcal{O})$ *of terms which generate the monoideal $\{X_1, \ldots, X_n\}^* \backslash \mathcal{O}$. The set of terms in* $\mathrm{Min}(\mathcal{O})$*, which are not among the leading terms of the canonical polynomials of $\mathcal{D}$ is called CutOut($\mathcal{O}$).*

Let us illustrate the situation with an Example.

Example 3.1

Let $\mathcal{D}$ be the 3^2 full design. Then $I(\mathcal{D}) = (f_1, f_2)$, where $f_1 = X(X-1)(X-2)$ and $f_2 = Y(Y-1)(Y-2)$. We have already said that $\{f_1, f_2\}$ is the reduced Gröbner basis of $I(\mathcal{D})$ with respect to every term-order.
So the unique standard set of power-products associated to $\mathcal{D}$ is $\mathcal{O}(\mathcal{D}) := \{1, X, Y, X^2, XY, Y^2, X^2Y, XY^2, X^2Y^2\}$. Let $\mathcal{O} := \{1, X, Y, X^2, XY, Y^2\}$.
It is easy to see that $\mathrm{CutOut}(\mathcal{O}) = \{X^2Y, XY^2\}$.
The meaning is that $\mathcal{O}$ is the standard set of all the power-products not divided by any element of the set $\{X^3, Y^3, X^2Y, XY^2\}$. The cardinality of $\mathcal{O}$ is 6, therefore we are looking for a fraction $\mathcal{F}$, whose elements are 6 out of the 9 points of $\mathcal{D}$ and a term-order σ such that $\mathcal{O} = \mathcal{O}_\sigma(\mathcal{F})$. We must have two σ-canonical confounding polynomials g_1, g_2, whose leading terms are X^2Y, XY^2 respectively. For instance it is easy to check that $g_1 := XY(X-1), g_2 := XY(Y-1)$ do the trick with respect to every term-order.

In order to see this example in the right perspective we recall some facts from Carrà and Robbiano (1990).

Definition 3.2 *Let k be a field, $T := X_1^{a_1} X_2^{a_2} \cdots X_n^{a_n}$ and let $\alpha_1, \ldots, \alpha_n$ be n sequences of elements of k, where $\alpha_r = (\alpha_{r,i})_{i \in \mathbf{N}}$ for $r := 1, \ldots, n$ and $\alpha_{r,i} \neq \alpha_{r,j}$ if $i \neq j$. Then we call the polynomial*

$$D(T) := \prod_{i=1}^{a_1} (X_1 - \alpha_{1,i}) \cdots \prod_{i=1}^{a_n} (X_n - \alpha_{n,i})$$

the **distraction** *of T with respect to $\alpha_1, \ldots, \alpha_n$.*

It is clear that, in order to get a distraction of T, it suffices to have vectors of sufficiently many elements of k. So sometimes we speak of a distraction associated to vectors, not to sequences.

Proposition 3.3 *Let $T_1, \ldots, T_h \in \{X_1, \ldots, X_n\}^*$, let $(\alpha_1, \ldots, \alpha_n)$ be n sequences of natural numbers and $I := (D(T_1), \ldots, D(T_h))$ the ideal generated by the distractions with respect to $(\alpha_1, \ldots, \alpha_n)$. Then $\{D(T_1), \ldots, D(T_h)\}$ is the reduced Gröbner basis of I with respect to every term-order.*

PROOF. See Carrà and Robbiano (1990), Theorem 1.4 and Corollary 2.6. □

We are ready to present a solution to our PROBLEM.

Theorem 3.4 *Let $\mathcal{D}$ be a full design with levels $(l_1, l_2, \ldots, l_n)$ and whose values of the coordinates are $0, 1, \ldots, l_i - 1$, for $i := 1, \ldots, n$. Let $I(\mathcal{D}) := (f_1, \ldots, f_n)$ be its generating ideal, where $f_j := X_j\ (X_j - 1) \cdots (X_j - l_j + 1)$, $j = 1, \ldots, n$; let $\mathcal{O}(\mathcal{D})$ be its corresponding standard set of power-products and $\mathcal{O}$ a standard set of power-products contained in $\mathcal{O}(\mathcal{D})$. Assume that* $\mathrm{CutOut}(\mathcal{O}) = \{T_1, \ldots, T_h\}$. *Let $\alpha_1, \ldots, \alpha_n$ be permutations of $(0, 1, \ldots, l_1 - 1), \ldots, (0, 1, \ldots, l_n - 1)$ and $D(T_1), \ldots D(T_h)$ the distractions of $T_1, \ldots, T_h$ with respect to $(\alpha_1, \ldots, \alpha_n)$.*

Then, for every term-order σ, we have: $\{f_1, \ldots, f_n, D(T_1), \ldots D(T_h)\}$ is a Gröbner basis and $\{D(T_1), \ldots D(T_h)\}$ is the σ-canonical set of confounding polynomials of a fraction such that $\mathcal{O}_\sigma(\mathcal{F}) = \mathcal{O}$.

PROOF. Let $T_i := X_1^{a_{1i}} X_2^{a_{2i}} \cdots X_n^{a_{ni}}$, $i = 1, \ldots, h$. By assumption $\mathcal{O} \subset \mathcal{O}(\mathcal{D})$, hence it follows that $a_{ri} < l_r$ for every $r := 1, \ldots, n$ and every $i := 1, \ldots, h$. Now $f_1, \ldots, f_n$ are the distractions of $X_1^{l_1}, \ldots, X_n^{l_n}$ with respect to $(\alpha_1, \ldots, \alpha_n)$, *independently* of $(\alpha_1, \ldots, \alpha_n)$. Then

$$\{f_1, \ldots, f_n, D(T_1), \ldots D(T_h)\} = \{D(X_1^{l_1}), \ldots, D(X_n^{l_n}), D(T_1), \ldots, D(T_h)\}$$

It is clear that $Lt_\sigma(D(T_i)) = T_i$ for every i and every σ, since all the power products in the support of $D(T_i)$, which are different from T_i, are divisors of T_i. Now the conclusion follows from Proposition 3.3. □

Example 3.2

Let $\mathcal{D}$ be the 3^2 full design; we consider the standard set of power-products $\mathcal{O} := \{1,\ Y,\ Y^2,\ X,\ XY,\ X^2\}$ as we did in Example 1. We see that $\mathrm{CutOut}(\mathcal{O}) = \{XY^2, X^2Y\}$. Now $l_1 = l_2 = 3$ and we choose $\alpha_1 := (1, 2, 0)$, $\alpha_2 := (2, 0, 1)$. The associated distractions are $((X-1)(Y-2)Y, (X-1)(X-2)(Y-2)) = (XY^2 - 2XY - Y^2 + 2Y, X^2Y - 2X^2 - 3XY + 6X + 2Y - 4)$. A solution to this problem is the fraction: $\{(0,2), (1,0), (1,1), (1,2), (2,0), (2,2)\}$, whose defining ideal is generated by $\{X^3 - 3X^2 + 2X,\ Y^3 - 3Y^2 + 2Y,\ XY^2 - 2XY - Y^2 + 2Y,\ X^2Y - 2X^2 - 3XY + 6X + 2Y - 4\}$.

Definition 3.5 *Given a full design $\mathcal{D}$, a term-order σ, a standard set of power-products $\mathcal{O} \subset \mathcal{O}(\mathcal{D})$, let $\mathcal{F}$ be a fraction of $\mathcal{D}$, whose σ-canonical set of confounding polynomials is given by distractions as indicated in Theorem 3.4. Then the fraction is called a* **distracted fraction**.

Corollary 3.6 *Given a full design $\mathcal{D}$ and a standard set of power-products $\mathcal{O} \subset \mathcal{O}(\mathcal{D})$, then the set of the points whose coordinates are the exponents of the elements of $\mathcal{O}$ is the distracted fraction associated to the* identical permutations.

Corollary 3.7 *Given a design $\mathcal{D}$ and a standard set of power-products $\mathcal{O} \subset \mathcal{O}(\mathcal{D})$, let* $\mathrm{Sol}(\mathcal{O}, \forall\sigma)$ *be the set of those fractions $\mathcal{F}$ such that $\mathcal{O}_\sigma(\mathcal{F}) = \mathcal{O}$ for* every *term-order σ; let* $\mathrm{Distr}(\mathcal{O})$ *be the set of distracted fractions, whose associated standard set of power-products is $\mathcal{O}$ for some term-order σ. Then* $\mathrm{Distr}(\mathcal{O}) \subseteq \mathrm{Sol}(\mathcal{O}, \forall\sigma)$.

PROOF. We have already seen (see Theorem 3.4) that the leading term of a distraction is independent of the term-order. Therefore the distractions that, added to the canonical polynomials of the design, yield the polynomials of the fraction, are the σ-canonical set of confounding polynomials with respect to every term-order. □

The next question is: how many distracted fractions are there?

Definition 3.8 *Let $\mathcal{D}$ be a full design with levels $(l_1, l_2, \ldots, l_n)$. Let $\mathcal{F} \subset \mathcal{D}$ be a fraction and assume that* $\mathrm{C} := \mathrm{CutOut}(\mathcal{F}) = \{T_1, T_2, \ldots, T_h\}$. *We denote by* $\mathrm{Log}_{X_i}(T_j)$ *the exponent of X_i in the power product T_j. Then we define* $\mathrm{Log}_{X_i}(\mathrm{C})$ *to be the h-tuple of the elements* $\mathrm{Log}_{X_i}(T_j)$*, $j := 1, \ldots, h$, listed in non decreasing order.*

Example 3.3

Let $\mathrm{C} = \{T_1, T_2, T_3\} := \{X^2YZ^3,\ XY^2Z^3,\ X^5Z\}$.
Then $\mathrm{Log}_X(\mathrm{C}) = (1,2,5)$, $\mathrm{Log}_Y(\mathrm{C}) = (0,1,2)$, $\mathrm{Log}_Z(\mathrm{C}) = (1,3,3)$.

Definition 3.9 *With the above notation, let* $\mathrm{Log}_{X_i}(\mathrm{C}) := (a_{1,i}, \ldots, a_{h,i})$. *Then we define*

$$\Delta(\mathrm{Log}_{X_i}(\mathrm{C})) := \binom{l_i}{a_{1,i}} \binom{l_i - a_{1,i}}{a_{2,i} - a_{1,i}} \cdots \binom{l_i - a_{h-1,i}}{a_{h,i} - a_{h-1,i}}$$

Example 3.4

For instance in the above Example 3 let us assume that $l_1 = 6$. Then we get
$\Delta(\mathrm{Log}_X(\mathrm{C})) = \binom{6}{1}\binom{6-1}{2-1}\binom{6-2}{5-2} = 120$

Theorem 3.10 *Let $\mathcal{D}$ be a full design with levels $(l_1, l_2, \ldots, l_n)$. Let $\mathcal{O} \subset \mathcal{O}(\mathcal{D})$ be a standard set of power-products and* $\mathrm{C} := \mathrm{CutOut}(\mathcal{O})$. *Then*

$$\#(\mathrm{Distr}(\mathcal{O})) = \prod_{i=1}^{n} \Delta(\mathrm{Log}_{X_i}(\mathrm{C}))$$

PROOF. Given n permutations $(\alpha_1, \ldots, \alpha_n)$ of $(0, 1, \ldots, l_1 - 1), \ldots, (0, 1, \ldots, l_n - 1)$ respectively, and given a power product T, we get the corresponding distraction $D(T)$. But it is clearly possible that, given other permutations, we get *the same* polynomial $D(T)$. Namely, if $\mathrm{Log}_{X_i}(\mathrm{C}) = (a_1, \ldots, a_h)$, it is clear that we have $\binom{l_i}{a_1}$ choices for distracting $X_i^{a_1}$. By definition there exists another power product in the minimal set of generators of C, where X_i has exponent $a_2 \geq a_1$. Having fixed a_1 coordinates, there are still $\binom{l_i - a_1}{a_2 - a_1}$ choices for distracting $X_i^{a_2}$, and so on. Therefore if $C = \{T_1, \ldots, T_h\}$, then we have $\Delta(\mathrm{Log}_{X_i}(\mathrm{C}))$ choices for distracting the sequence $(X_i^{a_1}, \ldots, X_i^{a_h})$. This applies to every indeterminate, hence $\prod_{i=1}^n \Delta(\mathrm{Log}_{X_i}(\mathrm{C}))$ is exactly the number of different representations of $\mathrm{D}(T_1), \ldots, \mathrm{D}(T_h)$ as products of linear polynomials. But the polynomial ring is a unique factorization domain, hence $\prod_{i=1}^n \Delta(\mathrm{Log}_{X_i}(\mathrm{C}))$ is the number of different h-tuples $(\mathrm{D}(T_1), \ldots, \mathrm{D}(T_h))$. To conclude the proof, we need to show that different h-tuples define different fractions. Now, if $(\mathrm{D}(T_1), \ldots, \mathrm{D}(T_h))$ defines a fraction $\mathcal{F}$ and we choose an arbitrary term-order σ, then Proposition 3.3 implies that $(\mathrm{D}(T_1), \ldots, \mathrm{D}(T_h))$ is the reduced Gröbner basis of the defining ideal of $\mathcal{F}$ with respect to σ. But the reduced Gröbner basis is unique and this concludes the proof. □

Example 3.5

Let us go back to Example 3. Assume that $l_1 = 6$, $l_2 = 5$, $l_3 = 4$. Then we get $\Delta(\mathrm{Log}_X(\mathrm{C})) = \binom{6}{1}\binom{6-1}{2-1}\binom{6-2}{5-2} = 120$; $\Delta(\mathrm{Log}_Y(\mathrm{C})) = \binom{5}{0}\binom{5-0}{1-0}\binom{5-1}{2-1} = 20$ and finally $\Delta(\mathrm{Log}_Z(\mathrm{C})) = \binom{4}{1}\binom{4-1}{3-1}\binom{4-3}{3-3} = 12$..
Therefore $\#(\mathrm{Distr}(\mathcal{O})) = 120 \cdot 20 \cdot 12 = 28,800$.

Remark 3.11 *It is easy to see that Theorem 3.4 does not yield the general solution. For instance, with the same data as in Example 2, an answer to the problem is also given by the following fraction:*
$\{(0,0), (0,1), (0,2), (1,0), (1,1), (2,2)\}$. *The DegRevLex-canonical set of confounding polynomials is* $\{X^2Y - 2X^2 - XY + 2X,\ XY^2 - 2X^2 - XY + 2X\}$ *and so it solves the problem. On the other hand, the Lex-canonical set of confounding polynomials is* $\{X^2 - \frac{1}{2}XY^2 + \frac{1}{2}XY - X\}$. *This shows in particular that the given fraction is not a distracted fraction.*

Remark 3.12 *The discussion on how to compute more general solutions to the* PROBLEM *is carried on in the paper of Caboara and Robbiano (1997).*

After Corollary 3.7 and the subsequent discussion it is natural to ask the following question: is it true that $\mathrm{Distr}(\mathcal{O}) = \mathrm{Sol}(\mathcal{O}, \forall\sigma)$? We are going to see that the answer is negative. To this end we prove the following

Theorem 3.13 *Given a Design $\mathcal{D}$ and a standard set of power-products $\mathcal{O} \subset \mathcal{O}(\mathcal{D})$, let $\mathcal{F}$ be a fraction. Let σ be a term-order and $G_\sigma := \{g_1, \ldots, g_h\}$ the*

σ-canonical set of confounding polynomials of $\mathcal{F}$. If we write $g_i := T_i - r_i$, with $T_i := Lt_\sigma(g_i)$ for $i := 1, \ldots, h$, then the following conditions are equivalent
a) T divides T_i for every T in the support of r_i and every $i := 1, \ldots, h$.
b) G_σ is the τ-canonical set of confounding polynomials of $\mathcal{F}$ for every term-order τ.
c) $\mathcal{F} \in \mathrm{Sol}(\mathcal{O}, \forall\sigma)$.

PROOF. Let us see that a) implies b). The assumption a) implies that $T_i = \mathrm{Lt}_\tau(g_i)$ for every i and every term-order τ. Since the T_i's already cut out $\mathcal{O}$ from $\mathcal{O}(\mathcal{F})$, another element in the reduced Gröbner basis of $I(\mathcal{F})$ with respect to τ would necessarily have support contained in $\mathcal{O}$. This would lower the dimension of $R/I(\mathcal{F})$ as a k-vector space, a contradiction. It is obvious that b) implies c). Let us assume that c) holds and suppose, for contradiction, that there exist i and T in the support of g_i, such that T does not divide T_i. Then there exists a term-order τ such that $T >_\tau T_i$. But $T \in \mathcal{O}$, hence $\mathcal{O}_\tau(\mathcal{F}) \neq \mathcal{O}$, a contradiction. □

Example 3.6

Let $\mathcal{D}$ be the 2^3 full design, $\mathcal{O} := \{1, X, Y, Z\} \subset \mathcal{O}(\mathcal{D})$ and consider the fraction $\mathcal{F} := \{(0,0,0), (0,0,1), (1,0,0), (1,1,0)\}$. Then it is easy to show that $I(\mathcal{F}) = (X^2 - X,\ Y^2 - Y,\ Z^2 - Z,\ XY - X,\ XZ,\ YZ)$ and that $\{XY - X,\ XZ,\ YZ\}$ is the DegRevLex-canonical set of confounding polynomials of $\mathcal{F}$. Therefore $\mathcal{O} = \mathcal{O}_{DegRevLex}(\mathcal{F})$ and we deduce from Theorem 3.13 that $\{XY - X,\ XZ,\ YZ\}$ is the τ-canonical set of confounding polynomials of $\mathcal{F}$ for every term-order τ and that $\mathcal{F} \in \mathrm{Sol}(\mathcal{O}, \forall\sigma)$. On the other hand, we are going to show that $\mathcal{F} \notin \mathrm{Distr}(\mathcal{O})$. Suppose the contrary; then it follows from Carrà and Robbiano (1990) that every pair of polynomials in $\{X^2 - X,\ Y^2 - Y,\ Z^2 - Z,\ XY - X,\ XZ,\ YZ\}$ should be a Gröbner basis. But clearly $\{XY - X,\ XZ\}$ is not such. With respect to the 2^3 full design and the standard set of power-products $\mathcal{O} := \{1,\ X,\ Y,\ Z\}$, we conclude that $\mathrm{Distr}(\mathcal{O}) \subset \mathrm{Sol}(\mathcal{O}, \forall\sigma)$. We were able to compute exactly the cardinality of the two sets. They turned out to be 8 and 32 respectively.

References

Box, G. E. P., Hunter, W. G., Hunter, J. S. (1978) *Statistics for Experimenters.* John Wyley & Sons, New York

Capani, A., Niesi, G., Robbiano, L. (1996) Some Features of CoCoA 3. *Comput. Sci. J. of Moldova* **4**, No 3: 296–314

Caboara, M., Robbiano, L. (1997) Families of Ideals in Statistics. In *Proceedings of the 1997 ISSAC*, Küchlin Ed., pp. 404–409

Caboara, M., Pistone, Riccomagno, E., G., Wynn, H.P. (1997) *The Fan of an Experimental Design.* Preprint

Carrà, G., Robbiano, L. (1990) On SuperG-Bases. *J. Pure Appl. Algebra* **68**: 279–292

Collombier, D. (1996) *Plans D' Expérience Factoriels. Construction et propriétés des fractions de plans.* Springer-Verlag, Collection Mathématiques et Applications **21**, Heidelberg

Cox, D., Little, J., O'Shea, D. (1992) *Ideal, Varieties, and Algorithms.* Springer-Verlag, New York

Eisenbud, D. (1995) *Commutative Algebra with a View Toward Algebraic Geometry.* Springer Graduate Texts in Mathematics **150**

Fontana, R., Pistone, G., Rogantin, M. P. (1997) Algebraic analysis and generation of two-levels designs. *Statistica Applicata* **9**: 15-29

Holliday, T., Riccomagno, E., Wynn, H. P., Pistone, G. (1997) The Application of Computational Algebraic Geometry to the Analysis of Designed Experiments. A Case Study. *Computational Statistics.* To appear.

Mora, F., Robbiano, L. (1989) The Gröbner Fan of an Ideal. In L. Robbiano (ed.): *Computational Aspects of Commutative Algebra.* Academic Press. London, pp. 183–208

Pistone, G., Wynn, H. P. (1996) Generalized confounding and Gröbner bases. *Biometrika* **83**: 653-666.

Raktoe, B. L., Hedayat, A., Federer, W. T. (1981) *Factorial Designs.* John Wiley.

Robbiano, L. (1985) Term orderings in the polynomial ring. In *Proceedings of EUROCAL-85.* Lecture Notes in Computer Science **203** II.

Robbiano, L. (1986) On the Theory of Graded Structures. *J. Symb. Comput.* **2**: 139–170.

Robbiano, L. (1998) Gröbner bases and Statistics. This volume.

L. Robbiano and M.P. Rogantin
Dipartimento di Matematica, Università di Genova (Italy)
E-mail: {robbiano,rogantin}@dima.unige.it

Polynomial interpolation of minimal degree and Gröbner bases

Thomas Sauer

Mathematisches Institut, Universität Erlangen
Bismarckstr. $1\frac{1}{2}$, D–91054 Erlangen, Germany
e-mail: sauer@mi.uni-erlangen.de

Abstract

This paper investigates polynomial interpolation with respect to a finite set of appropriate linear functionals and the close relations to the Gröbner basis of the associated finite dimensional ideal.

1 Introduction

In the 33 years since their introduction by Buchberger (1965, 1970), Gröbner bases have been applied successfully in various fields of Mathematics and to many types of problems. This paper wants to go the opposite way by presenting a different approach to Gröbner bases for zero dimensional ideals from the quite recent theory of polynomial interpolation of minimal degree. The latter one is an approach introduced by de Boor and Ron (1990, 1992) to solve interpolation problems defined by a finite number of linear functionals using appropriate polynomial spaces with certain useful properties.

Let me briefly explain this with the example of Lagrange interpolation in $\mathbb{R}^d$. Suppose that a finite set of pairwise disjoint points $\{x_0, \ldots, x_N\} \in \mathbb{R}^d$ is given. The *Lagrange interpolation problem* consists of finding, for any $y_0, \ldots, y_N$, a polynomial p such that $p(x_j) = y_j$, $j = 0, \ldots, N$. Clearly, this problem is always solvable and even has infinitely many solutions. The "real" question, however, is to find a *polynomial subspace* $\mathcal{P}$ such that for any given data the Lagrange interpolation problem has a unique solution in $\mathcal{P}$ and to choose $\mathcal{P}$ "as simple as possible". For $d = 1$ the generic choice for $\mathcal{P}$ is obvious: one takes the polynomials of degree less than or equal to N. For $d \geq 2$ the situation is more complicated since it may now happen that the points lie on an algebraic hypersurface of sufficiently low degree, for example when there are ≥ 6 points on a circle in $\mathbb{R}^2$. In order to deal with this type of interpolation problem, one chooses the space $\mathcal{P}$, depending on the data points, such that it satisfies additional minimality constraints.

In the same sense as Gröbner bases associated to a term order are called standard bases for the ideal, one may ask for "standard" interpolation spaces associated to a certain term order. It will turn out that a natural choice of interpolation spaces provides a very close relationship to Gröbner bases; indeed, we will show that interpolating a polynomial from this interpolation space is equivalent to reducing the polynomial modulo the Gröbner basis.

2 Polynomial interpolation of minimal degree

Let $\mathbb{K}$ be a field and denote by $\Pi = \mathbb{K}[\xi_1, \ldots, \xi_d]$ the ring of polynomials in d variables over $\mathbb{K}$. A finite set $\Theta \subset \Pi'$ of linear functionals mapping Π to $\mathbb{K}$ is said to *admit an ideal interpolation scheme* if the set

$$\mathcal{I}_\Theta := \ker \Theta = \{p \in \Pi : \Theta(p) = 0\}, \qquad \Theta(p) = (\theta(p) : \theta \in \Theta),$$

is an ideal in Π. This terminology has been introduced by G. Birkhoff (1979). The *interpolation problem* associated to Θ then consists of finding a $|\Theta|$–dimensional subspace $\mathcal{P} \subset \Pi$ and a projection $L_{(\mathcal{P},\Theta)} : \Pi \to \mathcal{P}$ such that

$$\Theta\left(L_{(\mathcal{P},\Theta)} q\right) = \Theta(q), \qquad q \in \Pi. \tag{2.1}$$

Note that if Θ is linearly independent this problem is equivalent to finding, for any $y \in \mathbb{K}^\Theta$, a $p = p_y \in \mathcal{P}$ such that $\Theta(p) = y$. Whenever (2.1) holds true, we call the pair $(\mathcal{P}, \Theta)$ an *interpolation system.*

The classical examples of ideal interpolation schemes are Lagrange interpolation (i.e., all functionals θ_j are the point evaluation functionals δ_{x_j} for pairwise disjoint points in $\mathbb{R}^d$) or certain types of Hermite interpolation (i.e., interpolation of partial differential operators at pairwise disjoint points, cf. (Lorentz 1992) and (Sauer and Xu 1995), for example). As pointed out by Marinari et al. (1991) as well as de Boor and Ron (1992), ideal interpolation schemes can even be characterized as Hermite interpolation schemes with an additional closedness condition. The key notion for this result is that of a *D–invariant polynomial subspace* which means a subspace $Q \subset \Pi$ with the property that

$$p \in \Pi, q \in Q \qquad \Rightarrow \qquad p(D)q \in Q,$$

where

$$p(D) = \sum_{\alpha \in \mathbb{N}_0^d} c_\alpha \frac{\partial^{|\alpha|}}{x^\alpha}, \qquad p = \sum_{\alpha \in \mathbb{N}_0^d} c_\alpha x^\alpha \in \Pi.$$

Theorem 1 *Let $\Theta \subset \Pi'$ be a finite set of linearly independent functionals. Then $\mathcal{I}_\Theta$ is an ideal if and only if there are points $x_1, \ldots, x_m \in \mathbb{K}^d$ and D–invariant subspaces $Q_1, \ldots, Q_m \subset \Pi$ such that*

$$\operatorname{span}_{\mathbb{K}} \Theta = \operatorname{span}_{\mathbb{K}} \left\{ \delta_{x_j} \circ q_j(D) : q_j \in Q_j, \, j = 1, \ldots, m \right\}.$$

It is also known, cf. (de Boor and Ron 1991), that the "local" Hermite interpolation conditions connected to some point x_j correspond to the primary decomposition of the ideal $\mathcal{I}_\Theta$. Examining the structure of the spaces Q_j, $j = 1, \ldots, m$, leads to a refined notion of the multiplicity of a zero, cf. (Marinari et al. 1996).

Let $\prec$ be any total order of the multiindices $\mathbb{N}_0^d$ which is compatible with addition and satisfies $0 \preceq \alpha$ for any $\alpha \in \mathbb{N}_0^d$. Clearly, this *well-ordering* induces a term order on the monomials x^α, $\alpha \in \mathbb{N}_0^d$, and therefore the notion of a *leading term* $\Lambda(p) = \Lambda_\prec(p)$ of a polynomial $p \in \Pi$ defined by

$$\Lambda(p) = \max_{\prec} \{c_\alpha x^\alpha : c_\alpha \neq 0\}, \qquad p = \sum_{\alpha \in \mathbb{N}_0^d} c_\alpha x^\alpha.$$

Finally, we write $p \prec q$ for $p, q \in \Pi$ if $\Lambda(p) \prec \Lambda(q)$ and interpret this as p being of lower degree than q. Let Θ define an ideal interpolation scheme. Then we call a subspace $\mathcal{P} \subset \Pi$ a *minimal degree interpolation space for* Θ *with respect to the term order* $\prec$ if

1. $(\mathcal{P}, \Theta)$ is an *interpolation system*, i.e., for any $y \in \mathbb{K}^\Theta$ there is a *unique* $p \in \mathcal{P}$ such that $\Theta(p) = y$,

2. $\mathcal{P}$ is $\prec$*-minimal* with this property, i.e., there exists no interpolation system $(\mathcal{Q}, \Theta)$ with $\mathcal{Q} \prec \mathcal{P}$. The latter means that for any $q \in \mathcal{Q}$ there exists a $p \in \mathcal{P}$ such that $q \prec p$.

3. $\mathcal{P}$ is $\prec$*-reducing*, i.e., when $L_{(\mathcal{P},\Theta)}$ denotes the projection on $\mathcal{P}$ (the interpolation polynomial), then

$$L_{(\mathcal{P},\Theta)} q \preceq q, \qquad q \in \Pi.$$

In general, there is no unique minimal interpolation space for a set of interpolation conditions Θ. More precisely, we have the following result.

Proposition 2 *Let Θ be a linearly independent set of linear functionals admitting an ideal interpolation scheme. The the following statements are equivalent:*

1. *There exists a unique minimal degree interpolation space $\mathcal{P}$ for Θ with respect to the term order $\prec$.*

2. *Let $A_{|\Theta|} \subset \mathbb{N}_0^d$ denote the first $|\Theta|$ multiindices with respect to the ordering $\prec$, then the Vandermonde matrix*

$$\left[\theta\left((\cdot)^\alpha\right)\right]_{\theta \in \Theta, \alpha \in A_{|\Theta|}}$$

has full rank.

3. $\mathcal{P} = \operatorname{span}_{\mathbb{K}} \{x^\alpha : \alpha \in A_{|\Theta|}\}$.

Proof: We first note that if $\operatorname{span}_{\mathbb{K}} \{x^\alpha : \alpha \in A_{|\Theta|}\}$ is an interpolation space, then it is clearly a minimal degree interpolation space, since

$$\dim \operatorname{span}_{\mathbb{K}} \left\{ x^\alpha : \alpha \prec \max_{\prec} A_{|\Theta|} \right\} = |\Theta| - 1.$$

Consequently, any other interpolation space must contain a polynomial p such that $p \succ x^\alpha$, $\alpha \in A_{|\Theta|}$, and therefore it cannot be minimal. Hence, conditions 2. and 3. imply uniqueness.

Conversely, suppose that $\mathcal{P}$ is a minimal degree interpolation space, let $p_0 \prec \cdots \prec p_N$ be a basis of $\mathcal{P}$ and suppose that there exist a polynomial $q \in \Pi \setminus \mathcal{P}$ such that $q \prec p_N$. Since $q \notin \mathcal{P}$ and since the projection $L_{(\mathcal{P},\Theta)}$ is degree reducing we conclude that

$$0 \neq q - L_{(\mathcal{P},\Theta)} q \preceq q \prec p_N.$$

But then, for any $0 \neq c \in \mathbb{K}$, we have that

$$\operatorname{span}_{\mathbb{K}} \{p_0, \ldots, p_{N-1}, p_N + c\left(q - L_{(\mathcal{P},\Theta)} q\right)\} \neq \mathcal{P}$$

is another minimal degree interpolation space. □

The above result indicates that a "good" or "natural" choice for a interpolation space would be one consisting of exactly $N+1$ monomials.

3 Newton basis and Gröbner basis

Our next goal is to derive a suitable basis for the minimal degree interpolation space and relate that to Gröbner bases. To simplify notation, we set $N := |\Theta| - 1$ and write $\Theta = \{\theta_0, \ldots, \theta_N\}$, where we assume the indexing to be chosen in a proper way which will be specified later. We will also use the notation $\Theta_j = (\theta_0, \ldots, \theta_j)$, $j = 0, \ldots, N$, for interpolation subproblems. Also for any polynomial $p \in \Pi$ we define the linear subspace of polynomials of the same or lower degree as

$$\Pi_{\preceq p} = \{q \in \Pi : q \preceq p\} \qquad \text{and} \qquad \Pi_{\prec p} = \{q \in \Pi : q \prec p\}.$$

Clearly, $p \in \Pi_{\preceq p}$.

A *Newton basis* of a polynomial subspace $\mathcal{P}$ with respect to Θ is a set of polynomials $p_0 \prec \cdots \prec p_N \in \mathcal{P}$ such that, after numbering Θ properly, the following properties are satisfied:

1. $\mathcal{P} = \operatorname{span}_{\mathbb{K}} \{p_0, \ldots, p_N\}$,
2. $\theta_i(p_j) = \delta_{ij}$, $0 \leq i \leq j \leq N$,

3. there exist polynomials $q_0, \dots, q_M \in \mathcal{I}_\Theta$ such that

$$\Pi_{\preceq p_N} = \operatorname{span}_{\mathbb{K}} \{p_0, \dots, p_N\} \oplus \operatorname{span}_{\mathbb{K}} \{q_0, \dots, q_M\} . \tag{3.1}$$

Recalling the terminology of Marinari et al. (1991), the polynomials $p_0, \dots, p_N$ are a *triangular sequence* with respect to Θ. The additional condition 3. will correspond to minimality. The name *Newton basis* stems from the fact that the interpolation polynomial can now be written in the iterative "Newton form"

$$L_{(\mathcal{P},\Theta)} q = \sum_{j=0}^{N} \theta_j \left(q - L_{(\mathcal{P}_{j-1},\Theta_{j-1})} q\right) p_j, \qquad \mathcal{P}_j = \operatorname{span}_{\mathbb{K}} \{p_0, \dots, p_j\} .$$

Indeed we have the following connection between minimal degree interpolation spaces and Newton bases.

Theorem 3 *Let Θ be a finite linearly independent set of linear functionals admitting an ideal interpolation scheme. Then $\mathcal{P}$ is a minimal degree interpolation space with respect to Θ if and only if it has a Newton basis with respect to Θ.*

Proof: The proof is similar to the one for the total degree situation (Sauer 1997). Let us first assume that $\mathcal{P}$ has a Newton basis, then the interpolation property is immediate and from equation (3.1) we observe that

$$\begin{aligned} &\operatorname{rank} \left[\theta\left((\cdot)^\alpha\right)\right]_{\theta\in\Theta,\, x^\alpha \prec p_N} \\ &= \operatorname{rank} \left[\theta(q_0), \dots, \theta(q_M), \theta(p_0), \dots, \theta(p_{N-1})\right]_{\theta\in\Theta} \\ &= \operatorname{rank} \left[\theta(p_0), \dots, \theta(p_{N-1})\right]_{\theta\in\Theta} = |\Theta| - 1, \end{aligned}$$

yielding that $\mathcal{P}$ is $\prec$–minimal. Since $\Pi = \mathcal{P} \oplus \mathcal{I}_\Theta$, we can write any $p \in \Pi$ as

$$p = \sum_{j=0}^{N} a_j p_j + \sum_{j=0}^{M} b_j q_j + q, \qquad q \in \mathcal{I}_\Theta,$$

and therefore we also conclude that

$$L_{(\mathcal{P},\Theta)} p = \sum_{j=0}^{N} a_j p_j \preceq p,$$

hence $\mathcal{P}$ is also $\prec$–reducing and therefore it is indeed a minimal degree interpolation space for Θ.

Conversely, if $\mathcal{P}$ is a minimal degree interpolation space for Θ, then $\mathcal{P}$ has dimension $|\Theta| = N+1$ and therefore there exists a "graded" basis $\tilde{p}_0 \prec \cdots \prec$

$\tilde{p}_N$ of $\mathcal{P}$. Gauß elimination on the Vandermonde matrix then immediately yields that, after numbering Θ properly, there exist a triangular basis $p_0 \prec \cdots \prec p_N$ for $\mathcal{P}$. Now, let $\tilde{q}_0 \prec \cdots \prec \tilde{q}_M$ be linearly independent polynomials which complete $p_0, \ldots, p_N$ to a basis of $\Pi_{\preceq p_N}$. Clearly, $\tilde{q}_j \prec p_N$. Since $\mathcal{P}$ is $\prec$–reducing, we have for any $j = 0, \ldots, M$ that $\tilde{q}_j \succ L_{(\mathcal{P},\Theta)}\tilde{q}_j$, hence $\tilde{q}_j \preceq q_j \preceq \tilde{q}_j$, where

$$q_j := \tilde{q}_j - L_{(\mathcal{P},\Theta)}\tilde{q}_j \in \mathcal{I}_\Theta, \qquad j = 0, \ldots, M,$$

while still

$$\Pi_{\preceq p_N} = \operatorname{span}_{\mathbb{K}} \{p_0, \ldots, p_N\} \oplus \operatorname{span}_{\mathbb{K}} \{q_0, \ldots, q_M\}.$$

This verifies that $p_0, \ldots, p_N$ is a Newton basis. □

Let us now recall that a basis of a polynomial ideal $\mathcal{I}$, i.e., a finite set of polynomials $f_1, \ldots, f_n$ such that

$$\mathcal{I} = \langle f_1, \ldots, f_n \rangle = \left\{ \sum_{j=1}^{n} q_j f_j : q_j \in \Pi, \, j = 1, \ldots, n \right\},$$

is called a *Gröbner basis* of $\mathcal{I}$ if

$$\langle \Lambda(f_1), \ldots, \Lambda(f_n) \rangle = \langle \Lambda(\mathcal{I}) \rangle := \{ \Lambda(q) : q \in \mathcal{I} \}.$$

It is well–known, cf. (Buchberger and Möller 1982), (Marinari et al. 1991), as well as (de Boor 1994), that a triangular system for Θ and a Gröbner basis for the ideal $\mathcal{I}_\Theta$ can be constructed *simultaneously* by properly applying Gauß elimination to the Vandermonde matrix. An equivalent approach is to use a Gram–Schmidt orthogonalization process for the monomials and successively factor out the subideals generated by the Gröbner basis elements obtained during this process; this has been described explicitly for the case of the graded lexicographical ordering (Sauer 1997)

Let me briefly outline how the latter approach works for finding the Newton basis (and therefore the "natural" interpolation space $\mathcal{P}_\Theta$ announced earlier in this paper) and the reduced Gröbner basis for $\mathcal{I}_\Theta$ simultaneously.

For that purpose we first remark that Θ defining an ideal interpolation scheme yields that any minimal degree interpolation space $\mathcal{P}$ is a subspace of Π_N, the space of all polynomials of total degree at most N and that therefore all elements of the reduced Gröbner basis can be found in Π_{N+1}. This elementary remark gives us an a priori bound on the total degree of the Gröbner and Newton basis elements which enables us to formulate the algorithm in a "finite environment".

We first initialize the polynomials $\phi_\alpha = x^\alpha$, $|\alpha| \leq N+1$, let $\Theta' = \Theta$ be the set of all "free" linear functionals from Θ and set $j = 0$. Now we loop over all multiindices α of absolute value $\leq N+1$ according to the ordering $\prec$ and check if $\Theta'(\phi_\alpha) = 0$. If this is the case and $\phi_\alpha \neq 0$, then we mark ϕ_α as a member of the Gröbner basis and set $\phi_{\alpha+\beta} = 0$, $\beta \in \mathbb{N}_0^d$, $|\alpha+\beta| \le N+1$. Otherwise, we pick $\theta \in \Theta'$ (which leaves room for a pivoting strategy) such that $\theta(\phi_\alpha) \neq 0$, set $\theta_j = \theta$, $p_j = \phi_\alpha/\theta_j(\phi_\alpha)$, do the orthogonalization $\phi_\beta \leftarrow \phi_\beta - \theta_j(\phi_\beta)p_j$, $\beta \succ \alpha$, $|\beta| \le N+1$ and continue with $\Theta' \leftarrow \Theta' \setminus \{\theta\}$ and $j \leftarrow j+1$.

It is easy to see that the interpolation space generated this way is spanned by exactly $|\Theta|$ monomials and that the Gröbner basis elements are of the form $x^\alpha + p$, $x^\alpha \succ p \in \mathcal{P}$. In particular, no leading term of a Gröbner basis element divides any term of another one which verifies the claim that the Gröbner basis generated by this algorithm is the *reduced* Gröbner basis.

Conversely, let $\mathcal{G}_\Theta = \{g_1, \dots, g_n\}$ denote the unique reduced Gröbner basis for $\mathcal{I}_\Theta$, i.e., the leading terms $\Lambda(g_j)$, $j = 1, \dots, n$, have coefficient 1 and no leading term of one of these polynomials divides any term of another one. Since the minimal degree interpolation space generated by the above construction satisfies

$$\mathcal{P} = \Pi \ominus \Lambda(\mathcal{I}_\Theta) = \Pi \ominus \langle \Lambda(\mathcal{G}_\Theta) \rangle, \tag{3.2}$$

there are exactly $N+1$ monomials not contained in $\Lambda(\mathcal{I}_\Theta)$. Hence we have the following result.

Theorem 4 *Let Θ be a finite linearly independent set of linear functionals admitting an ideal interpolation scheme. Then there exists a unique minimal degree interpolation space $\mathcal{P}_\Theta$ which is spanned by $|\Theta|$ monomials.*

Clearly, the ideal $\mathcal{I}_\Theta$ can also be written as

$$\mathcal{I}_\Theta = \left\{ p - L_{(\mathcal{P}_\Theta,\Theta)} p : p \in \Pi \right\} = \left\{ x^\alpha - L_{(\mathcal{P}_\Theta,\Theta)} (\cdot)^\alpha : \alpha \in \mathbb{N}_0 \right\}.$$

Set $q_\alpha = x^\alpha - L_{(\mathcal{P}_\Theta,\Theta)}(\cdot)^\alpha$, $\alpha \in \mathbb{N}_0^d$. Since $\mathcal{I}_\Theta$ is an ideal, the set $A_\Theta := \left\{ \alpha \in \mathbb{N}_0^d : q_\alpha \neq 0 \right\}$ is an *upper set*, i.e., $A_\Theta + \mathbb{N}_0^d \subset A_\Theta$. Then it is easily observed that the *corners* $C_\Theta \subset A_\Theta$, i.e., those elements $\alpha \in A_\Theta$ which cannot be written as $\alpha = \alpha' + \beta$, $\alpha' \in A_\Theta$, $0 \neq \beta \in \mathbb{N}_0^d$, point to the elements of the Gröbner basis. In other words,

$$\mathcal{G}_\Theta = \left\{ x^\alpha - L_{(\mathcal{P}_\Theta,\Theta)} (\cdot)^\alpha : \alpha \in C_\Theta \right\}. \tag{3.3}$$

A careful investigation of these relationships has been performed by Marinari et al. (1991).

Let us briefly comment on some algorithmic aspects of the above algorithm: of course, "sifting" the full Π_{N+1} in order to obtain the Newton basis and the Gröbner basis, may usually cause a lot of unnecessary effort. A more effective approach will proceed by degree and if, for some $k \in \mathbb{N}$, the space Π_k is not sufficient for the Newton basis (in terms of the above procedure this means that $\Theta' \neq \emptyset$), then one might add the polynomials $\phi_\alpha = x^\alpha - L_{\mathcal{P}_k,\Theta\setminus\Theta'}$, $|\alpha| = k+1$, where $\mathcal{P}_k \subset \Pi_k$ is the interpolation space spanned by the already constructed Newton polynomial which interpolates at $\Theta \setminus \Theta'$. Moreover, in the above "addition technique" those monomials can be omitted which are divisible by the leading term on an already found element of the Gröbner basis. Information on how to practically compute interpolation polynomials by means of the Newton method, including a triangular scheme for the coefficients of the Newton representation, can be found in (Sauer 1995).

4 Interpolation and reduction

Gröbner bases are closely connected to (and most frequently even defined by) the notion of reduction. We say the a polynomial $p \in \Pi$ reduces to $q \in \Pi$ modulo a finite set $\mathcal{F} = \{f_1, \ldots, f_n\} \subset \Pi$ if there exist $g_1, \ldots, g_n \in \Pi$ such that

$$p = \sum_{j=1}^{n} f_j g_j + q$$

and none of $\Lambda(f_1), \ldots, \Lambda(f_n)$ divides any term of q. It is well-known that if $\mathcal{G}$ is a Gröbner basis, then the remainder of reduction is unique and therefore reduction with respect to a Gröbner basis $\mathcal{G}$ can be understood as a mapping $\to_{\mathcal{G}}: \Pi \to \Pi$. Moreover, we call a polynomial $p \in \Pi$ *reduced* with respect to a Gröbner basis $\mathcal{G}$ if $p = p \to_{\mathcal{G}}$. Reduced polynomials can easily be described in the following way.

Lemma 5 *Let $\mathcal{G}$ be a Gröbner basis for the ideal $\mathcal{I}$. A polynomial $p \in \Pi$ is reduced modulo $\mathcal{G}$ if and only if*

$$\Lambda(g)(D)p = 0, \qquad \begin{cases} g \in \mathcal{I}, \\ g \in \mathcal{G}. \end{cases} \tag{4.1}$$

Proof: Equation (4.1) with $g \in \mathcal{G}$ is only a reformulation of the assumption that no term of p is divisible by any $\Lambda(g)$, $g \in \mathcal{G}$. Also, the validity of (4.1) for all $g \in \mathcal{I}$ trivially implies the validity for $g \in \mathcal{G}$. Conversely, since $\mathcal{G}$ is a Gröbner basis, there exist, for any $q \in \mathcal{I}$, monic polynomials q_g, $g \in \mathcal{G}$, such that

$$\Lambda(q) = \sum_{g \in \mathcal{G}} \Lambda(q_g) \Lambda(g)$$

and therefore, for any reduced polynomial $p \in \Pi$

$$\Lambda(q)(D)p = \sum_{g \in \mathcal{G}} \Lambda(q_g)(D) \underbrace{(\Lambda(g)(D)p)}_{=0} = 0.$$

□

From this lemma it immediately follows that the reduced polynomials form a D-invariant finite dimensional subspace which has the form $\Pi \ominus \Lambda(\mathcal{I}_\Theta)$. Recalling equation (3.2) we therefore obtain the representation

$$\mathcal{P}_\Theta = \bigcap_{q \in \mathcal{I}_\Theta} \ker \Lambda(q)(D) = \bigcap_{g \in \mathcal{G}_\Theta} \ker \Lambda(g)(D), \tag{4.2}$$

which implies the following result, cf. (Möller 1998).

Theorem 6 *Let $\Theta \subset \Pi'$ be a finite linearly independent set which admits an ideal interpolation scheme. A polynomial $p \in \Pi$ is reduced modulo $\mathcal{G}_\Theta$ if and only if $p \in \mathcal{P}_\Theta$. Moreover, for any $p \in \Pi$ we have*

$$L_{(\mathcal{P}_\Theta,\Theta)} p = p \rightarrow_{\mathcal{G}_\Theta} . \tag{4.3}$$

Proof: The first statement follows from Lemma 5 and (4.2). To prove (4.3) we first note that on $\mathcal{P}_\Theta$ both mappings act as the identity while for general $p \in \Pi$ we write

$$p = p \rightarrow_{\mathcal{G}} + \sum_{g \in \mathcal{G}} p_g g$$

and then

$$L_{(\mathcal{P}_\Theta,\Theta)} p = L_{(\mathcal{P}_\Theta,\Theta)} \left(p \rightarrow_{\mathcal{G}}\right) + \sum_{g \in \mathcal{G}} L_{(\mathcal{P}_\Theta,\Theta)} \left(p_g g\right) = p \rightarrow_{\mathcal{G}},$$

since $\Theta(p_g g) = 0$, $g \in \mathcal{G}$. □

Consequently, by switching between interpolation and reduction it is possible by simple and efficient algorithms to compute either $p \rightarrow_{\mathcal{G}_\Theta}$ from the data $\Theta(p)$ or $L_{(\mathcal{P}_\Theta,\Theta)} p$ if only $\mathcal{G}_\Theta$ (but not Θ itself) is known.

5 Least Interpolation

There is a way to write the space $\mathcal{P}_\Theta$ in a "closed form" which has been used by de Boor and Ron to introduce their concept of *least interpolation* (de Boor and Ron 1990). Following their approach (de Boor and Ron 1992), we identify Π' with the formal power series $\mathbb{K}[\![\xi_1, \ldots, \xi_d]\!]$ and use the pairing $\langle \cdot, \cdot \rangle : \mathbb{K}[\![\xi_1, \ldots, \xi_d]\!] \times \mathbb{K}[\xi_1, \ldots, \xi_d] \to \mathbb{K}$ defined by

$$\langle f, p \rangle = \sum_{\alpha \in \mathbb{N}_0^d} \frac{D^\alpha f(0) D^\alpha p(0)}{\alpha!} = p(D) f(0).$$

We can identify any $\theta \in \Theta$ with an $f_\theta \in \mathbb{K}[\![\xi_1, \ldots, \xi_d]\!]$ (for example, if $\theta = \delta_x$, then $f_\theta(y) = e^{x \cdot y}$). Then the following variant of Theorem 1 is valid; cf. (de Boor and Ron 1992), where also the proof is taken from which we give for the sake of completeness.

Theorem 7 *The finite set $\Theta \subset \Pi'$ admits an ideal interpolation scheme if and only if the subspace $f_\Theta := \operatorname{span}_{\mathbb{K}} \{f_\theta : \theta \in \Theta\}$ is closed under formal differentiation.*

Proof: The proof is based on the simple observation that, for $\alpha \in \mathbb{N}_0^d$,

$$\langle f, (\cdot)^\alpha p \rangle = (p(D) D^\alpha f)(0) = \langle D^\alpha f, p \rangle, \tag{5.1}$$

which holds for any

$$p \in \mathbb{K}[\xi_1, \ldots, \xi_d], \; f \in \mathbb{K}[[\xi_1, \ldots, \xi_d]].$$

Hence, if f_Θ is closed under differentiation, then $\ker\Theta$ is obviously an ideal by (5.1).

Conversely, since f_Θ is of finite dimension, it is closed and therefore

$$f_\Theta = \{ f \in \mathbb{K}[[\xi_1, \ldots, \xi_d]] : \langle f, p \rangle = 0, \; p \in \ker\Theta \},$$

we can again use (5.1) to conclude that f_Θ is closed under differentiation. □

For any $f \in \mathbb{K}[[\xi_1, \ldots, \xi_d]]$ we denote by $\lambda(f) = \lambda_\prec(f)$ the *least term* of f with respect to the term order $\prec$ as the $\prec$-minimal nonzero term of the power series f, i.e.,

$$\lambda(f) = \min_{\prec} \{ f_\alpha x^\alpha : c_\alpha \neq 0 \}, \qquad f = \sum_{\alpha \in \mathbb{N}_0^d} f_\alpha x^\alpha.$$

This notion gives the desired representation of $\mathcal{P}_\Theta$.

Theorem 8 *Assume that the finite and linearly independent set $\Theta \subset \Pi'$ admits an ideal interpolation scheme. Then*

$$\mathcal{P}_\Theta = \lambda(f_\Theta) = \{ \lambda(f) : f \in f_\Theta \}.$$

Proof: We first prove that $\lambda(f_\Theta) \subset \mathcal{P}_\Theta$. For that purpose we assume that there exist $f = \sum c_\theta f_\theta$, $c_\theta \in \mathbb{K}$, $\theta \in \Theta$, and $q \in \mathcal{I}_\Theta$ such that $\Lambda(q)(D)\lambda(f) \neq 0$. This implies that $q \preceq \lambda(f) \preceq q$ and therefore

$$0 \neq \Lambda(q)(D)\lambda(f) = q(D)f(0) = \langle f, q \rangle = \sum_{\theta \in \Theta} c_\theta \langle f_\theta, q \rangle = \sum_{\theta \in \Theta} c_\theta \theta(q) = 0,$$

which is a contradiction. Hence, $\Lambda(q)(D)\lambda(f) = 0$, $q \in \mathcal{I}_\Theta$, $f \in \lambda(\Theta)$ and therefore

$$\lambda(f_\Theta) \subset \bigcap_{q \in \mathcal{I}_\Theta} \ker \Lambda(q)(D) = \mathcal{P}_\Theta.$$

On the other hand, the functionals Θ were assumed to be linearly independent which implies that $\dim \lambda(f_\Theta) = \dim f_\Theta = \dim \Theta = |\Theta| = \dim \mathcal{P}_\Theta$, hence $\lambda(f_\Theta) = \mathcal{P}_\Theta$. □

References

Birkhoff, G. (1979) 'The algebra of multivariate interpolation', In: Coffman, C. V., Fix G. J. (eds.), *Constructive Approaches to Mathematical Models*, Academic Press Inc., 345–363.

de Boor, C. (1994) 'Gauss elimination by segments and multivariate polynomial interpolation', In: Zahar, R. V. M. (ed.), *Approximation and Computation: A Festschrift in Honor of Walter Gautschi*, Birkhäuser Verlag, 87–96.

de Boor, C., Ron, A. (1990) 'On multivariate polynomial interpolation', *Constr. Approx.* **6**, 287–302.

de Boor, C., Ron, A. (1991) 'On polynomial ideals of finite codimension with applications to box spline theory', *J. Math. Anal. and Appl.* **158**, 168–193.

de Boor, C., Ron, A. (1992) 'The least solution for the polynomial interpolation problem', *Math. Z.* **210**, 347–378.

Buchberger, B. (1965) 'Ein Algorithmus zum Auffinden der Basiselemente des Restklassenrings nach einem nulldimensionalen Polynomideal', Dissertation, Universität Innsbruck.

Buchberger, B. (1970) 'An algorithmic criterion for the solvability of algebraic systems of equations' (German), *Aequationes Math.* **4**, 374–383.

Buchberger, B. (1985) 'Gröbner bases: an algorithmic method in polynomial ideal theory', In: Bose, N. K. (ed.), *Multidimensional System Theory*, Reidel, Dordrecht Boston Lancaster, 184–232.

Buchberger, B., Möller, H. M. (1982) 'The construction of multivariate polynomials with preassigned zeros', In: Goos, G., Hartmanis, J. (eds.), *Computer Algebra, EUROCAM '82, European Computer Algebra Conference*, Springer Lecture Notes in Computer Science **144**, 24–31.

Lorentz, R. A. (1992) 'Multivariate Birkhoff Interpolation', Springer Lecture Notes in Mathematics **1516**.

Marinari, M. G., Möller, H. M., Mora, T. (1991) 'Gröbner bases of ideals given by dual bases', In: *Proceedings of ISAAC 1991.*

Marinari, M. G., Möller, H. M., Mora, T. (1996) 'On multiplicities in polynomial system solving', *Trans. Amer. Math. Soc.* **348**, 3283–3321.

Möller, H. M. (1998) 'Gröbner Bases and Numerical Analysis', This volume.

Sauer, T. (1995) 'Computational aspects of multivariate polynomial interpolation', *Advances Comput. Math.* **3**, 219–238.

Sauer, T., Xu, Yuan (1995) 'On multivariate Hermite interpolation', *Advances Comput. Math.* **4**, 207–259.

Sauer, T. (1997) 'Polynomial interpolation of minimal degree', *Numer. Math.*, to appear.

Inversion of Birational Maps with Gröbner Bases

Josef Schicho

Research Institute for Symbolic Computation
Johannes Kepler University
A-4040 Linz, Austria
Josef.Schicho@risc.uni-linz.ac.at

Abstract

We present an efficient algorithm for the inversion of a given birational map. The problem is reduced to finding the unique solution of a maximal ideal defined over an algebraic function field.

1 Introduction

The problem of inverting a birational maps arises in several contexts in algorithmic algebraic geometry, and an efficient solution is useful in many situations.

For instance, consider the parameterization problem, which is useful for numerous applications in CAD/CAM (see section 3). A closer look to the existing parameterization algorithms reveals that a parameterization is often obtained via inversion of some birational map.

In this paper, we present a new efficient algorithm for the inversion of birational maps, based on the method of Gröbner bases (Buchberger 1965, Buchberger 1979, Buchberger 1983, Beckers and Weispfenning 1993).

For a special case, the inversion problem has also been investigated in (Essen 1990, Audoly *et al.* 1991) (see also (Ollivier 1989) for related work), in the context of the Jacobian conjecture (Keller 1939). The general problem was solved in (Sweedler 1993) (see also Shannon and Sweedler 1988). However, Sweedler's method depends on computing the components of the inverse map one by one, i.e. it requires the computation of a Gröbner basis with lexicographical termorder. It is known (see Faugère *et al.* 1993) that such a Gröbner basis is much harder to compute than Gröbner bases with respect to other term orders (e.g. the reverse lexicographical termorder).

In this paper, we reduce the problem to the computation of the unique solution of a maximal ideal defined over some function field. Since this ideal is zero-dimensional, Gröbner bases with respect to any termorders can be computed fast (Lakshman and Lazard 1991, Faugère *et al.* 1993). This overweighs by far the disadvantage that the field arithmetic has become is more difficult. This holds especially when the image variety is simple.

2 The problem

Let k be a field. An affine variety is the zero set of an absolutely prime ideal (i.e. a prime ideal which remains prime over the algebraic closure) of $k[x_1, \ldots, x_m]$. Whenever we consider zero sets, we mean zeroes over the algebraic closure. By Hilbert's Nullstellensatz, the correspondence between varieties and absolutely prime ideals is one-to-one.

Problem 1:
Input: An affine variety $V \subset \mathbf{A}^m$,
where $\mathbf{A}^m$ is the affine space of dimension m.
A birational map $f : V \to f(V) \subset \mathbf{A}^n$.
Output: The inverse $f^{-1} : W \to V$, where $W := f(V)$.

The variety V is represented by a set of polynomials in $k[x_1, \ldots, x_n]$ that generates Ideal(V).

The image of a rational map is the Zariski-closure of the set-theoretic image of the partial function f. It is again a variety (see Safarevic 1974).

3 Some applications

There are several situations in algorithmic algebraic geometry, in which one has a variety and a rational map, and one knows that the map is birational but does not know the inverse. Here is a simple example.

Example 1 Let $V = \mathbf{A}^1$, and let $f = (x^3 - x, x^5 - x)$. We show that the map f is birational to its image (a plane curve).

Let $K := k(F_1, F_2) \subset k(x)$. By Lüroth's theorem (see Winter 1974), we know that $K = k(R)$ for some $R \in k(x)$. Degree considerations reveal that R has degree 1. (The degree of a rational function is the maximum of the degrees of numerator and generator. The degree of any rational function in $k(R)$ is a multiple of the degree of R.) All rational functions of degree 1 generate the same field, so we have $k(F_1, F_2) = k(x)$. Thus, x can be expressed as a rational function in F_1 and F_2. This rational function is the inverse of f.

The equation of the image and the inverse map can be found by computing the resultant and the first subresultant of $x^3 - x - y_1$ and $x^5 - x - y_2$ with respect to x (see Hong and Schicho 1997). The equation of the image curve is

$$y_1^5 - 8y_1^2y_2 + 5y_1y_2^2 - y_2^3 + 4y_2^3 = 0,$$

and the inverse map is

$$(y_1, y_2) \mapsto x = \frac{y_1^2y_2}{y_2^2 - 2y_1y_2}.$$

In the above example, the image variety W is now known a priori. In other situations, W is known, for instance we can have $W = \mathbf{A}^n$. This case is particularily important, because it is related to the rational parameterization problem.

If $n = 2$, then we are dealing with parameterization of surfaces, which has many applications in CAD/CAM: for example, reliable surface plotting and display (see Farin 1988), motion display and computing transformations (see Hoffmann 1989), computing cutter offset curves (see Qiulin and Davies 1987, Arrondro *et al.* 1997), and many others. The reason why parameterization is important to these algorithms is that they use the NURBSs (see Piegl and Tiller 1997), i.e. certain parametric representations.

The parameterization problem asks for a parameterization of an implicitly given surface, in terms of rational functions in two parameters. Not every algebraic surface has a parameterization, but those who have form an interesting and important subclass (the class of rational surfaces).

In the parameterization algorithm (Schicho 1995, Schicho 1997) (which is the only one working for arbitrary rational surfaces), a birational map f is constructed from the given surface to the plane. Then, a parameterization is obtained by inverting f.

Example 2 Let V be the surface in $\mathbf{A}^3$ with equation

$$x_1(x_2^2 - x_1x_3)(x_3^2 - x_2) + (x_1 - x_2x_3)^2 = 0.$$

Let $f : V \to \mathbf{A}^2$ be the map

$$(x_1, x_2, x_3) \mapsto (y_1, y_2) = \left(\frac{x_2^2 - x_1x_3}{x_1 - x_2x_3}, \frac{x_3^2 - x_2}{x_1 - x_2x_3}\right).$$

This map is birational, and an inverse will be constructed below. The inverse map is a parameterization of V in terms of rational functions in y_1, y_2.

What is true in the surface case can also be said about the curve case (i.e. $W = \mathbf{A}^1$). Almost any known curve parameterization algorithm, e.g. those in (Walker 1978, Abhyankar and Bajaj 1987, Sendra and Winkler 1991, Schicho 1992, Hoeij 1994, Mnuk *et al.* 1996, Mnuk 1996, Sendra and Winkler 1997), first construct a certain birational function f from the given curve (even if it is not always given a name). Then, the parameterization is found by inverting f. This could be done by our general algorithm. However, experiments showed that the ad hoc methods used in the algorithms mentioned above (based on resultant techniques) outperform the general inversion algorithm.

4 The algorithm

First, we observe that the problem can be reduced to the following easier problem.

Problem 2:
Input: An affine variety $V \subset \mathbf{A}^m$.
A birational *polynomial* map $f : V \to f(V) \subset \mathbf{A}^n$.
Output: The image $W := f(V) \in \mathbf{A}^n$.
The inverse $f^{-1} : W \to V$.

To reduce problem 1 to problem 2, we use a sort of 'Rabinowitsch trick' to invert the denominator by introducing a new variable.

Reduction of problem 1 to problem 2:
Reduce the rational functions defining f
to the greatest common denominator D;
[Now, $f = (F_1/D, \ldots, F_n/D)$.]
$S' :=$ add $x_{m+1}D - 1$ to S;
$f' := (F_1 x_{m+1}, \ldots, F_n x_{m+1})$;
$Y, (G_1, \ldots, G_{m+1}) :=$ solution of problem 2 for S', f';
return $Y, (G_1, \ldots, G_m)$;

Example 3 Let V and f be as in example 2. Then V' is defined as the surface in $\mathbf{A}^4$ defined by the equations

$$x_1(x_2^2 - x_1x_3)(x_3^2 - x_2) + (x_1 - x_2x_3)^2 = x_4(x_1 - x_2x_3) - 1 = 0,$$

and $f' : V' \to \mathbf{A}^2$ is the polynomial map

$$(x_1, x_2, x_3, x_4) \mapsto (y_1, y_2) = ((x_2^2 - x_1x_3)x_4, (x_3^2 - x_2)x_4).$$

Lemma 1 *Let $V \subset \mathrm{A}^m$ be a variety. Let $S \subset k[x_1, \ldots, x_m]$ be a set of generators for Ideal(V). Let $D \in k[x_1, \ldots, x_m]$, such that $D \notin$ Ideal(V). Then the ideal generated by S and $x_{m+1}D - 1$ in $k[x_1, \ldots, x_{m+1}]$ is absolutely prime.*

Proof: We may assume that k is algebraically closed, and replace 'absolutely prime' by 'prime'.

The function ring $k[V] := k[x_1, \ldots, x_m]/(S)$ is an integral domain, and the canonical image $\overline{D}$ of D in A is not zero. Obviously, we have

$$k[x_1, \ldots, x_{m+1}]/(S, x_{m+1}D - 1) \cong k[V][x_{m+1}]/(x_{m+1}\overline{D} - 1).$$

This is isomorphic to the localization $D^{-1}k[V]$, which is again an integral domain.

Theorem 1 *The above reduction of problem 1 to problem 2 is correct.*

Proof: Let V' be the variety defined by S', and let $p : V' \to V$ be the projection on the first m coordinates. Then p is birational and $f \circ p = f'$. Hence $f(V) = f'(V')$. If g is the inverse of f', then $p \circ g$ is the inverse of f.

If the equations of W are not known, then they can be computed in the following way. Suppose that $f : V \to f(V) \in \mathbf{A}^n$ is given by the polynomials $F_1, \ldots, F_n \in k[x_1, \ldots, x_n]$. We have to eliminate the variables $x_1, \ldots, x_m$ in the system

$$S \cup \{y_1 - F_1, \ldots, y_n - F_n\}.$$

(Here $y_1, \ldots, y_n$ are new variables, forming a system of coordinates for $\mathbf{A}^n$.) This can be done by computing a Gröbner base with lexicographical term order (any y_i must be smaller than any x_j), and intersecting the result with $k[y_1, \ldots, y_n]$. See (Cox *et al.* 1992) for details. (The essential point of the proof is the elimination property of Gröbner bases (see Buchberger 1985).) However, computation of such a Gröbner basis can be expensive. In general, computing the image is the most expensive step in the map inversion problem. Fortunately, we have some knowledge about the image in most applications.

To compute the inverse map, we do arithmetic in the function field $k(W)$. Its elements are represented by fractions (F/G) of polynomials in $k[y_1, \ldots, y_n]$, such that $G \notin \mathrm{Ideal}(V)$. Since we are equipped with a Gröbner basis of Ideal(V), we can test for zero, by deciding whether $F \in \mathrm{Ideal}(V)$. (It will never be necessary to decide ideal membership of denominators, because all fractions in our computation arise by addition, multiplication, and division by nonzero elements, starting with a set of integral elements.)

We have a canonical homomorphism

$$\phi : k[x_1, \ldots, x_m] \to k(W)$$

with kernel Ideal(W). In the following, we will denote the canonical images by overlining.

Lemma 2 *If f is birational, then the ideal I generated by S and $\overline{y_1} - F_1, \ldots, \overline{y_n} - F_n$ is an absolutely maximal ideal in $k(W)[x_1, \ldots, x_m]$ (i.e. a maximal ideal which remains maximal over the algebraic closure of $k(W)$). The inverse of f is represented by the unique zero in $(G_1, \ldots, G_m) \in k(W)^m$.*

Proof: Note that $P_g := (\overline{y_1}, \ldots, \overline{y_n})$ is a generic point of W, i.e. a point defined over some field extension with the property that $F(P_g) = 0$ implies $F \in \mathrm{Ideal}(W)$ for all $F \in k[y_1, \ldots, y_m]$. The zeroes of the ideal I form the inverse image of the generic point. Since f is birational, the inverse image of a generic point consists of exactly one point, which lies a priori in the algebraic closure of $k(W)$.

Let $g : W \to V$ be the inverse of f, represented by $(G_1, \ldots, G_m) \in k(W)^m$. Then $f \circ g = \mathrm{id}_W$ implies $f(G_1, \ldots, G_m) = P_g$.

Now, we have to show that I is absolutely maximal. Since I has exactly one zero over the algebraic closure, it is zero-dimensional. Since the unique zero is defined over the field $k(W)$, the ideal I is a primary ideal, and its associated prime ideal is absolutely maximal. It remains to show that I is prime.

Let I' be the ideal in $k[W][x_1,\ldots,x_m] = k[x_1,\ldots,x_m,\overline{y_1},\ldots,\overline{y_n}]$ generated by S and $\overline{y_1} - F_1,\ldots,\overline{y_n} - F_n$. It is a prime ideal, because the quotient is an integral domain (isomorphic to $k[V]$). Let S' be the set of all nonzero elements in $k[W]$. Then $k(W)[x_1,\ldots,x_m]$ is the localization of $k[W][x_1,\ldots,x_n]$ at S', and I is the extension of I'. Moreover, S' is disjoint from I'. Then, primality of I follows from the lemma below.

Lemma 3 *Let A be a Noetherian integral domain, let I be a prime ideal, and let S be a multiplicative set disjoint from I. Then $S^{-1}I$ is prime in $S^{-1}A$.*

Proof: Let a/s_1, b/s_2 be two fractions such that their product is in $S^{-1}I$. Because I is finitely generated, there is an element $s_3 \in S$ such that $s_3 ab \in I$. Now, I is prime, and s_3 is not in I, hence either a or b is in I, and the lemma follows.

If I is a maximal ideal in $k[x_1,\ldots,x_m]$ with a solution defined over k, then a reduced Gröbner bases of I has the shape

$$\{a_1x_1 + b_1,\ldots,a_mx_m + b_m\}$$

for some $a_1,\ldots,a_m \in k - \{0\}$, $b_1,\ldots,b_m \in k$, regardless of the term order (see Becker and Weispfenning 1993). Therefore, it is trivial to find the solution once we have the Gröbner basis.

Algorithm *InvertMap*:

Input: A finite set $S \subset k[x_1,\ldots,x_m]$, representing a variety V in A^m.

Input: A finite set $T \subset k[y_1,\ldots,y_n]$, representing a variety W in A^n.
An n-tuple $(F_1,\ldots,F_n) \in k[x_1,\ldots,x_m]^n$,
representing a birational polynomial map f to W.

Output: An m-tuple $(G_1,\ldots,G_m) \in k[y_1,\ldots,y_n]^m$,
representing the inverse f^{-1}.

$S' := S \cup \{y_1 - F_1,\ldots,y_n - F_n\}$;
$S'' :=$ a Gröbner basis of S' over $k(W)$;
$(G_1,\ldots,G_n) :=$ solution of S'' (trivial to compute);
return $(G_1,\ldots,G_n)$

Example 4 Let V and f be as V' and f' in example 3. We compute a Gröbner basis of the set

$S' := \{x_1(x_2^2-x_1x_3)(x_3^2-x_2)+(x_1-x_2x_3)^2, x_4(x_1-x_2x_3)-1, (x_2^2-x_1x_3)x_4-y_1,$

$$(x_3^2 - x_2)x_4 - y_2\}$$

with total degree order, where y_1 and y_2 are regarded as constants. The result is

$$S'' = \{(y_2y_1^2 - y_1)x_2 + y_1^3 + 1, y_1y_2x_1 + 1, (y_2y_1^2 - y_1)x_3 - y_2 - y_1^2,$$
$$(y_2y_1^5 - y_1 + 3y_2y_1^2 + y_2^2)x_4 - y_2^3y_1^4 + 2y_2^2y_1^3 - y_2y_1^2\}.$$

Therefore, the solution is

$$(G_1, G_2, G_3, G_4) = (\frac{-1}{y_1y_2}, \frac{-y_1^3 - 1}{y_2y_1^2 - y_1}, \frac{y_2 + y_1^2}{y_2y_1^2 - y_1}, \frac{y_2^3y_1^4 - 2y_2^2y_1^3 + y_2y_1^2}{y_2y_1^5 - y_1 + 3y_2y_1^2 + y_2^2}).$$

This is a representation of the inverse map $f^{-1} : \mathbf{A}^2 \rightarrow V$. The first three components form a parameterization of the surface in example 2 (see also example 3).

Theorem 2 *The algorithm Invertmap is correct.*

Proof: Obvious from lemma 2.

In order to test the performance of the algorithm, we took 5 examples from (Schicho 1995, Schicho 1997) and did the inversion step using the algorithm *InvertMap*. In all cases except for one small example, our method outperforms Sweedler's method. The Gröbner bases computation depends on the order of the variables, so we tested several orders for each example. It turned out that the computing time of *InvertMap* is relatively stable, whereas with Sweedler's method one can have a bad chance to choose orders where the computing time is much larger. (This is not surprising, because lexicographical Gröbner bases depend very sensitively on the order of variables.)

The following table shows the computing time in CPU seconds. All computations were done with the Gröbner basis implemention of MAPLE on SGI Indigo with a 100 MHz IP20 processor.

Example	Input terms	*InvertMap*	Sweedler's method
1	13	5	65–106
2	12	0.07–1.5	3.5
3	9	1.5	11–12
4	10	2.5	1.5
5	6	5.5–6	32–1650
6	19	390–520	> 5400 (interrupted)

References

S.S. Abhyankar and B.Bajaj. Automatic parametrization of curves and surfaces III. *Computer Aided Geometric Design*, 5-4:309–323, 1987.

E. Arrondo, J. Sendra, and J. R. Sendra. Parametric generalized offsets to hypersurfaces. *Journal of Symbolic Computation*, 23:267–285, 1997.

S. Audoly and G. Bellu and A. Buttu and L. d'Angio. Procedures to investigate injectivity of polynomial maps and compute the inverse. *Applicable Algebra in Engineering, Communication and Computing*, 2:91–103, 1991.

T. Becker and V. Weispfenning. *Gröbner bases - a computational approach to commutative algebra.* Graduate Texts in Mathematics. Springer, 1993.

B. Buchberger. *An Algorithm for Finding a Basis for the Residue Class Ring of a Zero-Dimensional Polynomial Ideal.* PhD thesis, Universitat Innsbruck, Institut fur Mathematik, 1965. German.

B. Buchberger. A criterion for detecting unnecessary reductions in the construction of Grobner bases. In *Proceedings of the EUROSAM 79 Symposium on Symbolic and Algebraic Manipulation.* Springer, 1979.

B. Buchberger. A note on the complexity of constructing Grobner-bases. In *Lecture Notes in Computer Science - Proc. EUROCAL '83.* Springer, 1983.

B. Buchberger. Groebner bases: An algorithmic method in polynomial ideal theory. In N. K. Bose, editor, *Recent Trends in Multidimensional Systems Theory*, chapter 6. D. Riedel Publ. Comp., 1985.

D. Cox, J. Little, and D. O'Shea. *Ideals, Varieties, and Algorithms.* Undergraduate Texts in Mathematics. Springer, 1992.

G. F. Farin. *Curves and Surfaces for Computer Aided Geometric Design: A practical guide.* Academic Press, New York, 1988.

J.C. Faugère, P. Gianni, D. Lazard, and T. Mora. Efficient computation of zero-dimensional Groebner bases by change of ordering. *Journal of Symbolic Computation*, 16:329–344, 1993.

C. M. Hoffmann. *Geometric and Solid Modelling – an Introduction.* Morgan Kauffmann Publisher, San Mateo, California, 1989.

H. Hong and J. Schicho. Algorithms for trigonometric curves. Submitted to *Journal of Symbolic Computation*, 1997.

O. Keller. Ganze cremona transformationen. *Monatsh. Math. Phys.*, 47:299–306, 1939.

Y. N. Lakshman and D. Lazard. On the complexity of zero-dimensional algebraic systems. In *Proceedings MEGA '90*, pages 217–225. Birkhäuser, 1991.

M. Mñuk. *Algebraic parametrization of rational curves.* PhD thesis, RISC Linz, 1996.

M. Mñuk, J. R. Sendra, and F. Winkler. On the complexity of parametrizing curves. *Beitr. Algebra und Geometrie*, 37/2, 1996.

F. Ollivier. Inversibility of rational mappings and structural identifiability in automatics. *ISSAC-89*, pages 43–54, 1989.

L. Piegl and W. Tiller. *The NURBS book*. Springer, 1997.

D. Qiulin and B. J. Davies. *Surface engineering geometry for computer-aided design and manufacture*. John Wiley, 1987.

J. Schicho. On the choice of pencils in the parametrization of curves. *Journal of Symbolic Computation*, 14(6):557–576, 1992.

J. Schicho. *Rational parametrization of surfaces*. PhD thesis, RISC Linz, 1995.

J. Schicho. Rational parameterization of surfaces. Submitted to *Journal of Symbolic Computation*, 1997.

J.R. Sendra and F. Winkler. Symbolic parametrization of curves. *Journal of Symbolic Computation*, 12(6):607–632, 1991.

J.R. Sendra and F. Winkler. Parametrization of algebraic curves over optimal field extensions. *Journal of Symbolic Computation*, 23(2/3):191–208, 1997.

I. R. Shafarevich. *Algebraic geometry*. Springer, 1974.

D. Shannon and M. Sweedler. Using gröbner bases to determine algebra membership, split surjective algebra homomorphisms and determine birational equivalence. *Journal of Symbolic Computation*, 6(2/3):267–273, 1988.

M. Sweedler. Using Gröbner bases to determine the algebraic and transcendental nature of field extensions. *Applicable Algebra in Engineering, Communication and Computing*, 10:66–75, 1993.

A. van den Essen. A criterion to decide if a polynomial map is invertible and to compute the inverse. *Comm. Alg.*, 18(10):3183–3186, 1990.

M. van Hoeij. Computing parametrizations of rational algebraic curves. In *ISSAC-94*, pages 187–190. ACM Press, 1994.

R. J. Walker. *Algebraic Curves*. Springer, 1978.

D. J. Winter. *The structure of fields*. Springer, 1974.

Reverse lexicographic initial ideals of generic ideals are finitely generated

Jan Snellman

Department of Mathematics, Stockholm University,
106 91 Stockholm, Sweden;
jans@matematik.su.se

Abstract

This article generalizes the well-known notion of *generic forms* to the algebra R', introduced in [12]. For the total degree, then reverse lexicographic order, we prove that the initial ideal of an ideal generated by finitely many generic forms (in countably infinitely many variables) is finitely generated. This contrasts to the lexicographic order, for which initial ideals of generic ideals in general are non-finitely generated.

The natural question, "is the reverse lexicographic initial ideal of an homogeneous, finitely generated ideal in R' finitely generated" is posed, but not answered; we do, however, point out one direction of investigation that might provide the answer: namely to view such an ideal as the "specialization" of a generic ideal.

1 Introduction

In this article, we study the initial ideals of generic and "almost generic" ideals with respect to the (total degree, then) reverse lexicographic term order. For a generic ideal $I \subset K[x_1, \ldots, x_n]$, generated by $r \leq n$ forms, there is a well-known conjecture on how $\mathrm{gr}(I)$ looks like. In particular, $\mathrm{gr}(I)$ is minimally generated in $K[x_1, \ldots, x_r]$. We interpret this result in the setting of the ring R', introduced in [12]: this ring, which is a proper subring on the power series ring on countably many variables, and which properly contains the polynomial ring on the same set of indeterminates, is the habitat of "generic forms in countably many indeterminates". In the non-noetherian ring R', finitely generated, homogeneous ideals need not have finitely generated initial ideals; in fact, there are many finitely generated, generic ideals that have non-finitely generated initial ideals, with respect to the pure lexicographic term order. However, we show that the result above implies that finitely generated, generic ideals in R' have finitely generated initial ideals with respect to the reverse lexicographic term order. The key property of the degrevlex order that we use is the fact that the forming of initial ideals with respect to this order commutes with the truncation homomorphisms ρ_n, so that $\mathrm{gr}(\rho_n(I)) = \rho_n(\mathrm{gr}(I))$, whereas for arbitrary term orderings we only have an inclusion.

We also study variants of generic ideals, where the coefficients of the monomials of the forms lie not in the field, but in some other polynomial ring, which is mapped onto the ground field by a *specialization* map. We call such ideals *pure generic ideals.* At first, we study them in the polynomial ring $K[y_1, \ldots, y_{t_n}; x_1, \ldots, x_n]$, where we show that their initial ideals, with respect to the "twisted" product order of degrevlex on the two groups of variables, is minimally generated in $K[y_1, \ldots, y_{t_r}; x_1, \ldots, x_r]$, if the pure generic ideal is generated by r pure generic forms.

This construction can be generalized to the ring $K[Y][[X]]'$. We prove similar results on the initial ideals of pure generic ideals. In particular, we show that they are finitely generated.

Finally, we study *specialization maps* from this ring to R', that is, maps which fix the X-variables and map $K[Y]$ onto K. Since every finitely generated, homogeneous ideal in R' may be regarded as the specialization of a generic ideal, it is natural to ask if the initial ideal (with respect to the reverse lexicographic term order) of a finitely generated, homogeneous ideal in R' is finitely generated. We are unable to answer this question, but we present some ideas that might be used to tackle it.

An expanded version of this paper, containing some of the proofs that was left out due to lack of space, and some hopefully illuminating examples, can be found at the authors homepage, **http://www.matematik.su.se/users/jans/**.

2 Preliminaries

The rings, algebras, semi-groups and other devices used below are defined in greater detail in [12, 13], to which we refer the reader.

Let K be a field, and let Q be its prime field. For any positive integer n, we denote by $\mathcal{M}^n$ the free commutative semigroup on the letters $\{x_1, \ldots, x_n\}$, and by $\mathcal{M}^n_d$ the subset of elements of total degree d. Since the polynomial ring $K[x_1, \ldots, x_n]$ is the monoid ring of $\mathcal{M}^n$ over K, we can identify it with the set of all finitely supported maps from $\mathcal{M}^n$ to K. For an arbitrary element $h \in K[x_1, \ldots, x_n]$, we denote by $\mathrm{Coeff}(m, h)$ the value of the corresponding map at $m \in \mathcal{M}^n$, and by $\mathrm{Mon}(h)$ the support of the map.

We mean by a *form* of degree $|f| = d$ a homogeneous element. This element is said to be a *generic form* if, in addition, the set of its coefficients, that is, the set $\{\, \mathrm{Coeff}(m, f) \mid m \in \mathrm{Mon}(f) \,\}$ is algebraically independent over Q, if no two coefficients are equal, and if every monomial of appropriate total degree occur in the set of monomials: $\mathrm{Mon}(f) = \mathcal{M}^n_d$. An ideal I of $K[x_1, \ldots, x_n]$

is said to be *generic* if we can find a (finite) generating set, whose members are generic forms, and furthermore the union of the sets of coefficients of the generators is algebraically independent over Q; we also demand that no two occuring coefficients are equal.

These concepts are well-known and well-studied by algebraists ([6, 5, 14]). We now generalize them to (countably) infinitely many variables. For this purpose, we first introduce $R = K[[x_1, x_2, x_3, \dots]]$, the power series ring on countably many variables, and then define the K-algebra R' as the sub-algebra of R that is generated by all homogeneous elements. We denote by $\mathcal{M}$ the free commutative monoid on the x_i's (in other words, the direct limit of the $\mathcal{M}^n$'s) and by $\mathcal{M}_d$ the subset of all elements of degree d. Then, elements of R may be viewed as maps from $\mathcal{M}$ to K, and we can define $\mathrm{Coeff}(m, h)$ and $\mathrm{Mon}(h)$ analogously to the polynomial case. We remark that similar rings have been studied extensively in the litterature; see for instance [10].

We mean by a *form in* R' a homogeneous element f in R'. A *generic form in* R', is a form f in R' such that $\{ \mathrm{Coeff}(m, f) \mid m \in \mathrm{Mon}(f) \}$ is algebraically independent over Q, such that no two coefficients occuring are equal, and such that $\mathrm{Mon}(f) = \mathcal{M}_{|f|}$. By a *generic ideal in* R' we mean an ideal I for which a finite set of generators, which are generalized forms, can be found, such that the union of the sets of coefficients for the generators is algebraically independent over Q, and such that no two coefficients occuring are equal. In particular, such an ideal is homogeneous and finitely generated.

We assume that K contains infinitely many elements that are transcendental over Q, and algebraically independent over Q.

In this article, except where otherwise stated, $>$ will denote the total-degree, then reverse lexicographic order on the semi-group $\mathcal{M}$ of monomials in the variables $x_1, x_2, x_3, \dots$, as well as its restriction to the subsemigroups $\mathcal{M}^n$. It is enough to define $>$ on each $\mathcal{M}^n$, where $x_1^{\alpha_1} x_2^{\alpha_2} \cdots x_n^{\alpha_n} > x_1^{\beta_1} x_2^{\beta_2} \cdots x_n^{\beta_n}$ if $\sum_{i=1}^n \alpha_i > \sum_{i=1}^n \beta_i$ or the total degrees are equal but $\exists r \in \{1, \dots, n\} : (\alpha_r < \beta_r) \wedge (i > r \implies \alpha_i = \beta_i)$. We say that $>$ is an *admissible order* on $\mathcal{M}$, by which we mean that it is a total order with 1 as the smallest element and such that $x_i > x_j \iff i < j, p > q \implies pt > pq, p, q, t \in \mathcal{M}$. It was showed in [12] that if $f \in R'$ and $>$ is an admissible order on M, then $\mathrm{Mon}(f) \subset \mathcal{M}$ has a maximal element (with respect to $<$) $\mathrm{Lpp}(f)$, which we call the *leading power product* of f. Therefore, we can associate to any ideal I in R' its *initial ideal* $\mathrm{gr}(I)$, the monomial ideal generated by all leading power products of elements in I. It was also showed that if I is locally finitely generated, that is, homogeneous and posesses a homogeneous generating set with only finitely many elements of any given total degree, then the initial ideal share that property. In order to show this, a Gröbner basis theory for locally finitely generated ideals in R' was developed. Since the polynomial rings $K[x_1, \dots, x_n]$ are embedded in R', this theory extends the classical theory pioneered by Buchberger [3, 4] (see also [2, 11]). In fact, most of the well-known results carry over to this case, and the proofs are either trivial modifications of the ordinary proofs, or reductions to the polynomial ring case. There are however some dissimilarities, due to the fact that R' is non-noetherian.

If n is any positive integer, denote by B_n the ideal generated in R by all power series in $K[[x_{n+1}, x_{n+2}, x_{n+3}, \dots]]$ with zero constant term. Then the *n'th truncation homomorphism* is defined by $\rho_n : R \twoheadrightarrow \frac{R}{B_n} \simeq K[[x_1, \dots, x_n]]$; restricted to R' it has image $K[x_1, \dots, x_n]$, and when restricted to $K[x_1, \dots ; x_m]$, $m \geq n$, it coincides with the homomorphism defined by $K[x_1, \dots, x_m] \twoheadrightarrow \frac{K[x_1,\dots,x_m]}{(x_{n+1},\dots,x_m)} \simeq K[x_1, \dots, x_n]$. We will abuse notations and let ρ_n denote both the function itself, and its restrictions to R' and $K[x_1, \dots, x_n]$.

The homomorphism ρ_n is the "linear extension" of its restriction to the monoid $\mathcal{M}$ (which is not a K-vector space basis) in the sense that $\rho_n \left(\sum_{m \in \mathcal{M}} c_m m\right) = \sum_{m \in \mathcal{M}^n} c_m m$. This is certainly not true for all homomorphisms from R; for instance, the quotient epimorphism $R \twoheadrightarrow R/(x_1, x_2, x_3, \dots)$ vanishes at every monomial, but is not identically zero.

3 Initial ideals of generic ideals in $K[x_1, \dots, x_n]$

In this section, we concern ourselves with the generic ideal $I = (f_1, \dots, f_r) \subset K[x_1, \dots, x_n]$ generated by generic forms f_i with total degree d_i. We note that the initial ideal is determined by the d_i; if $I' = (g_1, \dots, g_r)$ is another generic ideal, generated by generic forms g_i with $|g_i| = d_i$, then $\mathrm{gr}(I') = \mathrm{gr}(I)$. This holds for any admissible order, but, as stated above, we are interested in the case of the graded reverse lexicographic order.

To start, we establish two basic properties of the reverse lexicographic order:

Lemma 3.1. *If $h \in R'$ is homogeneous, and if v is any positive integer, then either $\rho_v(h) = 0$, or $\mathrm{Lpp}(h) = \mathrm{Lpp}(\rho_v(h))$.*

In particular, the result holds for $h \in K[x_1, \dots, x_n]$.

Lemma 3.2. *For any homogeneous ideal $J \subset R'$, and any positive integer v, we have that $\rho_v(\mathrm{gr}(J)) = \mathrm{gr}(\rho_v(J))$. The same formula holds for homogeneous ideals in $K[x_1, \dots, x_n]$.*

Proof. It is enough to prove the assertion about ideals in R'. By [13, Lemma 3.3], we have that $\rho_v(\mathrm{gr}(J)) \subset \mathrm{gr}(\rho_v(J))$. It remains to prove the reverse inclusion. Let $0 \neq m \in \mathrm{gr}(\rho_v(J)) \cap \mathcal{M}^n$, then there exists a homogeneous element $h \in R'$ such that $m = \mathrm{Lpp}(\rho_v(h))$. By Lemma 3.1, $\mathrm{Lpp}(h) = m$. Clearly, $m \in \mathcal{M}^v$, so that $\rho_v(m) = m$. Therefore, $m \in \rho_v(\mathrm{gr}(J))$. □

We also need

Lemma 3.3. *The image of I under the epimorphism ρ_r is a generic ideal in $K[x_1, \ldots, x_r]$.*

In [9, Section I.3], Moreno defines the *stairs* (with respect to $>$) of I as $E(I) = \mathcal{M}^n \setminus (\mathrm{gr}(I) \cap \mathcal{M}^n)$. In passing, he notes:

Proposition 3.4. *If $r < n$, then the stairs are* cylindrical, *that is,* $E(I) = \tilde{E}^0 \times \mathbb{N}$ *where* $\tilde{E}^0 = E(I) \cap \mathcal{M}^{n-1}$.

Corollary 3.5. *If $r < n$, then the minimal generators of* $\mathrm{gr}(I)$ *are contained in* $\mathcal{M}^{n-1}$.

In fact, we have

Proposition 3.6. *If $r < n$, then the minimal generators of* $\mathrm{gr}(I)$ *are contained in* $\mathcal{M}^r$, *and furthermore* $\mathrm{gr}(I) = \mathrm{gr}(\rho_r(I))^e$, *where the extension is to* $K[x_1, \ldots, x_n]$.

Moreno discusses in [9, Conjecture I.4.1] a conjecture, which, if it holds true (and the computational evidence for its veracity is owerwhelming) completely determines the structure of the $\mathrm{gr}(I)$. The claim of the conjecture is as follows: by definition, $\mathrm{gr}(I)$ has minimal monomial generators $m_1, \ldots, m_v$. Denote by $\mathrm{gr}(I)_{<d}$ the monomial ideal generated by those m_j's that have total degree $< d$. Then, the conjecture claims that the minimal monomial generators of degree d are those monomials of $\mathcal{M}_d^n \setminus (\mathrm{gr}(I)_{<d} \cap \mathcal{M}_d^n)$ that occupy the first $w(d)$ available spots, where $w(d)$ is determined by the difference of the Hilbert series of $\frac{K[x_1,\ldots,x_n]}{I}$ and the Hilbert series of $\frac{K[x_1,\ldots,x_n]}{\mathrm{gr}(I)_{<d}}$ in degree d.

4 Initial ideals of generic ideals in R'

We now generalize the results of the previous section to the ring R'. To that purpose, let $I = (f_1, \ldots, f_r) \subset R'$ be a generic ideal, generated by generic forms f_i with $\deg f_i = d_i$. As before, we note that the initial ideal is determined by the d_i's.

From Proposition 3.6 we can determine the structure of (almost) all generic initial ideals $\mathrm{gr}(\rho_n(I))$:

Proposition 4.1. *For all non-negative integers s,* $\mathrm{gr}(\rho_r(I))^e = \mathrm{gr}(\rho_{r+s}(I))^e$, *where the extension is to* R'.

We now use the theorem of degree-wise approximation from [13], which state that for all total degrees d, there exists an integer $N(d)$ such that, for any $n \geq N(d)$ we have that $\mathrm{gr}(I)_d = \mathrm{gr}(\rho_n(I))_d^e$ where the right-hand side is extended to R' using the natural inclusion. Since $\mathrm{gr}(\rho_n(I))_d^e$ stabilizes for $n \geq r$, for any d, we conclude:

Theorem 4.2. *For $n \geq r$,* $\mathrm{gr}(\rho_n(I))^e = \mathrm{gr}(I)$. *Thus,* $\mathrm{gr}(I)$ *is generated in* $\mathcal{M}^r$, *and is finitely generated.*

5 Initial ideals of "almost" generic ideals in $K[x_1, \ldots, x_n]$

5.1 The associated homogeneous ideal

For any f in the ring $K[x_1, \ldots, x_n]$ we denote by $c(f)$ the homogeneous component of f of maximal degree. If $I \subset K[x_1, \ldots, x_n]$ is an ideal, we denote by $c(I)$ the homogeneous ideal generated by all $c(f)$ for $f \in I$. This homogeneous ideal is the graded associated ideal with respect to the total degree filtration; since the initial ideal $\mathrm{gr}(I)$ is the graded associated ideal to the filtration induced by Lpp, and since this latter filtration is a refinement of the total-degree filtration, we have that $\mathrm{gr}(I) = \mathrm{gr}(c(I))$. We can also see this directly: for any $f \in I$, we have that $\mathrm{Lpp}(f) = \mathrm{Lpp}(c(f))$.

It is well know that not every generating set of an ideal is a Gröbner Basis. Similarly, not every generating set F of I has the property that $\{\, c(f) \mid f \in F \,\}$ generates $c(I)$. In the generic case, however, we have the following:

Lemma 5.1. *Let $J = (f_1, \ldots, f_r) \subset K[x_1, \ldots, x_n]$, with $r \leq n$. Suppose that all $c(f_i)$'s are generic, as is $I = (c(f_1), \ldots, c(f_r))$. Then $c(J) = I$.*

Proof. Assume, towards a contradiction, that there exists an $f \in J$ such that $c(f) \in c(J) \setminus I$. Let $d = |f|$. Since $f \in J$ we can write $f = \sum_{i=1}^{r} q_i f_i$. Furthermore, since $c(f) \notin I$, we must have that $\max_i |q_i f_i| > d$. Put $\mathcal{S} = \left\{ S = (a_1, \ldots, a_r) | a_i \in K[x_1, \ldots, x_n],\ f = \sum_{i=1}^{r} a_i f_i,\ \max_i |a_i f_i| > d \right\}$. For $S \in \mathcal{S}$, put $\delta_S = \max_i |a_i f_i|$. By assumptions, $\delta_S > d$, and $\mathcal{S}$ is non-empty, containing the element $(q_1, \ldots, q_r)$. Since the set $\{\, \delta_S \mid S \in \mathcal{S} \,\}$ is a non-void subset of the natural numbers, it contains a minimum. Choose an $S = (a_1, \ldots, a_r)$ where that minimum is obtained.

Now, the $c(f_i)$'s form a regular sequence, so all syzygies involving them are trivial (see [8, Theorem 16.5]) . We apply this to the homogeneous component of maximal (δ_S) degree in $f = \sum_{i=1}^{r} a_i f_i$. Denoting by $V \subset \{1, \ldots, r\}$ the set of the indices for which $|a_i f_i| = \delta_S$, we get that $0 = \sum_{v \in V} c(a_v) c(f_v)$. This is a syzygy, and must be a trivial one, that is, it must be generated by vectors $(0, \ldots, c(f_w), 0, \ldots, -c(f_v), 0, \ldots)$ with non-zero entries in positions v and w. This is a vector of length equal to the cardinality of V. Summing up, we have that $\forall v \in V: \quad c(a_v) = \sum_{w \in V} e_{vw} f_w$, where the homogeneous e_{vw}'s fulfill $e_{vw} = -e_{wv}$, $e_{vv} = 0$. By defining $e_{ij} = 0$ whenever $(i,j) \notin V \times V$, we get an $r \times r$ skew-symmetric matrix $E = (e_{ij})$ such that $\forall 1 \leq i \leq r$: $c(a_i) = \sum_{j=1}^{r} e_{ij} f_j$. Since E is skew-symmetric, for all vectors $x = (x_1, \ldots, x_r)$ in the r-fold cartesian product $K[x_1, \ldots, x_n]^r$ we have that $x E x^t = 0$. We apply this to the vector $(f_1, \ldots, f_r)$, and get that $\sum_{i=1}^{r} \sum_{j=i}^{r} e_{ij} f_i f_j = 0$. The conclusion draws near. Using the above, we write $f = \sum_{i=1}^{r} a_i f_i = \sum_{i=1}^{r} a_i f_i -$

$\sum_{i=1}^{r}\sum_{j=i}^{r} e_{ij}f_if_j = \sum_{i=1}^{r}\left(a_i - \sum_{j=i}^{r} e_{ij}f_j\right) f_i$. Now put $b_i = a_i - \sum_{j=i}^{r} e_{ij}f_j$ for $1 \leq i \leq r$. Since $c(a_i) = \sum_{j=i}^{r} e_{ij}c(f_j)$ we get that $|b_i| < |a_i|$ hence that $|b_if_i| < \delta_S$. But then $f = \sum_{i=1}^{r} b_if_i$ and $T = (b_1, \ldots, b_r) \in \mathcal{S}$ with a $\delta_T < \delta_S$. This contradicts the minimality of δ_S. □

Remark 5.2. It follows from our discussion above that $\mathrm{gr}(J) = \mathrm{gr}(c(I)) = \mathrm{gr}(I)$.

5.2 "Almost generic" ideals

What happens if we start with a generic ideal, generated by generic forms, and then replace some of the coefficients of the monomials in the forms with non-generic values?

In the first lemma, we study what happens when we leave the coefficients of monomials in $\mathcal{M}^r$ as they are, but manipulate the others:

Lemma 5.3. *Let I be a generic ideal in $K[x_1, \ldots, x_n]$ generated by generic forms f_1 to f_r, with $r < n$. For $1 \leq i \leq r$, $d_i = |f_i|$, let $g_i \in K[x_1, \ldots, x_n]$ be homogeneous of degree d_i, and such that each monomial in* $\mathrm{Mon}(g_i)$ *is divisible by at least one of the variables $x_{r+1}, \ldots, x_n$; put $h_i = f_i + g_i$. Denote by J the ideal $(h_1, \ldots, h_r) \subset K[x_1, \ldots, x_n]$. Then* $\mathrm{gr}(I) = \mathrm{gr}(J)$.

Proof. We know from previous results that $\mathrm{gr}(I) = \mathrm{gr}(\rho_r(I)^e)$, where the extension is to $K[x_1, \ldots, x_n]$. Since it will simplify our proof, we henceforth assume that the f_i's are generic forms in $K[x_1, \ldots, x_r]$. Let $m \in \mathrm{gr}(I) \cap \mathcal{M}^n$, $m \neq 0$. Then there exists a $g \in I$ with $\mathrm{Lpp}(g) = m$. We can write $g = \sum_{i=1}^{r} e_if_i$, where the e_i's are homogeneous. Put $h = \sum_{i=1}^{r} e_ih_i = \sum_{i=1}^{r} e_if_i + \sum_{i=1}^{r} e_ig_i$. Clearly, each monomial in $\mathrm{Mon}(\sum_{i=1}^{r} e_if_i)$ is greater than any monomial in $\mathrm{Mon}(\sum_{i=1}^{r} e_ig_i)$. It follows that $\mathrm{Lpp}(h) = \mathrm{Lpp}(g) = m$. Therefore, $m \in \mathrm{gr}(J)$.

We have showed that $\mathrm{gr}(I) \subset \mathrm{gr}(J)$. Since I is generic, the quotient $\frac{K[x_1,\ldots,x_n]}{I}$ has (lexicographically) minimal Hilbert series among all quotients of $K[x_1, \ldots, x_n]$ by a homogeneous ideal generated by forms of degree d_1 to d_r. This useful property was shown by Fröberg in [5], and is to be interpreted in the following way: if we write the Hilbert series of the generic quotient as $\sum_{k=0}^{\infty} v_kt^k$ and the Hilbert series of the other algebra as $\sum_{k=0}^{\infty} w_kt^k$, then if the set $\{ v_k - w_k \mid k \in \mathbb{N} \} \setminus \{0\}$ is non-empty, and if k is the smallest k such that $v_k \neq w_k$, then $v_k < w_k$.

The ideal J belongs to the prescribed class of homogeneous ideals. Therefore, $\frac{K[x_1,\ldots,x_n]}{J}$ have a Hilbert series that is no smaller than the Hilbert series of $\frac{K[x_1,\ldots,x_n]}{I}$, hence $\frac{K[x_1,\ldots,x_n]}{\mathrm{gr}(J)}$ have a Hilbert series that is no smaller than that of $\frac{K[x_1,\ldots,x_n]}{\mathrm{gr}(I)}$. This shows that the inclusion $\mathrm{gr}(I) \subset \mathrm{gr}(J)$ can not be strict. □

If on the other hand J is obtained from I by destroying the genericity of the *highest* variables, then we can not hope to get the same initial ideal. We believe, however, that the initial ideal is generated in the $r+v$ first variables, where v denotes the index of the last variable that is manipulated:

Conjecture 5.4. *Let $J \subset K[x_1,\ldots,x_v,x_{v+1},\ldots,x_{v+s}]$ be a homogeneous ideal generated by homogeneous $f_1,\ldots,f_r$ with $r \leq s$. Assume that the polynomials $\bar{f}_1,\ldots,\bar{f}_r$ generate a generic ideal in $K[x_{v+1},\ldots,x_{v+s}] \simeq \frac{K[x_1,\ldots,x_v,x_{v+1},\ldots,x_{v+s}]}{(x_1,\ldots,x_v)}$, where $\bar{f}_i$ denotes the image of f_i. Then the monoid ideal $\mathrm{gr}(J) \cap \mathcal{M}^{v+s}$ is minimally generated in $\mathcal{M}^{v+r}$.*

Remark 5.5. The conjecture is easily seen to be true in the two "extreme cases": when the f_i's are generic forms in $K[x_1,\ldots,x_v,x_{v+1},\ldots,x_{v+s}]$, we have that $\mathrm{gr}(J) \cap \mathcal{M}^{v+s}$ is generated in $\mathcal{M}^r$; when $f_i = \bar{f}_i$ for all i, clearly $\mathrm{gr}(J) \cap \mathcal{M}^{v+s}$ is generated in the commutative monoid on the letters $x_{v+1},\ldots,x_{v+r}$ and in particular in $\mathcal{M}^{v+r}$. The author has checked several other examples by computer.

5.3 Initial ideal generic ideals with "ordered coefficients"

Let n, r and $d_1,\ldots,d_r$ be positive integers, and define $t_n = \sum_{i=1}^r \binom{n+d_i-1}{n-1}$. Then t_n is the cardinality of the disjoint union of the set of monomials of degree d_i in $K[x_1,\ldots,x_n]$, for $i \leq i \leq r$. We can therefore define $f_i := \sum_{m \in \mathcal{M}^n_{d_i}} y_{i,m} m \in S_n$, where $S_n = K[\{y_{i,m}\}][x_1,\ldots,x_n] \simeq K[y_1,\ldots,y_{t_n}][x_1,\ldots,x_n]$, and put $I = (f_1,\ldots,f_r)$. The ordering of the y_i's is such that $y_1,\ldots,y_{t_1}$ are the variables that occur together with monomials in $\mathcal{M}^1$, $y_{t_1+1},\ldots,y_{t_2}$ together with monomials in $\mathcal{M}^2 \setminus \mathcal{M}^1$, and so forth. We say that the f_i's are *pure generic forms*, and that I is a *pure generic ideal.*

Example 5.6. If $r = n = d_1 = d_2 = 2$ then $f_1 = y_1x_1^2 + y_3x_1x_2 + y_4x_2^2$, and $f_2 = y_2x_1^2 + y_5x_1x_2 + y_6x_2^2$.

Let $>$ be the total degree, then reverse lexicographic order on S_n, when the Y-variables are given weight 0. This is the same as taking the "twisted" product order of revlex on the two submonoids on the y's and on the x's. That is, when comparing two monomials tm and $t'm'$, with $t,t' \in [y_1,\ldots,y_{t_n}]$, $m,m' \in [x_1,\ldots,x_n]$, we first compare m and m', and only if they are equal do we compare t and t'. Here, $[x_1,\ldots,x_n]$ denotes the free commutative monoid on the letters $x_1,\ldots,x_n$, and similarly for the y_j's.

The following lemma is obvious:

Lemma 5.7. *Let $>_{rev}$ be the ordinary degrevlex order on S_n (that is, when the y-variables are given weight 1), let $f \in S_n$ be bi-homogeneous with respect the two groups of variables, and let $J \subset S_n$ be generated by such bi-homogeneous elements. Then $\mathrm{Lpp}_>(f) = \mathrm{Lpp}_{>_{rev}}(f)$, and $\mathrm{gr}_>(J) = \mathrm{gr}_{>_{rev}}(J)$.*

In particular, this holds for the pure generic ideal I.

For any $1 \leq v < n$, we denote by $\rho_{*,v}$ the epimorphism $S_n \twoheadrightarrow S_n/(x_{v+1},\dots,x_n) \simeq K[y_1,\dots y_{t_n}][x_1,\dots,x_v]$.
We need "bi-graded" counterparts of Lemma 3.1 and Lemma 3.2. Since these results hold in a more general setting, namely in the ring $K[Y][[X]]'$, defined in Section 6, we do not give proofs here, but refer to the proofs of the more general Lemma 6.2 and Lemma 6.3.

Lemma 5.8. *If $h \in S_n$ is $\mathcal{M}$-homogeneous (that is, homogeneous when the Y-variables are given weight zero), and if $1 \leq v \leq n$, then either $\rho_{*,v}(h) = 0$, or $\text{Lpp}(h) = \text{Lpp}(\rho_{*,v}(h))$.*

Lemma 5.9. *For any $\mathcal{M}$-homogeneous ideal $J \subset S_n$, and for $1 \leq v \leq n$, we have that $\rho_{*,v}(\text{gr}(J)) = \text{gr}(\rho_{*,v}(J))$.*

The following lemma is a key ingredient in the proof of the generalization of Proposition 3.6:

Lemma 5.10. *If $r \leq n$ then S_n/I is a complete intersection.*

Proof. Let $V \subset \{y_1,\dots,y_{t_n}\}$ be the set of all variables y_v except those that occur as the coefficient of $x_i^{d_i}$ in f_i, and let J be the ideal generated by V. If we re-order the Y-variables so that $V = \{y_{r+1},\dots,y_{t_n}\}$ and $\{y_1,\dots,y_r\}$ are those Y-variables not in V, then the image $\bar{f}_i$ of f_i in S_n/J is $y_i x_i^{d_i}$. Therefore, $\frac{S_n}{I+J} \simeq \frac{K[y_1,\dots,y_r;x_1,\dots,x_n]}{(y_1 x_1^{d_1},\dots,y_r x_r^{d_r})}$, which is a complete intersection because the support of the monomials are disjoint; so it has Hilbert series $\frac{\prod_{i=1}^r (1-z^{(1+d_i)})}{(1-z)^{n+r}}$. We now get that $I+J = (f_1,\dots,f_r,y_{r+1},\dots,y_{t_n})$. Since $S_n/(I+J)$ has Hilbert series $\frac{\prod_{i=1}^r (1-z^{(1+d_i)})}{(1-z)^{(n+r)}} = \frac{(1-z)^{(t_n-r)}\prod_{i=1}^r (1-z^{(1+d_i)})}{(1-z)^{(n+t_n)}}$, it follows that $f_1,\dots,f_r,y_{r+1},\dots,y_{t_n}$ must be a regular sequence in S_n. Therefore, $f_1,\dots,f_r$ is also a regular sequence, hence S_n/I is a complete intersection. □

Proposition 5.11. *If $r \leq n$, then the minimal monomial generators of $\text{gr}(I)$ are contained in $[y_1,\dots,y_{t_n}] \oplus \mathcal{M}^r$, and furthermore $\text{gr}(I) = \text{gr}(\rho_{*,r}(I))^e$, where the extension is to S_n.*

Proof. By Lemma 5.10, S_n/I is a complete intersection; it follows from this that so is $K[y_1,\dots,y_{t_n}][x_1,\dots,x_r]/\rho_{*,r}(I)$. Therefore, their bi-graded Hilbert series are, respectively, $(1-u)^{-t_n}(1-v)^{-n}\prod_{i=1}^r(1-uv^{d_i})$ and $(1-u)^{-t_n}(1-v)^{-r}\prod_{i=1}^r(1-uv^{d_i})$; this is also the bi-graded Hilbert series of $S_n/\text{gr}(I)$ and $K[y_1,\dots,y_{t_n}][x_1,\dots,x_r]/\text{gr}(\rho_{*,v}(I))$. By Lemma 5.9, we can regard $\text{gr}(\rho_{*,v}(I))^e$ as a subideal of $\text{gr}(I)$ (the extension is to S_n). Since these ideals have the same bi-graded Hilbert series, they are equal. □

Since $\rho_{*,r}(I)$ is generated in S_r, we must have that a minimal Gröbner basis of the ideal is contained in that subring of S_n. Therefore:

Corollary 5.12. *If $r < n$, then the minimal monomial generators of $\text{gr}(I)$ are contained in $[y_1,\dots,y_{t_r}] \oplus \mathcal{M}^r$.*

6 Initial ideals of generic ideals in $K[Y][[X]]'$

If X is any set, and C is a commutative ring, we denote by $[X]$ the free commutative monoid on X, by $C[X]$ the polynomial ring on X with coefficients in C, with $C[[X]]$ the power series ring on X with coefficients in C, and by $C[[X]]'$ the subset of $C[[X]]$ consisting of all elements with bounded total degree with respect to $[X]$.

We now let $X = \{x_1, x_2, x_3, \dots\}$ and $Y = \{y_1, y_2, y_3, \dots\}$. Then $\mathcal{M} = [X]$ and $R' = K[[X]]'$. To generalize the results of Section 5.3, we consider the ring $S := K[Y][[X]]'$. The underlying monoid is $[Y] \oplus [X]$, which we order by the (graded) reverse lexicographic order where the Y-variables are given weight 0, that is, by the "twisted" product order of revlex on the two subsemigroups. There is no problem in finding leading power products in this ring.

For any $f \in K[Y][[X]]'$, we denote by $\mathrm{Mon}(f) \subset [Y] \oplus [X]$ the set of XY-monomials occuring with non-zero coefficient in f.

Remark 6.1. Note that $R'[Y] = K[[X]]'[Y] \subsetneq K[Y][[X]]'$, since $\sum_{i=1}^{\infty} y_i x_i \in K[Y][[X]]' \setminus R'[Y]$.

Let $r, d_1, \dots, d_r, n$ be positive integers, let t_n be as above, and set $f_{i,n} := \sum_{m \in \mathcal{M}^n_{d_i}} y_{i,m} m \in K[\{y_{i,m}\}][x_1, \dots, x_n] \simeq K[y_1, \dots, y_{t_n}][x_1, \dots, x_n] = S_n$. Then there exists $f_1, \dots, f_r \in K[Y][[X]]'$ such that for each i, v, $\rho_{*,v}(f_i) = f_{i,v}$. We have here generalized the definition of $\rho_{*,v}$ given in (5.3), so that $\rho_{*,v}$ is the quotient epimorphism $S \twoheadrightarrow S/C_v \simeq K[Y][x_1, \dots, x_v]$, where C_v is the ideal generated by all power series in $K[[x_{v+1}, x_{v+2}, x_{v+3}, \dots]]$ with zero constant term. This coincides with the former definition on S_n. Note that the orderering of the Y-variables is defined so that $S_n \subset S_{n+1} \subset S$ for all n, and that the S_n's form an increasing, exhaustive filtration on S.

Now put $I = (f_1, \dots, f_r)$. We say that I is a *pure generic ideal* in $K[Y][[X]]'$. If $n > r$, then $\rho_{*,n}(I)$ is an ideal in $K[Y][x_1, \dots, x_n]$ but generated in S_n, so that, by Proposition 5.11, we have that $\mathrm{gr}(\rho_{*,n}(I))$ is generated in S_r, and is in fact the extension of $\mathrm{gr}(\rho_{*,r}(I))$. Note also that Lemma 5.7 holds also in this more general situation, and that I is bi-homogeneous, so that $\mathrm{gr}(I)$ is in fact the initial ideal with respect to the graded revlex order on $K[Y][[X]]'$ (when the Y-variables have weight 1).

The following two lemmas generalize Lemma 5.8 and Lemma 5.9:

Lemma 6.2. *If $h \in K[Y][[X]'$ is $\mathcal{M}$-homogeneous (that is, homogeneous when the Y-variables are given weight zero), and if v is any positive integer, then either $\rho_{*,v}(h) = 0$, or $\mathrm{Lpp}(h) = \mathrm{Lpp}(\rho_{*,v}(h))$.*

Lemma 6.3. *For any $\mathcal{M}$-homogeneous ideal $J \subset K[Y][[X]]'$, and any positive integer v, we have that $\rho_{*,v}(\mathrm{gr}(J)) = \mathrm{gr}(\rho_{*,v}(J))$.*

Theorem 6.4. *The minimal monomial generators of $\mathrm{gr}(I)$ are contained in $[y_1, \dots, y_{t_r}] \oplus \mathcal{M}^r$, and furthermore $\mathrm{gr}(I) = \mathrm{gr}(\rho_{*,r}(I))^e$, where the extension is to S.*

The coefficient ideals $(\mathrm{gr}(I) : m) \cap K[Y]$, $m \in \mathcal{M}$, *are finitely generated monomial ideals generated in* $K[y_1, \ldots, y_{t_r}]$.

Proof. As observed above, for any $n > r$ we have that $\mathrm{gr}(\rho_{*,n}(I)) = \mathrm{gr}(\rho_{*,r}(I))^e$, where the extension is to S_n. Since, by Lemma 6.3, $\mathrm{gr}(\rho_{*,n}(I)) = \rho_{*,n}(\mathrm{gr}(I))$, and since for any monomial $m \in \mathrm{gr}(I) \cap ([Y] \oplus [X])$ there is an N such that $n > N \implies m \in \rho_{*,n}(\mathrm{gr}(I))$, we get that $\mathrm{gr}(I)$ is generated in $K[y_1, \ldots, y_{t_r}][x_1, \ldots, x_r]$.

We then have that $\mathrm{gr}(I) = (t_1 m_1, \ldots, t_s m_s)$, $t_i \in [y_1, \ldots, y_{t_r}]$, $m_i \in [x_1, \ldots, m_r]$. For any $m \in \mathcal{M}$, we have that $(\mathrm{gr}(I) : m) \cap K[Y]$ is generated by those t_i for which $m_i | m$. □

6.1 Regarding the Y-variables as coefficients in a domain

The admissible order $>$ on $[Y] \oplus [X]$ restricts to the ordinary reverse lexicographic order on $[X]$. We can also view it (the restriction) as a linear quasi order (using the terminology of [2]) $>_x$ on $[Y] \oplus [X]$. Adopting this view, we regard $K[Y][[X]]'$ as having underlying monoid $[X]$, ordered by $>_x$, and having coefficients in the domain $K[Y]$. We define the *initial term* of an element $f \in K[Y][[X]]'$ as $\mathrm{in}_{>_x}(f) = \mathrm{lc}_{>_x}(f)\,\mathrm{Lpp}_{>_x}(f)$, where the leading coefficient $\mathrm{lc}_{>_x}(f) \in K[Y]$, and the leading power product $\mathrm{Lpp}_{>_x}(f) \in [X]$, and the initial ideal with respect to $>_x$ of an ideal $J \subset K[Y][[X]]'$ as $\mathrm{in}_{>_x}(J) = \{\, \mathrm{in}_{>_x}(f) \mid f \in J \,\}$. Since $K[Y]$ is a domain, but not a field, the so called *coefficient ideals* $(\mathrm{in}_{>_x}(J) : m) \cap K[Y]$, $m \in [X]$, may be different from 0 and $K[Y]$; in fact, they can even be non-finitely generated.

Localizing $K[Y]$ in the multiplicatively closed set $K[Y] \setminus \{0\}$, we get its field of fractions $K(Y)$. There is a canonical inclusion $\alpha : K[Y][[X]]' \hookrightarrow K(Y)[[X]]'$. The ring $K(Y)[[X]]$ is similar to R'; we have simply replaced K with an overfield $K(Y)$. Therefore, leading power products and initial ideals in $K(Y)[[X]]'$ are defined in the usual fashion.

Remark 6.5. We do *not*, as in the polynomial ring case, get the ring $K(Y)[[X]]'$ by localizing $K[Y][[X]]'$ in $K[Y] \setminus \{0\}$, since the resulting ring does not contain i.e. the element $\sum_{j=1}^{\infty} x_j / y_j$.

Lemma 6.6. *Let $J \subset K[Y][[X]]'$ be an ideal, and denote by J^e its extension to $K(Y)[[X]]'$. Then a Gröbner basis of J is a Gröbner basis of J^e, hence* $\mathrm{in}_{>_x}(J)^e = \mathrm{gr}(J^e)$.

Proof. Similar to [1, Corollary 3.7] and [7, Prop. 3.4]. □

Lemma 6.7. *For any ideal $J \subset K[Y][[X]]'$, we have that* $\mathrm{gr}(J) = \mathrm{gr}(\mathrm{in}_{>_x}(J))$.

Proof. The filtration induced by $>$ is a refinement of the filtration induced by $>_x$. □

Lemma 6.8. *If $J \subset K[Y][[X]]'$ is a bi-homogeneous, $\mathcal{M}$-locally finitely generated ideal, then a $\mathcal{M}$-homogeneous Gröbner basis G of J with respect to $>$ is also a Gröbner basis of J with respect to $>_x$.*

Theorem 6.9. *For the pure generic ideal $I \subset K[Y][[X]]'$, the following assertions hold:*

(i) $\mathrm{gr}(I^e)$ *is generated by a finite number of monomials in* $\mathcal{M}^r$,

(ii) $\mathrm{gr}(I^e)^c$ *is generated by a finite number of monomials in* $\mathcal{M}^r$,

(iii) $\mathrm{in}_{>_x}(I)$ *is generated by a finite number of elements of the form* pm, $p \in K[y_1, \ldots, y_{t_r}]$, $m \in \mathrm{gr}(I)^e \cap \mathcal{M}^r$,

(iv) The coefficient ideal $(\mathrm{in}_{>_x}(I) : m) \cap K[Y]$, $m \in \mathcal{M}$, *is zero if* $m \notin \mathrm{gr}(I^e)$, *and generated by finitely many* $p_i \subset K[y_1, \ldots, y_{t_r}]$ *otherwise.*

Proof. The first assertion follows from Theorem 4.2, since I^e is a generic ideal in $K(Y)[[X]]'$.

The second assertion follows trivially from the first.

To prove the third assertion, we note that I has a $>_x$-Gröbner basis consisting of elements f which have $\mathrm{Lpp}_{>_x}(f) \in \mathcal{M}^r$ and $\mathrm{lc}(\mathrm{lc}_{>_x}(f)) \in K[y_1, \ldots, y_{t_r}]$. But by the construction of I, we must in fact have that $\mathrm{lc}_{>_x}(f) \in K[y_1, \ldots, y_{t_r}]$.

The fourth assertion follows from the preceding ones: we know that $\mathrm{in}_{>_x}(I) = (p_1 m_1, \ldots, p_v m_v)$, $p_i \in K[y_1, \ldots, y_{t_r}]$, $m_v \in \mathrm{gr}(I^e) \cap \mathcal{M}^r$. Therefore, for any $m \in \mathcal{M}$, $(\mathrm{in}_{>_x}(I) : m) \cap K[Y] = \langle \{ p_i \mid m_i | m \} \rangle_{K[Y]}$. □

7 Initial ideals of finitely generated, homogeneous ideals in $K[Y][[X]]'$

Having examined the initial ideals of generic homogeneous, finitely generated ideals in R', we are ready to turn to the study of the initial ideals of quite arbitrary homogeneous, finitely generated ideals in R'. In particular, the following question is of great interest:

Question 7.1. *Let J be a homogeneous, finitely generated ideal in R', and let $\mathrm{gr}(J)$ denote its initial ideal, with respect to the graded reverse lexicographic order. Is $\mathrm{gr}(J)$ finitely generated?*

Definition 7.2. We say that a $\mathcal{M}$-homogeneous, finitely generated ideal, in R' or in $K[Y][[X]]'$, is of type $d_1, \ldots, d_r$ if it can be generated by forms of these degrees. It is of minimal type $d_1, \ldots, d_r$ if it can be generated minimally by forms of these degrees.

Definition 7.3. A K-algebra homomorphism $\phi : K[Y][[X]]' \to R'$ is called a *specialization* if $\phi(K[Y]) \subset K$, and if $\phi(x_i) = x_i$ for all i. The specialization is said to be *good* if $\phi(T)$ is algebraically independent over Q for any $T \subset Y$.

Clearly, if θ is a good specialization of I, then $\theta(I)$ is a generic ideal in R'. In this case, we also have that $\mathrm{gr}(\theta(I)) = \theta(\mathrm{in}_{>_x}(I)) = \mathrm{gr}(I^e) \cap R'$, where the extension is to $K(Y)[[X]]'$.

Lemma 7.4. *If $I \subset K[Y][[X]]'$ is a pure generic ideal of minimal type $d_1, \ldots, d_r$, and if $J \subset R'$ is a finitely generated, homogeneous ideal of type $d_1, \ldots, d_r$, then there exists some specialization ϕ such that $\phi(I) = J$.*

We can now reformulate Question 7.1:

Question 7.5. *Let I be a pure generic ideal in $K[Y][[X]]'$, and let ϕ be a specialization. Is the monomial ideal $\mathrm{gr}(\phi(I)) \subset R'$ finitely generated?*

The following result is a straightforward generalization of the corresponding result in [1, Proposition 3.4] (it also bears some resemblance to [13, Lemma 10.3]). The short proof of that proposition works *mutatis mutandis.*

Lemma 7.6. *Let J be an ideal in $K[Y][[X]]'$. For any admissible order $>_x$ on $[X]$, $>_y$ on $[Y]$ and any specialization ϕ, we have that $\phi(\mathrm{in}_{>_x}(J)) \subset \mathrm{gr}_{>_x}(\phi(J))$. Similarly, $\phi(\mathrm{gr}_{>}(J)) \subset \mathrm{gr}_{>_x}(\phi(J))$. where $>$ is the "twisted" product of $>_x$ and $>_y$ on $[Y] \oplus [X]$.*

Remark 7.7. The reader should keep in mind that $\mathrm{gr}_{>_x}(J) \subset K[Y][[X]]'$, whereas $\mathrm{gr}_{>_x}(\phi(J)) \subset K[[X]]'$. In the first case, $>_x$ is regarded as a linear quasi order on $[Y] \oplus [X]$, in the second case, as a total order on $[X]$. We have also that $\mathrm{gr}_{>}(J) \subset K[Y][[X]]'$, and that $>$ is a total order on $[Y] \oplus [X]$.

From now on, we once again assume that $>_x$ and $>_y$ are degrevlex, and that $>$ is their twisted product.

Theorem 7.8. *Let $\phi : K[Y][[X]]' \to R'$ be a specialization, and let I be a pure generic ideal in $K[Y][[X]]'$. Then $\phi(\mathrm{in}_{>_x}(I))$ is a finitely generated monomial ideal, as is $\phi(\mathrm{gr}_{>}(I))$.*

Proof. The monomial ideals $\mathrm{in}_{>_x}(I)$ and $\mathrm{gr}_{>}(I)$ are finitely generated, hence any specialization of them is a finitely generated monomial ideal. □

By Theorem 7.8 and Lemma 7.6, we know that $\mathrm{gr}_{>_x}(\phi(I))$ contains the finitely generated monomial ideals $\phi(\mathrm{in}_{>_x}(I))$ and $\phi(\mathrm{gr}_{>}(I))$. R'-ideals may be regarded as R'-modules; therefore, we can form the quotient modules $\frac{\mathrm{gr}_{>_x}(\phi(I))}{\phi(\mathrm{in}_{>_x}(I))}$ and $\frac{\mathrm{gr}_{>_x}(\phi(I))}{\phi(\mathrm{gr}_{>}(I))}$. Now, an ideal is a finitely generated module iff it is a finitely generated ideal, and the quotient of a module A with a finitely generated module B is finitely generated iff A is finitely generated. Therefore, we conclude:

Proposition 7.9. *Let I be a pure generic ideal of $K[Y][[X]]'$, and let ϕ : $K[Y][[X]]' \to R'$ be a specialization. Then the following assertions are equivalent:*

(i) The R'-ideal $\mathrm{gr}_{>_x}(\phi(I))$ is a finitely generated ideal,

(ii) The R'-module $\frac{\mathrm{gr}_{>_x}(\phi(I))}{\phi(\mathrm{in}_{>_x}(I))}$ is finitely generated,

(iii) The R'-module $\frac{\mathrm{gr}_{>_x}(\phi(I))}{\phi(\mathrm{gr}_{>}(I))}$ is finitely generated.

Question 7.10. *If $I \subset K[Y][[X]]'$ is a pure generic ideal, and ϕ : $K[Y][[X]]' \to R'$ is a specialization, when are the quotient modules $\frac{\mathrm{gr}_{>_x}(\phi(I))}{\phi(\mathrm{in}_{>_x}(I))}$ and $\frac{\mathrm{gr}_{>_x}(\phi(I))}{\phi(\mathrm{gr}_{>}(I))}$ non-zero modules?*

Conjecture 7.11. *Let $I \subset K[Y][[X]]'$ be a pure generic ideal, and let ϕ be a specialization such that there exists a finite subset U of Y with the property that $\phi(W)$ is algebraically independent over Q for any $W \subset Y \setminus U$. Then the graded reverse lexicographic initial ideal $\mathrm{gr}(\phi(I))$ is finitely generated.*

7.1 Acknowledgements

I thank Jörgen Backelin, Ralf Fröberg and Torsten Ekedahl for useful advice and patient tutoring. Many of the proofs are due to them as is many of the ideas used in this paper.

References

[1] Dave Bayer, André Galligo, and Mike Stillman, *Gröbner Bases and extension of scalars*, Computational Algebraic Geometry and Commutative Algebra (David Eisenbud and Lorenzo Robbiano, eds.), Symposia Mathematica, vol. 24, 1991.

[2] Thomas Becker and Volker Weispfenning, *Gröbner bases: a computational approach to commutative algebra*, Graduate texts in mathematics, Springer Verlag, 1993.

[3] Bruno Buchberger, *Gröbner bases: An algorithmic method in polynomial ideal theory.*, Multidimensional systems theory, Progress, directions and open problems, Math. Appl. **16** (1985), 184–232.

[4] ———, *A survey on the method of Gröbner bases for solving problems in connection with systems of multivariate polynomials*, Symbolic and Algebraic Computation by Computers (1985), 69–83.

[5] Ralf Fröberg, *An inequality for Hilbert series of graded algebras*, Mathematica Scandinavica **56** (1985), 117–144.

[6] Ralf Fröberg and Joachim Hollman, *Hilbert series for ideals generated by generic forms*, Journal of Symbolic Computation **17** (1994), 149–157.

[7] P Gianni and B Trager, *Gröbner bases and primary decomposition in polynomial ideals*, Journal of Symbolic Computation **6** (1988), 148–166.

[8] Hideyuki Matsumura, *Commutative Ring Theory*, Cambridge Studies in Advanced Mathematics, vol. 8, Cambridge University Press, 1986, Translated from the Japanese by M. Reid.

[9] Guillermo Moreno Socias, *Autour de la fonction de Hilbert-Samuel (escaliers d'idéaux polynomiaux)*, Ph.D. thesis, École Polytechnique, 1991.

[10] Paulo Ribenboim, *Generalized power series rings*, Lattices, semigroups and universal algebra (J Almeida, G Bordalo, and P Dwinger, eds.), Plenum, New York, 1990, pp. 15–33.

[11] Lorenzo Robbiano, *Introduction to the Theory of Gröbner Bases*, Queen's papers in pure and applied math **5** (1988), no. 80.

[12] Jan Snellman, *Gröbner bases and normal forms in a subring of the power series ring on countably many variables over a field*, Journal of Symbolic Computation (1997), To appear.

[13] ———, *Initial ideals of truncated homogeneous ideals*, Communications in algebra **26** (1998), no. 3, To appear.

[14] B. L. Van der Waerden, *Moderne Algebra*, Die Grundlehren der Mathematischen Wissenschaften, Verlag von Julius Springer, 1930.

Parallel Computation and Gröbner Bases: an Application for Converting Bases with the Gröbner Walk

Quôc-Nam Trân[1]

Research Institute for Symbolic Computation (RISC–Linz),
Johannes Kepler University, A–4040 Linz, Austria
E-mail: tqnam@risc.uni-linz.ac.at

Abstract

Basis conversion arises in many parts of computational mathematics and computer science such as solving algebraic equations, implicitization of algebraic sets, elimination theory, etc. In this paper we discuss the Gröbner walk method of Collart et al. to convert a given Gröbner basis of a multivariate polynomial ideal of arbitrary dimension into a Gröbner basis of the ideal with respect to another term order. We describe some improvements and a parallel implementation in parallel Maple, where we can still utilize the whole sequential library of the popular computer algebra system Maple. The system supports a variety of virtual machines that differ in the manner in which nodes are connected. Therefore, it is independent of the devices and easy to program. The programs may run on different hardware ranging from shared-memory machines over distributed memory architectures up to networks of workstations without any modification or re-compilation. Moreover, the programs are scalable in that they may be written to execute on many thousands of nodes. We show that our best implementation achieves a speed up of up to six over a sequential implementation. We also outline further applications of parallel computation in the Gröbner bases method.

1 Introduction

Buchberger's algorithm (Buchberger, 1965; Buchberger, 1985) for the computation of Gröbner bases has became one of the most important algorithms in providing exact solutions of scientific problems in multivariate polynomial

[1]Supported by the Austrian Science Foundation (FWF) project HySaX, Proj. No. P11160-TEC

ideal theory, elimination theory and so on. Even though Gröbner bases are a giant step forward in so far as implementations have became feasible and have actually provided answers to scientists, many problems of no more than moderate input size still defy computation. Since the algorithm operates with exact arithmetic, they are therefore more expensive with respect to time and memory than numerical methods.

In order to improve the situation to the satisfaction of users, one may use some strategies to speed up the algorithm (Giovini et al., 1991; Becker and Weispfenning, 1993; Faugère et al., 1993; Faugère, 1994b; Noro and Yokoyama, 1995) or acquire fast computers with a huge working memory. However, in spite of the fact that the performance of the fastest computers has grown exponentially from 1945 to the present, clock cycle times are decreasing slowly and appear to be approaching physical limits such as the speed of light. On the other hand, a fundamental result in Very Large Scale Integration (VLSI) complexity theory says that one should not utilize internal concurrency in a chip, for example by operating simultaneously on all 64 bits of two numbers that are to be multiplied, because this strategy is expensive. Actually, surveys of trends in computer architecture and networking suggest a future in which parallelism pervades not only supercomputers but also workstations, personal computers, and networks. In this future, programs will be required to exploit the multiple processors located inside each computer and the additional processors available across the network. Because most computer algebra systems and most existing algorithms are specialized for a single processor, this situation implies a need for new parallel algorithms on the one hand and the reuse of huge libraries in existing computer algebra systems on the other.

It is well known that the size and form of a Gröbner basis of a polynomial ideal and also the complexity of its computation strongly depend on the chosen term order. The choice of the term order depends on the problem itself, in that elimination orders are cumbersome to compute. Therefore, it may be more efficient to compute first a Gröbner basis with respect to a total degree order and then convert the basis since total degree bases are generally much faster to compute. Basis conversion arises in many parts of computational mathematics and computer science such as solving algebraic equations, implicitization of algebraic sets, elimination theory, etc. In this paper we discuss the Gröbner walk method of Collart et al. (Collart et al., 1996; Kalkbrener, 1996; Amrhein et al., 1996) to convert a given Gröbner basis of a multivariate polynomial ideal of arbitrary dimension into a Gröbner basis of the ideal with respect to another term order. We describe some improvements and a parallel implementation in parallel Maple, where we can still utilize the whole sequential library of the popular computer algebra system Maple. The system supports a variety of virtual machines that differ in the manner in which nodes are connected. Therefore, it is independent of the devices and easy to program. The programs may run on different hardware ranging from shared-memory machines over distributed memory architectures up to networks of workstations without any modification or re-compilation. Moreover, the programs are scalable in that they may be written to execute on many thousands of nodes. We show that our best implementation achieves a speed up of up to six over a sequential implementation, which in turn achieves a speed up of up to ten over the direct application of Buchberger's algorithm (Aigner, 1997). We also outline further applications of parallel computation in the Gröbner bases method.

The structure of the paper is as follows. In Section 2, we give a short introduction to basic facts and analyze the performance of sequential algorithms for the Gröbner walk method. The main aspects of parallelization of the algorithms as well as their performance and comparisons are reported in Section 3.

2 Gröbner Walk

Before describing the parallel algorithms, we give a short introduction to basic facts and analyze the performance of the sequential algorithms for the Gröbner walk method. We refer to (Collart et al., 1996) for missing details.

Given the reduced Gröbner basis of an ideal $I \in \mathbb{K}[x_1, \ldots, x_n]$ with respect to an admissible term order $\prec_1$, where $\mathbb{K}$ is a computable field. Our goal is to compute the reduced Gröbner basis of I with respect to another admissible order $\prec_2$ without applying Buchberger's algorithm. The term orders $\prec_1$ and $\prec_2$ can be expressed by sequences of rational vectors $S_{\prec_1}$ and $S_{\prec_2}$ where their first elements, denoted by σ and τ, are weight vectors refined by $\prec_1$ and $\prec_2$, respectively.

For σ and τ in the set of weight vectors Ω, we denote the line segment in Ω between σ and τ by $\overline{\sigma\tau}$, i.e.

$$\overline{\sigma\tau} = \{(1-t)\sigma + t\tau : 0 \leq t \leq 1\}$$

There exists finitely many weight vectors $\sigma = \omega_0, \omega_1, \ldots, \omega_m = \tau$ in $\overline{\sigma\tau}$ and pairwise different Gröbner cones $C_{\prec_1}(I) = C_0(I)$, $C_1(I) = C_{(\omega_1|\prec_2)}(I), \ldots,$ $C_m(I) = C_{\prec_2}(I)$ in the Gröbner fan of I such that for every $k \in \{1, \ldots, m\}$, ω_k is the weight vector with

$$\overline{\omega_{k-1}\omega_k} = \overline{\omega_{k-1}\tau} \cap C_{k-1}(I)$$

We denote the reduced Gröbner basis of I over the Gröbner cone $C_k(I)$ by G_k.

We perform the Gröbner walk method by moving on the line segment $\overline{\sigma\tau}$ from σ to τ, i.e., we compute $\omega_1, \ldots, \omega_{m-1}$ and $G_1, \ldots, G_m$ successively. The crucial point is that this conversion can be done efficiently without applying

Buchberger's algorithm. We first check if $C_{k-1}(I)$ is equal to $C_{\prec_2}(I)$ for a given Gröbner basis $G_{k-1} = \{g_1, \ldots, g_r\}$. If it is a case then G_{k-1} is already the reduced Gröbner basis of I with respect to $\prec_2$. Otherwise, we have to determine the next weight vector ω_k which is the point on the segment $\overline{\sigma\tau}$ where we leave the Gröbner cone $C_{k-1}(I)$. The weight ω_k can be easily computed from ω_{k-1}, τ and G_{k-1} as $\omega_k = \omega(\bar{t}) = \omega_{k-1} + \bar{t}(\tau - \omega_{k-1})$, $0 < t \leq 1$, where

$$\bar{t} = \min(\{t \in \mathbb{Q} \cap (0,1] : \deg_{\omega(t)} p_1 = \deg_{\omega(t)} p_i, \text{ for some } g = p_1 + \cdots + p_n \in G_{k-1}, 2 \leq i \leq n\})$$

From our experiments, this computation costs an average of 15.02 percent of the total time needed for the walk.

After leaving $C_{k-1}(I)$ we enter $C_k = C_{(\omega_k|\prec_2)}(I)$. We now have to transform G_{k-1} into G_k. Note that there exists a term order $\prec$ which refines ω_k such that $C_\prec(I) = C_{k-1}(I)$. Therefore $in_\prec(f) = in_\prec(in_{\omega_k}(f))$ for all $f \in I$ and

$$\langle (I_{\omega_k})_\prec \rangle = \langle I_\prec \rangle = \langle (G_{k-1})_\prec \rangle = \langle ((G_{k-1})_{\omega_k})_\prec \rangle$$

hence $(G_{k-1})_{\omega_k}$ is the reduced Gröbner basis of I_{ω_k} with respect to $\prec$. We now convert $(G_{k-1})_{\omega_k}$ into the the reduced Gröbner basis $M = \{m_1, \ldots, m_s\}$ of $\langle I_{\omega_k} \rangle$ with respect to $(\omega_k| \prec_2)$. Note that this conversion itself can be done with any basic conversion, for example by using Hilbert-Poincaré series (Traverso, 1996) since we have a complete intersection case here. However, we may want to use a specialized Buchberger algorithm in this case because we can perturb the weight vectors when needed. In a perturbed walk (Amrhein et al., 1996), most of the initials are monomial. Beside of that, critical pairs of two monomials are unnecessary since its S-polynomial is always zero. Furthermore, experiments show that Gröbner bases of two adjacent Gröbner cones are very similar, i.e., the computation of the initial Gröbner basis usually adds just few polynomials to the basis. As most of the S-polynomials reduce to zero in one step, there is no need of sophisticated selection strategies. From our experiments, this computation costs an average of 16.54 percent of the total time needed for the walk.

Since $m_1, \ldots, m_s$ are ω_k-homogeneous, we can compute ω_k-homogeneous polynomials $h_{i1}, \ldots, h_{ir}$ with

$$m_i = \sum_{j=1}^{r} h_{ij} in_{\omega_k}(g_j) \quad \text{and} \quad \deg_{\omega_k}(m_i) = \deg_{\omega_k}(h_{ij} in_{\omega_k}(g_j)),$$

for $j = 1 \ldots r$ with $h_{ij} \neq 0$. From our experiments, this computation costs an average of 3.20 percent of the total time needed for the walk.

Replacing $in_{\omega_k}(g_j)$ by g_j, we obtain

$$f_i = \sum_{j=1}^{r} h_{ij} g_j \quad \text{and} \quad G = \{f_1, \cdots, f_s\}$$

It immediately follows that that $in_{\omega_k}(f_i) = m_i$ and therefore

$$\langle I_{(\omega_k|\prec_2)}\rangle = \langle\langle I_{\omega_k}\rangle_{(\omega_k|\prec_2)}\rangle = \langle M_{(\omega_k|\prec_2)}\rangle = \langle G_{(\omega_k|\prec_2)}\rangle$$

Hence G is a Gröbner basis of I with respect to $(\omega_k| \prec_2)$ which we reduce to G_k. From our experiments, this computation costs an average of 58.56 percent of the total time needed for the walk.

3 Parallel Implementation

As we have seen in Section 2, the most time consuming parts in the sequential algorithm are:

1. the computation of the borders ω_k,
2. the computation of the reduced Gröbner basis of $\langle I_{\omega_k}\rangle$ with respect to $(\omega_k| \prec_2)$,
3. the computation of $G = \{f_1, \ldots, f_s\}$, where $f_i = \sum_{j=1}^{r} h_{ij}g_j$, and the reduction of G into the reduced Gröbner basis of I with respect to $(\omega_k| \prec_2)$.

They cost an average of 15.02%, 16.54% and 58.56% of the total time needed for the walk respectively. For this reason, the main goal is to parallelize these parts of the algorithm.

Beside of that, we will avoid to walk through the intersection of several Gröbner cones since with a weight on the intersection, the initial polynomials become larger thus making the problem harder for the consequent steps. We can guarantee to stay within the right Gröbner cone by controlling the initial polynomials. Especially, at the end weight vectors it often happens that we are on an intersection. By slightly perturbing these weight vectors, we can improve the situation. We will also use integer weight vectors instead of rational ones since comparison of two monomials involving rational arithmetic is rather costly.

We have implemented the algorithms using $\|MAPLE\|$ (Siegl, 1993) and basing upon a sequential implementation of the Gröbner walk method in CASA (Tran and Winkler, 1997b; Tran and Winkler, 1997a; Aigner, 1997). Both systems, parallel Maple and CASA, have been developed at RISC-Linz over the last several years, where CASA is a system for computational algebra and constructive geometry while parallel Maple is a portable system for parallel symbolic computation. The core of the system is built on top of an interface between the parallel declarative programming language Strand (Foster and Taylor, 1989) and the sequential computer algebra system Maple (Monagan et al., 1996), thus providing the elegance of Strand and the power of the existing sequential algorithms in Maple. The programs may run on different hardware ranging from shared-memory machines over distributed memory architectures up to networks of workstations without any modification or re-compilation. All necessary communications are done automatically by the system without any additional programming effort.

3.1 Determining the Borders

We partition the problem into concurrent components where each component is corresponding with a polynomial in G_{k-1}. We than apply the procedure on each of the components in parallel. The procedure `possible_t` runs through

Table 1: Parallel Algorithm for Determining Borders

```
border := proc(omega, tau, G)
  components := seq('possible_t'(omega, tau, G[i]), i=1..nops(G));
  peval([components],min(seq(parg[i], i=1..nops(G))))
end:

possible_t := proc(omega, tau, g)
  find t such that deg_omega_t(p_1) = deg_omega_t(p_i)
                   for some  2 <= i <= n
                   where g = p_1 + ... + p_n
end:
```

all monomials of g and determines the possible parameter t. The procedure `peval` sends each function in the list of unevaluated functions `[components]` to a different node for evaluation in parallel while the composition function `min(seq(parg[i], i=1..nops(G)))` takes as input the results produced by the parallel evaluations and produces the final output.

Note that we may partition the problem into concurrent components corresponding with a monomial of a polynomial in G_{k-1} but in this case the granularity is too small that the program is unlikely to benefit from parallel execution. We may also utilize divide-and-conquer principle here but it is inefficient is this case since in parallel Maple we have a higher overhead for communication and a bigger memory overhead due to the need of copying the required data between the Maple and Strand heaps. The structure of the problem for determining the borders is regular, i.e., we can allocate the components into a particular virtual machine, for example a ring virtual machine or a star-shaped virtual machine which utilizes the manager-worker load balancing strategy.

3.2 Computation of the Reduced Gröbner Basis of $\langle I_{\omega_k} \rangle$

There have been several attempts to parallelize Buchberger's algorithm (Buchberger, 1987; Vidal, 1989; Bradford, 1990; Pascale, 1990; Neun and Melenk, 1990; Hawley, 1991; Schwab, 1992; Siegl, 1993; Faugère, 1994a; Gräber and Lassner, 1994; Siegl, 1994). We need to compute the reduced Gröbner basis $M = \{m_1, \ldots, m_s\}$ of $\langle I_{\omega_k} \rangle$ with respect to $(\omega_k | \prec_2)$ from $(G_{k-1})_{\omega_k}$. This computation itself can be done, for example by using Hilbert-Poincaré series (Traverso, 1996) since we have a complete intersection case here. However, we will perturb the weight vectors when needed. In a perturbed walk (Amrhein et al., 1996), most of the initials are monomial. Beside of that, critical pairs of two monomials are unnecessary since its S-polynomial is always zero. Furthermore, experiments show that Gröbner bases of two adjacent Gröbner cones are very similar, i.e., the computation of the initial Gröbner basis usually adds just few polynomials to the basis. As most of the S-polynomials reduce to zero in one step, there is no need of sophisticated selection strategies. Therefore, we will use a specialized Buchberger algorithm in this case. Additionally, for applications where we are only interested with the algebraic set of the ideal (Gräber and Lassner, 1994), we can use factorization within a parallel Gröbner basis algorithm, which will allow the cancellation of superfluous computational branches in an early stage of the computation. Using factorization, one can split the problem of computation of a Gröbner basis into several subproblems which can be run in parallel then. The sequential Buchberger's algorithm is a Critical Pair Completion algorithm. In that given a set of input polynomials as input basis, one computes S-polynomials and reduces them with respect to the basis polynomials. The resulting polynomials are added to the basis and new S-polynomials are formed.

In the parallel version, similar to (Siegl, 1993) we create a pipeline of processes reducing the input polynomials with respect to an increasing set of basis polynomials. The irreducible intermediate values are appended to the reduction process. For generating new S-polynomials as input, the new polynomials which is jointed to the new basis are sent back through the pipe building the S-polynomials with the earlier computed basis polynomials. The process at the front of the pipe needs to collect the computed S-polynomials and sends them to the reduction processes again. As a result we have a bidirectional pipe forming a dynamically expandable ring. We will combine the advantages of the pipeline method with the factorization method by splitting up the pipe into a factorization tree. As long as no polynomials are factored, we will reduce them with the conventional method to cover the demand. This will be internally parallelized using the manager-worker approach to guarantee that faster reducible polynomials will show up earlier in the basis. But as soon as a polynomial can be factored (in case we are only interested with the algebraic set of the ideal), we add a new pipeline process for each factor. In a parallelized Buchberger's algorithm, the most important special feature is the intermediate growth, that usually cannot be predicted by the input data. Thus an efficient automatic load balancing scheme is needed. Fortunately, this can be achieved by using the manager-worker strategy.

Table 2: Parallel Algorithm for a Specialized Buchberger's Algorithm

```
special_gb := proc(F,order)
  local irrSet,G;
  irrSet := {op(F)};
  G := F;
  while irrSet <> {}
    SpolSet := spol(G,order);
    components := seq('reduce_poly'(SpolSet[i], G, order),
                       i=1..nops(SpolSet));
    peval([components],irrSet:={seq(parg[i],i=1..nops(G))}\{0})
end:

reduce_poly := proc(p,G,order)
  pp := convert(p, list);
  while there is a monomial in pp which is reducible in peval
    components:=seq('reduce_mon'(pp[i],G,order),i=2..nops(pp));
    peval([components], pp:=[seq(parg[i],i=1..nops(G))])
  convert(pp,'+')
end:
```

3.3 Lifting and Reducing

This part is the most time consuming part of the sequential algorithm which costs an average of 58.56% of the total time needed for the walk. Thus it needs to be carefully investigated. While lifting algorithm is easy to be parallelized

Table 3: Parallel Algorithm for Lifting

```
lift_gb := proc(G,H)
  local F;
  components:=seq('f_i'(i,G,H), i=1..nops(H));
  peval([components], F:=[seq(parg[i], i=1..nops(H))])
end:

f_i := proc(i,G,H)
  convert(seq(H[i][j]*G[j], j=1..nops(G)), '+')
end:
```

(see Table 3), the parallelization of the reduction process is not straightforward since algorithms for reduction found in the literature (Becker and Weispfenning, 1993; Winkler, 1996) are typically designed for sequential computation.

As mentioned in (Collart et al., 1996), the Gröbner walk method also works by using minimal Gröbner bases instead of reduced Gröbner bases. Since reduction is expensive with respect to time and memory on the one hand and it is not easy to be parallelized on the other, one may think that it would be better if one can avoid the reduction process of the lifted Gröbner bases. However, experiments show that it is actually not helpful since if we do not have reduced Gröbner bases, we may have to walk several steps within the same Gröbner cone. Beside of that it can lead to lots of additional steps and lager bases thus making the problem harder for the consequent steps.

We now proposed a new parallel algorithm for reduction as follows. We

Table 4: Parallel Algorithm for Reduction

```
reduce_basis := proc(G,order)
  local Gp, reduced_G
  Gp := minimal(G);
  while there is p in Gp which is reducible in peval
    components := seq('reduce_poly'(p, subsop(i=NULL, Gp), order),
                      i=1..nops(Gp));
    peval([components], Gp:=[seq(parg[i], i=1..nops(Gp))])
  reduced_G:=Gp
end:

reduce_poly := proc(p,Gp,order)
  pp := convert(p, list);
  while there is a monomial in pp which is reducible in peval
    components := seq('reduce_mon'(pp[i],Gp,order), i=2..nops(pp));
    peval([components], pp:=[seq(parg[i], i=1..nops(Gp))])
  convert(pp,`+`)
end:
```

partition the problem into concurrent components where each component is corresponding with a polynomial p in `Gp` and `Gp\{p}`. We than apply the procedure on each of the components in parallel. The procedure **`reduce_poly`** reduces p with respect to `Gp\{p}`. The procedure **`peval`** sends each function in the list of unevaluated functions **`[components]`** to a different node for evaluation in parallel while the composition function `Gp:=[seq(parg[i], i=1..nops(Gp))]` takes as input the results produced by the parallel evaluations and produces the final output. The old basis is replaced by the set of new reduced polynomials. The procedure **`reduce_poly`** in turn is partitioned into concurrent components corresponding with a monomial of the polynomial since in this case the granularity is coarse enough for the benefit from parallel execution. We also use the manager-worker strategy to achieve an efficient automatic load balancing for this case.

Lemma 3.1 *The algorithm* **`reduce_basis`** *computes the reduced Gröbner basis of the ideal* $I = \langle G \rangle$.

PROOF. We may assume that G is minimal. $\langle Gp \rangle = \langle G \rangle$ and $\langle in(Gp) \rangle = \langle in(G) \rangle$; therefore `reduced_G` is the reduced Gröbner basis of the ideal $I = \langle G \rangle$. □

We tested our algorithms with several examples found in the literature. We summarize the timings using a network of six and twelve Silicon Graphics (SGI) workstations. All computation times are given in second and including times for garbage collections. The orders of variables and the equations of the examples are as follows.

Table 5: Time Consuming

Example	Seq	6 Procs	12 Procs	Speedup
Example 1	23.400	9.449	4.799	4.88
Example 2	108.629	40.909	18.337	5.93
Canny 1	146.490	56.110	25.983	5.64
Canny 2	85.680	33.487	16.089	5.33
Trinks 1	7.910	3.503	2.034	3.89

1. Example 1: $x \succ y \succ z$,
   $F = [x^3 + y + xy^2 + y^3 + z^2, 2x^3y + y^2 + xy^2 + x^3z + xz^3]$.
2. Example 2: $x \succ y \succ z$,
   $F = [2x^2y + x^3y + 2xy^2z, xy^3 + y^4 + yz^2 - z^3 - 2xz^3, 2 - 3x^2y + 2x^3y + yz^3]$.
3. Canny 1: $x \succ y \succ z \succ t \succ u$,
   $F = [-x^3 - xy + z, -y^3 + t - 2, -x^2y - xy^2 + u]$.
4. Canny 2: $x \succ y \succ z \succ t \succ u$,
   $F = [x^3 - xy^2 - z, x^2y - t - 2, y^3 - x^2 - u]$.
5. Trinks 1: $v \succ u \succ t \succ z \succ y \succ x$,
   $F = [45y + 35u - 165v - 36, 35y + 25z + 40t - 27u, 25yu - 165v^2 + 15x - 18z + 30t, 15yz + 20tu - 9x, -11v^3 + xy + 2zt, -11uv + 3v^2 + 99x]$.

4 Conclusion

We discussed the Gröbner walk method of Collart et al. to convert a given Gröbner basis of a multivariate polynomial ideal of arbitrary dimension into a Gröbner basis of the ideal with respect to another term order. We described some improvements and a parallel implementation in parallel Maple, where we can still utilize the whole sequential library of the popular computer algebra system Maple. The system supports a variety of virtual machines that differ in the manner in which nodes are connected. Therefore, it is independent of the devices and easy to program. The programs may run on different hardware ranging from shared-memory machines over distributed memory architectures up to networks of workstations without any modification or re-compilation. Moreover, the programs are scalable in that they may be written to execute on many thousands of nodes. We showed that our best implementation achieves a speed up of up to six over a sequential implementation. We also outlined further applications of parallel computation in the Gröbner bases method.

References

Aigner, K. (1997). Maple V procedure for computation of Gröbner bases by the Gröbner walk. Technical Report 97–14, RISC-Linz, The University of Linz, Linz, Austria.

Amrhein, B., Gloor, O., and Küchlin, W. (1996). Walking faster. In *Proceedings of the DISCO'96*, Karlsruhe, Germany.

Becker, T. and Weispfenning, V. (1993). *Gröbner Bases: a Computational Approach to Commutative Algebra.* Graduate Texts in Mathematics. Springer–Verlag.

Bradford, R. (1990). A parallelization of the Buchberger algorithm. In *Proceedings of the ISSAC'90*, pages 296–298, Tokyo, Japan.

Buchberger, B. (1965). *An Algorithm for Finding a Basis for the Residue Class Ring of a Zero-dimensional Polynomial Ideal (in German).* PhD thesis, Institute of Mathematics, Univ. Innsbruck, Innsbruck, Austria.

Buchberger, B. (1985). Gröbner Bases: An algorithmic method in polynomial ideal theory. In Bose, N. K., editor, *Multidimensional Systems Theory*, chapter 6, pages 184–232. Reidel Publishing Company, Dodrecht.

Buchberger, B. (1987). The Parallelization of Critical-Pair/Completion Procedures on the L-Machine. In *Proc. Japanese Symposium on Functional Programming*, pages 54–61, Japan.

Collart, S., Kalkbrener, M., and Mall, D. (1996). Converting bases with the Gröbner walk. *J. Symb. Comp.*

Faugère, F. C., Gianni, P., Lazard, D., and Mora, T. (1993). Efficient Computation of Zero-dimensional Gröbner Bases by Change of Ordering. *Journal of Symbolic Computation*, 16:329–344.

Faugère, J. C. (1994a). Parallelization of Gröbner basis. In *Proceedings of the First International Symposium on Parallel Symbolic Computation PASCO-94*, pages 124–132, Hagenberg, Austria. World Scientific.

Faugère, J. C. (1994b). *Résolution des systèmes d'équations algégriques.* PhD thesis, Univ. Paris VI, paris, France.

Foster, I. and Taylor, S. (1989). *Starnd – New concepts in Parallel Programming.* Prentice-Hall.

Giovini, A., Mora, T., Niesi, G., Robbiano, T., and Traverso, C. (1991). "One sugar cube, please" or selection strategy in Buchberger's algorithm. In *Proceeding of ISSAC-91*, pages 49–54.

Gräber, H.-G. and Lassner, W. (1994). A parallel Gröbner factorizer. In *Proceedings of the First International Symposium on Parallel Symbolic Computation PASCO-94*, pages 174–180, Hagenberg, Austria. World Scientific.

Hawley, D. (1991). A Buchberger algorithm for distributed memory multiprocessors. In *Proceedings of the First International ACPC conference*, pages 385–390, Salzburg, Austria. Springer Verlag.

Kalkbrener, M. (1996). On the complexity of Gröbner bases conversion. Preprint.

Monagan, M. B., Geddes, K. O., Heal, K. M., Labahn, G., and Vorkoetter, S. M. (1996). *Maple V Programming Guide.* Springer Verlag.

Neun, W. and Melenk, H. (1990). Very large Gröbner bases calculations. In *Computer Algebra and Parallelism. Proceedings of the Second International Workshop*, pages 89–99, Ithaca, USA. Springer Verlag.

Noro, M. and Yokoyama, K. (1995). New Method for the Change-Of-Ordering in Gröbner Basis Computation. Technical report, International Institute for Advanced Study of Social Information Science (IIAS-SIS), Fujitsu Laboratories, Numazu-shi, Shizuoka, Japan.

Pascale, S. (1990). Boolean Gröbner bases and their MIMD implementation. In *Computer Algebra and Parallelism. Proceedings of the Second International Workshop*, pages 101–114, Ithaca, USA. Springer Verlag.

Schwab, S. A. (1992). Extended parallelism in the Gröbner basis algorithm. *Int. Journal of Parallel Programming*, 21:39–66.

Siegl, K. (1993). Parallelizing algorithms for symbolic computation using $\|MAPLE\|$. In *Proceedings of the Fourth ACM SIGPLAN Symposium on Principles and Practice of Parallel Programming*, pages 179–186, San Diego, California.

Siegl, K. (1994). A parallel factorization tree Gröbner basis algorithm. In *Proceedings of the First International Symposium on Parallel Symbolic Computation PASCO-94*, pages 356–362, Hagenberg, Austria. World Scientific.

Tran, Q.-N. and Winkler, F. (1997a). Casa reference manual (version 2.3). Technical report, RISC-Linz, The University of Linz, Austria.

Tran, Q.-N. and Winkler, F. (1997b). An overview of CASA – a system for computational algebra and constructive algebraic geometry. In Effelterre, T. V., Racio, T., and Winkler, F., editors, *The Symbolic and Algebraic Computation (SAC) Newsletter*, volume 2. Stichting CAN, Computer Algebra Nederland, Universiteit van Amsterdam, the Netherlands.

Traverso, C. (1996). Hilbert functions and the Buchberger algorithm. *Journal of Symbolic Computation*, 22:355–376.

Vidal, J.-P. (1989). The computation of Gröbner bases on a shared memory multiprocessor. Technical report, School of Computer Science, Carnegie-Mellon University.

Winkler, F. (1996). *Polynomial Algorithms in Computer Algebra.* Texts and Monographs in Symbolic Computation. Springer-Verlag.

Appendix

An Algorithmic Criterion for the Solvability of a System of Algebraic Equations[1]

Bruno Buchberger

1 Problem Statement

We start from the polynomial ring $K[x_1, \ldots, x_n]$ over a commutative field K (abbreviated $K[x_i]$) and an arbitrary polynomial ideal $\mathcal{A} = (f_1, \ldots, f_s)$ ($f_j \in K[x_i]$ for $j = 1, \ldots, s$). (Notions not explicitly defined here will be used in precisely the sense defined in Gröbner (1949) and van der Waerden (1937).)

The residue class ring $\mathcal{O} = K[x_i]/\mathcal{A}$ is well known to be an (in general, infinite dimensional) algebra over K. If $\mathcal{A}$ is zero dimensional, then $\mathcal{O}$ has finite dimension over K, and conversely. The residue classes of the power products (abbreviated PP) $x_1^{i_1} \cdots x_n^{i_n}$ form a basis of the algebra, which is linearly dependent in the case $\mathcal{A} \neq (0)$. (Since in what follows we will speak often of the basis of an ideal and of the basis of the corresponding residue class algebra, we adopt the following convention to avoid ambiguity: we say simply *basis* if we mean the basis of the residue class algebra and *ideal basis* otherwise.)

The goal of the present study is to develop an algorithm which, given an ideal basis $(f_1, \ldots, f_s)$, extracts a linearly independent basis for $\mathcal{O}$ from the set of all residue classes of PPs, and, for two arbitrary elements of this basis, allows a representation of its product to be computed as a linear combination of basis elements (i.e. produces the complete multiplication table in the finite dimensional case). For zero dimensional ideals, Gröbner (1964a) suggested such an algorithm, for which it was still undecided, when it could be terminated in concrete cases. The justification that the algorithm suggested here can be applied to arbitrary polynomial ideals emerged during the investigation of this last question. By applying the algorithm, an assertion is possible about the existence of a zero for the ideal as well as the dimension of the ideal.

[1]Original article appeared as 'Ein algorithmisches Kriterium für die Lösbarkeit eines algebraischen Gleichungssystems', *Aeq. Math.* **4** (1970), 374-383. Received 27 March 1969, and in revised form 11 November 1969. Translation by Michael Abramson and Robert Lumbert.

2 Description of the Algorithm

2.1 Definitions

For the purposes of our study, we order the PPs by increasing degree and within the same degree lexicographically in the sense that $x_1^{i_1} \cdots x_n^{i_n}$ precedes $x_1^{i'_1} \cdots x_n^{i'_n}$ if $i_1 > i'_1$ or $i_t = i'_t$ (for $t = 1, \ldots, k$ and $1 \leq k < n$) and $i_{k+1} > i'_{k+1}$. Beginning with $x_1^0 \cdots x_n^0$, we associate the integers $1, 2, 3, \ldots$ to these ordered PPs as the *indices of the corresponding PPs.* The PP having the highest index among those occuring in a polynomial, will be called the *LPP of this polynomial.*

Let the basis polynomials of the above polynomial ideal $\mathcal{A} = (f_1, \ldots, f_s)$ be

$$f_j = x_1^{I_{j,1}} \cdots x_n^{I_{j,n}} + \sum a_{i_1 \ldots i_n}^{(j)} x_1^{i_1} \cdots x_n^{i_n} \qquad (j = 1, \ldots, s) \tag{2.1}$$

Let $pp_j = x_1^{I_{j,1}} \cdots x_n^{I_{j,n}}$ be the LPP of f_j and let the other terms of f_j be collected under the summation sign in (1). Without loss of generality, we have assumed the coefficient of pp_j to be 1. Because of (1) we have

$$x_1^{I_{j,1}+v_1} \cdots x_n^{I_{j,n}+v_n} \leftrightarrow -\sum a_{i_1 \ldots i_n}^{(j)} x_1^{i_1+v_1} \cdots x_n^{i_n+v_n} \tag{2.2}$$

($j = 1, \ldots, s$ and $v_t = 0, 1, 2, \ldots$ for $t = 1, \ldots, n$. We use "$\leftrightarrow$" as the symbol for *congruent modulo* $\mathcal{A}$).

A PP $x_1^{i_1} \cdots x_n^{i_n}$ is called a *multiple* of the PP $x_1^{k_1} \cdots x_n^{k_n}$ if $i_t \geq k_t$ for $t = 1, \ldots, n$. PPs which are multiples of at least one of the pp_j ($j = 1, \ldots, s$), i.e. occur at least once on the left-hand side of (2), will be called *MPPs relative to the ideal basis* $(f_1, \ldots, f_s)$. Otherwise a PP will be called *NPP relative to the ideal basis* $(f_1, \ldots, f_s)$. When no ambiguity is possible, we will omit the phrase "relative to the ideal basis $(f_1, \ldots, f_s)$". Polynomials, in which only NPPs occur and terms containing the same PPs are collected, are called *NPP-polynomials* (relative to the ideal basis $(f_1, \ldots, f_s)$).

A given polynomial (for which we do not wish to assume that terms containing the same PPs are collected) can now be transformed into a congruent NPP-polynomial, which is in general not necessarily unique, by successively replacing all MPPs by the corresponding right-hand sides of (2) and by collecting all like terms in arbitrary stages of this reduction procedure (which we call *M-reduction*). Thus the residue classes of the NPPs already form a basis for $\mathcal{O}$, although still linearly dependent in general.

2.2 A Lemma

The following now holds:

Lemma. *If an ideal basis $(f_1, \ldots, f_s)$ has the property that all possible M-reductions of a polynomial lead to the same result, then the residue classes of the NPPs relative to the ideal basis form a linear independent basis for $\mathcal{O}$.*

Proof. A linear dependency between residue classes of NPPs would correspond to an NPP-polynomial q in $\mathcal{A}$ which would possess a representation

$$q \equiv \sum_{j=1}^{s} h_j \cdot f_j \qquad \text{with} \quad h_j \in K[x_j] \quad \text{for } j = 1, \ldots, s. \tag{2.3}$$

Multiplying out the $h_j \cdot f_j$ ($j = 1, \ldots, s$) without collecting terms containing the same PPs produces a polynomial on the right-hand side of (3) that can be M-reduced in two different ways with two different results, contradicting the assumption of the lemma: It can be M-reduced to $q \not\equiv 0$ precisely by collecting like terms. And it can be trivially M-reduced to 0 by subtracting the polynomials $h_j \cdot f_j$ and afterwards collecting like terms. (Notice that both procedures satisfy the definition of M-reduction!) □

2.3 Description of One Step of the Algorithm

In one step of the algorithm, the following procedure is carried out: Form the *least common multiple* (abbreviated LCM) of the LPPs pp_j and pp_k of two distinct basis polynomials f_j and f_k, namely the PP

$$pp_{j,k} = x_1^{L_{j,k,1}} \cdots x_n^{L_{j,k,n}}, \qquad \text{where} \quad L_{j,k,t} = \max(I_{j,t}, I_{k,t}) \quad (t = 1, \ldots, n).$$

For $pp_{j,k}$, both of the following congruences arise from (2):

$$pp_{j,k} \leftrightarrow -\sum a^{(j)}_{i_1 \ldots i_n} x_1^{i_1 + d_{j,1}} \cdots x_n^{i_n + d_{j,n}} \tag{2.4}$$

and

$$pp_{j,k} \leftrightarrow -\sum a^{(k)}_{i_1 \ldots i_n} x_1^{i_1 + d_{k,1}} \cdots x_n^{i_n + d_{k,n}} \tag{2.5}$$

where

$$d_{j,t} = L_{j,k,t} - I_{j,t} \quad \text{and} \quad d_{k,t} = L_{j,k,t} - I_{k,t} \qquad (t = 1, \ldots, n).$$

We call the polynomial that results from forming the difference of the right-hand sides of (4) and (5) the *S-polynomial corresponding to* $pp_{j,k}$. Two situations can arise through M-reduction of this polynomial relative to the ideal basis in question:

1. The S-polynomial corresponding to $pp_{j,k}$ is M-reduced to zero. In this case, we go immediately to the next step, as indicated in 2.4.
2. The S-polynomial corresponding to $pp_{j,k}$ is M-reduced to a polynomial (6) which does not vanish:

$$a x_1^{I_1} \cdots x_n^{I_n} + \sum a_{i_1 \ldots i_n} x_1^{i_1} \cdots x_n^{i_n} \qquad (a \in K, \quad a \neq 0), \tag{2.6}$$

where again we have singled out the LPP $x_1^{I_1} \cdots x_n^{I_n}$ of this polynomial. In this case, we add the resulting polynomial in the form

$$x_1^{I_1} \cdots x_n^{I_n} + \frac{1}{a} \sum a_{i_1 \dots i_n} x_1^{i_1} \cdots x_n^{i_n} \tag{2.7}$$

to the ideal basis. By doing this, we also increase the set of relations appearing in (2) and thereby the number of possible M-reductions that can possibly be carried out in the future for any polynomial. In accordance with 2.4, we go to the next step.

2.4 Combination of Steps

The operations given in 2.3 will be carried out in succession for all $pp_{j,k}$, where $j = 1, \dots, s-1$ and $k = j+1, \dots, s$. Notice that during the execution of the algorithm, the number of generating polynomials changes in general. For practical computation, it is advisable to process the $pp_{j,k}$ with smallest index first, since this can reduce the computation time considerably.

For a specific combination of indices j and k, one of the following special cases can arise:

S1. $pp_{j,k} = pp_j$ (or $pp_{j,k} = pp_k$). In this case, the basis polynomial f_j (f_k resp.) can be deleted from the basis, after which the operations defined in 2.3 are carried out for $pp_{j,k}$.

S2. $pp_{j,k} = pp_j \cdot pp_k$. In this case, we need not carry out the operations defined in 2.3.

The algorithm is terminated when all $pp_{j,k}$ have been processed according to the instructions of 2.3. We will now prove the following two claims:

B1. *Let $(g_1, \dots, g_u)$ be the current ideal basis when the algorithm terminates. Then $(g_1, \dots, g_u)$ has the property required by the hypothesis of the lemma.*

B2. *The algorithm terminates for every ideal in finitely many steps.*

3 The Proofs

3.1 Proof of B1

Let

$$g_j = x_1^{S_{j,1}} \cdots x_n^{S_{j,n}} + \sum c^{(j)}_{i_1 \dots i_n} x_1^{i_1} \cdots x_n^{i_n} \quad (j = 1, \dots, u)$$

be the new basis polynomials.

For this special ideal basis, we have: If we were to process every $pp_{j,k}$ ($pp_{j,k}$ would now be the LCM of $x_1^{S_{j,1}} \cdots x_n^{S_{j,n}}$ and $x_1^{S_{k,1}} \cdots x_n^{S_{k,n}}$) once more according to the instructions of the algorithm ($j = 1, \dots, u-1$ and $k = j+1, \dots, u$), then we could now M-reduce every resulting S-polynomial to zero in at least one way, since we have just added the necessary polynomials of the form (7) to the ideal basis during the previous run of the algorithm.

In the case $pp_{j,k} = pp_j \cdot pp_k$ (pp_j is now the abbreviation for $x_1^{S_{j,1}} \cdots x_n^{S_{j,n}}$), the corresponding S-polynomial has the form

$$-\sum c_{i_1 \dots i_n}^{(j)} x_1^{i_1+S_{k,1}} \cdots x_n^{i_n+S_{k,n}} + \sum c_{k_1 \dots k_n}^{(k)} x_1^{k_1+S_{j,1}} \cdots x_n^{k_n+S_{j,n}}.$$

The special M-reduction, in which we replace every

$$x_1^{i_1+S_{k,1}} \cdots x_n^{i_n+S_{k,n}}$$

by

$$-\sum c_{k_1 \dots k_n}^{(k)} x_1^{k_1+i_1} \cdots x_n^{k_n+i_n}$$

and every

$$x_1^{k_1+S_{j,1}} \cdots x_n^{k_n+S_{j,n}}$$

by

$$-\sum c_{i_1 \dots i_n}^{(j)} x_1^{i_1+k_1} \cdots x_n^{i_n+k_n}$$

makes this polynomial zero immediately. Thus in this case, an M-reduction to zero of the corresponding S-polynomial always exists, even when the operations defined in 2.3 were not carried out for $pp_{j,k}$.

We now show that every M-reduction relative to the ideal basis $(g_1, \dots, g_u)$ of a given polynomial leads to the same result. We prove this by induction and begin with $x_1^0 \cdots x_n^0$. $x_1^0 \cdots x_n^0$ has a uniquely determined M-reduced form relative to every ideal basis, because it is either an NPP and therefore already (uniquely) M-reduced, or an MPP and therefore M-reducible to zero.

The induction hypothesis is: For every polynomial whose LPP does not have a greater index than a fixed PP pp_0, the M-reduction relative to $(g_1, \dots, g_u)$ produces a uniquely determined NPP-polynomial.

What must be shown is: For a polynomial

$$f = c_1 \cdot x_1^{V_1} \cdots x_n^{V_n} + \dots + c_p \cdot x_1^{V_1} \cdots x_n^{V_n} + \sum a_{v_1 \dots v_n} x_1^{v_1} \cdots x_n^{v_n},$$

whose LPP $x_1^{V_1} \cdots x_n^{V_n}$ has an index that is greater by one than pp_0, the M-reduction produces a uniquely determined NPP-polynomial. (Note that we must prove the claim for polynomials in which like terms are not yet collected because we used them in this form in 2.2.)

We distinguish between three cases:

Case A. $x_1^{V_1} \cdots x_n^{V_n}$ is NPP. Then M-reduction of f means: M-reduction of

$$\sum a_{v_1 \dots v_n} x_1^{v_1} \cdots x_n^{v_n}$$

which produces a unique result by the induction hypothesis, and collection of the terms $c_q \cdot x_1^{V_1} \cdots x_n^{V_n}$ ($q = 1, \ldots, p$) which leads similarly to a unique result.

Case B. $x_1^{V_1} \cdots x_n^{V_n}$ is a multiple of precisely one $x_1^{S_{j,1}} \cdots x_n^{S_{j,1}}$ ($1 \leq j \leq u$). At some stage of the M-reduction of f, every term $c_q \cdot x_1^{V_1} \cdots x_n^{V_n}$ ($q = 1, \ldots, p$) must first be replaced by $c_q \cdot \sum(j)$, where

$$\sum(j) = \sum c_{i_1 \ldots i_n}^{(j)} x_1^{i_1+d_{j,1}} \cdots x_n^{i_n+d_{j,n}} \quad (d_{j,t} = V_t - S_{j,t} \text{ for } t = 1, \ldots, n),$$

and then, together with the other terms of the polynomial, be further M-reduced, leading again to a unique result because of the induction hypothesis.

Case C. $x_1^{V_1} \cdots x_n^{V_n}$ is a multiple of several $x_1^{S_{j,1}} \cdots x_n^{S_{j,n}}$, e.g. $j = j_1, \ldots, j_z$ ($1 \leq z \leq u$). During the M-reduction of f, every $c_q \cdot x_1^{V_1} \cdots x_n^{V_n}$ ($q = 1, \ldots, p$) will be replaced by a polynomial $c_q \cdot \sum(j_{r_q})$ where $j_{r_q} \in \{j_1, \ldots, j_z\}$. Thus a fixed combination of indices $(j_{r_1}, \ldots, j_{r_p})$ characterizes a family of M-reductions of f, which all use the same "initial substitutions" for the terms $c_q \cdot x_1^{V_1} \cdots x_n^{V_n}$ and all produce the same result, as we can easily check by slightly modifying the argument in case B. In particular, every M-reduction characterized by the p-tuple $(j_1, j_1, \ldots, j_1)$ leads to the same NPP-polynomial. We are done if we can show that among the M-reductions characterized by a specific combination $(j_{r_1}, \ldots, j_{r_p})$ and those characterized by $(j_1, \ldots, j_1)$, one of each can be specified which leads to the same result. In any case, a suitable M-reduction from the class characterized by $(j_1, \ldots, j_1)$ will first replace $c_q \cdot x_1^{V_1} \cdots x_n^{V_n}$ in f ($q = 1, \ldots, p$) by

$$c_q \cdot \sum c_{i_1 \ldots i_n}^{(j_1)} x_1^{V_1'+i_1+d_{j_1,1}} \cdots x_n^{V_n'+i_n+d_{j_1,n}}, \tag{3.1}$$

where

$$V_t' = V_t - L_{j_1,j_{r_q},t} \qquad d_{j_1,t} = L_{j_1,j_{r_q},t} - S_{j_1,t}$$
$$L_{j_1,j_{r_q},t} = \max(S_{j_1,t}, S_{j_{r_q},t})$$
$$\text{(exponents of the LCM } pp_{j_1,j_{r_q}}), \quad t = 1, \ldots, n.$$

An appropriate M-reduction from the class characterized by $(j_{r_1}, \ldots, j_{r_p})$ will first replace $c_q \cdot x_1^{V_1} \cdots x_n^{V_n}$ ($q = 1, \ldots, p$) by

$$c_q \cdot \sum c_{i_1 \ldots i_n}^{(j_{r_q})} x_1^{V_1'+i_1+d_{j_{r_q},1}} \cdots x_n^{V_n'+i_n+d_{j_{r_q},n}} \tag{3.2}$$

where

$$d_{j_{r_q},t} = L_{j_1,j_{r_q},t} - S_{j_{r_q},t} \qquad (t = 1, \ldots, n).$$

The polynomials (8) and (9) are precisely the polynomials whose difference yielded the S-polynomial corresponding to $pp_{j_1,j_{r_q}}$, except both are multiplied by $x_1^{V_1'} \cdots x_n^{V_n'}$. From an M-reduction that reduces this S-polynomial to zero (and that always exists relative to $(g_1, \ldots, g_u)$ because of the argument at the beginning of this section), we can immediately obtain an M-reduction to zero of the polynomial consisting of the difference of (8) and (9) by multiplying all PPs resulting from the M-reduction of the S-polynomial by $x_1^{V_1'} \cdots x_n^{V_n'}$. However if the difference of two polynomials to M-reduces to zero, then it is easily seen that each of the two polynomials can be M-reduced to the same polynomial in at least one way. Hence, among the M-reductions of f characterized by $(j_1, \ldots, j_l)$ and those characterized by $(j_{r_1}, \ldots, j_{r_p})$, there is at least one of each which leads to the same NPP-polynomial. With this, B1 is proved. □

The justification of the treatment of the special case S1 still remains open. However, we can easily convince ourselves that by removing a generating polynomial which has the property specified in S1, the family of possible M-reductions will not be reduced, as long as we add the NPP-polynomial corresponding to the S-polynomial to the ideal basis, as the algorithm specifies. The claim that all of the S-polynomials corresponding to $pp_{j,k}$ can be M-reduced to zero relative to the ideal basis $(g_1, \ldots, g_u)$, which forms the basis for the proof of uniqueness of M-reduction, remains correct after removing a basis polynomial in step S1.

3.2 Proof of B2

During the processing of an LCM $pp_{j,k}$ according to the algorithm, the ideal basis will eventually be enlarged by the addition of a polynomial whose LPP is an NPP relative to the previous ideal basis. B2 is proved if we can prove the following theorem:

Theorem. *A sequence of PPs $x_1^{I_{j,1}} \cdots x_n^{I_{j,n}}$ $(j = 1, 2, 3, \ldots)$, which has the property that $x_1^{I_{k,1}} \cdots x_n^{I_{k,n}}$ $(k = 2, 3, \ldots)$ is not a multiple of any $x_1^{I_{m,1}} \cdots x_n^{I_{m,n}}$ $(m < k)$, has only finitely many elements. (A sequence of PPs with this property will be called an M-sequence).*

Proof. The theorem is easily seen for $n = 1$. We assume its correctness for all $n < N$ and consider an M-sequence which begins with $x_1^{I_1} \cdots x_N^{I_N}$. Let $x_1^{v_1} \cdots x_N^{v_N}$ be another element of the M-sequence, then $v_i < I_i$ for at least one i $(1 \leq i \leq N)$. For an arbitrary combination of indices $(i_1, \ldots, i_k)$ $(1 \leq k \leq N$, $1 \leq i_1 < \ldots < i_k \leq n)$, there are only finitely many k-tuples $(v_{i_1}, \ldots, v_{i_k})$ of positive integers (including zero) which satisfy $v_{i_j} < I_{i_j}$ for $j = 1, \ldots, k$. Therefore, the PPs of an M-sequence belong to finitely many types, each of which is characterized by a fixed combination of exponents $(v_{i_1}, \ldots, v_{i_k})$ of k fixed variables $x_{i_1}, \ldots, x_{i_k}$.

In an M-sequence with infinitely many elements, an infinite subsequence must exist whose elements all belong to the same type. By removing the variables $x_{i_1}, \ldots, x_{i_k}$ for which the PPs of this subsequence have fixed exponents

$(v_{i_1}, \ldots, v_{i_k})$, an infinite M-sequence in $n-k$ variables arises from these PPs, contradicting our induction hypothesis. □

Remark. This proof shows that the algorithm is finite in principal, independent of special properties of the ideal (e.g. the dimension), but says nothing about its practicality. In this setting, the following fact seems to be interesting: When applied to a system of linear equations, the algorithm becomes the Gaussian elimination procedure, as one can easily verify. It is therefore a generalization of the Gaussian algorithm in a certain sense.

4 Results about the Existence of a Zero and the Dimension

It is well known that a polynomial ideal $\mathcal{A}$ has no zeros if and only if $\mathcal{A} = (1)$, i.e. the corresponding residue class ring $\mathcal{O}$ consists of only one residue class. This is the case if and only if during the course of the algorithm, the polynomial $x_1^0, \ldots, x_n^0$ must be adjoined to the ideal basis. (If, after termination of the algorithm, $x_1^0, \ldots, x_n^0$ is not present in the ideal basis, then there is more than one NPP and therefore more than one residue class in $\mathcal{O}$!) Hence, we have:

Criterion 4.1. *A polynomial ideal has no zeros if and only if, during the course of the algorithm, the polynomial* $x_1^0, \ldots, x_n^0$ *must be adjoined to the ideal basis.*

Moreover, we arrive at the following criterion about the dimension of the ideal:

Criterion 4.2. *A polynomial ideal has dimension greater than zero if and only if the ideal basis* $(g_1, \ldots, g_u)$ *produced by the algorithm has the following property: There is an* i $(1 \leq i \leq n)$*, such that no PP of the form* x_i^h $(h \geq 0)$ *occurs among the LPPs of the basis polynomials.*

Proof. Applying the definition of the dimension of a polynomial ideal ([1, p. 98]), we easily see that a polynomial ideal has dimension zero if and only if the residue class algebra is finite dimensional. But the number of basis elements of the residue class algebra (= the number of NPPs relative to $(g_1, \ldots, g_u)$) is infinite if and only if the condition expressed in Criterion 4.2 holds. □

5 Calculating the Multiplication Table of the Residue Class Algebra

Let $\overline{x_1^{i_1} \cdots x_n^{i_n}}$ and $\overline{x_1^{k_1} \cdots x_n^{k_n}}$ be quantities which were constructed as basis elements for $\mathcal{O}$ by the algorithm. (By the bar, we denote the the corresponding residue class.) Then we obtain a representation of their product $\overline{x_1^{i_1+k_1} \cdots x_n^{i_n+k_n}}$ as a linear combination of the residue classes of the NPPs through the M-reduction of $x_1^{i_1+k_1} \cdots x_n^{i_n+k_n}$ relative to the ideal basis produced by the algorithm.

6 Remark on the Construction of Zeros

We suggest here a method for finding the zeros of the polynomial ideal using the knowledge about the structure of the residue class algebra. First we consider the zero dimensional case: Let $u_1, \ldots, u_m$ be the PPs whose residue classes form a basis for $\mathcal{O}$. Using the multiplication table, we can successively find representations for all $\overline{x_1^k}$ $(k = 0, 1, \ldots)$ as linear combinations of the $\overline{u_j}$ $(j = 1, \ldots, m)$, and for each $\overline{x_1^k}$, we can test whether it already depends linearly on $1, \overline{x_1}, \ldots, \overline{x_1^{k-1}}$. Let $k = m_1$ $(m_1 \leq m + 1)$ be the first time this is the case, then $p_1(\overline{x_1}) = 0$ holds, where $p_1(x_1)$ is a polynomial of degree m_1 in $K[x_1]$.

Next, we form the representations of the residue classes of the PPs which arise by multiplying $1, x_1, \ldots, x_1^{m_1-1}$ by $x_2, x_2^2, \ldots,$ until for a fixed residue class $\overline{x_1^{i_1} x_2^{i_2}}$, a linear dependency with previous PP-residue classes is found which can be written in the form

$$p_2(\overline{x_1}, \overline{x_2}) = 0 \qquad \text{with} \qquad p_2(x_1, x_2) \in K[x_1, x_2].$$

We continue similarly and obtain a sequence of polynomials $p_k(x_1, \ldots, x_k) \in \mathcal{A}$ $(k = 1, \ldots, n)$, *which we can solve successively.* Certainly, every zero of the ideal occurs among the set of zeros of the p_k, but the converse is not the case in general. Precautions must still be taken (by taking additional polynomials from $\mathcal{A}$) in order to eliminate extraneous zeros. A detailed study addressing questions in this context is however beyond the scope of the present work.

The d dimensional case $(d > 0)$ can be reduced to the zero dimensional case by substituting numerical values for d independent variables relative to $\mathcal{A}$. Thereby we can obtain at least finitely many solutions. For every solution point, under certain assumptions, we can construct the family of solutions in the neighborhood of these points, e.g. with the help of Lie series (Gröbner (1964b), p. 72ff). However many questions remain to be studied in detail.

7 An Example and a Remark on Programming

In order to illustrate the steps of the algorithm, we give a simple example:

$$\mathcal{A} = (x_3^2 - \tfrac{1}{2}x_1^2 - \tfrac{1}{2}x_2^2, \quad x_1x_3 - 2x_3 + x_1x_2, \quad x_1^2 - x_2).$$

We consider first the $pp_{1,2}$, namely $x_1x_3^2$. The corresponding S-polynomial is

$$S_{1,2} = \tfrac{1}{2}x_1^3 + \tfrac{1}{2}x_1x_2^2 - 2x_3^2 + x_1x_2x_3.$$

M-reduction of $S_{1,2}$ produces

$$x_1x_2^2 - x_1x_2 + 2x_2 + 2x_2^2 - 4x_2x_3.$$

We adjoin this polynomial to the ideal basis. We must treat the other $pp_{j,k}$, where $j = 1, \ldots, s-1$ and $k = 2, \ldots, s$, in an analogous manner. s is 3 at the start, but becomes 4 by addition of the new basis polynomial and changes again several times during the course of the algorithm. For example, $pp_{1,3}$ is a PP for which the arguments of the special case S2 apply, so the M-reduction of $S_{1,3}$ can be skipped.

Finally we obtain the new ideal basis

$$\begin{aligned} \mathcal{A} \;=\; & (x_3^2 - \tfrac{1}{2}x_2^2 - \tfrac{1}{2}x_2, \quad x_1x_3 - 2x_3 + x_1x_2, \quad x_1^2 - x_2, \\ & x_1x_2^2 + 7x_1x_2 + 2x_2 + 6x_2^2 - 16x_3, \\ & x_2x_3 + x_2^2 + 2x_1x_2 - 4x_3, \quad x_2^3 - 12x_2 - 29x_2^2 + 64x_3 - 24x_1x_2). \end{aligned}$$

The LPPs of the six basis polynomials are x_3^2, x_1x_3, x_1^2, $x_1x_2^2$, x_2x_3, x_2^3. Therefore, Criterion 4.2 shows that $\mathcal{A}$ is zero dimensional. The residue classes of the PPs 1, x_1, x_2, x_3, x_1x_2, x_2^2 form a basis of the residue class algebra. The representation of the product of the fourth and sixth basis elements, for example, is formed from the the M-reduction of $x_2^2x_3$. It produces

$$x_2^2x_3 \leftrightarrow -18x_1x_2 + 48x_3 - 8x_2 - 21x_2^2.$$

The remaining elements of the multiplication table can be found similarly.

Of course, the computation time increases very quickly with increased number of variables and increased degree of the polynomials. The programming of the algorithm is very easy because of its simple structure. It is advantageous to use list processing concepts. Programs exist in Freiburg Code and machine code of the ZUSE Z 23.

For other results on questions studied in the present paper, Hermann (1926) should be consulted.

I would like to use this opportunity to express sincere thanks to my distinguished teacher, Professor W. Gröbner.

References

Gröbner, W. (1949). *Moderne algebraische Geometrie.* Springer-Verlag, Vienna, Innsbruck.

Gröbner, W. (1964a). *Teoria degli ideali e geometria algebrica.* Seminari dell'Istituto Nazionale di Alta Matematica 1962-1963. Edizione Cremonese, Rome.

Gröbner, W. (1964b). Oral communication in the Math. Seminar of the University of Innsbruck.

Hermann, G. (1926). 'Die Frage der endlich vielen Schritte in der Theorie der Polynomideale'. *Math. Ann.* **95**, 736-788.

van der Waerden, B. L. (1939). *Moderne Algebra*, Volume 2, 2nd Edition. Springer-Verlag, Berlin.

Universität Innsbruck

Index of Tutorials

Printed in the United States
By Bookmasters